Fundamentals of Stochastic Models

Stochastic modeling is a set of quantitative techniques for analyzing practical systems with random factors. This area is highly technical and mainly developed by mathematicians. Most existing books are for those with extensive mathematical training; this book minimizes that need and makes the topics easily understandable.

Fundamentals of Stochastic Models offers many practical examples and applications and bridges the gap between elementary stochastics process theory and advanced process theory. It addresses both performance evaluation and optimization of stochastic systems and covers different modern analysis techniques such as matrix analytical methods and diffusion and fluid limit methods. It goes on to explore the linkage between stochastic models, machine learning, and artificial intelligence, and discusses how to make use of intuitive approaches instead of traditional theoretical approaches.

The goal is to minimize the mathematical background of readers that is required to understand the topics covered in this book. Thus, the book is appropriate for professionals and students in industrial engineering, business and economics, computer science, and applied mathematics.

Operations Research Series

Series Editors:
Natarajan Gautam
A. Ravi Ravindran

Operations Research
A Practical Introduction
Michael W. Carter, Camille C. Price

Operations Research and Management Science Handbook
Edited by A. Ravi Ravindran

Operations Research Calculations Handbook, Second Edition
Dennis Blumenfeld

Introduction to Linear Optimization and Extensions with MATLAB®
Roy H. Kwon

Multiple Criteria Decision Making in Supply Chain Management
Edited By A. Ravi Ravindran

Multiple Criteria Decision Analysis for Industrial Engineering
Methodology and Applications
Gerald William Evans

Supply Chain Engineering
Models and Applications
A. Ravi Ravindran, Donald P. Warsing, Jr.

Analysis of Queues
Methods and Applications
Natarajan Gautam

Operations Planning
Mixed Integer Optimization Models
Joseph Geunes

Big Data Analytics Using Multiple Criteria Decision-Making Models
Edited By Ramakrishnan Ramanathan, Muthu Mathirajan, A. Ravi Ravindran

Service Systems Engineering and Management
A. Ravi Ravindran, Paul M. Griffin, Vittaldas V. Prabhu

Probability Models in Operations Research
C. Richard Cassady, Joel A. Nachlas

Fundamentals of Stochastic Models
Zhe George Zhang

Fundamentals of Stochastic Models

Zhe George Zhang

CRC Press
Taylor & Francis Group
Boca Raton London New York

CRC Press is an imprint of the
Taylor & Francis Group, an **informa** business

Designed cover image: © Shutterstock

MATLAB® is a trademark of The MathWorks, Inc. and is used with permission. The MathWorks does not warrant the accuracy of the text or exercises in this book. This book's use or discussion of MATLAB® software or related products does not constitute endorsement or sponsorship by The MathWorks of a particular pedagogical approach or particular use of the MATLAB® software.

First edition published 2023
by CRC Press
6000 Broken Sound Parkway NW, Suite 300, Boca Raton, FL 33487-2742

and by CRC Press
4 Park Square, Milton Park, Abingdon, Oxon, OX14 4RN

CRC Press is an imprint of Taylor & Francis Group, LLC

© 2023 Zhe George Zhang

ISBN: 978-0-367-71261-7 (hbk)
ISBN: 978-0-367-71262-4 (pbk)
ISBN: 978-1-003-15006-0 (ebk)

DOI: 10.1201/9781003150060

Typeset in Nimbus font
by KnowledgeWorks Global Ltd.

Publisher's note: This book has been prepared from camera-ready copy provided by the authors.

Contents

List of Figures xv

List of Tables xix

Preface xxi

Acknowledgments xxv

Chapter 1   Introduction 1

    1.1   Stochastic Process Classification 2
    1.2   Organization of the Book 3

Part I   ***Fundamentals of Stochastic Models***

Chapter 2   Discrete-Time Markov Chains 9

    2.1   Dynamics of Probability Measures 10
    2.2   Formulation of DTMC 13
    2.3   Performance Analysis of DTMC 16
        2.3.1   Classification of States 16
        2.3.2   Steady State Analysis 22
        2.3.3   Positive Recurrence for DTMC 30
        2.3.4   Transient Analysis 36
        2.3.5   Branching Processes 41
    References 45

Chapter 3   Continuous-Time Markov Chains 47

    3.1   Formulation of CTMC 47
    3.2   Analyzing the First CTMC: A Birth-and-Death
        Process 50
    3.3   Transition Probability Functions for CTMC 53
        3.3.1   Uniformization 56
    3.4   Stationary Distribution of CTMC 58

3.4.1   Open Jackson Networks.................................. 60
3.5   Using Transforms 65
3.6   Using Time Reversibility.............................. 67
3.7   Some Useful Continuous-time Markov Chains......... 74
       3.7.1   Poisson Process and Its Extensions.............. 74
               3.7.1.1   Non-Homogeneous Poisson
                          Process.............................. 78
               3.7.1.2   Pure Birth Process 79
               3.7.1.3   The Yule Process 81
       3.7.2   Pure Death Processes 86
               3.7.2.1   Transient Analysis on CTMC........ 88
References.. 89

Chapter 4   Structured Markov Chains.................................... 91

4.1   Phase-Type Distributions............................. 91
4.2   Properties of PH Distribution 97
       4.2.1   Closure Properties..................................... 97
       4.2.2   Dense Property of PH Distribution 103
       4.2.3   Non-Uniqueness of Representation of PH
               Distribution 104
4.3   Fitting PH Distribution to Empirical Data or
       a Theoretical Distribution 105
       4.3.1   The EM Approach.................................... 105
4.4   The EM Algorithm 107
       4.4.1   Convergence of EM Algorithm.................... 113
       4.4.2   EM Algorithm for PH Distribution.............. 116
4.5   Markovian Arrival Processes..................... 122
       4.5.1   From PH Renewal Processes to
               Markovian Arrival Processes 122
       4.5.2   Transition Probability Function Matrix 125
4.6   Fitting MAP to Empirical Data..................... 132
       4.6.1   Grouped Data for MAP Fitting................... 132
4.7   Quasi-Birth-and-Death Process (QBD) – Analysis
       of MAP/PH/1 Queue................................... 136
       4.7.1   QBD – A Structured Markov Chain 137
               4.7.1.1   Matrix-Geometric Solution –
                          R-Matrix 141
               4.7.1.2   Fundamental Period – **G**-Matrix . 147
       4.7.2 $MAP/PH/1$ Queue – A Continuous-Time
               QBD Process................................... 151
4.8 $GI/M/1$ Type and $M/G/1$ Type Markov Chains..... 153
       4.8.1 $GI/M/1$ Type Markov Chains..................... 153

4.8.2 $M/G/1$ Type Markov Chains....................... 156
4.8.3   Continuous-Time Counterpart of
          Discrete-Time QBD Process 162
References.. 168

Chapter 5   Renewal Processes and Embedded Markov Chains 171

5.1   Renewal Processes 171
       5.1.1   Basic Results of Renewal Processes 172
       5.1.2   More on Renewal Equations 177
       5.1.3   Limit Theorems for Renewal Processes....... 181
       5.1.4   Blackwell's Theorem and Key Renewal
                Theorem 187
       5.1.5   Inspection Paradox...................................... 195
       5.1.6   Some Variants of Renewal Processes........... 202
       5.1.7   Renewal Reward Processes........................... 205
       5.1.8   Regenerative Processes 207
5.2   Markov Renewal Processes 211
       5.2.1   Basic Results for Markov Renewal
                Processes 211
       5.2.2   Results for Semi-Markov Processes 222
       5.2.3   Semi-Regenerative Processes...................... 224
       5.2.4 $M/G/1$ and $GI/M/1$ Queues 226
                5.2.4.1 $M/G/1$ Queue 226
                5.2.4.2 $GI/M/1$ Queue............................. 231
       5.2.5   An Inventory Model..................................... 235
       5.2.6   Supplementary Variable Method.................. 238
References.. 242

Chapter 6   Random Walks and Brownian Motions............................. 245

6.1   Random Walk Processes... 245
       6.1.1   Simple Random Walk – Basics.................... 245
       6.1.2   Spitzer's Identity Linking Random Walks
                to Queueing Systems.................................... 256
       6.1.3   System Point Level Crossing Method.......... 264
       6.1.4   Change of Measures in Random Walks 270
       6.1.5   The Binomial Securities Market Model....... 272
       6.1.6   The Arc Since Law 275
       6.1.7   The Gambler's Ruin Problem 279
       6.1.8   General Random Walk 280
                6.1.8.1   A Brief Introduction to DTMP.... 280
       6.1.9   Basic Properties of GRW 284
                6.1.9.1   Central Limit Theorem for GRW 287

6.2   Brownian Motion 288
     6.2.1   Brownian Motion as a Limit of Random
           Walks 289
     6.2.2   Gaussian Processes 291
     6.2.3   Sample Path Properties 294
           6.2.3.1   Infinite Zeros and
                    Non-Differentiability 294
           6.2.3.2   Re-Scaling a Process 295
           6.2.3.3   The Reflection Principle – Hit-
                    ting Times and Maximum
                    of BM 297
           6.2.3.4   Conditional Distribution of BM .. 302
           6.2.3.5   BM as a Martingale 304
     6.2.4   Transition Probability Function of BM 309
     6.2.5   The Black-Scholes Formula 313
  References 318

Chapter 7   Reflected Brownian Motion Approximations to
Simple Stochastic Systems 319

7.1   Approximations to $G/G/1$ Queue 319
7.2   Queue Length as Reflection Mapping 321
7.3   Functional Strong Law of Large Numbers
     (FSLLN) – Fluid Limit 327
7.4   Functional Central Limit Theorem (FCLT) –
     Diffusion Limit 329
7.5   Heavy Traffic Approximation to $G/G/1$ Queue 331
7.6   Bounds for Fluid and Diffusion Limit
     Approximations 334
7.7   Applications of RBM Approach 337
     7.7.1   A Two-Station Tandem Queue 337
     7.7.2   A Production-Inventory Model 341
  References 349

Chapter 8   Large Queueing Systems 351

8.1   Multi-Dimensional Reflected Brownian Motion
     Approximation to Queueing Networks 351
     8.1.1   Oblique Reflection Mapping 353
     8.1.2   A Fluid Network 355
     8.1.3   A Brownian Motion Network 358
     8.1.4   Fluid and Diffusion Approximations to
           Queueing Networks 361
8.2   Decomposition Approach 365

8.2.1 Superposition of Flows 365
8.2.2 Flowing through a Queue 366
8.2.3 Splitting a Flow 369
8.2.4 Decomposition of a Queueing Network 370
8.3 One-Stage Queueing System with Many Servers 373
8.3.1 Multi-Server Queues without Customer
Abandonments 374
8.3.1.1 Increasing ρ with fixed s and μ ... 375
8.3.1.2 Increasing λ and s with fixed ρ ... 375
8.3.1.3 Increasing λ and s with an
Increasing ρ 375
8.3.2 Multi-Server Queues with Customer
Abandonments 377
8.4 Queues with Time-Varying Parameters 381
8.4.1 Fluid Approximation 382
8.4.2 Diffusion Approximation 384
8.5 Mean Field Method for a Large System with
Many Identical Interacting Parts 386
References 393

Chapter 9 Static Optimization in Stochastic Models 395

9.1 Optimization Based on Regenerative Cycles 395
9.1.1 Optimal Age Replacement Policy for a
Multi-State System 395
9.1.2 Optimal Threshold Policy for a $M/G/1$
Queue 402
9.2 Optimization Based on Stationary Performance
Measures – Economic Analysis of Stable
Queueing Systems 406
9.2.1 Individual Optimization 407
9.2.2 Social Optimization 408
9.2.3 Service Provider Optimization 411
9.3 Optimal Service-Order Policy for a Multi-Class
Queue 416
9.3.1 Preliminary Results for a Multi-Class
M/G/1 Queue 416
9.3.2 Optimal Service-Order Policy for a Multi-
Class Queue with Nonpreemtive Priority
– $c\mu$ Rule 420
9.4 Customer Assignment Problem in a Queue
Attended by Heterogeneous Servers 423
9.4.1 Problem Description 423

9.4.2   Characterization of Optimal Policy 426
9.4.3   Optimal Multi-Threshold Policy – c/μ Rule434
9.5   Performance Measures in Optimization of
      Stochastic Models 438
      9.5.1   Value at Risk and Conditional Value
              at Risk 438
      9.5.2   A Newsvendor Problem 442
References 446

Chapter 10   Dynamic Optimization in Stochastic Models 449

10.1   Discrete-Time Finite Markov Decision Process 450
10.2   Computational Approach to DTMDP 471
       10.2.1   Value Iteration Method 471
       10.2.2   Policy Iteration Method 475
       10.2.3   Computational Complexity 479
       10.2.4   Average Cost MDP with Infinite Horizon 480
10.3   Semi-Markov Decision Process 495
       10.3.1   Characterizing the Structure of Optimal
                Policy 495
       10.3.2   Computational Approach to SMDP 504
10.4   Stochastic Games – An Extension of MDP 508
References 513

Chapter 11   Learning in Stochastic Models 515

11.1   Multi-Arm Bandits Problem 515
       11.1.1   Sample Average Methods 517
       11.1.2   Effect of Initial Values 519
       11.1.3   Upper Confidence Bounds 520
       11.1.4   Action Preference Method 527
11.2   Monte Carlo-Based MDP Models 530
       11.2.1   Model-Based Learning 531
       11.2.2   Model-Free Learning 533
       11.2.3   Model-Free Learning with Bootstrapping ... 535
       11.2.4 Q-Learning 536
       11.2.5   Temporal-Difference Learning 540
       11.2.6   Convergence of Learning Algorithms 547
       11.2.7   Learning in Stochastic Games – An
                Extension of Q-Learning 549
11.3   Hidden Markov Models 558
11.4   Partially Observable Markov Decision Processes 566
References 582

Part II *Appendices: Elements of Probability and Stochastics*

Chapter A   Basics of Probability Theory 587

   A.1   Probability Space 587
   A.2   Basic Probability Rules.. 589
         A.2.0.1   Bayesian Belief Networks 590
   A.3   Random Variables 592
   A.4   Probability Distribution Function 594
         A.4.1   Multivariate Distribution and Copulas......... 596
         A.4.2   Transforming Distribution Functions........... 601
   A.5   Independent Random Variables 602
   A.6   Transforms for Random Variables 602
   A.7   Popular Distributions in Stochastic Models 604
         A.7.1   Chi-Square Distribution 606
         A.7.2   *F* Distribution 607
         A.7.3   *t* Distribution 608
         A.7.4   Derivation of Probability Density Function. 609
         A.7.5   Some Comments on Degrees of Freedom ... 614
   A.8   Limits of Sets 615
   A.9   Borel-Cantelli Lemmas.. 618
   A.10 A Fundamental Probability Model 619
   A.11 Sets of Measure Zero 624
   A.12 Cantor Set 626
   A.13 Integration in Probability Measure 630
         A.13.1 Radon-Nikodym Theorem 631
   References.. 638

Chapter B   Conditional Expectation and Martingales 639

   B.1 σ-Algebra Representing Amount of Information..... 639
   B.2   Conditional Expectation in Discrete Time................ 642
   B.3   Conditional Expectation in Continuous-Time 646
   B.4   Martingales 647
         B.4.1   Optional Sampling 649
   References.. 651

Chapter C   Some Useful Bounds, Inequalities, and Limit Laws 653

   C.1   Markov Inequality.. 653
   C.2   Jensen's Inequality... 658
   C.3   Cauchy–Schwarz Inequality 659

List of Figures

2.1 A three-state DTMC for a computer system with two processors...... 14
2.2 A four-state DTMC for a computer system with two processors....... 14
2.3 A five-state DTMC for a computer system with two processors........ 15
2.4 An irreducible DTMC. 17
2.5 A reducible DTMC. 17
2.6 A three-state DTMC with period $d = 2$. 33
2.7 A DTMC model for liver disease treatment process. 38
2.8 Two cases of $\phi_{Y_1}(z)$. 44

3.1 A multi-server queue. 59
3.2 Expected total cost per unit time vs. total number of RNs available.. 73

4.1 Convolution of two independent PH distributions............................. 99
4.2 A service facility serving three types of customers. 165
4.3 A service facility serving three types of customers with fast
    removing type 2 customers. 166
4.4 A service facility serving three types of customers with fast
    adding type 2 customers. 166

5.1 Renewal process. 176
5.2 Noise function $h(t)$. 192
5.3 A graphic representation of $H(t)$. 192
5.4 Age of a renewal process (the sloping line is 45 degrees). 198
5.5 Residual life of a renewal process. 199
5.6 Cycle life of a renewal process. 199
5.7 Busy cycles of a single server system.. 207
5.8 A fixed point ξ satisfying $\xi = \sum_{n=0}^{\infty} \xi^n a_n$. 233

6.1 Reflection principle in random walk.. 247
6.2 Mirroring for maximum... 249
6.3 Transforming from the original random walk to its dual.................. 251
6.4 Two-dimensional random walk. 255
6.5 The maze problem for a mouse. 256
6.6 Sample path of $W(t), t \geq 0$ with a level x. 265
6.7 Arcsine law. 277
6.8 Brownian motion. 295
6.9 Mirroring the maximum. 298

6.10   A simple binomial model.. 306

7.1   One-dimensional reflection mapping of a fluid queue..................... 323

8.1   A six-node network with two external arrivals........................... 355
8.2   A three-node network with two external arrivals......................... 364
8.3   A six-node network with one external arrival stream for an ER in
       a hospital.. 373
8.4   State transition in an $M/M/1/1$ queue................................... 388
8.5   State transition in a system with two $M/M/1/1$ queues.................. 388
8.6   Transition rates in the first $M/M/1/1$ queue depending on the
       second $M/M/1/1$ queue.. 389
8.7   Markov chain with lumping states from the two $M/M/1/1$ queues. 389
8.8   Overall view of the Markov chain.. 390
8.9   Transition rates in the tagged queue depending on the Markov
       chain with lumped states... 390
8.10   Markov chain with identical entities.................................... 391
8.11   Markov chain of one of identical entities............................... 391
8.12   Probability of queue length as function of N. Source: Adapted
       from A. Bobbio, M. Gribaudo, M. Telek, Analysis of large scale
       interacting systems by mean field method. *Quantitative Evalua-
       tion of Systems, IEEE Computer Society*, 215-224, 2008. 392
8.13   The coefficient of variation of queue length 0 and 1 as functions
       of N. Source: Adapted from A. Bobbio, M. Gribaudo, and M.
       Telek, Analysis of large scale interacting systems by mean field
       method. *Quantitative Evaluation of Systems, IEEE Computer So-
       ciety*, 215-224, 2008. 393

9.1   Probabilities that the one element system is in state $i =
       5, 4, 3, 2, 1, 0$ at time t. Source: Adapted from S.H. Sheu and Z.G.
       Zhang, An optimal age replacement policy for multi-state sys-
       tems, *IEEE Transactions on Reliability*, 62, (3), 722-723, 2013...... 401
9.2 $C_{ARP}(T, w)$ and $J_{ARP}(T, w)$ functions for $\theta = 0.5$. Source:
       Adapted from S.H. Sheu and Z.G. Zhang, An optimal age re-
       placement policy for multi-state systems, *IEEE Transactions on
       Reliability*, 62, (3), 722-723, 2013................................... 402
9.3   Work process over a service cycle in an $M/G/1$ queue with server
       vacations... 403
9.4   VaR definition.. 439
9.5   CVaR definition... 441

10.1   Base-stock policy is optimal.. 460
10.2   Base-stock policy is not optimal.. 460

10.3 Case i: $[x, z]$ is to the right of S_t. 462

10.4 Case ii: $[x, z]$ is to the left of S_t. 462

10.5 Case iii: q is to the right of S_t. 462

10.6 Case iii: q is to the left of S_t 463

10.7 One-step reward function Gridworld problem.................................. 472

10.8 Gridworld – An agent tries to move in the maze to a terminal state to achieve the maximum rewards 473

10.9 Gridworld – Initial Value function – also the immediate reward function 474

10.10 Gridworld – Value function after the first itearation......................... 475

10.11 Gridworld – Value function after the fifth itearation 476

10.12 Gridworld – Value function after the tenth itearation 476

10.13 Gridworld – Value function after the 1000th itearation.................... 477

10.14 Gridworld – Value function and Optimal policy after the 100th itearation 477

10.15 Gridworld – Value function values for a given initial policy (Go North for all states) and the updated greedy policy with respect to these values – the first iteration. 478

10.16 Gridworld – Value function values under the policy from the first iteration and the updated greedy policy with respect to these values – the second iteration.. 478

10.17 Gridworld – Value function values under the policy from the second iteration and the updated greedy policy with respect to these values – the third iteration. 479

10.18 Customer allocation in a two queue system. 496

10.19 Structure of the optimal policy in a two queue system.................... 497

10.20 Structure of the optimal policy in two queue system....................... 508

10.21 Transition probabilities and one-step reward function for the two-player stochastic game... 510

11.1 10-arm MAB problem simulations under different algorithms. 529

11.2 Gridworld problem without transition probability and one-step reward information. 531

11.3 Training the MDP model by four episodes (replications) via model-based estimation. 532

11.4 Training the MDP model by four episodes (replications) via model-free estimation. 534

11.5 A robot trying to exit to the house. 537

11.6 Model-free Monte Carlo versus TD learning. 542

11.7 The progress of an episode in Gridworld problem. 543

11.8 TD learning based on an episode in Gridworld problem.................. 544

11.9 Optimal value function and policy determined by MDP for Gridworld problem.. 544

11.10 The value function estimated for the optimal policy by TD for Gridworld problem. 545

11.11 The Gridworld with cliff problem. 546

11.12 Two Gridworld games.. 550

11.13 A pair of NE strategies for two players. 551

11.14 Other possible NE policy pairs for two players in Grid Game 1...... 551

11.15 Possible NE policy pairs for two players in Grid Game 2............... 552

11.16 Q-values for three NEs in Grid Game 2. 554

11.17 Q-values updating during the learning process for Grid Games 1 and 2... 557

11.18 Q-values updating during the learning process for Grid Games 1 and 2... 557

11.19 A hidden Markov process with T periods. 559

11.20 Horizon 1 value function. 571

11.21 Horizon 1 value function. 573

11.22 Transformed value function for a given $a1$ and $o1$........................... 574

11.23 Transformed belief states under $a1$.. 575

11.24 Transformed value functions for all observations under $a1$. 576

11.25 Optimal Strategy at the second stage under $a1$ for the first stage..... 576

11.26 Value function for stage 2 under $a_0^1 = a1$. 577

11.27 Value function for stage 2 under $a_0^2 = a2$. 577

11.28 Value functions for both $a1$ and $a2$ as the first stage action, respectively... 578

11.29 Value function for horizon 2 problem.. 578

11.30 Transformed Value functions for horizon 2 problem given $a1$ at the beginning and different outcomes... 579

11.31 Value function for horizon 2 problem given $a1$ at the beginning. 579

11.32 Value function for horizon 2 problem given $a2$ at the beginning. 580

11.33 Optimal starting action for a horizon 3 problem. 580

11.34 Value function for a horizon 3 problem.. 580

A.1 Bayes' belief network and probability information......................... 591

A.2 Measurability of a random variable. 595

A.3 Distribution function of a random variable...................................... 595

A.4 A typical set in Chebyshev's inequality... 623

B.1 Two period stock price tree.. 640

B.2 Three period stock price tree. 642

B.3 Atom structure under different σ algebras for the three period stock price tree.. 644

C.1 Markov Inequality... 653

C.2 Jensen's inequality. 659

H.1 Construction of a standard Brownian motion. 722

List of Tables

1.1   Classification of Stochastic Processes 2

4.1   Results for Case 1: Knowing the coin selected each time 109
4.2   Expected value of unobservable variables Y_{i1} and Y_{i2} 112
4.3   Results for Case 2: Not knowing the coin selected each time 112

9.1   Impact of θ on T^*, $C_{ARP}(T^*, w)$, and $J_{ARP}(T^*, w)$ for a single element multiple-state system with $r_3 = 1000, r_4 = 1100, r_5 = 1200, c_u = 1500, c_p = 1000$ 401

11.1   Nash Q-values in state (0,2) of Grid Game 1. 553
11.2   Nash Q-values in state (0,2) for Grid Game 2. 554
11.3   State sequence probabilities... 561
11.4   HMM probabilities 561

A.1   Major Discrete Distributions 604
A.2   Major Continuous Distributions 605
A.3   Distributions Derived from the Normal Distributions 606

D.1   A Sample of Size n for Estimating an MLR Model 674

Preface

The aim of this book is to present a comprehensive collection of modeling techniques for analyzing stochastic systems for service, manufacturing, and computer network / telecommunication applications. The book contains two parts: the main chapters and the appendices. The main chapters cover the fundamentals of stochastic models, and the appendices cover the elements of probability and stochastics, which are relevant to stochastic modeling.

Main Chapters:

In the first four main chapters (Chapters 2–5), we present the foundation of stochastic modeling by focusing on the discrete-time and continuous-time Markov chains and extending them to the structured Markov chains and matrix-analytic methods (MAMs). Then, we study the renewal processes and extend them to the Markov renewal processes. The main applications of models in this part are usually for small-scale systems with discrete-state space, called microscopic models. The topic of fitting the phase-type distribution and Markov arrival process to empirical data is also discussed.

The next two chapters (Chapters 6 and 7) focus on the random walk and Brownian motion, the two fundamental stochastic processes of the discrete-time and continuous-time versions, respectively. These two processes are not only important in themselves, but also are the building blocks for more complex stochastic models. Since Brownian motions are the limits for the corresponding random walks, the relation between these two types of models revealed in Chapter 6 lays the foundation of developing stochastic process limits for approximating the stochastic processes. In Chapter 7, the reflected Brownian motion is utilized to approximate the $G/G/1$ queue, which is an important waiting line model. Through this simple example, we establish the theoretical foundation of using the stochastic process limits. A simple queueing network and a production-inventory system are presented to demonstrate the use of the reflected Brownian motion as a building block for a larger stochastic system.

Chapter 8 focuses on large queueing systems, called macroscopic models. Under heavy traffic conditions, fluid and Brownian motion networks are utilized to approximate the general queueing networks via oblique reflection mapping. A decomposition approach is presented for the cases where the first moment performance measures are sufficient for system evaluations. Other types of large queueing systems such as service systems with many servers, time-varying parameters, and many interacting parts are also discussed in this chapter.

In the last three chapters (Chapters 9–11), we discuss optimization and learning in stochastic models. Chapter 9 is focused on the static optimization problems. In this class of problems, the optimal policy parameters do not depend on time. The most popular approach, which is based on regenerative cycles, is demonstrated by solving a multi-state system's maintenance problem and analyzing a queueing system with a threshold service policy. Static optimization problems often occur when a system reaches steady state as a result of an economic equilibrium. This class of problems is presented by analyzing queueing systems with customer choice. Moreover, a service order policy for a multi-class queueing system is presented to show a situation where the optimal policy does not depend on every state of the system. Finally, more general performance measures are discussed for optimizing the operations of stochastic systems. In Chapter 10, we consider dynamic optimization problems for stochastic models. The objective of this class of problems is to find the optimal policy that depends on system state and/or time under a certain decision criterion. The solution is obtained by formulating the system dynamics as a Markov decision process (MDP) or a semi-Markov decision process (SMDP). We address two issues in MDPs and SMDPs: characterization of the optimal policy structures and development of computational algorithms. Chapter 11 is devoted to learning in stochastic models. Multi-arm bandit problems and Monte Carlo-based MDP models are utilized to demonstrate the learning processes in stochastic models. The hidden Markov model (HMM) and the partially observable Markov decision process (POMDP) are introduced to solve a class of reinforcement learning problems. Challenging issues such as the high computational complexity in stochastic models with learning processes are demonstrated through numerical examples.

Appendices: Since the foundation of stochastic models is the theory of probability and stochastics, we provide a series of appendices, labeled as Chapters A–J, to accommodate readers with different mathematical backgrounds who wish to obtain quick refreshers or reviews on certain topics. We start with a review of basic probability theory in Chapter A. The definitions, concepts, propositions, and theorems summarized in this chapter are relevant in developing the basic stochastic models. Chapter B is focused on the conditional expectation and martingales, which are useful in analyzing some stochastic processes such as random walks and Brownian motions. Chapter C provides a review of some useful bounds, inequalities, and limit laws, which are foundation topics for several chapters of the book. For example, the law of the iterated logarithm (LIL) is presented as it gives readers a flavor of the theoretical derivation of major functional law of the iterated logarithms (FLILs). These FLILs are useful in bounding the approximation errors generated by the stochastic process limits. Chapter D presents a brief review of non-linear programming techniques and examples of typical applications in probability and statistical models. A technique called change of probability measure is reviewed in Chapter E. Convergences of random variables and stochastic processes are presented in Chapter F and Chapter G, respectively. A brief review of stochastic calculus is provided in Chapter H. Chapter I discusses the stochastic orders, which are used to compare stochastic processes. Finally, in Chapter J, we give a review on matrix algebra, which is fundamental in developing the theory of multi-dimensional Markov chains. Internet Supplement (online chapters) to the book may be created to provide supplementary materials related to the subject of the book (e.g., emerging topics or errata) after this book has been published.

This book contains several features. First, in terms of presentation style, I try to maintain a balance between intuitive interpretation and mathematical rigor for major results in stochastic models. To achieve this balance, detailed derivations of the key results are provided for simplified cases in which more elementary mathematical treatments are sufficient and easy to understand. Most of these results also apply for more general cases where more advanced mathematical machinery is required in proofs. These more theoretical proofs are omitted, and interested readers are referred to advanced or specialized books or research papers. Second, in terms of methodologies, I try to cover most major analytical methods used in stochastic models. For the small-scale stochastic systems, the classical Markov chains and their extensions, structured Markov chains, are formulated and treated with the matrix-analytic method (MAM). For large-scale systems, to overcome the "the curse of dimensionality", the stochastic process limits (SPLs) are established to study the system dynamics or behaviors. Fluid and diffusion limits are utilized to approximate the system performance measures. The connection between these two approaches is illustrated via the relation between the random walk process (RWP) and Brownian motion (BM). Third, to show the wide applicability of stochastic models, I have addressed three aspects of stochastic modeling in this book: the performance evaluation, static and dynamic optimization, and learning. Each aspect has major applications in certain areas or disciplines. For example, performance evaluation is needed for many stochastic service systems such as computer and telecommunication networks; static optimization is a major issue in maintenance engineering, manufacturing, and economic systems; dynamic optimization is appropriate for some production and inventory control systems; and finally learning in stochastic models is relevant to the areas of artificial intelligence and machine learning.

There are many possible uses of this book for different groups of students or different courses. For example, for an introductory course in stochastic processes for undergraduate students in business or engineering, the following chapters could be used: Chapter A (part of it), Chapter 2, Chapter 3, and Chapter 5 (part of it). For a course on computational probability, one could use Chapter 2, Chapter 3, Chapter 4, and Chapter J. A selection of Chapter 5,

Chapter 6, Chapter 7, and Chapter 8 supplemented by Chapters B to G can be used for an introduction course on stochastic process limits. Chapter 3, Chapter 4, Chapter 5, Chapter 6, and Chapter 7 can be selected to use for a course on queueing systems. Finally, some chapters such as Chapter 9, Chapter 10, and Chapter 11 can be selected as supplementary materials for courses on stochastic optimization or machine learning.

Acknowledgments

This book would not have been possible without the support of many people. I would like to begin by thanking Dr. Raymond G. Vickson, my PhD supervisor. As my teacher and mentor, he has taught me more than I could ever give him credit for here. He has shown me, by his example, what a good scientist (and person) should be. I am especially indebted to Dr. Craig Tyran, Chairman of the Department of Decision Sciences at Western Washington University, who has been supportive of my career goals and who worked actively to provide me with the protected academic time to pursue those goals.

I am grateful to my research collaborators – Professors Ernie Love, Shey-Huei Sheu, Youhua Chen, Minghui Xu, Li Xia, and Hui Shao as some of our collaborative research has appeared in example problems of this book. In addition, I have been fortunate to have had the opportunity to work with Professors Naishuo Tian, Hsing Paul Luh, Yu-Hung Chien, Jauchuan Ke, Zhenghua Long, Hailun Zhang, Jiheng Zhang, Jinting Wang, Daniel Ding, Mahesh Nagarajan, Pengfei Guo, and Robin Lindsey on a variety of research projects about stochastic modeling and applications. The use of different methodological approaches in these studies motivated me to develop this book. During the process of writing this book, I benefited from my research projects sponsored by Simon Fraser University. I would like to express my gratitude to Professor Sudheer Gupta for his support to my research activities at Beedie School of Business. In addition, my long-term collaboration with Sauder School of Business at the University of British Columbia has enriched the presentation of some topics in the book. I am indebted to Professors Mahesh Nagarajan and Tim Huh for their support to my research collaboration at Sauder. I want to thank Dr. Qi-Ming He for reading all chapters of the book and providing constructive comments. I would also like to thank Dr. Natarajan Gautam, the editor of the Operations Research Series at CRC Press, for encouragement, guidance, and advice for developing this book.

Nobody has been more important to me in the pursuit of writing this book than the members of my family. I would like to thank my parents, Maoxi and Yuwen, whose love and patience are with me in whatever I pursue. Most importantly, I wish to thank my loving and supportive wife, Siping, who has provided unending inspiration and moral/emotional support, and my two wonderful daughters, Nancy and Lucy, who have provided encouragement and compassion during the entire book writing process.

1 Introduction

This book treats both microscopic and macroscopic models for stochastic systems. A stochastic model is a mathematical description of the dynamic behavior of a system involving uncertain or random factors. There are many examples, from a simple daily demand model for a product, which can be a single normally distributed random variable with a mean of a linear decreasing function of price $\mu(p) = a - bp$ and a constant variance σ^2, to a complex queueing network model for a large theme park with many random variables. Microscopic models are formulated for small-scale systems such as a single server system or an inventory system with a single type of discrete items. The system state can be described by fewer discrete random variables such as the number of customers in the system and status of the server. On the other hand, macroscopic models are formulated for large-scale systems such as a call center with a huge number of servers or a large network with many service stations connected in a complex way. The system state can be described by many discrete and/or continuous random variables. We study stochastic processes in these models, which represent how the system state changes with time. A standard definition for a stochastic process is a collection of random variables with a time index set. In this book, we focus on stochastic processes with a conditional independence property called Markov property. Let $X(t_k)$ denote a stochastic process of our interest. Then $X(t_k)$ represents the system state at a time point t_k, where $k = 0, 1, 2, \ldots$. The actual state at the kth point in time is denoted by x_k.

Markov Property Definition: *The future of the process does not depend on its past, only on its present (the kth point in time). In terms of conditional probability distribution function, that is*

$$P\{X(t_{k+1}) \leq x_{k+1} | X(t_k) = x_k, X(t_{k-1}) = x_{k-1}, \ldots, X(t_0) = x_0\}$$
$$= P(X(t_{k+1}) \leq x_{k+1} | X(t_k) = x_k).$$

This is also called a conditional independent property, meaning that given the present state, the future is independent from the past. Note that the kth time point is deterministically chosen. If it is a random point with a special property (called stopping time), this conditional independence can be called "strong Markov property", which will be discussed later in the book. A stochastic process with the Markov property is called a Markov process. Such a process can be analyzed in an elegant way. In theory, almost all stochastic processes in practice can be modeled as a Markov process at the expense of the size of the state space. However, the "curse of dimensionality" prevents us from

DOI: 10.1201/9781003150060-1

TABLE 1.1

Classification of Stochastic Processes

Time		Discrete	Continuous
State	Discrete	Discrete-time Markov chain (DTMC)	Continuous-time Markov chain (CTMC)
	Continuous	Discrete-time Markov process (DTMP)	Continuous-time Markov process (CTMP)

analyzing a Markov process with a super large state space in practice. We start with the classification of Markov processes.

Exercise 1.1 *In the Markov property definition, why the condition is in form of a set of equalities, while the future event is in form of inequality?*

1.1   STOCHASTIC PROCESS CLASSIFICATION

A stochastic process with the Markov property can be classified according to the state space and the time index as shown in Table 1.1. In this book, the process with discrete state space is called a *Markov chain*, and the process with continuous state space is called a *Markov process*.

We give an example for each class. (i) Discrete-time Markov chain (DTMC): Assume that weather condition for each day can be classified into three categories in summer. They are "sunny", "cloudy", and "rainy". We also assume that the weather condition for tomorrow only depends on the weather condition today. The system state is defined as the weather condition on each day. Since the state change happens each day (a fixed period), a three-state DTMC for the weather condition can be developed. (ii) Continuous-time Markov chain (CTMC): Consider a single server queue (a barber shop) where customer arrival follows a Poisson process and the service time is independently and exponentially distributed with a constant service rate. If we use the number of customers in the system as the state variable, then the queue size can be modeled as a CTMC (to be discussed in details later). This is because that the state change can happen any time. (iii) Discrete-time Markov process (DTMP): Observe the stock level in an inventory system at one time point per day (either at the beginning or at the ending of the business day) and assume that the daily demand is identically and independently distributed (i.i.d.) and continuous random variable. Since the state variable now is continuous, the stock process can be modeled as a DTMP. (iv) Continuous-time Markov process (CTMP): Observe the stock price continuously (taking the limit of process in a DTMC or DTMP appropriately). The process can be shown to be a Brownian motion, which is an example of CTMP.

It is worth pointing out that different types of models require different methodologies. For the microscopic models, we mainly perform exact analysis. Usually, the analysis in this category can be done by using transform techniques or developing computationally efficient algorithms. For the macroscopic models, due to the "curse of dimensionality", we often use the approximation approach to analyzing the system behaviors. A natural and effective method is to develop the approximations to the stochastic processes based on the stochastic process limits. These process limits can be either deterministic fluid limits or stochastic diffusion limits.

1.2   ORGANIZATION OF THE BOOK

This book has two parts. Part I contains ten chapters (Chapters 2-11) and Part II contains ten appendices (Chapters A–J). Part I is the focus of the book and Part II provides supplementary and technical materials, which are foundation to stochastic modeling. In Part I, we start with the simplest stochastic models – the discrete-time Markov chains in Chapter 2. In this class of models, the time index is deterministic. In other words, the state-change event can only happen at beginning or ending of each period. Thus, the inter-event time is deterministic and the uncertain factor is the random transitions among the discrete states. Due to its simplicity, we tend to provide proofs for the fundamental theorems and propositions to shed the light on the basic ideas in developing the main results in more general settings (e.g., continuous-time and/or continuous state space models), which require more advanced mathematical machinery. In Chapter 3, we consider the continuous-time Markov chains, which allow the state-change to happen at any time. Thus, the inter-event time is a continuous random variable. Compared with the discrete-time models, this class of models make Markov chains applicable in more practical situations. A restriction in this class of models is that all random variables must be exponentially distributed to preserve the Markov property. The advantage of the simplicity is kept and the analysis remains elementary for the CTMC. Chapters 2 and 3 provide the theory of elementary stochastic modeling. The next five chapters extend the basic theory to more advanced stochastic modeling theory in three directions.

To generalize these basic CTMCs to more general stochastic models without losing the Markov property, in Chapter 4, we introduce the phase-type (PH) distributed random variables and Markov arrival processes (MAP) and discuss the structured Markov chains. A powerful computational method called matrix analytic method is presented. Since the PH distribution, an extension of exponential distribution can be used to approximate any general distribution and MAP, an extension of Poisson process can be used to approximate more general arrival processes with non-independent inter-arrival times, we can still build the structured Markov chains with multi-dimensional state spaces to study more general stochastic systems. This direction leads to the computational approach to analyzing stochastic models. We provide the literature in this area for interested readers pursuing this direction.

In Chapter 5, we focus on the renewal processes and its generalization, Markov renewal processes, which form a class of important stochastic processes. Based on the Markov renewal process, we can define the semi-Markov processes and semi-regenerative processes, which can be utilized to study the stochastic systems with non-exponential random variables. For example, by identifying the semi-regenerative processes, the embedded Markov chains can be formulated for queueing systems with general service times or general inter-arrival times. The main tools utilized in this direction include a variety of transforms and generating functions such as moment generating functions or probability generating functions for random variables. Interested readers in this direction are referred to some specialized books and research papers for further exploration.

Chapters 6–8 are devoted to using the stochastic process limits to analyze the stochastic systems. We start with extending the random walk, a fundamental Markov chain with discrete state space, to the Brownian motion, a fundamental Markov process with continuous state space in Chapter 6. Since the Brownian motion is the building block of this approach, its basic theory is presented in this chapter. The relation between the Brownian motion and some practical stochastic models is demonstrated and the basic approach to using the stochastic process limit is established in Chapter 7. Chapter 8 is focused on analyzing large-scale stochastic models via stochastic process limits and reveals the advantages of this approximation approach. Some references on stochastic process limits are provided for interested readers in this direction.

The last three chapters focus on the optimization and learning in stochastic models. Specifically, Chapter 9 treats the static optimization problems. These problems are based on the stationary performance measures when systems reach the steady state. In these problems, the structures of the optimal policies are assumed to be given. Such an assumption can be justified based on the reasonableness and easy implementation of the policy considered. For some special inventory and queueing models, the structure of the optimal policy can be characterized theoretically. In Chapter 10, we discuss the dynamic optimization problems in stochastic models. The policies considered are state-dependent. Two issues are addressed. The first issue is to characterize the structure of the optimal policy and the second issue is to compute the parameters of the optimal policy. Markov decision process (MDP), as the model to solve these dynamic optimization problems, is the focus of Chapter 10. Chapter 11 discusses the situations where the model parameters may not be known, the states may not be observable, and even the model structure may not be available. These situations can be treated by learning processes based on data simulated or collected. The basic learning process in a stochastic situation is introduced by considering a multi-arm bandits problem. Then the MDPs with unknown parameters are discussed. Finally, hidden Markov models and partially observable MDP are presented for the situations where the system states are not observable.

The appendices in Part II provide readers with some quick refreshers and references on certain topics related to the chapters in Part I. Specifically, some topics are pre-requisites or foundations for formulating the stochastic models or developing the key results. Other topics present some similar results in a simplified setting to the key results in a more general setting with the purpose of providing intuitions about the theoretical results to readers.

Part I

Fundamentals of Stochastic Models

2 Discrete-Time Markov Chains

In this and next chapters, we focus on the basic Markov chains. The simplest stochastic process is the discrete-time Markov chain (DTMC) where the state space E is discrete and the time index is a non-negative integer. For a DTMC, the state at the kth time point (also called a period or step) is denoted by X_k and the Markov property in terms of conditional probability can be written as

$$P\{X_{k+1} = x_{k+1} | X_k = x_k, ..., X_0 = x_0\} = P\{X_{k+1} | X_k = x_k\}, \tag{2.1}$$

where x_k is the actual state of the DTMC at time k. We can define the DTMC as follows:

Definition 2.1 *A stochastic process $\{X_k, k = 0, 1, 2, ...\}$ with a discrete state space that satisfies the Markov property (2.1) is said to be a DTMC.*

We can define the one-step transition probability at kth point (current time $= k$) as

$$p_{ij}(k) = P\{X_{k+1} = j | X_k = i\}. \tag{2.2}$$

Since the transition probability is a conditional probability and forms a complete discrete distribution, for all i at any time k, we have

$$\sum_{j \in \Gamma(i)} p_{ij}(k) = 1, \tag{2.3}$$

where $\Gamma(i)$ is the set of states that the DTMC can reach from state i. Similarly, the n-step transition probability from period k to $k+n$ can be defined as

$$p_{ij}(k, k+n) = P\{X_{k+n} = j | X_k = i\}. \tag{2.4}$$

Clearly, the state variable of a DTMC X_k is a random variable. In the college statistics course, a random variable's behavior is characterized by its probability distribution. Similarly, the behavior of a DTMC is described by the probability distribution of its state variable at a particular time k or when time goes to infinity. The starting point of obtaining such kind of distribution, simply called the distribution of the DTMC, is the transition probability defined above as it contains all information about how the DTMC evolves with time.

DOI: 10.1201/9781003150060-2

9

2.1 DYNAMICS OF PROBABILITY MEASURES

The basis of deriving the distribution of DTMC is the two fundamental probability formulas in elementary probability theory. They are the total probability
and conditional probability formulas (see Appendix A. In this book, we also
call Appendix A Chapter A or Appendix Chapter A). First, we establish the
Chapman-Kolmogorov (C-K) equation as a fundamental relation for the multi-
step transition probabilities in a DTMC with a finite-state space. Suppose that
at a period m ($k < m < k+n$), the DTMC $\{X_m\}$ can take one of $R < \infty$ states,
which are numbered from 1 to R. Using the total probability formula and the
Markovian property, we can write the n-step transition probability from state i
at time k to state j at time $k+n$ as

$$p_{ij}(k,k+n) = \sum_{r=1}^{R} P\{X_{k+n} = j | X_m = r, X_k = i\} P\{X_m = r | X_k = i\}$$

$$= \sum_{r=1}^{R} P\{X_{k+n} = j | X_m = r\} P\{X_m = r | X_k = i\} \tag{2.5}$$

$$= \sum_{r=1}^{R} p_{ir}(k,m) p_{rj}(m,k+n),$$

which is called the C-K equation for the DTMC. The C-K equation reveals how
the probability measure for the DTMC evolves with time and can be expressed
in the matrix form. Define the n-step transition probability matrix $\mathbf{P}(k,k+n) =$
$[p_{ij}(k,k+n)]_{R \times R}$ (matrices in this book are in boldface). The C-K equation can
be written as

$$\mathbf{P}(k,k+n) = \mathbf{P}(k,m)\mathbf{P}(m,k+n). \tag{2.6}$$

To express the C-K equation in terms of one-step transition probability matrix,
we can choose $m = k+n-1$ (one period before ending period). Then we have
the so-called forward C-K equation as

$$\mathbf{P}(k,k+n) = \mathbf{P}(k,k+n-1)\mathbf{P}(k+n-1,k+n) = \mathbf{P}(k,k+n-1)\mathbf{P}(k+n-1), \tag{2.7}$$

where only the starting time index is used in the notation for the one-step transition probability matrix. Alternatively, we can choose $m = k+1$ and obtain
the backward C-K equation as

$$\mathbf{P}(k,k+n) = \mathbf{P}(k,k+1)\mathbf{P}(k+1,k+n) = \mathbf{P}(k)\mathbf{P}(k+1,k+n). \tag{2.8}$$

Note that we have kept the time index k for all transition probabilities so
far. This is because the transition probability may change with absolute time
point. A DTMC with time-dependent transition probability is said to be time-
nonhomogeneous. If the transition probability is independent of time index,

the DTMC is said to be time-homogeneous and its analysis can be significantly simplified. In this book, we mainly discuss the time-homogeneous DTMC and CTMC because in theory a time-nonhomogeneous DTMC or CTMC can be homogenized at the expense of increasing the state space. A straightforward way is to include the time index as a state variable which progresses from k to $k+1$ deterministically (with probability one). Then the multi-dimensional DTMC (with one dimension as the time index) can be treated as being time-homogeneous.

Exercise 2.1 *The weather condition for a region changes from day to day and can be modeled as a DTMC if the weather condition tomorrow only depends on today's condition. However, such a Markovian property also depends on the season of the year. Thus, the DTMC should be time-nonhomogeneous. How can such a DTMC be converted into a time-homogeneous DTMC?*

For a time-homogeneous DTMC, the one-step transition probabilities are independent of time k. That is $\mathbf{P}(k) = \mathbf{P}$ or $[p_{ij}] = [P\{X_{k+1} = j | X_k = i\}] = [P\{X_1 = j | X_0 = i\}]$. This property makes it easy to compute the multi-step transition probability. The n-step transition probability from state i to state j can be denoted by p_{ij}^n. Thus, the $(n+m)$-step transition probability can be written as

$$
\begin{aligned}
p_{ij}^{n+m} &= P\{X_{n+m} = j | X_0 = i\} \\
&= \sum_{k \in \Gamma(i)} P\{X_{n+m} = j | X_n = k, X_0 = i\} P\{X_n = k | X_0 = i\} \\
&= \sum_k p_{kj}^m p_{ik}^n.
\end{aligned}
\tag{2.9}
$$

Let $\mathbf{P}^{(n)}$ be the matrix of the n-step transition probabilities. Then we have

$$
\mathbf{P}^{(n+m)} = \mathbf{P}^{(n)} \mathbf{P}^{(m)}.
\tag{2.10}
$$

By induction, we have $\mathbf{P}^{(n)} = \mathbf{P}^n$. Therefore, with the one-step transition probability matrix, we could compute the probability of a future state from the current state.

Example 2.1 *There are three product brands in a consumer market. Consumers switch between brands from week to week. Assume that the consumer's brand-switching behavior has reached the steady state. The state of the system X is the brand used in the week. That is $X = \{1, 2, 3\}$. The transition probability matrix is estimated as*

$$
\mathbf{P} = \begin{bmatrix} 0.90 & 0.07 & 0.03 \\ 0.02 & 0.82 & 0.16 \\ 0.20 & 0.12 & 0.68 \end{bmatrix}.
\tag{2.11}
$$

For a current brand 1 user, what is the probability that after two weeks, he or she becomes a brand 3 user?

Note: In this book, there are many examples. Some examples are presented with formal solutions followed. Other examples are given for general discussions. The former type is clearly labeled with the "**Solution**" and the example number, and the latter is without such a label.

Solution 2.1 *Raising the transition probability matrix to power of 2, we obtain*

$$
\mathbf{P} = \begin{bmatrix} 0.90 & 0.07 & 0.03 \\ 0.02 & 0.82 & 0.16 \\ 0.20 & 0.12 & 0.68 \end{bmatrix}^2 = \begin{bmatrix} 0.8174 & 0.1240 & 0.0586 \\ 0.0664 & 0.6930 & 0.2406 \\ 0.3184 & 0.1940 & 0.4876 \end{bmatrix}. \quad (2.12)
$$

The probability we want is the entry at row 1 and column 3 in the matrix of (2.12), which is 5.86%.

Exercise 2.2 *Let L_n be the amount of water in a regional reservoir at the beginning of the nth month and R_n be the amount of new water (rain fall) that is added to the reservoir during the nth month. Each month, exactly 5 units of water is removed at the end of each month by the region (or whatever part of 5 unit amount is available). R_n can be treated as a discrete i.i.d. random variable, denoted by R, and with a distribution of $P(R = 5) = 0.50, P(R = 4) = P(R = 6) = 0.20$ and $P(R = 3) = P(R = 7) = 0.05$. (a) Show that L_n process is a DTMC and write the one-step transition probability matrix. (b) Suppose that the water amount of the current month is 5 units (i.e., $L_n = 5$). Then what is the probability that there will be a drought after 2 months? (i.e., what is $P(L_{n+2} = 0)$?)*

Another quantity of interest for a DTMC is the probability distribution (unconditional) at a time k. Define $\pi_i(k) = P\{X_k = i\}$. For all possible states, we define the probability vector

$$
\pi(k) = [\pi_0(k), \pi_1(k)., ..., \pi_R(k)]. \quad (2.13)
$$

Using the total probability formula, we have the recursive relation

$$
\pi_i(k) = \sum_j P\{X_k = i | X_{k-1} = j\} P\{X_{k-1} = j\} = \sum_j p_{ji}(k-1)\pi_j(k-1), \quad (2.14)
$$

which can be written in matrix form as $\pi(k) = \pi(k-1)\mathbf{P}(k-1)$ or $\pi(k) = \pi(k-1)\mathbf{P}$ for the time-homogeneous DTMC. Although we introduce the multi-step transition probabilities and unconditional distribution for DTMCs with an finite state space, the definitions and computational formulas can be extended to DTMCs with an infinite state space under certain conditions (to be discussed).

Exercise 2.3 *In Exercise 2.2, if at the beginning of January of a particular year, the distribution of L_0 is $P(L_0 = 3) = P(L_0 = 4) = P(L_0 = 5) = 0.10$ and $P(L_0 = 6) = P(L_0 = 7) = 0.35$. (a) Compute the probability distribution of the water amount in the reservoir at the beginning of April of the year (i.e., the distribution of L_4)? (b) What are the weaknesses of this DTMC for modeling the dynamic process of the water amount in the reservoir.*

2.2 FORMULATION OF DTMC

There are several steps in formulating a DTMC. We formalize this process as follows:

1. Define the state space
2. Identify the events that lead to the state transitions
3. Make a state transition diagram (optional)
4. Determine the transition probability matrix

We demonstrate these steps by developing a DTMC for a simple computer system.

Example 2.2 *Consider a two processor computer system where time is divided into time slots. The system operates as follows: At most one job can arrive during any time slot and this can happen with probability α (arrival process). Jobs are served by whichever processor is available. If both are available, then the job is given to processor 1 (priority rule). If both processors are busy, then the job will wait to see if any processor completes a job during the current time slot. Whenever a service is completed during the current time slot, the job is accepted; otherwise, it is lost (i.e., one-time slot patience). When a processor is busy, it can complete the job with probability β during any one time slot (service process).*

Solution 2.2 *The system can be considered as a discrete-time two-server queue without waiting room. Although it is simple, it contains the main characteristics of a queueing system. They are (i) an arrival process; (ii) a service process; and (iii) customer balking/reneging. We first define the state space. Let the number of jobs in the system at the end of the time slot be the state variable. Then the state space is $X = \{0, 1, 2\}$. The events for state change in a time slot will be $\{a$ (job arrival), d (job departure), $-$ (nothing happening)$\}$. The feasible event sets, $\Gamma(X)$, for the three states are $\Gamma(0) = \{a\}$, $\Gamma(1) = \{a, d\}$, and $\Gamma(2) = \{a, d\}$. Based on the system description, we can make a state transition diagram in Figure 2.1 (Note: we use shaded circles to represent states for DTMCs and non-shaded circles to represent states in state-transition diagrams for continuous-time Markov chains in the next chapter). The transition probability matrix is given by*

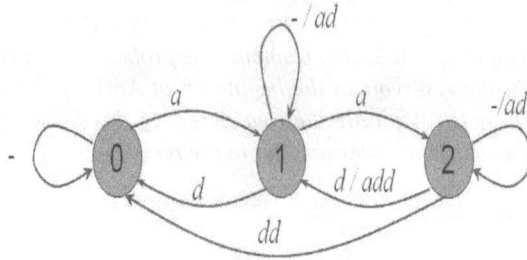

Figure 2.1   A three-state DTMC for a computer system with two processors

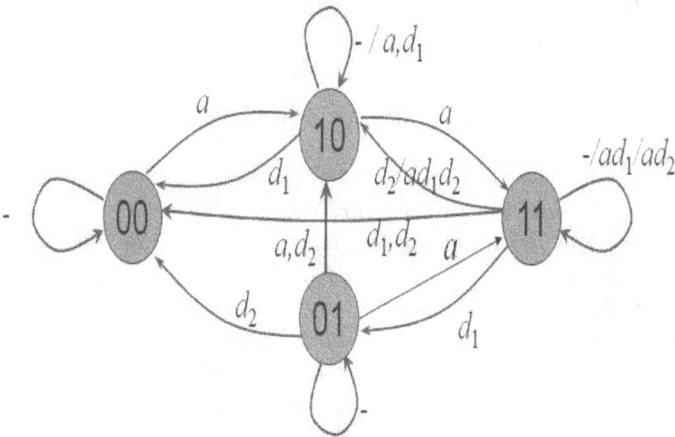

Figure 2.2   A four-state DTMC for a computer system with two processors

$$\mathbf{P} = \begin{bmatrix} 1-\alpha & \alpha & 0 \\ \beta(1-\alpha) & (1-\beta)(1-\alpha)+\alpha\beta & \alpha(1-\beta) \\ \beta^2(1-\alpha) & \beta^2\alpha+2\beta(1-\beta)(1-\alpha) & (1-\beta)^2+2\alpha\beta(1-\beta) \end{bmatrix}. \tag{2.15}$$

Note that there may be multiple ways of formulating a DTMC model for the same system. As an example, we show an alternative solution by using two-dimensional state space for this two-processor system. Let the two-server status, denoted by (X_1, X_2), be the state of the system with $X_i = \{0,1\}$, where $0(1)$ means idle (busy). Thus, there will be four possible states for the system. The event set is $\{a, d_i, -\}$, where d_i represents a customer departure from server i. Then the feasible event sets are $\Gamma((0,0)) = \{a\}$, $\Gamma((0,1)) = \{a, d_2\}$, $\Gamma((1,0)) = \{a, d_1\}$, and $\Gamma((1,1)) = \{a, d_1, d_2\}$. The transition diagram for this four state DTMC is shown in Figure 2.2. The transition probability matrix is

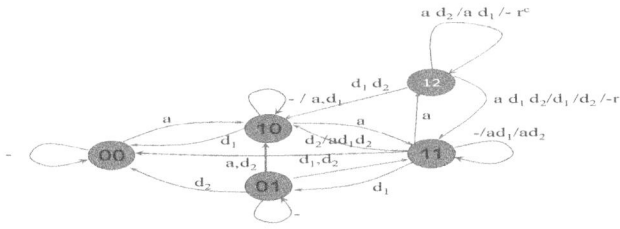

Figure 2.3 A five-state DTMC for a computer system with two processors.

given by

$$
\mathbf{P} = \begin{bmatrix}
1-\alpha & \alpha & 0 & 0 \\
\beta(1-\alpha) & (1-\alpha)(1-\beta)+\alpha\beta & 0 & \alpha(1-\beta) \\
\beta(1-\alpha) & \alpha\beta & (1-\alpha)(1-\beta) & \alpha(1-\beta) \\
\beta^2(1-\alpha) & (1-\beta)\beta(1-\alpha)+\beta^2\alpha & (1-\beta)\beta(1-\alpha) & (1-\beta)^2+2\beta(1-\beta)\alpha
\end{bmatrix},
\tag{2.16}
$$

Since $\mathbf{P}$ *is stochastic matrix, the row sum is 1. The advantage of using two-dimensional state model is that more information about the system can be obtained although the state space is a little larger. For example, if we want to know the proportion of time each server is idle. Then it can be obtained in the second model but not in the first model.*

Exercise 2.4 *In the example above, suppose that* $\alpha = 0.5$ *and* $\beta = 0.7$ *and the system starts with an empty state at the beginning. Find the probability that in period 2, there is only one job in the system. Examining the probability distribution of the number of jobs in the system as time increases. What is going to happen to this distribution?*

Example 2.3 *For the two processor system considered in the previous example, we assume that there is a waiting spot (a buffer) for holding a job. For each time slot, the job leaves without being processed with probability of* γ *at the end of the time slot. Formulate a DTMC for this system.*

Solution 2.3 *This is an example that a queue is formed in a service system. An extra state is needed and labeled as (1,2), which represents that two processors are busy and one job is waiting in the buffer. Again, the waiting job is patient enough during the time slot and only reneges with probability* γ *at the end of the time slot. Based on the state transition diagram shown in Figure 2.3, the transition probability matrix is given by*

$$
\mathbf{P} = \begin{bmatrix}
1-\alpha & \alpha & 0 & 0 & 0 \\
\beta(1-\alpha) & (1-\alpha)(1-\beta)+\alpha\beta & 0 & \alpha(1-\beta) & 0 \\
\beta(1-\alpha) & \alpha\beta & (1-\alpha)(1-\beta) & \alpha(1-\beta) & 0 \\
\beta^2(1-\alpha) & (1-\beta)\beta(1-\alpha)+\beta^2\alpha & (1-\beta)\beta(1-\alpha) & (1-\beta)^2(1-\alpha)+2\beta(1-\beta)\alpha & (1-\beta)^2\alpha \\
0 & \beta^2(1-\alpha) & 0 & (1-\beta)^2\gamma+2\beta(1-\beta)(1-\alpha)+\beta^2\alpha & (1-\beta)^2(1-\gamma)+2\beta(1-\beta)\alpha)
\end{bmatrix}.
\tag{2.17}
$$

It is worth noting that the inter-arrival time, service time, and customer patient time in the example above are all geometrically distributed random variables.

The advantage of using geometric random variables is that we do not have to record the time elapsed. This property is also called "memoryless property", which makes the process to possess the Markovian property. A continuous-time counterpart to the geometric distribution is the exponential distribution, which is discussed later in this book.

Exercise 2.5 *In Example 2.2, it is assumed that the patience time is two-point distributed. That is the patient time is one-period with probability p and two-period with probability* $1 - p$. *Formulate a DTMC for the system and make some comments about the state space of DTMC if the patience time is multi-point distributed and/or the waiting buffer is getting bigger.*

2.3 PERFORMANCE ANALYSIS OF DTMC

If a DTMC system is in operation indefinitely (no ending point), then we are interested in the system behavior when time goes to infinity. Thus, an important question is whether or not the steady-state exists for the DTMC. For finite state-space DTMCs, it is relatively easier to answer this question. However, for a DTMC with infinite state space, we need to develop the conditions for the existence of the steady state. Only under these conditions, the stationary analysis makes sense. To develop these conditions, we start with the classification of states for a DTMC.

2.3.1 CLASSIFICATION OF STATES

In a DTMC, a state j is said to be *reachable* if the probability to go from i to j in $n > 0$ steps is greater than zero (state j is reachable from state i if in the state transition diagram there is a path from i to j). Two states i and j that are reachable to each other are said to *communicate*, and we write $i \leftrightarrow j$. A subset S of the state space X is *closed* if $p_{ij} = 0$ for every $i \in S$ and $j \notin S$. A state i is said to be *absorbing* if it is a single element closed set. A closed set S of states is *irreducible* if any state $j \in S$ is reachable from every state $i \in S$. A Markov chain is said to be irreducible if the state space X is irreducible .

Example 2.4 *A five-state DTMC has the following transition probability matrix*

$$\mathbf{P} = \begin{bmatrix} p_{00} & p_{01} & 0 & 0 & 0 \\ p_{10} & 0 & p_{12} & 0 & p_{14} \\ 0 & p_{21} & p_{22} & p_{23} & 0 \\ 0 & 0 & p_{32} & p_{33} & 0 \\ p_{40} & 0 & 0 & 0 & 0 \end{bmatrix}. \tag{2.18}$$

The transition diagram in Figure 2.4 clearly shows that the DTMC is irreducible as any state is reachable from every other state by a direct or an

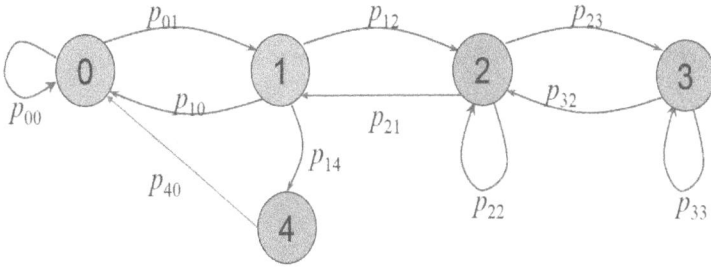

Figure 2.4 An irreducible DTMC.

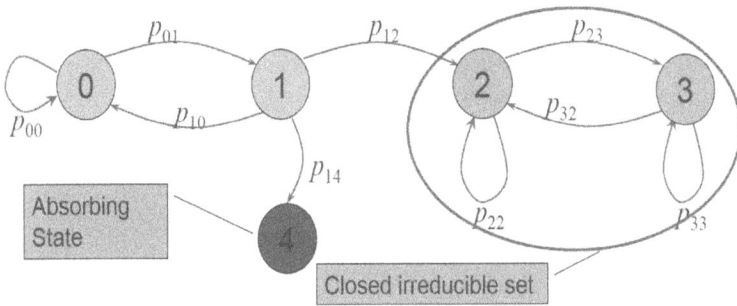

Figure 2.5 A reducible DTMC.

indirect link. However, if $p_{40} = 0$ and $p_{21} = 0$, then the DTMC becomes reducible as shown in Figure 2.5.

Exercise 2.6 *Each of the following probability transition matrices is for a DTMC. Please identify the communication classes for each DTMC.*

(a) $\mathbf{P} = \begin{pmatrix} \frac{1}{4} & \frac{1}{4} & \frac{1}{4} & \frac{1}{4} \\ \frac{1}{2} & \frac{1}{2} & 0 & 0 \\ 0 & 0 & 0 & 1 \\ 0 & \frac{1}{2} & 0 & \frac{1}{2} \end{pmatrix}$, (b) $\mathbf{P} = \begin{pmatrix} 0 & 0 & \frac{1}{2} & \frac{1}{2} \\ 0 & 0 & \frac{1}{2} & \frac{1}{2} \\ 0 & 0 & \frac{1}{2} & \frac{1}{2} \\ 0 & 0 & \frac{1}{2} & \frac{1}{2} \end{pmatrix}$,

(c) $\mathbf{P} = \begin{pmatrix} \frac{1}{3} & 0 & \frac{1}{3} & 0 & \frac{1}{3} \\ 0 & \frac{1}{2} & 0 & \frac{1}{2} & 0 \\ \frac{1}{5} & 0 & \frac{1}{5} & 0 & \frac{3}{5} \\ 0 & \frac{1}{2} & 0 & \frac{1}{2} & 0 \\ \frac{1}{4} & 0 & \frac{1}{4} & 0 & \frac{1}{2} \end{pmatrix}$, (d) $\mathbf{P} = \begin{pmatrix} \frac{1}{5} & \frac{1}{5} & \frac{1}{5} & \frac{1}{5} & \frac{1}{5} \\ \frac{1}{3} & \frac{1}{3} & \frac{1}{3} & 0 & 0 \\ \frac{1}{2} & 0 & 0 & \frac{1}{2} & 0 \\ 0 & \frac{1}{2} & 0 & \frac{1}{2} & 0 \\ \frac{1}{3} & 0 & \frac{1}{3} & 0 & \frac{1}{3} \end{pmatrix}$.

To further classify the states, we define the following.

Definition 2.2 *Hitting time T_{ij}: Given the initial state i, the first time the DTMC reaches state j is called the hitting time to state j, that is $T_{ij} = \min\{k > 0 : X_k = j | X_0 = i\}$.*

We can establish some theorems and propositions about the properties for DTMCs below (Note: the proofs of some results if omitted are left as exercises).

Theorem 2.1 *For a DTMC with state space E and transition matrix $\mathbf{P}$, let $C = \{c_n : n \in I\} \subset E$ with $I \subset \mathbb{N}$ be a closed communication class and $\mathbf{P}' = [p'_{ij}]$ be the transition matrix for this class with $p'_{ij} := p_{c_i,c_j}$ for all $i, j \in I$. Then $\mathbf{P}'$ is stochastic (i.e., the row sum is equal to 1).*

Exercise 2.7 *Prove Theorem 2.1.*

A DTMC may have multiple closed communication classes C_n for $n = 1, 2, \ldots$ with the following transition matrix

$$
\mathbf{P} = \begin{pmatrix}
\mathbf{Q} & \mathbf{Q}_1 & \mathbf{Q}_2 & \mathbf{Q}_3 & \mathbf{Q}_4 & \cdots \\
\mathbf{0} & \mathbf{P}_1 & \mathbf{0} & \mathbf{0} & \mathbf{0} & \cdots \\
\mathbf{0} & \mathbf{0} & \mathbf{P}_2 & \mathbf{0} & \mathbf{0} & \cdots \\
\mathbf{0} & \mathbf{0} & \mathbf{0} & \mathbf{P}_3 & \mathbf{0} & \cdots \\
\vdots & \vdots & \ddots & \ddots & \ddots &
\end{pmatrix}
\tag{2.19}
$$

with $\mathbf{P}_n$ being stochastic matrices on the closed communication classes C_n. The first row contains the transition probabilities from open communication classes.

Exercise 2.8 *Suppose that a DTMC has 10 states with state space $E = \{1, \ldots, 10\}$. The transition probability matrix is given by*

$$
\begin{pmatrix}
1/3 & 0 & 2/3 & 0 & 0 & 0 & 0 & 0 & 0 & 0 \\
0 & 1/2 & 0 & 0 & 0 & 0 & 1/2 & 0 & 0 & 0 \\
1 & 0 & 0 & 0 & 0 & 0 & 0 & 0 & 0 & 0 \\
0 & 0 & 0 & 0 & 1 & 0 & 0 & 0 & 0 & 0 \\
0 & 0 & 0 & 1/4 & 1/4 & 0 & 0 & 0 & 1/2 & 0 \\
0 & 0 & 0 & 0 & 0 & 1 & 0 & 0 & 0 & 0 \\
0 & 0 & 0 & 0 & 0 & 0 & 1/2 & 0 & 1/2 & 0 \\
0 & 0 & 1/4 & 1/4 & 0 & 0 & 0 & 1/4 & 0 & 1/4 \\
0 & 1 & 0 & 0 & 0 & 0 & 0 & 0 & 0 & 0 \\
0 & 1/4 & 0 & 0 & 1/4 & 0 & 0 & 0 & 0 & 1/2
\end{pmatrix}.
$$

Reorder the states according to their communication classes and determine the resulting form of the transition matrix as in the form of (2.19).

Denote the distribution of $T_{i,j}$ by $F_k(i,j) := P(T_{i,j} = k)$ for all $i, j \in E$ and $k \in \mathbb{N}$.

Proposition 2.1 *The hitting time distribution can be computed recursively by*

$$F_k(i,j) = \begin{cases} p_{i,j}, & k = 1, \\ \sum_{h \neq j} p_{ih} F_{k-1}(h,j), & k \geq 2, \end{cases} \quad (2.20)$$

for all $i, j \in E$.

Proof. The case of $k = 1$ is obvious. For $k \geq 2$, conditioning on X_1 yields

$$\begin{aligned} F_k(i,j) &= P(X_1 \neq j, ..., X_{k-1} \neq j, X_k = j | X_0 = i) \\ &= \sum_{k \neq j} P(X_1 = h | X_0 = i) P(X_2 \neq j, ..., X_{k-1} \neq j, X_k = j | X_0 = i, X_1 = h) \\ &= \sum_{k \neq j} p_{ih} P(X_1 \neq j, ..., X_{k-2} \neq j, X_{k-1} = j | X_0 = h) \end{aligned}$$

due to Markov property. ∎

Exercise 2.9 *Show that the relation $\mathbf{P}^n(i,j) = \sum_{k=1}^m F_k(i,j) \mathbf{P}^{n-k}(j,j)$ for all $n \in \mathbb{N}$ and $i, j \in E$.*

Define

$$f_{ij} := P(T_{ij} < \infty) = \sum_{k=1}^{\infty} F_k(i,j) \quad \text{for } i, j \in E, \quad (2.21)$$

which is the probability of ever visiting j given the DTMC starts at state i. Conditioning on the state of the first transition, we obtain

$$f_{ij} = p_{i,j} + \sum_{h \neq j} p_{ih} f_{hj} \quad \text{for } i, j \in E. \quad (2.22)$$

Define N_j as the total number of visits to the state $j \in E$ for an infinite time horizon. We can determine the conditional distribution of N_j given an initial state in terms of f_{ij}.

Theorem 2.2 *For a DTMC with state space E, the distribution of $(N_j | X_0)$ is given by*

$$P(N_j = m | X_0 = j) = f_{jj}^{m-1}(1 - f_{jj})$$

and for $i \neq j$

$$P(N_j = m | X_0 = i) = \begin{cases} 1 - f_{ij}, & m = 0, \\ f_{ij} f_{jj}^{m-1}(1 - f_{jj}), & m \geq 1. \end{cases}$$

Proof. Define $\tau_j^{(1)} := T_{ij}$ and $\tau_j^{(k+1)} := \min\{n > \tau_j^{(k)} : X_n = j\}$ for all $k \in \mathbb{N}$, with the convention that $\min\emptyset = \infty$. Note that $\tau_j^{(k)} = \infty$ implies $\tau_j^{(l)} = \infty$ for all $l > k$. Then, we have a sequence of stopping times of $\{\tau_j^{(k)} : k \in \mathbb{N}\}$. (For the definition of stopping time, see Appendix Chapter Section B.4.1). The event $\{N_j = m | X_0 = i\}$ is the same as the intersection of the events $\{\tau_j^{(k)} < \infty\}$ for $k = 1, ..., M$ and $\{\tau_j^{(M+1)} = \infty\}$, with $M = m$ if $i \neq j$ and $M = m - 1$ if $i = j$. Note that this event is equivalent to the intersection of the events $\{\tau_j^{(k+1)} - \tau_j^{(k)} < \infty\}$ for $k = 1, ..., M - 1$ and $\{\tau_j^{(M+1)} - \tau_j^{(M)} = \infty\}$, with M as defined above and the convention $\tau_j^{(0)} := 0$. The event $\{\tau_j^{(k+1)} - \tau_j^{(k)} < \infty\}$ has probability of f_{ij} for $k = 0$ and probability of f_{jj} for $k > 0$. The event $\{\tau_j^{(M+1)} - \tau_j^{(M)} = \infty\}$ has probability of $1 - f_{ij}$ for $M = 0$ and $1 - f_{jj}$ for $M > 0$. It follows from the strong Markov property (i.e., the Markovian property holds at a stopping time instant) that these events are independent. Thus, multiplying these probabilities of sub-events yields the results of the theorem. ∎

Note that It is possible to prove the results by utilizing (2.21) without using the stopping time.

Corollary 2.1 *For all $j \in E$, the zero-one law*

$$P(N_j < \infty | X_0 = j) = \begin{cases} 1, & f_{jj} < 1, \\ 0, & f_{jj} = 1. \end{cases}$$

holds.

Exercise 2.10 *Prove Corollary 2.1.*

Note that the recurrence time to a state i is T_{ii}, the first time that the DTMC returns to state i. Then, $f_{ii} = \sum_{k=1}^{\infty} P\{T_{ii} = k\} = P\{T_{ii} < \infty\}$.

Definition 2.3 *A state is recurrent if $f_{ii} = 1$ and transient if $f_{ii} < 1$.*

Theorem 2.3 *If a Markov chain has finite state space, then at least one of the states is recurrent.*

Exercise 2.11 *Prove Theorem 2.3 using contradiction argument.*

Now we define the so-called potential matrix $\mathbf{R} = [r_{ij}]_{i,j \in E}$ of the DTMC with

$$r_{ij} := E(N_j | X_0 = i), \quad \text{for } i, j \in E.$$

Thus, r_{ij} represents the expected number of visits to state $j \in E$ given the starting state is $i \in E$ in an infinite time horizon. It can be computed by noting

$$(N_j | X_0 = i) = \sum_{k=0}^{\infty} \mathbf{1}_{\{X_k = j | X_0 = i\}}.$$

Taking the expectation of the above equation yields

$$r_{ij} = E(N_j|X_0 = i) = \sum_{k=0}^{\infty} E(1_{\{X_k=j|X_0=i\}}) = \sum_{k=0}^{\infty} P(X_k = j|X_0 = i) = \sum_{k=0}^{\infty} P^k(i,j),$$

where $P^k(i,j)$ denotes the k-step transition probability from state i to state j. Such a notation is consistent with the power operations of the transition probability matrix for computing multi-step transition probabilities. From the mean formula of geometric distribution (and modified geometric distribution), we have

$$r_{jj} = (1 - f_{jj})^{-1} \quad \text{and} \quad r_{ij} = f_{ij}r_{jj}, \text{ for } i,j,\in E.$$

Theorem 2.4 *Recurrence and transience of states in a DTMC are class properties. Furthermore, a recurrent communication class is always closed.*

Proof. Assume $i \in E$ is transient and $i \leftrightarrow j$ (communication). Then there are numbers $m,n \in \mathbb{N}$ with $0 < P^m(i,j) \leq 1$ and $0 < P^n(j,i) \leq 1$. Then, the following inequalities hold

$$\infty > r_{ii} = \sum_{k=0}^{\infty} P^k(i,i) \geq \sum_{h=0}^{\infty} P^{m+h+n}(i,i) \geq P^m(i,j)P^n(j,i) \sum_{h=0}^{\infty} P^h(j,j)$$
$$= P^m(i,j)P^n(j,i)r_{jj}.$$

Here, the first strict inequality is from the definition of transient state (i.e., state i); the second inequality is due to fewer positive terms in the sum of a convergent series on the right-hand side; and the third inequality is due to a subset of all paths considered for the right-hand side. The above relation implies that $r_{jj} = \sum_{h=0}^{\infty} P^h(j,j) < \infty$, i.e., state j is transient as well. If j is recurrent, then the same inequalities result in

$$r_{ii} \geq P^m(i,j)P^n(j,i)r_{jj} = \infty,$$

which implies that i is recurrent too. Since above arguments are symmetric in i and j, the proof of the first statement is complete.

For the second statement, we can prove it by showing that if a class is not closed, then the states in the class are transient. Assume that $i \in E$ belongs to a communication class $C \subset E$ and $p_{ij} > 0$ for some state $j \in E \setminus C$ (i.e., an open class). Then

$$f_{ii} = p_{ii} + \sum_{h \neq i} p_{ih}f_{hi} \leq 1 - p_{ij} < 1,$$

where the inequality holds because the left-hand side cannot have the term p_{ij} (since $f_{ji} = 0$) and hence is upper bounded by $1 - p_{ij}$, which implies i is transient. This completes the proof. ∎

Based on the mean formulas of the geometric distribution (modified), i.e., r_{jj} (r_{ij}), we can easily verify $\lim_{n \to \infty} P^n(i,j) = 0$ regardless of the starting state i.

Recurrent states can be further classified into two types. Let M_i be the mean recurrence time of state i. That is $M_i \equiv E[T_{ii}] = \sum_{k=1}^{\infty} kP\{T_{ii} = k\}$. A state is said to be *positive recurrent* if $M_i < \infty$. If $M_i = \infty$ then the state is said to be *null-recurrent*. The theorems above apply to the positive recurrent states. Intuitively, the steady-state probability is non-zero only for positive recurrent state. Another property for the recurrence time of a DTMC is called *period*. Suppose that the structure of the DTMC is such that state i is visited after a number of steps that is an integer multiple of an integer $d > 1$. Then the state is called periodic with period d. If no such an integer exists (i.e., $d = 1$), then the state is called *aperiodic*. An irreducible, positive recurrent, and aperiodic DTMC is called ergodic.

Exercise 2.12 *Answer the following questions: (1) Is it possible for a DTMC with a finite state space to have the null-recurrent states? Discuss. (2) For a positive recurrent state $i \in E$ in a DTMC, what is the meaning of $1/M_i$?*

2.3.2 STEADY STATE ANALYSIS

Recall that the probability of finding the DTMC in state i at the kth step is given by $\pi_i(k) \equiv P\{X_k = i\}$, $i \in E$, where E can be either finite or infinite. Then the distribution vector at the kth step can be written as $\pi(k) = [\pi_0(k), \pi_1(k), ...]$. An interesting question is what happens to this probability in the "long run"? We define the *limiting probability* for state i (also called the equilibrium or steady state probability) as $\pi_i \equiv \lim_{k \to \infty} \pi_i(k)$ and the time-independent probability vector is written as $\pi = [\pi_0, \pi_1, ...]$. We need to answer the following questions: (i) Do these probability limits exist? (ii) If they exist, do they converge to a legitimate probability distribution (i.e., $\pi_i \geq 0$ for $i \in E$ and $\sum \pi_i = 1$)? and (iii) How to evaluate π_j for all j. Recall the recursive relation $\pi(k+1) = \pi(k)\mathbf{P}$. If steady state exists, then $\pi(k+1) \approx \pi(k)$ when k is getting large, and therefore the steady state probabilities are given by the solution to the equations

$$\pi = \pi\mathbf{P}, \quad \sum_i \pi_i = 1 \tag{2.23}$$

and is called *stationary distribution*. Note that it is possible that the limiting distribution does not exist while the system still reaches the steady state, hence the stationary distribution can be solved via (2.23). A simple example is the two-state DTMC with the following transition matrix

$$P = \begin{pmatrix} 0 & 1 \\ 1 & 0 \end{pmatrix},$$

which does not have the limiting distribution but the stationary distribution can be determined as $(1/2, 1/2)$. However, as long as the limiting distribution exists, it is the same as the stationary distribution. This answers the third question. We can also prove that a probability distribution satisfying the above relation is a stationary distribution for the DTMC.

Theorem 2.5 *For a DTMC with state space E and transition probability $\mathbf{P}$, if $\pi\mathbf{P} = \pi$ and $\sum_{j\in E}\pi_j = 1$, then π is a stationary distribution for the DTMC. If π is a stationary distribution for the DTMC, then $\pi\mathbf{P} = \pi$ holds.*

Proof. Let $P(X_0 = i) = \pi_i(0)$ for all $i \in E$. We need to show $P(X_n = i) = \pi_i(n) = \pi_i(0) = P(X_0 = i) = \pi_i$ for all $n \in \mathbb{N}$ and $i \in E$. This can be done by induction. The case of $n = 1$ holds as $\pi_i(1) = \sum_{j\in E}\pi_j(0)p_{ji} = \pi_i(0)$. Now assume that such a relation holds for $n = k$, i.e., $\pi_i(k) = \pi_i(0)$. Then for $n = k+1$, we have

$$\pi_i(k+1) = \sum_{j\in E}\pi_j(k)p_{ji} = \pi_i(k) = \pi_i(0).$$

where the first equality is due to the Markov property, the second equality is due to the assumption, and the third equality is by induction hypothesis. The last statement is obvious. ∎

For the first two questions, we need to consider more properties of the DTMC. An simple example of a DTMC without a stationary distribution is the Bernoulli process.

Example 2.5 *The transition probability matrix of a Bernoulli process has the following structure*

$$\mathbf{P} = \begin{pmatrix} 1-p & p & 0 & 0 & \cdots \\ 0 & 1-p & p & 0 & \ddots \\ 0 & 0 & 1-p & p & \cdots \\ \vdots & \ddots & \ddots & \ddots & \ddots \end{pmatrix}.$$

Thus, $\pi\mathbf{P} = \pi$ leads to

$$\pi_0 \cdot (1-p) = \pi_0 \Rightarrow \pi_0 = 0,$$

due to $0 < p < 1$. Assume that $\pi_n = 0$ for any $n \in \mathbb{N}_0$. This and the condition of $\pi\mathbf{P} = \pi$ further imply that

$$\pi_n \cdot p + \pi_{n+1} \cdot (1-p) = \pi_{n+1} \Rightarrow \pi_{n+1} = 0.$$

This completes the induction argument that proves $\pi_n = 0$ for all $n \in \mathbb{N}_0$. Therefore, the Bernoulli process does not have a stationary distribution or all states are transient. Such an observation can be formalized as a theorem.

Theorem 2.6 *A DTMC with all transient states has no stationary distributions.*

Proof. Suppose that $\pi \mathbf{P} = \pi$ holds for some stationary distribution π, which is for a DTMC with state space $E = (s_n : n \in \mathbb{N})$. Choose any index $m \in \mathbb{N}$ with $\pi_{s_m} > 0$. Since $\sum_{n=1}^{\infty} \pi_{s_n} = 1$ is bounded, there exists an index $M > m$ such that $\sum_{n=M}^{\infty} \pi_{s_n} < \pi_{s_m}$ (a tail probability for a state above s_m becomes smaller than π_{s_m}). Set $\varepsilon := \pi_{s_m} - \sum_{n=M}^{\infty} \pi_{s_n}$. If state s_m is transient, then there exists an index N such that $P^n(s_i, s_m) < \varepsilon$ for all $i \leq M$ (states with index smaller than M or non-tail states) and $n \geq N$. Then the stationarity of π indicates

$$\pi_{s_m} = \sum_{i=1}^{\infty} \pi_{s_i} P^N(s_i, s_m) = \sum_{i=1}^{M-1} \pi_{s_i} P^N(s_i, s_m) + \sum_{i=M}^{\infty} \pi_{s_i} P^N(s_i, s_m)$$

$$< \varepsilon + \sum_{i=M}^{\infty} \pi_{s_i} = \pi_{s_m}.$$

which is a contradiction. Note that the last inequality holds because of $P^N(s_i, s_m) < \varepsilon$ for all $i \leq M$, the partial sum of probability distribution $\sum_{i=1}^{M-1} \pi_{s_i} < 1$, and $P^N(s_i, s_m) < 1$ for all $i \geq M$. ∎

Define $N_i(n)$ as the number of visits to state i until time n. Recalling the definition of the recurrent time for state $i \in E$ and using the elementary renewal theorem (to be discussed in Chapter 5), we have

$$\lim_{n \to \infty} \frac{E(N_i(n) | X_0 = j)}{n} = \frac{1}{M_i},$$

for all recurrent state $i \in E$ and independently of $j \in E$ if $j \leftrightarrow i$ with the convention of $1/\infty := 0$, where M_i denotes the mean recurrence time of state i. While this relation is rigorously justified in Chapter 5, it is quite intuitive from the perspective of the law of large numbers. Thus, the limiting rate of visits to a recurrent state, which is also the stationary probability, is the inverse of the mean recurrence time of this state. Positive recurrent state with a finite recurrence time has a positive limiting rate of visits while null recurrent state with an infinite recurrence time has a zero limiting rate of visits. These two types of recurrence are class properties.

Theorem 2.7 *Positive recurrence and null recurrence are class properties with respect to the communication relation between states.*

Proof. For two states in the same class, assume that i is null recurrent. Because of $i \leftrightarrow j$, there exist numbers $m, n \in \mathbb{N}$ with $P^n(i, j) > 0$ and $P^m(j, i) > 0$.

Based on $E(N_i(n)|X_0 = i) = \sum_{l=0}^{k} P^l(i,i)$ and $M_i = \infty$, we have

$$
\begin{aligned}
0 &= \lim_{k \to \infty} \frac{\sum_{l=0}^{k} P^l(i,i)}{k} \\
&\geq \lim_{k \to \infty} \frac{\sum_{l=0}^{k-m-n} P^l(j,j)}{k} \cdot P^n(i,j)P^m(j,i) \\
&= \lim_{k \to \infty} \frac{k-m-n}{k} \cdot \frac{\sum_{l=0}^{k-m-n} P^l(j,j)}{k-m-n} \cdot P^n(i,j)P^m(j,i) \\
&= \lim_{k \to \infty} \frac{\sum_{l=0}^{k} P^l(j,j)}{k} \cdot P^n(i,j)P^m(j,i) \\
&= \frac{P^n(i,j)P^m(j,i)}{M_j}.
\end{aligned}
$$

which implies that $M_j = \infty$ or state j is also null recurrent. Similarly, assume that state j is positive recurrent, then $M_j < \infty$, that is the limiting rate of visits to state j is positive. Using the relation above again, we conclude that the limiting rate of visits to state i is also positive. Thus state i is also positive recurrent. ∎

For a positive recurrent class, we can construct the stationary distribution for a DTMC.

Theorem 2.8 *Based on a positive recurrent state i in a DTMC, a stationary distribution can be constructed as*

$$
\pi_j := M_i^{-1} \sum_{n=0}^{\infty} P(X_n = j, T_{ii} > n | X_0 = i) \quad \text{for all } j \in E.
$$

Proof. We first verify that π is a probability distribution as

$$
\sum_{j \in E} \sum_{n=0}^{\infty} P(X_n = j, T_{ii} > n | X_0 = i) = \sum_{n=0}^{\infty} \sum_{j \in E} P(X_n = j, T_{ii} > n | X_0 = i)
$$

$$
= \sum_{n=0}^{\infty} P(T_{ii} > n | X_0 = i) = M_i.
$$

To show the stationarity of π, we need to verify $\pi = \pi P$. Note that

$$
\begin{aligned}
\pi_j &= M_i^{-1} \sum_{n=0}^{\infty} P(X_n = j, T_{ii} > n | X_0 = i) \\
&= M_i^{-1} \sum_{n=1}^{\infty} P(X_n = j, T_{ii} \geq n | X_0 = i) \\
&= M_i^{-1} \sum_{n=1}^{\infty} P(X_n = j, T_{ii} > n - 1 | X_0 = i).
\end{aligned}
$$

due to $X_0 = X_{T_{ii}} = i$ in the conditioning set $\{X_0 = i\}$. Note that the summation in the above equation represents the expected number of visits to state j over a recurrence time for state i (a renewal cycle) and using time index $n-1$ is for introducing the one-step transition probability into the expression. Now

$$P(X_n = j, T_{ii} > n-1 | X_0 = i) = \frac{P(X_n = j, T_{ii} > n-1, X_0 = i)}{P(X_0 = i)}$$

$$= \sum_{k \in E} \frac{P(X_n = j, X_{n-1} = k, T_{ii} > n-1, X_0 = i)}{P(X_0 = i)}$$

$$= \sum_{k \in E \setminus \{i\}} \frac{P(X_n = j, X_{n-1} = k, T_{ii} > n-1, X_0 = i)}{P(X_{n-1} = k, T_{ii} > n-1, X_0 = i)}$$

$$\times \frac{P(X_{n-1} = k, T_{ii} > n-1, X_0 = i)}{P(X_0 = i)}$$

$$= \sum_{k \in E} p_{kj} P(X_{n-1} = k, T_{ii} > n-1 | X_0 = i).$$

Substituting the above expression into π constructed yields

$$\pi_j = M_i^{-1} \sum_{n=1}^{\infty} \sum_{k \in E} p_{kj} P(X_{n-1} = k, T_{ii} > n-1 | X_0 = i)$$

$$= \sum_{k \in E} p_{kj} \cdot M_i^{-1} \sum_{n=0}^{\infty} P(X_n = k, T_{ii} > n | X_0 = i) = \sum_{k \in E} \pi_k p_{kj},$$

which completes the proof. ∎

Next, we show that there is a unique stationary distribution for an irreducible and positive recurrent DTMC.

Theorem 2.9 *An irreducible and positive recurrent DTMC has a unique stationary distribution.*

Proof. Assume π is the stationary distribution as constructed in Theorem 2.8 and let i be the positive recurrent state that is utilized as recurrence point for π. Moreover, let v be another stationary distribution for the DTMC. Then there is a state $j \in E$ with $v_j > 0$ and a number n such that $P^n(j, i) > 0$ due to the irreducibility. Thus, we have

$$v_i = \sum_{k \in E} v_k P^n(k, i) \geq v_j P^n(j, i) > 0.$$

Then we can multiply v_i by a constant c such that $c \cdot v_i = \pi_i = 1/M_i$. Using this c as a scaler factor, define a vector $\hat{v} := c \cdot \hat{v}$. Let $\tilde{P}_i$ denote the transition probability matrix P with the ith column replaced with the column of zeros. That is $\tilde{p}_{jk} = p_{jk}$ if $k \neq i$ and zero otherwise. This matrix $\tilde{P}_i$ ensures that the number of visits to the reference state i is 1 during the recurrence cycle. Also define the Dirac measure vector on i as δ^i. That is $\delta^i_j = 1$ if $i = j$ and zero otherwise. This row vector δ^i ensures that state i is the reference point for the recurrence cycle and the number of visits to state $j \neq i$ is accumulated during the cycle. Then the stationary distribution constructed in Theorem 2.8 can be expressed as $\pi = M_i^{-1} \delta^i \sum_{n=0}^{\infty} \tilde{P}_i^n = M_i^{-1} \delta^i (1 - \tilde{P}_i)^{-1}$. Using $\hat{v} := c \cdot v = \pi$, we re-write this relation as $M_i \hat{v} = \delta^i + M_i \hat{v} \tilde{P}_i$. This relation is clearly true for the entry $\hat{v}_i$ and also easily verifiable for $\hat{v}_j$ for $j \neq i$ as $(\hat{v} \tilde{P}_i)_j = c \cdot (vP)_j = \hat{v}_j$ Then we can proceed with the same argument so that

$$M_i \hat{v} = \delta^i + (\delta^i + M_i \hat{v} \tilde{P}_i) \tilde{P}_i = \delta^i + \delta^i \tilde{P}_i + M_i \hat{v} \tilde{P}_i^2 = \dots$$

$$= \delta^i \sum_{n=0}^{\infty} \tilde{P}_i^n = M_i \pi.$$

Thus, $\hat{v}$ is a vector of probability measure and the scalar factor must be $c = 1$. This implies $v = \hat{v} = \pi$. ∎

Exercise 2.13 *Suppose that the condition of a production facility can be classified into four states. Let X be the state variable. Then $X = 1, 2$, representing two "In control" states and $X = 3, 4$, representing two "Out-of-control" states. Its state changes according to a Markov process with the transition probability matrix given by*

$$\mathbf{P} = \begin{pmatrix} 0.5 & 0.3 & 0.2 & 0 \\ 0.2 & 0.5 & 0.2 & 0.1 \\ 0.1 & 0.3 & 0.4 & 0.2 \\ 0.1 & 0.2 & 0.3 & 0.4 \end{pmatrix}.$$

(a) In the long run, what proportion of time is the facility in out-of-control state?

(b) If every period that the facility spends in state 1, 2, 3, and 4 incurs a cost of £5, £10, £15, and £20, respectively, What is the long run operating cost per period?

(c) In the long run, what fraction of transitions are from in-control state to out-of-control state?

Exercise 2.14 *For a transition probability matrix $\mathbf{P}$ for a DTMC, if the rows sum to 1 as well as the columns, then $\mathbf{P}$ is called doubly stochastic. That is*

$$p_{ij} \geq 0, \quad and \quad \sum_k p_{ik} = \sum_k p_{kj} = 1 \quad for \; all \; i, j.$$

Assume that the DTMC with a doubly stochastic transition probability matrix has N states $0, 1, ..., N-1$. Show that the unique limiting distribution is the discrete uniform distribution $\pi = (1/N, ..., 1/N)$.

Note that the irreducibility assumption can be relaxed to the situation where there are a set of transient states and a single closed positive recurrent class. It follows from Theorem 2.8 that the stationary probability of a transient state k is zero, i.e. $\pi_k = 0$. In addition, the constructed stationary probability for a positive recurrent state can be obtained by setting $j = i$. That is $\pi_i = M_i^{-1}$ which is also the stationary distribution due to the fact of $\sum_{n=0}^{\infty} P(X_n = i, T_{ii} > n|X_0 = i) = P(X_0 = i, T_{ii} > 0|X_0 = i) + 0 + \ldots = 1$ and Theorem 2.9. Furthermore, if the limiting probability $p_j = \lim_{n \to \infty} = P(X_n = j)$ exists, then it is the same as the stationary probability π_j since

$$p_j = \lim_{n \to \infty} (vP^n)_j = \sum_{i \in E} v_i \lim_{n \to \infty} P^n(i, j)$$

$$= \sum_{i \in E} v_i \lim_{n \to \infty} \frac{E(N_j(n)|X_0 = i)}{n} = \sum_{i \in E} v_i \lim_{n \to \infty} \frac{\sum_{l=0}^{n} P^l(i, j)}{n}$$

$$= \sum_{i \in E} v_i \pi_j = \pi_j.$$

If we are interested in a subset of the state space of the DTMC, we may find the stationary distribution restricted to these states in the subset by analyzing a restricted Markov chain. Consider the process X_n of a positive recurrent DTMC on state space E. Suppose that we are interested in a subset $G \subset E$. Define the kth visit of the DTMC to G as

$$\tau_G(k) := \min\{n > \tau_G(k-1) : X_n \in G\}, \quad k = 1, 2, \ldots$$

with $\tau_G(0) := 0$. Clearly, $\tau_G(k)$ is a stopping time. Due to the strong Markov property the process restricted to the subset G is also a positive recurrent Markov chain denoted by $(X_{\tau_G(n)} : n \in \mathbb{N})$. For the definition of and discussion on the stopping times and strong Markov property, see Appendix B. This process, which is observed only at a state in G, is called the Markov chain restricted to G. It is easy to verify that the stationary distribution for this restricted Markov chain is given by

$$\pi_j^G = \frac{\pi_j}{\sum_{k \in G} \pi_k} \quad \text{for all } j \in G.$$

Using the restricted Markov chain, we can extend the property from a finite subset of a DTMC to the states outside the finite subset. The following theorem is an example.

Theorem 2.10 *For an irreducible and positive recurrent DTMC with state space E, a measure v on E is stationary for the DTMC if and only if this measure on a subset $G \in E$, denoted by $v' = (v_i : i \in G)$, is stationary and*

$$v_j = \sum_{k \in G} v_k \sum_{n=0}^{\infty} P(X_n = j, \tau_G > n|X_0 = k), \quad \text{for all } j \in E \setminus G, \qquad (2.24)$$

where $\tau_G := \min\{n \in \mathbb{N} : X_n \in G\}$.

Proof. Since the stationary distribution on a finite subset of the positive recurrent Markov chain exists, if we can show that the relation (2.24) holds, the theorem is proved. Define

$$\tau_i := \min\{n \in \mathbb{N} : X_n = i\} \quad \text{for all } i \in G.$$

Using $i \in G$ as a reference state (the state of the DTMC at time 0), it follows from the renewal reward theorem (to be discussed in Chapter 5) that a component of the stationary measure vector v (outside subset G) is given by

$$v_j = \frac{\sum_{n=0}^{\infty} P(X_n = j, \tau_i > n | X_0 = i)}{M_i} = v_i \cdot \sum_{n=0}^{\infty} P(X_n = j, \tau_i > n | X_0 = i)$$

$$= v_i \cdot E_i \left(\sum_{n=0}^{\tau_i - 1} \mathbf{1}_{X_n = j} \right) \quad \text{for } j \in E \setminus G,$$

where E_i represents the conditional expectation given $X_0 = i$. Define

$$\tau_i^G := \min\{n \in \mathbb{N} : X_n^G = i\},$$

which is the first time to reach a particular state i in the finite subset G. Due to the strong Markov property, we have

$$v_j = v_i \cdot E_i \left(\sum_{n=0}^{\tau_i^G - 1} E_{X_n^G} \sum_{m=0}^{\tau_G - 1} \mathbf{1}_{X_m = j} \right)$$

$$= v_i \cdot \sum_{k \in G} E_i \left(\sum_{n=0}^{\tau_i^G - 1} \mathbf{1}_{X_n^G = k} \right) \cdot E_k \left(\sum_{m=0}^{\tau_G - 1} \mathbf{1}_{X_m = j} \right). \tag{2.25}$$

The RHS of the first equality, which is the total expected number of visits to state j during a cycle between two consecutive visits to state i, can be explained as follows: the cycle time, τ^G, is decomposed into the $\tau_i^G - 1$ transitions to states in subset G except for i. For nth inter-transition time (the interval between nth and $(n+1)$st visits to G) given a starting state X_n^G, we compute the expected number of visits to state $j \in E \setminus G$. The RHS of the second equality holds by conditioning on the starting state of the nth inter-transition time to G during the cycle time. For the restricted Markov chain, we know

$$E_i \left(\sum_{n=0}^{\tau_i^G - 1} \mathbf{1}_{X_n^G = k} \right) = \sum_{n=0}^{\infty} P(X_n^G = k, \tau_i^G > n | X_0^G = i) = v_k / v_i,$$

for all $k \in G$. Substituting this equation into (2.25) yields

$$v_j = \sum_{k \in G} v_k \sum_{n=0}^{\infty} P(X_n = j, \tau^G > n | X_0 = k).$$

This completes the proof. ∎

Exercise 2.15 *Let* $\mathbf{P}$ *be the transition matrix of an irreducible DTMC with state space E which can be decomposed into two disjoint subsets. That is* $E = F \cup F^c$. *Write* $\mathbf{P}$ *in block form as*

$$\mathbf{P} = \left(\begin{array}{cc} \mathbf{P}_{FF} & \mathbf{P}_{FF^c} \\ \mathbf{P}_{F^cF} & \mathbf{P}_{F^cF^c} \end{array} \right).$$

Show that the DTMC restricted to the subset F has the following transition probability matrix

$$\mathbf{P}_F = \mathbf{P}_{FF} + \mathbf{P}_{FF^c}(\mathbf{I} - \mathbf{P}_{F^cF^c})^{-1}\mathbf{P}_{F^cF},$$

where $\mathbf{I}$ *is the identity matrix on* $\mathbf{F}^c$.

2.3.3 POSITIVE RECURRENCE FOR DTMC

To perform the stationary analysis, we need to have the conditions for positive recurrence satisfied. For the DTMC with finite-state space, this can be easily confirmed.

Theorem 2.11 *An irreducible Markov chain with finite state space G is positive recurrent.*

Proof. Due to the finiteness of the state space, for all $n \in \mathbb{N}$, we have $\sum_{j \in G} P^n(i, j) = 1$. Thus, it is not possible that $\lim_{n \to \infty} P^n(i, j) = 0$ for all $j \in G$. Then, there is a state s such that it will be visited infinite number of times as $n \to \infty$, i.e., $\sum_{n=0}^{\infty} P^n(s, s) = \infty$. This means that s is recurrent and by irreducibility that the Markov chain is recurrent. Assume that the chain is null recurrent. Then we have

$$\lim_{n \to \infty} \frac{1}{n} \sum_{k=1}^{n} P^k(i, j) = \frac{1}{M_j} = 0,$$

which holds for all $j \in G$ due to the irreducibility. However, this implies that $\lim_{n \to \infty} P^n(i, j) = 0$ for all $j \in G$, which contradicts the observation made in the proof. Therefore, the chain is positive recurrent. ∎

The finite expected recurrence time for a state as a condition for the positive recurrence can be extended to the finite expected recurrence time for a subset of the DTMC.

Theorem 2.12 *A DTMC with state space E is positive recurrent if and only if* $E(\tau_G | X_0 = i) < \infty$ *for all* $i \in G \subset E$.

Proof. If the chain is positive recurrent, then $E(\tau_G | X_0 = i) \leq E(\tau_i | X_0 = i) < \infty$ for all $i \in G$.

Now assume that $E(\tau_G|X_0 = i) < \infty$ for all $i \in G$. Define the stopping time $\phi(i) := \min\{k \in \mathbb{N} : X_k^G = i\}$, which is the first time to visit a particular state in subset G, and random variables $Y_k := \tau_G(k) - \tau_G(k-1)$, the inter-visit time between kth visit and $k+1$st visit to subset G. Since G is finite, we have the maximum $m := \max_{j \in G} E(\tau_G|X_0 = j) < \infty$. Denote the conditional expectation given $X_0 = i$ by E_i. For $i \in G$, we have

$$E_i(\tau_i|X_0 = i) = E\left(\sum_{k=1}^{\phi(i)} Y_k\right) = \sum_{k=1}^{\infty} E_i(E(Y_k|X_{\tau_G(k-1)}) \cdot \mathbf{1}_{k \le \phi(i)})$$

$$\le m \cdot \sum_{k=1}^{\infty} P(\phi(i) \ge k|X_0 = i) = m \cdot E(\phi(i)|X_0 = i).$$

Since G is finite, the restricted Markov chain is positive recurrent. Thus, we have $E(\phi(i)|X_0 = i) < \infty$ which shows that the DTMC is positive recurrent. ∎

An important criteria for the existence of stationary distribution of a DTMC with infinite state space in queueing theory is the Foster's criterion as stated in the following theorem.

Theorem 2.13 *For an irreducible Markov chain with countable state space E and transition probability matrix* **P**, *if there exists a function $h : E \to \mathbb{R}$ with $\inf\{h(i) : i \in E\} > -\infty$, such that the conditions*

$$\sum_{k \in E} p_{ik} h(k) < \infty \quad and \quad \sum_{k \in E} p_{jk} h(k) < h(j) - \varepsilon$$

hold for some $\varepsilon > 0$ and all $i \in G \subset E$ and $j \in E \setminus G$, then the DTMC is positive recurrent.

Proof. We assume $h(i) \ge 0$ for all $i \in E$ and define the stopping time $\tau_G := \min\{n \in \mathbb{N}_0 : X_n \in G\}$, the first time to reach a state in subset G. First, note that

$$E(h(X_{n+1}) \cdot \mathbf{1}_{\tau_G > n+1}|X_0, ..., X_n) \le E(h(X_{n+1}) \cdot \mathbf{1}_{\tau_G > n}|X_0, ..., X_n)$$
$$= \mathbf{1}_{\tau_G > n} \cdot \sum_{k \in E} p_{X_n,k} h(k) \qquad (2.26)$$
$$\le \mathbf{1}_{\tau_G > n} \cdot (h(X_n) - \varepsilon)$$
$$= h(X_n) \cdot \mathbf{1}_{\tau_G > n} - \varepsilon \cdot \mathbf{1}_{\tau_G > n}$$

holds for all $n \in \mathbb{N}_0$, where the first inequality is due to the fact of $\{\tau_G > n+1\} \subset \{\tau_G > n\}$ and the first equality is due to the fact that the random variable $\mathbf{1}_{\tau_G > n}$ is measurable with respect to the sigma algebra $\sigma(X_0, ..., X_n)$

(see Appendix B.2). Furthermore, we have

$$
\begin{aligned}
0 &\le E(h(X_{n+1}) \cdot \mathbf{1}_{\tau_G > n+1} | X_0 = i) \\
&= E(E(h(X_{n+1}) \cdot \mathbf{1}_{\tau_G > n+1} | X_0, ..., X_n) | X_0 = i) \\
&\le E(h(X_n) \cdot \mathbf{1}_{\tau_G > n} | X_0 = i) - \varepsilon P(\tau_G > n | X_0 = i) \\
&\le \cdots \\
&\le E(h(X_0) \cdot \mathbf{1}_{\tau_G > n} | X_0 = i) - \varepsilon \sum_{k=0}^{n} P(\tau_G > k | X_0 = i),
\end{aligned}
\tag{2.27}
$$

which holds for all $i \in E \setminus G$ and $n \in \mathbb{N}_0$. Note that the equality in (2.27) follows from the tower property of conditional expectation (see Appendix B.2) and the inequalities following hold by using (2.26) and the tower property repetitively. For $n \to \infty$, this implies

$$
E(\tau_G | X_0 = i) = \sum_{k=0}^{\infty} P(\tau_G > k | X_0 = i) \le h(i)/\varepsilon < \infty
$$

for $i \in E \setminus G$. Hence, the mean return time to the state set G is bounded by

$$
\begin{aligned}
E(\tau_G | X_0 = i) &= \sum_{j \in G} p_{ij} + \sum_{j \in E \setminus G} p_{ij} E(\tau_G + 1 | X_0 = j) \\
&\le 1 + \varepsilon^{-1} \sum_{j \in E} p_{ij} h(j) < \infty
\end{aligned}
$$

for all $i \in G$, which completes the proof. ∎

Exercise 2.16 *Assume that $\mathbf{P}$ is the transition probability matrix of a positive recurrent Markov chain with discrete state space E. Show that there exists a function $h : E \to \mathbb{R}$ and a finite subset $F \subset E$ such that*

$$
\sum_{k \in E} p_{ik} h(k) < \infty \quad \text{for all } i \in F,
$$

$$
\sum_{k \in E} p_{ik} h(k) \le h(j) - 1 \quad \text{for all } j \in E \setminus F.
$$

Hint: Consider the conditional expectation of the remaining time until returning to a fixed set F of states.

Exercise 2.17 *Suppose that the DTMC has the following transition probability matrix*

$$
\mathbf{P} = \begin{pmatrix}
p_{00} & p_{01} & & & \\
p_{10} & 0 & p_{12} & & \\
& p_{10} & 0 & p_{12} & \\
& & \ddots & \ddots & \ddots \\
& & & \ddots & \ddots
\end{pmatrix}.
$$

Find the criterion of positive recurrence for this DTMC by applying Theorem 2.13.

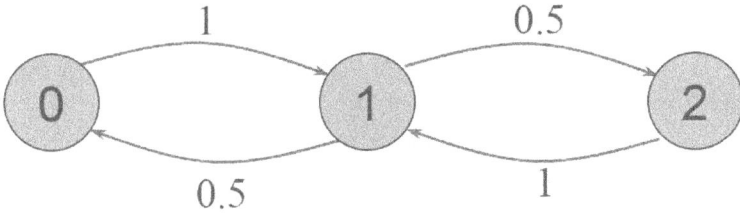

Figure 2.6   A three-state DTMC with period $d = 2$.

In addition, we notice that in an irreducible DTMC, the presence of periodic states prevents the existence of a limiting (steady state) distribution. Here is an example of this type of DTMC.

Example 2.6 *Suppose that a three-state DTMC has the state transition diagram shown in Figure 2.6.*
The transision probability matrix is

$$\mathbf{P} = \begin{bmatrix} 0 & 1 & 0 \\ 0.5 & 0 & 0.5 \\ 0 & 1 & 0 \end{bmatrix}. \tag{2.28}$$

Note that this DTMC has a period of 2. If the DTMC starts with the initial probability vector $\pi(0) = [1,0,0]$, then we have $\pi(1) = \pi(0)\mathbf{P} = [0,1,0]$, $\pi(2) = \pi(1)\mathbf{P} = [0.5,0,0.5]$, $\pi(3) = \pi(2)\mathbf{P} = [0,1,0]$, $\pi(4) = \pi(3)\mathbf{P} = [0.5,0,0.5]$, and so forth. Thus, π oscillates between $[0,1,0]$ and $[0.5,0,0.5]$ and the limiting probability does not exist.

From this example, we find that the existence of limiting probability requires the DTMC to be aperiodic.

Exercise 2.18 *Consider a DTMC with n states. That is $E = \{1,2,...,n\}$. The transition probability matrix is given by*

$$\mathbf{P} = \begin{pmatrix} 0 & 1 & 0 & 0 & \cdots & 0 \\ 0 & 0 & 1 & 0 & \cdots & 0 \\ 0 & 0 & 0 & 1 & \cdots & 0 \\ \vdots & \vdots & \ddots & \ddots & \cdots & \vdots \\ 0 & 0 & & & \cdots & 1 \\ 1 & 0 & 0 & & \cdots & 0 \end{pmatrix}.$$

(a) Determine the period of this DTMC. (b) For $n = 4$, evaluate $p_{00}, p_{00}^{(2)}, p_{00}^{(3)}, p_{00}^{(4)}, p_{00}^{(5)}$, and $p_{00}^{(6)}$. Based on these probabilities, justify that the period of this DTMC is 2.

Theorem 2.14 *In an irreducible aperiodic DTMC consisting of positive recurrent states, a unique steady state probability vector π exists such that $\pi_j > 0$ and $\pi_j = \lim_{k\to\infty} \pi_j(k) = 1/M_j$. Such a DTMC is called an ergodic Markov chain.*

Exercise 2.19 *Prove Theorem 2.14.*

The steady state (also called stationary) distribution vector can be solved by using (2.23). Note that although the stationary distribution does not exist for a periodic DTMC, the proportion of time that the DTMC spends for each state can still be computed by using (2.23). The condition for the existence of the steady state distribution for a DTMC can be more complex when the state space becomes infinite. For an irreducible DTMC with infinite state space, we can find the situations in which all states are transient, positive recurrent, and null recurrent, respectively. We show these cases by the following example.

Example 2.7 *Consider a discrete-time birth-and-death process (DTBDP), also called a discrete-time single server queue. In such a system, for each period, if the system is non-empty, then with a probability 1-p there is a birth and with a probability p there is a death; If the system is empty, there is "no-death" and with a probability 1-p there is a birth and with a probability p there is no birth. This is a DTMC with the transition probability matrix as follows:*

$$
\mathbf{P} = \begin{bmatrix} p & 1-p & 0 & \cdots \\ p & 0 & 1-p & \cdots \\ 0 & p & 0 & \cdots \\ \vdots & \vdots & \vdots & \ddots \end{bmatrix}. \tag{2.29}
$$

Due to the tri-diagonal structure of $\mathbf{P}$, $\pi = \pi\mathbf{P}$ can be written as the following recursive relations:

$$
\begin{aligned} \pi_0 &= \pi_0 p + \pi_1 p, \\ \pi_j &= \pi_{j-1}(1-p) + \pi_{j+1} p, \quad j = 1, 2, \dots. \end{aligned} \tag{2.30}
$$

Solving these equations yields $\pi_j = \left(\frac{1-p}{p}\right)^j \pi_0$, for $j = 0, 1, 2, \dots$. Using the normalization condition, we obtain

$$
\pi_0 \sum_{i=0}^{\infty} \left(\frac{1-p}{p}\right)^i = 1 \Rightarrow \pi_0 = 1 \Big/ \sum_{i=0}^{\infty} \left(\frac{1-p}{p}\right)^i,
$$

$$
\pi_j = \left(\frac{1-p}{p}\right)^j \Big/ \sum_{i=0}^{\infty} \left(\frac{1-p}{p}\right)^i. \tag{2.31}
$$

Now, if $p < 1/2$, then $\sum_{i=0}^{\infty} \left(\frac{1-p}{p}\right)^i = \infty$, thus $\pi_j = 0$ for all j. This means all states are transient; if $p > 1/2$, then $\sum_{i=0}^{\infty} \left(\frac{1-p}{p}\right)^i = \frac{p}{2p-1}$, thus $\pi_j = \frac{2p-1}{p}\left(\frac{1-p}{p}\right)^j$, for all j. This means all states are positive recurrent; finally, if $p = 1/2$, then $\sum_{i=0}^{\infty} \left(\frac{1-p}{p}\right)^i = \infty$, thus $\pi_j = 0$ for all j. This is the case in which all states are null recurrent. The last case is a little involved. Recall that the null recurrent state has two properties: (a) $\rho_i = \sum_{k=1}^{\infty} P\{T_{ii} = k\} = P\{T_{ii} < \infty\} = 1$; and (b) the mean recurrence time $M_i \equiv E[T_{ii}] = \sum_{k=1}^{\infty} kP\{T_{ii} = k\} = \infty$ for all i. Here, $\pi_i = 0$ for all i implies (b). Hence, we need to show $\rho_i = 1$ for all i. This can be shown by considering a symmetric random walk (SRW) process. To do that, we first develop an alternative criterion for identifying a recurrent state based on the expected number of time periods that the process is in state i, denoted by $E(N_i)$. Obviously, for a recurrent state, $E(N_i) = \infty$. For a transient state, N_i, the actual number of times of visiting state i, is finite and follows a geometric distribution with parameter $\rho_i < 1$. It is easy to show that state i is recurrent (transient) if $\sum_{n=1}^{\infty} P_{ii}^n = \infty (< \infty)$ as it is equivalent to $E(N_i)$. Therefore, to show (a), we can simply show $\sum_{n=1}^{\infty} P_{00}^n = \infty$ for a SRW process. Note that the DTBDP can be considered as the SRW with a reflecting barrier at origin and its $\sum_{n=1}^{\infty} P_{00}^n$ is the same as in the SRW without the reflecting barrier at origin. For the SRW, only possible time instants for returning to the origin are at even number of steps. Thus, $\sum_{n=1}^{\infty} P_{00}^n = \sum_{k=1}^{\infty} P_{00}^{2k}$ and

$$P_{00}^{2k} = \binom{2k}{k} q^k p^k, \quad q = 1 - p. \tag{2.32}$$

Denote by $a_n \sim b_n$ if $\lim_{n \to \infty} a_n/b_n = 1$. Using Stirling formula $n! \sim \sqrt{2\pi n}(n/e)^n = \sqrt{2\pi n} n^n e^{-n}$, we have

$$\binom{2k}{k} = \frac{(2k)!}{k!(2k-k)!} = \frac{(2k)!}{k!k!} \sim \frac{\sqrt{2\pi 2k}(2k)^{2k} e^{-2k}}{\sqrt{2\pi k}(k)^k e^{-k}\sqrt{2\pi k}(k)^k e^{-k}} = \frac{1}{\sqrt{\pi k}} 4^k, \tag{2.33}$$

and hence

$$\sum_{n=1}^{\infty} P_{00}^n = \sum_{k=1}^{\infty} P_{00}^{2k} \sim \sum_{k=1}^{\infty} \frac{1}{\sqrt{\pi k}} 4^k q^k p^k = \frac{1}{\sqrt{\pi}} \sum_{k=1}^{\infty} \frac{1}{\sqrt{k}} (4qp)^k. \tag{2.34}$$

For a SRW with $p = q = 1/2$, the RHS of the above becomes $(1/\sqrt{\pi}) \sum_{k=1}^{\infty} (1/\sqrt{k}) = \infty$ by using the harmonic series of $\sum_{k=1}^{\infty} (1/k^{\alpha}) = \infty$ for $\alpha \leq 1$. This shows (a). Thus, all states are null recurrent for the DTBDP when $p = 1/2$. We will return to study the random walk processes in more details in Chapter 6.

Exercise 2.20 *Let $\{\alpha_i : i = 1, 2, ...\}$ be a probability distribution. Consider a DTMC with the transition probability matrix as follows:*

$$\mathbf{P} = \begin{pmatrix} \alpha_1 & \alpha_2 & \alpha_3 & \alpha_4 & \alpha_5 & \alpha_6 & \cdots \\ 1 & 0 & 0 & 0 & 0 & 0 & \cdots \\ 0 & 1 & 0 & 0 & 0 & 0 & \cdots \\ 0 & 0 & 1 & 0 & 0 & 0 & \cdots \\ 0 & 0 & 0 & 1 & 0 & 0 & \cdots \\ \vdots & \vdots & \vdots & \vdots & \vdots & \vdots & \ddots \end{pmatrix}.$$

Find the condition on the probability distribution $\{\alpha_i : i = 1, 2...\}$ that is necessary and sufficient for the DTMC to have a limiting distribution. Determine this limiting distribution if it exists.

2.3.4 TRANSIENT ANALYSIS

If a DTMC has some transient states, it is meaningful to perform a transient analysis that mainly answers two questions: (i) If the process starts in a transient state, how long, on average, does it spend among the transient states? (ii) What is the probability that the process will enter a given single recurrent state (absorbing state) or a given subset of the recurrent states? To answer these questions, we consider a finite m-state DTMC with some transient states. Assume all states are numbered so that the first t states, denoted by $T = \{1, 2, ..., t\}$, are transient and the remaining $m - t$ states, denoted by $R = \{t + 1, t + 2, ..., m\}$, are recurrent. Let $\mathbf{P_T}$ be the matrix of one-step transition probabilities among these transient states. Let $\mathbf{P_{TR}}$ be the $t \times (m - t)$ matrix of one-step transition probabilities from transient states to the recurrent states and $\mathbf{P_R}$ be the $(m - t) \times (m - t)$ matrix of one-step transition probabilities among the recurrent states. Then the entire one-step transition probability matrix can be written in block form as

$$\mathbf{P} = \begin{bmatrix} \mathbf{P_T} & \mathbf{P_{TR}} \\ \mathbf{0} & \mathbf{P_R} \end{bmatrix}. \tag{2.35}$$

If the recurrent states are all absorbing, then $\mathbf{P_R} = \mathbf{I}$.

To answer the first question, define $\delta_{ij} = 1$ if $i = j$ and 0 otherwise. For states i and j in T, let s_{ij} be the expected number of periods (expected sojourn time) that the process is in state j given that it started in start i. Conditioning on the state of the first transition and noting that transitions from recurrent states to transient states should not be counted, we obtain

$$s_{ij} = \delta_{ij} + \sum_{k=1}^{t} p_{ik} s_{kj}. \tag{2.36}$$

Introducing matrix $\mathbf{S} = [s_{ij}]_{t \times t}$, the above can be written as $\mathbf{S} = \mathbf{I} + \mathbf{P_T} \mathbf{S}$. Rewriting it as $(\mathbf{I} - \mathbf{P_T})\mathbf{S} = \mathbf{I}$, we have $\mathbf{S} = (\mathbf{I} - \mathbf{P_T})^{-1}$. If the initial distribution vector is π_0, the total expected time that the DTMC spends before reaching one of the recurrent states is given by $\pi_0 \mathbf{S} \mathbf{e}$ where $\mathbf{e}$ is a column vector of 1's with dimension of t. Clearly, the total time spent in these transient states is a random sum of non-identical and geometrically distributed random variables. This random period is said to follow a discrete-time phase-type (PH) distribution with a representation of $(\pi_0, \mathbf{S})$. Discrete-time PH distribution is a generalization of geometric distribution and can be utilized to characterize more general discrete random variables in stochastic modeling. We will study both discrete- and continuous-time PH distributions in details later. From the discussion above, we understand that a discrete-time PH distribution can be completely determined by its representation and has the mean of $\pi_0 Se$.

To answer the second question, we consider a simple case in which all recurrent states are absorbing states. Define the conditional probability $f_{ij} = P$ –DTMC absorbed at state j | DTMC started at state i″, where $i \in T, j \in R$. Conditioning on the state of the first transition, we have

$$f_{ij} = p_{ij} + \sum_{k \in T} p_{ik} f_{kj}. \tag{2.37}$$

Define the matrix $\mathbf{F} = [f_{ij}]_{t \times (m-t)}$. We can write the equation (2.37) in matrix form as $\mathbf{F} = \mathbf{P_{TR}} + \mathbf{P_T F}$. Hence, $\mathbf{F} - \mathbf{P_T F} = \mathbf{P_{TR}}$ or $(\mathbf{I} - \mathbf{P_T})\mathbf{F} = \mathbf{P_{TR}}$. Left multiplying the inverse of $(\mathbf{I} - \mathbf{P_T})$, we obtain $\mathbf{F} = (\mathbf{I} - \mathbf{P_T})^{-1}\mathbf{P_{TR}}$ which gives the probability that the process will eventually be absorbed at any recurrent state given it started from any transient state.

Example 2.8 *Suppose a liver disease can be diagnosed as in one of the three different stages. For each stage, the patient can be either completely treated or dead. It is also possible to change from one stage to another stage. The process can be modeled as a DTMC with five states (three transient states and two absorbing states) with the transition probability matrix according to the state numbering rule.*

$$\mathbf{P} = \begin{bmatrix} p_{11} & p_{12} & p_{13} & p_{14} & p_{15} \\ p_{21} & p_{22} & p_{23} & p_{24} & p_{25} \\ p_{31} & p_{32} & p_{33} & p_{34} & p_{35} \\ 0 & 0 & 0 & 1 & 0 \\ 0 & 0 & 0 & 0 & 1 \end{bmatrix}. \tag{2.38}$$

Here, the first three states represent three stages of the disease and state 4 represents completed treated or recovered and state 5 represents death. The generic state transition is shown in Figure 2.7. To illustrate the use of the formulas developed, we give a numerical example as follows:

$$\mathbf{P} = \begin{bmatrix} 0.30 & 0.20 & 0 & 0.40 & 0.10 \\ 0.10 & 0.30 & 0.20 & 0.30 & 0.10 \\ 0 & 0.05 & 0.40 & 0.20 & 0.35 \\ 0 & 0 & 0 & 1 & 0 \\ 0 & 0 & 0 & 0 & 1 \end{bmatrix}. \tag{2.39}$$

Clearly, the two sub-matrices are

$$\mathbf{P_T} = \begin{bmatrix} 0.30 & 0.20 & 0 \\ 0.10 & 0.30 & 0.20 \\ 0 & 0.05 & 0.40 \end{bmatrix}, \mathbf{P_{TR}} = \begin{bmatrix} 0.40 & 0.10 \\ 0.30 & 0.10 \\ 0.20 & 0.35 \end{bmatrix}. \tag{2.40}$$

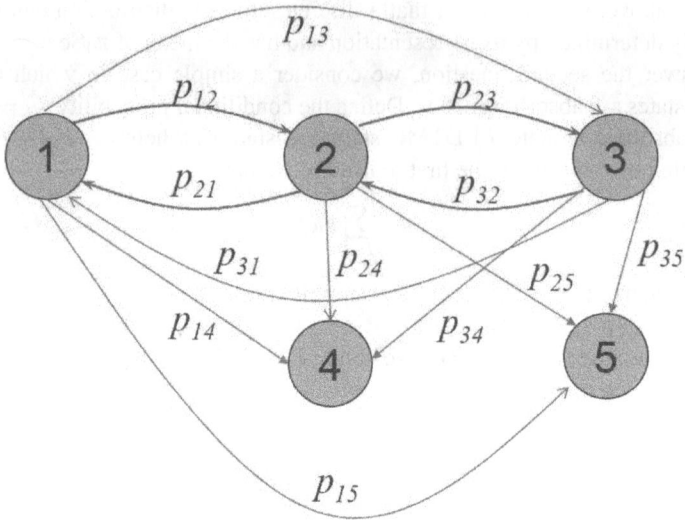

Figure 2.7 A DTMC model for liver disease treatment process.

The sojourn time matrix for the transient states can be computed as

$$
\mathbf{S} = (\mathbf{I} - \mathbf{P_T})^{-1} = \begin{bmatrix} 0.70 & -0.20 & 0 \\ -0,10 & 0.70 & -0.20 \\ 0 & -0.05 & 0.60 \end{bmatrix}^{-1}
$$

$$
= \begin{bmatrix} 1.490909 & 0.436364 & 0.145455 \\ 0.218182 & 1.527273 & 0.509091 \\ 0.018182 & 0.127273 & 1.709091 \end{bmatrix}.
$$

(2.41)

If the process starts at state 1 or $\pi(0) = [1,0,0]$, the total expected time spent in the transient set is $\pi(0)\mathbf{S}e = 2.072727$. Similarly, the absorbing probability matrix can be computed as

$$
\mathbf{F} = (\mathbf{I} - \mathbf{P_T})^{-1}\mathbf{P_{TR}} = \begin{bmatrix} 1.490909 & 0.436364 & 0.145455 \\ 0.218182 & 1.527273 & 0.509091 \\ 0.018182 & 0.127273 & 1.709091 \end{bmatrix} \begin{bmatrix} 0.40 & 0.10 \\ 0.30 & 0.10 \\ 0.20 & 0.35 \end{bmatrix}
$$

$$
= \begin{bmatrix} 0.756364 & 0.243636 \\ 0.647273 & 0.352727 \\ 0.387273 & 0.612727 \end{bmatrix}.
$$

(2.42)

These probabilities are meaningful. For example, if the disease is diagnosed in stage 2, then the probability of treatment success is about 64.73% and the probability of death is about 35.27%. A theoretical question is whether or not the matrix $\mathbf{I} - \mathbf{P_T}$ is invertible. Note that $\mathbf{P_T}$ is a substochastic matrix in which the row sum is strictly less than 1. Thus, for a finite-state DTMC with absorbing states, this matrix is always invertible (see Appendix J for more details).

Exercise 2.21 *Assume that the number of traffic accidents in a urban highway system per day can be considered as an i.i.d. discrete random variable, denoted by u_n for day n, where $n = 1, 2, \ldots$. This non-negative discrete distribution is given by $a_i = P(u = i)$ where $i = 0, 1, \ldots$, and u is the generic notation for these i.i.d. random variables u_n. We are interested in the maximum number of accidents per day since day 1 (record starting day). Denote by v_n for this maximum on day n. Then we have $v_n = \max\{u_1, \ldots, u_n\}$. Clearly, v_n is a DTMC. Let $b_i = \sum_{k=0}^{i} a_i$. The transition probability matrix for v_n is given by*

$$P = \begin{pmatrix} b_0 & a_1 & a_2 & a_3 & \cdots \\ 0 & b_1 & a_2 & a_3 & \cdots \\ 0 & 0 & b_2 & a_3 & \cdots \\ 0 & 0 & 0 & b_3 & \cdots \\ \vdots & \vdots & \vdots & \vdots & \ddots \end{pmatrix}.$$

If the probability distribution of u is on a finite support $\{0, 1, 2, \ldots, M\}$, where M is the maximum number of accidents that can occur in one day, then v_n becomes a DTMC with an absorbing state.

(a) Write the transition probability matrix for the case of $M = 3$ and show that the mean time until absorbtion is $1/a_3$.

(b) For the $M = \infty$ case (i.e., no finite upper bound on u), let N be the prescribed level and define $T = \min\{n \geq 1; v_n \geq N\}$. Then show that $E[T] = 1 + E[T]P(u < N)$. (Note: this model can also be used to study the bidding process on a certain asset.)

A more general case is that the reducible DTMC has several recurrent classes that are not single states (absorbing states). While the computations for S and F matrices are the same, we may need an extra step to compute the probability that the DTMC eventually enters a recurrent class or an absorbing state given it started from a transient state. Assume that a DTMC has t transient states, denoted by set T, and m classes of recurrent states, denoted by $R_1, R_2, \ldots R_m$. Recurrent class R_j has n_j states, $j = 1, 2, \ldots, m$. We number these states as before. That is $T = \{1, 2, \ldots, t\}, R_1 = \{t+1, t+2, \ldots, t+n_1\}, R_2 = \{t+n_1+1, \ldots, t+n_1+n_2\}, \ldots, R_m = \{t+\sum_{k=1}^{m-1} n_k + 1, \ldots, t+\sum_{k=1}^{m} n_k\}$. Then, the transition probability matrix can be written as

$$P = \begin{bmatrix} P_T & P_{TR_1} & P_{TR_2} & \cdots & P_{TR_m} \\ & P_{R_1 R_1} & & & \\ & & P_{R_2 R_2} & & \\ & & & \ddots & \\ & & & & P_{R_m R_m} \end{bmatrix}, \qquad (2.43)$$

where

$$
\mathbf{P_T} = \begin{bmatrix} p_{11} & p_{12} & \cdots & p_{1t} \\ p_{21} & p_{22} & \cdots & p_{2t} \\ \vdots & \vdots & \vdots & \vdots \\ p_{t1} & p_{t2} & \cdots & p_{tt} \end{bmatrix}, \mathbf{P_{TR_1}} = \begin{bmatrix} p_{1t+1} & p_{1t+2} & \cdots & p_{1t+n_1} \\ p_{2t+1} & p_{2t+2} & \cdots & p_{2t+n_1} \\ \vdots & \vdots & \vdots & \vdots \\ p_{tt+1} & p_{tt+2} & \cdots & p_{tt+n_1} \end{bmatrix},
$$

$$
\mathbf{P_{TR_2}} = \begin{bmatrix} p_{1t+n_1+1} & p_{1t+n_1+2} & \cdots & p_{1t+n_1+n_2} \\ p_{2t+n_1+1} & p_{2t+n_1+2} & \cdots & p_{2t+n_1+n_2} \\ \vdots & \vdots & \vdots & \vdots \\ p_{tt+n_1+1} & p_{tt+n_2+2} & \cdots & p_{tt+n_1+n_2} \end{bmatrix}, \dots,
$$

$$
\mathbf{P_{TR_m}} = \begin{bmatrix} p_{1t+\sum_{k=1}^{m-1} n_k+1} & p_{1t+\sum_{k=1}^{m-1} n_k+2} & \cdots & p_{1t+\sum_{k=1}^{m} n_k} \\ p_{2t+\sum_{k=1}^{m-1} n_k+1} & p_{2t+\sum_{k=1}^{m-1} n_k+2} & \cdots & p_{2t+\sum_{k=1}^{m} n_k} \\ \vdots & \vdots & \vdots & \vdots \\ p_{tt+\sum_{k=1}^{m-1} n_k+1} & p_{tt+\sum_{k=1}^{m-1} n_k+2} & \cdots & p_{tt+\sum_{k=1}^{m} n_k} \end{bmatrix}.
$$

$$(2.44)$$

To compute the probability that the DTMC enters each recurrent class, we need to multiply the $\mathbf{F}$ matrix by the $\mathbf{R}$ matrix defined as follows:

$$
\mathbf{R} = \begin{bmatrix} 1 & \cdots & & \\ 1 & & & \\ \vdots & & & \\ 1 & & & \\ & 1 & & \\ & \vdots & & \\ & 1 & \ddots & \\ & & 1 & \\ & & 1 & \\ & & \vdots & \\ & & & 1 \end{bmatrix}_{(n_1+\dots+n_m)\times m.}
$$

$$(2.45)$$

This $\mathbf{R}$ matrix has $\sum_{k=1}^{m} n_k$ rows and m columns. Column k has n_k 1's in appropriate rows and other elements are zeros. Thus, the recurrent class entrance probability matrix can be computed by $\mathbf{F'} = \mathbf{FR}$. Note that when all recurrent classes are absorbing states, then $\mathbf{R} = \mathbf{I}$.

Exercise 2.22 *Analyze the limiting behavior of the DTMC with the following probability transition matrix:*

$$
P = \begin{pmatrix}
0.1 & 0.2 & 0.1 & 0.1 & 0.2 & 0.1 & 0.1 & 0.1 \\
0 & 0.1 & 0.2 & 0.1 & 0 & 0.3 & 0.1 & 0.2 \\
0.5 & 0 & 0 & 0.2 & 0.1 & 0.1 & 0.1 & 0 \\
0 & 0 & 0.5 & 0.5 & 0 & 0 & 0 & 0 \\
0 & 0 & 0.7 & 0.3 & 0 & 0 & 0 & 0 \\
0 & 0 & 0 & 0 & 0 & 0.3 & 0.3 & 0.4 \\
0 & 0 & 0 & 0 & 0 & 0.1 & 0.2 & 0.7 \\
0 & 0 & 0 & 0 & 0 & 0.4 & 0.4 & 0.2
\end{pmatrix}.
$$

2.3.5 BRANCHING PROCESSES

While the transient analysis in the previous subsection is focused on the DTMC with a finite state space, we now discuss the absorbing probability issue in a DTMC with an infinite state space by presenting a classical DTMC model called a Branching process. Suppose each individual produces a random number of offspring, which is i.i.d., and the system starts with only one individual, called 0 generation. Denote by X_n the population of the nth generation where $n = 0, 1, \ldots$. Thus, $X_0 = 1$ and $X_n = \sum_{i=1}^{X_{n-1}} Y_i$, where Y_i denoted the number of offspring of the ith individual of the $(n-1)$st generation. Since Y_i is i.i.d. with a given distribution $P(Y_1 = k) = p_k$ where $k = 0, 1, \ldots$, we can compute the mean and variance of Y_1 as $\mu = \sum_{i=0}^{\infty} i p_i$ and $\sigma^2 = \sum_{i=0}^{\infty} (i - \mu)^2 p_i$. Clearly, the process $\{X_n, n = 0, 1, \ldots\}$ is a DTMC with the set of non-negative integers as the state space. We are interested in a few important quantities for this DTMC. These include the mean and variance for the population size of the nth generation and the extinction probabilities (absorbing probabilities). To compute the mean of X_n, we condition on X_{n-1} and obtain

$$
E[X_n] = E[E[X_n|X_{n-1}]] = E\left[E\left[\sum_{i=1}^{X_{n-1}} Y_i \Big| X_{n-1}\right]\right]
$$
$$
= E[X_{n-1}\mu] = \mu E[X_{n-1}].
$$

For the rule on conditional expectations, readers are referred to Chapters A and B. Using this recursion n times with $E[X_0] = 1$, we have

$$
E[X_n] = \mu E[X_{n-1}] = \mu^2 E[X_{n-2}] = \cdots = \mu^{n-1} E[X_1] = \mu^n.
$$

Similarly, it follows from the conditional variance formula that

$$
Var(X_n) = E[Var(X_n|X_{n-1})] + Var(E[X_n|X_{n-1}]). \tag{2.46}
$$

Note that $E[X_n|X_{n-1}] = X_{n-1}\mu$ and $Var(X_n|X_{n-1}) = X_{n-1}\sigma^2$. Using these expressions, recursively, in (2.46) yields

$$
\begin{aligned}
Var(X_n) &= E[X_{n-1}\sigma^2] + Var(X_{n-1}\mu) \\
&= \sigma^2\mu^{n-1} + \mu^2 Var(X_{n-1}) \\
&= \sigma^2\mu^{n-1} + \mu^2(\sigma^2\mu^{n-2} + \mu^2 Var(X_{n-2})) \\
&= \sigma^2(\mu^{n-1} + \mu^n) + \mu^4 Var(X_{n-2}) \\
&= \sigma^2(\mu^{n-1} + \mu^n) + \mu^4(\sigma^2\mu^{n-3} + \mu^2 Var(X_{n-3})) \\
&= \sigma^2(\mu^{n-1} + \mu^n + \mu^{n+1}) + \mu^6 Var(X_{n-3}) \\
&= \cdots \\
&= \sigma^2(\mu^{n-1} + \mu^n + \cdots + \mu^{2n-2}) + \mu^{2n} Var(X_0) \\
&= \sigma^2(\mu^{n-1} + \mu^n + \cdots + \mu^{2n-2}).
\end{aligned}
$$

Thus, we have

$$
Var(X_n) = \begin{cases} \sigma^2\mu^{n-1}\left(\frac{1-\mu^n}{1-\mu}\right), & \text{if } \mu \neq 1, \\ n\sigma^2, & \text{if } \mu = 1. \end{cases} \tag{2.47}
$$

Note that the population size variance increases geometrically if $\mu > 1$, decreases geometrically if $\mu < 1$, and increase linearly if $\mu = 1$.

Now we compute the probability of population extinction. This event happens when the population size is reduced to 0. Thus, we can treat state 0 as the absorbing state for the DTMC $\{X_n, n \geq 1\}$. Denote the extinction time by $N = \min\{n \geq 0 : X_n = 0|X_0 = 1\}$. Then, the probability of extinction at or before the nth generation can be defined as

$$
v_n = P(N \leq n) = P(X_n = 0), \quad n \geq 1.
$$

With this probability, we can also determine the probability that the population v will eventually die out by taking $\lim_{n \to \infty} v_n = v$. By using the generating function of discrete probability distribution, we can derive a recursive functional relation between the extinction probabilities of successive generations. Denote by $\phi_{Y_1}(z) = \sum_{k=0}^{\infty} p_k z^k$ the generating function of the individual's offspring distribution. Then, the following result can be established.

Proposition 2.2 *For a branching process, the extinction probability of the nth generation satisfies the following functional recursion:*

$$
v_n = \phi_{Y_1}(v_{n-1}) \text{ for } n \geq 2, \tag{2.48}
$$

with $v_1 = p_0$. Further, the limiting extinction probability $v \in (p_0, 1]$ is the fixed point of $v = \phi_{Y_1}(v)$.

Proof. Due to $X_0 = 1$, by conditioning on X_1, we have

$$v_n = \sum_{k=0}^{\infty} P(X_n = 0 | X_1 = k) p_k, \quad n \geq 1. \tag{2.49}$$

Clearly, $v_1 = p_0$, which is the case of $X_1 = 0$, is a direct result of this expression. For $X_1 = k \geq 1$, each of these k individuals produces a separate subpopulation that follows the same branching process as the original one starting at time 1. Thus, the event of the population extinction at time n is the union of the same events for k i.i.d. subpopulations that have evolved for $n - 1$ time units. Hence, we have

$$P(X_n = 0 | X_1 = k) = v_{n-1}^k. \tag{2.50}$$

Substituting (2.50) into (2.49) yields (2.48). Moreover, since $\phi_{Y_1}(z)$ is continuous and $v_n \to v$, then taking the limit $n \to \infty$ in (2.48) gives $v = \phi_{Y_1}(v)$. ∎

Since the limiting extinction probability is a fixed point of ϕ_{Y_1}, it can be computed by the recursion (2.48) numerically. Further, we can characterize this probability as follows.

Theorem 2.15 *For a branching process, if $\mu \leq 1$, then $v = 1$. If $\mu > 1$, then v is the unique root in $(0, 1)$ for equation $z = \phi_{Y_1}(z)$.*

Proof. If $\mu < 1$, then by the Markov inequality, we have

$$P(X_n \geq 1) \leq E[X_n] = \mu^n \to 0,$$

as $n \to \infty$. Thus, $v = \lim_{n \to \infty} (1 - P(X_n \geq 1)) = 1$. To prove the cases of $\mu > 1$ and $\mu = 1$, we need to establish the strict convexity of the generating function $\phi_{Y_1}(z)$ in two cases. One case is for $\mu > 1$ and the other is for $\mu = 1$ and $p_0 + p_1 < 1$.

First, for the both cases, to show the strict convexity of $\phi_{Y_1}(z)$ over $[0, 1]$, we need to verify the second-order derivative condition

$$\phi_{Y_1}''(z) = \sum_{k=2}^{\infty} k(k-1) p_k z^{k-2} > 0, \quad z \in [0, 1].$$

This condition holds only when $p_j > 0$ for some $j \geq 2$. If $\mu = \sum_{k=1}^{\infty} k p_k > 1$, which implies $p_0 + p_1 < 1$, hence, $p_j > 0$ for some $j \geq 2$. If $\mu = 1$, then we need the additional condition of $p_0 + p_1 < 1$ to make the condition satisfied.

With the strict convexity of ϕ_{Y_1}, to show the solution to $\phi_{Y_1}(z) = z$ in $[0, 1]$, we graph $\phi_{Y_1}(z)$ with boundary points $(0, p_0 > 0)$ and $(1, 1)$ for the two cases in Figure 2.8.

Note that the solution is the intersection of $\phi_{Y_1}(z)$ and the 45 degree line. For the case of $\mu > 1$, it follows from the strict convexity of ϕ_{Y_1} and $\phi_{Y_1}'(1) = \mu > 1$ that there exists exactly one $z < 1$ for which $\phi_{Y_1}(z) = z$. That is ϕ_{Y_1} has a unique

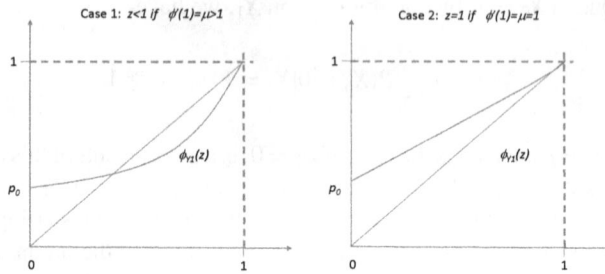

Figure 2.8 Two cases of $\phi_{Y_1}(z)$.

root $z = v$ in $(p_0, 1)$. For the case of $\mu = 1$, the slope of ϕ_{Y_1} coincides with that of the 45 degree line, indicating that the root to $\phi_{Y_1}(z) = z$ is $z = v = 1$. ∎

For the more interesting case with $\mu > 1$, if Y_1 is a distribution over a finite value range, then the limiting extinction probability is a unique solution $v = z \in (0, 1)$ to a polynomial equation $z = \phi_{Y_1}(z)$. For a special case where the maximum number of offspring produced by an individual is 2, the limiting extinction probability v is the unique solution to the quadratic equation $z = p_0 + p_1 z + p_2 z^2$ which is equivalent to

$$p_2 z^2 - (1 - p_1)z + p_0 = 0.$$

Using the root formula for the quadratic equation and the normalization condition $p_0 + p_1 + p_2 = 1$, we have a nice result $v = z = \min\{p_0/p_1, 1\}$. In general, we can obtain the numerical solution by using recursion based on $z_n = \phi_{Y_1}(z_{n-1})$ with $z_0 = 0$ and $n \uparrow$. Such a procedure is justified by the fixed point theorem (see Chapter G). Starting with $z_0 = 0$, we continue the iteration until $z_{n+1} - z_n \leq \varepsilon$ where ε is a pre-specified accuracy parameter. The solution is given by

$$z^* = \min\{z_{n+1} : \phi_{Y_1}(z_n) \leq z_n + \varepsilon\}.$$

Note that if $\mu \leq 1$, we can conclude immediately that $v = 1$, which implies that the population extinction is a sure thing.

For a finite Y_1 case, we may compute the time-dependent extinction probabilities recursively based on $v_n = \phi_{Y_1}(v_{n-1}) = \sum_{k=0}^{N} p_k (v_{n-1})^k$ for $n = 1, 2....$

Exercise 2.23 *Families in a rural region of a country determine the number of children that they will have according to the following rule: If the first child is a boy, they will have exactly one more child. If the first child is a girl, they continue to have children until the first boy, and then stop childbearing.*

(a) Find the probability that a particular family will have exactly k children in total, where $k = 0, 1, 2, ...$

(b) Find the probability that a particular family will have exactly k female children among their offspring, where $k = 0, 1, 2, ...$.

REFERENCE NOTES

This chapter presents the elementary stochastic models, namely, the discrete-time Markov chains (DTMCs). Due to its applications in computer and communication networks, we introduce the formulation of a DTMC for a small computer system with two processors. The steady-state and transient performance evaluations for DTMCs are discussed in this chapter. By taking the advantage of the simplicity of DTMC, we present the terminology, definitions, important concepts, and major properties for this class of stochastic models with both intuitive explanations and rigorous proofs. Although most of these results hold in more general settings, the proofs can be more technical and involved and require more mathematical background. For example, the proof of the Foster's criterion in a DTMC setting is relatively easy and taken from Breuer and Baum [2]. The branch processes are mainly based on Ross [3]. There are many good textbooks on introduction to DTMCs such as [1].

REFERENCES

1. A.S. Alfa, "Applied Discrete-Time Queues", Springer, New York, 2016.
2. L. Breuer and D. Baum, "An Introduction to Queueing Theory and Matrix-Analytic Methods", Springer, The Netherlands, 2005.
3. S. Ross, "Introduction to Probability Models", Academic Press, 11th ed., 2014.

3 Continuous-Time Markov Chains

In this Chapter, we consider the continuous-time Markov chain (CTMC). A stochastic process $\{X(t), t \geq 0\}$ is a CTMC if for all $s, t \geq 0$ and non-negative integers i, j (discrete states for MC) with state space Ω, and process history $x(u)$, where $0 \leq u < s$, the following Markovian property holds

$$P\{X(s+t) = j | X(s) = i, X(u) = x(u), 0 \leq u < s\} = P\{X(s+t) = j | X(s) = i\}. \tag{3.1}$$

If (3.1) is independent of s, then the CTMC is called time-homogeneous CTMC. That is $P\{X(s+t) | X(s) = i\} = P\{X(t) = j | X(0) = i\} = P_{ij}(t)$, which is called the transition probability function. Writing it in matrix form, we obtain the probability transition matrix for the CTMC as follows:

$$\mathbf{P}(t) = \begin{bmatrix} P_{00}(t) & P_{01}(t) & \cdots \\ P_{10}(t) & P_{11}(t) & \cdots \\ \vdots & \vdots & \ddots \end{bmatrix}. \tag{3.2}$$

The most significant difference between CTMC and DTMC is the extension from the deterministic time slots in which a state transition can occur to any random time point at which a state transition can occur. Such an extension may require the information about the time elapsed since the last state transition instant for predicting the future behavior of the process. This requirement may make the process non-Markovian unless the time elapsed since the last state transition does not affect the prediction of the process's future state. Fortunately, if this inter-state transition time follows an exponential distribution, this time elapsed information is not required due to the "memoryless property". In fact, the exponential distribution is the only continuous probability distribution with such a property and its discrete-time counterpart is the geometric distribution.

3.1 FORMULATION OF CTMC

We start with the simplest CTMC which is the arrival process to a service system. From time $t = 0$ in a store (when its business starts), we count the number of customers arriving at the store with rate λ as time goes on. If we assume that times between any two consecutive arrivals are independent, identical, and exponentially distributed random variables with probability density function

DOI: 10.1201/9781003150060-3

$f(x) = \lambda e^{-\lambda x}$, then the cumulative number of arrivals over $[0, t]$, defined as the time-dependent state variable, is a CTMC. This is because at any point in time, the number of customers arrived can be used to completely predict the number of customers at any future time point (the next state) with a given probability due to the memoryless property of the inter-arrival time. Clearly, this CTMC is the well-known Poisson process (see Ross [5] and the discussion later in this chapter). Due to its simplicity, we know the transition probability function and the marginal probability distribution of this CTMC at any finite time point. However, it is not much interesting to consider this CTMC's limiting behavior. This is because as time goes to infinity it surely does not have a stationary distribution (i.e., the number of customers arrived will go to infinity as well). Poisson process plays a fundamental role in stochastic modeling theory. As a basic stochastic process, it can be extended to general CTMCs from the state variable perspective such as a queueing process and to general point processes from the inter-event time perspective such as a renewal process. In this chapter, we focus on general CTMCs.

As a typical CTMC, we consider a simple queueing system. Suppose that a service station has a single server and customers arrive individually at the station according to a Poisson process with rate λ. Furthermore, service times are independent and identically distributed (i.i.d.) exponential random variables with mean of $1/\mu$. The state variable is the number of customers in the system (including the one in service), also called queue length, at a time point, denoted by $X(t)$. This state space is discrete. Let's look at the queue length process at an arbitrary time instant, call it present or $t = 0$, and observe $X(0) = i > 0$ customers in the system. Then the next state change will occur either when a customer arrives or the customer in service departs. Since the state change is either +1 or -1, the current queue length $X(0)$ and the occurrence of the next event type (arrival or departure) can predict the future state. The second part, in general, requires the information of the time elapsed since the last arrival and service time elapsed for the customer in service if the inter-arrival times and service times are i.i.d. random variables. Since both inter-arrival times and service times are exponentially distributed, the time elapsed information is not needed due to the memoryless property. Specifically, we can show that an exponentially distributed random variable, denoted by Y, has the following memoryless property: $P\{Y > s + t \mid Y > t\} = P\{Y > s\}$ (left as an exercise). If Y is an inter-arrival time between two consecutive customer arrivals (or a service time), this property indicates that the time elapsed since the last arrival (or the time elapsed since the latest service starts) will not affect the time to the next arrival (or the time to the next departure). Thus, the future queue length $X(s)$ can be completely predicted by the current queue length $X(t)$ where $s > t$ and hence the queueing process become a CTMC. Such a single server queue is denoted by $M/M/1$ in Kendall notation, where the first M (for "Markovian") indicates the exponential inter-arrival time, the second M indicates the exponential service time, and 1 indicates a single server.

Exercise 3.1 *(a) Show the memoryless property of an exponentially distributed random variable. (b) For two exponentially distributed random variables X_1 and X_2 with rates λ_1 and λ_2, respectively, show that the probability of $P(X_1 < X_2) = \lambda_1/(\lambda_1 + \lambda_2)$. (c) For the two random variables in (b), show that the minimum of them is also an exponentially distributed random variable with rate $\lambda_1 + \lambda_2$. Generalize this two random variable result to the minimum of n independent exponentially distributed random variables.*

In a CTMC, a state transition can happen at any time. Thus, there are two aspects about each transition from the current state: when and where? For the $M/M/1$ queue, the second question is easy for a non-empty state ($X(0) = i > 0$). It is either $X(0) + 1$ (arrival) or $X(0) - 1$ (departure). The answer to the first question is that the time to the next state transition is the minimum of the two independent exponential random variables (i.e., the minimum of the remaining service time and the time to the next arrival). It is well-known that the minimum of two exponential random variables is still an exponential random variable with the rate equal to the sum of the two rates (Exercise 3.1). For the $M/M/1$ queue, if the current state is $i > 0$, then the time until next state change is exponentially distributed with rate of $\lambda + \mu$. Furthermore, we know that if an arrival occurs before a departure, then state $i+1$ is reached, otherwise, state $i - 1$ is reached. The former case occurs with probability $\lambda/(\lambda + \mu)$ and the latter case occurs with probability $\mu/(\lambda + \mu)$ (Exercise 3.1). Therefore, for state $i > 0$ in the $M/M/1$ queue, the amount of time it spends before making a transition into a different state is exponentially distributed with rate $\lambda + \mu$. When the process leaves state i, it enters state j with some probability, denoted by P_{ij} (also called jump-to probability). Then $P_{ij} = \lambda/(\lambda + \mu)$ if $j = i + 1$ or $P_{ij} = \mu/(\lambda + \mu)$ if $j = i - 1$; otherwise $P_{ij} = 0$. For state $i = 0$, the amount of time it spends before making a transition into a different state is exponentially distributed with rate λ and $P_{01} = 1$. A more general process with this type of transitions (but state-dependent) is called a birth-and-death process which will be discussed below. Now we give a constructive definition of a CTMC.

Definition 3.1 *If a stochastic process, $\{X(t), t \geq 0\}$ on a discrete state space Ω, has the following properties that each time it enters state i: (a) the amount of time it spends in that state before making a transition into a different state is exponentially distributed with rate v_i (i.e., the mean sojourn time is $1/v_i$); and (b) when the process leaves state i, it enters state j with probability P_{ij}. For all i, P_{ij} must satisfy*

$$P_{ii} = 0, \quad \sum_{j \in \Omega} P_{ij} = 1, \tag{3.3}$$

it is said to be a CTMC.

Exercise 3.2 *In the lobby of a college building, there are three public computers for students to use. Users do not wait when no computer is idle. The computer use time for each user is exponentially distributed with rate μ_i for $i = 1, 2, 3$. The inter-arrival time between two consecutive potential users is exponentially distributed with rate λ. Show that the number of users at any time can be modeled as a CTMC.*

3.2 ANALYZING THE FIRST CTMC: A BIRTH-AND-DEATH PROCESS

If the state transition of a CTMC is limited to either an increase by 1 or a decrease by 1 (for the non-empty state), it is called a birth-and-death process (BDP). Clearly, the $M/M/1$ queue introduced above is a special BDP. Now we consider a more general case where the birth rate and death rate depend on the states. It is considered as one of the most popular CTMC's. We may consider the BDP as a population changing with time. Let $X(t)$ be the population size at time t and the state space be $E = \{0, 1, 2, ...\}$. Assume that, in state $X(t) = i$, the time to the next birth (or death) is exponentially distributed with rate λ_i (or μ_i), denoted by B_i (or D_i). That is $B_i \sim Exp(\lambda_i), i \geq 0$ and $D_i \sim Exp(\mu_i), i \geq 1$, respectively, and they are independent. Then it follows from the properties of exponential distributions that this BDP is a CTMC with the following sojourn times in different states, denoted by $T_i : T_i = min\{B_i, D_i\} \sim Exp((\lambda_i + \mu_i) = v_i), i \geq 1$, and $T_0 \sim Exp(\lambda_0)$. The transition probabilities are as follows:

$$P_{i,i+1} = \frac{\lambda_i}{\lambda_i + \mu_i}, P_{i,i-1} = \frac{\mu_i}{\lambda_i + \mu_i}, P_{ij} = 0, \; j \neq i \pm 1, i \geq 1, \; P_{01} = 1. \quad (3.4)$$

Due to the continuous-time process, we cannot develop the recursion-based formulas to compute the time-dependent performance measures as in the DTMC cases discussed in Chapter 2. Instead, we have to utilize a different approach based on differential equations to analyze CTMCs. The common performance measures for CTMCs are the expected values of some random variables related the state of the process (i.e., state variable), which can be either time-dependent (for transient state) or time-independent (for stationary or steady state). As an example, we start with developing a time-dependent performance measure for a special BDP with immigration, which is taken from Ross [5], where α is the immigration rate, λ is the birth rate, and μ is the death rate. Suppose we are interested in the expected population size at time t, denoted by $M(t) = E[X(t)]$, given that the system starts with initial population size $X(0) = i$. For a small time increment h, from the property of exponential distribution (see Ross [5]), we have

$$X(t+h) = \begin{cases} X(t)+1 & \text{with probability}[\alpha + X(t)\lambda]h + o(h), \\ X(t)-1 & \text{with probability} X(t)\mu h + o(h), \\ X(t) & \text{otherwise,} \end{cases} \quad (3.5)$$

where $o(h)/h \to 0$ as $h \to 0$. Conditioning on $X(t)$ yields the conditional expectation at time $t + h$

$$E[X(t+h)|X(t)] = X(t) + \alpha h + X(t)\lambda h - X(t)\mu h + o(h). \quad (3.6)$$

Taking the expectation on both sides of the equation above, we obtain

$$E[E[X(t+h)|X(t)]] = M(t+h) = M(t) + \alpha h + M(t)\lambda h - M(t)\mu h + o(h), \quad (3.7)$$

which is equivalent to

$$\frac{M(t+h) - M(t)}{h} = \alpha + (\lambda - \mu)M(t) + \frac{o(h)}{h}. \quad (3.8)$$

Taking the limit of $h \to 0$, a differential equation is established for $M(t)$ as

$$M'(t) = (\lambda - \mu)M(t) + \alpha. \quad (3.9)$$

with a solution of

$$M(t) = \left(\frac{\alpha}{\lambda - \mu} + i \right) e^{(\lambda - \mu)t} - \frac{\alpha}{\lambda - \mu}. \quad (3.10)$$

However, for the BDP, some expected time intervals of interest, rather than the time-dependent performance measures, may be obtained by using the recursive relations. For example, let T_{ij} be the time to reach j first time starting from i. The expect time of this interval can be computed by using the recursions. Due to the simple structure of the transitions, it is easy to write

$$E[T_{ij}] = \frac{1}{\lambda_i + \mu_i} + P_{i,i+1} \times E[T_{i+1,j}] + P_{i,i-1} \times E[T_{i-1,j}], \ i,j \geq 1. \quad (3.11)$$

This expected first passage time can be solved by starting with $E[T_{0,1}] = 1/\lambda_0$. As an example, we consider an $M/M/1$ queue where $\lambda_i = \lambda$, and $\mu_i = \mu$. Clearly, $E[T_{0,1}] = 1/\lambda$. Note that $E[T_{1,2}] = 1/(\lambda + \mu) + P_{1,2} \times 0 + P_{1,0} \times E[T_{0,2}] = 1/(\lambda + \mu) + [\mu/(\lambda + \mu)](E[T_{0,1}] + E[T_{1,2}])$. With this recursion, we obtain $E[T_{1,2}] = \frac{1}{\lambda}\left[1 + \frac{\mu}{\lambda}\right]$. Repeatedly, we can derive $E[T_{i,i+1}] = \frac{1}{\lambda}\left[1 + \frac{\mu}{\lambda} + \frac{\mu^2}{\lambda^2} + \cdots + \frac{\mu^i}{\lambda^i}\right] = \frac{1}{\lambda - \mu}[1 - (\mu/\lambda)^{i+1}]$ (more analysis on this problem can be found in Ross [5]).

Furthermore, the distribution of T_{ij} can be obtained by using Laplace–Stieltjes transform (LST) (see A.6 of Chapter A). For a random variable Y, denote its LST by $\widetilde{L}_Y(s) = E[e^{-sY}]$. Based on the random variable relation in a BDP

$$T_{ij} = T_i + I_{i,i+1}T_{i+1,j} + I_{i,i-1}T_{i-1,j}, \tag{3.12}$$

where I_{ij} is an indicator variable with $I_{ij} = 1$ if $i = j$ and 0 otherwise. Taking the LST on both sides, we have

$$\widetilde{L}_{T_{ij}}(s) = E[e^{-sT_{ij}}] = \widetilde{L}_{T_i}(s) \times \{P_{i,i+1}E[e^{-sT_{i+1,j}}] + P_{i,i-1}E[e^{-sT_{i-1,j}}]\}, \tag{3.13}$$

where

$$\widetilde{L}_{T_i}(s) = \frac{\lambda_i + \mu_i}{\lambda_i + \mu_i + s}. \tag{3.14}$$

The LST for the first passage time can be solved by starting from $E[e^{-sT_{0,1}}] = \lambda_0/(\lambda_0 + s)$. Using the same example of $M/M/1$ queue, equation (3.13) with $i \geq 1$ and $j = 0$ becomes

$$E[e^{-sT_{i,0}}] = \frac{1}{\lambda + \mu + s}[\lambda E[e^{-sT_{i+1,0}}] + \mu E[e^{-sT_{i-1,0}}]], \ i \geq 1, \tag{3.15}$$

with $E[e^{-sT_{0,0}}] = 1$. Note that $T_{i,0} = T_{i,i-1} + T_{i-1,i-2} + ... + T_{1,0}$ (time from a non-empty state to the empty state) and these i one-step down first passage times are i.i.d.'s due to the transition structure of BDP with constant λ and μ. Thus $\widetilde{L}_{T_{i,0}}(s) = (\widetilde{L}_{T_{1,0}}(s))^i$. Thus, for a fixed s, (3.15) can be considered as a second order difference equation with the solution

$$E[e^{-sT_{i,0}}] = \frac{1}{(2\lambda)^i}[\lambda + \mu + s - \sqrt{(\lambda + \mu + s)^2 - 4\lambda\mu}]^i. \tag{3.16}$$

Using the LST, we could obtain the higher moments of $T_{i,0}$. For example, the variance of this time period can be computed.

Transition probability as a time function in a CTMC is more complex than the multi-step transition probability in a DTMC. Similar to the analysis above, the close form expression is only possible for some very simple CTMCs. For example, as mentioned earlier, the Poisson process with constant rate λ is a simple CTMC (a special case of a pure birth process) with the cumulative number of arrivals in the system as the state variable. Assume that the system starts with i customers present or $X(0) = i$, then we can write the transition probability function as

$$P_{ij}(t) = P\{j - i \text{ arrivals in } (0,t]|X(0) = i\} = P\{j - i \text{ arrivals in } (0,t]\}$$
$$= e^{-\lambda t}\frac{(\lambda t)^{j-i}}{(j-i)!}. \tag{3.17}$$

For more general CTMCs, we usually cannot get the explicit expressions for transition probability functions. However, a set of differential equations satisfied by the transition probability function can be obtained as indicated in the next section.

Exercise 3.3 *For the $M/M/1$ queue as a BDP, develop the recursion for computing the variance of T_i. Then compute the variance for the time to go from 0 to 3 in a case of $\lambda = 2$ and $\mu = 3$.*

Exercise 3.4 *For the $M/M/1$ queue as a BDP, find the closed form expression for $E[T_i]$ and $E[T_{kj}]$ with $k < j$ for the $\lambda = \mu$ case (Note: this is an unstable queue case).*

3.3 TRANSITION PROBABILITY FUNCTIONS FOR CTMC

For a time-homogeneous CTMC, $\{X(t), t \geq 0\}$ on a discrete state space Ω, the transition probability function defined as

$$P_{ij}(t) = P\{X(s+t) = j | X(s) = i\} \text{ for any } s \text{ and } i, j \in \Omega. \tag{3.18}$$

Since the state transition can occur at any point in time, we need to analyze the process dynamics in a infinitesimal time interval. Thus, we introduce the instantaneous transition rates and jump-to transition probabilities. The jump-to transition probability from state i to state j is defined as

$P_{ij} = P\{\text{CTMC makes a transition from } i \text{ to } j \text{ at a state transition instant}\}$.

Obviously, $\sum_j P_{ij} = 1$ and $P_{ii} = 0$. Note that the jump-to transition probability does not have the time argument. Since the duration time in state i follows an exponential distribution with rate v_i, we can define the instantaneous transition rate from state i to j as $q_{ij} = v_i P_{ij}$. q_{ij} can be interpreted as the rate, when in state i, at which the process makes a transition into state j. It is easy to get $P_{ij} = q_{ij}/v_i = q_{ij}/\sum_j q_{ij}$. Based on the property of exponential distribution, we can give an alternative definition of the instantaneous transition rate. Suppose that the CTMC is at state i at time zero. For a very small time increment h, the probability that the CTMC makes a transition (or leaves the current state i) is $v_i h + o(h) = 1 - P_{ii}(h)$, where $\lim_{h \to 0} o(h)/h \to 0$. From this probability, we obtain

$$v_i = \lim_{h \to 0} \frac{1 - P_{ii}(h)}{h}. \tag{3.19}$$

Now we consider the transition probability function at time h. Clearly, $P_{ij}(h) = P-(\text{Transitions occur during } h) \cap (\text{the transition is to } j \text{ from } i)'' = (hv_i + o(h)) \times P_{ij} = hv_i P_{ij} + o(h)$ from which it follows

$$q_{ij} = v_i P_{ij} = \lim_{h \to 0} \frac{P_{ij}(h)}{h} \equiv \frac{dP_{ij}(t)}{dt}\Big|_{t=0+}, \tag{3.20}$$

which denote the right derivative at 0 because $P_{i,j}(t)$ is not defined for $t < 0$. The expressions of (3.19) and (3.20) will be used to establish differential equations that transition probability functions satisfy.

Similar to the DTMC, the C-K equations can be written as

$$P_{ij}(t+s) = \sum_{k \in \Omega} P_{ik}(t)P_{kj}(s). \tag{3.21}$$

The proof can be easily done by conditioning on the state visited at time t and using the total probability formula. Using (3.19), (3.20), and (3.21), we can establish the differential equations that $P_{ij}(t)$ has to satisfy. Consider a small time increment h followed by the time interval of t. We can write $P_{ij}(h+t) = \sum_{k \in \Omega} P_{ik}(h)P_{kj}(t)$. Subtracting $P_{ij}(t)$ from both sides of the CK equation, we obtain

$$\begin{aligned}
&P_{ij}(h+t) - P_{ij}(t) \\
&= \sum_{k \in \Omega} P_{ik}(h)P_{kj}(t) - P_{ij}(t) = \sum_{k \in \Omega, k \neq i} P_{ik}(h)P_{kj}(t) - P_{ij}(t) + P_{ii}(h)P_{ij}(t) \\
&= \sum_{k=\Omega, k \neq i} P_{ik}(h)P_{kj}(t) - (1 - P_{ii}(h))P_{ij}(t).
\end{aligned}$$
$$\tag{3.22}$$

Dividing both sides of the above equation by h and taking the limit $h \to 0$, we have

$$\lim_{h \to 0} \frac{P_{ij}(h+t) - P_{ij}(t)}{h} = \lim_{h \to 0} \left\{ \sum_{k \neq i} \frac{P_{ik}(h)}{h} P_{kj}(t) - \left[\frac{1 - P_{ii}(h)}{h} \right] P_{ij}(t) \right\}, \tag{3.23}$$

which is

$$dP_{ij}(t)/dt = \sum_{k \neq i} q_{ik}P_{kj}(t) - v_i P_{ij}(t). \tag{3.24}$$

Note that for the infinite state space CTMC the interchange the limit and summation in above derivation can be justified in this set of differential equations. An intuitive explanation is that the summation of transition probability over the destination state should be 1 or upper bounded (a complete distribution) so that the dominance convergence theorem applies to justify the interchange of the limit and summation. Equations (3.24) are called Kolmogorov's backward equations as the first time increment is approaching zero and period t follows it. If the time increment approaching zero is taken after time period t, we can obtain another set of differential equations called Kolmogorov's forward equations. Following the similar development as above, we start with

$$P_{ij}(t+h) - P_{ij.}(t) = \sum_{k \neq j} P_{ik}(t)P_{kj}(h) - [1 - P_{jj}(h)]P_{ij}(t), \tag{3.25}$$

and end up with

$$dP_{ij}(t)/dt = \sum_{k \neq j} q_{kj}P_{ik}(t) - v_j P_{ij}(t). \tag{3.26}$$

Note that the interchange of limit and summation cannot be justified in all cases as the similar explanation for the backward equations cannot be applied here (see Feller [2]). With an infinite state space, there could be infinitely many transitions in a finite time period. This can cause a technical problem that the interchange of limit and summation cannot be justified since for a fixed period of time t, infinite many transitions may happen, before doing the asymptotic analysis. In contrast, for the backward equations, the many possible transitions are considered for a small time internal h and when $h \to 0$, infinitely many transitions cannot happen and a finite state space behavior follows. Thus we can interchange the limit and summation. However, the forward equations hold for most practical models including DBP and all finite state space models. In matrix form, the forward or backward equations can be expressed as

$$\mathbf{P}'(t) = \mathbf{P}(t)\mathbf{Q}, \quad \text{or} \quad \mathbf{P}'(t) = \mathbf{Q}\mathbf{P}(t). \tag{3.27}$$

where

$$\mathbf{Q} = \begin{bmatrix} -v_0 & q_{01} & q_{02} & \cdots \\ q_{10} & -v_1 & q_{12} & \cdots \\ \vdots & \vdots & \vdots & \ddots \end{bmatrix}, \tag{3.28}$$

which is called infinitesimal generator (also called rate matrix) of the CTMC. The proofs of these equations by using matrix notation are simple (left as exercises).

Exercise 3.5 *Verify (3.27) and (3.28).*

Exercise 3.6 *(a) Verify if the following matrix*

$$\mathbf{Q} = \begin{pmatrix} -5 & 3 & 2 \\ 0 & -4 & 4 \\ 2 & 5 & -7 \end{pmatrix},$$

is a valid infinitesimal generator of a three-state CTMC. (b) Find the distribution of the sojourn time in each state and the jump-to transition probability matrix. (c) Like a DTMC, draw a state transition diagram for this CTMC with the transition rate on each directed arc.

Recall that the scalar ordinary differential equation (ODE) $f'(t) = cf(t)$ has an exponential function solution as $f(t) = f(0)e^{ct}, t > 0$. Similarly, the matrix ODEs (forward and backward) have an exponential form solution.

Theorem 3.1 *(matrix exponential representation) The transition probability function can be expressed as a matrix-exponential function of the rate matrix $\mathbf{Q}$, i.e.*

$$\mathbf{P}(t) = e^{\mathbf{Q}t} \equiv \sum_{n=0}^{\infty} \frac{\mathbf{Q}^n t^n}{n!}. \tag{3.29}$$

This matrix exponential is the unique solution to the two ODEs with initial condition $\mathbf{P}(0) = \mathbf{I}$.

If the interchange of the differentiation and summation is valid, we can verify that (3.29) satisfies the two ODEs. That is

$$
\frac{d\mathbf{P}(t)}{dt} = \frac{de^{\mathbf{Q}t}}{dt} = \frac{d}{dt} \sum_{n=0}^{\infty} \frac{\mathbf{Q}^n t^n}{n!} = \sum_{n=0}^{\infty} \frac{d}{dt} \frac{\mathbf{Q}^n t^n}{n!}
$$
$$
= \sum_{n=0}^{\infty} \frac{n\mathbf{Q}^n t^{n-1}}{n!} = \mathbf{Q} \sum_{n=0}^{\infty} \frac{\mathbf{Q}^n t^n}{n!} = \mathbf{Q}e^{\mathbf{Q}t} = \mathbf{Q}\mathbf{P}(t). \tag{3.30}
$$

In general, it is hard to solve these differential equations to get the explicit expressions for the transition probability functions. Only for some extremely simple CTMCs, we may obtain the solution to these equations. For example, if the BDP has only two states of 0 and 1 with birth rate of λ and death rate of μ, the backward equation becomes

$$
P'_{00}(t) = \lambda \left[P_{10}(t) - P_{00}(t) \right], \quad P'_{10}(t) = \mu P_{00}(t) - \mu P_{10}(t). \tag{3.31}
$$

These ordinary differential equations can be solved as

$$
P_{00}(t) = \frac{\lambda}{\lambda + \mu} e^{-(\lambda+\mu)t} + \frac{\mu}{\lambda + \mu}, \quad P_{10}(t) = \frac{\mu}{\lambda + \mu} - \frac{\mu}{\lambda + \mu} e^{-(\lambda+\mu)t}. \tag{3.32}
$$

If $t \to \infty$, $P_{00}(\infty) = P_{10}(\infty) = \mu/(\lambda + \mu)$ which is the stationary probability for state 0.

Exercise 3.7 *(a) Write the differential equations (3.31) in matrix form. (b) Solve these two differential equations for (3.32).*

Exercise 3.8 *Consider a pure birth process which can be considered as a special case of BDP. That is, for state $i(\geq 0)$, the birth rate is λ_i and the death rate $\mu_i = 0$. Show that the transition probability functions are given by*

$$
P_{ii}(t) = e^{-\lambda_i t}, \quad i \geq 0
$$
$$
P_{ij}(t) = \lambda_{j-1} e^{-\lambda_j t} \int_0^t e^{\lambda_j s} P_{i,j-1}(s)\,ds, \quad j \geq i+1.
$$

3.3.1   UNIFORMIZATION

To better link the CTMC to its corresponding DTMC, we introduce the uniformization technique. The unique feature of the CTMC is that the time between two consecutive transitions is exponentially distributed with the state-dependent rate v_i (which is also the minimum of a set of exponential random variables corresponding to transition times to all possible future states T_{ij}, $j \in \Gamma(i)$). If we ignore the sojourn time in each state and only focus on the state transitions at transition instants, then we have a corresponding DTMC. The inter-transition times of the CTMC can be treated as the inter-arrival times of an independent Poisson process. However, a big difference between the CTMC transition process and the standard Poisson process is that former's transition rate (or mean sojourn time) is state-dependent and the latter's is state-independent (i.e., constant). To overcome this difference, we introduce a uniform transition rate, denoted by λ, which is not smaller than the maximum transition rate of the CTMC. That is $\lambda \geq v_i \equiv -q_{ii} = \sum_{j \neq i} q_{ij}$ for all i. We then use the Poisson process with rate λ to model the state transitions of the CTMC over a time interval. Since this Poisson process with the rate greater than or equal to the maximum transition rate of all states, it makes more frequent transitions than any feasible state does. To keep the same probability law of the original CTMC process, some of these transitions will not represent real (actual) state transitions and are considered as fictitious transitions to the current state itself (i.e., no real transitions occur). Clearly, the proportion of transitions of the Poisson process for state i that represent the real state transition will be v_i/λ. This makes common sense as for smaller v_i, the mean sojourn time in state i (i.e., $1/v_i$) is longer or the transition frequency is lower, thus the lower proportion of transitions of the Poisson process will be the real state transitions. The proportion of fictitious transitions (transitions to the state itself) is $1 - v_i/\lambda$. Accordingly, at the transition instants of the Poisson process, the transition probability of the corresponding DTMC will have non-zero probability for the self-transition. This is different from the DTMC embedded at the real state transition instants considered earlier where the probability of self-transition is zero or $P_{ii} = 0$. Since the transition rate for every state is the same with value λ, this approach to modeling a CTMC is called the uniformization technique. The transition probabilities can be determined as follows:

$$\widetilde{P}_{i,j} = \frac{v_i}{\lambda} P_{ij} = \frac{q_{ij}}{\lambda}, \; j \neq i; \; \widetilde{P}_{i,i} = 1 - \sum_{j \neq i} \widetilde{P}_{i,j} = 1 - \frac{v_i}{\lambda} = 1 + \frac{q_{i,i}}{\lambda} = 1 - \frac{\sum\limits_{j \neq i} q_{i,j}}{\lambda}.$$

$$(3.33)$$

In matrix form, we have

$$\widetilde{\mathbf{P}} = \mathbf{I} + \lambda^{-1} \mathbf{Q}. \qquad (3.34)$$

Uniformization allows us to apply properties of DTMC's to analyze CTMC's. For a general CTMC characterized by the rate matrix $\mathbf{Q}$, we can express its transition probability function in terms of the transition probability of the corresponding DTMC where the transitions are governed by the Poisson process. That is

$$P_{i,j}(t) \equiv P(X(t) = j | X(0) = i) = \sum_{k=0}^{\infty} \widetilde{P}_{i,j}^{k} P(N(t) = k) = \sum_{k=0}^{\infty} \widetilde{P}_{i,j}^{k} \frac{e^{-\lambda t}(\lambda t)^k}{k!}. \quad (3.35)$$

If we denote the DTMC by $\{Y_n : n \geq 0\}$ and the associated Poisson process by $\{N(t) : t \geq 0\}$, the CTMC $\{X(t) : t \geq 0\}$ can be constructed as random time change of DTMC by the Poisson process. That is

$$X(t) = Y_{N(t)}, \ t \geq 0. \tag{3.36}$$

Theorem 3.2 *The CTMC constructed by (3.33) and (3.35) follows the same probability law of the original CTMC.*

Proof. We can verify that the two CTMC's have the same transition rates. Using (3.35), for $i \neq j$ and a small time interval h, we have

$$\begin{aligned}
P_{i,j}(h) &= \sum_k \widetilde{P}_{i,j}^k \frac{e^{-\lambda h}(\lambda h)^k}{k!} = \lambda h e^{-\lambda h} P_{i,j}^1 + o(h) \\
&= \lambda h e^{-\lambda h} \frac{q_{i,j}}{\lambda} + o(h) = q_{ij} h + o(h).
\end{aligned} \tag{3.37}$$

∎

Exercise 3.9 *Find the transition probability function for the two-state $(0,1)$ BDP by using the uniformization technique.*

3.4 STATIONARY DISTRIBUTION OF CTMC

Suppose that the CTMC is ergodic. Then it has a limiting distribution. The existence of the limiting (stationary) distribution can be shown by using the same arguments as in the DTMC. This is because that any CTMC can be related to a corresponding DTMC called embedded Markov chain. Define $\alpha_j = \lim_{t \to \infty} P_{ij}(t)$ as the stationary probability for state i. . Taking the limit of $t \to \infty$ on both sides of (3.26) and using the fact of $\lim_{t \to \infty} P'_{ij}(t) = 0$, we obtain

$$0 = \sum_{k \neq j} q_{kj} \alpha_k - v_j \alpha_j, \ \text{ or } \ v_j \alpha_j = \sum_{k \neq j} q_{kj} \alpha_k. \tag{3.38}$$

This set of equations are called rate balance equations. Each equation means that the total rate leaving a state (node in the state transition diagram) equals the total rate entering this state when the system has reached steady state. This rate balance condition holds for any set of states and its complement. That is the total rate out of a set A equals the total rate into the set A. Such a rate balance equation between A and A^c is called a cut balance equation in contrast to the node balance equation and can be more convenient in solving for the stationary distribution. Combining the set of rate balance equations with the normalization condition $\sum_j \alpha_j = 1$, we can solve for the stationary distribution

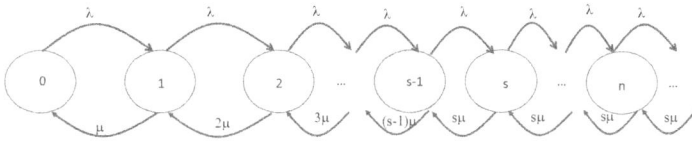

Figure 3.1 A multi-server queue.

of the CTMC. Define the transition rate matrix as

$$\mathbf{Q} = \begin{bmatrix} -v_0 & q_{01} & q_{02} & q_{03} & \cdots \\ q_{10} & -v_1 & q_{12} & q_{13} & \cdots \\ q_{20} & q_{21} & -v_2 & q_{23} & \cdots \\ q_{30} & q_{31} & q_{32} & -v_3 & \cdots \\ \vdots & \vdots & \vdots & \vdots & \ddots \end{bmatrix}, \tag{3.39}$$

and stationary probability vector as $\alpha = [\alpha_0, \alpha_1, ...]$. Then the stationary distribution satisfies the equations in matrix form

$$\alpha \mathbf{Q} = 0. \tag{3.40}$$

The normalization condition can be written as $\alpha \mathbf{e} = 1$, where $\mathbf{e}$ is a column vector of 1's with appropriate dimension. For the $\mathbf{Q}$ matrix, the sum of each row is zero. For ergodic CTMCs with finite state space, solving a set of rate balance equations for the stationary distribution is straightforward. However, for CTMCs with infinite state space, solving for the stationary distribution can be tricky. In the case of birth-and-death (BD) type CTMC, we can utilize the recursions to solve for the stationary distribution as shown in the example below.

Example 3.1 *Suppose a queue with Poisson arrivals of rate λ and s identical servers. The service time is exponentially distributed with rate μ. Such a queue is denoted by $M/M/s$ and the state transition diagram is shown in Figure 3.1. Solve for the stationary distribution of the queue length.*

Let state variable, $X(t)$, be the number of customers in the system. Then the process $\{X(t), t \geq 0\}$ is a CTMC with transition rates as

$$q_{i,i+1} = \lambda, \quad q_{i,i-1} = \min(i, s)\mu. \tag{3.41}$$

Denote by π_i the limiting probability of the queue length. Using the cut balance condition, we have $\lambda \pi_0 = \mu \pi_1, \lambda \pi_1 = 2\mu \pi_2, ...$ In general, we can write

$$\begin{aligned} \lambda \pi_j &= (j+1)\mu \pi_{j+1}, & 0 \leq j \leq s-1, \\ \lambda \pi_j &= s\mu \pi_{j+1}, & j \geq s. \end{aligned} \tag{3.42}$$

Using these recursions repeatedly and defining $\rho = \lambda/(s\mu)$, we obtain

$$\pi_j = \begin{cases} \pi_0 \frac{(s\rho)^j}{j!}, & \text{if } 0 < j < s, \\ \pi_0 \frac{\rho^j s^s}{s!}, & \text{if } s \leq j, \end{cases} \tag{3.43}$$

where

$$\pi_0 = \left[\left(\sum_j^{s-1} \frac{(s\rho)^j}{j!} \right) + \frac{(s\rho)^s}{s!} \frac{1}{1-\rho} \right]^{-1}. \tag{3.44}$$

For non-BDP type CTMCs, using cut balance or node balance conditions may not work in solving for the stationary distributions. However, we can apply other methods such as using transforms or time-reversibility property to solve for the stationary distributions.

3.4.1 OPEN JACKSON NETWORKS

As an important example of CTMC, we consider a special type of queueing network, called open Jackson network, where external customer arrivals are according to Poisson processes and service times are all exponentially distributed. Specifically, the following conditions are met:

1. The network is made of N work stations (or nodes). Each node i has s_i identical servers with infinite buffers, where $1 \leq s_i \leq \infty$ and $i = 1,...,N$. The service times at node i are exponentially distributed with rate μ_i.

2. External customers arrive at each node i according to a Poisson process with rate $0 \leq \lambda_i < \infty$. All arrival processes are independent of each other and are also independent of the service processes. At least one node has no external arrivals, i.e., $\lambda_i = 0$ (this condition is optional).

3. When a customer completes service at node i, he or she exits the network with probability q_i or joins the queue at node j with probability p_{ij}. It is assumed that $q_i + \sum_{j=1}^{N} p_{ij} = 1$. Define the $N \times N$ routing matrix $\mathbf{P} = [p_{ij}]$, which is independent of the state of the network. The $\mathbf{I} - \mathbf{P}$ matrix is invertible if every customer would eventually exit the network after a finite sojourn.

Since we are interested in the stationary distribution of the network, we first derive the stability condition. Denote by a_j the effective input rate to station j (external arriving customers plus customers sent from other stations after services). Then, it follows from the flow balance in terms of average rates that

$$a_j = \lambda_j + \sum_{i=1}^{N} a_i p_{ij} \quad \text{for } j = 1, 2, ..., N. \tag{3.45}$$

Writing this equation in matrix form, we have

$$\mathbf{a} = \lambda + \mathbf{a}\mathbf{P},$$

where $\mathbf{a} = (a_1, a_2, ..., a_N)$ and $\lambda = (\lambda_1, \lambda_2, ..., \lambda_N)$. Clearly, we have

$$\mathbf{a} = \lambda (\mathbf{I} - \mathbf{P})^{-1}. \tag{3.46}$$

For any node i, the stability condition is $a_i < s_i \mu_i$ with $i = 1, 2, ..., N$. The stability condition in matrix form can be written as

$$\mathbf{a} < \mathbf{s}\mu,$$

where $\mathbf{s} = (s_1, s_2, ..., s_N)$ and $\mu = \text{diag}(\mu_1, \mu_2, ..., \mu_N)$. To analyze a Jackson network, we formulate the N-dimensional CTMC with state vector $\mathbf{X}(t) = (X_1(t), X_2(t), ..., X_N(t))$, where $X_i(t)$ is the number of customers in node i of the network at time t with $i = 1, ..., N$. Under the stability condition, we denote the stationary distribution of the CTMC state $\mathbf{x} = (x_1, x_2, ..., x_N)$ by

$$p(\mathbf{x}) = \lim_{t \to \infty} P(\mathbf{X}(t) = (x_1, x_2, ..., x_N)).$$

To write the flow balance equation, we utilize the unit vector $\mathbf{e}_i$ which has one as the ith element and zeros everywhere else. With this notation, we can write the generic balance equation for state $\mathbf{x}$ as follows:

$$p(\mathbf{x}) \left(\sum_{i=1}^{N} \lambda_i + \sum_{i=1}^{N} \min(x_i, s_i) \mu_i \right)$$
$$= \sum_{i=1}^{N} p(\mathbf{x} - \mathbf{e}_i) \lambda_i + \sum_{i=1}^{N} p(\mathbf{x} + \mathbf{e}_i) q_i \min(x_i + 1, s_i) \mu_i$$
$$+ \sum_{j=1}^{N} \sum_{i=1}^{N} p(\mathbf{x} + \mathbf{e}_i - \mathbf{e}_j) p_{ij} \min(x_i + 1, s_i) \mu_i.$$

The LHS of the equation above represents all transitions out of state $\mathbf{x}$, which includes any external arrivals or service completions. On the other hand, the RHS includes all transitions into state $\mathbf{x}$. Specifically, the first term represents that an arrival will make the process to reach state $\mathbf{x}$; the second term represents that a customer's exiting the network after a service will bring the process to state $\mathbf{x}$; and finally, the third term represents that a customer's joining other node after a service will lead the process to state $\mathbf{x}$. To solve this balance equation, we consider a "guessed form" of the solution and verify its satisfying the balance equation. Then it is the solution to the balance equation due to its uniqueness. An extreme case of the open Jackson network is one station system without feedback. That is the $M/M/s$ queue treated earlier which has the stationary distribution (3.43). A simple extension is a two station tandem queue which turns out to be the multiplication of two stationary distributions of (3.43) form (as an exercise). Thus, we guess that such a product form still holds for the stationary distribution of the Jackson network. Following the distribution form (3.43), we define

$$\phi_j(n) = \begin{cases} \frac{1}{n!}\left(\frac{a_j}{\mu_j}\right)^n \phi_j(0) & \text{if } 0 \le n \le s_j - 1, \\ \frac{1}{s_j! s_j^{n-s_j}}\left(\frac{a_j}{\mu_j}\right)^n \phi_j(0) & \text{if } n \ge s_j, \end{cases} \tag{3.47}$$

where

$$\phi_j(0) = \left[\sum_{n=0}^{s_j-1} \frac{1}{n!}\left(\frac{a_j}{\mu_j}\right)^n + \left(\frac{a_j}{\mu_j}\right)^{s_j}\left(\frac{1}{s_j!}\right)\frac{1}{1 - a_j/(s_j\mu_j)}\right]^{-1}.$$

We assume that the solution to the balance equation of the Jackson network has the product form as

$$p(\mathbf{x}) = \phi_1(x_1)\phi_2(x_2)\cdots\phi_N(x_N),$$

where $\phi_j(n)$ is given by (3.47). Now we need to verify that this product-form solution does satisfy the balance equation. To do that, due to the product form of $p(\mathbf{x})$, we have the following relations:

$$\frac{p(\mathbf{x})}{p(\mathbf{x} \pm \mathbf{e}_i)} = \frac{\phi_i(x_i)}{\phi_i(x_i \pm 1)},$$

$$\frac{p(\mathbf{x})}{p(\mathbf{x} + \mathbf{e}_i - \mathbf{e}_j)} = \frac{\phi_i(x_i)\phi_j(x_j)}{\phi_i(x_i + 1)\phi_j(x_j - 1)},$$

for all i and j. Also it follows from (3.47) that

$$a_i\phi_i(x_i - 1) = \min(x_i, s_i)\mu_i\phi_i(x_i),$$
$$a_i\phi_i(x_i) = \min(x_i + 1, s_i)\mu_i\phi_i(x_i + 1),$$

with an additional condition of $\phi_i(n) = 0$, if $n < 0$. By dividing both sides of the balance equation by $p(\mathbf{x})$ and using these relations, we have

$$\sum_{i=1}^{N} \lambda_i + \sum_{i=1}^{N} \min(x_i, s_i)\mu_i$$

$$= \sum_{i=1}^{N} \frac{\min(x_i, s_i)\mu_i\lambda_i}{a_i} + \sum_{i=1}^{N} a_i q_i + \sum_{j=1}^{N}\sum_{i=1}^{N} \frac{a_i}{a_j}\min(x_j, s_j)\mu_j p_{ij}$$

$$= \sum_{i=1}^{N} \frac{\min(x_i, s_i)\mu_i\lambda_i}{a_i} + \sum_{i=1}^{N} a_i q_i + \sum_{j=1}^{N} \frac{\min(x_j, s_j)\mu_j}{a_j}\sum_{i=1}^{N} a_i p_{ij}$$

$$= \sum_{i=1}^{N} \frac{\min(x_i, s_i)\mu_i \lambda_i}{a_i} + \sum_{i=1}^{N} a_i q_i + \sum_{j=1}^{N} \frac{\min(x_j, s_j)\mu_j}{a_j}(a_j - \lambda_j)$$

$$= \sum_{i=1}^{N} a_i q_i + \sum_{j=1}^{N} \min(x_j, s_j)\mu_j,$$

where the third equality on the RHS follows from $\sum_{i=1}^{N} a_i p_{ij} = a_j - \lambda_j$, which is the average rate balance equation (3.45). Clearly, the "guessed solution" $p(\mathbf{x}) = \phi_1(x_1)\phi_2(x_2)\cdots\phi_N(x_N)$ satisfies the balance equation if $\sum_{i=1}^{N} \lambda_i = \sum_{i=1}^{N} a_i q_i$ holds. It follows from (3.46) that

$$\lambda \mathbf{e} = \mathbf{a}(\mathbf{I} - \mathbf{P})\mathbf{e},$$

$$\Rightarrow \sum_{i=1}^{N} \lambda_i = \sum_{i=1}^{N} a_i \left(1 - \sum_{j=1}^{N} p_{ij}\right),$$

$$\Rightarrow \sum_{i=1}^{N} \lambda_i = \sum_{i=1}^{N} a_i q_i.$$

Thus, we have verified that the product form $p(\mathbf{x})$ satisfies the balance equation and concluded that it is the stationary distribution of the Jackson network. With the simple product-form distributions, we can analyze the queueing network with a large number of stations. However, the assumptions of exponential service times for all servers and infinite waiting buffers can be too strong for evaluating and optimizing practical systems. In Chapter 8, we will develop the stochastic process limits as the approximations to large practical stochastic systems including the queueing network with general service times.

Exercise 3.10 *Consider a CTMC with three states modeling a system that has the following conditions: (1) working state, (2) testing state, and (3) repair state. The infinitesimal generator is given by*

$$\mathbf{Q} = \begin{pmatrix} -1 & 1 & 0 \\ 5 & -10 & 5 \\ 5 & 0 & -5 \end{pmatrix}.$$

Determine the stationary distribution of this CTMC. Verifying that $\alpha\mathbf{Q} = \mathbf{0}$ is equivalent to the rate balance equations.

Exercise 3.11 *(Facility Repair Problem) Consider a production system with N facilities (e.g., machines). Assume that at most $W \leq N$ facilities can be in operating (i.e., working status) at any one time and the rest are put as spares (i.e., cold standby). For an operating facility, the time to failure is exponentially distributed with rate λ. Upon failure, the facility will be repaired by one of R repairmen. That means at most R failure facilities can be in repair at any one* *time. If all R repairmen are busy, the failure facility has to wait until one of the repairmen becomes available. The repair time is also exponentially distributed with rate μ. Clearly, a facility can be in one of the four states: (i) operating, (ii) cold standby (or spare), (iii) in repair, and (iv) waiting for repair. While the first two states, (i) and (ii), are functional facilities (also called "up") and the remaining states, (iii) and (iv), are failure facilities (also called "down").*

(a) Develop a CTMC for this system by specifying the transition rate matrix. (Hint: Use the number of "up" machines as the state variable and formulate a BDP).

(b) Assume that λ = 1, μ = 2, N = 4, W = 3, and R = 2. Compute the stationary distribution of the CTMC and evaluate the average number of facilities in operation and the average idle repair capacity.

(c) If W = N = R, prove that the stationary distribution of the CTMC is a binomial distribution.

Exercise 3.12 *Consider a simple two station tandem queue where each station has a single server with infinite buffer and the service times are exponentially distributed with rates μ_1 and μ_2 for the two stations, respectively. Customers arrive at the system according to a Poisson process with rate λ. Write the flow balance equations and verify that the joint stationary distribution for the two queues is the product of the two stationary distributions of the M/M/1 queues.*

Exercise 3.13 *Consider a serial system with N stations (also called tandem queueing system) with an external arrival stream following a Poisson process with rate λ. All the service times are exponentially distributed with rate μ. Show that the mean sojourn time in the system is given by*

$$E[W]_{series} = \frac{N}{\mu - \lambda}.$$

Exercise 3.14 *(Closed Jackson Network) A closed Jackson network is a queueing network with constant C customers in the system without external arrivals to the system and departures from the system. All other assumptions are the same as those in the open Jackson network. A difference from the open Jackson network is that the stability condition is not required. Use the same approach as the one used for open Jackson network, derive the stationary distribution of the closed Jackson network. (Hint: the product-form stationary distribution still holds in closed Jackson network with $\phi_j(n) = \prod_{k=1}^{n} \left(\frac{a_j}{\mu_j(k)} \right)$ for $n \geq 1$ and $\phi_j(0) = 1$. In addition, the effective input rate vector satisfies $\mathbf{a} = \mathbf{aP}$ due to no external arrivals, i.e., $\lambda = \mathbf{0}$. Note that the normalizing constant does not have closed form expression and is not computationally trivial.)*

3.5 USING TRANSFORMS

For the non-BDP type CTMCs, we may need to use z-transforms to solve the infinite number of rate balance equations. The z-transform is one of transforms for probability distributions of random variables (for more discussion on transforms, see Appendix A.6). It is for a discrete random variable X and is defined as the expected value of a variable z to the power of the random variable X. That is, for a discrete distribution of $a_i = P(X = i)$ where $i = 0, 1, ...$, its z-transform is

$$G(z) = E[z^X] = \sum_{i=0}^{\infty} P(X = i)z^i = \sum_{i=0}^{\infty} a_i z^i. \tag{3.48}$$

It follows from the definition that z-transform contains the same information as the probability distribution. In theory, we can obtain the probability distribution from the z-transform by the inversion process. It is usually not easy, if not impossible, to get the closed-form inverse of z-transform. However, getting the moments of the random variable is not hard. For example, the mean can be computed as $E[X] = G'(1)$. In general, the kth factorial moment can be obtained as $E[X(X - 1) \cdot (X - k + 1)] = G^{(k)}(1)$ by taking the kth derivative and setting it at 1. Thus, a z-transform is also called a probability generating function. A queueing system with batch arrivals is a good example of using z-transforms.

Example 3.2 *Consider a food kiosk on a university campus where customers arrive in groups according to a Poisson process with rate λ per unit time on average. The size of each group, denoted by Y, is independent and identically distributed with a probability g_i of having a batch of size $Y = i$, i.e., $P(Y = i) = g_i$, (with generating function $\phi(z)$ described earlier). Customers arriving in batches join a queue and are served one by one. It is assumed that the service time follows an exponential distribution with service rate μ. Such a model can be denoted as $M^Y/M/1$ queue.*

We model this service system as a CTMC and determine its stationary distribution by solving the balance equations. Note that this is a CTMC with infinite state space and the number of flow balance equations is inifinite. Since it is not a BDP, we cannot use the recursive relations to solve these equations for the stationary distribution. Thus, we have to utilize the z-transforms to solve the infinite number of equations.

Define the state variable $X(t)$ as the number of customers in the system at time t. Due to the exponentially distributed customer inter-arrival times and service times, the stochastic process $\{X(t), t \geq 0\}$ is a CTMC. Assume that the stability condition is satisfied and let $p_i = \lim_{t \to \infty} P(X(t) = i)$ be the stationary probability. Using the node balance, we have the following equations:

$$p_0 \lambda = \mu p_1$$
$$p_1(\lambda + \mu) = \mu p_2 + \lambda g_1 p_0$$
$$p_2(\lambda + \mu) = \mu p_3 + \lambda g_1 p_1 + \lambda g_2 p_0$$
$$p_3(\lambda + \mu) = \mu p_4 + \lambda g_1 p_2 + \lambda g_2 p_1 + \lambda g_3 p_0$$
$$p_4(\lambda + \mu) = \mu p_5 + \lambda g_1 p_3 + \lambda g_2 p_2 + \lambda g_3 p_1 + \lambda g_4 p_0$$

$$\vdots \quad \vdots \quad \vdots$$

where the left hand side is the total rate out of a node and the right hand side is the total rate into the node if a transition diagram is drawn. To solve these equations, we need to make the z-transforms by multiplying both sides of each equation by an appropriate power term of z and summing them up. In this example, we first define two z-transforms as $\Psi(z) = \sum_{i=0}^{\infty} p_i z^i$ and $\phi(z) = \sum_{j=1}^{\infty} g_j z^j$. Then, we multiply the first equation by z^0, the second by z^1, the third by z^2, the fourth by z^3, and so on. Next, by adding all these equations, we obtain

$$\lambda \Psi(z) + \mu \Psi(z) - \mu p_0 = \frac{\mu}{z} \Psi(z) - \frac{\mu}{z} p_0 + \lambda g_1 z \Psi(z)$$
$$+ \lambda g_2 z^2 \Psi(z) + \lambda + g_3 z^3 \Psi(z) + \cdots$$
$$= \frac{\mu}{z} \Psi(z) - \frac{\mu}{z} p_0 + \lambda \Psi(z) \phi(z).$$

It follows from the above expression that

$$\Psi(z) = \frac{\mu p_0 (1-z)}{\mu(1-z) + \lambda z (\phi(z) - 1)}. \tag{3.49}$$

Using the normalization condition $\Psi(1) = 1$, we can determine the unknown p_0. After some routine algebra, we can get $p_0 = (1 - \lambda E[Y])/\mu$, which implies that the necessary condition for stability is $\lambda E[Y] < \mu$ (i.e., $p_0 > 0$). It is a good exercise to verify the average queue length denoted by L as follows:

$$L = \Psi'(1) = \frac{\lambda E[Y] + \lambda E[Y^2]}{2(\mu - \lambda E[Y])},$$

It is interesting to see that to obtain the first moment performance measures, we may not need the entire distribution of some random variables. Here, to get the mean queue length or mean waiting time, we only need the first two moments of the batch size. This is another advantage of utilizing z-transform or other types of transforms. However, whether or not you can manipulate these equations to convert the infinite series into the z-transforms depends the structure of the state transitions of the specific CTMC. Some other examples, in which this approach works, include BDPs such as $M/M/s$ queue, $M/M/1/1$ queue with retrials, and $M/M/1$ queue with catastrophic breakdown (see Gautam [3]). It is worth noting that using z-transforms may become more complex in solving the balance equations for stationary distribution in some CTMCs with multi-dimensional and/or finite state space. Some special techniques of solving difference equations may be used (see Elaydi [1]) on a case by case basis.

Exercise 3.15 *Consider a single server queue where customers arrive according to a Poisson process with rate λ. Service times are i.i.d. exponential random variables with rate μ. The server is subject to failures. When the server is in operation, the time to failure is exponentially distributed with rate α. Each failure is catastrophic and all customers in the system are clear (i.e., rejected without service). The failure server is repaired and the repair time is exponentially distributed with rate β. Furthermore, it is assumed that the server can fail for either non-empty (busy) or empty (idle) system. No customers can enter the system is in repair state. Develop a CTMC and obtain the stationary performance measures such as the average number of customers in the system, the average waiting time in the system, and the fraction of customers lost (rejected). (Hint: Use the z-transform to solve the rate balance equations).*

Exercise 3.16 *Extend the analysis on $M^Y/M/1$ queue in Example 3.2 to the analysis on $M^Y/M/s$ queue (a multi-server Markovian queue with batch arrivals). (Hint: First try to analyze $M^Y/M/2$ queue and then generalize the results to the s server case.)*

3.6 USING TIME REVERSIBILITY

At time t, let $X(t)$ be the state of the system described earlier that is continuously observed, for all $t \in (-\infty, \infty)$. If stochastic process $\{X(t), -\infty < t < \infty\}$ is stochastically identical to the process $\{X(\tau - t), -\infty < t < \infty\}$ for all $\tau \in (-\infty, \infty)$, then $\{X(t), -\infty < t < \infty\}$ is a reversible process. The process $\{X(\tau - t), -\infty < t < \infty\}$ for any $\tau \in (-\infty, \infty)$ is known as the reversed process at τ. Although this is the definition of a reversed process, it is usually hard to show a CTMC is reversible based on that directly. Instead we resort to one of the properties of reversible processes that are especially applied to CTMCs. The first step is to check and see if a CTMC is not reversible. In the rate diagram if there is an arc from node i to node j of the CTMC, then there must be an arc from j to i as well for the CTMC to be reversible. This is straightforward because only if you can go from j to i in the forward video, you can go from i to j in the reversed video. Note that this is necessary but not sufficient as we need to verify if the same probability law is followed by both the process and its reversed process (the requirement of being stochastically identical). For an ergodic CTMC that has reached the steady state, we study the probability structure of the reverse process. Tracing the process, denoted by $X(t)$, going backward in time, we first look at the time spent in each state. Given that the CTMC is in state i at some time t, the probability that the reverse process has been in this state for an amount of time greater than s is just $e^{-v_i s}$. This is because

$$P(X(\tau) = i, \tau \in [t-s,t] | X(t) = i) = \frac{P(X(\tau) = i, \tau \in [t-s,t])}{P(X(t) = i)}$$

$$= \frac{P(X(t-s) = i)e^{-v_i s}}{P(X(t) = i)} = e^{-v_i s},$$

where $P(X(t-s) = i) = P(X(t) = i)$ due to the steady state. Thus, going backward in time, the amount of time spent in state i follows the same exponential distribution as that in the original process. Next, we need to find the condition under which the jump-to probability distribution of the reverse process is the same as that in the original process. Again, we assume that the CTMC has reached the steady state and consider a sequence of state transition instants going backward in time. That is, starting at time instant n (nth transition instant), consider the sequence of states reached at these instants, denoted by $X_n, X_{n-1}, X_{n-2}, \ldots$. It turns out that this sequence of states is itself a Markov chain process with the transition probabilities, denoted by Q_{ij}. According to the definition, we have

$$Q_{ij} = P(X_m = j | X_{m+1} = i) = \frac{P(X_m = j, X_{m+1} = i)}{P(X_{m+1} = i)}$$

$$= \frac{P(X_m = j)P(X_{m+1} = i | X_m = j)}{P(X_{m+1} = i)} = \frac{p_j P_{ji}}{p_i}.$$

To prove that the reversed process is indeed a Markov chain, we must verify that

$$P(X_m = j | X_{m+1} = i, X_{m+2}, X_{m+3}, \ldots) = P(X_m = j | X_{m+1} = i).$$

To confirm this property, we can use the fact that the Markov property implies the conditional independence between the past and future given the current state and the independence is a symmetric relationship (can you make such an argument?). Thus, to ensure that the reverse process has the same probability structure as the original process, we need $Q_{ij} = P_{ij}$ which results in the condition for the reversible Markov process. That is $p_j P_{ji} = p_i P_{ij}$. Such a relation can simplify solving the balance equations for stationary distributions. The next two properties will enhance the power of using the time reversibility to find stationary distributions of more complex CTMCs.

Property 1: Joint processes of independent reversible processes are reversible.

Suppose that we have n independent reversible processes, denoted by $\{X_1(t), -\infty < t < \infty\}, \{X_2(t), -\infty < t < \infty\}, ..., \{X_n(t), -\infty < t < \infty\}$. If we are interested in the joint process $\{X_1(t), X_2(t), ..., X_n(t), -\infty < t < \infty\}$, then it is also reversible and its steady-state probabilities would just be the product of those of the corresponding states of the individual reversible processes. As an example, if each process is a one-dimensional BDP, then the joint process is an n-dimensional BDP which is also reversible. You can verify the reversible condition in terms of the stationary probabilities and transition rates. A truncated process is the process of a finite-state space that results from cutting off part of the infinite state space.

Property 2: Truncated processes of reversible processes are reversible.

Consider a reversible and ergodic CTMC $\{X(t), -\infty < t < \infty\}$ with infinitesimal generator $Q = [q_{ij}]$ defined on state space S and steady-state probabilities p_j that the CTMC is in state j for all $j \in S$. Now consider another CTMC $\{Y(t), -\infty < t < \infty\}$ which is a truncated version of $\{X(t), -\infty < t < \infty\}$ defined on state space A such that $A \in S$. By truncation, we can keep the inter-state transition rates of the truncated process $Y(t)$ the same as those in the original process $X(t)$ and adjust the diagonal elements of the infinitesimal generator of $Y(t)$ by letting its negative value equal to the sum of the off-diagonal transition rates in the row. For more discussion about these properties, readers are referred to Kelly (1979). An example that demonstrates these properties is to study the relation between the $M/M/s$ queue and the $M/M/s/K$ queue. Next we provide a stochastic service system with a limited capacity and random demand. Analysis of multimedia traffic in Internet is a good example.

Example 3.3 *The link between the Internet and a web server farm has a limited capacity measured as "kilobits per second" (Kbps) or "megabits per second" (Mbps). Such a capacity is also called bandwidth capacity. A file to be uploaded needs to take a certain amount of capacity of this transmission link for a random time period. There are different types of files that request different numbers of Kpbs of bandwidth. For example, video and audio files take more bandwidth capacities than image and text files in transmission. We can formulate this resource sharing problem as a stochastic model. Suppose that the capacity of the link (service resource) is C Kbps. There are N classes of files (i.e., N classes of customers) to be uploaded by accessing the link. Class i files arrive according to a Poisson process with rate λ_i. The requested connection (transmission) time for class i file is exponentially distributed with rate μ_i. During the entire transmission time, each class i file uses c_i Kpbs of bandwidth. Furthermore, we assume that whenever the maximum capacity C of the link is taken, new connection requests are denied. That means no buffering takes place. We can use the time reversibility to figure out the stationary distribution of the files in transmission.*

Let $X_i(t)$ be the number of files of class i in transmission at time t. The capacity constraint can be written as

$$c_1 X_1(t) + c_2 X_2(t) + \cdots + c_N X_N(t) \leq C.$$

This constraint makes the state space finite as it controls the admission of transmission requests. For example, if a transmission request of a class i file arrives and $c_i + c_1 X_1(t) + c_2 X_2(t) + \cdots + c_N X_N(t) > C$ holds at that instant, this request is rejected. Due to the finite state space, the multi-dimensional CTMC, $\{X_1(t), X_2(t), ..., X_N(t)\}$, has the stationary distribution denoted by

$$p_{x_1, x_2, ..., x_N} = \lim_{t \to \infty} P(X_1(t) = x_1, X_2(t) = x_2, ..., X_N(t) = x_N).$$

To obtain this stationary distribution, we consider an N-dimensional CTMC with the same state variables and $C = \infty$. Such a CTMC has the state space $S = \mathbf{Z}_+^N$. Obviously, the CTMC of our model is a truncated process of this unconstrained CTMC due to the constraint. Note that $C = \infty$ implies that the system can be considered as N independent $M/M/\infty$ queues since the arrival processes are independent. Each process is reversible and the joint process is also reversible according to Property 1. Moreover, the stationary probability of the number of jobs (files in transmission) for process i, denoted by $p_i^\infty(j), j = 0, 1, ...,$ has a closed-form expression (i.e., Poisson process with parameter λ_i/μ_i) as follows:

$$p_i^\infty(j) = e^{-\lambda_i/\mu_i} \left(\frac{\lambda_i}{\mu_i} \right)^j \frac{1}{j!},$$

where the superscript ∞ indicates $C = \infty$ (omitting this superscript implies $C < \infty$, the truncated or constrained case). Clearly, the stationary distribution for the joint process is given by

$$\lim_{t \to \infty} P(X_1^\infty(t) = x_1, X_2^\infty(t) = x_2, ..., X_N^\infty(t) = x_N)$$

$$= p_1^\infty(x_1) p_2^\infty(x_2) \cdots p_N^\infty(x_N) = \left(e^{-\sum_{i=1}^N \lambda_i/\mu_i} \right) \prod_{i=1}^N \left(\frac{\lambda_i}{\mu_i} \right)^{x_i} \frac{1}{x_i!}.$$

It follows from Property 2 that the stationary probability for our system with bandwidth capacity constraint, denoted by $p_{x_1, x_2, ..., x_N}$, can be written as

$$p_{x_1, x_2, ..., x_N} = M \prod_{i=1}^N \left(\frac{\lambda_i}{\mu_i} \right)^{x_i} \frac{1}{x_i!}, \tag{3.50}$$

subject to $c_1 x_1 + c_2 x_2 + \cdots + c_N x_N \leq C$. Here M is a constant and can be determined by the normalization condition as

$$M = \left[\sum_{x_1, x_2, ..., x_N : c_1 x_1 + c_2 x_2 + \cdots + c_N x_N \leq C} \prod_{i=1}^N \left(\frac{\lambda_i}{\mu_i} \right)^{x_i} \frac{1}{x_i!} \right]^{-1}.$$

It is worth noting that this model is general enough to analyze a variety of stochastic service systems besides the multi-media traffic problem in the Internet. For example, the service resource can be number of hospital beds in a hospital. Different types of patients require different types of beds. Each type of bed has a specific operating cost. Then the capacity constraint can be replaced by an operating cost budget constraint (maximum dollars per time unit). If different types of customers arrive according to different independent Poisson processes and the bed-stay times are exponentially distributed. Then the model can be utilized to analyze the relation between the blocking probabilities of different classes of customers and the total budget. However, determining the constant M is not trivial under the capacity constraint for a large N case. We can use a two-class customer system (i.e., $N = 2$) to demonstrate the application in resource management as M can be computed easily.

Example 3.4 *Inpatient units in a hospital includes intensive care patients, surgery patients, and rehab patients. Consider a cardiology/General Internal Medicine Inpatient Unit that admits two types of patients, cardiology acute care (CAC) called type 1, and general internal medicine (GIM) called type 2. A key constraint for admitting a patient is the number of acute care nurse practitioners (denoted by RN – registered nurses with advanced training) available. Let C be the total number of RNs available. Each CAC patient admitted requires c_1 RNs and each GIM patient admitted requires c_2 RNs. Assume that (a) patients of type n arrive at the unit according to an independent Poisson process with rate λ_n; and (b) each admitted patient type n spends an exponentially distributed time with rate μ_n for $n = 1, 2$. Let $X_1(t)$ and $X_2(t)$ be the number of type-1 and type-2 patients at time t. Then the two-dimensional CTMC $\{(X_1(t), X_2(t)), t = 0\}$ will reach steady-state.*

We now write down the stationary probabilities and related performance measures of this model based on the structure of the solution for the general model. Due to the linear capacity constraint $c_1 X_1(t) + c_2 X_2(t) \leq C$, the state space is finite and the boundary states can be determined in two directions. These two directions will determine two types of boundary states, called X_1 boundary and X_2 boundary states. For a given $x_2 = j$ with $j = 1, ..., \lfloor \frac{C}{c_2} \rfloor$, the feasible state is (x_1, j) where $x_1 = 0, 1, ..., \lfloor \frac{C - jc_2}{c_1} \rfloor^+$. Here $\lfloor x \rfloor$ is the lower floor function that gives the greatest integer less than or equal to x and $\lfloor x \rfloor^+ = max(\lfloor x \rfloor, 0)$. Thus, X_1 boundary state is $(\lfloor \frac{C - jc_2}{c_1} \rfloor^+, j)$ where $j = 0, 1, ..., \lfloor \frac{C}{c_2} \rfloor$. This boundary state set is denoted by $S_{X_1}^B$. Similarly, For a given $x_1 = i$, the feasible state is (i, x_2) where $x_2 = 0, 1, ..., \lfloor \frac{C - ic_1}{c_2} \rfloor^+$. Thus, X_2 boundary state is $(i, \lfloor \frac{C - ic_1}{c_2} \rfloor)^+$ where $i = 0, 1, ..., \lfloor \frac{C}{c_1} \rfloor$. This boundary state set is denoted by $S_{X_2}^B$. Then the double boundary state set is $S_{X_1, X_2}^B = \{(X_1(t) = i, X_2(t) = j : (i, j) \in S_{X_1}^B \cap S_{X_2}^B\}$; X_1 only boundary state set is $\{(X_1(t) = i, X_2(t) = j : (i, j) \in S_{X_1, X_2}^B \setminus S_{X_2}^B\}$;

and X_2 only boundary state set is $\{(X_1(t) = i, X_2(t) = j : (i,j) \in S^B_{X_1,X_2} \; S^B_{X_1}\}$. According to the solution of the general model, we have the stationary probabilities as

$$p_{ij} = \lim_{t \to \infty} P(X_1(t) = i, X_2(t) = j) = M \left(\frac{\lambda_1}{\mu_1}\right)^i \left(\frac{\lambda_2}{\mu_2}\right)^j \frac{1}{i!j!},$$

$$i = 0, 1, ..., \lfloor \frac{C}{c_1} \rfloor; j = 0, 1, ..., \lfloor \frac{C - ic_1}{c_2} \rfloor^+,$$

where

$$M = \left[\sum_{i=0}^{\lfloor \frac{C}{c_1} \rfloor} \sum_{j=0}^{\lfloor \frac{C - ic_1}{c_2} \rfloor^+} p_{ij} \right]^{-1}.$$

The performance measures, such as the expected number of patients in the system and the blocking probability of each type, can be obtained based on the stationary probabilities. For example, under a certain cost structure, we can construct a cost function for the system. Assume that there are two types of costs that are critical from the customer service and operating cost perspectives. The first type is the cost of rejecting a patient of either type due to the system blocking (i.e., no room for admitting a patient). Let L_n be the cost of rejecting a type n patient, $n = 1, 2$. and let h be the cost of hiring one RN per time unit. Then the total expected cost per time unit, denoted by g, is given by

$$g = \lambda_1 L_1 \sum_{j=0}^{\lfloor \frac{C}{c_2} \rfloor} P_{(\lfloor \frac{C - jc_2}{c_1} \rfloor, j)} + \lambda_2 L_2 \sum_{i=0}^{\lfloor \frac{C}{c_1} \rfloor} P_{(i, \lfloor \frac{C - ic_1}{c_2} \rfloor)} + hC.$$

Clearly the first two terms are decreasing in C and the last term is increasing in C. Numerically, we can determine the optimal C that minimizes the expected cost rate g. As a numerical example, consider a system with $c_1 = 2, c_2 = 1, \lambda_1 = 2, \lambda_2 = 3, \mu_1 = 0.5$, and $\mu_2 = 1$ and varying C. The normalization constant is

$$M = \left[\sum_{i=0}^{\lfloor C/2 \rfloor} \sum_{j=0}^{C-2i} \left(\frac{\lambda_1}{\mu_1}\right)^i \left(\frac{\lambda_2}{\mu_2}\right)^j \frac{1}{i!j!} \right]^{-1}.$$

Figure 3.2 shows g as a function of C for $h = 16, L_1 = 100$ and $L_2 = 20$. The optimal number of RNs in this example is 8. Based on this performance measure and Erlang B formulas, we can also evaluate the benefit of completely sharing admission policy compared with the two dedicated systems for the two types of patients.

Exercise 3.17 *For the two station tandem queue in Exercise 3.12, use the time reversibility of an M/M/1 queue to argue that the output process is identical to the input process, i.e. the Poisson process with rate λ, then the product-form joint stationary distribution can be justified.*

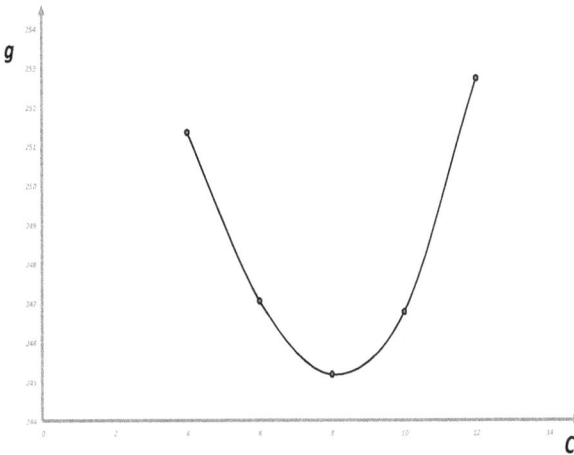

Figure 3.2 Expected total cost per unit time vs. total number of RNs available.

Exercise 3.18 *Argue that the departure processes of the following queueing systems are also Poisson processes: (1) $M/M/s$ queue; (2) $M/G/\infty$ queue; and (3) $M/G/s/s$ queue.*

Exercise 3.19 *A queueing system with n heterogeneous servers has no waiting room. The service times of server i are independent and exponentially distributed with rate $\mu_i, i = 1, ..., n$. Customers arrive at the system according to a Poisson process with rate λ and balk if they find all servers are busy. An arriving customer who finds k idle severs will be served by any of these k servers with probability $1/k$. (a) Formulate a CTMC for the number of customers in the system. (b) Show that this CTMC is time reversible. (c) Determine the stationary distribution for this CTMC.*

Exercise 3.20 *A clinic accepts k different types of patients, where type i patients arrive according to a Poisson process with rate λ_i. These k Poisson processes are independent. The length of stay (LOS) of each type i patient is exponentially distributed with rate $\mu_i, i = 1, ..., k$. Assume that each type i patient consumes c_i units of resources for each unit time of stay. The clinic will not accept a new patient if it would result in the total of all patient's resources needs exceeding the amount C per unit time. Let n_i be the number of patients of type i in the clinic. Then this constraint can be written as $\sum_{i=1}^{k} n_i c_i \leq C$.*

(a) Formulate a CTMC for this clinic.
(b) If $C = \infty$ and $N_i(t)$ represents the number of type i customers in the

system at time t, what type of process is $\{N_i(t), t \geq 0\}$? Is it time reversible?

(c) *Find the limiting probabilities of the k-dimensional CTMC*
$\{(N_1(t), ..., N_k(t)), t \geq 0\}$ *if $C = \infty$.*

(d) *Find the limiting probabilities of the k-dimensional CTMC*
$\{(N_1(t), ..., N_k(t)), t \geq 0\}$ *if $C < \infty$.*

(e) *What fraction of patients are lost if $C < \infty$. What fraction of patients of type i are lost if $C < \infty$, where $i = 1, ..., k$.*

3.7 SOME USEFUL CONTINUOUS-TIME MARKOV CHAINS

3.7.1 POISSON PROCESS AND ITS EXTENSIONS

We start with the introduction to Poisson distribution. This distribution can be explained by the law of rare events. Consider a large number N of independent Bernoulli trials where the probability of success at each trial (e.g., the chance that an arrival occurs in one of N fixed time intervals) is a constant and small value p. Denote by $X_{(N,p)}$ the total number of successes in N trials which follows the binomial distribution for $k = 0, 1, ..., N$. That is

$$P(X_{(N,p)} = k) = \binom{N}{k} p^k (1-p)^{N-k}.$$

If taking the limit of $N \to \infty$ and $p \to 0$ such that $Np = \mu$, which is a constant, then the distribution for $X_{(N,p)}$ becomes the so-called Poisson process with the probability mass function as

$$P(X_\mu = k) = \frac{e^{-\mu} \mu^k}{k!}, \quad \text{for } k = 0, 1, ...$$

The result of this limiting process is called the law of rare events which is used to suggest the situations under which the Poisson distribution might be expected to prevail, at least approximately. The Poisson process entails notions of Poisson distribution with independence and can be used for modeling a random arrival process. In fact, a Poisson process with rate $\lambda > 0$ (i.e., the expected number of arrivals per unit of time) is a CTMC denoted by $\{X(t), t \geq 0\}$, where $X(t)$ represents the number of arrivals already occurred in $(0, t]$, often called a counting process (It is also often denoted by $N(t)$). This process satisfies the following properties:

1. for a set of time points $t_0 < t_1 < \cdots < t_n$ and $t_0 = 0$, the number of arrivals occurred in the disjoint intervals

$$X(t_1) - X(t_0), X(t_2) - X(t_1), ..., X(t_n) - X(t_{n-1})$$

are independent random variables.

2. for $s \geq 0$ and $t \geq 0$, the random variable $X(s+t) - X(s)$ follows the Poisson distribution

$$P(X(s+t) - X(s) = k) = \frac{(\lambda t)^k e^{-\lambda t}}{k!}.$$

3. $X(0) = 0$.

For a given $t > 0$, it is easy to derive $E[X(t)] = \lambda t$ and $Var[X(t)] = \lambda t$. Another property of Poisson process is that for a small time interval h, the probability of a single arrival can be written in terms of $o(h)$

$$P(X(t+h) - X(t) = 1) = \frac{(\lambda h)e^{-\lambda h}}{1!} = (\lambda h) \sum_{n=0}^{\infty} \frac{(-\lambda h)^n}{n!}$$

$$= (\lambda h) \left(\frac{(-\lambda h)^0}{0!} + \frac{(-\lambda h)^1}{1!} + \frac{(-\lambda h)^2}{2!} + \cdots \right)$$

$$= (\lambda h) \left(1 - \lambda h + \frac{1}{2}\lambda^2 h^2 - \cdots \right)$$

$$= \lambda h + o(h).$$

In a Poisson process, the time between two consecutive arrivals is called an inter-arrival time, denoted by $T_i = S_{i+1} - S_i$, where S_i is the arrival instant of the ith customer. The inter-arrival times $T_1, ..., T_{n-1} ...$ are independent random variables, each having the exponential probability density function $f_{T_k}(s) = \lambda e^{-\lambda s}$.

Exercise 3.21 *Prove that S_n, the arrival instant of the nth customer follows a gamma distribution with parameter n and λ. That is the p.d.f. of S_n is as follows:*

$$f_{S_n}(t) = \lambda e^{-\lambda t} \frac{(\lambda t)^{n-1}}{(n-1)!}, \quad t \geq 0.$$

To understand more properties of Poisson process, readers are either referred to some textbook on stochastic processes (e.g., Ross [5]) or encouraged to practice the following exercises.

Exercise 3.22 *Suppose that $\{N_1(t), t \geq 0\}$ and $\{N_2(t), t \geq 0\}$ represent respectively the number of type 1 and type 2 events occurring in $[0,t]$. Let $N(t) = N_1(t) + N_2(t)$. If $N(t)$ is a Poisson process with rate λ, then $\{N_1(t), t \geq 0\}$ and $\{N_2(t), t \geq 0\}$ are two independent Poisson processes having respective rates λp and $\lambda(1-p)$, where p is the proportion of type 1 customers.*

Exercise 3.23 *Suppose that $\{N_1(t), t \geq 0\}$ and $\{N_2(t), t \geq 0\}$ are two independent Poisson processes having respective rates λ_1 and λ_2. Denote by S_n^i the time of the nth event of the ith process, $i = 1, 2$.*

(a) Show that

$$P(S_1^1 < S_1^2) = \frac{\lambda_1}{\lambda_1 + \lambda_2}.$$

(b) Show that

$$P(S_n^1 < S_m^2) = \sum_{k=n}^{n+m-1} \binom{n+m-1}{k} \left(\frac{\lambda_1}{\lambda_1 + \lambda_2}\right)^k \left(\frac{\lambda_2}{\lambda_1 + \lambda_2}\right)^{n+m-1-k}.$$

Exercise 3.24 *(Conditional distribution of the Poisson arrival times) For a Poisson process $\{N(t), t \geq 0\}$ with rate λ, given that $N(t) = n$, the n arrival times $S_1, ..., S_n$ have the same distribution as the order statistics corresponding to n independent random variables uniformly distributed on the interval $(0,t)$. That is the conditional p.d.f. is given by*

$$f(s_1, ..., s_n | n) = \frac{n!}{t^n}, \quad 0 < s_1 < \cdots < s_n < t.$$

Exercise 3.25 *Suppose that $\{N(t), t \geq 0\}$ is a Poisson process with rate λ. Then for $0 < s < t$ and $0 \leq j \leq n$, show that*

$$P(N(s) = j | N(t) = n) = \binom{n}{j} \left(\frac{s}{t}\right)^j \left(1 - \frac{s}{t}\right)^{n-j}.$$

For a Poisson process $\{X(t), t \geq 0\}$ with rate $\lambda > 0$ representing an arrival process of customers, assume that each arrival has an associated i.i.d. random variable, denoted by $Y_i, i = 1, 2, ...$, which is also independent of the Poisson process $\{X(t), t \geq 0\}$. Denote the distribution function of Y_i by $F_Y(y) = P(Y_i \leq y)$. Then we can define a *compound Poisson process* $Z(t)$ by

$$Z(t) = \sum_{i=1}^{X(t)} Y_i \quad t \geq 0,$$

which is a cumulative value process. For example, if each arrival is a customer order for a particular product and Y_i is the number of items ordered, then $Z(t)$ represents the total number of items ordered by time t. With the associated sequence of random variables Y_i, we also define a *marked Poisson process*. A marked Poisson process is the sequence of pairs $(S_1, Y_1), (S_2, Y_2), ...$, where $S_1, S_2, ...$ are the arrival times of the Poisson process. With the marked Poisson process, we can split a Poisson process into two independent Poisson processes. For example, suppose that $\{X(t), t \geq 0\}$ represents a Poisson arrival process for customers entering a store. Let Y_i denote the gender of the ith arrival. That is $Y_i = 1$ with probability p, representing a male arrival, and $Y_i = 0$ with probability $1 - p$, representing a female arrival. Now consider separately the processes of points marked with 1's and of points marked with 0's. Then we can define the relevant Poisson processes explicitly by

$$X_m(t) = \sum_{i=1}^{X(t)} Y_i \quad X_f(t) = X(t) - X_m(t), \tag{3.51}$$

for male and female arrivals, respectively.

Exercise 3.26 *Consider the compound Poisson process defined above.*
(a)Show that the distribution function of $Z(t)$ is given by

$$P(Z(t) \leq z) = \sum_{i=0}^{\infty} \frac{(\lambda t)^i e^{-\lambda t}}{n!} F_Y^{(i)}(z),$$

where $F_Y^{(i)}(z)$ is the convolution of Y to itself. That is $F_Y^{(i)} = P(Y_1 + \cdots + Y_i \leq z)$.
(b) Denote by $\mu = E[Y_1]$ and $\sigma^2 = Var[Y_1]$ the mean and variance, respectively. Show that

$$E[Z(t)] = \lambda \mu t, \quad Var[Z(t)] = \lambda(\sigma^2 + \mu^2)t.$$

Exercise 3.27 *Suppose that claims arrive at an insurance company according to a Poisson process with rate λ and claim i's amount follows an exponential distribution with mean μ. Let T be the time that the cumulative claim exceeds a threshold h. Show that the expected value $E[T] = (1 + \mu h)/\lambda$.*

Exercise 3.28 *Prove that the two Poisson processes in (3.51) are independent.*

Exercise 3.29 *For a marked Poisson process, if Y_i's are i.i.d. continuous random variables with a p.d.f. of $h(y)$, then $(S_1, Y_1), (S_2, Y_2), \ldots$ form a 2-dimensional non-homogeneous Poisson point process in the (t, y) plane. Consider a region A in the (t, y) plane, argue that the mean number of points in A is given by*

$$\mu(A) = \iint_A \lambda h(y) dy dt.$$

The two-dimensional Poisson process in Exercise 3.29 is a special case of spatial Poisson process which can be considered as a generalization of the regular Poisson process. For more details about spatial Poisson processes, interested readers are referred to Taylor and Karlin [7].

Note that in the Poisson process defined, the arrival rate λ is constant. Such a Poisson process is called a homogeneous Poisson process. There are several ways of generalizing or extending this process. We present a few of them below:

3.7.1.1   Non-Homogeneous Poisson Process

If the arrival rate is a function of time, i.e. $\lambda = \lambda(t)$, then the process is said to be a non-homogeneous process. A formal definition is as follows: Let $\lambda(t):$ $[0,\infty) \mapsto [0,\infty)$ be an integrable function. The counting process $\{X(t), t \geq 0\}$ is called a non-homogeneous Poisson process with rate $\lambda(t)$ if the following conditions are satisfied:

1. $X(t)$ has independent increments.
2. for any $t \in [0,\infty)$, the following holds

$$P(X(t+h) - X(t) = 1) = \lambda(t)h + o(h),$$
$$P(X(t+h) - X(t) = 0) = 1 - \lambda(t)h + o(h),$$
$$P(X(t+h) - X(t) \geq 2) = o(h).$$

3. $X(0) = 0$.

For a non-homogeneous Poisson process with rate $\lambda(t)$, the number of arrivals in any interval is a Poisson random variable; however, its parameter can depend on the location of the interval. Thus, for any time interval $(t, t+s]$, the number of arrivals $X(t+s) - X(t)$ follows a Poisson process with rate $\Lambda((t, t+s]) = \int_t^{t+s} \lambda(\tau)d\tau$. Non-homogeneous Poisson process is often used in modeling the maintenance and replacement systems subject to failures. An interesting case of the non-homogeneous Poisson process is when the rate function is a stochastic process. Such processes, introduced by Cox in 1955, are called doubly stochastic processes (also called Cox processes). The next exercise develops some interesting results for this type of non-homogeneous Poisson processes.

Exercise 3.30 *Suppose that* $\{N(t), t \geq 0\}$ *is a Poisson process with rate* Λ *which is a continuous random variable with the p.d.f.* $f(\lambda)$. *Then, given a value of* Λ, *this non-homogeneous Poisson process becomes a Poisson process with constant rate* $\lambda = \Lambda$. *Thus, the* $\{N(t), t \geq 0\}$ *is also called a mixed Poisson process. Now, if* $f(\lambda) = e^{-\lambda}$ *for* $\lambda > 0$.
(a) Show that

$$P(N(t) = i) = \left(\frac{t}{1+t}\right)^i \left(\frac{1}{1+t}\right), \quad for \ i = 0, 1, \dots.$$

(b) Show that

$$P(N(t) = i, N(t+s) = i+k) = \binom{i+k}{i} t^i s^k \left(\frac{1}{1+s+t}\right)^{i+k+1}.$$

Based on this formula, what is the main difference between the mixed Poisson process and a homogeneous Poisson process?

3.7.1.2 Pure Birth Process

Another generalization of the homogeneous Poisson process is to allow the arrival rate to depend on the number of arrivals already occurred $X(t)$. Such a process is called a pure birth process, which can model the reproduction of living organisms without mortality and migration out of the system. Let $X(t)$ denote the number of births in the time interval $(0,t]$. In terms of the transition probability in a small time interval h, the pure birth process can be defined as a CTMC satisfying the following:

$$P_{i,i+j}(h) = P(X(t+h) - X(t) = j | X(t) = i)$$
$$= \begin{cases} \lambda_i h + o(h) & j = 1, \\ 1 - \lambda_i h + o(h) & j = 0, \\ 0 & j < 0. \end{cases} \tag{3.52}$$

Define the probability distribution function $P_n(t) = P(X(t) = n)$. This function satisfies the differential equations:

$$\frac{dP_0(t)}{dt} = -\lambda_0 P_0(t),$$
$$\frac{dP_n(t)}{dt} = -\lambda_n P_n(t) + \lambda_{n-1} P_{n-1}(t). \tag{3.53}$$

with initial conditions

$$P_0(0) = 1, \quad P_n(0) = 0, \ n > 0.$$

The first equation can be solved easily with $P_0(0) = 1$ as

$$P_0(t) = e^{-\lambda_0 t}, \quad \text{for } t > 0.$$

To solve the second equation, we utilize the recursive approach. Since the number of births increases with time t, it is natural to guess that the $P_n(t)$ has the following form

$$P_n(t) = e^{-\lambda_n t} R_n(t), \quad \text{for } n = 0, 1, 2, \cdots,$$

which is expected to be a decreasing function of t for any given n (i.e., no birth probability given that n is decreasing in t). Equivalently, it can be written as

$$R_n(t) = e^{\lambda_n t} P_n(t).$$

Rewrite the second equation of (8.56) as

$$\lambda_{n-1} P_{n-1}(t) = \frac{dP_n(t)}{dt} + \lambda_n P_n(t), \tag{3.54}$$

which is used for the following. Differentiating R_n with respect to t yields

$$
\begin{aligned}
\frac{dR_n(t)}{dt} &= \lambda_n e^{\lambda_n t} P_n(t) + e^{\lambda_n t} \frac{P_n(t)}{dt} \\
&= e^{\lambda_n t} \left[\lambda_n P_n(t) + \frac{P_n(t)}{dt} \right] \\
&= e^{\lambda_n t} \lambda_{n-1} P_{n-1}(t),
\end{aligned}
\tag{3.55}
$$

where the last equality follows from (3.54). Integrating both sides of (3.55) and using the initial condition $R_n(0) = 0$ yields

$$
R_n(t) = \int_0^t e^{\lambda_n x} \lambda_{n-1} P_{n-1}(x) dx, \quad \text{for } n \geq 1.
$$

Substituting $R_n(t)$ into the expression for $P_n(t)$, we have

$$
\begin{aligned}
P_n(t) &= e^{-\lambda_n t} \int_0^t e^{\lambda_n x} \lambda_{n-1} P_{n-1}(x) dx \\
&= \lambda_{n-1} e^{-\lambda_n t} \int_0^t e^{\lambda_n x} P_{n-1}(x) dx, \quad \text{for } n = 1, 2, \dots.
\end{aligned}
\tag{3.56}
$$

To make the process reasonable, we require $\sum_{n=0}^{\infty} P_n(t) = 1$ for a finite $t > 0$, which implies that during a finite time period, the number of births occurred cannot be infinite. A sufficient and necessary condition for this requirement is $\sum_{n=0}^{\infty}(1/\lambda_n) = \infty$ which represents the expected time before the population becomes infinite. In other words, this condition indicates that the probability that the process becomes infinite for a finite time t, $1 - \sum_{n=0}^{\infty} P_n(t)$, is zero (i.e., $P(X(t) = \infty) = 0$ for $t > 0$). The recursive integral equation (3.56) can be solved for $n = 1$ and $n = 2$ as follows:

$$
\begin{aligned}
P_1(t) &= \lambda_0 e^{-\lambda_1 t} \int_0^t e^{\lambda_1 x} P_0(x) dx \\
&= \lambda_0 \left(\frac{e^{-\lambda_0 t}}{\lambda_1 - \lambda_0} + \frac{e^{-\lambda_1 t}}{\lambda_0 - \lambda_1} \right),
\end{aligned}
$$

$$
\begin{aligned}
P_2(t) &= \lambda_1 e^{-\lambda_2 t} \int_0^t e^{\lambda_2 x} P_1(x) dx \\
&= \lambda_0 \lambda_1 \left(\frac{e^{-\lambda_0 t}}{(\lambda_1 - \lambda_0)(\lambda_2 - \lambda_0)} + \frac{e^{-\lambda_1 t}}{(\lambda_0 - \lambda_1)(\lambda_2 - \lambda_1)} + \frac{e^{-\lambda_2 t}}{(\lambda_0 - \lambda_2)(\lambda_1 - \lambda_2)} \right).
\end{aligned}
$$

Following the pattern, we can obtain $n \geq 3$ case with $P_n(0) = 0$ as follows:

$$
\begin{aligned}
P_n(t) &= P(X(t) = n | X(0) = 0) \\
&= \lambda_0 \cdots \lambda_{n-1} \left(A_{0,n} e^{-\lambda_0 t} + \cdots + A_{n,n} e^{-\lambda_n t} \right),
\end{aligned}
\tag{3.57}
$$

where

$$A_{0,n} = \frac{1}{(\lambda_1 - \lambda_0)(\lambda_2 - \lambda_0) \cdots (\lambda_n - \lambda_0)},$$

$$A_{k,n} = \frac{1}{(\lambda_0 - \lambda_k) \cdots (\lambda_{k-1} - \lambda_k) \cdots (\lambda_n - \lambda_k)} \quad \text{for } 0 < k < n,$$

$$A_{n,n} = \frac{1}{(\lambda_0 - \lambda_n) \cdots (\lambda_1 - \lambda_n) \cdots (\lambda_{n-1} - \lambda_n)}.$$

The proof of this general expression can be done by the induction via tedious algebra and hence is omitted.

Exercise 3.31 *Consider a pure birth process starting from $X(0) = 0$. It has the following birth parameters for state $n \geq 0$: $\lambda_0 = 1$ and $\lambda_n = n\lambda_{n-1}$. Find $P_n(t)$ for $n = 0, 1, 2, 3$.*

Exercise 3.32 *A new product with an environment protection feature (also called "green feature") is introduced to the market. The sales are expected to be determined by both media (newspaper, radio, and television) advertising and word-of-month advertising (i.e., satisfied customers tell friends about the product). Assume that media advertising generates new customers according to a Poisson process with rate $\lambda = 50$ per month. For the word-of-month advertising, assume that each customer buying a product will generate sales to new customers with a rate of $\alpha = 100$ per month. Denote by $X(t)$ the total number of customers of green products during $[0, t)$.*
(a) Develop a pure birth process for $X(t)$.
(b) Find the probability that exact 100 customers will buy the green products.

3.7.1.3 The Yule Process

The Yule process is an example of a pure birth process that commonly arises in physics and biology. Let the state variable $X(t)$ denote the population size at time t and the initial population is $X(0) = N$ so that $P_i(0) = \delta_{iN}$. The population size can only increase in size because there are only births. In this process, every individual has a probability $\lambda h + o(h)$ of giving birth to a new member during a small time interval h, where $\lambda > 0$. Then, given a population of size n, using the independence among members and binomial formula, we have

$$P(X(t+h) - X(t) = 1 | X(t) = n) = \binom{n}{1} [\lambda h + o(h)][1 - \lambda h + o(h)]^{n-1}$$

$$= n\lambda h + o_n(h).$$

With this property, the transition probabilities for sufficiently small h are given by

$$P_{i,i+j} = P(X(t+h) - X(t) = j | X(t) = i)$$

$$= \begin{cases} 1 - i\lambda h + o(h) & j = 0, \\ i\lambda h + o(h) & j = 1, \\ o(h) & j \geq 2, \\ 0 & j < 0. \end{cases}$$

Using these transition probabilities, we can develop differential equations for the probability distribution function. Note that

$$\begin{aligned}
P_i(t+h) =& P(X(t+h) = i) \\
=& P(X(t) = i, X(t+h) - X(t) = 0) \\
&+ P(X(t) = i-1, X(t+h) - X(t) = 1) \\
&+ P(X(t) = i-2, X(t+h) - X(t) \geq 2) \\
=& P(X(t) = i-1) \cdot P(X(t+h) - X(t) = 1) \\
&+ P(X(t) = i-2) \cdot P(X(t+h) - X(t) \geq 2) \\
=& P_i(t)[1 - i\lambda h + o(h)] + P_{i-1}[(i-1)\lambda h + o(h)] \\
&+ P_{i-2}(t) \cdot o(h),
\end{aligned}$$

which can be re-written as

$$P_i(t+h) - P_i(t) = (i-1)\lambda h P_{i-1}(t) - i\lambda h P_i(t) + o(h). \tag{3.58}$$

Dividing both sides of the equation above and taking the limit $h \to 0$ lead to the system of forward Kolmogorov equations

$$\begin{aligned}
\frac{dP_i(t)}{dt} &= (i-1)\lambda P_{i-1}(t) - i\lambda P_i(t), \quad i = N, N+1, N+2, \cdots \\
\frac{dP_i(t)}{dt} &= 0, \quad i = 0, 1, 2, \cdots, N-1.
\end{aligned} \tag{3.59}$$

The solution is given by

$$P_i(t) = e^{-\lambda t}(1 - e^{-\lambda t})^{i-1}, \quad i \geq 1. \tag{3.60}$$

There are multiple ways of obtaining this solution. For example, we can apply (3.57) to this process to get the solution. This approach is left as an exercise to readers. Another way of solving the differential equations is to utilize the transform approach that requires to solve a partial differential equation (PDE). Due to its importance in analysis of stochastic models, we present the details of obtaining the probability distribution function for the Yule process via this approach. The z-transform (also called probability generating function, p.g.f.) of $P_i(t)$, denoted by $\mathbb{P}(z,t)$, is given by

$$\mathbb{P}(z,t) = \sum_{i=0}^{\infty} P_i(t)z^i.$$

With initial condition $P_N(0) = 1$ and $P_i(0) = 0$ for $i \neq N$, we derive the PDE for the z-transform by multiplying the differential equations in (3.59) by z^i and sum over i

$$\frac{\partial \mathbb{P}(z,t)}{\partial t} = \sum_{i=0}^{\infty} \frac{dP_i(t)}{dt} z^i = \sum_{i=0}^{\infty} (i-1)\lambda P_{i-1}(t)z^i - \sum_{i=0}^{\infty} i\lambda P_i(t)z^i$$

$$= \sum_{i=N+1}^{\infty} (i-1)\lambda P_{i-1}(t)z^i - \sum_{i=N}^{\infty} i\lambda P_i(t)z^i$$

$$= \lambda z^2 \sum_{i=N}^{\infty} iP_i(t)z^{i-1} - \lambda z \sum_{i=N}^{\infty} iP_i(t)z^{i-1}$$

$$= \lambda z(z-1) \sum_{i=N}^{\infty} iP_i(t)z^{i-1}.$$

Note that $\partial \mathbb{P}(z,t)/\partial z = \sum_{i=1}^{\infty} iP_i(t)z^{i-1}$. Thus, we have

$$\frac{\partial \mathbb{P}(z,t)}{\partial t} = \lambda z(z-1)\frac{\partial \mathbb{P}(z,t)}{\partial z} \tag{3.61}$$

with initial condition $\mathbb{P}(z,0) = z^N$. The first order PDE for the moment generating function (m.g.f) can be obtained by a change of variable. Let $z = e^\theta$. Then we have $d\theta/dz = 1/e^\theta = 1/z$. Denote the m.g.f. by $\mathbb{M}(\theta,t) = E[e^{\theta X(t)}] = \mathbb{P}(e^\theta,t)$. It follows from the chain rule that

$$\frac{\partial \mathbb{P}}{\partial z} = \frac{\partial \mathbb{M}}{\partial \theta}\frac{d\theta}{dz} = \frac{1}{z}\frac{\partial \mathbb{M}}{\partial \theta}.$$

Using this expression, we can re-write the PDE for the p.g.f. in terms of the m.g.f. as follows:

$$\frac{\partial \mathbb{M}(\theta,t)}{\partial t} = \lambda(e^\theta - 1)\frac{\partial \mathbb{M}(\theta,t)}{\partial \theta} \tag{3.62}$$

with initial condition $\mathbb{M}(\theta,0) = e^{N\theta}$. To solve this PDE by using the characteristics method, we re-write it as

$$\frac{\partial \mathbb{M}(\theta,t)}{\partial t} + \lambda(1 - e^\theta)\frac{\partial \mathbb{M}(\theta,t)}{\partial \theta} = 0.$$

We introduce the characteristics equations to parameterize the PDE as follows:

$$\frac{dt}{d\tau} = 1, \quad \frac{d\theta}{d\tau} = \lambda(1 - e^{\theta}), \text{ and } \frac{dv}{d\tau} = 0$$

with initial conditions

$$\theta(s,0) = s, \quad t(s,0) = 0, \text{ and } v(s,0) = e^{Ns}.$$

The first characteristic equation can be solved as

$$t = \tau + c_1,$$

and $c_1 = 0$ by using the initial condition $t(s,0) = 0$. Thus the solution is simply $t = \tau$. Solving the second characteristic equation is more involved. We first separate the variables

$$\frac{d\theta}{1 - e^{\theta}} = \lambda d\tau \quad \Rightarrow \quad \frac{e^{-\theta} d\theta}{e^{-\theta} - 1} = \lambda d\tau.$$

Let $u = e^{-\theta} - 1$. Then $du = -e^{-\theta} d\theta$. The equation above can be re-written as

$$-\frac{du}{u} = \lambda d\tau,$$

which yields a general solution after integrating both sides

$$\ln(u) = -\lambda\tau + c_2 \text{ or } \ln(e^{-\theta} - 1) = -\lambda\tau + c_2.$$

This solution can be written as

$$e^{-\theta} - 1 = e^{-\lambda\tau + c_2} = C_2 e^{-\lambda\tau}.$$

Using the initial condition $\theta(s,0) = s$, we can determine $C_2 = e^{-s} - 1$. Hence, the particular solution for the second characteristic equation is given by

$$e^{-\theta} - 1 = (e^{-s} - 1)e^{-\lambda\tau}.$$

The solution to the third characteristic equation is

$$v(s, \tau) = e^{Ns}.$$

Letting $v(s, \tau) = \mathbb{M}(s, \tau)$, we have the m.g.f. as

$$\mathbb{M}(s, \tau) = e^{Ns}.$$

The solution $\mathbb{M}$ must be expressed in terms of θ and t. Based the solution to the second characteristic equation, we have

$$e^{-s} = 1 - e^{\lambda\tau}(1 - e^{-\theta}).$$

It follows from $e^{Ns} = (e^{-s})^{-N}$ that the m.g.f. for the Yule process is given by

$$\mathbb{M}(\theta,t) = (1 - e^{\lambda\tau}(1 - e^{-\theta}))^{-N} = (1 - e^{\lambda t}(1 - e^{-\theta}))^{-N},$$

where the second equality follows from the solution to the first characteristic equation $t = \tau$. Using $e^{-\theta} = z^{-1}$, we obtain the p.g.f. as

$$\mathbb{P}(z,t) = \left(1 - e^{\lambda t}(1 - z^{-1})\right)^{-N}$$

$$= z^N e^{-\lambda Nt}\left(ze^{-\lambda t} - (z-1)\right)^{-N}$$

$$= \frac{z^N e^{-\lambda Nt}}{\left(1 - z(1 - e^{-\lambda t})\right)^N}.$$

Letting $p = e^{-\lambda t}$ and $q = 1 - p$, we can write

$$\mathbb{P}(z,t) = \frac{(pz)^N}{(1 - qz)^N}. \qquad (3.63)$$

For the case of $N = 1$ (i.e., the system starts with one individual), we have

$$\mathbb{P}(z,t) = \frac{pz}{1 - qz}.$$

Then, we take a series of partial derivatives of $\mathbb{P}(z,t)$ with respect to z as follows:

$$\frac{\partial \mathbb{P}(z,t)}{\partial z} = \frac{p}{(1 - qz)^2}\bigg|_{z=0} = p,$$

$$\frac{\partial^2 \mathbb{P}(z,t)}{\partial z^2} = \frac{2pq}{(1 - qz)^3}\bigg|_{z=0} = 2pq,$$

$$\frac{\partial^3 \mathbb{P}(z,t)}{\partial z^3} = \frac{6pq^2}{(1 - qz)^4}\bigg|_{z=0} = 6pq^2,$$

$$\frac{\partial^4 \mathbb{P}(z,t)}{\partial z^4} = \frac{24pq^3}{(1 - qz)^5}\bigg|_{z=0} = 24pq^3,$$

$$\vdots$$

It follows from

$$P_k(t) = \frac{1}{k!}\frac{\partial^k \mathbb{P}(z,t)}{\partial z^k}\bigg|_{z=0}$$

and the pattern of the partial derivative series that

$$P_j(t) = \binom{j-1}{j-N}p^N q^{j-N},$$

where $j = N + i$. Clearly, when $N = 1$, this formula reduces to $P_j(t) = pq^{j-1}$ which is equivalent to

$$P_{i+1}(t) = e^{-\lambda t}(1 - e^{-\lambda t})^i \text{ or } P_i(t) = e^{-\lambda t}(1 - e^{-\lambda t})^{i-1}.$$

For the general case of $N \geq 1$ (initial population size), we have

$$P_{i+N}(t) = \binom{N+i-1}{i} e^{-\lambda N t}(1 - e^{-\lambda t})^i, \quad i = 0, 1, 2, \dots.$$

Due to the negative binomial distribution, the mean and variance for the Yule process are given by

$$\mu(t) = \frac{N}{p} = Ne^{\lambda t} \text{ and } \sigma^2(t) = \frac{Nq}{p} = Ne^{2\lambda t}(1 - e^{-\lambda t}).$$

The pure birth process is often used in modeling the failure process of a machine due to aging (i.e., the number of failures during a finite time period) where the machine state at time t can be described by the total number of failures occurred in $[0, t]$. The Yule process can be considered as a special case of the pure birth process and a branching process in continuous time. The analysis presented above demonstrates the use of generating function approach to stochastic process analysis.

Exercise 3.33 *Consider a Yule process $X(t)$ with parameter λ. For a random period U, which is uniformly distributed over $[0, 1)$, show that $P(X(U) = k) = (1 - e^{-\beta})^k / (\beta k)$ for $k = 1, 2, \dots.$*

Exercise 3.34 *A Yule process $X(t)$ with immigration has birth parameters $\beta_n = \alpha + n\lambda$ for $n = 0, 1, 2, \dots$, where α is the immigration rate and λ is the individual birth rate. Find $P_n(t)$ for $n = 0, 1, 2, \dots$ given $X(0) = 0$.*

3.7.2 PURE DEATH PROCESSES

Complementing the increasing processes such as the Poisson process and its extension, we now consider a decreasing stochastic process called pure death process. Such a process starts with a fix population size N and decreases successively through states $N, N-1, \dots, 2, 1$ and eventually is absorbed in state 0 (extinction state). Let $X(t)$ be the population size at time t, which is a CTMC if the time between two consecutive deaths is exponentially distributed. Denote by μ_k the death rate for state k where $k = N, N-1, \dots, 1$. The infinitesimal transition probabilities for the pure death process are given by

$$P_{i,i+j}(h) = P(X(t+h) - X(t) = j | X(t) = i)$$

$$= \begin{cases} \mu_i h + o(h) & j = -1, \\ 1 - \mu_i h + o(h) & j = 0, \\ 0 & j > 0. \end{cases} \quad (3.64)$$

for $i = 1, 2, ..., N$ in the first and second equations and $i = 0, 1, ..., N$ in the third equation with a convention of $\mu_0 = 0$. With distinct $\mu_1, \mu_2, ..., \mu_N$ (i.e., $\mu_j \neq \mu_j$ if $i \neq j$), based on the forward Kolmogorov equation, we can obtain

$$\frac{dP_N(t)}{dt} = -\mu_N P_N(t).$$

Using the initial condition $X(0) = N$ or $P_N(0) = 1$, we have the explicit solution

$$P_N(t) = e^{-\mu_N t}.$$

For i, N, $P_i(t)$ satisfies

$$\frac{dP_N(t)}{dt} = \mu_{i+1} P_{i+1}(t) - \mu_i P_i(t),$$

which can be solved with

$$P_i(t) = P(X(t) = i | X(0) = N) \\ = \mu_{i+1} \mu_{i+2} \cdots \mu_N (A_{i,i} e^{-\mu_i t} + \cdots + A_{N,i} e^{-\mu_N t}),$$ (3.65)

where

$$A_{k,i} = \frac{1}{(\mu_N - \mu_k) \cdots (\mu_{k+1} - \mu_k)(\mu_{k-1} - \mu_k) \cdots (\mu_i - \mu_k)}.$$

Complementing the Yule process, we can consider a linear death process where $\mu_i = i\mu$. The infinitesimal transition probabilities are given by

$$P_{i,i+j} = P(X(t+h) - X(t) = j | X(t) = i)$$
$$= \begin{cases} i\mu h + o(h) & j = -1, \\ 1 - i\mu h + o(h) & j = 0, \\ o(h) & j \leq -2, \\ 0 & j > 0. \end{cases}$$

Using these transition probabilities, we can develop differential equations for the probability distribution function $P_i(t)$. Note that

$$\begin{aligned} P_i(t+h) &= P(X(t+h) = i) \\ &= P(X(t) = i, X(t+h) - X(t) = 0) \\ &\quad + P(X(t) = i+1, X(t+h) - X(t) = -1) \\ &\quad + P(X(t) = i+2, X(t+h) - X(t) \leq -2) \\ &= P(X(t) = i) \cdot P(X(t+h) - X(t) = 0) \\ &\quad + P(X(t) = i+1) \cdot P(X(t+h) - X(t) = -1) \\ &\quad + P(X(t) = i+2) \cdot P(X(t+h) - X(t) \leq -2) \\ &= P_i(t)[1 - i\mu h + o(h)] + P_{i+1}[(i+1)\mu h + o(h)] \\ &\quad + P_{i+2}(t) \cdot o(h), \end{aligned}$$

which can be written as

$$P_i(t+h) - P_i(t) = (i+1)\mu h P_{i+1}(t) + o(h).$$

Dividing the equation above by h and taking the limit of $h \to 0$, we obtain the forward Kolmogorov equations

$$\frac{dP_i(t)}{dt} = (i+1)\mu P_{i+1}(t) - i\mu P_i(t)$$

$$\frac{dP_N(t)}{dt} = -N\mu P_N(t)$$

for $i = 0, 1, 2, ..., N-1$ and with initial conditions $P_N(0) = 1$ and $P_i(0) = 0$ for $i \neq N$. Similar to the Yule process, the probability distribution function can be obtained by using the generating function approach. Since the derivation is almost identical to that for the Yule process, we omit it and only present the solution:

$$P_i(t) = \binom{N}{i} e^{-i\mu t}(1 - e^{-\mu t})^{N-i} \tag{3.66}$$

for $i = 0, 1, 2, ..., N$. Clearly, this is a binomial distribution with parameters N and $e^{-\mu t}$. Thus, the mean and the variance are given by

$$\mu(t) = Ne^{-\mu t} \quad \text{and} \quad \sigma^2(t) = Ne^{-\mu t}(1 - e^{-\mu t}).$$

Exercise 3.35 *A pure death process starting with $X(0) = 3$ has death rates $\mu_3 = 5, \mu_2 = 3, \mu_1 = 2$, and $\mu_0 = 0$.*
(a) Find $P_n(t)$ for $n = 0, 1, 2, 3$.
(b) Let T_3 be the random time that it takes this pure death process to reach 0. Find the mean and variance of T_3.

The CTMCs presented in this section are often used as a building block of more complex stochastic systems such as an arrival process for queueing systems, a demand process for inventory systems, or a machine failure process for maintenance systems. Note that these CTMCs do not have interesting or meaningful limiting behaviors, either infinite state for pure birth processes or zero for pure death processes. Therefore, we are more interested in the transient or time-dependent behavior of these processes.

3.7.2.1   Transient Analysis on CTMC

Similar to the DTMC with absorbing states, we can also perform the transient analysis for the CTMC with absorbing states. Deriving the transition probability function or time-dependent performance measures usually requires solving some differential equations. Although we have demonstrated this type of analysis in those useful CTMCs above, in more general cases, the transient analysis can be very hard, if not impossible, to obtain the closed-form solution. However, an interesting quantity for a CTMC with an absorbing state is the time interval from a starting transient state to the absorbing state. This time interval follows a so-called phase-type distribution and becomes a building block for extending one-dimensional CTMCs to multi-dimensional CTMCs. This is the topic to be covered in the next chapter.

REFERENCE NOTES

This chapter is focused on the continuous-time Markov chains (CTMCs). To ensure the Markovian property to hold at any point in time, the random time intervals in a stochastic system are modeled as exponentially distributed random variables. The relation between the CTMC and DTMC can be established by the uniformization technique. Thus, the major concepts and properties presented in DTMCs also hold in CTMCs. However, compared with the DTMC treated in the previous chapter, the rigorous proofs of the major properties for CTMC are more technical. We refer interested readers to some classical textbooks such as Feller [2], Taylor and Karlin [7], and Ross [5]. for more detailed discussions on the technical aspects of CTMCs. Several queueing model examples in this chapter are based on Guatam [3]. The nurse-staffing problem in Example 3.4 is based on a working paper [6].

REFERENCES

1. S. Elaydi, "An Introduction to Difference Equations", Springer Science, New York, 3rd ed., 2005.
2. W. Feller, "An Introduction to Probability Theory and Its Applications", Vol. II. John Wiley & Sons Inc., New York, 2nd ed., 1971.
3. N. Gautam, "Analysis of Queues: Methods and Applications", CRC Press, London, 2012.
4. F.P. Kelly, "Reversibility and Stochastic Networks", Chichester, UK: Wiley, 1979.
5. S. Ross, "Introduction to Probability Models", Academic Press, 11th ed., 2014.
6. S. Su, S.H. Sheu, K. Wang, and Z.G. Zhang, Staffing problem for inpatient units with multi-type patients in a hospital, working paper CORMS 2022-01, Western Washington University, 2022.
7. H.M. Taylor and S. Karlin, "An Introduction to Stochastic Modeling", Academic Press, 3rd ed., 1998.

4 Structured Markov Chains

In this Chapter, we extend the exponential distribution to the phase-type (PH) distribution and the Poisson process to the Markovian arrival process (MAP). This extension can overcome the limitation of "memoryless property" of exponential random variables in modeling practical stochastic systems. As a typical example of structured Markov chains, we present the theory of quasi-birth-and-death (QBD) process that extends the birth-and-death process (BDP) discussed in the previous chapter. Furthermore, we introduce the $GI/M/1$ type and $M/G/1$ type Markov chains, which both generalize the QBD process.

4.1 PHASE-TYPE DISTRIBUTIONS

Consider a finite-state CTMC with m transient states and one absorbing state which is numbered as $m+1$. If the CTMC starts from a transient state, the random time interval to the absorbing state, denoted by T, is of our interest and its distribution can be determined by its rate matrix (i.e., infinite generator matrix)

$$\mathbf{Q} = \begin{pmatrix} \mathbf{T} & \eta \\ \mathbf{0} & 0 \end{pmatrix}, \tag{4.1}$$

where $\mathbf{T}$ is an $m \times m$ matrix of transition rates among transient states, $\eta = -\mathbf{T1}$ is a column vector, representing the transition rates from transient states to the absorbing state. Here, $\mathbf{1}$ is a column vector of one's of size $m \times 1$ (Note: this column vector is also denoted by $\mathbf{e}$ in this book). If the starting state is determined by a distribution, denoted by a row vector $\alpha = (\alpha_1, \alpha_2, ..., \alpha_m)$ and α_{m+1} with $\alpha_{m+1} = 1 - \alpha\mathbf{1}$. Then T is said to follow a phase-type (PH) distribution with representation $(\alpha, \mathbf{T})$. We can use the notation $T \sim PH(\alpha, \mathbf{T})$.

Theorem 4.1 *A PH distributed random variable T with representation $(\alpha, \mathbf{T})$ has the cumulative distribution function (cdf)*

$$F_T(t) = P(T \le t) = 1 - \alpha e^{\mathbf{T}t}\mathbf{1}. \tag{4.2}$$

Proof. Using the transition probability function matrix for the CTMC governed by $\mathbf{Q}$, we have

$$\mathbf{P}(t) = e^{\mathbf{Q}t} = \sum_{n=0}^{\infty} \frac{(\mathbf{Q}t)^n}{n!} = \sum_{n=0}^{\infty} \begin{pmatrix} \mathbf{T} & \eta \\ \mathbf{0} & 0 \end{pmatrix}^n \frac{t^n}{n!}$$

$$= \begin{pmatrix} \mathbf{I} & \mathbf{0} \\ \mathbf{0} & 1 \end{pmatrix} + \begin{pmatrix} \mathbf{T} & \eta \\ \mathbf{0} & 0 \end{pmatrix} t + \begin{pmatrix} \mathbf{T} & \eta \\ \mathbf{0} & 0 \end{pmatrix}^2 \frac{t^2}{2!} + \begin{pmatrix} \mathbf{T} & \eta \\ \mathbf{0} & 0 \end{pmatrix}^3 \frac{t^3}{3!} + \cdots$$

$$= \begin{pmatrix} \mathbf{I} & \mathbf{0} \\ \mathbf{0} & 1 \end{pmatrix} + \begin{pmatrix} \mathbf{T} & \eta \\ \mathbf{0} & 0 \end{pmatrix} t + \begin{pmatrix} \mathbf{T}^2 & \mathbf{T}\eta \\ \mathbf{0} & 0 \end{pmatrix} \frac{t^2}{2!} + \begin{pmatrix} \mathbf{T}^3 & \mathbf{T}^2\eta \\ \mathbf{0} & 0 \end{pmatrix} \frac{t^3}{3!} + \cdots$$

$$= \begin{pmatrix} e^{\mathbf{T}t} & 1 - e^{\mathbf{T}t}\mathbf{1} \\ \mathbf{0} & 1 \end{pmatrix}.$$

$$(4.3)$$

The third equality is due to the structure of the matrices and the last equality holds because of power series expression of matrix exponential. Note that

$$\mathbf{0} + \eta t + \mathbf{T}\eta \frac{t^2}{2!} + \mathbf{T}^2\eta \frac{t^3}{3!} + \cdots$$

$$= 1 - 1 - \mathbf{T}\mathbf{1}t - \mathbf{T}^2\mathbf{1}\frac{t^2}{2!} - \mathbf{T}^2\mathbf{1}\frac{t^3}{3!} - \cdots$$

$$= 1 - \left(1 + \mathbf{T}\mathbf{1}t + \mathbf{T}^2\mathbf{1}\frac{t^2}{2!} + \mathbf{T}^3\mathbf{1}\frac{t^3}{3!} + \cdots \right)$$

$$= 1 - e^{\mathbf{T}t}\mathbf{1}.$$

$$(4.4)$$

For $T > 0$, using the fact of $\alpha\mathbf{1} = 1$, we have

$$F_T(t) = P(T \le t) = \alpha \left(1 - e^{\mathbf{T}t}\mathbf{1} \right) = 1 - \alpha e^{\mathbf{T}t}\mathbf{1}. \qquad (4.5)$$

■

The probability density function (pdf) of T can be obtained by taking the first-order derivative of $F_T(t)$ with respect to t as follows:

$$f_T(t) = \frac{dF_T(t)}{dt} = \frac{d}{dt}\left(1 - \alpha e^{\mathbf{T}t}\mathbf{1} \right) = \frac{d}{dt}\left(1 - \alpha \sum_{n=0}^{\infty} \frac{(\mathbf{T}t)^n}{n!}\mathbf{1} \right)$$

$$= -\alpha \sum_{n=0}^{\infty} \frac{n(\mathbf{T}t)^{n-1}\mathbf{T}}{n!}\mathbf{1} = -\alpha \sum_{n=1}^{\infty} \frac{(\mathbf{T}t)^{n-1}}{(n-1)!}\mathbf{T}\mathbf{1} = \alpha \sum_{n=0}^{\infty} \frac{(\mathbf{T}t)^n}{n!}\eta = \alpha e^{\mathbf{T}t}\eta.$$

$$(4.6)$$

With the pdf, we can also obtain the LST, denoted by $L_T(s)$, as follows:

$$L_T(s) = \int_0^{\infty} e^{-st} \alpha e^{\mathbf{T}t}\eta \, dt + \alpha_{m+1} = \alpha \int_0^{\infty} e^{-st+\mathbf{T}t} dt\eta + \alpha_{m+1}$$

$$= \alpha \Delta\eta + \alpha_{m+1},$$

$$(4.7)$$

where $\Delta = \int_0^\infty e^{-st + \mathbf{T}t} dt$ which can be obtained by the integration by parts.

$$\Delta = \int_0^\infty e^{-st + \mathbf{T}t} dt = \int_0^\infty \left(-\frac{1}{s}\right) e^{\mathbf{T}t} d(e^{-st})$$

$$= -\frac{1}{s}\left(e^{-st} e^{\mathbf{T}t}\big|_0^\infty - \int_0^\infty e^{-st} d(e^{\mathbf{T}t})\right) \tag{4.8}$$

$$= -\frac{1}{s}[\mathbf{0} - \mathbf{I}] + \frac{1}{s}\int_0^\infty e^{-st} e^{\mathbf{T}t}\mathbf{T} dt = \frac{1}{s}\mathbf{I} + \frac{1}{s}\Delta\mathbf{T}.$$

Thus we obtain $\Delta = (s\mathbf{I} - \mathbf{T})^{-1}$ and hence

$$L_T(s) = \alpha(s\mathbf{I} - \mathbf{T})^{-1}\eta + \alpha_{m+1}. \tag{4.9}$$

With the LST, we can get the nth moment of T by taking the nth derivative of $L_T(s)$ with respect to s and then set $s = 0$. That is

$$\frac{d^n L_T(s)}{ds^n} = \alpha(-1)^n n!(s\mathbf{I} - \mathbf{T})^{-(n+1)}\eta. \tag{4.10}$$

Thus, we have $E(T^n) = (-1)^n n!\alpha\mathbf{T}^{-n}\mathbf{1}$ by using $\eta = -\mathbf{T}\mathbf{1}$. Here we have to utilize the basic rules of matrix derivatives as stated in the following lemma.

Lemma 4.1 *If $\mathbf{A}$ is non-singular square matrix as a function of scalar s (i.e., $\mathbf{A}^{-1}$ exists), then*

$$\frac{d\mathbf{A}^{-1}}{ds} = -\mathbf{A}^{-1}\frac{d\mathbf{A}}{ds}\mathbf{A}^{-1}. \tag{4.11}$$

Proof. Taking the first-order derivative with respect to s on both sides of $\mathbf{A}^{-1}\mathbf{A} = \mathbf{I}$ yields

$$\frac{d}{ds}(\mathbf{A}^{-1}\mathbf{A}) = \frac{d\mathbf{A}^{-1}}{ds}\mathbf{A} + \mathbf{A}^{-1}\frac{d\mathbf{A}}{ds} = \frac{d\mathbf{I}}{ds} = \mathbf{0}$$

$$\Rightarrow \frac{d\mathbf{A}^{-1}}{ds}\mathbf{A} = -\mathbf{A}^{-1}\frac{d\mathbf{A}}{ds}.$$

By right-multiplying by $\mathbf{A}^{-1}$, we have the result (4.11). ∎

Exercise 4.1 *Prove $E(T^n) = (-1)^n n!\alpha\mathbf{T}^{-n}\mathbf{1}$ and find the mean and variance of the PH distributed random variable.*

Example 4.1 *Suppose that an operating system is subject to three types of failures, which are minor, moderate, and major in nature. While the first two types (minor and moderate) can be repaired, the third (major) one is not repairable and the system must be replaced. When the system is in operation,* *it can be in one of the two conditions: good and average. The time to fail-ure for a good (average) condition system follows an exponential distribution with mean of 100 hours (50 hours). When the system of good condition fails, the probabilities for minor, moderate, and major types are 0.5, 0.4, and 0.1, respectively. For the system of average condition, these probabilities are 0.15, 0.50, and 0.35, respectively. Further it is assumed that the repair time for type 1 (type 2) is also exponentially distributed with mean of 1 hour. The repair for a type 1 failure will bring the system to good condition and the repair for a type 2 failure will result in average condition. Argue that the life time of a good con-dition system is a PH distributed random variable and give the representation of this PH distribution.*

Solution 4.1 *Based on the problem description, we know that the system can be in one of five states. Four of these states are the transient states and one is the absorbing state. These states are*

- *State 1: Good condition system in operation*
- *State 2: System under repair after a type 1 failure*
- *State 3: System under repair after a type 2 failure*
- *State 4: Average condition system in operation*
- *State 5: System in type 3 failure*

Note that the first four states are transient and state 5 is absorbing. Since all random variables are exponentially distributed, the life time of a brand new system (good condition), the duration from the state 1 to the first time of reaching state 5, is a PH distributed random variable, denoted by D. The representation is given by

$$\alpha = (1,0,0,0)$$

$$\mathbf{T} = \begin{pmatrix} -0.010 & 0.005 & 0.004 & 0 \\ 1 & -1 & 0 & 0 \\ 0 & 0 & -1 & 1 \\ 0 & 0.003 & 0.01 & -0.02 \end{pmatrix}.$$

Thus, $D \sim PH(\alpha, \mathbf{T})$.

Exercise 4.2 *In Example 4.1, suppose that a type 1 repair results in a working system of good condition with 50% chance and of average condition with 50% chance; and a type 2 repair results in a working system of good condition with 10% chance and of average condition with 90% chance. Compute the mean and vairance of the life time of the system and compare them with the mean and variance of the system with the parameters of Example 4.1.*

One special type of PH distributions are called acyclic PH distributions and have a $\mathbf{T}$ as an upper (or lower) triangular matrix. Many well-known distributions such as Erlang or Hyper-exponential distributions belong to this class.

Exercise 4.3 *Write down the PH representation for the following distributions: (i) Exponential distribution; (ii) Erlang-k distribution; (iii) Hypoexponential distribution; (iv) Hyper-exponential distribution; (v) A certain type of Coxian distribution.*

We can also develop the discrete PH distribution. Consider a DTMC denoted by $\{X_n : n \in \mathbb{N}\}$ on state space $S = \{1, 2, ..., m, m+1\}$ with state $m+1$ being absorbing. Then, the one-step transition probability matrix can be written as

$$\mathbf{P} = \begin{pmatrix} \mathbf{T} & \eta \\ \mathbf{0} & 1 \end{pmatrix}, \tag{4.12}$$

where $\mathbf{T}\mathbf{1} + \eta = \mathbf{1}$. Similar to the CTMC case, starting from an initial transient state, we define

$$K = \inf(n \in \mathbb{N} : X_n = m+1), \tag{4.13}$$

as the discrete-time PH distributed random variable. Its representation is denoted by $PH_d(\alpha, \mathbf{T})$. Clearly, $\eta = (\mathbf{I} - \mathbf{T})\mathbf{1}$, and $\alpha_{m+1} = 1 - \alpha\mathbf{1}$. It is easy to get the cdf and probability mass function (pmf) as follows:

$$F_K(k) = P(K \leq k) = 1 - \alpha\mathbf{T}^k\mathbf{1},$$
$$f_K(0) = P(K = 0) = \alpha_{m+1}, \text{ and } f_K(k) = P(K = k) = \alpha\mathbf{T}^{k-1}\eta. \tag{4.14}$$

Recall that the cdf and pmf for a geometrically distributed random variable are $F_X(x) = P(X \leq x) = 1 - q^x$ and $f_X(x) = q^{x-1}p$, respectively, where $p = 1 - q$. It is noted that the discrete-time PH distribution is the extension of the geometric distribution and the matrix $\mathbf{T}$ plays the same role as the parameter q. Similarly, we can work out the z-transform of K (moment generating function), denoted by $G_K(z)$, as follows:

$$G_K(z) = E(z^K) = \sum_{k=0}^{\infty} z^k f_K(k) = \alpha_{m+1} + \sum_{k=1}^{\infty} z^k \alpha\mathbf{T}^{k-1}\eta$$
$$= \alpha_{m+1} + z\alpha(\mathbf{I} - z\mathbf{T})^{-1}\eta. \tag{4.15}$$

Here we have used the geometric series $\sum_{i=0}^{\infty} x^i = 1/(1 - x)$ for matrices. That is $\sum_{i=0}^{\infty} \mathbf{A}^i = (\mathbf{I} - \mathbf{A})^{-1}$ whenever $|\lambda_i| < 1$ for all i, where λ_i's are the eigenvalues of $\mathbf{A}$ (see Appendix J). The factorial moments for a discrete PH random variable can be obtained by successive differentiation of the generating function. The matrix $\mathbf{S} = (\mathbf{I} - \mathbf{T})^{-1}$ is of special importance as the i, j'th element has an important probabilistic interpretation as the expected time spent in state j before absorption conditioned on starting in state i (see Chapter 2). Specifically, we can establish the following theorem.

Theorem 4.2 *A $PH_d(\alpha, \mathbf{T})$ distribution is nondefective if and only if the matrix $\mathbf{I} - \mathbf{T}$ is invertible. In this case, the expected number S_{ij} of visits to state j before absorption, given that the initial state is i, is $S_{ij} = [(\mathbf{I} - \mathbf{T})]_{ij}^{-1}$.*

Proof. $\Leftarrow$ can be proved by using Section 2.3.4 of Chapter 2. To show $\Rightarrow$, we assume that $\mathbf{I} - \mathbf{T}$ is invertible. Based on the multiple-step transition probability matrix given by

$$\mathbf{P}^{(k)} = \mathbf{P}^k = \begin{pmatrix} \mathbf{T}^k & (\mathbf{I} - \mathbf{T}^k)\mathbf{1} \\ \mathbf{0} & 1 \end{pmatrix}, \tag{4.16}$$

we have the probability vector $\Theta(k) = (\mathbf{I} - \mathbf{T}^k)\mathbf{1}$ in which all elements $\Phi_i(k)$ with $1 \le i \le m$ and $k \in \mathbb{N}$ are between 0 and 1. Consider the equation

$$(\mathbf{I} - \mathbf{T}) \sum_{k=0}^{n-1} \mathbf{T}^k = \sum_{k=0}^{n-1} \mathbf{T}^k - \sum_{k=0}^{n} \mathbf{T}^k = \mathbf{I} - \mathbf{T}^n. \tag{4.17}$$

Multiplying both sides of the equation above by $(\mathbf{I} - \mathbf{T})^{-1}$ from the left side and by $\mathbf{1}$ from the right side, we have

$$\sum_{k=0}^{n-1} \mathbf{T}^k \mathbf{1} = (\mathbf{I} - \mathbf{T})^{-1}(\mathbf{I} - \mathbf{T}^n)\mathbf{1} = (\mathbf{I} - \mathbf{T})^{-1}\Theta(n), \tag{4.18}$$

which implies

$$\lim_{n \to \infty} \sum_{k=0}^{n-1} \mathbf{T}^k = \sum_{k=0}^{\infty} \mathbf{T}^k < \infty \tag{4.19}$$

in an entry-wise meaning. Note that T_{ij}^n is the probability of being in phase j at time n, given that the initial phase is i. Hence we have

$$\begin{aligned} S_{ij} &= E\left(\sum_{n=0}^{\infty} I(X_n = j) | X_0 = i \right) \\ &= \sum_{n=0}^{\infty} E(I(X_n = j) | X_0 = i) \\ &= \sum_{k=0}^{\infty} T_{ij}^k < \infty \quad \forall 1 \le i, j \le m, \end{aligned} \tag{4.20}$$

which means that all states are transient and absorption in state $m + 1$ is certain from any initial state, i.e. $F_K(1) = 1$. $\blacksquare$

Now we can explain how to get the expression of the geometric matrix series. Taking the limit $n \to \infty$ of

$$(\mathbf{I} - s\mathbf{T}) \sum_{k=0}^{n-1} (s\mathbf{T})^k = \mathbf{I} - (s\mathbf{T})^n \quad \text{for } |s| < 1, \tag{4.21}$$

results in

$$(\mathbf{I} - s\mathbf{T}) \sum_{k=0}^{\infty} (s\mathbf{T})^k = \mathbf{I} \quad \text{for } |s| < 1. \tag{4.22}$$

Exercise 4.4 *Prove the cdf and pmf of a discrete PH random variable with representation $PH_d(\alpha, \mathbf{T})$ are given by (4.14).*

Exercise 4.5 *Write down the representation of discrete PH distribution for (i) Geometric distribution with parameter p; (ii) Negative Binomial distribution with parameter (k, p).*

Exercise 4.6 *Prove that the z-transform for K in terms of matrix $\mathbf{S}$ is $G_K(z) = \alpha_{m+1} + \alpha \left(\mathbf{S} \frac{1-z}{z} + \mathbf{I} \right)^{-1} \mathbf{1}$.*

4.2 PROPERTIES OF PH DISTRIBUTION

4.2.1 CLOSURE PROPERTIES

The family of discrete or continuous PH distributions is closed under certain operations such as addition, convex mixtures, minimum, and maximum. These properties make PH distributions a powerful tool for probabilistic modeling. From the definition of PH distribution, we know that PH random variable is a time period and certainly non-negative. Thus, these operations with closure properties are expected. For example, one time interval plus another interval should be a time interval; a weighted sum of two time intervals should be a time interval; and minimum or maximum of several time intervals should be also a time interval. This intuition also explains why other operations such as subtraction, multiplication, or division do not have the closure properties. Clearly, a time interval is multiplied (or divided) by another time interval is not going to be a time interval. Subtracting one time interval from another time interval may result in a negative number which is not a time interval. Thus, PH distributions are not closed under these operations. We show these closure properties in either discrete or continuous PH distribution as one case can be proved in the same way as the other case. Thus, for each property, we only provide the proof for one of the discrete- and continuous-time cases and leave the other case as an exercise for readers. We start with the discrete time case for the sum of two independent PH random variables.

Theorem 4.3 *(Convolution of PH_d) Let $Z_i \sim PH_d(\alpha^{(i)}, \mathbf{T}^{(i)})$ be two independent discrete PH distributed random variables of order m_i for $i = 1, 2$. Then their sum $Z = Z_1 + Z_2 \sim PH_d(\alpha, \mathbf{T})$ has a PH distribution of order $m = m_1 + m_2$ with representation*

$$\alpha = (\alpha^{(1)}, \alpha_{m_1+1}^{(1)} \alpha^{(2)})$$

and

$$\mathbf{T} = \begin{pmatrix} \mathbf{T}^{(1)} & \eta^{(1)}\alpha^{(2)} \\ \mathbf{0} & \mathbf{T}^{(2)} \end{pmatrix}.$$

Proof. Consider two independent DTMCs $X^{(i)}$ with m_i transient states and one absorbing state for $i = 1,2$. Let Z_i, for $i = 1,2$, be the random variable expressing the time until absorption of the $X^{(i)}$, the corresponding DTMCs with an absorbing state. The sum $Z = Z_1 + Z_2$ represents the total time until absorption in the concatenated Markov chain, denoted by X, with $m = m_1 + m_2 + 1$ states where the first m_1 states refer to the transient states of $X^{(1)}$, the next m_2 states refer to the transient states of $X^{(2)}$ and the final state is the absorbing state. The chain X first moves along the transient states of $X^{(1)}$ defined by $(\alpha^{(1)}, \mathbf{T}^{(1)})$ until it reaches the absorbing state of $X^{(1)}$, but upon absorption, the concatenated process X immediately starts to run through the second Markov chain $X^{(2)}$ defined by $(\alpha^{(2)}, \mathbf{T}^{(2)})$ until absorption in this second chain.

With probability $\alpha^{(1)}$ this concatenated chain starts in any of the m_1 transient states of $X^{(1)}$ and with probability $\alpha^{(1)}_{m_1+1}\alpha^{(2)}$ in any of the m_2 transient states of $X^{(2)}$ since that would imply that the first chain starts in the absorbing state and we immediately move on to the second chain. The probability that the total time until absorption in the concatenated Markov chain is 0 hence equals $\alpha^{(1)}_{m_1+1}\alpha^{(2)}_{m_2+1}$.

The transitions within the first chain are defined by $\mathbf{T}^{(1)}$ and transitions between the first chain and the second are determined by the product of the exit probabilities from the first chain and the initial distribution $\alpha^{(2)}$ of the second chain $X^{(2)}$. As Z_1 and Z_2 are independent, the first state in the path of the second chain $X^{(2)}$ is always determined by its initial distribution $\alpha^{(2)}$. Once the concatenated chain enters the transient states of $X^{(2)}$ there is no possibility of returning to a state of $X^{(1)}$ and transitions are governed by $\mathbf{T}^{(2)}$. This illustrates that the sum of Z_1 and Z_2 can be represented as a PH distribution with representation $(\alpha, \mathbf{T})$ as defined in the formulation of the theorem. Note that for the sum Z we have

$$\eta = 1 - \mathbf{T1} = \begin{pmatrix} \eta^{(1)}\alpha^{(2)}_{m_2+1} \\ \eta^{(2)} \end{pmatrix}.$$

Starting in any of the transient states of $X^{(1)}$, the exit probabilities of X are the product of the exit probabilities of $X^{(1)}$ and the initial probability of starting in the absorbing state of $X^{(2)}$. Starting in any of the transient states of $X^{(2)}$, the exit probabilities of X are simply the exit probabilities of $X^{(2)}$. ∎

This proof can be clearly illustrated by making a transition diagram similar to those for CTMCs. Such a diagram is shown in Figure 4.1. Other closure properties can be shown by similar transition diagrams (left as exercises).

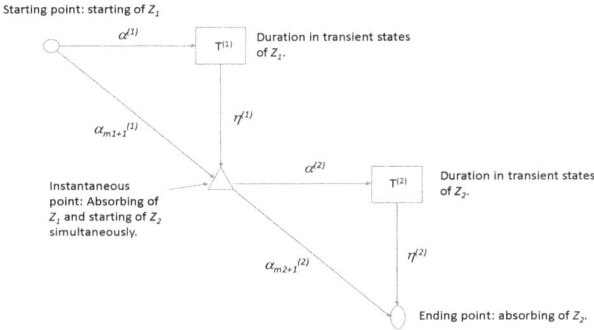

Figure 4.1 Convolution of two independent PH distributions.

Remark 4.1 : *Using the probabilistic interpretation of the PH representation to prove the closure properties is intuitive. Alternatively, these properties can be proved analytically by using the moment generating function (or Laplace transforms for the continuous-time case). In this case of sum of two independent PH-distributed random variable, we have*

$$G_{Z_1+Z_2}(z) = (\alpha^{(1)}_{m_1+1} + z\alpha^{(1)}(\mathbf{I} - z\mathbf{T}^{(1)})^{-1}\eta^{(1)})(\alpha^{(2)}_{m_2+1} + z\alpha^{(2)}(\mathbf{I} - z\mathbf{T}^{(2)})^{-1}\eta^{(2)})$$

$$= \alpha_{m+1} + z\alpha(\mathbf{I} - z\mathbf{T})^{-1}\eta,$$

where the second equality follows after some straightforward but tedious calculations (left as an exercise).

Exercise 4.7 *For two independent geometrically distributed random variables X with parameter p_x and Y with p_y, write down the PH representation of $Z = X + Y$.*

Exercise 4.8 *Prove the closure property for the sum of two independent continuous PH random variables.*

Exercise 4.9 *Let Z be the sum of k independent exponentially distributed random variables with parameter λ_i with $i = 1, 2, ...k$. Write down the PH representation of Z.*

The next closure property is regarding the finite mixtures of PH distributions. We use the continuous PH distribution case to show this property.

Theorem 4.4 *A finite convex mixture of PH-distributions is a PH-distribution.*

Proof. Suppose $X_i \sim PH(\alpha_i, \mathbf{T}_i)$ with $i = 1, 2, ..., k$. Let $Z = \sum_{i=1}^{k} p_i X_i$ where $0 \leq p_i \leq 1$ and $\sum_{i=1}^{k} p_i = 1$. Then we show that $Z \sim PH(\gamma, \mathbf{L})$ where

$$\gamma = (p_1 \alpha_1, p_2 \alpha_2, ..., p_k \alpha_k)$$

$$\mathbf{L} = \begin{pmatrix} T_1 & 0 & \cdots & 0 \\ 0 & T_2 & \cdots & 0 \\ \vdots & \vdots & \vdots & \vdots \\ 0 & 0 & \cdots & T_k \end{pmatrix}.$$

The proof is completed by using the probabilistic interpretation of $(\gamma, \mathbf{L})$. ■

Exercise 4.10 *Make a transition diagram similar to Figure 4.1 for Theorem 4.4.*

Exercise 4.11 *Prove the closure property for a finite convex mixture of discrete PH distributions.*

Exercise 4.12 *Consider k exponentially distributed random variables $X_i \sim \exp(\lambda_i)$ with $i = 1, ..., k$. Assume that Z is a convex mixture of these exponential random variables. Write down the PH-representation of Z.*

In addition to the sum and the convex mixture of PH random variables, the order statistic of a finite number of independent PH-distributed random variables is itself PH-distributed. Again, we use the continuous PH-distribution case to state this property. First, we use an example to demonstrate the basic idea. Before we analyze such an example, we need to introduce two useful matrix operations. The first one is the Kronecker product of two matrices, denoted by $\otimes$. Let $\mathbf{A}$ be a matrix with dimension $l \times k$ and $\mathbf{B}$ be a matrix with dimension $n \times m$. Then, the Kronecker product of $\mathbf{A} = [a_{ij}]$ and $\mathbf{B} = [b_{ij}]$ is defined by

$$\mathbf{A} \otimes \mathbf{B} = \begin{pmatrix} a_{11}\mathbf{B} & a_{12}\mathbf{B} & \cdots & a_{1k}\mathbf{B} \\ a_{21}\mathbf{B} & a_{22}\mathbf{B} & \cdots & a_{2k}\mathbf{B} \\ \vdots & \vdots & \vdots & \vdots \\ a_{l1}\mathbf{B} & a_{l2}\mathbf{B} & \cdots & a_{lk}\mathbf{B} \end{pmatrix},$$

with dimension $ln \times km$. Thus, it is valid to perform Kronecker product with any two matrices of different dimensions.

Another useful operation is the Kronecker sum of two matrices $\mathbf{A}$ and $\mathbf{B}$ and is defined by

$$\mathbf{A} \oplus \mathbf{B} = \mathbf{A} \otimes \mathbf{I} + \mathbf{I} \otimes \mathbf{B}.$$

Example 4.2 *Let X be a two-stage hypo-exponential random variable (also called generalized Erlang-2) with parameters $\lambda_{x,1}, \lambda_{x,2}$ and Y be a hyperexponential of mixture 2 random variable with parameters $p_y, \lambda_{y,1}, \lambda_{y,2}$. We want to investigate the distributions of $\min(X, Y)$ and $\max(X, Y)$, respectively.*

Clearly, we can write the PH-representations of these two random variables as

$$\alpha_x = (1,0), \qquad \mathbf{T_x} = \begin{pmatrix} -\lambda_{x,1} & \lambda_{x,1} \\ 0 & -\lambda_{x,2} \end{pmatrix},$$

$$\alpha_y = (p_y, 1 - p_y), \quad \mathbf{T_y} = \begin{pmatrix} -\lambda_{y,1} & 0 \\ 0 & -\lambda_{y,2} \end{pmatrix}.$$

First, let's consider $\min(X,Y)$. We construct a two-dimensional CTMC (denoted by 2-D CTMC) that simultaneously describes the evolution of the two independent CTMCs corresponding to the two PH-distributed random variables. The transient state of this 2-D CTMC is denoted by (i,j) with $i = 1,2$ representing the stage of X and $j = 1,2$ representing the parameter index of Y. The only single absorbing state is denoted by A, representing the ending of X or Y, whichever is smaller. Then, we have four transient states for the 2-D CTMC: $(1,1)$, $(1,2)$, $(2,1)$, $(2,2)$, a convenient sequence of the states. For example, state $(1,2)$ indicates that X is in the first stage with parameter $\lambda_{x,1}$ and Y is an exponential random variable with parameter $\lambda_{y,2}$. Clearly, the time until the 2-D CTMC get absorbed is $\min(X,Y)$, which is also a PH distributed random variable. The PH-representation of $\min(X,Y)$ is given by

$$\alpha_{\min(X,Y)} = \alpha_x \otimes \alpha_y = (p_y, 1 - p_y, 0, 0),$$

$$\mathbf{T}_{\min(X,Y)} = \mathbf{T_x} \otimes \mathbf{I_y} + \mathbf{I_x} \otimes \mathbf{T_y}$$

$$= \begin{pmatrix} -(\lambda_{x,1} + \lambda_{y,1}) & 0 & \lambda_{x,1} & 0 \\ 0 & -(\lambda_{x,1} + \lambda_{y,2}) & 0 & \lambda_{x,1} \\ 0 & 0 & -(\lambda_{x,2} + \lambda_{y,1}) & 0 \\ 0 & 0 & 0 & -(\lambda_{x,2} + \lambda_{y,2}) \end{pmatrix}.$$

Note that we have defined the Kronecker product and Kronecker sum of two matrices and know how to perform the calculations. Here, we provide some intuition about these operations. For example, in $\mathbf{T}_{\min(X,Y)} = \mathbf{T_x} \otimes \mathbf{I_y} + \mathbf{I_x} \otimes \mathbf{T_y}$, the first term $\mathbf{T_x} \otimes \mathbf{I_y}$ represents that the CTMC corresponding to X makes a transition to the 2nd stage while the CTMC corresponding to Y does not make any transition. Similarly, the second term $\mathbf{I_x} \otimes \mathbf{T_y}$ represents that the CTMC corresponding to Y makes a transition while the CTMC corresponding to X does not make any transition. Therefore, the identity matrix for one variable in Kronecker product can be interpreted as the CTMC corresponding that variable does not make any transition.

Next, we consider $\max(X,Y)$. To obtain the distribution of $\max(X,Y)$ by analyzing a 2-D CTMC, we need to add another value 3 to each state variable (i.e., the ending of each variable). For example, state $(3,1)$ represents that X has reached the absorbing state of its corresponding CTMC while Y is in the state with exponential duration of parameter $\lambda_{y,1}$. Hence, we will have eight transient states in the 2-D CTMC sequenced as $(1,1),(1,2),(2,1),$

(2,2),(1,3),(2,3),(3,1),(3,2). The first four states are those in which neither the CTMC for X nor the CTMC for Y has been absorbed and the last four states are those in which one of the two CTMCs has been absorbed. The only absorbing state, denoted by A, is the case in which both CTMCs are absorbed. Clearly, the time until this 2-D CTMC with eight transient states get absorbed is $\max(X,Y)$, which is also a PH distributed random variable. Its PH-representation is given by

$$\alpha_{\max(X,Y)} = (\alpha_x \otimes \alpha_y, \alpha_x \alpha_{y,m+1}, \alpha_{x,k+1} \alpha_y)$$
$$= (p_y, 1 - p_y, 0, 0, \alpha_{y,m+1}, 0, \alpha_{x,k+1} p_y, \alpha_{x,k+1}(1 - p_y)),$$

$$\mathbf{T}_{\max(X,Y)} = \begin{pmatrix} \mathbf{T}_x \otimes \mathbf{I}_y + \mathbf{I}_x \otimes \mathbf{T}_y & \mathbf{I}_x \otimes \eta_y & \eta_x \otimes \mathbf{I}_y \\ \mathbf{0} & \mathbf{T}_x & \mathbf{0} \\ \mathbf{0} & \mathbf{0} & \mathbf{T}_y \end{pmatrix}$$

$$= \left(\begin{array}{cccc|cc|cc} -(\lambda_{x,1}+\lambda_{y,1}) & 0 & \lambda_{x,1} & 0 & \lambda_{y,1} & 0 & 0 & 0 \\ 0 & -(\lambda_{x,1}+\lambda_{y,2}) & 0 & \lambda_{x,1} & \lambda_{y,2} & 0 & 0 & 0 \\ 0 & 0 & -(\lambda_{x,2}+\lambda_{y,1}) & 0 & 0 & \lambda_{y,1} & \lambda_{x,2} & 0 \\ 0 & 0 & 0 & -(\lambda_{x,2}+\lambda_{y,2}) & 0 & \lambda_{y,2} & 0 & \lambda_{x,2} \\ \hline 0 & 0 & 0 & 0 & -\lambda_{x,1} & \lambda_{x,1} & 0 & 0 \\ 0 & 0 & 0 & 0 & 0 & -\lambda_{x,2} & 0 & 0 \\ \hline 0 & 0 & 0 & 0 & 0 & 0 & -\lambda_{y,1} & 0 \\ 0 & 0 & 0 & 0 & 0 & 0 & 0 & -\lambda_{y,2} \end{array} \right).$$

Now we can summarize the closure property under the min and max operations in the following theorem.

Theorem 4.5 *Suppose that $X \sim PH(\alpha_x, T_x)$ and $Y \sim PH(\alpha_y, T_y)$ are two independent PH random variables. Let $Z_{min} = \min(X,Y)$ and $Z_{max} = \max(X,Y)$. Then Z_{min} and Z_{max} are also PH-distributed. The PH representation for Z_{min}, denoted by $PH(\gamma_{min}, \mathbf{L}_{min})$, is given by*

$$\gamma_{min} = \alpha_x \otimes \alpha_y, \quad \mathbf{L}_{min} = \mathbf{T}_x \otimes \mathbf{I}_y + \mathbf{I}_x \otimes \mathbf{T}_y.$$

The PH representation for Z_{max}, denoted by $PH(\gamma_{max}, \mathbf{L}_{max})$, is given by

$$\gamma_{max} = (\alpha_x \otimes \alpha_y, \alpha_x \alpha_{y,m+1}, \alpha_{x,k+1} \alpha_y)$$
$$\mathbf{L}_{\max(X,Y)} = \begin{pmatrix} \mathbf{T}_x \otimes \mathbf{I}_y + \mathbf{I}_x \otimes \mathbf{T}_y & \mathbf{I}_x \otimes \eta_y & \eta_x \otimes \mathbf{I}_y \\ \mathbf{0} & \mathbf{T}_x & \mathbf{0} \\ \mathbf{0} & \mathbf{0} & \mathbf{T}_y \end{pmatrix},$$

and

$$\eta_{max} = \begin{pmatrix} \mathbf{0} \\ \eta_x \\ \eta_y \end{pmatrix}.$$

Proof. The proof can be completed directly by using the probabilistic interpretation of $(\gamma, \mathbf{L})$. ∎

Exercise 4.13 *Make a transition diagram similar to Figure 4.1 for Theorem 4.5.*

Exercise 4.14 *Let X be a negative binomial random variable with parameters $k_x = 2$ and p_x and let Y be a negative binomial random variable with parameters $k_y = 2$ and p_y. Write down the PH representations for the minimum and maximum of these two discrete PH random variables.*

4.2.2 DENSE PROPERTY OF PH DISTRIBUTION

We prove the following theorem.

Theorem 4.6 *The class of phase-type distributions is dense (in terms of weak convergence) within the class of all distributions on $\mathbb{R}_0^+$*

Proof. Let $F : \mathbb{R}_0^+ \to [0,1]$ denote any non-negative distribution function. This proof has three parts: (1) show that F can be approximated by a step function bounded by ε; (2) argue that such a step function can be approximated by a combination of Dirac functions bounded by ε; and (3) show that these Dirac functions can be approximated by appropriate Erlang distributions bounded by ε.

Since F is bounded, monotone and right-continuous, we can approximate F by a step function G with countably many jumps at $(t_n : n \in \mathbb{N}_0)$, where $t_n < t_{n+1}$ for all $n \in \mathbb{N}$. The error ε of approximation can be chosen arbitrarily small such that $|F(t) - G(t)| < \varepsilon$ holds for all $t \geq 0$. If $t_0 = 0$, i.e. if there is a jump of G at zero, we can write $G(t) = p_0 \delta_0 + (1 - p_0)\tilde{G}(t)$ with $p_0 = G(0)$ and $\tilde{G}(t) = (G(t) - G(0))/(1 - p_0)$. Note that the Dirac distribution function δ_0 can be considered as a phase-type distribution with $m = 0$ and $\alpha_{m+1} = 1$. Next we show that $\tilde{G}$ can be approximated by a finite mixture of Dirac functions. This is done by first finding a truncation point T of $\tilde{G}$ such that $G(T) > 1 - \varepsilon$. Then there is a number $N \in \mathbb{N}$ such that T has been reached no later than the Nth jump, i.e. $t_n > T$ for all $n > N$. Thus, $\tilde{G}$, an increasing step function, can be approximated by

$$H = \sum_{n=1}^{N-1} (\tilde{G}(t_n) - \tilde{G}(t_{n+1}))\delta_{t_n} + (1 - \tilde{G}(t_N))\delta_{t_N}, \qquad (4.23)$$

with an error bounded by ε. Finally, for every $n = 1, ..., N$, we approximate the Dirac distribution δ_{t_n} by a suitable Erlang distribution. This is feasible due to the following argument: The variance of an Erlang distribution $E_k^{k\lambda}$ of order k with parameter $k \cdot \lambda$ is given by $(k \cdot \lambda^2)^{-1}$ (left as an exercise) and thus tends to zero as k grows larger. It follows from the mean of such an Erlang distribution equal to $k \cdot \frac{1}{k\lambda} = \frac{1}{\lambda}$ and Chebyshev's inequality that the sequence $(E_k^{k\lambda} : k \in \mathbb{N})$ converges in probability (or called weakly) toward δ_{t_n} if $\lambda = 1/t_n$ is chosen. This means that there is a number $K_n \in \mathbb{N}$ such that the distribution function H_n of an $E_{K_n}^{K_n/t_n}$ distribution satisfies $|H_n(t) - \delta_{t_n}(t)| < \varepsilon$ for all $t \geq 0$. If we pursue the above approximation method for every $n = 1, ..., N$ and define

$$\tilde{H} = \sum_{n=1}^{N-1} (\tilde{G}(t_n) - \tilde{G}(t_{n+1})) \cdot H_n + (1 - \tilde{G}(t_N)) \cdot H_N, \qquad (4.24)$$

then we obtain an approximation bound $|H - \tilde{H}| < \varepsilon$. It follows from Theorem 4.4 and Erlang distribution property that the distribution $\tilde{H}$ is phase type. In summary, we have approximated F by $p_0 \cdot \delta_0 + (1 - p_0)\tilde{H}$, which is a PH distribution, with an approximation bound of 3ε. This completes the proof. ∎

This theorem indicates in theory that any non-negative distribution can be approximated by a PH distribution at a pre-specified accuracy level. Such a property makes the PH distribution a powerful modeling tool. However, the "curse of dimensionality" can become an issue in utilizing the PH distribution.

4.2.3 NON-UNIQUENESS OF REPRESENTATION OF PH DISTRIBUTION

The representation of a PH distribution is not unique. In general, there exists another representation $(\beta, \mathbf{S})$ of size either $m = n$ or $m \neq n$ that represents the same phase-type distribution with representation $(\alpha, \mathbf{T})$ of size n. For example, if a PH distribution has a representation $(\alpha, \mathbf{T})$, then we can find a matrix $\mathbf{B}$ which is invertible and $\mathbf{B1} = \mathbf{1}$ (hence, $\mathbf{B}^{-1}\mathbf{1} = \mathbf{1}$). It is easy to verify that another PH distribution of the same size with the representation $(\beta = \alpha\mathbf{B}, \mathbf{S} = \mathbf{B}^{-1}\mathbf{TB})$ has the same PH distribution with representation $(\alpha, \mathbf{T})$. This is because that its cdf is given by

$$1 - \alpha\mathbf{B}e^{\mathbf{B}^{-1}\mathbf{TB}x}\mathbf{1} = 1 - \alpha\mathbf{B}\sum_{k=0}^{\infty}\frac{(\mathbf{B}^{-1}\mathbf{TB}x)^k}{k!}\mathbf{1} = 1 - \alpha\mathbf{B}\sum_{k=0}^{\infty}\frac{(\mathbf{B}^{-k})(\mathbf{T}x)^k(\mathbf{B})^k}{k!}\mathbf{1}$$

$$= 1 - \alpha\mathbf{B}\mathbf{B}^{-1}\sum_{k=0}^{\infty}\frac{(\mathbf{T}x)^k}{k!}\mathbf{B1} = 1 - \alpha\mathbf{B}\mathbf{B}^{-1}e^{\mathbf{T}x}\mathbf{B1} = 1 - \alpha e^{\mathbf{T}x}\mathbf{1}.$$

$$(4.25)$$

Exercise 4.15 *Verify that the following three PH representations describe the same PH distribution.*

$$\alpha^{(1)} = (0.3, 0.3, 0.4), \quad \mathbf{T}^{(1)} = \begin{pmatrix} -4 & 1 & 2 \\ 1 & -4 & 2 \\ 0 & 0 & -1 \end{pmatrix},$$

$$\alpha^{(2)} = (0.21, 0.39, 0.4), \quad \mathbf{T}^{(2)} = \begin{pmatrix} -4.3 & 1.3 & 2 \\ 0.7 & -3.7 & 2 \\ 0 & 0 & -1 \end{pmatrix},$$

$$\alpha^{(3)} = (0.6, 0.4), \quad \mathbf{T}^{(3)} = \begin{pmatrix} -3 & 2 \\ 0 & -1 \end{pmatrix}.$$

4.3 FITTING PH DISTRIBUTION TO EMPIRICAL DATA OR A THEORETICAL DISTRIBUTION

There are several ways of fitting a PH distribution to either an empirical data or a theoretical distribution. In this section, we introduce the Expectation Maximization (EM) algorithm developed by Asmussen et al. (1996). It is an iterative method for calculating maximum likelihood estimates for probability density functions in cases with missing data or where a direct evaluation of the observed data likelihood function is difficult. As the name indicates, the algorithm contains two steps. The expectation (E) step replaces the missing (unobserved) data with their conditional expectation given the current parameter estimates as well as the observed data. In the maximization (M) step, maximum likelihood estimates of the parameters are calculated based on the observed data as well as the expectations obtained through the E step.

When using the EM algorithm, it is assumed that two sets of data exist. The observed data x which is incomplete and the complete data z which contains x and an unobserved part y. If the underlying probability distribution function of the data belongs to the family of exponential distributions, e.g., a CTMC, it is possible to substitute z with its sufficient statistic (Asmussen et al., 1996). The general procedure for the EM algorithm is described below.

4.3.1 THE EM APPROACH

The EM approach to fitting a probability distribution to a sample of data for a random variable belongs to the well-known maximum likelihood method. We first describe it in general. Suppose we have a random i.i.d. sample $\mathbf{X} = (X_1, ..., X_n)$ generated from a probability density function $f_{\mathbf{X}}(\cdot; \theta)$ of which the parameter vector θ belongs to the parameter space $\Theta \subset \mathbb{R}^d$ and is unknown. Our goal is to estimate θ with an observed sample $\mathbf{x} = (x_1, ..., x_n)$. We here use the maximum likelihood method that has asymptotic optimal properties (asymptotically unbiased, efficient, and normally distributed under certain regularity conditions). The idea is to estimate θ that makes the observed data, $\mathbf{x}$, most likely to occur. Since the data is independent and identically distributed with probability density function $f_{\mathbf{X}}(\cdot; \theta)$, the density of the observed sample $\mathbf{x} = (x_1, ..., x_n)$ is given by

$$f_{\mathbf{X}}(\mathbf{x}; \theta) = \prod_{i=1}^{n} f_{\mathbf{X}}(x_i; \theta). \tag{4.26}$$

This function is called the likelihood function of θ given the sample $\mathbf{x}$ and denoted as $\mathscr{L}(\theta;\mathbf{x})$. The maximum likelihood method obtains the estimate of θ in the parameter space which maximizes the likelihood function:

$$\hat{\theta} = \arg\max_{\theta \in \Theta} \mathscr{L}(\theta;\mathbf{x}). \tag{4.27}$$

Taking the natural logarithm of the likelihood function, we obtain the log likelihood function as

$$l(\theta;\mathbf{x}) = \ln \mathscr{L}(\theta;\mathbf{x}) = \sum_{i=1}^{n} \ln(f_{\mathbf{X}}(x_i;\theta)). \tag{4.28}$$

Since the logarithm is a monotone function, we can just as well maximize the log likelihood,

$$\hat{\theta} = \arg\max_{\theta \in \Theta} l(\theta;\mathbf{x}), \tag{4.29}$$

which is more convenient in analysis. For some distributions with fewer parameters, such as the Gaussian distribution, the maximization problem can be solved analytically. Solving this maximization problem may however not always be possible analytically or even numerically. Direct maximization of the likelihood using a standard numerical optimization method such as Newton–Raphson might require heavy analytical preparatory work to obtain the gradient and, possibly, the Hessian of the likelihood function. Moreover, the implementation of these methods may present numerical challenges (memory requirements, convergence, instabilities, etc.), particularly when the number of parameters to be estimated is large. For a PH distribution, the number of parameters depends on the order of the representation (i.e., the number of transient states of the corresponding CTMC) and the structure of the $\mathbf{T}$. For a general PH distribution with representation $(\alpha, \mathbf{T})$ of order n, the number of parameters should be $n + n^2$ which can be large. Moreover, a sample of data about the PH distributed random variable are not complete with regard to the underlying CTMC dynamics. For example, if we sample the service time of a queueing system which is PH distributed, a time interval observed (or recorded) is the period starting from an initial transient state to the absorbing state. However, such a time interval does not contain the detailed path information which specifies which states the CTMC visited and the duration in each state before its absorption. The incompleteness of data and large number of parameters to be estimated motivates us to develop a general iterative method for finding the maximum likelihood estimates of parameters of a PH distribution. That is the EM algorithm to be described next.

Exercise 4.16 *Explain for a general PH distribution with representation* $(\alpha, \mathbf{T})$ *of order n, the number of parameters should be* $n + n^2$.

4.4 THE EM ALGORITHM

Denote the complete data vector by $\mathbf{Z}$ as opposed to the observed or incomplete data vector. Thus, $\mathbf{Z}$ can be decomposed into the "observed part", $\mathbf{X}$, and "unobserved part" (also called "hidden" or "latent" part), $\mathbf{Y}$. That is $\mathbf{Z} = (\mathbf{X}, \mathbf{Y})$ which is called a "completion" of the data given by $\mathbf{X}$. For the complete data vector $\mathbf{Z}$ we can write the probability density function as

$$f_{\mathbf{Z}}(\mathbf{z}; \theta) = f_{\mathbf{Z}}(\mathbf{x}, \mathbf{y}; \theta) = f_{\mathbf{Z}|\mathbf{X}}(\mathbf{x}, \mathbf{y}|\mathbf{x}; \theta) f_{\mathbf{X}}(\mathbf{x}; \theta) = f_{\mathbf{Z}|\mathbf{X}}(\mathbf{y}|\mathbf{x}; \theta) f_{\mathbf{X}}(\mathbf{x}; \theta).$$

Note that the last equality indicates $f_{\mathbf{Z}|\mathbf{X}}(\mathbf{z}|\mathbf{x}; \theta) f_{\mathbf{X}}(\mathbf{x}; \theta) = f_{\mathbf{Z}|\mathbf{X}}(\mathbf{y}|\mathbf{x}; \theta) f_{\mathbf{X}}(\mathbf{x}; \theta)$ which can be used interchangeably. The form of this joint density function differs depending on the situation and establishes the relationship between the missing $\mathbf{y}$ and observed $\mathbf{x}$ values. For the partially classified data or grouped version of the data, this means that we have to specify the distribution at the most refined level. For random effects or latent variables, the form of the joint density will depend on the distributional assumption for these unobservables. The density function of $\mathbf{Z}$ can then be used to define the complete data likelihood $\mathscr{L}(\theta; \mathbf{Z}) = f_{\mathbf{Z}}(\mathbf{Z}; \theta)$, and the corresponding complete data log likelihood

$$l(\theta; \mathbf{Z}) = \ln(\mathscr{L}(\theta; \mathbf{Z})) = \ln(f_{\mathbf{Z}}(\mathbf{Z}; \theta)). \tag{4.30}$$

Note however that $\mathbf{Z}$ is not fully observed and the function above is in fact a random variable. Hence, we will take the expectation on it in the EM algorithm. As opposed to the complete data log likelihood, we refer to the log likelihood in (4.28) as the observed or incomplete data log likelihood. Some textbooks utilize different notations for the complete data likelihood such as $\mathscr{L}(\theta|\mathbf{Z}) = \mathscr{L}(\theta|\mathbf{X}, \mathbf{Y}) = f_{\mathbf{Z}}(\mathbf{X}, \mathbf{Y}|\theta)$, where the last notation apparently indicates considering the parameter vector θ as given for the convenience. Similarly, the incomplete data likelihood can be denoted by $\mathscr{L}(\theta|\mathbf{X}) = f_{\mathbf{X}}(\mathbf{X}|\theta)$. The observed data log likelihood can be maximized by iteratively maximizing the complete data log likelihood, using the EM algorithm. The EM algorithm is tailored to the situation where it is difficult to maximize the observed data log likelihood (4.28), but maximizing the complete data likelihood (4.30) is simple. In other words, computation of the maximum likelihood estimates given $\mathbf{x}$ alone is difficult, while it is easy to find the maximum likelihood estimates if $\mathbf{Z}$ were known. Since the complete data vector is unavailable, it is not possible to directly optimize the complete data log likelihood. The intuitive idea for obtaining parameter estimates in case of incomplete data is to estimate the complete data log likelihood and then use this estimated likelihood function to estimate the parameters. However, estimating the complete data log likelihood requires the knowledge of the parameter vector θ, so the algorithm uses an iterative approach. This technique is much used when the data has missing values, due to problems or limitations in the observation process. As a result, the incomplete data log likelihood might not be easily available while the form of the complete data log likelihood is known. Here we utilize a simple example to demonstrate this EM process.

Example 4.3 *Consider two biased coins, call them Coin 1 and Coin 2. Let θ_k be the probability of getting "heads" (H) for Coin k with $k = 1, 2$ if tossed. We want to estimate θ_1 and θ_2 by taking a random sample $\mathbf{X} = (\mathbf{X}_1, ..., \mathbf{X}_n)$ of size n. Each sample point is a vector of m values. That is $\mathbf{X}_i = (X_{i1}, ..., X_{i,m})$ with $i = 1, ..., n$. Such a sample is obtained by (i) selecting a coin at random and flipping that one coin m times; (ii) repeating this process in (i) n times. Assume that for each random selecting the coin, we do not know if it is Coin 1 or Coin 2. Then we obtain the observed data $\mathbf{X}$ and unobserved data $\mathbf{Y} = (Y_1, ..., Y_n)$ as follows:*

$$
\begin{array}{ccccc}
X_{11} & X_{12} & \cdots & X_{1m} & Y_1 \\
X_{21} & X_{22} & \cdots & X_{2m} & Y_2 \\
\vdots & \vdots & \vdots & \vdots & \vdots \\
X_{n1} & X_{n2} & \cdots & X_{nm} & Y_n
\end{array} ,
$$

where

$$
X_{ij} = \begin{cases} 1, & \text{if the jth flip for the ith selected coin is H} \\ 0, & \text{if the jth flip for the ith selected coin is T} \end{cases} ,
$$

$$
Y_i = \begin{cases} 1, & \text{if the ith coin is Coin 1} \\ 2, & \text{if the ith coin is Coin 2} \end{cases}
$$

for $i = 1, 2, ..., n$ and $j = 1, ..., m$. Note that $(X_{ij} | Y_i = k) \sim Bernoulli(p_k)$ with $k = 1, 2$ are i.i.d.. Given a sample shown in Table 4.1 below, show the EM process. To numerically experience this process, we consider a small-scale problem with $n = 5$ and $m = 10$ (i.e., selecting one of the coins 5 times and flipping the selected coin 10 times). The data observed are given by

$$
\mathbf{X} = \begin{pmatrix} \mathbf{X}_1 \\ \mathbf{X}_2 \\ \mathbf{X}_3 \\ \mathbf{X}_4 \\ \mathbf{X}_5 \end{pmatrix} = \begin{pmatrix} 1 & 0 & 0 & 0 & 1 & 1 & 0 & 1 & 0 & 1 \\ 1 & 1 & 1 & 1 & 0 & 1 & 1 & 1 & 1 & 1 \\ 1 & 0 & 1 & 1 & 1 & 1 & 1 & 0 & 1 & 1 \\ 1 & 0 & 1 & 0 & 0 & 0 & 1 & 1 & 0 & 0 \\ 0 & 1 & 1 & 1 & 0 & 1 & 1 & 1 & 0 & 1 \end{pmatrix}, \quad \mathbf{Y} = \begin{pmatrix} Y_1 \\ Y_2 \\ Y_3 \\ Y_4 \\ Y_5 \end{pmatrix} .
$$

We want to estimate θ_1 and θ_2 in two cases: (1) $\mathbf{Y}$ is observable (i.e., knowing which coin is selected each time); and (2) $\mathbf{Y}$ is unobservable (i.e., not knowing which coin is selected each time).

The first case is trivial. Suppose that $\mathbf{Y} = (2, 1, 1, 2, 1)^T$. Then, we have the results in Table 4.1. The second case is what we are really interested in. In such a case, we must start with some initial guess of θ_i. Assume the initial guessed

TABLE 4.1
Results for Case 1: Knowing the coin selected each time

Coin 1	Coin 2
	5H, 5T
9H, 1T	
8H, 2T	
	4H 6T
7H, 3T	

$$\hat{\theta}_1 = \frac{24}{24+6} = 0.80, \quad \hat{\theta}_2 = \frac{9}{9+11} = 0.45.$$

values are $\hat{\theta}_1^{(0)} = 0.60, \hat{\theta}_2^{(0)} = 0.50$. We start with the pdf of the complete data vector

$$
\begin{aligned}
f(\mathbf{X}_1,...,\mathbf{X}_5,Y_1,...,Y_5|\theta) &= f(\{X_{11},...,X_{1,10}\},...,\{X_{51},...,X_{5,10}\},Y_1,...,Y_5|\theta) \\
&= f(\{X_{11},...,X_{1,10}\},...,\{X_{51},...,X_{5,10}\}|Y_1,...,Y_5,\theta) \\
&\quad \times f(Y_1,...,Y_5) \\
&= \prod_{i=1}^{5} f(\{X_{i1},...,X_{i,10}\}|Y_i,\theta) \prod_{i=1}^{5} f(Y_i),
\end{aligned}
$$

(4.31)

where $f(Y_i) = \prod_{k=1}^{2} \pi_k^{Y_{ik}}$ and π_k is the probability of selecting Coin $k \in \{1,2\}$. Here Y_i is defined as an indicator variable

$$Y_i = \begin{pmatrix} Y_{i1} \\ Y_{i2} \end{pmatrix} \in \left\{ \begin{pmatrix} 1 \\ 0 \end{pmatrix}, \begin{pmatrix} 0 \\ 1 \end{pmatrix} \right\}.$$

Note that

$$f(\{X_{i1},...,X_{i,10}\}|Y_i,\theta) = \prod_{j=1}^{10} f(X_{ij}|Y_i,\theta),$$

(4.32)

where $X_{ij} = 1(0)$ if the jth toss of the ith run is a head (tail). Thus, we have

$$f(X_{ij}|Y_i,\theta) = \prod_{k=1}^{2} [\theta_k^{X_{ij}}(1-\theta_k)^{1-X_{ij}}]^{Y_{ik}}.$$

(4.33)

Substituting (4.33) and (4.32) into (4.31) yields

$$f(\mathbf{X}_1,...,\mathbf{X}_5,Y_1,...,Y_5|\theta) = \prod_{i=1}^{5}\prod_{j=1}^{10}\prod_{k=1}^{2}[\theta_k^{X_{ij}}(1-\theta_k)^{1-X_{ij}}]^{Y_{ik}} \prod_{i=1}^{5}\prod_{k=1}^{2}\pi_k^{Y_{ik}}, \quad (4.34)$$

which is the complete data likelihood. Taking the logarithm of the equation above, we obtain the complete data log likelihood as

$$\ln f(\mathbf{X}_1,...,\mathbf{X}_5,Y_1,...,Y_5|\boldsymbol{\theta}) = \sum_{i=1}^{5}\sum_{j=1}^{10}\sum_{k=1}^{2} Y_{ik} \ln \theta_k^{X_{ij}}(1-\theta_k)^{1-X_{ij}} + \sum_{i=1}^{5}\sum_{k=1}^{2} Y_{ik}\ln\pi_k.$$

$$(4.35)$$

Since we have an observed sample $\mathbf{X}$, we can take the expectation of (4.35) of $\mathbf{Y}$ given $\mathbf{X}$. That is

$$E_{f(\mathbf{Y}|\mathbf{X})}[\ln f(\mathbf{X}_1,...,\mathbf{X}_5,Y_1,...,Y_5|\boldsymbol{\theta})] = \sum_{i=1}^{5}\sum_{j=1}^{10}\sum_{k=1}^{2} E_{f(\mathbf{Y}|\mathbf{X})}[Y_{ik}]\ln\theta_k^{X_{ij}}(1-\theta_k)^{1-X_{ij}}$$

$$+ \sum_{i=1}^{5}\sum_{k=1}^{2} E_{f(\mathbf{Y}|\mathbf{X})}[Y_{ik}]\ln\pi_k.$$

$$(4.36)$$

Note that $f(\mathbf{Y}|\mathbf{X},\boldsymbol{\theta}) = f(Y_1,...,Y_5|\mathbf{X}_1,...,\mathbf{X}_5,\boldsymbol{\theta})$ and

$$E_{f(\mathbf{Y}|\mathbf{X})}[Y_{ik}] = \sum_{Y_1}\cdots\sum_{Y_5} Y_{ik}f(\mathbf{Y}|\mathbf{X},\boldsymbol{\theta}) = \sum_{Y_i} Y_{ik}f(Y_i|\{X_{i1},...,X_{i,10}\},\boldsymbol{\theta}),$$

where

$$f(Y_i|\{X_{i1},...,X_{i,10},\boldsymbol{\theta}\}) = \frac{f(Y_i\cap\{X_{i1},...,X_{i,10}\}|\boldsymbol{\theta})}{f(\{X_{i1},...,X_{i,10}\}|\boldsymbol{\theta})} = \frac{f(\{X_{i1},...,X_{i,10}\}|Y_i,\boldsymbol{\theta})f(Y_i)}{f(\{X_{i1},...,X_{i,10}\}|\boldsymbol{\theta})}$$

$$= \frac{\prod_{j=1}^{10}\prod_{k=1}^{2}[\theta_k^{X_{ij}}(1-\theta_k)^{1-X_{ij}}]^{Y_{ik}}\pi_k^{Y_{ik}}}{\sum_{Y_i}\prod_{j=1}^{10}\prod_{k=1}^{2}[\theta_k^{X_{ij}}(1-\theta_k)^{1-X_{ij}}]^{Y_{ik}}\pi_k^{Y_{ik}}}$$

$$(4.37)$$

Simplifying the above yields

$$E_{f(\mathbf{Y}|\mathbf{X})}[Y_{ik}] = \sum_{Y_i} Y_{ik}\frac{\prod_{j=1}^{10}\prod_{k=1}^{2}[\theta_k^{X_{ij}}(1-\theta_k)^{1-X_{ij}}]^{Y_{ik}}\pi_k^{Y_{ik}}}{\sum_{Y_i}\prod_{j=1}^{10}\prod_{k=1}^{2}[\theta_k^{X_{ij}}(1-\theta_k)^{1-X_{ij}}]^{Y_{ik}}\pi_k^{Y_{ik}}}$$

$$= \frac{\sum_{Y_i} Y_{ik}\prod_{j=1}^{10}\prod_{k=1}^{2}[\theta_k^{X_{ij}}(1-\theta_k)^{1-X_{ij}}]^{Y_{ik}}\pi_k^{Y_{ik}}}{\sum_{Y_i}\prod_{j=1}^{10}\prod_{k=1}^{2}[\theta_k^{X_{ij}}(1-\theta_k)^{1-X_{ij}}]^{Y_{ik}}\pi_k^{Y_{ik}}}$$

$$= \frac{\pi_k\prod_{j=1}^{10}\theta_k^{X_{ij}}(1-\theta_k)^{1-X_{ij}}}{\pi_1\prod_{j=1}^{10}\theta_1^{X_{ij}}(1-\theta_1)^{1-X_{ij}}+\pi_2\prod_{j=1}^{10}\theta_2^{X_{ij}}(1-\theta_2)^{1-X_{ij}}}.$$

Assume $\pi_1 = \pi_2 = 1/2$. Then, we have

$$E_{f(\mathbf{Y}|\mathbf{X})}[Y_{ik}] = \frac{\prod_{j=1}^{10}\theta_k^{X_{ij}}(1-\theta_k)^{1-X_{ij}}}{\prod_{j=1}^{10}\theta_1^{X_{ij}}(1-\theta_1)^{1-X_{ij}}+\prod_{j=1}^{10}\theta_2^{X_{ij}}(1-\theta_2)^{1-X_{ij}}}. \qquad (4.38)$$

This is the end of the expectation step. Next, for a given fixed $E_{f(\mathbf{Y}|\mathbf{X})}[Y_{ik}]$, we complete the maximization step. That is to maximize the complete data log likelihood as follows:

$$\max l(\boldsymbol{\theta}, \mathbf{Z}) = E_{f(\mathbf{Y}|\mathbf{X})}[\ln f(\mathbf{X}_1, ..., \mathbf{X}_5, Y_1, ..., Y_5 | \boldsymbol{\theta})]$$

$$= \sum_{i=1}^{5} \sum_{j=1}^{10} \sum_{k=1}^{2} E_{f(\mathbf{Y}|\mathbf{X})}[Y_{ik}] \ln \theta_k^{X_{ij}} (1 - \theta_k)^{1 - X_{ij}} + \sum_{i=1}^{5} \sum_{k=1}^{2} E_{f(\mathbf{Y}|\mathbf{X})}[Y_{ik}] \ln \pi_k.$$

$$= \sum_{i=1}^{5} \sum_{j=1}^{10} \sum_{k=1}^{2} E_{f(\mathbf{Y}|\mathbf{X})}[Y_{ik}] (X_{ij} \ln \theta_k + (1 - X_{ij}) \ln(1 - \theta_k)) + constant.$$

$$(4.39)$$

Taking the first-derivative of $l(\boldsymbol{\theta}, \mathbf{Z})$ in (4.39) with respect to θ_1 and setting it to zero yields

$$\frac{dl(\boldsymbol{\theta}, \mathbf{Z})}{d\theta_1} = \sum_{i=1}^{5} \sum_{j=1}^{10} E_{f(\mathbf{Y}|\mathbf{X})}[Y_{i1}] \left(X_{ij} \frac{1}{\theta_1} - (1 - X_{ij}) \frac{1}{1 - \theta_1} \right) = 0,$$

which can be re-written as $\sum_{i=1}^{5} \sum_{j=1}^{10} E_{f(\mathbf{Y}|\mathbf{X})}[Y_{i1}] (X_{ij}(1 - \theta_1) - (1 - X_{ij})\theta_1) = 0$. Solving this equation for θ_1, we get θ_1. Similarly, we can obtain θ_2. They are given by

$$\theta_1 = \frac{\sum_{i=1}^{5} \sum_{j=1}^{10} E_{f(\mathbf{Y}|\mathbf{X})}[Y_{i1}] X_{ij}}{\sum_{i=1}^{5} 10 E_{f(\mathbf{Y}|\mathbf{X})}[Y_{i1}]}, \quad \theta_2 = \frac{\sum_{i=1}^{5} \sum_{j=1}^{10} E_{f(\mathbf{Y}|\mathbf{X})}[Y_{i2}] X_{ij}}{\sum_{i=1}^{5} 10 E_{f(\mathbf{Y}|\mathbf{X})}[Y_{i2}]}. \quad (4.40)$$

These formulas are used to update the two parameters θ_1 and θ_2 until the convergence. That is the end of maximization step. Now we apply these two steps to our numerical example. With the initial values $\hat{\theta}_1^{(0)} = 0.60, \hat{\theta}_2^{(0)} = 0.50$ and the given observed data $\mathbf{X}$, we first use (4.38) to compute the expected value of Y_{i1} and Y_{i2} for $i = 1, ..., 5$. They are given in Table 4.38. Here is an example. The expected value of Y_{11} is computed as

$$E_{f(\mathbf{Y}|\mathbf{X})}[Y_{11}] = \frac{\prod_{j=1}^{10} \theta_1^{X_{ij}} (1 - \theta_1)^{1 - X_{ij}}}{\prod_{j=1}^{10} \theta_1^{X_{ij}} (1 - \theta_1)^{1 - X_{ij}} + \prod_{j=1}^{10} \theta_2^{X_{ij}} (1 - \theta_2)^{1 - X_{ij}}}$$

$$= \frac{(0.6)^5 \cdot (1 - 0.6)^5}{(0.6)^5 \cdot (1 - 0.6)^5 + (0.5)^5 \cdot (1 - 0.5)^5} = 0.45.$$

Other values in Table 4.2 can be computed similarly. Using $E_{f(\mathbf{Y}|\mathbf{X})}[Y_{ik}]$ values in Table 4.38 and (4.40), we can obtain the updated $\hat{\theta}_1^{(1)}$ and $\hat{\theta}_2^{(1)}$ in Table 4.3. With these updated $\hat{\theta}_i^{(1)}$ values and the observed data $\mathbf{X}$, we can start the next iteration (E-step and M-step). After 10 iterations, these values converge to

TABLE 4.2

Expected value of unobservable variables Y_{i1} and Y_{i2}

i	1	2	3	4	5
$E_{f(\mathbf{Y}\mid\mathbf{X})}[Y_{i1}]$	0.45	0.80	0.73	0.35	0.65
$E_{f(\mathbf{Y}\mid\mathbf{X})}[Y_{i2}]$	0.55	0.20	0.27	0.65	0.35

TABLE 4.3

Results for Case 2: Not knowing the coin selected each time

Coin 1	Coin 2
2.2H, 2.2 T	2.8H, 2.8T
7.2H, 0.8T	1.8H, 0.2T
5.9H, 1.5T	2.1H, 0.5T
1.4H, 2.1T	2.6H, 3.9T
4.5H, 1.9T	2.5H, 1.1T

$$\hat{\theta}_1^{(1)} = \frac{2.2+7.2+5.9+1.4+4.5}{(2.2+7.2+5.9+1.4+4.5)+(2.2+0.8+1.5+2.1+1.9)} = \frac{21.2}{21.2+8.5} = 0.7138,$$

$$\hat{\theta}_2^{(1)} = \frac{2.8+1.8+2.1+2.6+2.5}{(2.8+1.8+2.1+2.6+2.5)+(2.8+0.2+0.5+3.9+1.1)} = \frac{11.8}{11.8+8.5} = 0.5813 \frac{11.7}{11.7+8.4} =$$

0.58.

$\hat{\theta}_1^{(10)} = 0.80, \hat{\theta}_2^{(10)} = 0.52$ which are quite close to the values $\hat{\theta}_1 = 0.80, \hat{\theta}_2 = 0.45$ in Case 1 with knowing $\mathbf{Y}$ and the same observed data $\mathbf{X}$. This fictitious example demonstrates the EM algorithm in details. EM algorithms provide a general approach to learning in presence of unobservable variables. Learning in stochastic models will be discussed in Chapter 11.

The reason why we will use the EM algorithm in our context (i.e., estimating the PH presentation) however is that maximizing the likelihood function itself is analytically intractable, but when we assume the existence of additional but latent variables, the optimization problem can be simplified. The EM algorithm iterates between two steps. After initializing the starting values of the algorithm, we compute the expected augmented-data log likelihood in the E-step and maximize this likelihood in the M-step. Both steps are iterated until convergence.

Initial step. An initial guess $\theta^{(0)}$ for θ is needed to start the algorithm. The closer the starting value is to the true maximum likelihood estimator, the faster the algorithm will converge.

E-step. In the kth iteration of the E-step, we compute the expected value of the complete data log likelihood (4.30) with respect to the unknown data $\mathbf{Z}$ given the observed data $\mathbf{x}$ and using the current estimate of the parameter vector $\theta^{(k-1)}$ as true values:

$$
\begin{aligned}
Q(\theta; \theta^{(k-1)}) &= E\left[l(\theta; \mathbf{Z})|\mathbf{x}; \theta^{(k-1)}\right] \\
&= E\left[\ln(f_{\mathbf{Z}}(\theta; \mathbf{Z})|\mathbf{x}; \theta^{(k-1)}\right] \quad\quad (4.41) \\
&= \int \ln(f_{\mathbf{Z}}(\mathbf{z}; \theta) f_{\mathbf{Z}|\mathbf{X}}(\mathbf{z}|\mathbf{x}; \theta^{(k-1)}) dz.
\end{aligned}
$$

It is important to realize that the current parameter estimates $\theta^{(k-1)}$ are used to compute this conditional expectation through the density of the complete data given the observed data, whereas the expectation itself is still a function of the unknown parameter vector θ as it appears in the complete data log likelihood. The notation $Q(\theta; \theta^{(k-1)})$ indicates that it is a function of the parameter vector θ that depends on the current parameter estimates $\theta^{(k-1)}$. Functions through which the incomplete data enter the complete data likelihood are replaced by their expectations, given the observed part of the data and the current value of the parameters. In case the log likelihood is linear in sufficient statistics (e.g., the exponential family), the computation of the E-step simplifies a great deal as it is reduced to calculating the conditional expectations of the sufficient statistics.

M-step In the M-step, we maximize the expected value of the complete data log likelihood (4.41) obtained in the E-step with respect to the parameter vector:

$$
\theta^{(k)} = \arg\max_{\theta \in \Theta} Q(\theta; \theta^{(k-1)}). \quad\quad (4.42)
$$

The new estimate $\theta^{(k)}$ hence satisfies $Q(\theta^{(k)}; \theta^{(k-1)}) \geq Q(\theta; \theta^{(k-1)}) \; \forall \theta \in \Theta$. The E-step and the M-step are iterated until the difference $l(\theta^{(k)}; \mathbf{x}) - l(\theta^{(k-1)}; \mathbf{x})$ is sufficiently small. Other convergence criteria can be used as well.

4.4.1 CONVERGENCE OF EM ALGORITHM

If we can show that the likelihood increases at every iteration, then the convergence (to a local maximum) of the EM algorithm holds since the likelihood is bounded from above. The proof is based on the following well-known lemma.

Lemma 4.2 *(Jensen's inequality) Let X be a random variable with $E[X] < \infty$. If g is a convex function, then*

$$
E[g(X)] \geq g(E[X]). \qu\quad (4.43)
$$

If g is a concave function, then

$$E[g(X)] \leq g(E[X]). \qquad (4.44)$$

The proof can be found in Appendix C.

Now we can prove the non-decreasing property of expected conditional complete data log likelihood given the observed sample values and the previous estimates of the parameter vector.

Theorem 4.7 *If for $\theta \in \Theta$, we have*

$$Q(\theta;\theta^{(k-1)}) \geq Q(\theta^{(k-1)};\theta^{(k-1)}),$$

then

$$l(\theta;\mathbf{x}) \geq l(\theta^{(k-1)};\mathbf{x}).$$

Proof. The observed data likelihood and the complete data likelihood are connected as follows

$$
\begin{aligned}
l(\theta;\mathbf{X}) &= l(\theta;\mathbf{Z}) - l(\theta;\mathbf{Z}) + l(\theta;\mathbf{X}) \\
&= l(\theta;\mathbf{Z}) - \ln(\mathscr{L}(\theta;\mathbf{Z}) + \ln(\mathscr{L}(\theta;\mathbf{X})) \\
&= l(\theta;\mathbf{Z}) - \ln\left(\frac{f_{\mathbf{Z}}(\mathbf{Z};\theta)}{f_{\mathbf{X}}(\mathbf{X};\theta)}\right).
\end{aligned}
\qquad (4.45)
$$

Taking the expectation on both sides with respect to the conditional distribution of the complete data Z given the observed data x and using the current parameter estimates $\theta^{(k-1)}$ results in

$$l(\theta;\mathbf{x}) = Q(\theta;\theta^{(k-1)}) - H(\theta;\theta^{(k-1)}), \qquad (4.46)$$

where

$$
\begin{aligned}
H(\theta;\theta^{(k-1)}) &= E\left[\ln\left(\frac{f_{\mathbf{Z}}(\mathbf{Z};\theta)}{f_{\mathbf{X}}(\mathbf{x};\theta)}\right)|\mathbf{x};\theta^{(k-1)}\right] \\
&= E\left[\ln(f_{\mathbf{Z}|\mathbf{X}}(\mathbf{Z}|\mathbf{x};\theta))|\mathbf{x};\theta^{(k-1)}\right]
\end{aligned}
$$

Evaluating (4.46) with $\theta^{(k-1)}$ and subtracting it from (4.46) results in

$$
\begin{aligned}
l(\theta;\mathbf{x}) - l(\theta^{(k-1)};\mathbf{x}) &= Q(\theta;\theta^{(k-1)}) - Q(\theta^{(k-1)};\theta^{(k-1)}) \\
&\quad - \left[H(\theta;\theta^{(k-1)}) - H(\theta^{(k-1)};\theta^{(k-1)})\right].
\end{aligned}
\qquad (4.47)
$$

Note that $H(\theta;\theta^{(k-1)})$, as a function of θ, reaches its maximum at $\theta = \theta^{(k-1)}$. That is

$$H(\theta;\theta^{(k-1)}) \leq H(\theta^{(k-1)};\theta^{(k-1)}) \quad \forall \theta \in \Theta, \qquad (4.48)$$

which follows because of

$$
\begin{aligned}
H(\theta;\theta^{(k-1)}) - H(\theta^{(k-1)};\theta^{(k-1)}) &= E\left[\ln(f_{\mathbf{Z}|\mathbf{X}}(\mathbf{Z}|\mathbf{x};\theta))|\mathbf{x};\theta^{(k-1)}\right] \\
&\quad - E\left[\ln(f_{\mathbf{Z}|\mathbf{X}}(\mathbf{Z}|\mathbf{x};\theta^{(k-1)}))|\mathbf{x};\theta^{(k-1)}\right] \\
&= E\left[\ln\left(\frac{f_{\mathbf{Z}|\mathbf{X}}(\mathbf{Z}|\mathbf{x};\theta)}{f_{\mathbf{Z}|\mathbf{X}}(\mathbf{Z}|\mathbf{x};\theta^{(k-1)})}\right)|\mathbf{x};\theta^{(k-1)}\right] \\
&\leq \ln\left(E\left[\frac{f_{\mathbf{Z}|\mathbf{X}}(\mathbf{Z}|\mathbf{x};\theta)}{f_{\mathbf{Z}|\mathbf{X}}(\mathbf{Z}|\mathbf{x};\theta^{(k-1)})}|\mathbf{x};\theta^{(k-1)}\right]\right) \\
&= \ln\left(\int \frac{f_{\mathbf{Z}|\mathbf{X}}(\mathbf{Z}|\mathbf{x};\theta)}{f_{\mathbf{Z}|\mathbf{X}}(\mathbf{Z}|\mathbf{x};\theta^{(k-1)})} f_{\mathbf{Z}|\mathbf{X}}(\mathbf{Z}|\mathbf{x};\theta^{(k-1)})dz\right) \\
&= \ln\left(\int f_{\mathbf{Z}|\mathbf{X}}(\mathbf{Z}|\mathbf{x};\theta)dz\right) \\
&= \ln 1 = 0.
\end{aligned}
$$

The inequality above holds by applying Jensen's inequality to the concave natural logarithm. Thus, if for $\theta \in \Theta$,

$$
Q(\theta;\theta^{(k-1)}) \geq Q(\theta^{(k-1)};\theta^{(k-1)}),
$$

then based on (4.47) and (4.48) we have

$$
l(\theta;\mathbf{x}) \geq l(\theta^{(k-1)};\mathbf{x}).
$$

■

This theorem ensures that the likelihood function is non-decreasing after each iteration in EM algorithm. Combining the kth iteration in the M-step and the theorem, we obtain

$$
Q(\theta^{(k)};\theta^{(k-1)}) \geq Q(\theta;\theta^{(k-1)}) \geq Q(\theta^{(k-1)};\theta^{(k-1)}).
$$

In many situations, the maximum likelihood method does not have closed-form solutions for the optimal estimated parameters. The EM approach offers an alternative to numerically estimate the parameters of a distribution. PH distribution is an example of such a situation. Note however that the EM algorithm is not guaranteed to converge to a global maximizer of the likelihood in all cases. Under some regularity conditions on the likelihood $\mathscr{L}(\theta;\mathbf{X})$ and on the parameter set Θ, it is possible, however, to show that the sequence $\{\theta^{(k)}|k \in \mathbb{N}\}$ obtained by the EM algorithm converges to a local maximizer of $\mathscr{L}(\theta;\mathbf{X})$ or at least to a stationary point of $\mathscr{L}(\theta;\mathbf{X})$. See Wu (1983) for a further discussion on the convergence properties of the EM algorithm in its generality. In case the observed log likelihood has multiple local maxima, the quality of the initial estimate $\theta^{(0)}$ can greatly affect result and should hence be chosen carefully. It is often wise to try and compare the results of several initial starting points. It can be shown, again under regularity conditions, that the EM algorithm converges linearly near the solution. The rate of convergence however can be slow, depending on the amount of missing information (Dempster et al. (1977)). The (sometimes very) slow convergence is a disadvantage of the EM algorithm as well as the fact that there is no automatic provision of precision estimates as the observed information matrix, and hence the co-variance matrix of the parameter vector estimator, is not directly accessible. Alternatives have been suggested to speed up the convergence, but for these extensions the simplicity of the EM algorithm is lost and the implementation becomes more complex. These modifications also bring about the appropriate methodology on how to arrive at precision estimates for the estimated parameter vector. See McLachlan and Krishnan (2007) for a thorough treatment on how to calculate standard errors and an overview of variants of the EM algorithm with improved convergence speed. Often, the computations in both the E-step as well as the M-step are much simpler than in the corresponding direct likelihood method, such as Newton-Raphson or Fisher scoring. The EM algorithm is therefore in general easy to implement, both analytically and computationally. The cost per iteration is generally low, which can offset the larger number of iterations needed for the EM algorithm compared to other competing procedures. For some problems however, the E- or M-steps may be analytically intractable. In such cases, we need to use iterative methods to compute the expectation in the E-step and/or the maximization in the M-step, which can be computationally expensive and may cause the EM algorithm to be slow. Variants of the EM algorithm that alter the E- and/or M-step, while preserving the convergence properties of the algorithm, have been suggested to alleviate this problem (McLachlan and Krishnan (2007)).

4.4.2   EM ALGORITHM FOR PH DISTRIBUTION

Let $\mathbf{X} = (X_1, ..., X_n)$ be an i.i.d. random sample from the continuous PH distribution $PH_c(\alpha, \mathbf{T})$ of order m with probability density function given in (4.6). With loss of generality, we assume that the data set does not contain zeros, i.e. $\alpha_{m+1} = 1 - \alpha\mathbf{1} = 0$. The parameter vector $\theta = (\alpha, \mathbf{T}, \eta)$ must satisfy $\alpha\mathbf{1} = 1$ and $\eta = -\mathbf{T1}$. We can write the likelihood of an observed sample $\mathbf{x} = (x_1, ..., x_n)$ as

$$\mathscr{L}(\theta; \mathbf{x}) = \prod_{v=1}^{n} \alpha e^{\mathbf{T}x_v} \eta,$$

and the corresponding log likelihood as

$$l(\theta; \mathbf{x}) = \sum_{v=1}^{n} \ln\left(\alpha e^{\mathbf{T}x_v} \eta\right).$$

To find the maximum likelihood estimates, we make use of the connection between a PH distribution and its underlying CTMC $\{X^{[v]}(t)|t \geq 0\}$. The random variable X_v is the absorption time of the CTMC and can be regarded as an incomplete version of that Markov process. This is because that we only observe the time until the Markov chain reaches the absorbing state and not the embedded jump chain and the sojourn times in each state. In the context of the EM algorithm we can therefore define the complete data vector consisting of the embedded jump Markov Chain (JMC) $Y_0^{[v]}, Y_1^{[v]}, ..., Y_{L^{[v]}-1}^{[v]}, (Y_{L^{[v]}}^{[v]} = m + 1)$ and sojourn times $S_0^{[v]}, S_1^{[v]}, ..., S_{L^{[v]}-1}^{[v]}, (S_{L^{[v]}}^{[v]} = \infty)$ where $L^{[v]}$ is the number of jumps until $X^{[v]}(t)$ hits the absorbing state $m+1$ and $X_v = S_0^{[v]} + S_1^{[v]} + ... + S_{L^{[v]}-1}^{[v]}$, where $v = 1, ..., n$. From the definition of PH distribution, we have the transition probabilities for the JMC as

$$
p_{ij} = P(Y_{u+1}^{[v]} = j | Y_u^{[v]} = i) = \begin{cases} \frac{t_{ij}}{\lambda_i} & \text{for } i, j = 1, ..., m, \\ \frac{t_i}{\lambda_i} & \text{for } i = 1, ..., m \text{ and } j = m + 1, \end{cases} \quad (4.49)
$$
$$
\text{for } u = 0, 1, ..., L^{[v]} - 1.
$$

and the holding times are exponentially distributed with parameter $\lambda_i = -t_{ii}$. The density of a complete observation $z_v = (y_0^{[v]}, y_1^{[v]}, ..., y_{l^{[v]}-1}^{[v]}, s_0^{[v]}, s_1^{[v]}, ..., s_{l^{[v]}-1}^{[v]})$ corresponding to the observed absorption time $x_v = s_0^{[v]} + s_1^{[v]} + ... + s_{l^{[v]}-1}^{[v]}$ is then given by

$$
f(z_v; \theta) = \alpha_{y_0^{[v]}} \lambda_{y_0^{[v]}} \exp\left(-\lambda_{y_0^{[v]}} s_0^{[v]}\right) p_{y_0^{[v]}, y_1^{[v]}} ... \lambda_{y_{l^{[v]}-1}^{[v]}} \exp\left(-\lambda_{y_{l^{[v]}-1}^{[v]}} s_{l^{[v]}-1}^{[v]}\right) p_{y_{l^{[v]}-1}^{[v]}, m+1}
$$
$$
= \alpha_{y_0^{[v]}} \exp\left(-\lambda_{y_0^{[v]}} s_0^{[v]}\right) t_{y_0^{[v]}, y_1^{[v]}} ... \exp\left(-\lambda_{y_{l^{[v]}-1}^{[v]}} s_{l^{[v]}-1}^{[v]}\right) t_{y_{l^{[v]}-1}^{[v]}}.
$$

The density of the complete sample $\mathbf{z} = (z_1, ..., z_n)$ results in the complete data likelihood

$$
\mathcal{L}(\theta; \mathbf{z}) = \prod_{v=1}^{n} \left(\alpha_{y_0^{[v]}} \exp\left(-\lambda_{y_0^{[v]}} s_0^{[v]}\right) t_{y_0^{[v]}, y_1^{[v]}} ... \exp\left(-\lambda_{y_{l^{[v]}-1}^{[v]}} s_{l^{[v]}-1}^{[v]}\right) t_{y_{l^{[v]}-1}^{[v]}} \right)
$$
$$
= \prod_{i=1}^{m} \alpha_i^{B_i} \prod_{i=1}^{m} \exp(t_{ii} Z_i) \prod_{i=1}^{m} \prod_{j=1, j\neq i}^{m} t_{ij}^{N_{ij}} \prod_{i=1}^{m} t_i^{N_i},
$$

$$(4.50)$$

where B_i is the number of processes starting in state i, Z_i the total time spent in state i for all processes, N_{ij} the number of jumps from state i to state j among all the processes and N_i the number of processes reaching the absorbing state

$m+1$ from state i. The log likelihood for the complete data is given by

$$l(\theta;\mathbf{z}) = \sum_{i=1}^{m} B_i \ln(\alpha_i) + \sum_{i=1}^{m} t_{ii} Z_i + \sum_{i=1}^{m} \sum_{j=1,j\neq i}^{m} N_{ij} \ln(t_{ij}) + \sum_{i=1}^{m} N_i \ln(t_i). \quad (4.51)$$

Making explicit use of the fact that the parameters must satisfy the constraints $\alpha\mathbf{1} = 1$ and $\mathbf{T1} + \eta = 0$ where η is a column vector with element t_i. Setting

$$\alpha_m = 1 - \sum_{i=1}^{m-1} \alpha_i \quad \text{and} \quad t_{ii} = -\left(t_i + \sum_{j=1,j\neq i}^{m} t_{ij} \right),$$

for $i = 1,...,m$, it is easy to derive that the maximum likelihood parameters for the complete data vector z are given by

$$\hat{\alpha}_i = \frac{B_i}{n}, \quad \hat{t}_{ij} = \frac{N_{ij}}{Z_i}, \quad \hat{t}_i = \frac{N_i}{Z_i}, \quad \hat{t}_{ii} = -\left(\hat{t}_i + \sum_{j=1,j\neq i}^{m} \hat{t}_{ij} \right), \quad (4.52)$$

for $i,j = 1,...,m, i \neq j$. We are now ready to describe the E- and M-steps of the EM algorithm for fitting continuous PH distributions.

E-step In the E-step of the kth iteration of the algorithm, we calculate the expected value of the complete data log likelihood in (4.51) with respect to the unknown data $\mathbf{Z}$ given the observed data $\mathbf{x}$ and using the current estimate of the parameter vector $\theta^{(k-1)}$ as true value. Inspecting the form of the complete data log likelihood in (4.51), we see that the log likelihood is linear in the sufficient statistics B_i, Z_i, N_{ij} and N_i for $i,j = 1,...,m; i \neq j$ and hence the calculation of the conditional expectation of the log likelihood boils down to computing the conditional expectations of the sufficient statistics. By writing the sufficient statistics as a sum over the individual contributions of the n observations in the sample, i.e.

$$B_i = 1 - \sum_{v=1}^{n} B_i^{[v]}, \quad Z_i = \sum_{v=1}^{n} Z_i^{[v]}, \quad N_{ij} = \sum_{v=1}^{n} N_{ij}^{[v]}, \quad N_{ij} = \sum_{v=1}^{n} N_{ij}^{[v]}$$

for $i,j = 1,...,m, i \neq j$. For the kth iteration, conditioning on the sample $\mathbf{x}$ reduces to conditioning on each observation, we have

$$B_i^{(k)} = E(B_i|\mathbf{x};\theta^{(k-1)}) = \sum_{v=1}^{n} E(B_i^{[v]}|x_v;\theta^{(k-1)}),$$

$$Z_i^{(k)} = E(Z_i|\mathbf{x};\theta^{(k-1)}) = \sum_{v=1}^{n} E(Z_i^{[v]}|x_v;\theta^{(k-1)}),$$

$$N_{ij}^{(k)} = E(N_{ij}|\mathbf{x};\theta^{(k-1)}) = \sum_{v=1}^{n} E(N_{ij}^{[v]}|x_v;\theta^{(k-1)}),$$

$$N_i^{(k)} = E(N_{ij}|\mathbf{x};\theta^{(k-1)}) = \sum_{v=1}^{n} E(N_{ij}^{[v]}|x_v;\theta^{(k-1)}),$$

for $i, j = 1, ..., m, i \neq j$. Each expected conditionals can be obtained as follows:

$$E(B_i^{[v]}|x_v; \theta) = \frac{\alpha_i \mathbf{e}_i' \exp(\mathbf{T}x_v)\eta}{\alpha \exp(\mathbf{T}x_v)\eta},$$

$$E(Z_i^{[v]}|x_v; \theta) = \frac{\int_0^{x_v} \alpha \exp(\mathbf{T}u)\mathbf{e}_i \mathbf{e}_i' \exp(\mathbf{T}(x_v - u))\eta du}{\alpha \exp(\mathbf{T}x_v)\eta},$$

$$E(N_{ij}^{[v]}|x_v; \theta) = \frac{t_{ij} \int_0^{x_v} \alpha \exp(\mathbf{T}u)\mathbf{e}_i \mathbf{e}_i' \exp(\mathbf{T}(x_v - u))\eta du}{\alpha \exp(\mathbf{T}x_v)\eta},$$

$$E(N_i^{[v]}|x_v; \theta) = \frac{t_i \alpha \exp(\mathbf{T}x_v)\mathbf{e}_i}{\alpha \exp(\mathbf{T}x_v)\eta},$$

These conditional expectations can be derived by using the basic conditional probability formula. We show them given the parameter vector (estimates in EM algorithm) below. To simplify the notation, we assume a single observation sample (i.e., $n = 1$) so that the superscript and subscript can be suppressed. That is $B_i^{[1]} = B_i$ and $x_1 = x$. Thus, we have

$$E(B_i|x; \theta) = \frac{\mathbf{P}(Y_0 = i, X \in dx)}{\mathbf{P}(X \in dx)}$$

$$= \frac{\mathbf{P}(Y_0 = i)\mathbf{P}(X \in dx|Y_0 = i)}{\mathbf{P}(X \in dx)}$$

$$= \frac{\alpha_i \mathbf{e}_i' \exp(\mathbf{T}x)\eta}{\alpha \exp(\mathbf{T}x)\eta},$$

Similarly, we can get the conditional expectation of the total time spent in state i for the observation of the CTMC as follows:

$$E(Z_i|x; \theta) = E\left[\int_0^\infty 1_{Y_u=i} du | X = x\right]$$

$$= \int_0^\infty \mathbf{P}(Y_u = i|X = x) du$$

$$= \int_0^\infty \frac{\mathbf{P}(Y_u = i, X \in dx)}{\mathbf{P}(X \in dx)} du$$

$$= \frac{\int_0^x \mathbf{P}(Y_u = i)\mathbf{P}(X \in dx|Y_u = i) du}{\mathbf{P}(X \in dx)}$$

$$= \frac{\int_0^x \alpha \exp(\mathbf{T}u)\mathbf{e}_i \mathbf{e}_i' \exp(\mathbf{T}(x - u))\eta du}{\alpha \exp(\mathbf{T}x)\eta}.$$

The exchange of the order of integration and conditional expectation is justified by positivity of integrand. (The forth equality follows due to the fact that $Y_u = 0$ for $u > x$.) Next, we show the third conditional expectation which is a little more involved and based on Asmussen et al. (1996). First note that the expectation of the total number of jumps $E(\sum_{i\neq j} N_{ij})$ is finite. We introduce a discrete approximation to N_{ij} as

$$N_{ij}^{\varepsilon} = \sum_{k=0}^{\infty} \mathbf{1}_{\{Y_{k\varepsilon}=i, Y_{(k+1)\varepsilon}=i\}} \quad \varepsilon > 0, i \neq j,$$

which are all dominated by $\sum_{i\neq j} N_{ij}$ and converge to N_{ij} as $\varepsilon \to 0$. Moreover, we have

$$E(N_{ij}^{\varepsilon}|x;\theta) = \sum_{k=0}^{[x/\varepsilon]-1} \frac{\mathbf{P}(Y_{k\varepsilon}=i, Y_{(k+1)\varepsilon}=j, X \in dx)}{\mathbf{P}(X \in dx)}$$

$$= \sum_{k=0}^{[x/\varepsilon]-1} \frac{\mathbf{P}(Y_{k\varepsilon}=i)\mathbf{P}(Y_{(k+1)\varepsilon}=j|Y_{k\varepsilon}=i)\mathbf{P}(X \in dx|Y_{(k+1)\varepsilon}=j)}{\mathbf{P}(X \in dx)}$$

$$= \frac{\sum_{k=0}^{[x/\varepsilon]-1}(\alpha\exp(\mathbf{T}k\varepsilon)\mathbf{e}_i)(\mathbf{e}_i'\exp(\mathbf{T}\varepsilon)\mathbf{e}_j)(\mathbf{e}_j'\exp(\mathbf{T}(x-(k+1)\varepsilon))\eta)}{\alpha\exp(\mathbf{T}x)\eta}$$

$$\to \frac{\int_0^x \alpha\exp(\mathbf{T}u)\mathbf{e}_i t_{ij}\mathbf{e}_i'\exp(\mathbf{T}(x-u))\eta du}{\alpha\exp(\mathbf{T}x)\eta}$$

due to the continuity of $\exp(\mathbf{T}\varepsilon)$ and the fact that

$$\frac{\exp(\mathbf{T}\varepsilon)-I}{\varepsilon} \to \mathbf{T} \quad \text{as} \quad \varepsilon \to 0.$$

The last limit makes

$$(\mathbf{e}_i'\exp(\mathbf{T}\varepsilon)\mathbf{e}_j) = \frac{(\mathbf{e}_i'\exp(\mathbf{T}\varepsilon)\mathbf{e}_j)}{\varepsilon}\varepsilon \to \frac{(\mathbf{e}_i'(I+\mathbf{T}\varepsilon)\mathbf{e}_j)du}{\varepsilon} = (\mathbf{e}_i'\mathbf{T}\mathbf{e}_j)du = t_{ij}du$$

$$\text{as} \quad \varepsilon \to 0, i \neq j.$$

Using the dominated convergency theorem (for conditional expectations), we obtain

$$E(N_{ij}|x;\theta) = \frac{\int_0^x \alpha\exp(\mathbf{T}u)\mathbf{e}_i t_{ij}\mathbf{e}_i'\exp(\mathbf{T}(x-u))\eta du}{\alpha\exp(\mathbf{T}x)\eta},$$

Finally, the conditional expectations of the number of observations (CTMC) reaching the absorbing state $m+1$ from transient state i can be derived as follows. For the single observation sample, it can be interpreted as the conditional probability that the final absorbing jump at time x came from state i. We use the ε-argument again and have

$$E(N_i|x;\theta) = \frac{\mathbf{P}(Y_{x-\varepsilon}=i)\mathbf{P}(X \in dx|Y_{x-\varepsilon}=i)}{\mathbf{P}(X \in dx)}$$

$$= \frac{\alpha\exp(\mathbf{T}(x-\varepsilon))\mathbf{e}_i\mathbf{e}_i'\exp(\mathbf{T}\varepsilon)\eta}{\alpha\exp(\mathbf{T}x)\eta} \quad i = 1,...,m, x > \varepsilon > 0,$$

$$\to \frac{\alpha\exp(\mathbf{T}x)\mathbf{e}_i t_i}{\alpha\exp(\mathbf{T}x)\eta} \quad \text{as} \quad \varepsilon \to 0.$$

The expected value of the complete log likelihood in (4.51), with respect to the unknown path $\mathbf{Z}$ of the underlying CTMC, given the absorption times x and using the current value $\theta^{(k-1)}$ of the parameter vector, then becomes

$$
\begin{aligned}
Q(\theta;\theta^{(k-1)}) = E\left[l(\theta;\mathbf{Z})|\mathbf{x};\theta^{(k-1)}\right] \\
= \sum_{i=1}^{m} B_i^{(k)} \ln(\alpha_i) + \sum_{i=1}^{m} t_{ii} Z_i^{(k)} \\
+ \sum_{i=1}^{m} \sum_{j=1,j\neq i}^{m} N_{ij}^{(k)} \ln(t_{ij}) + \sum_{i=1}^{m} N_i^{(k)} \ln(t_i).
\end{aligned}
\tag{4.53}
$$

The E-step of the EM algorithm is the difficult part as it requires the computation of matrix exponentials and integrals of matrix exponentials that cannot be reduced any further. Asmussen et al. (1996b) consider a $p(p+2)$ dimensional linear system of homogeneous differential equations in order to calculate these functions and solve the system numerically with high precision. In the EMpht-program, the Runge-Kutta method of fourth order is implemented for this purpose.

M-step In the M-step, the expected complete log likelihood in (4.53) is maximized over θ. The new estimates are obtained by simply replacing the sufficient statistics in (4.52) with their conditional expectations evaluated in the E-step:

$$
\alpha_i^{(k)} = \frac{B_i^{(k)}}{n}, \quad t_{ij}^{(k)} = \frac{N_{ij}^{(k)}}{Z_i^{(k)}}, \quad t_i^{(k)} = \frac{N_i^{(k)}}{Z_i^{(k)}}, \quad t_{ii}^{(k)} = -\left(t_i^{(k)} + \sum_{j=1,j\neq i}^{m} t_{ij}^{(k)}\right).
$$

for $i,j = 1,...,m, i \neq j$. The E- and M-step can then be repeated until convergence of the maximum likelihood estimates, i.e. until the changes in the successive parameter estimates have become negligible. This stopping criterion has however not been implemented in some popular program such as the EMpht-program, where you have to specify the number of iterations of the EM algorithm to be performed and the starting values of the parameters of the PH distribution (Olsson (1998)). A good way to choose the starting values of the parameter vector in the initial step of the algorithm is not described in Asmussen et al. (1996b). In recent work of Bladt et al. (2011), the method of uniformization is applied for the evaluation of the matrix exponentials and related integrals appearing in the E-step. This alternative has a higher numerical precision and speeds up the execution of the EM algorithm considerably for small to medium sized data sets, while the Runge-Kutta method may outperform this method for large amounts of data. In case the data does contain zeros, we must allow for $\alpha_{m+1} > 0$. The estimation procedure that produces the maximum likelihood estimator for the full model can then be split up into the following components: (i) let $\hat{\alpha}_{m+1} > 0$ denote the proportion of zeros in the data set, (ii) eliminate the zeros from the data, (iii) fit a PH distribution to the remaining data (Bladt et al. (2011)). Bladt et al. (2011) also show how the estimation of discrete PH distributions can be treated analogously using the EM algorithm.

4.5 MARKOVIAN ARRIVAL PROCESSES

To model more general arrival processes, we study the Markovian arrival processes in this section. We start with the PH-renewal process.

4.5.1 FROM PH RENEWAL PROCESSES TO MARKOVIAN ARRIVAL PROCESSES

We consider a renewal process with PH distributed inter-arrival times, $X_i \in PH(\alpha, \mathbf{T}), i > 1$. Let $J(t)$ be the phase or state in the Markov chain related to X_i. Thus, $E(X_i) = \mu_1 = \alpha(-\mathbf{T})^{-1}\mathbf{1}$. The probabilistic interpretation of $(-\mathbf{T})^{-1}$ can be made similarly as in the discrete PH distribution. Define the stochastic process

$$I_{ij}(t) = \begin{cases} 1 & \text{if } J(t) = j, \\ 0 & \text{otherwise.} \end{cases} \tag{4.54}$$

Denote by τ_{ij} the time spent in phase j given that $J(0) = i$. Then we have $\tau_{ij} = \int_0^\infty I_{ij}(t)dt$ and

$$E(\tau_{ij}) = E\left(\int_0^\infty I_{ij}(t)dt\right) = \int_0^\infty p_{ij}(t)dt.$$

Recall that the transition probability function matrix for the related CTMC is $\mathbf{P}(t) = e^{\mathbf{T}t}$.

$$\int_0^K e^{\mathbf{T}t}dt = \int_0^K \sum_{i=0}^\infty \frac{(\mathbf{T}t)^i}{i!}dt = \sum_{i=0}^\infty \mathbf{T}^i \int_0^K \frac{t^i}{i!}dt = \sum_{i=0}^\infty \mathbf{T}^i \left[\frac{t^{i+1}}{(i+1)!}\right]_0^K =$$

$$= \sum_{i=0}^\infty \mathbf{T}^i \frac{K^{i+1}}{(i+1)!} = \mathbf{T}^{-1}\left(e^{\mathbf{T}K} - \mathbf{I}\right) \to (-\mathbf{T})^{-1}, \text{ as } K \to \infty. \tag{4.55}$$

Thus the element (i, j) in $(-\mathbf{T})^{-1}$ is the expected time spent in phase j before absorption given that the chain was started in phase i. It is clear from this probabilistic interpretation that $(-\mathbf{T})^{-1} \geq 0$. We can now get the mean time before absorption conditioning on the starting state. At the end of one inter-arrival time, another inter-arrival time starts again with the initial probability vector α. Thus, $\mathbf{T} + \eta\alpha$ is the generator matrix of an ergodic CTMC. The steady-state distribution for this CTMC is given by $\pi = (1/\mu_1)\alpha(-\mathbf{T})^{-1}$. We use a renewal process with Erlang-2 inter-arrival times as an example of the PH renewal process.

$$\mathbf{T} = \begin{pmatrix} -\lambda & \lambda \\ 0 & \lambda \end{pmatrix}, \quad \eta = \begin{pmatrix} 0 \\ \lambda \end{pmatrix}, \quad \alpha = (1, 0).$$

Exercise 4.17 *Consider the renewal process with Hyper-exponential distributed inter-arrival times. Find the expressions for T, η, α and the generator matrix of the CTMC for the phase change.*

If we keep track of the current phase of the PH distribution during renewal intervals, then we obtain a Markovian description $(N(t), J(t))$ of the PH renewal process $N(t)$. Clearly, $(N(t), J(t))$ is a two-dimensional CTMC with the state space of $\Omega = \mathbb{N}_0 \times \{1, ..., m\}$, with $N(t)$ denoting the number of arrivals until time t and $J(t)$ denoting the current phase of the PH inter-arrival time X_i at time t. Thus there are state transitions from (n, i) to $(n + 1, j)$ or to (n, j), which are called transitions with or without arrivals, respectively. For transitions without arrivals, there is no renewal event and hence no absorption for the PH distribution, which means that these transitions are described by the parameter matrix $\mathbf{T}$ of the PH distribution of the renewal intervals. For transitions with arrivals, we observe the following dynamics. Being in phase i, there is an absorption (hence a renewal event) with rate η_i. After that, a new PH distributed renewal interval begins and a phase is chosen according to the initial phase distribution α. Hence transitions with arrivals are determined by the rate matrix $\mathbf{A} := \eta \alpha$. After ordering the state space $\Omega = \mathbb{N}_0 \times \{1, ..., m\}$ lexicographically, we can write the infinitesimal generator matrix $\mathbf{Q}$ of the CTMC $(N(t), J(t))$ as a block matrix

$$\mathbf{Q} = \begin{pmatrix} \mathbf{T} & \mathbf{A} & & & \\ & \mathbf{T} & \mathbf{A} & & \\ & & \mathbf{T} & \mathbf{A} & \\ & & & \mathbf{T} & \mathbf{A} \\ & & \ddots & \ddots & \ddots \end{pmatrix}, \tag{4.56}$$

with the non-specified blocks being zero matrices. An essential feature of the PH renewal process is that immediately after an arrival (i.e., a renewal event) the phase distribution always is α. This makes it a real renewal process with i.i.d. renewal intervals. However, in modern communication systems like the Internet or other computer networks there may be strong correlations between subsequent inter-arrival times. Thus it is a natural idea to introduce a dependence between the subsequent inter-arrival intervals. This can be done without changing the block structure of the generator. Note that the row vectors of the matrix $\mathbf{A}$ in a PH renewal process has the following structure

$$\mathbf{A} = \begin{pmatrix} \eta_1 \cdot \alpha \\ \vdots \\ \eta_m \cdot \alpha \end{pmatrix},$$

meaning that the row entries differ only by a scalar η_i and thus the new phase after an arrival is chosen independently of the phase immediately before that arrival. If we relax this restriction, we arrive at a new matrix

$$\mathbf{A}' = \begin{pmatrix} \eta_1 \cdot \alpha_1 \\ \vdots \\ \eta_m \cdot \alpha_m \end{pmatrix},$$

with the only requirement that $\alpha_i \mathbf{1} = 1$ for all $i = 1, ..., m$. Here the phase distribution after an arrival depends on the phase immediately before that arrival. However, the requirement $\alpha_i \mathbf{1} = 1$ in connection with the fact that $\eta = -\mathbf{T1}$ simply restates the observation that the row entries of a generator matrix sum up to zero. Thus there is no real restriction in choosing $\mathbf{A}'$. Denoting $\mathbf{D}_0 := \mathbf{T}$ and $\mathbf{D}_1 := \mathbf{A}'$ as usually done in the literature, we arrive at a generator

$$\mathbf{Q} = \begin{pmatrix} \mathbf{D}_0 & \mathbf{D}_1 & & & \\ & \mathbf{D}_0 & \mathbf{D}_1 & & \\ & & \mathbf{D}_0 & \mathbf{D}_1 & \\ & & & \mathbf{D}_0 & \mathbf{D}_1 \\ & & & & \ddots & \ddots & \ddots \end{pmatrix}. \tag{4.57}$$

The matrix $\mathbf{D}_0$, associated with transitions without arrivals (i.e., phase change only), is non-singular with negative diagonal elements and non-negative off diagonal elements. The matrix $\mathbf{D}_1$, associated with transitions with both phase changes and arrivals, is non-negative. A Markov process with such a generator is called Markovian arrival process (or shortly MAP). This type of arrival processes, as an important class of point processes, play a critical role in queueing models. Differing from PH renewal processes as a special case, MAPs are not renewal but semi-Markov processes. The methods we will employ can be further generalized in analyzing the traffic in modern communication networks. Here are a few simple examples.

Example 4.4 *If the arrival rate for a Poisson process depends on the environment which can be modeled as the state of an underlying CTMC with a transition rate matrix $\mathbf{R}$, then the process is called the Markov-Modulated Poisson Process (MMPP). Such a process is a special case of MAP with*

$\mathbf{D_1} = diag\{\lambda_1, ..., \lambda_m\}$ and $\mathbf{D_0} = \mathbf{R} - \mathbf{D_1}$. *A special case of MMPP is a two-rate switched Poisson process (SPP) with the following rate matrices:*

$$\mathbf{D_0} = \begin{pmatrix} -(\sigma_1 + \gamma_1) & \sigma_1 \\ \sigma_2 & -(\sigma_2 + \gamma_2) \end{pmatrix}, \quad \mathbf{D_1} = \begin{pmatrix} \gamma_1 & 0 \\ 0 & \gamma_2 \end{pmatrix},$$

$$\mathbf{D} = \begin{pmatrix} -\sigma_1 & \sigma_1 \\ \sigma_2 & -\sigma_2 \end{pmatrix}.$$

In such a process, whenever the process is in state i, the arrivals occur according to a Poisson process with rate γ_i. The sojourn time in each of the two states is exponential with mean $(1/\sigma_i)$. If either $\gamma_1 = 0$ or $\gamma_2 = 0$ holds, the process is called an Interrupted Poisson Process (IPP).

Exercise 4.18 *If the inter-arrival times follow an IPP, the arrival process is a renewal process. Why? Further, show that this renewal process is stochastically equivalent to a renewal process with hyper-exponentially distributed inter-arrival times. Specify the parameters of the hyper-exponential distribution.*

4.5.2 TRANSITION PROBABILITY FUNCTION MATRIX

Define the transition probability function as

$$P_{ij}(n,t) = P(N(s+t) = n, J(s+t) = j | N(s) = 0, J(s) = i) \quad \text{for any } s. \quad (4.58)$$

Then, assuming $s = 0$, we can introduce the transition probability function matrix $\mathscr{P}(t) = (\mathbf{P}(0,t), \mathbf{P}(1,t), ...)$ where $\mathbf{P}(n,t) = (P_{ij}(n,t))$, which is an $m \times m$ matrix. According to the forward C-K equation for CTMC, we have $\mathscr{P}'(t) = \mathscr{P}(t)\mathbf{Q}$, which can be written as

$$(\mathbf{P}'(0,t), \mathbf{P}'(1,t), ...) = (\mathbf{P}(0,t), \mathbf{P}(1,t), ...)\mathbf{Q}. \quad (4.59)$$

Therefore, the transition probability function matrices satisfy a set of linear differential equations that can be effectively solved by using Laplace transforms. Writing them as

$$\mathbf{P}'(0,t) = \mathbf{P}(0,t)\mathbf{D_0}$$
$$\mathbf{P}'(1,t) = \mathbf{P}(0,t)\mathbf{D_1} + \mathbf{P}(1,t)\mathbf{D_0}$$

$$\vdots \qquad\qquad (4.60)$$

$$\mathbf{P}'(n+1,t) = \mathbf{P}(n,t)\mathbf{D_1} + \mathbf{P}(n+1,t)\mathbf{D_0}$$

$$\vdots$$

Multiplying the ith equation by z^i yields

$$\mathbf{P}'(0,t) = \mathbf{P}(0,t)\mathbf{D}_0$$
$$z\mathbf{P}'(1,t) = z\mathbf{P}(0,t)\mathbf{D}_1 + z\mathbf{P}(1,t)\mathbf{D}_0$$
$$\vdots \qquad\qquad\qquad\qquad\qquad (4.61)$$
$$z^{n+1}\mathbf{P}'(n+1,t) = z^{n+1}\mathbf{P}(n,t)\mathbf{D}_1 + z^{n+1}\mathbf{P}(n+1,t)\mathbf{D}_0.$$
$$\vdots$$

Summing these equations, we have

$$\sum_{i=0}^{\infty} z^i \mathbf{P}'(i,t) = z\sum_{i=0}^{\infty} z^i \mathbf{P}(i,t)\mathbf{D}_1 + \sum_{i=0}^{\infty} z^i \mathbf{P}(i,t)\mathbf{D}_0. \qquad (4.62)$$

Defining $\mathbf{P}(z,t) = \sum_{i=0}^{\infty} z^i \mathbf{P}(i,t)$ and $\mathbf{D}(z) = \mathbf{D}_0 + z\mathbf{D}_1$ and noting that $\sum_{i=0}^{\infty} z^i \mathbf{P}'(i,t) = \mathbf{P}'(z,t)$, we obtain

$$\mathbf{P}'(z,t) = \mathbf{P}(z,t)\mathbf{D}(z). \qquad (4.63)$$

The solution to the above equation is

$$\mathbf{P}(z,t) = e^{\mathbf{D}(z)t} = e^{(\mathbf{D}_0 + z\mathbf{D}_1)t}. \qquad (4.64)$$

A special case is the Poisson process with intensity parameter λ where $\mathbf{D}_0 = -\lambda$, $\mathbf{D}_1 = \lambda$ and $E(z^{N(t)}) = e^{-\lambda(1-z)t}$. Thus in the transform domain we see that the MAP is a very natural generalization of the Poisson process. It should be stressed however, that the distribution $\mathbf{P}(n,t)$ generally is very complicated. This is true even for the most simple generalizations of the Poisson process as the IPP or the Erlang 2 renewal process. Next, we can obtain the moments of the number of arrivals during an interval of length t for the MAP by taking derivatives in (4.64).

$$\frac{d}{dz}\mathbf{P}(z,t) = \frac{d}{dz}\sum_{i=0}^{\infty} \frac{(\mathbf{D}(z)t)^i}{i!} = \frac{d}{dz}\sum_{i=0}^{\infty} \frac{((\mathbf{D}_0 + z\mathbf{D}_1)t)^i}{i!}$$
$$= \sum_{i=1}^{\infty} \frac{t^i}{i!}\sum_{j=0}^{i-1}(\mathbf{D}_0 + z\mathbf{D}_1)^j \mathbf{D}_1(\mathbf{D}_0 + z\mathbf{D}_1)^{i-1-j}. \qquad (4.65)$$

Here, we utilize the formula of taking derivative of power of matrix function $\mathbf{A}(z)$ as

$$\frac{d}{dz}(\mathbf{A}(z))^n = \sum_{j=1}^{n-1}(\mathbf{A}(z))^j \frac{d(\mathbf{A})}{dz}(\mathbf{A}(z))^{k-1-j},$$

which can be easily proved by using induction. Then we have

$$\frac{d}{dz}\mathbf{P}(z,t)|_{z=1} = \sum_{i=1}^{\infty} \frac{t^i}{i!} \sum_{j=0}^{i-1} \mathbf{D}^j \mathbf{D}_1 \mathbf{D}^{i-1-j}. \tag{4.66}$$

The (i,j)-th element of this matrix is the expected number of $N(t)I_{ij}(t)$, i.e. the expected number of events in the interval $(0,t]$ if the phase at time t is j and the phase at time 0 is i. Taking row sums by post-multiplying with $\mathbf{e}$ (also denoted by $\mathbf{1}$) using $\mathbf{De} = \mathbf{0}$, we have

$$\frac{d}{dz}\mathbf{P}(z,t)|_{z=1}\mathbf{e} = \mu(t) = \sum_{i=1}^{\infty} \frac{t^i}{i!} \mathbf{D}^{i-1}\mathbf{D}_1\mathbf{e}, \tag{4.67}$$

where $\mu(t)$ is a column vector. Here we use $\mathbf{De} = \mathbf{0}$ and note that the only non-zero matrix term is the one when $j = i-1$ in (4.66). The kth element of $\mu(t)$ is the expected number of events occurred in $(0,t]$ conditioning on phase k at time 0. For the stationary MAP, we assume that the phase process distribution is represented by probability vector θ, then $\theta\mathbf{D} = 0$ and $\theta\mathbf{e} = 1$. Thus, the average event rate is

$$\theta\mu(t) = \theta\mathbf{D}_1\mathbf{e}t = \lambda^* t. \tag{4.68}$$

Note that pre-multiplying $\mu(t)$ by θ will result in the only non-zero matrix term for $i = 1$. The value of $\lambda^* = \theta\mathbf{D}_1\mathbf{e}$ is called fundamental rate of the MAP. After some algebraic manipulation, the expression has some flavor of the matrix exponential. Consider a relation

$$\sum_{i=1}^{\infty} \frac{t^i}{i!} \mathbf{D}^{i-1}(\mathbf{D} - \mathbf{e}\theta) = \sum_{i=0}^{\infty} \frac{t^i}{i!} \mathbf{D}^i - \mathbf{I} - t\mathbf{e}\theta = e^{\mathbf{D}t} - \mathbf{I} - t\mathbf{e}\theta.$$

Post-multiplying the above relation by $(\mathbf{D} - \mathbf{e}\theta)^{-1}$, we obtain

$$\sum_{i=1}^{\infty} \frac{t^i}{i!} \mathbf{D}^{i-1} = (e^{\mathbf{D}t} - I - t\mathbf{e}\theta)(\mathbf{D} - \mathbf{e}\theta)^{-1}. \tag{4.69}$$

Substituting (4.69) into (4.67), we obtain

$$\begin{aligned}
\frac{d}{dz}\mathbf{P}(z,t)|_{z=1}\mathbf{e} = \mu(t) &= (e^{\mathbf{D}t} - I - t\mathbf{e}\theta)(\mathbf{D} - \mathbf{e}\theta)^{-1}\mathbf{D}_1\mathbf{e} \\
&= \left[-t\mathbf{e}\theta(\mathbf{D} - \mathbf{e}\theta)^{-1} + (e^{\mathbf{D}t} - I)(\mathbf{D} - \mathbf{e}\theta)^{-1}\right]\mathbf{D}_1\mathbf{e} \\
&= t\mathbf{e}\theta\mathbf{D}_1\mathbf{e} + (e^{\mathbf{D}t} - I)(\mathbf{D} - \mathbf{e}\theta)^{-1}\mathbf{D}_1\mathbf{e}.
\end{aligned} \tag{4.70}$$

Here we use the relation $\theta(\mathbf{D} - \mathbf{e}\theta)^{-1} = -\theta$ which is justified by multiplying both sides from right by $(\mathbf{D} - \mathbf{e}\theta)$ and noting that $\theta = -\theta(\mathbf{D} - \mathbf{e}\theta) = \theta$. Note that $\lambda = \theta \mathbf{D}_1 \mathbf{e}$. The mean number of arrivals over time period $(0,t]$ given an initial phase distribution vector α is given by

$$
\begin{aligned}
E(N(t)) &= \alpha \frac{d}{dz} \mathbf{P}(z,t)|_{z=1} \mathbf{e} = \alpha \lambda t \mathbf{e} + \alpha (e^{\mathbf{D}t} - I)(\mathbf{D} - \mathbf{e}\theta)^{-1} \mathbf{D}_1 \mathbf{e} \\
&= \lambda t + \alpha (e^{\mathbf{D}t} - I)(\mathbf{D} - \mathbf{e}\theta)^{-1} \mathbf{D}_1 \mathbf{e}.
\end{aligned}
\tag{4.71}
$$

In principle, by taking second-order derivative of $\mathbf{P}(z,t)$ and applying the matrix operations, we can obtain the expressions for the second moment of $N(t)$ or $Var(N(t))$. The calculations are routine but tedious. Interested readers are referred to Neuts (1979) for the details about getting higher moments of $N(t)$ for MAPs.

Example 4.5 *Consider an operating system with two identical units which is maintained by a repairman. One of the two units is put in operation and the other is on cold standby. If the unit in operation fails, the repairman starts repairing it right away and the standby unit is put in operation. If the repair is completed before the failure of the operating unit, the repaired unit is on cold standby; otherwise, the failed unit has to wait for repair. A repaired unit is put in operation right away if the other unit is failed (waiting for repair). The life time and repair time are exponentially distributed with parameters λ and μ, respectively. Formulate an MAP for the number of repairs completed in $[0,t]$ and find the average time between two consecutive repairs if $\lambda = 0.01$ and $\mu = 0.5$.*

We first number the four status of each unit:

- *0: in repair*
- *1: waiting for repair*
- *2: on cold standby*
- *3: in operation*

Since the two units are identical, we can identify three states for the system. They are (a) 2 functioning units: one is in operation and the other is on cold standby; (b) 1 functioning unit: one is in operation and the other is in repair; (c) 0 functioning unit: one is in repair and the other is waiting for repair. These three states can be denoted by a pair of numbers showing the status of two units. Thus the state space is $\{(3,2),(3,0),(1,0)\}$. Since the life time and repair time are all exponentially distributed, the system evolves as a three-state CTMC, denoted by $\{J(t), t \geq 0\}$. The process of the "repair" occurrences is govern by the evolution of the underlying 3-state CTMC in which the transitions among these three states (also called phases of the MAP) are as follows:

(i) For phase (3,2), the time to the next state transition is the time to failure of the operating unit which is exponentially distributed with rate λ, and the transition is to (3,0) (i.e., one unit in operation and one unit in repair).

(ii) For phase (3,0), the time to the next state transition is the minimum of the time to failure of the operating unit and the time to repair completion of the unit in repair, which is exponentially distributed with rate $\lambda + \mu$. If the time to failure is smaller, the transition is to (1,0) and the probability of this event is $\lambda/(\lambda + \mu)$; otherwise, it is to (3,2) and the probability is this event of $\mu/(\lambda + \mu)$.

(iii) For phase (1,0), the time to the next state transition is the time to repair completion, which is exponentially distributed with rate μ, and this transition is to (3,0).

Let $N(t)$ be the number of repairs completed in $[0,t]$. Then the process $\{N(t), J(t)\}$ is an MAP with the following matrix representation:

$$\mathbf{D} = \begin{matrix} & \begin{matrix} (3,2) & (3,0) & (1,0) \end{matrix} \\ \begin{matrix} (3,2) \\ (3,0) \\ (1,0) \end{matrix} & \begin{pmatrix} -\lambda & \lambda & 0 \\ \mu & -\lambda-\mu & \lambda \\ 0 & \mu & -\mu \end{pmatrix} \end{matrix},$$

$$\mathbf{D_0} = \begin{matrix} & \begin{matrix} (3,2) & (3,0) & (1,0) \end{matrix} \\ \begin{matrix} (3,2) \\ (3,0) \\ (1,0) \end{matrix} & \begin{pmatrix} -\lambda & \lambda & 0 \\ 0 & -\lambda-\mu & \lambda \\ 0 & 0 & -\mu \end{pmatrix} \end{matrix},$$

$$\mathbf{D_1} = \begin{matrix} & \begin{matrix} (3,2) & (3,0) & (1,0) \end{matrix} \\ \begin{matrix} (3,2) \\ (3,0) \\ (1,0) \end{matrix} & \begin{pmatrix} 0 & 0 & 0 \\ \mu & 0 & 0 \\ 0 & \mu & 0 \end{pmatrix} \end{matrix},$$

where $\mathbf{D_0}$ is the rate matrix for these transitions without repair completions and $\mathbf{D_1}$ is the rate matrix for these transitions with repair completions. If $\lambda = 0.01$ and $\mu = 0.5$, the stationary distribution of $\{J(t), t \geq 0\}$ is determined by solving $\theta\mathbf{D} = 0$ and $\theta\mathbf{e} = 1$. The solution is $\theta = (0.9800, 0.0196, 0.0004)$ and the arrival rate (repair completion rate) is $\theta\mathbf{D_1}\mathbf{e} = 0.009996$. Then the average time between two consecutive repair completions is $1/0.009996 = 100.04$ time units.

Exercise 4.19 *The MAP can be extended to batch Marovian arrival process (BMAP) if each arrival may represent $k \geq 1$ simultaneously arriving customers. Let $\mathbf{D_k}$ with $k \geq 1$ be the k-arriving matrix. Its entry $[\mathbf{D_k}]_{ij}$ with $k \geq 1$ and $i, j \leq m$ represents the transition rate for a batch arrival of size k occurring in connection with a phase transition from i to j. Write the infinitesimal generator of this BMAP. What if the batch size is upper bounded by M?*

To study the BMAP, we need to define the convolutions of matrix sequences and the convolutional powers of a matrix sequence. Denote two sequences of $m \times m$ matrices by $\mathscr{G} = \{\mathbf{G}_n : n \in \mathbb{N}_0\}$ and $\mathscr{H} = \{\mathbf{H}_n : n \in \mathbb{N}_0\}$, respectively. The convolution of these two sequences is defined as the sequence $\mathscr{L} = \mathscr{G} \star \mathscr{H} = \{\mathbf{L}_n : n \in \mathbb{N}_0\}$ of $m \times m$ matrices with $\mathbf{L}_n := \sum_{k=0}^{n} \mathbf{G}_k \mathbf{H}_{n-k}$ for all $n \in \mathbb{N}_0$.

The convolution powers of a matrix sequence $\mathscr{G}$ is defined recursively as follows: $\mathscr{G}^{*(n+1)} := \mathscr{G}^{*n} \star \mathscr{G}$ for all $n \in \mathbb{N}_0$ with the initial sequence $\mathscr{G}^{*0} := \{\mathbf{I}, 0, 0...\}$. If we denote the kth matrix of a sequence $\mathscr{G}^{*n}$ by $\mathscr{G}_k^{*n}$, then we can write $\mathscr{G}_k^{*0} = \delta_{k0} \cdot \mathbf{I}$ for all $k \in \mathbb{N}_0$, where δ_{k0} denotes the Kronecker function and $\mathbf{I}$ denotes $m \times m$ identity matrix. Clearly, we have $\mathscr{G}_k^{*1} = \mathbf{G}_k$ for $k \in \mathbb{N}_0$. We can perform the same analysis on the BMAP as on the MAP. For the CTMC $(N(t), J(t))$ associated with BMAP, we can develop the transition probability matrix function and the expected number of arrivals for a given period of time in terms of z-transforms. We again provide a set of exercises to illustrate these fundamental results. Readers who are not interested in the derivations and proofs can use these results without doing these exercises.

Exercise 4.20 *Show that the transition probability matrix for the CTMC $(N(t), J(t))$ associated with a BMAP is given by*

$$\mathbf{P}(t) = e^{\mathbf{Q}t} := \sum_{n=0}^{\infty} \frac{t^n}{n!} \mathbf{Q}^n,$$

for all $t \geq 0$, where $\mathbf{Q}^n$ is the nth power of the matrix $\mathbf{Q}$. Here $\mathbf{Q}$ has the following structure

$$\mathbf{Q} = \begin{pmatrix} \mathbf{D}_0 & \mathbf{D}_1 & \mathbf{D}_2 & \mathbf{D}_3 & \cdots \\ 0 & \mathbf{D}_0 & \mathbf{D}_1 & \mathbf{D}_2 & \ddots \\ 0 & 0 & \mathbf{D}_0 & \mathbf{D}_1 & \ddots \\ \vdots & \vdots & \vdots & \ddots & \ddots \end{pmatrix}$$

and $[\mathbf{P}]_{(ki)(nj)} = P(N(t) = n, J(t) = j | N(0) = k, J(0) = i)$.

Exercise 4.21 *Denote the matrix sequence by $\Delta = \{\mathbf{D}_n : n \in \mathbb{N}_0\}$, called the characterizing sequence of a BMAP. For $n \in \mathbb{N}_0$, show that the nth power of the infinitesimal generator matrix $\mathbf{Q}$ of a BMAP has block entries as*

$$[\mathbf{Q}^n]_{kl} = \Delta_{l-k}^{*n},$$

for all $k \leq l \in \mathbb{N}_0$, and $[\mathbf{Q}^n]_{kl} = \mathbf{0}$ for all $k < l \in \mathbb{N}_0$.

Exercise 4.22 *Show that the (k, l)th block entry of the transition probability matrix of a BMAP with characterizing sequence Δ is given by*

$$[\mathbf{P}(t)]_{kl} = [e^{*\Delta t}]_{l-k} := \sum_{n=0}^{\infty} \frac{t^n}{n!} \Delta_{l-k}^{*n},$$

for all $k \leq l \in \mathbb{N}_0$, and $[\mathbf{P}(t)]_{kl} = \mathbf{0}$ for all $k > l \in \mathbb{N}_0$.

Exercise 4.23 *Show that the matrix containing the probabilities that there are no arrivals over time interval $[0,t)$, denoted by $\mathbf{P}_0(t)$ is given by*

$$\mathbf{P}_0(t) = e^{\mathbf{D}_0 t},$$

for all $t \geq 0$ given an initial empty system.

Exercise 4.24 *Denote by $\mathbf{P}_k(t) := \mathbf{P}_{0,k}(t)$ the probability transition matrix for phases given the starting state is 0 and k customers have arrived during $[0,t)$. Define $\mathbf{D} := \sum_{n=0}^{\infty} \mathbf{D}_n$. Show that the transition probability matrix for the phase process $\{J(t), t \geq 0\}$ is given by*

$$\mathbf{P}_J(t) = \sum_{n=0}^{\infty} \mathbf{P}_k(t) = e^{\mathbf{D}t},$$

for all $t \geq 0$ and $1 \leq i, j \leq m$. Note that $[\mathbf{P}_J(t)]_{ij} := P(J(t) = j | J(0) = i)$.

To find the expected number of arrivals over a finite time interval $E[N(t)]$, we need to derive the z-transform of $N(t)$. That is $\mathbf{N}_t(z) := \sum_{n=0}^{\infty} \mathbf{P}_n(t) z^n$ for $z \in \mathbb{C}$ with $|z| \leq 1$. We also define the z-transform of the matrices $\{\mathbf{D}_n : n \in \mathbb{N}_0\}$ as $\mathbf{D}(z) := \sum_{n=0}^{\infty} \mathbf{D}_n z^n$.

Exercise 4.25 *Show that the z-transform of $N(t)$ is given by*

$$\mathbf{N}_t(z) = e^{\mathbf{D}(z)t},$$

for all $z \in \mathbb{C}$ with $|z| \leq 1$.

Exercise 4.26 *Given the initial phase distribution π, show that the expected number of arrivals over time interval $[0,t)$ is given by*

$$E_\pi[N(t)] = t \cdot \pi \sum_{k=1}^{\infty} k \cdot \mathbf{D}_k \mathbf{e}.$$

The following two exercises are for the PH renewal process, a special case of MAP.

Exercise 4.27 *For a PH renewal process with parameters $(\alpha, \mathbf{T})$ starting in phase equilibrium, show that the expected number of arrivals over $[0,t)$ is given by*

$$E_\pi[N(t)] = t \cdot \pi \eta,$$

with $\eta = -\mathbf{T}\mathbf{e}$. Here the stationary phase distribution π is given by

$$\pi = \frac{1}{-\alpha \mathbf{T}^{-1}\mathbf{e}} \int_0^{\infty} \alpha e^{\mathbf{T}t} dt.$$

Exercise 4.28 *The PH renewal process in the exercise above is a delayed renewal process with initial inter-renewal time $X_0 \sim PH(\pi, \mathbf{T})$ and following renewal intervals $X_n \sim PH(\alpha, \mathbf{T})$ for $n \in \mathbb{N}$. Show that $E[X_1] = (\pi \eta)^{-1}$.*

4.6 FITTING MAP TO EMPIRICAL DATA

If the phase process of a MAP is govern by the embedded Markov chain, a discrete-time Markov chain (DTMC), with the transition probability matrix $\mathbf{P} = (-\mathbf{D}_0)^{-1}\mathbf{D}_1$, then the inter-arrival time follows a PH random variable. Let π be the probability vector for initial phase at time $t = 0$. Then, the time to the first arrival occurrence becomes a PH random variable with PH parameters $(\pi, \mathbf{D}_0, \mathbf{D}_1\mathbf{e})$. Generally, let $X_0, X_1, ... X_k$ be the k-successive inter-arrival times of MAP with with $(\mathbf{D}_0, \mathbf{D}_1)$. Note that the inter-arrival times are non-i.i.d. PH distributed random variables. Some important moment-based characteristics of the MAP are

- the lag-l joint moments:

$$E(X_0^i X_l^j) = i!\, j!\, \pi_s(-\mathbf{D}_0)^{-i}\mathbf{P}^l(-\mathbf{D}_0)^{-j}\mathbf{e}, \qquad (4.72)$$

- the lag-l autocorrelation

$$\rho_l = \frac{E(X_0 X_l) - E(X^2)}{E(X^2) - E(X)^2}, \qquad (4.73)$$

where $E(X^k) = k!\,\pi_s(-\mathbf{D}_0)^{-k}\mathbf{e}$ is the k-th moment of the PH distribution with π_s to be the stationary vector of $\mathbf{P}$ satisfying $\pi_s = \pi_s\mathbf{P}$. To fit MAP to empirical data set, there are some different methods such as Matching Moments (MM), Maximum Likelihood Estimation (MLE), and Bayes estimation (BM). Below, we only briefly present the basic idea of MLE method which is again the EM algorithm as used in fitting PH distribution to empirical data set. For more details, interested readers are referred to the work by Okamura and Dohi (2016) on which the following is based.

4.6.1 GROUPED DATA FOR MAP FITTING

The grouped data for MAP fitting is defined by time intervals and the number of arrivals in those intervals. Consider the grouped observed data $\mathscr{D} = \{n_1, n_2, ..., n_K\}$ with break points $0 = t_0 < t_1 < ... < t_K$, where n_k is the number of arrivals in $[t_{k-1}, t_k)$. With such an observed data set, we show how to estimate the MAP parameters $(\pi, \mathbf{D}_0, \mathbf{D}_1)$, where π is the probability vector for the initial phase at $t = 0$. In this section, π_i, $\lambda_{i,j}$, and $\mu_{i,j}$ are the i-th element of π, (i, j) elements of $\mathbf{D}_0$ and $\mathbf{D}_1$, respectively. Furthermore, we define the following unobserved variables:

- B_i: an indicator random variable for the event that the phase process of MAP begins with phase i.
- $Z_i^{(k)}$: the total time spent in phase i in the interval $[t_{k-1}, t_k)$.

- $Y_{ij}^{(k)}$: the total number of arrivals leading to phase transitions from phase i to phase j in the time interval $[t_{k-1}, t_k)$.
- $M_{ij}^{(k)}$: the total number of phase transitions from phase i to phase j without arrivals in the time interval $[t_{k-1}, t_k)$.

Based on the expected values of the unobserved variables, the M-step formulas for MAP parameters can be given by

$$\pi_i \leftarrow E(B_i|\mathscr{D}), \quad \lambda_{i,j} \leftarrow \frac{\sum_{k=1}^{K} E(M_{i,j}^{(k)}|\mathscr{D})}{\sum_{k=1}^{K} E(Z_i^{(k)}|\mathscr{D})}, \quad \mu_{i,j} \leftarrow \frac{\sum_{k=1}^{K} E(Y_{i,j}^{(k)}|\mathscr{D})}{\sum_{k=1}^{K} E(Z_i^{(k)}|\mathscr{D})}. \quad (4.74)$$

To derive the expected values of the unobserved variables in the E-step, we need to define the following row and column vectors:

$$[\mathbf{f}_k(n,u)]_i = P(\Delta N_1 = n_1, ..., \Delta N_{k-1} = n_{k-1}, N_{u+t_{k-1}} - N_{t_{k-1}} = n, J_{u+t_{k-1}} = i),$$
$$\text{for } 0 \le u \le \Delta t_k,$$
$$(4.75)$$

$$[\mathbf{b}_k(n,u)]_i = P(N_{t_k} - N_{t_k - u} = n, \Delta N_{k+1} = n_{k+1}, ..., \Delta N_K = n_K | J_{t_k - u} = i),$$
$$\text{for } 0 \le u \le \Delta t_k,$$
$$(4.76)$$

where $\Delta N_k = N_{t_k} - N_{t_{k-1}}$ and $\Delta t_k = t_k - t_{k-1}$. The sum of row vector $\mathbf{f}_k(n,u)$ represents the likelihood of observed data in $[0, t_{k-1} + u]$, and the i-th element of $\mathbf{b}_k(n,u)$ denotes the conditional likelihood of observed data in $[t_k - u, t_k]$ provided that $J_{t_k - u} = i$. By using $\mathbf{f}_k(n,u)$ and $\mathbf{b}_k(n,u)$, the expected $E(B_i|\mathscr{D})$ is given by

$$
\begin{aligned}
E(B_i|\mathscr{D}) &= \frac{P(J_0 = i, \Delta N_1 = n_1, ..., \Delta N_K = n_K)}{P(\Delta N_1 = n_1, ..., \Delta N_K = n_K)} \\
&= \frac{P(J_0 = i)P(\Delta N_1 = n_1, ..., \Delta N_K = n_K | J_0 = i)}{P(\Delta N_1 = n_1, ..., \Delta N_K = n_K)} \\
&= \frac{\pi_i [\mathbf{b}_1(n_1, \Delta t_1)]_i}{\pi \mathbf{b}_1(n_1, \Delta t_1)},
\end{aligned}
\quad (4.77)
$$

where $\pi \mathbf{b}_1(n_1, \Delta t_1)$ is the likelihood of $\mathscr{D}$. Similarly, the expected value $E(Z_i^{(k)}|\mathscr{D})$ is given by

$$E(Z_i^{(k)}|\mathscr{D}) = \frac{\int_0^{\Delta t_k} P(\Delta N_1 = n_1, ..., \Delta N_k = n_k, J_{t_{k-1}+u} = i, ..., \Delta N_K = n_K) du}{P(\Delta N_1 = n_1, ..., \Delta N_K = n_K)}, \quad (4.78)$$

where

$$P(\Delta N_1 = n_1, ..., \Delta N_k, J_{t_{k-1}+u} = i, ...\Delta N_K = n_K)$$

$$= \sum_{l=0}^{n_k} P(\Delta N_1 = n_1, ..., N_{t_{k-1}+u} - N_{t_{k-1}} = l, J_{t_{k-1}+u} = i)$$

$$\times P(N_{t_k} - N_{t_k-u} = n_k - l, ..., \Delta N_K = n_K | J_{t_{k-1}+u} = i) \tag{4.79}$$

$$= \sum_{l=0}^{n_k} [\mathbf{f}_k(l, u)]_i [\mathbf{b}_k(n_k - l, \Delta t_k - u)]_i.$$

With this relation, we have

$$E(Z_i^{(k)} | \mathscr{D}) = \frac{\int_0^{\Delta t_k} \sum_{l=0}^{n_k} [\mathbf{f}_k(l, u)]_i [\mathbf{b}_k(n_k - l, \Delta t_k - u)]_i du}{\pi \mathbf{b}_1(n_1, \Delta t_1)}. \tag{4.80}$$

Similarly, the expected value $E(M_{i,j}^{(k)} | \mathscr{D})$ is given by

$$E(M_{i,j}^{(k)} | \mathscr{D}) = \frac{1}{P(\Delta N_1 = n_1, ..., \Delta N_K = n_K)}$$

$$\times \int_0^{\Delta t_k} P(\Delta N_1 = n_1, ..., \Delta N_k = n_k, J_{t_{k-1}+u}^- = i, J_{t_{k-1}+u} = j, ..., \Delta N_K = n_K) du$$

$$= \frac{1}{\pi \mathbf{b}_1(n_1, \Delta t_1)} \int_0^{\Delta t_k} \sum_{l=0}^{n_k} P(\Delta N_1 = n_1, ..., \Delta N_{t_{k-1}+u} - N_{t_{k-1}} = l, J_{t_{k-1}+u}^- = i)$$

$$\times P(J_{t_{k-1}+u} = j | J_{t_{k-1}+u}^- = i P(N_{t_k} - N_{t_k-u} = n_k - l, ..., \Delta N_K = n_K | J_{t_{k-1}+u} = j) du$$

$$= \frac{\int_0^{\Delta t_k} \sum_{l=0}^{n_k} [\mathbf{f}_k(l, u)]_i \lambda_{i,j} [\mathbf{b}_k(n_k - l, \Delta t_k - u)]_j du}{\pi \mathbf{b}_1(n_1, \Delta t_1)}, \tag{4.81}$$

and the expected value of $E(Y_{i,j}^{(k)} | \mathscr{D})$ is given by

$$E(Y_{i,j}^{(k)} | \mathscr{D}) = \frac{1}{P(\Delta N_1 = n_1, ..., \Delta N_K = n_K)}$$

$$\times \int_0^{\Delta t_k} P(\Delta N_1 = n_1, ..., \Delta N_k = n_k, N_{t_{k-1}+u} - N_{t_{k-1}+u}^- = l, J_{t_{k-1}+u}^- = i, J_{t_{k-1}+u} = j,$$

$$..., \Delta N_K = n_K) du = \frac{1}{\pi \mathbf{b}_1(n_1, \Delta t_1)} \int_0^{\Delta t_k} \sum_{l=0}^{n_k-1} P(\Delta N_1 = n_1, ...,$$

$$\Delta N_{t_{k-1}+u}^- - N_{t_{k-1}} = l, J_{t_{k-1}+u}^- = i)$$

$$\times P(N_{t_{k-1}+u} - N_{t_{k-1}+u}^- = l, J_{t_{k-1}+u} = j | J_{t_{k-1}+u}^- = j)$$

$$\times P(N_{t_k} - N_{t_k-u} = n_k - l - 1, ..., \Delta N_K = n_K | J_{t_{k-1}+u} = j) du$$

$$= \frac{\int_0^{\Delta t_k} \sum_{l=0}^{n_k-1} [\mathbf{f}_k(l, u)]_i \mu_{i,j} [\mathbf{b}_k(n_k - l - 1, \Delta t_k - u)]_j du}{\pi \mathbf{b}_1(n_1, \Delta t_1)}, \tag{4.82}$$

where $N_t^- = \lim_{t' \to 0^+} N_{t-t'}$.

Clearly, in the E-step of the EM algorithm, we need to compute $\mathbf{f}_k(n, u)$, $\mathbf{f}_k(n, u)$, and their convolutions. To facilitate the computations, we introduce the following row and column vectors

$$\hat{\mathbf{f}}_k(u) = (\mathbf{f}_k(0, u), ..., \mathbf{f}_k(n_k, u)), \quad \hat{\mathbf{b}}_k(u) = \begin{pmatrix} \mathbf{b}_k(n_k, u) \\ \vdots \\ \mathbf{b}_k(0, u) \end{pmatrix}. \tag{4.83}$$

Then, we can write

$$\hat{\mathbf{f}}(u) = \hat{\pi} e^{(\hat{\mathbf{D}}_0(1)\Delta t_1)} \hat{\mathbf{I}}(1) \times ... \times e^{(\hat{\mathbf{D}}_0(k-1)\Delta t_{k-1})} \hat{\mathbf{I}}(k-1) e^{\hat{\mathbf{D}}_0(k)u},$$
$$\hat{\mathbf{b}}(u) = e^{\hat{\mathbf{D}}_0(k)u} \hat{\mathbf{I}}(k) e^{\hat{\mathbf{D}}_0(k+1)\Delta t_{k+1}} \hat{\mathbf{I}}(k+1) \times ... \times e^{\hat{\mathbf{D}}_0(K)\Delta t_K} \hat{\mathbf{I}}(K) \hat{\mathbf{I}}, \tag{4.84}$$

where $\hat{\mathbf{D}}_0(k)$ and $\hat{\mathbf{I}}(k)$ are the following block matrices:

$$\hat{\mathbf{D}}_0(k) = \begin{pmatrix} \mathbf{D}_0 & \mathbf{D}_1 & & \\ & \mathbf{D}_0 & \mathbf{D}_1 & \\ & & \ddots & \ddots \\ & & & \mathbf{D}_0 \end{pmatrix}_{m(n_k+1) \times m(n_k+1)},$$

$$\hat{\mathbf{I}}(k) = \begin{pmatrix} \mathbf{O} & \cdots\cdots & & \mathbf{O} \\ \vdots & \cdots & \cdots & \vdots \\ \mathbf{O} & \cdots & \ddots & \mathbf{O} \\ \mathbf{I} & \mathbf{O} & \cdots & \mathbf{O} \end{pmatrix}_{m(n_k+1) \times m(n_k+1)} \tag{4.85}$$

The row and column vectors $\hat{\pi}$ and $\hat{\mathbf{I}}$ are defined as follows:

$$\hat{\pi} = \underbrace{(\pi \, \mathbf{0} \cdots \mathbf{0})}_{m(n_1+1)}, \quad \hat{\mathbf{I}} = \left. \begin{pmatrix} \mathbf{0} \\ \vdots \\ \mathbf{0} \\ 1 \end{pmatrix} \right\} m(n_K+1). \tag{4.86}$$

Now we define the convolution integral of $\hat{\mathbf{b}}_k(u)$ and $\hat{\mathbf{f}}_k(u)$:

$$\hat{\mathbf{H}}_k(\Delta t_k) = \int_0^{\Delta t_k} \hat{\mathbf{b}}_k(\Delta t_k - s) \hat{\mathbf{f}}_k(s) ds. \tag{4.87}$$

From the definition of $\hat{\mathbf{b}}_k(u)$ and $\hat{\mathbf{f}}_k(u)$, the sum of diagonal matrices is an $m \times m$ matrix representing the convolution of $\mathbf{f}_k(n, u)$ and $\mathbf{b}_k(n, u)$. That is

$$\sum_{v=0}^{n_k} \hat{\mathbf{H}}_k(u)[v, v] = \int_0^u \sum_{l=0}^{n_k} \mathbf{b}_k(n_k - l, u - s) \mathbf{f}_k(l, s) ds, \tag{4.88}$$

where $\hat{\mathbf{H}}_k(u)[v, v]$ is the $(v+1)$-st diagonal submatrix of $\hat{\mathbf{H}}_k(u)$. Also, the sum of subdiagonal matrices becomes

$$\sum_{v=0}^{n_k} \hat{\mathbf{H}}_k(u)[v, v-1] = \int_0^u \sum_{l=0}^{n_k-1} \mathbf{b}_k(n_k - l - 1, u - s)\mathbf{f}_k(l, s)ds, \qquad (4.89)$$

To simplify the notations, we define $\hat{\mathbf{f}}_k = \hat{\mathbf{f}}_k(\Delta t_k)$ and $\hat{\mathbf{b}}_k = \hat{\mathbf{b}}_k(\Delta t_k)$. Let $\hat{\mathbf{f}}_k[v]$ and $\hat{\mathbf{b}}_k[v]$, $v = 0, 1, ..., n_k$ be the v-th subvectors of $\hat{\mathbf{f}}_k$ and $\hat{\mathbf{b}}_k$, respectively. Then the expected values given the observed sample are given by

$$E(B_i|\mathscr{D}) = \frac{\pi_i[\hat{\mathbf{b}}_1[0]]_i}{\pi \hat{\mathbf{b}}_i[0]}, \quad E(Z_i^{(k)}|\mathscr{D}) = \frac{[\sum_{v=0}^{n_k} \hat{\mathbf{H}}_k[v, v]]_{i,i}}{\pi \hat{\mathbf{b}}_i[0]}, \qquad (4.90)$$

$$E(M_{i,j}^{(k)}|\mathscr{D}) = \frac{\lambda_{i,j}[\sum_{v=0}^{n_k} \hat{\mathbf{H}}_k[v, v]]_{j,i}}{\pi \hat{\mathbf{b}}_i[0]}, \quad E(Y_{i,j}^{(k)}|\mathscr{D}) = \frac{\mu_{i,j}[\sum_{v=0}^{n_k} \hat{\mathbf{H}}_k[v, v-1]]_{j,i}}{\pi \hat{\mathbf{b}}_i[0]}. \qquad (4.91)$$

From Equations (4.84), $\hat{\mathbf{f}}_k$ and $\hat{\mathbf{b}}_k$ can be obtained by computing the matrix exponential in the forward–backward manner. We introduce the convolution integral of matrix exponential $\mathbf{T}$ as follows:

$$\Gamma(x; \mathbf{v}_1, \mathbf{v}_2) = \int_0^x e^{\mathbf{T}(x-u)} \mathbf{v}_1 \mathbf{v}_2 e^{\mathbf{T}u} du,$$

where $\mathbf{v}_1$ and $\mathbf{v}_2$ are arbitrary column and row vectors, respectively. Then we have $\hat{\mathbf{H}}_k(\Delta t_k)$ as the convolution integral of matrix exponential $\hat{\mathbf{D}}_0(k)$

$$\begin{aligned}
\hat{\mathbf{H}}_k(\Delta t_k) &= \int_0^{\Delta t_k} e^{\hat{\mathbf{D}}_0(k)(\Delta t_k - s)} \hat{\mathbf{I}}(k)\hat{\mathbf{b}}_{k+1}\hat{\mathbf{f}}_{k-1}\hat{\mathbf{I}}(k-1)e^{\hat{\mathbf{D}}_0(k)s} ds \\
&= \Gamma(\Delta t_k; \hat{\mathbf{I}}(k)\hat{\mathbf{b}}_{k+1}, \hat{\mathbf{f}}_{k-1}\hat{\mathbf{I}}(k-1)).
\end{aligned} \qquad (4.92)$$

Based on these equations, we can develop the algorithm and the parameters are updated with $\pi_i \leftarrow \pi_i[\mathbf{B}]_i, \lambda_{i,j} \leftarrow \lambda_{i,j}[\mathbf{H}]_{i,i}$, and $\mu_{i,j} \leftarrow \mu_{i,j}[\mathbf{Y}]_{j,i}/[\mathbf{H}]_{i,i}$.

Exercise 4.29 *Design a flowchart for developing an algorithm of implementing E-step procedure for MAP fitting with grouped data.*

4.7 QUASI-BIRTH-AND-DEATH PROCESS (QBD) – ANALYSIS OF MAP/PH/1 QUEUE

Since, under certain conditions, a CTMC can be transformed into a DTMC using the so-called uniformization technique (see Chapter 3), we use the discrete time case to introduce the quasi-birth-and-death process (QBD), a structured Markov chain that extends the birth-and-death process into a two-dimensional case. Since the theory for the continuous-time QBD processes can be developed similarly, we only present some results and leave the rest as exercises. We start with a motivational example.

4.7.1 QBD – A STRUCTURED MARKOV CHAIN

Consider a discrete-time queue-inventory model which can be used to introduce the discrete-time QBD process. Customers arrive at a service system according to a geometric distribution. As a discrete-time system, the time line is divided into discrete equal time slots. The events may occur at the boundary of each slot. At the boundary of each slot, a customer may arrive with probability p and his service requires one unit of inventory. There are many practical situations where each service requires an inventory item. For example, some auto-service that involves replacement of certain part fits such a model. It is assumed that if the stock-out occurs (i.e., the inventory level becomes zero), customers are not allowed to join the system. Such an assumption is to ensure a certain customer service level can be achieved. The service time process is modeled as a geometric distribution with parameter q. The inventory is controlled by an $(s\,S)$ policy. That is whenever the inventor level falls to s, an order is placed to bring the level to S. The lead time from the placing the order to receiving the order follows a geometric distribution with parameter l and takes at least one time slot to complete. Thus, an order cannot be received at the time slot when it is placed. We assume that a departure or replenishment occurs in slightly prior to the boundary of the slot (k^-,k) and an arrival in slightly after the boundary of the slot (k,k^+). We denote the complement probabilities as $p^c = 1 - p$, $q^c = 1 - q$, and $l^c = 1 - l$. Let X_k denote the number of customers in the system and J_k the inventory level at epoch k^+. The joint queue length and inventory level process, denoted by $\{(X_k, J_k) : k \in \mathbb{N}\}$, is a two-dimensional DTMC with state space $\Omega = \{0,1,2,...\} \times \{0,1,2,...,s,s+1,...,S\}$. The state space of this DTMC can be partitioned into levels with the first state variable as the "level variable" and the second variable as the "phase variable". That is the level $\hat{i}$ is a finite subset of the state space $\hat{i} = \{(i,0),(i,1),...,(i,s),(i,s+1),...,(i,S)\}$. The one step transition probability matrix for this DTMC is given by

$$
\mathbf{P} =
\begin{pmatrix}
\mathbf{B}_{00} & \mathbf{B}_{01} & & & \\
\mathbf{A}_2 & \mathbf{A}_1 & \mathbf{A}_0 & & \\
& \mathbf{A}_2 & \mathbf{A}_1 & \mathbf{A}_0 & \\
& & \ddots & \ddots & \ddots \\
& & & \ddots & \ddots
\end{pmatrix},
$$

where each entry is a square matrix of order $S+1$. This two-dimensional DTMC $\{X_k, J_k\}$ with the transition probability matrix of the tridiagonal structure is called a QBD process. Note that the matrices $\mathbf{B}_{00}$ and $\mathbf{B}_{01}$ contain transition probabilities from the states in level 0, called boundary states. Some transitions are among the states within level 0 and their probabilities are in $\mathbf{B}_{00}$. Other transitions are to the states in level 1 and their probabilities are in $\mathbf{B}_{01}$.

From level 1 and up, the three non-zero matrices, containing transition probabilities among the interior states, remain the same and this is the characteristics
of the level-independent QBD process. The tridiagonal structure is called "skip
free" property which means that transitions among the interior states can only
be to one level up with probabilities in $\mathbf{A}_0$ or to one level down with probabilities in $\mathbf{A}_2$. According to the problem description and the inventory policy, we
can obtain the entries of these matrices as follows:

$$[\mathbf{B}_{00}]_{ij} = \begin{cases} l^c & j=i, & i=0, \\ p^c l^c & j=i, & i=1,2,...,s, \\ p^c & j=i, & i=s+1,s+2,...,S, \\ p^c l & j=S, & i=0,1,...,s, \\ 0, & \text{otherwise.} \end{cases}$$

$$[\mathbf{B}_{01}]_{ij} = \begin{cases} pl^c & j=i, & i=1,2,...,s, \\ p & j=i, & i=s+1,s+2,...,S, \\ pl & j=S, & i=0,1,...,s, \\ 0 & \text{otherwise.} \end{cases}$$

$$[\mathbf{A}_2]_{ij} = \begin{cases} ql^c & j=i-1, & i=1, \\ p^c ql^c & j=i-1, & i=2,3,...,s, \\ p^c q & j=i-1, & i=s+1,s+2,...,S, \\ p^c ql & j=S-1, & i=1,2,...,s, \\ 0 & \text{otherwise.} \end{cases}$$

$$[\mathbf{A}_1]_{ij} = \begin{cases} l^c & j=i, & i=0, \\ p^c q^c r^c & j=i, & i=1,2,...,s, \\ p^c l^c & j=i, & i=s+1,s+2,...,S, \\ pql^c & j=i-1, & i=2,3,...s, \\ pql & j=S-1, & i=1,2,...s, \\ p^c l & j=S, & i=0, \\ p^c q^c l & j=S, & i=1,2,...s, \\ 0 & \text{otherwise.} \end{cases}$$

$$[\mathbf{A}_0]_{ij} = \begin{cases} pq^c l^c & j=i, & i=1,2,...,s, \\ p & j=i, & i=s+1,s+2,...,S, \\ pl & j=S, & i=0, \\ pq^c l & j=S, & i=1,2,...s, \\ 0 & \text{otherwise.} \end{cases}$$

This model can be denoted by $Geo/Geo/1/(s\ S)$, which is a typical discrete-
time QBD process. We will use this example throughout this section.

Remark 4.2 *In this queue-inventory model, the number of phases at level 0 is
the same as that in higher levels and only level 0 is the boundary level. In general, the number of phases at level 0 may be different from that in higher levels.*

Thus, both level 0 and level 1 can be boundary levels. This fact is reflected in the following QBD definition.

Definition 4.1 *A discrete-time QBD* $\{(X_k, J_k), k \in \mathbb{N}_0\}$ *is a DTMC with state space* $\{(0,1), (0,2), ..., (0, m_0)\} \cup \{\{1, 2, ...\} \times \{1, 2, ..., m\}\}$ *with the transition probability matrxi given by*

$$
\mathbf{P} = \begin{pmatrix}
\mathbf{B}_{00} & \mathbf{B}_{01} & & & \\
\mathbf{B}_{10} & \mathbf{B}_{11} & \mathbf{A}_0 & & \\
& \mathbf{A}_2 & \mathbf{A}_1 & \mathbf{A}_0 & \\
& & \ddots & \ddots & \ddots \\
& & & \ddots & \ddots
\end{pmatrix}, \tag{4.93}
$$

where $\mathbf{B}_{00}$ *is an* $m_0 \times m_0$ *matrix,* $\mathbf{B}_{01}$ *is an* $m_0 \times m$ *matrix,* $\mathbf{B}_{10}$ *is an* $m \times m_0$ *matrix, and* $\{\mathbf{B}_{11}, \mathbf{A}_0, \mathbf{A}_1, \mathbf{A}_2\}$ *are non-negative matrices of order m.*

Since $\mathbf{P}$ is a stochastic matrix, its entry matrices satisfy

$$
(\mathbf{A}_0 + \mathbf{A}_1 + \mathbf{A}_2)\mathbf{e} = \mathbf{e}
$$
$$
\mathbf{B}_{10}\mathbf{e} + (\mathbf{B}_{11} + \mathbf{A}_0)\mathbf{e} = \mathbf{e}
$$
$$
\mathbf{B}_{00}\mathbf{e} + \mathbf{B}_{01}\mathbf{e} = \mathbf{e}.
$$

If we denote $\mathbf{A} = \mathbf{A}_0 + \mathbf{A}_1 + \mathbf{A}_2$, then $\mathbf{A}$ is a stochastic matrix of order m, which governs the transitions of the phase variable J_k for level $X_k \geq 2$. If the QBD process defined by (4.93) is irreducible and positive recurrent (i.e., ergodic), then the limiting probabilities exist based on the DTMC theory in Chapter 2. Define

$$
\pi_{nj} = \lim_{k \to \infty} P(X_k = n, J_k = j | (X_0, J_0)), \quad n = 0, 1, 2, ..., j = 1, ..., m.
$$
$$
\pi_0 = (\pi_{01}, \cdots, \pi_{0m_0}); \tag{4.94}
$$
$$
\pi_n = (\pi_{n1}, \cdots, \pi_{nm}), \quad n = 1, 2, ...;
$$
$$
\pi = (\pi_0, \pi_1 \cdots).
$$

To ensure the existence of the limiting distribution, we first present the condition for ergodicity.

Theorem 4.8 *The irreducible and aperiodic discrete-time QBD process* $\{(X_k, J_k), k \in \mathbb{N}_0\}$ *is ergodic if and only if* $\theta\mathbf{A}_0\mathbf{e} < \theta\mathbf{A}_2\mathbf{e}$, *where* θ *satisfies* $\theta\mathbf{A} = \theta$ *and* $\theta\mathbf{e} = 1$.

The detailed proofs of this theorem can be found in Neuts (1981) and Latouche and Ramaswami (1999). To give readers some basic idea about the proof of ergodicity of Markov chains, we provide a brief proof of sufficiency following He (2014). Such a proof reveals the important role of the Perron–Frobenius theorem in linear algebra played in probability and stochastic process theory (see Appendix J).

Proof. (sufficiency of Theorem 4.8) The proof is based on Foster's criterion mentioned in Chapter 2 (more details about this criterion can be found in Cohen (1982)). The basic idea is to find the mean-drift of some Lyapunov function of states of the Markov chain and to show that this mean-drift is negative under the condition. First, we define a matrix Lyapunov function

$$\mathbf{A}^*(z) = \mathbf{A}_0 + z\mathbf{A}_1 + z^2\mathbf{A}_2, \ z \geq 0.$$

Note that $\mathbf{A}^*(1) = \mathbf{A}$. Let $\lambda(z)$ be the Perron–Frobenius eigenvalue of the non-negative matrix $\mathbf{A}^*(z)$ (i.e., the eigenvalue of $\mathbf{A}^*(z)$ with the largest real part). Let $\mathbf{u}(z)$ and $\mathbf{v}(z)$ be the left and right eigenvectors of $\mathbf{A}^*(z)$. That is

$$\mathbf{u}(z)\mathbf{A}^*(z) = \lambda(z)\mathbf{u}(z), \ \ \mathbf{A}^*(z)\mathbf{v}(z) = \lambda(z)\mathbf{v}(z),$$

normalized by $\mathbf{u}(z)\mathbf{v}(z) = 1$ and $\mathbf{u}(z)\mathbf{e} = 1$. It follows from the Perron–Frobenius theorem (Seneta (2006)) that both $\mathbf{u}(z)$ and $\mathbf{v}(z)$ are non-negative. Taking the first-order derivative with respect to z on both sides of $\mathbf{u}(z)\mathbf{A}^*(z) = \lambda(z)\mathbf{u}(z)$ and right-multiplying by $\mathbf{v}(z)$ on both sides, we have

$$\mathbf{u}(z)\mathbf{A}^{*(1)}(z)\mathbf{v}(z) + \mathbf{u}^{(1)}(z)\mathbf{A}^*(z)\mathbf{v}(z) = \lambda^{(1)}(z)\mathbf{u}(z)\mathbf{v}(z) + \lambda(z)\mathbf{u}^{(1)}(z)\mathbf{v}(z),$$

where $\mathbf{A}^{*(1)}(z), \mathbf{u}^{(1)}(z)$, and $\lambda^{(1)}(z)$ are the first-order derivatives. Note that $\mathbf{u}(1) = \theta$, $\mathbf{v}(1) = \mathbf{e}$, and hence, $\mathbf{A}^*(1)\mathbf{v}(1) = \mathbf{A}\mathbf{e} = \mathbf{e} = \mathbf{v}(1)$. Setting $z = 1$, we have

$$\mathbf{u}(1)\mathbf{A}^{*(1)}(1)\mathbf{v}(1) + \mathbf{u}^{(1)}(1)\mathbf{A}^*(1)\mathbf{v}(1) = \lambda^{(1)}(1)\mathbf{u}(1)\mathbf{v}(1) + \lambda(1)\mathbf{u}^{(1)}(1)\mathbf{v}(1),$$
$$\Rightarrow \mathbf{u}(1)\mathbf{A}^{*(1)}(1)\mathbf{e} = \lambda^{(1)}(1),$$

by using the normalization conditions $\mathbf{u}(1)\mathbf{v}(1) = 1$ and $\mathbf{u}^{(1)}(1)\mathbf{v}(1) = 0$. This relation implies

$$\lambda^{(1)}(1) = \theta\mathbf{A}^{*(1)}(1)\mathbf{e} = \theta(\mathbf{A}_1 + 2\mathbf{A}_2)\mathbf{e}$$
$$= \theta(\mathbf{A}_0 + \mathbf{A}_1 + \mathbf{A}_2 + \mathbf{A}_2 - \mathbf{A}_0)\mathbf{e} = \theta(\mathbf{A} + \mathbf{A}_2 - \mathbf{A}_0)\mathbf{e}$$
$$= 1 + \theta\mathbf{A}_2\mathbf{e} - \theta\mathbf{A}_0\mathbf{e} > 1.$$

Therefore, there exists a z such that $0 < \lambda(z) < z < 1$, and $\mathbf{v}(z)$ is positive elementwise. Define a vector Lyapunov function for states in level n as

$$\mathbf{f}^*(n) = z^{-n}\mathbf{v}(z), \ \ n = 0, 1, 2, \dots \tag{4.95}$$

Clearly, $\mathbf{f}^*(n) \to \infty$ elementwise as $n \to \infty$. The difference $\mathbf{f}^*(X_{k+1}) - \mathbf{f}^*(X_k)$ is called the drift of the Markov chain at X_k. Then the mean-drift can be calculated with a negative infinity limit as follows, for $n > 2$,

$$
\begin{aligned}
E[\mathbf{f}^*(X_{k+1}) &- \mathbf{f}^*(X_k)|X_k = n] \\
&= z^{-n-1}\mathbf{A}_0\mathbf{v}(z) + z^{-n}\mathbf{A}_1\mathbf{v}(z) + z^{-n+1}\mathbf{A}_2\mathbf{v}(z) - z^{-n}\mathbf{v}(z) \\
&= z^{-(n+1)}\mathbf{A}^*(z)\mathbf{v}(z) - z^{-n}\mathbf{v}(z) \\
&= z^{-(n+1)}(\lambda(z) - z)\mathbf{v}(z) \to -\infty
\end{aligned}
\tag{4.96}
$$

as $n \to \infty$. Note that the last expression in (4.96) is less than or equal to $(\lambda(z) - z_{\mathbf{v}}(z) < 0$. For $n \le 2$, the mean drift is clearly finite. By Foster's criteria, the DTMC is ergodic. This completes the proof of the sufficiency of the theorem. ∎

4.7.1.1 Matrix-Geometric Solution – R-Matrix

Assume that the mean-drift condition holds for the QBD process. Now we focus on finding the limiting distribution of the QBD process by solving the $\pi = \pi\mathbf{P}$ subject to the normalization condition $\pi\mathbf{e} = 1$. Using (4.93), this set of equations for the limiting probabilities can be written in vector form as

$$
\begin{aligned}
\pi_0 &= \pi_0\mathbf{B}_{00} + \pi_1\mathbf{B}_{10} \\
\pi_1 &= \pi_0\mathbf{B}_{01} + \pi_1\mathbf{B}_{11} + \pi_2\mathbf{A}_1 \\
\pi_n &= \pi_{n-1}\mathbf{A}_0 + \pi_n\mathbf{A}_1 + \pi_{n+1}\mathbf{A}_2, \quad n = 2, 3, \ldots
\end{aligned}
\tag{4.97}
$$

subject to $\pi\mathbf{e} = 1$. To solve this set of equations, we utilize a comparison approach. Since the transition probability matrix (4.93) has the same structure of the transition probability matrix of a discrete-time birth-and-death process (for the discrete-time single server queue), we conjecture that the limiting distribution for the discrete-time QBD process has the same form as that in the discrete-time single server queue. Recall that the limiting distribution for the discrete-time single server queue is $\pi_n = \pi_0\left(\frac{1-p}{p}\right)^n = \pi_1\left(\frac{1-p}{p}\right)^{n-1} = \pi_1\rho^{n-1}$ for $j = 1, 2\ldots$, which is a geometric distribution with parameter ρ. Then we guess that the solution to (4.97) has the *matrix-geometric* form $\pi_n = \pi_1\mathbf{R}^{n-1}$ for $n = 1, 2\ldots$. Substituting this matrix-geometric solution into (4.97), we obtain

$$
\pi_1\mathbf{R}^{n-1}(\mathbf{R} - \mathbf{A}_0 - \mathbf{R}\mathbf{A}_1 - \mathbf{R}^2\mathbf{A}_2) = 0,
$$

for $n = 2, 3, \ldots$. If we can find a non-negative matrix $\mathbf{R}$ satisfying $\mathbf{R} = \mathbf{A}_0 + \mathbf{R}\mathbf{A}_1 + \mathbf{R}^2\mathbf{A}_2$, and appropriate π_0 and π_1 satisfying the normalization condition, then we obtain the matrix-geometric solution π. This fundamental result is summarized in the following theorem.

Theorem 4.9 *If the discrete-time QBD process $\{(X_k, J_k), k \in \mathbb{N}\}$ is ergodic, its limiting probabilities are given by*

$$\pi_n = \pi_1 \mathbf{R}^{n-1}, \quad n = 1, 2, \ldots, \tag{4.98}$$

where $\mathbf{R}$, called rate matrix, is the minimum non-negative solution to the non-linear equation

$$\mathbf{R} = \mathbf{A}_0 + \mathbf{R}\mathbf{A}_1 + \mathbf{R}^2\mathbf{A}_2 \tag{4.99}$$

and the vector π_0 and π_1 are the unique positive solution to linear system

$$(\pi_0, \pi_1) = (\pi_0, \pi_1) \begin{pmatrix} \mathbf{B}_{00} & \mathbf{B}_{01} \\ \mathbf{B}_{10} & \mathbf{B}_{11} + \mathbf{R}\mathbf{A}_2 \end{pmatrix}, \tag{4.100}$$

$$1 = \pi_0 \mathbf{e} + \pi_1 (\mathbf{I} - \mathbf{R})^{-1} \mathbf{e}.$$

The detailed and rigorous proof can be found in Neuts (1981). In general, there is no explicit solution for $\mathbf{R}$. However, using a recursive algorithm, we can obtain $\mathbf{R}$ numerically. We present a simple recursion on which an algorithm can be developed. Starting with $\mathbf{R}[0] = \mathbf{0}$, we can compute

$$\mathbf{R}[k+1] = \mathbf{A}_0 + \mathbf{R}[k]\mathbf{A}_1 + (\mathbf{R}[k])^2 \mathbf{A}_2, \quad k = 0, 1, 2, \ldots, \tag{4.101}$$

recursively until $|\mathbf{R}[k+1] - \mathbf{R}[k]|_{ij} < \varepsilon$, where ε is a small error tolerance (convergent criterion). Other more efficient algorithms have been developed in the literature by using matrix analytic methods. Interested readers are referred to Alfa (2010). It is reasonable to assume that $\mathbf{R}$ exists and finite as the entry in $\mathbf{R}$, R_{jv}, can be interpreted later as the expected number of visits of the QBD to a particular state v in level $i+1$ (one level higher than the starting state level i) during the time interval starting from state (i, j) to the first return to level i. Clearly for a stable QBD, such a number should be finite. The following proposition justify the use of the recursion to compute $\mathbf{R}$ recursively.

Proposition 4.1 *Suppose that $\mathbf{R}$ exists and is finite. Then (i) the sequence $\{\mathbf{R}[k], k = 0, 1, 2, \ldots\}$ computed by (4.101) is non-decreasing; and (ii) $\mathbf{R}[k] \leq \mathbf{R}$ for $k = 0, 1, 2, \ldots$; and (iii) $\{\mathbf{R}[k], k = 0, 1, 2 \ldots\}$ converges to $\mathbf{R}$.*

Proof. We use the induction to prove $\mathbf{R}[k]$ is increasing in k. First, for $k = 0$ with $\mathbf{R}[0] = 0$, we have

$$\mathbf{R}[1] = \mathbf{A}_0 + \mathbf{R}[0]\mathbf{A}_1 + (\mathbf{R}[0])^2 \mathbf{A}_2$$
$$= \mathbf{A}_0 > 0 = \mathbf{R}[0].$$

Assume that $\mathbf{R}[k] - \mathbf{R}[k-1] \geq 0$. It follows from

$$\mathbf{R}[k+1] = \mathbf{A}_0 + \mathbf{R}[k]\mathbf{A}_1 + (\mathbf{R}[k])^2 \mathbf{A}_2 \text{ and}$$
$$\mathbf{R}[k] = \mathbf{A}_0 + \mathbf{R}[k-1]\mathbf{A}_1 + (\mathbf{R}[k-1])^2 \mathbf{A}_2,$$

that

$$\mathbf{R}[k+1] - \mathbf{R}[k] = (\mathbf{A}_0 - \mathbf{A}_0) + (\mathbf{R}[k] - \mathbf{R}[k-1])\mathbf{A}_1 + (\mathbf{R}^2[k] - \mathbf{R}^2[k-1])\mathbf{A}_2$$
$$= (\mathbf{R}[k] - \mathbf{R}[k-1])\mathbf{A}_1 + (\mathbf{R}^2[k] - \mathbf{R}^2[k-1])\mathbf{A}_2$$
$$\geq \mathbf{0},$$

where the last inequality holds from the assumption. This completes the proof of (i). Since $\mathbf{R}$ exists and is finite, it must satisfy (4.99). Then $\mathbf{R}$ must be the limit of $\mathbf{R}[k]$ as $k \to \infty$. This is because when $\mathbf{R}[k]$ converges to a limit (either finite or infinite), this limit must also satisfy (4.99). We only focus on the finite limit case as infinity limit is meaningless. For the finite limit case, we must have $\lim_{k \to \infty} \mathbf{R}[k] = \mathbf{R}$. It follows from the increasing sequence of $\mathbf{R}[k]$ that $\mathbf{R}[k] \leq \mathbf{R}$ for $k = 0, 1, 2, \dots$. This also implies that this sequence will converge to $\mathbf{R}$. ∎

Next, we provide an algorithm for computing the stationary distribution of a discrete-time QBD as follows:

- Step 1: Input transition blocks $\{\mathbf{B}_{00}, \mathbf{B}_{01}, \mathbf{B}_{10}, \mathbf{B}_{11}, \mathbf{A}_0, \mathbf{A}_1, \mathbf{A}_2\}$.
- Step 2: Compute θ by solving $\theta(\mathbf{A}_0 + \mathbf{A}_1 + \mathbf{A}_2) = \theta$ and $\theta \mathbf{e} = 1$. Then check the ergodicity condition.
- Step 3: Set $k = 0$ and $\mathbf{R}[0] = 0$. Apply (4.101) to compute $\mathbf{R}$ recursively. Stop the iteration if $\|\mathbf{R}[k+1] - \mathbf{R}[k]\|_1 < \varepsilon$, where ε is a sufficiently small number and go to Step 4. (Note: $\|\mathbf{A}\|_1 = \max_{1 \leq j \leq n} (\sum_{i=1}^{n} |a_{ij}|)$ for matrix $\mathbf{A} = [a_{ij}]$.)
- Step 4: Solve (4.100) for π_0 and π_1.
- Step 5: Use $\{\mathbf{R}, \pi_0, \pi_1\}$ to compute limiting probabilities of the QBD process.

Exercise 4.30 *Consider a discrete-time QBD with the transition probability matrix in the form of (4.93) with the following block entries:*

$$\mathbf{B}_{00} = \begin{pmatrix} 0.3 & 0.2 \\ 0 & 0.5 \end{pmatrix}, \quad \mathbf{B}_{01} = \begin{pmatrix} 0.4 & 0.1 \\ 0.3 & 0.2 \end{pmatrix},$$

$$\mathbf{B}_{10} = \begin{pmatrix} 0.4 & 0.1 \\ 0.2 & 0.3 \end{pmatrix}, \quad \mathbf{B}_{11} = \begin{pmatrix} 0 & 0.2 \\ 0.2 & 0 \end{pmatrix},$$

$$\mathbf{A}_2 = \begin{pmatrix} 0.3 & 0.2 \\ 0.2 & 0.3 \end{pmatrix}, \quad \mathbf{A}_1 = \mathbf{B}_{11}, \quad \mathbf{A}_0 = \begin{pmatrix} 0.1 & 0.2 \\ 0.1 & 0.2 \end{pmatrix}.$$

(a) Check the stability condition (the QBD process is ergodic).
(b) If it is ergodic, find $\{\mathbf{R}, \pi_0, \pi_1\}$.

To interpret $\mathbf{R}$, we introduce the *taboo probability*.

Definition 4.2 *The taboo probability denoted by* $_iP^{(k)}_{(i,j);(i+n,v)}$*: It is the probability that given starting in state* (i,j)*, the discrete-time QBD process reaches state* $(i+n,v)$ *at time* k *without visiting the levels* $\{0,1,...,i\}$ *in between, for* $k \geq 0, i \geq 0, j = 1,...,m, n \geq 1,$ *and* $v = 1,...,m.$

Note that the taboo probability is a special kind of transition probability and the corresponding transition can be called "*taboo transition*". Define an indicator variable $_iI^{(k)}_{(i,j);(i+n,v)}$. That is $_iI^{(k)}_{(i,j);(i+n,v)} = 1$ if the discrete-time QBD visits $(i+n,v)$ at time k without visiting the levels $\{0,1,...,i\}$ in between, given that it starts in state (i,j); otherwise, $_iI^{(k)}_{(i,j);(i+n,v)} = 0$. Then $\sum_{k=0}^{\infty} {_iI^{(k)}_{(i,j);(i+n,v)}}$ is the total number of visits to $(i+n,v)$ before returning to the level i, given that the discrete-time QBD starts in state (i,j). Then we can show that $\mathbf{R}^{(n)}_{j,v} = E\left[\sum_{k=0}^{\infty} {_iI^{(k)}_{(i,j);(i+n,v)}}\right] = \sum_{k=0}^{\infty} E[{_iI^{(k)}_{(i,j);(i+n,v)}}]$

$= \sum_{k=0}^{\infty} {_iP^{(k)}_{(i,j);(i+n,v)}}$ is the expected number of visits to state $(i+n,v)$ before returning to the level i, given that the QBD process starts in state (i,j). With this interpretation of $\mathbf{R}^{(n)}$, we can obtain the following properties.

Proposition 4.2 *(i)* $\mathbf{R}^{(n)} = (\mathbf{R}^{(1)})^n$ *for* $n = 1,2,...;$ *(ii)* $\mathbf{R}^{(1)} = \mathbf{A}_0 + \mathbf{R}^{(1)}\mathbf{A}_1 + \mathbf{R}^{(2)}\mathbf{A}_2.$

Proof. (i) Decompose the taboo transitions from state (i,j) to $(i+n,v)$ into two parts: taboo transition from (i,j) to $(i+1,w)$, and from taboo transition from $(i+1,w)$ to $(i+n,v)$. Since the end of the first part t is the last visit to level $i+1$, by conditioning on t and the phase w, we have

$$_iP^{(k)}_{(i,j);(i+n,v)} = \sum_{t=1}^{k}\sum_{w=1}^{m} {_iP^{(t)}_{(i,j);(i+1,w)}} \cdot {_{i+1}P^{(k-t)}_{(i+1,w);(i+n,v)}}.$$

Summing both sides of the equation above yields $\mathbf{R}^{(n)} = \mathbf{R}^{(1)}\mathbf{R}^{(n-1)}$. It follows from induction that $\mathbf{R}^{(n)} = (\mathbf{R}^{(1)})^n$ holds.

(ii) Since the last state before visiting state $(i+1,v)$ can only be in the level i (from lower level), $i+1$ (from the same level), or $i+2$ (from the higher level), by conditioning on this last state without returning to the level i, we have

$$_iP^{(k)}_{(i,j);(i+1,v)} = \begin{cases} [\mathbf{A}_0]_{j,v}, & \text{if } k = 1, \\ \sum_{u=1}^{m}\left({_iP^{(k-1)}_{(i,j);(i+1,u)}}[\mathbf{A}_1]_{u,v} + {_iP^{(k-1)}_{(i,j);(i+2,u)}}[\mathbf{A}_2]_{u,v}\right), & \text{if } k \geq 2. \end{cases}$$

$$(4.102)$$

It follows from the definition of $\mathbf{R}^n$ that

$$\mathbf{R}_{j,v}^{(1)} = \sum_{k=1}^{\infty} {}_iP_{(ij);(i+1,v)}^{(k)}$$

$$= [\mathbf{A}_0]_{j,v} + \sum_{k=2}^{\infty} {}_iP_{(i,j);(i+1,v)}^{(k)}$$

$$= [\mathbf{A}_0]_{j,v} + \sum_{k=2}^{\infty} \left(\sum_{u=1}^{m} {}_iP_{(i,j);(i+1,u)}^{(k-1)} [\mathbf{A}_1]_{u,v} + \sum_{u=1}^{m} {}_iP_{(i,j);(i+2,u)}^{(k-1)} [\mathbf{A}_2]_{u,v} \right)$$

$$= [\mathbf{A}_0]_{j,v} + \sum_{u=1}^{m} \left(\sum_{k=2}^{\infty} {}_iP_{(i,j);(i+1,u)}^{(k-1)} \right) [\mathbf{A}_1]_{u,v} + \sum_{u=1}^{m} \left(\sum_{k=2}^{\infty} {}_iP_{(i,j);(i+2,u)}^{(k-1)} \right) [\mathbf{A}_2]_{u,v}$$

$$= [\mathbf{A}_0]_{j,v} + [\mathbf{R}^{(1)} \mathbf{A}_1]_{j,v} + [\mathbf{R}^{(2)} \mathbf{A}_2]_{j,v}.$$

This completes the proof. ∎

Using this proposition, we can show

Proposition 4.3 *The rate matrix to one level up* $\mathbf{R}^{(1)} = \mathbf{R}$, *where* $\mathbf{R}$ *is the minimal non-negative solution to equation (4.99).*

Proof. Since $\mathbf{R}$ is defined as the minimal non-negative solution to equation (4.99), it follows from Proposition 4.2 that $\mathbf{R}^{(1)} \geq \mathbf{R}$. Based on the definition of $R^{(1)}$, we have $\mathbf{R}^{(1)} \leq \mathbf{R}$. This completes the proof. ∎

Based on the discussion about $\mathbf{R}$ above, it is clear that the (i, j)th entry of $\mathbf{R}$ can be interpreted as the mean number of visits to $(n+1, j)$ before the QBD returns to levels n or lower, given that it starts in (n, i). This "visits to one level up" event can be easily remembered by interpreting the notation of rate matrix $\mathbf{R}$ as "**R**ising" in addition to **R**ate matrix. For example, we can interpret the equation (4.99) as follows: the visits to states in level $n+1$, starting from the current level n, can be classified into three mutually exclusive categories: visits immediately from level n, represented by $\mathbf{A}_0$; visits from level $n+1$, represented by $\mathbf{R}\mathbf{A}_1$ (going one step up by $\mathbf{R}$ and then remaining in the level by $\mathbf{A}_1$); and visits from level $n+2$, represented by $\mathbf{R}^2\mathbf{A}_2$ (going two steps up by $\mathbf{R}^2$ and then going one step down by $\mathbf{A}_2$).

Furthermore, using the definition of $\mathbf{R}$, we can prove

Proposition 4.4 *For* $n = 1, 2, ...$, $\pi_{n+1} = \pi_n \mathbf{R}$ *and* $\pi_n = \pi_1 \mathbf{R}^{n-1}$ *hold.*

Proof. It follows from the definition of the limiting probability and the law of large numbers, we have

$$\pi_{n+1,j} = \lim_{k \to \infty} \frac{\sum_{t=1}^{k} P_{(0,1);(n+1,j)}^{(t)}}{k},$$

with an arbitrary starting state $(0,1)$. By conditioning on the last time instant s $(1 \leq s \leq t-1)$ the QBD is in level n and the corresponding phase v (i.e., the last state in level n is $(X_s, J_s) = (n,v)$), we have

$$
\begin{aligned}
\pi_{n+1,j} &= \lim_{k \to \infty} \frac{\sum_{t=1}^{k} \left(\sum_{s=1}^{t-1} \sum_{v=1}^{m} P^{(s)}_{(0,1);(n,v)} {}_n P^{(t-s)}_{(n,v);(n+1,j)} \right)}{k} \\
&= \lim_{k \to \infty} \sum_{v=1}^{m} \frac{1}{k} \sum_{s=1}^{k} P^{(s)}_{(0,1);(n,v)} \sum_{t=1}^{k-s} {}_n P^{(t)}_{(n,v);(n+1,j)} \\
&= \sum_{v=1}^{m} \left(\lim_{k \to \infty} \frac{1}{k} \sum_{s=1}^{k} P^{(s)}_{(0,1);(n,v)} \right) \mathbf{R}_{v,j} \\
&= \sum_{v=1}^{m} \pi_{n,v} \mathbf{R}_{v,j},
\end{aligned}
$$

which is equivalent to $\pi_{n+1} = \pi_n \mathbf{R}$. This completes the proof. ■

Based on its probabilistic interpretation, it is clear that $\mathbf{R}$ provides the information about the ergodicity of the QBD process. Although some entries of $\mathbf{R}$ can be large numbers, the Perron–Frobenius eigenvalue of $\mathbf{R}$ must be between 0 and 1 to ensure the stability of the QBD process.

Proposition 4.5 *The rate matrix $\mathbf{R}$ is the unique non-negative solution to (4.99) such that $sp(R) < 1$ if the QBD process is ergodic. If the QBD process is nonergodic, then $sp(R) = 1$.*

The proof of this proposition is provided in Chapter 1 of Neuts (1981). This proposition indicates that (4.99) always has a non-negative solution with $sp(R) \leq 1$. Note that if the QBD is ergodic, then the matrix $\mathbf{I} - \mathbf{R}$ is invertible. With the matrix-geometric solution for the stationary distribution, we can obtain the marginal distributions, moments, and tail asymptotics for the QBD process. For example, the mean of the limiting (steady state) variable X_∞ is given by

$$
E[X_\infty] = \sum_{n=1}^{\infty} n\pi_n \mathbf{e} = \sum_{n=1}^{\infty} n\pi_1 \mathbf{R}^{n-1} \mathbf{e} = \pi_1 \sum_{n=1}^{\infty} n\mathbf{R}^{n-1} \mathbf{e} = \pi_1 (\mathbf{I} - \mathbf{R})^{-2} \mathbf{e}.
$$

Exercise 4.31 *Consider a discrete-time QBD process $\{(X_k, J_k), k = 0, 1, 2, ...\}$ with $m_0 = m$. Determine the limiting marginal distributions of the level and phase variables ($X_\infty = \lim_{k \to \infty} X_k$ and $J_\infty = \lim_{k \to \infty} J_k$), respectively, by applying the matrix-geometric solution.*

Exercise 4.32 *For an irreducible rate matrix $\mathbf{R}$ for a QBD process, let $\mathbf{v}$ and $\mathbf{u}$ be left and right Perron–Frobenius eigenvectors of $\mathbf{R}$, which are normalized* by $\mathbf{ve} = \mathbf{vu} = 1$. *Show that the limiting probabilities have a geometric decay with decay rate $sp(\mathbf{R}) < 1$. Thats is*

$$\pi_n = \pi_1 \mathbf{R}^{n-1} \approx (\pi_1 \mathbf{u}) \mathbf{v} (sp(\mathbf{R}))^{n-1} + o((sp(\mathbf{R}))^{n-1})$$

if n is large. (Hint: See Neuts (1986) for more information and Appendix 0.10.)

Exercise 4.33 *Consider an ergodic continuous-time QBD process $\{(X(t), J(t)), t \geq 0\}$ with infinitesimal generator $\mathbf{Q}$ with the tri-diagonal structure and $\mathbf{Qe} = \mathbf{0}$. Assume that $\mathbf{A}_0 + \mathbf{A}_1 + \mathbf{A}$ is irreducible.*
(a)Develop the drift condition for the stability of the QBD process.
(b)Show that the stationary distribution is a matrix-geometric solution.

4.7.1.2 Fundamental Period – G-Matrix

While $\mathbf{R}$-matrix is about the expected number of visits to states in one level up via taboo transitions, now we consider a $\mathbf{G}$-matrix with the (i,j)th entry representing the probability that the QBD process goes down to the level $n-1$ for the first time by entering state $(n-1, j)$, given the process starts in state (n, i). In other words, we want to investigate the evolution of the QBD process in downward direction. To do that we define the fundamental period.

Definition 4.3 *A fundamental period for a discrete-time QBD process, denoted by τ, is defined as a period that begins at an epoch the Markov chain is in a state at level n and ends at the first epoch the Markov chain is in a state at level $n-1$.*

According to the definition of $\mathbf{G}$ matrix, its entry g_{ij} is the probability of the following event

$$\{(X_\tau, J_\tau) = (n-1, j) | (X_0, J_0) = (n, i)\}$$
$$= \bigcup_{k=1}^{\infty} \{(X_k, J_k) = (n-1, j), X_t \geq n, t = 1, 2, ..., k-1 | (X_0, J_0) = (n, i)\}.$$

Since the right-hand side of the above is the union of countably infinite number of disjoint events, we have

$$g_{ij} = \sum_{i=1}^{\infty} P((X_k, J_k) = (n-1, j), X_t \geq n, t = 1, 2, ..., k-1 | (X_0, J_0) = (n, i)).$$

Clearly, g_{ij} is independent of the transitions related to levels $n-1$ and lower. It follows from the structure of (4.93) with $\mathbf{B}_{11} = \mathbf{A}_1$ that $\mathbf{G}$ is related to transitions of the following sub-matrix of $\mathbf{P}$

$$
\begin{pmatrix}
\mathbf{A}_1 & \mathbf{A}_0 & & & \\
\mathbf{A}_2 & \mathbf{A}_1 & \mathbf{A}_0 & & \\
& \mathbf{A}_2 & \mathbf{A}_1 & \mathbf{A}_0 & \\
& & \ddots & \ddots & \ddots \\
& & & \ddots & \ddots
\end{pmatrix},
$$

by eliminating the boundary state row and column. Note that we can consider the boundary states as absorbing ones in computing $\mathbf{G}$. Such a Markov chain is called sub-Markov chain since $(\mathbf{A}_1 + \mathbf{A}_0)\mathbf{e} < \mathbf{e}$. Due to the *QBD* structure, $\mathbf{G}$ is independent on level n (≥ 2). Similar to $\mathbf{R}$, we can establish the properties and computational algorithms for $\mathbf{G}$. Since the proofs are similar to and much simpler than the case of $\mathbf{R}$, we present them as exercises for readers.

Exercise 4.34 *Define matrix $\mathbf{G}^{(2)}$ as the two-level down first passage probability. That is its (i, j)th entry $g_{ij}^{(2)}$ is the probability that given that the QBD process starts in $(n+2, i)$, it goes down to level n for the first time by entering state (n, j). Then we have $\mathbf{G}^{(2)} = \mathbf{G}^2$. In general, $\mathbf{G}^{(n)} = \mathbf{G}^n$ for $n \geq 2$.*

Hint: To show this property in the exercise, one needs to (1) decompose the two-level down first passage time into sum of two one-level down first time passage times; (2) note that the first passage time from level $n+2$ to level $n+1$ is conditionally independent of and probabilistic equivalent to the first passage time from level $n+1$ to level n; and (3) condition on the time instant and the phase that the QBD process first reaches level $n+1$ and interchange the summation orders to obtain the results.

Using the property in Exercise 4.34 and by using the conditioning argument similar to the $\mathbf{R}$ case, we can show that $\mathbf{G}$ satisfy the following matrix quadratic equation (both entry-wise form and matrix form):

$$
g_{ij} = [\mathbf{A}_2]_{ij} + \sum_{k=1}^{m} [\mathbf{A}_1]_{ik} g_{kj} + \sum_{k=1}^{m} [\mathbf{A}_0]_{ik} \sum_{u=1}^{m} g_{ku} g_{uj}, \tag{4.103}
$$

$$
\mathbf{G} = \mathbf{A}_2 + \mathbf{A}_1 \mathbf{G} + \mathbf{A}_0 \mathbf{G}^2.
$$

Again, we can explain the equation (4.103) intuitively based on the definition of $\mathbf{G}$ as follows. The one-level down first passage probabilities can be decomposed into three mutually exclusive groups: (i) The probabilities that the QBD goes down to level $n-1$ for the first time directly (one-step), represented by $\mathbf{A}_2$. (ii) The probabilities that the QBD goes down to level $n-1$ in two-steps (i.e., two transitions) represented by $\mathbf{A}_1 \mathbf{G}$. The event in this group is that in the first step, the transition occurs in level n and then in the second step, the transition from step n to $n - 1$ occurs for the first step. (iii) The probabilities that the QBD goes down to level $n - 1$ for the first time in three-steps, represented by $\mathbf{A}_0\mathbf{G}^2$. The event in this group is that in the first step, the transition to level $n + 1$ occurs, then in step 2, the transition from level $n + 1$ to level n occurs for the first time, and finally in step 3, the transition from level n to level $n - 1$ for the first time occurs. Again, the event of "transitions to one level down for the first time" can be easily remembered by interpreting the notation of $\mathbf{G}$ as "Grounded". The fundamental difference between $\mathbf{R}$ and $\mathbf{G}$ is that the former one is regarding the expected excursion times of the QBD spent in the states at one level up before returning to the starting level and the latter one is regarding the probabilities of the first passage to the states one level down. By introducing another matrix $\mathbf{U}$, the relations between these matrices can be established. The $\mathbf{U}$ matrix is defined as follows:

Definition 4.4 *The* $\mathbf{U}$ *matrix is a matrix with the entry* u_{ij} *defined as the probability that the QBD process returns to level* n *by visiting state* (n, j) *without visiting any state in levels* $n - 1, n - 2, ..., 0, 1$ *in between, given that it starts in state* (n, i).

Like $\mathbf{G}$ matrix, $\mathbf{U}$ matrix is also a stochastic matrix. Since it is regarding the taboo probabilities of reaching states of the same level, the meaning can be easily remembered by the notation $\mathbf{U}$ for "Unchanged". With this matrix, we have the relations among these three matrices.

Proposition 4.6 *For the discrete-time QBD process, the followings hold: (i)* $\mathbf{R} = \mathbf{A}_0 + \mathbf{RU}$; *(ii)* $\mathbf{G} = \mathbf{A}_2 + \mathbf{UG}$; *(iii)* $\mathbf{U} = \mathbf{A}_1 + \mathbf{RA}_2 = \mathbf{A}_1 + \mathbf{A}_0\mathbf{G}$.

Proof. The proof can be done by using the conditioning argument. We only present the proof of (i) and leave the proofs of (ii) and (iii) as exercises. Based on the definition of $\mathbf{R}$, by conditioning on the first time to reach a state in one level up, s, and its corresponding phase, u, we have

$$R_{j,v}^{(1)} = \sum_{k=1}^{\infty} {}_i P_{(i,j);(i+1,v)}^{(k)} = [\mathbf{A}_0]_{j,v} + \sum_{k=2}^{\infty} {}_i P_{(i,j);(j+1,v)}^{(k)}$$

$$= [\mathbf{A}_0]_{j,v} + \sum_{k=2}^{\infty} \sum_{s=1}^{k-1} \sum_{u=1}^{\infty} {}_i P_{(i,j);(i+1,u)}^{(s)} P(X_k = i+1, J_k = v, X_t \geq i+2, t = s+1,$$

$$..., k-1 | X_s = i+1, J_s = u)$$

$$= [\mathbf{A}_0]_{j,v} + \sum_{u=1}^{m} R_{j,u}^{(1)} u_{u,v},$$

which is (i). ∎

The relations in Proposition 4.6 provide another way of computing $\mathbf{R}$. That is: first compute $\mathbf{G}$ and then compute $\mathbf{U}$ and $\mathbf{R}$ by using the relations between these matrices. The advantage of this alternative approach is that the computation of $\mathbf{G}$ is more stable since both the entries and sums of rows are upper bounded by 1. The properties and computational algorithms similar to those for $\mathbf{R}$ can be developed for $\mathbf{G}$. For example, $\mathbf{G}$ can be computed numerically as a limit of the sequence $\{\mathbf{G}[k], k = 0, 1, 2, ...\}$ determined by the following recursion:

$$\mathbf{G}[k+1] = \mathbf{A}_2 + \mathbf{A}_1 \mathbf{G}[k] + \mathbf{A}_0 \mathbf{G}[k]^2 \tag{4.104}$$

with $\mathbf{G}[0] = \mathbf{0}$ until $|\mathbf{G}[k+1] - \mathbf{G}[k]|_{ij} < \varepsilon$, where ε is a small error tolerance (convergent criterion). Other more efficient algorithms have been developed in the literature, Interested readers are referred to Bini and Meini, Alfa (2010)).

Exercise 4.35 *An alternative algorithm can be developed for computing the* $\mathbf{G}$ *matrix based on (4.103). By solving (4.103) for* $\mathbf{G}$ *in terms of* $\mathbf{A}_2, \mathbf{A}_0$, $(\mathbf{I} - \mathbf{A}_1)^{-1}$, *and* $\mathbf{G}$ *itself, we can define the following sequence:*

$$\mathbf{G}[k+1] = (\mathbf{I} - \mathbf{A}_1)^{-1} \mathbf{A}_2 + (\mathbf{I} - \mathbf{A}_1)^{-1} \mathbf{A}_0 \mathbf{G}[k]^2$$

for $k = 1, 2, ...,$ *with* $\mathbf{G}[0] = \mathbf{0}$. *Show that this sequence converges monotonically to* $\mathbf{G}$.

Exercise 4.36 *An important application of Proposition 4.6 is to develop the Wiener–Hopf factorization (also called RG-factorization) as follows:*

$$\mathbf{I} - (z^{-1}\mathbf{A}_2 + \mathbf{A}_1 + z\mathbf{A}_0) = (\mathbf{I} - z\mathbf{R})(\mathbf{I} - \mathbf{U})(\mathbf{I} - z^{-1}\mathbf{G}).$$

Show this relation in which the three factors on the right-hand side involve matrices $\mathbf{R}, \mathbf{U},$ *and* $\mathbf{G}$, *respectively.*

Note that The RG-factorization is useful in analyzing $G/G/1$ type Markov chain. For more details of utilizing RG-factorization, interested readers are referred to Li (2010).

Proposition 4.7 *The matrix G is the minimal non-negative solution to equation (4.103).*

The proof of this proposition is left as an exercise. Another property of $\mathbf{G}$ is shown in Neuts (1981) as follows:

Proposition 4.8 *If the QBD process is recurrent, $\mathbf{G}$ is stochastic. That is* $\mathbf{G}\mathbf{e} = \mathbf{e}$. *If the QBD is trnasient, then at least one row sum of $\mathbf{G}$ is less than one.*

For more details about designing stable algorithms for computing **G**, interested readers are referred to Latouche and Ramaswami (1993, 1999). As shown later in the next section, the **G** matrix plays an important role in $M/G/1$ type structured Markov chains as the **R** does in $GI/M/1$ type Markov chains. This section is concluded with an example of continuous-time QBD process.

Exercise 4.37 *Develop the algorithm for computing* **G** *matrix for a continuous-time QBD process with an irreducible infinitesimal generator* **Q**.

4.7.2 *MAP/PH/*1 **QUEUE – A CONTINUOUS-TIME QBD PROCESS**

As an application of PH distribution, MAP and QBD process, we present an analysis on MAP/PH/1 queue. Other variants of queueing systems with PH distributed time intervals or MAPs can mostly be treated as QBD processes (see He 2016 for more models of this class). Let $(\mathbf{D}_0, \mathbf{D}_1)$ be the parameter matrix of the MAP of order m_a with stationary mean arrival rate of $\lambda = \theta \mathbf{D}_1 \mathbf{e}$ where θ is the invariant probability vector of the generator $\mathbf{D} = \mathbf{D}_0 + \mathbf{D}_1$. That is $\theta \mathbf{D} = \mathbf{0}$ and $\theta \mathbf{1} = 1$. Let $(\beta, \mathbf{S})$ be the representation of the PH service time distribution with order of m_b with mean $\mu^{-1} = -\beta \mathbf{S}^{-1} \mathbf{e}$. Define the state variables as follows:

- $x(t)$: the number of customers in the system at time t (also called queue length).
- $I_a(t)$: the phase of the MAP at time t.
- $I_b(t)$: the phase of the service process at time t, if the server is busy; 0, otherwise.

Clearly, the process $\{[x(t), I_a(t), I_b(t)], t \geq 0\}$ is a CTMC with state space $\{\{0\} \times \{1, 2, ..., m_a\}\} \cup \{\{1, 2, ...\} \times \{1.2..., m_a\} \times \{1, 2, ..., m_b\}\}$. Since when the system is empty, there is no need to keep track of the phase of service time (i.e., $I_b(t) = 0$ when $x(t) = 0$), the boundary states are two-dimensional. Accordingly the generator can be written as

$$\mathbf{Q} = \begin{pmatrix} \mathbf{B}_1 & \mathbf{B}_0 & 0 & 0 & 0 & \cdots \\ \mathbf{B}_2 & \mathbf{A}_1 & \mathbf{A}_0 & 0 & 0 & \cdots \\ 0 & \mathbf{A}_2 & \mathbf{A}_1 & \mathbf{A}_0 & 0 & \cdots \\ 0 & 0 & \mathbf{A}_2 & \mathbf{A}_1 & \mathbf{A}_0 & \cdots \\ 0 & 0 & 0 & \mathbf{A}_2 & \mathbf{A}_1 & \cdots \\ \vdots & \vdots & \vdots & \vdots & \vdots & \ddots \end{pmatrix}, \tag{4.105}$$

where

$$\mathbf{B}_0 = \mathbf{D}_1 \otimes \beta, \quad \mathbf{B}_1 = \mathbf{D}_0, \quad \mathbf{B}_2 = \mathbf{I}_{m_a} \otimes \mathbf{S}^0,$$

$$\mathbf{A}_0 = \mathbf{D}_1 \otimes \mathbf{I}_{m_b}, \quad \mathbf{A}_1 = \mathbf{I}_{m_a} \otimes \mathbf{S} + \mathbf{D}_0 \otimes \mathbf{I}_{m_b}, \quad \mathbf{A}_2 = \mathbf{I}_{m_a} \otimes \mathbf{S}^0 \beta.$$

Here $\mathbf{I}_k$ is the order k identity matrix, $\mathbf{S}^0 = -\mathbf{S}\mathbf{e}$, and $\otimes$ denotes the Kroneker product. For non-boundary state levels, i.e. $x(t) \geq 2$, the transition of the two-dimensional phase process $\{(I_a(t), I_b(t), t \geq 0\}$ is governed by generator matrix of $\mathbf{A}_2 + \mathbf{A}_1 + \mathbf{A}_0 = (\mathbf{D}_0 + \mathbf{D}_1) \otimes \mathbf{I}_{m_b} + \mathbf{I}_{m_a} \otimes (\mathbf{S}^0 \beta + \mathbf{S})$. Due to the CTMC property, state-transition events cannot occur simultaneously. Hence, a state transition for a given level in the QBD is caused by either an arrival process related event (arrival rate $\mathbf{D}_1 \otimes \mathbf{I}_{m_b}$), or a service process related event (service completion rate $\mathbf{I}_{m_a} \otimes (\mathbf{S}^0 \beta)$), or the phase change only event (only phase change rate $\mathbf{D}_0 \otimes \mathbf{I}_{m_b} + \mathbf{I}_{m_a} \otimes \mathbf{S}$).

Assume that the QBD process to be irreducible and positive recurrent (i.e., $\lambda < \mu$), then the stationary distribution of $\{[x(t), I_a(t), I_b(t)], t \geq 0\}$ exists and can be defined as follows:

$$\pi = (\pi_0, \pi_1, ...) \quad \text{where}$$
$$\pi_0 = (\pi_{0,1}, ... \pi_{0,m_a}), \pi_n = (\pi_{n,1}, \pi_{n,2}, ... \pi_{n,m_a}), \quad \text{for } n = 1, 2, ... \text{ and}$$
$$\pi_{n,j} = (\pi_{n,j,1}, \pi_{n,j,2}, ... \pi_{n,j,m_b}), \quad \text{for } n = 1, 2, ... \ j = 1, 2, ..., m_a.$$
$$\pi_{n,j,i} = \lim_{t \to \infty} P(x(t) = n, I_a(t) = j, I_b(t) = i).$$

Such a distribution has a matrix geometric solution form as follows:

$$(\pi_0, \pi_1) \begin{pmatrix} \mathbf{B}_{00} & \mathbf{B}_{01} \\ \mathbf{B}_{10} & \mathbf{B}_{11} + \mathbf{R}\mathbf{A}_2 \end{pmatrix} = 0,$$
$$\pi_{k+1} = \pi_1 \mathbf{R}^k, \quad k \in \mathbb{Z}_+, \tag{4.106}$$
$$\pi_0 \mathbf{e} + \pi_1 (\mathbf{I} - \mathbf{R})^{-1} \mathbf{e} = 1,$$

where $\mathbf{R}$ is the minimal non-negative solution of the equation

$$\mathbf{R}^2 \mathbf{A}_2 + \mathbf{R}\mathbf{A}_1 + \mathbf{A}_0 = 0. \tag{4.107}$$

Using the stationary distribution π, we can compute major performance measures such as the tail distribution of queue length, denoted by X as

$$P(X > x) = \sum_{k=x+1}^{\infty} \pi_k \mathbf{e} = \pi_1 \mathbf{R}^x (\mathbf{I} - \mathbf{R})^{-1} \mathbf{e}, \quad x \geq 0, \tag{4.108}$$

and the expected queue length as

$$E(X) = \sum_{k=0}^{\infty} (k+1)\pi_{k+1} \mathbf{e} = \sum_{k=0}^{\infty} (k+1)\pi_1 \mathbf{R}^k \mathbf{e},$$
$$= \pi_1 \sum_{k=0}^{\infty} \mathbf{R}^k \mathbf{e} + \pi_1 \sum_{k=0}^{\infty} k \mathbf{R}^k \mathbf{e} = \pi_1 (\mathbf{I} - \mathbf{R})^{-1} \mathbf{e} + \pi_1 \mathbf{R}(\mathbf{I} - \mathbf{R})^{-2} \mathbf{e}. \tag{4.109}$$

It follows from the Little's law that the expected waiting time, denoted by $E(W)$, is given by

$$
\begin{aligned}
E(W) &= \lambda^{-1}\left(\pi_1(\mathbf{I}-\mathbf{R})^{-1}\mathbf{e}+\pi_1\mathbf{R}(\mathbf{I}-\mathbf{R})^{-2}\mathbf{e}\right)\\
&= (\theta\mathbf{D}_1\mathbf{e})^{-1}\left(\pi_1(\mathbf{I}-\mathbf{R})^{-1}\mathbf{e}+\pi_1\mathbf{R}(\mathbf{I}-\mathbf{R})^{-2}\mathbf{e}\right).
\end{aligned}
\tag{4.110}
$$

Let η be the spectral radius of matrix R, which is also called the caudal characteristic and provides the stability condition for the MAP/PH/1 queue. Furthermore, η is indicative of the tail behavior of the stationary queue length distribution. Let $\mathbf{u}$ and $\mathbf{v}$ be the left and right eigenvectors corresponding to η normalized by $\mathbf{ue}=1$ and $\mathbf{uv}=1$. Then it is well known that

$$
\mathbf{R}^x = \eta^x \mathbf{v}\cdot\mathbf{u} + o(\eta^x), \quad \text{as } x\to\infty,
\tag{4.111}
$$

With this asymptotic relation, we can express the tail probability as

$$
P(X>x) = \frac{\pi_1\mathbf{v}}{1-\eta}\eta^x + o(\eta^x), \quad \text{as } x\to\infty,
\tag{4.112}
$$

and thus

$$
\lim_{x\to\infty}\frac{P(X>x)}{\eta^x} = \frac{\pi_1\mathbf{v}}{1-\eta},
\tag{4.113}
$$

which implies

$$
P(X>x) \sim \frac{\pi_1\mathbf{v}}{1-\eta}\eta^x, \quad \text{as } x\to\infty.
\tag{4.114}
$$

Exercise 4.38 *For the opening example of the discrete-time queue-inventory model in Section 4.7.1, given a set of parameter values $p,q,l,s,$ and S, compute the expected number of customers waiting in the system and the average inventory level.*

4.8 $GI/M/1$ TYPE AND $M/G/1$ TYPE MARKOV CHAINS

In this section, we discuss two other types of structured Markov chains that generalize the QBD processes.

4.8.1 $GI/M/1$ TYPE MARKOV CHAINS

The discrete-time QBD process can be extended to another class of structured Markov chains with the same state space if the transition probability matrix has the following structure

$$
\mathbf{P} = \begin{pmatrix}
\mathbf{B}_0 & \mathbf{A}_{01} & & & \\
\mathbf{B}_1 & \mathbf{A}_{11} & \mathbf{A}_0 & & \\
\mathbf{B}_2 & \mathbf{A}_2 & \mathbf{A}_1 & \mathbf{A}_0 & \\
\mathbf{B}_3 & \mathbf{A}_3 & \mathbf{A}_2 & \mathbf{A}_1 & \mathbf{A}_0 \\
\vdots & \vdots & \ddots & \ddots & \ddots & \ddots
\end{pmatrix},
\tag{4.115}
$$

where the matrix entries satisfy

$$\mathbf{B}_0\mathbf{e} + \mathbf{A}_{01}\mathbf{e} = \mathbf{e}$$
$$\mathbf{B}_1\mathbf{e} + (\mathbf{A}_{11} + \mathbf{A}_0)\mathbf{e} = \mathbf{e}$$
$$\mathbf{B}_n\mathbf{e} + (\mathbf{A}_n + \cdots + \mathbf{A}_0)\mathbf{e} = \mathbf{e}, \quad n \geq 2.$$

The feature of such a process is "skip-free to the right" as the DTMC can only go up by at most one level. Denote $\mathbf{A} = \sum_{n=0}^{\infty} \mathbf{A}_n$, which is a stochastic/substochastic matrix. The name "$GI/M/1$ type" for this class of Markov chains is due to the fact that the transition probability matrix of the queue length process embedded at arrival instants for the single server queue with i.i.d. inter-arrival times (renewal process) and exponential service times, denoted by $GI/M/1$ queue, has the same structure as in (4.115). Renewal processes and embedded Markov chains are discussed in Chapter 5 of this book.

Fortunately, almost all results obtained from the QBD process can be applied to the $GI/M/1$ type Markov chains with some modifications. Thus, we present the main results. First, we specify the mean-drift condition for the stability of the $GI/M/1$ type Markov chains. That is

$$\theta \mathbf{A}_0 \mathbf{e} < \theta \sum_{k=1}^{\infty} (k-1)\mathbf{A}_k \mathbf{e},$$

where θ satisfies $\theta = \theta \sum_{k=0}^{\infty} \mathbf{A}_k$ and $\theta \mathbf{e} = 1$. Note that the right-hand side of this condition is the average service rate and the left-hand side is the average arrival rate.

Exercise 4.39 *Verify that the mean-drift condition can be re-written as* $\theta \sum_{k=1}^{\infty} k\mathbf{A}_k\mathbf{e} > 1$.

An intuitive result presented in Neuts (1981) is presented below.

Theorem 4.10 *The $GI/M/1$ type Markov chain with an irreducible and stochastic $\mathbf{A}$ is ergodic if and only if* $\theta \sum_{k=1}^{\infty} k\mathbf{A}_k\mathbf{e} > 1$.

Under the stability condition, we can determine the limiting probabilities $\pi = (\pi_0, \pi_1, \dots)$ by solving $\pi\mathbf{P} = \pi$ and $\pi\mathbf{e} = 1$. This linear system can be written as

$$\pi_0 = \sum_{j=0}^{\infty} \pi_j \mathbf{B}_j$$

$$\pi_1 = \pi_0 \mathbf{A}_{01} + \pi_1 \mathbf{A}_{11} + \sum_{j=2}^{\infty} \pi_j \mathbf{A}_j \qquad (4.116)$$

$$\pi_n = \sum_{j=0}^{\infty} \pi_{n-1+j} \mathbf{A}_j, \quad n = 2, 3, \dots.$$

Following the same approach, it can be shown that the limiting probabilities still have the matrix-geometric solution form (Neuts (1981)).

Theorem 4.11 *Suppose that a DTMC $\{(X_k, J_k), k = 0, 1, 2, ...\}$ with the transition probability matrix (4.115) is irreducible and positive recurrent. The limiting probability distribution is given by*

$$\pi_n = \pi_1 \mathbf{R}^{n-1}, \quad n = 1, 2, ..., \tag{4.117}$$

where the rate matrix $\mathbf{R}$ is the minimal non-negative solution to the nonlinear matrix equation

$$\mathbf{R} = \sum_{n=0}^{\infty} \mathbf{R}^n \mathbf{A}_n, \tag{4.118}$$

and (π_0, π_0) is the unique solution to

$$(\pi_0, \pi_1) = (\pi_0, \pi_1) \begin{pmatrix} \mathbf{B}_0 & \mathbf{A}_{01} \\ \sum_{n=1}^{\infty} \mathbf{R}^{n-1} \mathbf{B}_n & \mathbf{A}_{11} + \sum_{n=2}^{\infty} \mathbf{R}^{n-1} \mathbf{A}_n \end{pmatrix}, \tag{4.119}$$

$$1 = \pi_0 \mathbf{e} + \pi_1 (\mathbf{I} - \mathbf{R})^{-1} \mathbf{e}.$$

The proofs and detailed derivations are omitted and can be left as exercises. Interested readers are referred to Neuts (1981) for details for this type of DTMCs and corresponding CTMCs.

Compared with the QBD processes, the key difference noted in the $GI/M/1$ type Markov chains is that $\mathbf{R}$ matrix is obtained by solving the nonlinear matrix equation involved a sum of infinite number of terms of form $\mathbf{R}^n \mathbf{A}_n$ rather than a quadratic matrix equation. However, the solution to this nonlinear equation remains to be in the same matrix-geometric form as in the QBD processes. The main reason for having this specific nonlinear equation to retain the matrix-geometric solution for the $GI/M/1$ type Markov chains is due to the special structure of the transition probability matrix (4.115). That is starting from column 3, every column has the same structure of non-zero matrix entries (i.e., $\mathbf{A}_0, \mathbf{A}_1, ...$) after an increasing but finite number of zero blocks. Furthermore, due to the "skip-free to the right" characteristic, the limiting probabilities at level n are related on probabilities in one level below, current, and higher levels.

Again, we can use iterative algorithms to compute $\mathbf{R}$ from the nonlinear equation (4.118). The simplest but not necessarily efficient one is based on the following recursion. Starting with $\mathbf{R}(0) = \mathbf{A}_0$, we can compute

$$\mathbf{R}[k+1] = \sum_{n=0}^{\infty} (\mathbf{R}[k])^n \mathbf{A}_n \tag{4.120}$$

for $k = 0, 1, 2, ...$ until $|\mathbf{R}[k+1] - \mathbf{R}[k]|_{ij} < \varepsilon$, where ε is a small error tolerance (convergent criterion). Other more efficient algorithms have been developed in the literature on matrix analytic methods. Interested readers are referred to Bini and Meini, Alfa (2010)).

4.8.2 $M/G/1$ **TYPE MARKOV CHAINS**

Now we consider a class of structured Markov chains $\{(X_k, J_k), k = 9, 1, ...\}$ with the state space $\{(0,1),(0,2),...,(0,m_0)\} \cup \{(1,2,...\} \times \{1,2,...,m\}\}$, where m_0 and m are positive integers, and transition probability matrix

$$
\mathbf{P} = \begin{pmatrix}
\mathbf{A}_{00} & \mathbf{A}_{01} & \mathbf{A}_{02} & \mathbf{A}_{03} & \cdots \\
\mathbf{A}_{10} & \mathbf{A}_1 & \mathbf{A}_2 & \mathbf{A}_3 & \cdots \\
 & \mathbf{A}_0 & \mathbf{A}_1 & \mathbf{A}_2 & \cdots \\
 & & \ddots & \ddots & \ddots \\
 & & & \ddots & \ddots
\end{pmatrix}.
\tag{4.121}
$$

As the stochastic matrix, the matrix entries of (4.121) satisfy the following conditions:

$$\mathbf{A}_{00}\mathbf{e} + (\mathbf{A}_{01} + \mathbf{A}_{02} + \cdots)\mathbf{e} = \mathbf{e},$$

$$\mathbf{A}_{10}\mathbf{e} + (\mathbf{A}_1 + \mathbf{A}_2 + \cdots)\mathbf{e} = \mathbf{e},$$

$$(\mathbf{A}_0 + \mathbf{A}_1 + \mathbf{A}_2 + \cdots)\mathbf{e} = \mathbf{e}.$$

Denote the stochastic matrix by $\mathbf{A} = \sum_{n=0}^{\infty} \mathbf{A}_n$.

Again, the name "$M/G/1$ type" for this class of Markov chains is due to the fact that the transition probability matrix of the queue length process embedded at departure instants for the single server queue with Poisson arrivals and general i.i.d. service times, denoted by $M/G/1$ queue, has the same structure as in (4.121). This class of Markov chains can be called "skip-free to the left" processes. The embedded Markov chains are discussed in the next chapter. The condition for the ergodicity based on the mean-drift method can be established as in the $GI/M/1$ type Markov chain and is stated in the following theorem (Neuts (1981)).

Theorem 4.12 *Suppose that an $M/G/1$ type Markov chain is irreducible and $\sum_{k=0}^{\infty} k\mathbf{A}_{0k}\mathbf{e}$ is finite componentwise. Then it is ergodic if and only if $\theta \sum_{k=0}^{\infty} k\mathbf{A}_k\mathbf{e} < 1$, where θ satisfies $\theta = \theta \sum_{k=0}^{\infty} \mathbf{A}_k$ and $\theta\mathbf{e} = 1$.*

Next, we develop the algorithm for computing the limiting probabilities $\pi = (\pi_0, \pi_1, ...)$. Using the transition probability matrix (4.121), $\pi = \pi\mathbf{P}$ can be written as

$$
\begin{aligned}
\pi_0 &= \pi_0 \mathbf{A}_{00} + \pi_1 \mathbf{A}_{10}, \\
\pi_n &= \pi_0 \mathbf{A}_{0n} + \sum_{j=1}^{n+1} \pi_j \mathbf{A}_{n+1-j}, \quad n = 1, 2, ...,
\end{aligned}
\tag{4.122}
$$

which will be solved along with the normalization condition $\pi\mathbf{e} = 1$. Compared linear equation system (4.122) with (4.116), we notice that the limiting probabilities in level n for the $M/G/1$ type Markov chain depend on limiting probabilities for all levels from 0 to $n+1$ while the limiting probabilities in level n for the $GI/M/1$ type Markov chain depend on only limiting probabilities in levels $n-1$ and higher. Such a difference implies that there is no explicit form solution to (4.122). However, we can make use of matrix $\mathbf{G}$ and fundamental period introduced earlier to develop the recursive algorithm for computing the limiting probabilities for the $M/G/1$ type Markov chains. First, we extend the $\mathbf{G}$ matrix to the one that includes the length of the fundamental period and is denoted by $\mathbf{G}(k)$ with the entry $g_{ij}(k)$ as

$$g_{ij}(k) = P((X_k, J_k) = (n-1, j), X_t \geq n, t = 1, ..., k-1 | (X_0, J_0) = (n, i)).$$
(4.123)

In words, $g_{ij}(k)$ is the probability that the Markov chain reaches level $n-1$ for the first time by visiting state $(n-1, j)$ in step k, given that it starts in state (n, i). Define the matrix of moment generating function (z-transform) $\mathbf{G}^*(z) = [g_{ij}^*(z)]$ with respect to the length of the fundamental period, where $g_{ij}^*(z) = \sum_{k=1}^{\infty} z^k g_{ij}(k)$. Note that similar definition can be made for $\mathbf{R}$ for the $GI/M/1$ type Markov chains (Ramaswami (1980)). Clearly, we have $\mathbf{G} = \mathbf{G}^*(1)$. Due to the structure of $\mathbf{P}$ in (4.121), for $n \geq 1$, the first passage time from level $n+1$ to level n is probabilistically identical. Thus, we can define the transition probability matrix for k-step "multi-level fundamental period" $\mathbf{G}^{(s)}(k)$. The entry $g_{ij}^{(s)}$ is the probability that given starting in state $(n+s, i)$, the $M/G/1$ type Markov chain goes down to level n for the first time by visiting state (n, j) after exactly k steps (transitions). The z-transform can be defined as $\mathbf{G}^{*(s)}(z) = \sum_{k=1}^{\infty} z^k \mathbf{G}^{(s)}(k)$, where $0 \leq z \leq 1$.

Exercise 4.40 *Show the relation* $\mathbf{G}^{*(s)}(z) = (\mathbf{G}^*(z))^s$ *for* $s = 1, 2, ...$

Some fundamental equations regarding $\mathbf{G}^*(z)$ and $\mathbf{G}$ can be established and they are useful in developing computational algorithms for the limiting probabilities of the $M/G/1$ type Markov chains.

Proposition 4.9 *For the $M/G/1$ type Markov chain, the following relations hold*

$$\mathbf{G}^*(z) = z \sum_{n=0}^{\infty} \mathbf{A}_n (\mathbf{G}^*(z))^n,$$

$$\mathbf{G} = \sum_{n=0}^{\infty} \mathbf{A}_n \mathbf{G}^n.$$
(4.124)

Proof. Using the relation in Exercise (4.40) and conditioning on the state of the next transition (i.e., level and phase), we have

$$
g_{ij}^*(z) = \sum_{k=1}^{\infty} z^k g_{ij}(k)
$$

$$
= z[\mathbf{A}_0]_{ij} + \sum_{k=1}^{\infty} z^k \sum_{n=1}^{\infty} \sum_{v=1}^{m} [\mathbf{A}_n]_{iv} g_{vj}^{(n)}(k-1)
$$

$$
= z[\mathbf{A}_0]_{ij} + z \sum_{n=1}^{\infty} \left(\sum_{v=1}^{m} [\mathbf{A}_n]_{iv} \sum_{k=2}^{\infty} z^{k-1} g_{vj}^{(n)}(k-1) \right)
$$

$$
= z[\mathbf{A}_0]_{ij} + z \sum_{n=1}^{\infty} \left(\sum_{v=1}^{m} [\mathbf{A}_n]_{iv} g_{vj}^{*(n)}(z) \right)
$$

$$
= z[\mathbf{A}_0]_{ij} + z \sum_{n=1}^{\infty} [\mathbf{A}_n (\mathbf{G}^*(z))^n]_{ij},
$$

which leads to $\mathbf{G}^*(z) = z \sum_{n=0}^{\infty} \mathbf{A}_n (\mathbf{G}^*(z))^n$. Setting $z = 1$ yields the second equation in (4.124). ∎

Furthermore, the following property holds for $\mathbf{G}$ matrix.

Proposition 4.10 $\mathbf{G}$ *is the minimal non-negative solution to the second equation in (4.124). If the $M/G/1$ type Markov chain is ergodic, then $\mathbf{Ge} = \mathbf{e}$ (i.e., $sp(\mathbf{G}) = 1$). Otherwise, at least one row sum of $\mathbf{G}$ is less than 1.*

From (4.124), we can obtain the conditional mean numbers of transitions in a fundamental period.

Proposition 4.11 *Suppose that the $M/G/1$ type Markov chain is ergodic and $\mathbf{G}$ is irreducible. Then the conditional mean numbers of transitions in a fundamental period are given by the entries of the following column vector*

$$
\left. \frac{d\mathbf{G}^*(z)}{dz} \right|_{z=1} \mathbf{e} = (\mathbf{I} - \mathbf{G} + \mathbf{eg}) \left[\mathbf{I} - \mathbf{A} + \mathbf{eg} - \left(\sum_{n=0}^{\infty} n\mathbf{A}_n \right) \mathbf{eg} \right]^{-1} \mathbf{e}, \qquad (4.125)
$$

where $\mathbf{e}$ is actually a right eigenvector of $\mathbf{G}$ with eigenvalue 1 (i.e., $\mathbf{Ge} = \mathbf{e}$), and $\mathbf{g}$ is the invariant measure of $\mathbf{G}$, a left eigenvector with eigenvalue 1 (i.e., $\mathbf{Ge} = \mathbf{g}$ and $\mathbf{ge} = 1$).

Proof. Taking the first-order derivative of (4.124) with respect to z, we obtain

$$
\frac{d\mathbf{G}^*(z)}{dz} = \sum_{n=0}^{\infty} \mathbf{A}_n \left((\mathbf{G}^*(z))^n + z \sum_{k=0}^{n-1} (\mathbf{G}^*(z))^k \frac{d\mathbf{G}^*(z)}{dz} (\mathbf{G}^*(z))^{n-k-1} \right).
$$

$$
(4.126)
$$

Multiplying both sides of the equation above from right by $\mathbf{e}$, setting $z = 1$, and using the fact $\mathbf{Ge} = \mathbf{e}$ and $\mathbf{Ae} = \mathbf{e}$, we have

$$\left.\frac{d\mathbf{G}^*(z)}{dz}\right|_{z=1}\mathbf{e} = \mathbf{e} + \left[\sum_{n=0}^{\infty}\mathbf{A}_n\sum_{k=0}^{n-1}(\mathbf{G}^*(1))^k\right]\left.\frac{d\mathbf{G}^*(z)}{dz}\right|_{z=1}\mathbf{e}, \qquad (4.127)$$

which can be written as

$$\left[\mathbf{I} - \sum_{n=0}^{\infty}\mathbf{A}_n\sum_{k=0}^{n-1}\mathbf{G}^k\right]\left.\frac{d\mathbf{G}^*(z)}{dz}\right|_{z=1}\mathbf{e} = \mathbf{e}. \qquad (4.128)$$

To find the inverse of $\left[\mathbf{I} - \sum_{n=0}^{\infty}\mathbf{A}_n\sum_{k=0}^{n-1}\mathbf{G}^k\right]$, right-multiplying this matrix by $\mathbf{I} - \mathbf{G} + \mathbf{eg}$ yields

$$\left[\mathbf{I} - \sum_{n=0}^{\infty}\mathbf{A}_n\sum_{k=0}^{n-1}\mathbf{G}^k\right](\mathbf{I} - \mathbf{G} + \mathbf{eg})$$

$$= \mathbf{I} - \mathbf{G} + \mathbf{eg} - \sum_{n=0}^{\infty}\mathbf{A}_n\left(\sum_{k=0}^{n-1}\mathbf{G}^k - \sum_{k=0}^{n-1}\mathbf{G}^{k+1} + \sum_{k=0}^{n-1}\mathbf{G}^k\mathbf{eg}\right)$$

$$= \mathbf{I} - \mathbf{G} + \mathbf{eg} - \sum_{n=0}^{\infty}\mathbf{A}_n(\mathbf{I} - \mathbf{G}^n + n\mathbf{eg})$$

$$= \mathbf{I} - \mathbf{G} + \mathbf{eg} - \mathbf{A} + \mathbf{G} - \sum_{n=0}^{\infty}n\mathbf{A}_n\mathbf{eg}$$

$$= \mathbf{I} - \mathbf{A} + \mathbf{eg} - \left(\sum_{n=0}^{\infty}n\mathbf{A}_n\right)\mathbf{eg}.$$

It can be shown that both $\mathbf{I} - \mathbf{G} + \mathbf{eg}$ and $\mathbf{I} - \mathbf{A} + \mathbf{eg} - (\sum_{n=0}^{\infty}n\mathbf{A}_n)\mathbf{eg}$ are invertible (left as an exercise). Then pre-multiplying both sides of the equation above by $\left[\mathbf{I} - \sum_{n=0}^{\infty}\mathbf{A}_n\sum_{k=0}^{n-1}\mathbf{G}^k\right]^{-1}$ (also invertible due to the invertibility of the two matrices above) yields

$$\mathbf{I} - \mathbf{G} + \mathbf{eg} = \left[\mathbf{I} - \sum_{n=0}^{\infty}\mathbf{A}_n\sum_{k=0}^{n-1}\mathbf{G}^k\right]^{-1}\left[\mathbf{I} - \mathbf{A} + \mathbf{eg} - \left(\sum_{n=0}^{\infty}n\mathbf{A}_n\right)\mathbf{eg}\right]$$

which results in

$$\left[\mathbf{I} - \sum_{n=0}^{\infty}\mathbf{A}_n\sum_{k=0}^{n-1}\mathbf{G}^k\right]^{-1} = (\mathbf{I} - \mathbf{G} + \mathbf{eg})\left[\mathbf{I} - \mathbf{A} + \mathbf{eg} - \left(\sum_{n=0}^{\infty}n\mathbf{A}_n\right)\mathbf{eg}\right]^{-1}. $$
$$(4.129)$$

Using (4.129) in (4.128) completes the proof. ∎

By some tedious calculations, we can also obtain higher moments of the fundamental period from $\mathbf{G}^*(z)$.

Exercise 4.41 *Prove that matrices* $\mathbf{I} - \mathbf{G} + \mathbf{eg}$ *and* $\mathbf{I} - \mathbf{A} + \mathbf{eg} - \left(\sum_{n=0}^{\infty} n\mathbf{A}_n\right)\mathbf{eg}$ *are invertible*

It is interesting to see the similarity of the nonlinear equation satisfied by $\mathbf{R}$ in $GI/M/1$ type Markov Chains, $\mathbf{R} = \sum_{n=0}^{\infty} \mathbf{R}^n \mathbf{A}_n$, and the nonlinear equation satisfied by $\mathbf{G}$ in $M/G/1$ type Markov chains, $\mathbf{G} = \sum_{n=0}^{\infty} \mathbf{A}_n \mathbf{G}^n$.

Like in the QBD process, the matrix $\mathbf{G}$ can be computed recursively. Starting with $\mathbf{G}[0] = 0$, we can compute

$$\mathbf{G}[k+1] = \sum_{n=0}^{\infty} \mathbf{A}_n (\mathbf{G}[k])^n \tag{4.130}$$

for $k = 0, 1, 2, \dots$ until $|\mathbf{G}[k+1] - \mathbf{G}[k]|_{ij} < \varepsilon$, where ε is a small error tolerance (convergent criterion). Other more efficient algorithms have been developed in the literature on matrix analytic methods. Interested readers are referred to Alfa (2010).

To develop the algorithm for computing the limiting probabilities for the $M/G/1$ type Markov chain, we need to apply the censoring technique which can truncate Markov chains with infinite number of states into ones with finite number of states. Using this technique, a finite state Markov chain with $n + 1$ levels, called a censored process of the $M/G/1$ type Markov chain with $\mathbf{P}$ in (4.121), can be formulated with the following transition probability matrix

$$\mathbf{P}_{\leq n} = \begin{pmatrix} \mathbf{A}_{00} & \mathbf{A}_{01} & \mathbf{A}_{02} & \mathbf{A}_{03} & \cdots & \mathbf{A}_{0,n-1} & \sum_{k=0}^{\infty} \mathbf{A}_{0,n+k} \mathbf{G}^k \\ \mathbf{A}_{10} & \mathbf{A}_1 & \mathbf{A}_2 & \mathbf{A}_3 & \cdots & \mathbf{A}_{n-1} & \sum_{k=0}^{\infty} \mathbf{A}_{n+k} \mathbf{G}^k \\ & \mathbf{A}_0 & \mathbf{A}_1 & \mathbf{A}_2 & \cdots & \mathbf{A}_{n-2} & \sum_{k=0}^{\infty} \mathbf{A}_{n-1+k} \mathbf{G}^k \\ & & \ddots & \ddots & \ddots & \vdots & \vdots \\ & & & \ddots & \ddots & \vdots & \vdots \\ & & & & \mathbf{A}_0 & \mathbf{A}_1 & \sum_{k=0}^{\infty} \mathbf{A}_{2+k} \mathbf{G}^k \\ & & & & & \mathbf{A}_0 & \sum_{k=0}^{\infty} \mathbf{A}_{1+k} \mathbf{G}^k \end{pmatrix} \tag{4.131}$$

by eliminating the rows and columns representing states in levels above n and putting the probabilities from the states at each level $0, 1, 2, \dots, n$ to the states at level n after an excursion into states at level $n + 1$ and higher. That is, the transition probabilities from states in level i to states in level n are contained in the matrix entry $\sum_{k=0}^{\infty} \mathbf{A}_{n-(i-1)+k} \mathbf{G}^k$ with $i = 1, 2, \dots n$. Note that $\mathbf{A}_{n-(i-1)+k}$ contains the transition probabilities from states in level i to states in level k above n (i.e., level $n + k$) and $\mathbf{G}^k$ gives the first passage transition probabilities from the states in level k above n back to the states in level n with $k = 0, 1, 2, \dots$. Clearly, the length of transition time from states in level i to the states in level n in this censored process is different from that to the states in level $n - 1$ or lower. Thus, this censored process can be considered as the Markov chain embedded at epochs of visits to the set of states $\{(i,v); 0 \leq i \leq n, 1 \leq v \leq m\}$ that also belongs to a class called Markov renewal process to be discussed in the next Chapter. For notational simplicity, denote

$$\hat{\mathbf{A}}_{0,n} = \sum_{k=0}^{\infty} \hat{\mathbf{A}}_{0,n+k} \mathbf{G}^k, \quad \hat{\mathbf{A}}_n = \sum_{k=0}^{\infty} \hat{\mathbf{A}}_{n+k} \mathbf{G}^k, \quad n = 1, 2, \dots.$$

With this censored Markov chain, we can develop a stable recursive algorithm for computing the limiting probabilities π for the $M/G/1$ type Markov chain (Ramwasmi (1988)).

Theorem 4.13 *Suppose that the $M/G/1$ type Markov chain with transition probability $\mathbf{P}$ of (4.121) is ergodic. Then the limiting probabilities are determined by*

$$\pi_n = \left(\pi_0 \hat{\mathbf{A}}_{0,n} + \sum_{k=1}^{n-1} \pi_k \hat{\mathbf{A}}_{n+1-k} \right) (\mathbf{I} - \hat{\mathbf{A}}_1)^{-1}, \quad n = 2, 3, \dots,$$

$$(4.132)$$

$$(\pi_0, \pi_1) = (\pi_0, \pi_1) \left(\begin{array}{cc} \mathbf{A}_{00} & \sum_{k=0}^{\infty} \mathbf{A}_{0,1+k} \mathbf{G}^k \\ \mathbf{A}_{10} & \sum_{k=0}^{\infty} \mathbf{A}_{1+k} \mathbf{G}^k \end{array} \right)$$

and the normalization condition $\pi \mathbf{e} = 1$.

Proof. Based on the theory of the Markov renewal processes (Chapter 5), the limiting probability vector $\pi_{\leq n} = (\pi_0, \pi_1, \dots, \pi_n)$ for the censored Markov chain with $\mathbf{P}_{\leq n}$ of (4.131) is an eigenvector of $\mathbf{P}_{\leq n}$ corresponding to eigenvalue 1. That is $\pi_{\leq n} = \pi_{\leq n} \mathbf{P}_{\leq n}$. For level n states, we have

$$\pi_n = \pi_0 \hat{\mathbf{A}}_{0,n} + \sum_{k=1}^{n} \pi_k \hat{\mathbf{A}}_{n+1-k}, \quad (4.133)$$

which can be written as

$$\pi_n (\mathbf{I} - \hat{\mathbf{A}}_1) = \pi_0 \hat{\mathbf{A}}_{0,n} + \sum_{k=1}^{n-1} \pi_k \hat{\mathbf{A}}_{n+1-k}. \quad (4.134)$$

For the ergodic $M/G/1$ type Markov chain, we know that $(\mathbf{I} - \hat{\mathbf{A}}_1)$ is invertible. Then, post-multiplying both sides of (4.134) by $(\mathbf{I} - \hat{\mathbf{A}}_1)^{-1}$ yields the first relation of (4.132). Next, we consider the censored process of $n = 1$ with

$$\mathbf{P}_{\leq 1} = \left(\begin{array}{cc} \mathbf{A}_{00} & \sum_{k=0}^{\infty} \mathbf{A}_{0,1+k} \mathbf{G}^k \\ \mathbf{A}_{10} & \sum_{k=0}^{\infty} \mathbf{A}_{0,1+k} \mathbf{G}^k \end{array} \right). \quad (4.135)$$

Then, it follows from $\pi_{\leq 1} = \pi_{\leq 1} \mathbf{P}_{\leq 1}$ that the second relation in (4.132) holds.

∎

Note that unlike the $GI/M/1$ type Markov chains or QBD processes, there is no matrix-geometric solution form for the limiting distribution of the $M/G/1$ type Markov chain. However, using the $\mathbf{G}$ matrix and the relations (4.132), some recursive computational algorithms can be developed. In fact, there are plenty of theoretical results on $M/G/1$ type Markov chains in the literature. Interested readers are referred to Neuts (1989) and Alfa (2010).

4.8.3 CONTINUOUS-TIME COUNTERPART OF DISCRETE-TIME QBD PROCESS

Although we have presented the theoretical results by focusing on the discrete-time structured Markov chains, the similar results can be obtained for continuous structured Markov chains in parallel for continuous structured Markov chains by using the same approach. The details of deriving the results have been put as exercises and examples throughout this chapter. A connection between these two classes of models is the uniformization technique introduced in Chapter 3. Here we summarize the changes needed to make to obtain the corresponding results for the continuous-time counterparts from the results for the discrete-time structured Markov chains.

- The transition probability matrix $\mathbf{P}$ is replaced by the infinitesimal generator matrix denoted by $\mathbf{Q}$ (rate matrix). Accordingly, $\mathbf{Pe} = \mathbf{1}$ is replaced by $\mathbf{Qe} = \mathbf{0}$. The limiting probability equations $\pi\mathbf{P} = \pi, \pi\mathbf{e} = 1$ are replaced by $\pi\mathbf{Q} = \mathbf{0}, \pi\mathbf{e} = 1$.
- The $\mathbf{Ae} = \sum_{n=0}^{\infty} \mathbf{A}_n\mathbf{e} = \mathbf{e}$ in the $GI/M/1$ type Markov chain is replaced by $\mathbf{Ae} = \sum_{n=0}^{\infty} \mathbf{A}_n\mathbf{e} = \mathbf{0}$; and $\theta\mathbf{A} = \theta, \theta\mathbf{e} = 1$ is replaced by $\theta\mathbf{A} = \mathbf{0}, \theta\mathbf{e} = 1$.
- The nonlinear rate matrix equation $\mathbf{R} = \sum_{n=0}^{\infty} \mathbf{R}^n\mathbf{A}_n$ in the $GI/M/1$ type Markov chain is replaced by $\mathbf{0} = \sum_{n=0}^{\infty} \mathbf{R}^n\mathbf{A}_n$.

In this chapter, we have only presented the basic fixed point iteration algorithms for computing matrices $\mathbf{R}$ and $\mathbf{G}$. This is the simplest successive substitution approach by expressing $\mathbf{R}$ from the nonlinear equation it satisfies. For the discrete-time case, the nonlinear equation can be used directly. For example, we can get $\mathbf{R}[k+1] = \sum_{n=0}^{\infty} (\mathbf{R}[k])^n\mathbf{A}_n$ directly based on $\mathbf{R} = \sum_{n=0}^{\infty} \mathbf{R}^n\mathbf{A}_n$. For the continuous-time case, we need to solve

$$\sum_{n=0}^{\infty} \mathbf{R}^n\mathbf{A}_n = \mathbf{A}_0 + \mathbf{R}\mathbf{A}_1 + \sum_{n=2}^{\infty} \mathbf{R}^n\mathbf{A}_n = \mathbf{0}$$

for $\mathbf{R}$ as

$$\mathbf{R} = \left(-\mathbf{A}_0\mathbf{A}_1^{-1}\right) + \sum_{n=2}^{\infty} \mathbf{R}^n\left(-\mathbf{A}_n\mathbf{A}_1^{-1}\right).$$

Based on this expression, the following recursion can be developed for computing $\mathbf{R}$ with $\mathbf{R}[0] = 0$

$$\mathbf{R}[k+1] = \left(-\mathbf{A}_0\mathbf{A}_1^{-1}\right) + \sum_{n=2}^{\infty} (\mathbf{R}[k])^n \left(-\mathbf{A}_n\mathbf{A}_1^{-1}\right), \quad k = 0,1,2,.... \quad (4.136)$$

As mentioned earlier, these successive substitution algorithms may not be efficient (i.e., quite time-consuming). In particular, for the $GI/M/1$ type Markov chain, the algorithm involves a sum of infinite terms. However, in many cases, the structure of the $\mathbf{P}$ or $\mathbf{Q}$ is such that the blocks $\mathbf{A}_n$ are zero for relative small values of n, which limits the computational effort needed in each iteration. To improve the computational efficiency, alternative algorithms with faster convergence can be developed. As a simple example, we consider matrix $\mathbf{G}$ computation. The simplest one is given by

$$\mathbf{G}[k+1] = \sum_{n=0}^{\infty} \mathbf{A}_n (\mathbf{G}[k])^n, \quad k = 0,1,...$$

with $\mathbf{G}[0] = \mathbf{0}$, which is quite slow. A faster variant suggested by Neuts (1981)

$$\mathbf{G}[k+1] = (\mathbf{I} - \mathbf{A}_1)^{-1} \left(\mathbf{A}_0 + \sum_{n=0}^{\infty} \mathbf{A}_n (\mathbf{G}[k])^n \right), \quad k = 0,1,...$$

with $\mathbf{G}[0] = \mathbf{0}$. Asmussen and Bladt (1996) developed the following algorithm with fast convergence

$$\mathbf{G}[k+1] = \left(\mathbf{I} - \sum_{n=1}^{\infty} \mathbf{A}_n (\mathbf{G}[k])^{n-1} \right)^{-1} \mathbf{A}_0, \quad k = 0,1,...$$

with $\mathbf{G}[0] = \mathbf{0}$. More advanced algorithms based cyclic reduction have been also developed and converge much faster (see Latouche and Ramaswami (1999)).

To numerically implement the algorithms for computing the stationary distributions for the QBD, $GI/M/1$ type, and $M/G/1$ type processes, we consider a continuous-time queueing system operating in a stochastic environment.

Example 4.6 *Consider a service system with a single service facility. The service facility can be in one of three states serving three types of customers (or doing three types of jobs). While type 2 customers are queueing customers who arrive according to Poisson processes, type 1 and type 3 customers are non-queueing customers. There are two ways of serving type 1 or type 3 customers, a reactive approach and a proactive approach. (1) When the system has type 2 customers (either waiting or in service), a type 1 (or type 3) customer arriving* *according to a Poisson process with rate α_1 (or α_3) has preemptive priority to be served. However, only one type 1 or type 3 customer is allowed in the system (in service). That is no waiting room for type 1 or type 3 customers if there is any type 2 customer in the system. (2) When the system has no type 2 customers (queue is zero), the service facility will request serving type 1 and type 3 customers alternatively until the first arrival of a type 2 customer occurs. Note that a departure leaving an empty system will trigger serving either a type 1 or type 2 with equal probability and assume that type 1 and type 3 customers are always available as long as there is no type 2 customer in the system. Type 2 customers arrival rate depends on the state of the service facility. When it is serving a type i customer, the arrival rate is λ_i with $i = 1,2,3$. Service times of type i customers are exponentially distributed with rate μ_i with $i = 1,2,3$. The transition rate diagram is shown in Figure 4.2. We can model this service system as a continuous-time QBD process $\{X(t),J(t)\}$ with $X(t)$ being the number of type 2 customers in the system and $J(t)$ being the type of customer in service. Thus, the state space is $\{(0,1),(0,3)\}\cup\{\{1,2,...\}\times\{1,2,3\}\}$. The infinitesimal generator is given by*

$$
\mathbf{Q} =
$$

$$
\begin{pmatrix}
-(\mu_1+\lambda_1) & \mu_1 & \lambda_1 & 0 & 0 & & & & \\
\mu_3 & -(\mu_3+\lambda_3) & 0 & 0 & \lambda_3 & & & & \\
0 & 0 & -(\mu_1+\lambda_1) & \mu_1 & 0 & \lambda_1 & 0 & 0 & \\
\mu_2/2 & \mu_2/2 & \alpha_1 & -\binom{\mu_2+\alpha_1}{+\alpha_3+\lambda_2} & \alpha_3 & 0 & \lambda_2 & 0 & \\
0 & 0 & 0 & \mu_2 & -(\mu_2+\lambda_3) & 0 & 0 & \lambda_3 & \\
0 & 0 & 0 & 0 & 0 & -(\mu_1+\lambda_1) & \mu_1 & 0 & \cdots \\
0 & 0 & 0 & \mu_2 & 0 & \alpha_1 & -\binom{\mu_2+\alpha_1}{+\alpha_3+\lambda_2} & \alpha_3 & \cdots \\
0 & 0 & 0 & 0 & 0 & 0 & \mu_3 & -(\mu_3+\lambda_3) & \cdots \\
\vdots & \vdots & \vdots & \vdots & \vdots & \vdots & \vdots & \vdots & \ddots
\end{pmatrix}.
$$

We can write the block entries in $\mathbf{Q}$ as follow:

$$
\mathbf{B}_{00} = \begin{pmatrix} -(\mu_1+\lambda_1) & \mu_1 \\ \mu_3 & -(\mu_3+\lambda_3) \end{pmatrix},
$$

$$
\mathbf{B}_{01} = \begin{pmatrix} \lambda_1 & 0 & 0 \\ 0 & 0 & \lambda_3 \end{pmatrix}, \quad
\mathbf{B}_{10} = \begin{pmatrix} 0 & 0 \\ \mu_2/2 & \mu_2/2 \\ 0 & 0 \end{pmatrix},
$$

$$
\mathbf{A}_1 = \begin{pmatrix} -(\mu_1+\lambda_1) & \mu_1 & 0 \\ \alpha_1 & -\binom{\mu_2+\alpha_1}{+\alpha_3+\lambda_2} & \alpha_3 \\ 0 & \mu_2 & -(\mu_2+\lambda_3) \end{pmatrix},
$$

$$
\mathbf{A}_0 = \begin{pmatrix} \lambda_1 & 0 & 0 \\ 0 & \lambda_2 & 0 \\ 0 & 0 & \lambda_3 \end{pmatrix}, \quad
\mathbf{A}_2 = \begin{pmatrix} 0 & 0 & 0 \\ 0 & \mu_2 & 0 \\ 0 & 0 & 0 \end{pmatrix},
$$

so that the infinitesimal generator $\mathbf{Q}$ can be written in a tri-diagonal form that indicates that the CTMC is a QBD process.

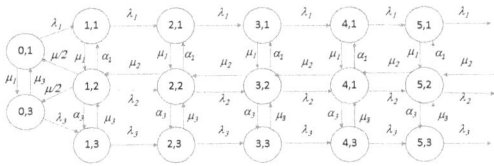

Figure 4.2   A service facility serving three types of customers.

Now we provide a series of exercises based on this system described.

Exercise 4.42 *For the service system in Example 4.6, write the stability condition in terms of the system parameters.*

Exercise 4.43 *For the service system in Example 4.6, suppose that a set of parameters are given as* $\lambda_1 = 0.5, \lambda_2 = 1, \lambda_3 = 0.2, \mu_2 = 2, \alpha_1 = 0.2, \mu_1 = 1, \alpha_3 = 0.1,$ *and* $\mu_3 = 0.5.$
   (a) Check the stability condition to see if the QBD process is ergodic.
   (b) Compute the rate matrix **R**, *stationary probability vectors* $\pi_0, \pi_1,$ *and the expected queue length for type 2 customers.*
   (c) Compute the **G** *and* **U** *matrices.*

Exercise 4.44 *In Example 4.6, if we assume that as long as there are at least two customers of type 2 are in the system and the service facility is in serving one type 1 or type 3 customer, two type 2 customers can be removed from the system and the removing time is exponentially distributed with rate* ξ_i *with* $i = 1, 3.$ *This is shown in Figure 4.3.*
   (a) Write the infinitesimal generator **Q** *for this two-dimensional CTMC that shows the GI/M/1 type process.*
   (b) Re-block the **Q** *into a QBD process.*
   (c) Use the set of parameters in the previous exercise and $\xi_1 = 0.25$ *and* $\xi_3 = 0.75,$ *compute the quantities in the previous example for this example.*

Exercise 4.45 *In Example 4.6, if we assume that as long as the service facility is in serving one type 1 or type 3 customer, two type 2 customers (couple arrivals) can be admitted in addition to the original individual type 2 arrivals with. The time between two consecutive couple arrivals is exponentially distributed with rate* ξ_i *with* $i = 1, 3.$ *This is shown in Figure 4.4.*
   (a) Write the infinitesimal generator **Q** *for this two-dimensional CTMC that shows the M/G/1 type process.*
   (b) Use the set of parameters in the previous exercise and $\xi_1 = 0.02$ *and* $\xi_3 = 0.06,$ *compute the* **G** *matrix and stationary distribution of the process.*

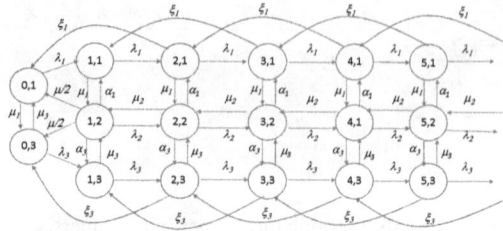

Figure 4.3 A service facility serving three types of customers with fast removing type 2 customers.

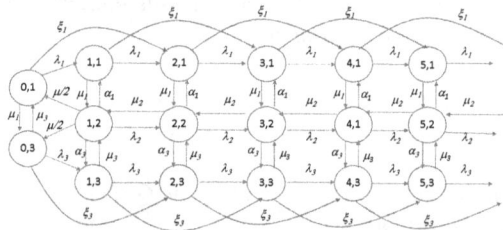

Figure 4.4 A service facility serving three types of customers with fast adding type 2 customers.

Next, we develop a discrete-time QBD process for a queueing model in which the arrival process depends on the environment.

Example 4.7 *Consider a clinic serving randomly arrival customers. Suppose that the time axis is discretized into time slots (hourly intervals). It is assumed that in each time slot (1-hour period), at most one of arrival and departure (service completion) events can occur. While in each time slot the service completion probability is constant, the probability of an arrival occurring depends on the weather condition and wait time. To simplify the model development, we assume that there are only two weather conditions for each time slot: $S =$ Sunny and $S^c =$ Non-sunny. Denote by A the customer arrival event and D the customer departure event. In each time slot, weather condition may or may not change. Define the following state variables:*

$$X_k = \text{the number of customers in the clinic at time slot } k,$$

$$J_k = \begin{cases} 1 & \text{the weather condition is sunny at time slot } k, \\ 2 & \text{the weather condition is non-sunny at time slot } k. \end{cases}$$

Then, $\{(X_k, J_k), k = 0, 1, ...\}$ is a discrete-time QBD with the state space given by $\Omega = \{0, 1, 2, ...\} \times \{1, 2\}$. The transition probability matrix for the weather *condition* $\mathbf{P}_W$ *is given by*

$$\mathbf{P}_W = \begin{array}{c} \\ S \\ S^c \end{array} \begin{pmatrix} S & S^c \\ 0.70 & 0.30 \\ 0.35 & 0.65 \end{pmatrix}.$$

Using events A, D, and S, we can specify the matrices in the transition probability matrix of the discrete-time QBD as follows:

$$\mathbf{A}_{00} = \begin{array}{c} \\ A^c \cap S \\ A^c \cap S^c \end{array} \begin{pmatrix} A^c \cap S & A^c \cap S^c \\ 0.30 & 0.20 \\ 0.25 & 0.40 \end{pmatrix}, \quad \mathbf{A}_{01} = \begin{array}{c} \\ A \cap S \\ A \cap S^c \end{array} \begin{pmatrix} A \cap S & A \cap S^c \\ 0.40 & 0.10 \\ 0.10 & 0.25 \end{pmatrix},$$

$$\mathbf{A}_{10} = \mathbf{A}_2 = \begin{array}{c} \\ D \cap S \\ D \cap S^c \end{array} \begin{pmatrix} D \cap S & D \cap S^c \\ 0.35 & 0.15 \\ 0.15 & 0.25 \end{pmatrix},$$

$$\mathbf{A}_{11} = \mathbf{A}_1 = \begin{array}{c} \\ A^c \cap D^c \cap S \\ A^c \cap D^c \cap S^c \end{array} \begin{pmatrix} A^c \cap D^c \cap S & A^c \cap D^c \cap S^c \\ 0.15 & 0.05 \\ 0.15 & 0.35 \end{pmatrix},$$

$$\mathbf{A}_0 = \begin{array}{c} \\ A \cap S \\ A \cap S^c \end{array} \begin{pmatrix} A \cap S & A \cap S^c \\ 0.20 & 0.10 \\ 0.05 & 0.05 \end{pmatrix}.$$

Clearly, these matrices imply several characteristics of the system. For example, (a) By examining $\mathbf{A}_{01}$ and $\mathbf{A}_0$, we see that the probability of an arrival occurring in a sunny period is higher than that in a non-sunny day and the probability of an arrival occurring in an empty system (no delay) is higher than that in a non-empty system. (b) It follows from $\mathbf{A}_2$ that the probability of completing a customer service in a sunny period is higher than that in a non-sunny period. The transition probability matrix of the QBD process is given by

$$\mathbf{P} = \begin{pmatrix} \mathbf{A}_{00} & \mathbf{A}_{01} & & & \\ \mathbf{A}_{10} & \mathbf{A}_{11} & \mathbf{A}_0 & & \\ & \mathbf{A}_2 & \mathbf{A}_1 & \mathbf{A}_0 & \\ & & \ddots & \ddots & \ddots \\ & & & \ddots & \ddots \end{pmatrix}.$$

With this transition probability matrix, one can obtain the matrix geometric solution for the stationary distribution and related major performance *measures provided the stability condition is satisfied (left as an exercise). Note that the relation between matrix $\mathbf{P}_W$ and $\mathbf{A}_{00}, \mathbf{A}_{01}$ or $\mathbf{A}_2, \mathbf{A}_1, \mathbf{A}_0$ (i.e., matrix elements of $\mathbf{P}$) is that each entry in $\mathbf{P}_W$ is the sum of corresponding entries in $\mathbf{A}_{00}, \mathbf{A}_{01}$ or $\mathbf{A}_2, \mathbf{A}_1, \mathbf{A}_0$.*

Exercise 4.46 *For the discrete-time QBD process in Example 4.7, please check the stability condition and compute the stationary distribution of the system, the average queue length, and average waiting time.*

In this chapter, we have presented the basic components, PH random variables and MAPs for formulating complex structured Markov chains and developed the fundamental theory for QBD processes, which is a typical class of structured Markov chain. We have also discussed two other classes of structured Markov chains – the $GI/M/1$ type and $M/G/1$ type Markov chains. It is worth noting that there are more theoretical developments on the structured Markov chains such as $GI/M/1$ type and $M/G/1$ type Markov chains, and Markov chains with tree structure. Furthermore, the state space can be countable or uncountable. Here, we only provide an introduction to this active research area with rich literature and refer interested readers to more specialized books such as Neuts (1981) (1989), Latouche and Ramaswami (1999), He (2014), and references therein for further study.

REFERENCE NOTES

This chapter covers a major methodology in stochastic modeling, namely, the matrix analytical method. The basic theory is developed by Neuts [11], [12] and further discussed in [9]. Recently, a book by He [7] provides an excellent introduction to this approach. Fitting PH distributions to empirical data or another theoretical distribution addressed in Section 4.3 is based on [2]. Fitting MAP to empirical data presented in Section 4.6 is taken from [13].

REFERENCES

1. A.S. Alfa, "Queueing Theory for Telecommunications: Discrete Time Modelling of a Single Node System", Springer, New York, 2010.
2. S. Asmussen, O. Nerman, and M. Olsson, Fitting phase-type distributions via the EM algorithm. *Scandinavian Journal of Statistics*, 23(4), 419–441, 1996.
3. S. Asmussen and M. Bladt, Renewal theory and queueing algorithms for matrix-exponential distributions. *Matrix-Analytic Methods in Stochastic Models* (eds A. S. Alfa and S. Chakravarty), Marcel Dekker, New York, 1996.
4. M. Bladt and B.F. Nielsen, "Matrix-Exponential Distributions in Applied Probability", Springer, New York, 2017.
5. J.W. Cohen, "The Single Server Queue", North-Holland series in applied mathematics and mechanics. North-Holland, Amsterdam, 1982.

6. A.P. Dempster, N.M. Laird, and D.B. Rubin, Maximum likelihood from incomplete data via the EM algorithm. *Journal of the Royal Statistical Society. Series B (Methodological)*, 39(1), 1–38, 1977.

7. Q. He, "Fundamentals of matrix-analytic methods", Springer, New York, 2014.

8. Q. Li, "Constructive Computation in Stochastic Models with Applications: The RG-Factorizations", Tsinghua University Press, 2010.

9. G. Latouche and V. Ramaswami, "Introduction to Matrix Analytic Methods in Stochastic Model", ASA/SIAM Series on Statistics and Applied Probability, 1999.

10. G.J. McLachlan and T. Krishnan, "The EM Algorithm and Extensions", New York: JOHN WILEY and SONS, INC., 2nd ed., 2008.

11. M.F. Neuts, "Matrix Geometric Solutions in Stochastic Models: An Algorithmic Approach", Johns Hopkins University Press, Baltimore, 1981.

12. M.F. Neuts, "Structured Stochastic Matrices of M/G/1 Type and Their Applications", Marcel Dekker, New York, 1989.

13. H. Okamura and T. Dohi, Fitting phase-type distributions and Markovian arrival processes: algorithms and tools, 49–75, "Principles of Performance and Reliability Modeling and Evaluation", 2016.

14. M. Olsson, Estimation of phase-type distributions from censored data. *Scandinavian Journal of Statistics* 23, 443–460, 1996.

15. V. Ramaswami, The N/G/1 queue and its detailed analysis. *Advances in Applied Probability*, 12(1), 222–261, 1980.

16. S. Ross, "Introduction to Probability Models", Academic Press, 11th ed., 2014.

17. E. Seneta, "Non-Negative Matrices and Markov chains", 2nd ed., Springer, New York, 2006.

18. C.F.J. Wu, On the Convergence Properties of the EM Algorithm. *The Annals of Statistics*, 11 (1), 95–103, 1983.

5 Renewal Processes and Embedded Markov Chains

In this chapter, we turn to the renewal processes and embedded Markov chains. For a continuous-time stochastic process, if we are only interested in whether or not a particular type of events will happen and the cumulative number of those events occurred in a finite time period, the state space can be considered as 0 (not happened) and 1 (happened), called a binary state space. For such a process, the next state to be visited is deterministic (i.e., the state-change pattern is certain with 0,1,0,1,0,1...) and we do not need the time elapsed since the last state change to predict the future state probabilistically. Thus, we do not need the memoryless property for the time between two consecutive events if we are only interested in the total number of events occurred in a fixed time period or the time when the Nth event occurs. This implies that the time between state-change events can be generally distributed. For example, a machine in operation has two states, working or broken. Denote the state at time t by $X(t)$, then $X(t) = 0$ ($X(t) = 1$) represents that the machine is working (broken) at time t. A working machine will incur some operating cost and a broken machine will cause some repair or replacement cost repair or replacement may or may not take time. If we are interested in the total cost of using the machine for a fixed period of time, then we are interested in the total number of failures occurred in this period. Since the state change is completely deterministic, we can focus on the time-related variables. In the machine-repair example, the time between two consecutive machine failures can be a generally distributed random variable. We will study a type of stochastic processes in which times between two consecutive events of interest are i.i.d. random variables. Such a process is called a renewal process, which is the first topic of this chapter. Then, we extend the renewal process with the binary state space to the stochastic process with the general discrete state space, which is called the Markov renewal process. Under such a framework, we introduce the embedded Markov chains that can model some queueing and stochastic inventory systems with non-exponential random variables.

5.1 RENEWAL PROCESSES

Consider a light bulb in a dark room. It is on all the time. As soon as the light bulb is burned out, it is replaced with a new and identical one either instantly

DOI: 10.1201/9781003150060-5

or by taking a random amount of time. If successive lifetimes of light bulbs are i.i.d. random variables, denoted by $\{X_k, k \geq 1\}$, and the corresponding replacement times are also i.i.d. random variables, denoted by $\{Y_k, k \geq 1\}$, then we can define a process cycle time as $Z_k = X_k + Y_k$ where $k = 1, 2, ...$, which forms a sequence of i.i.d. random variables. Since the process cycle time repeats itself in a probabilistically identical way, a renewal process can be defined. We start with a simpler case for the light bulb replacement problem where $Y_k \equiv 0$ or the burned-out light bulb is replaced instantly. In this case, we may be interested in two quantities. The first one is the time instant when the nth replacement occurs and the second one is the total number of failures (or replacements) occurred in a fixed time period t. Clearly, these two quantities are two stochastic processes which can be defined as renewal processes.

Definition 5.1 *For a sequence of i.i.d. random variables $\{X_k, k \geq 1\}$, called inter-renewal times, let $S_n = \sum_{k=1}^{n} X_k, n \geq 0$, be the time instant of the nth event, and $N(t) = \max\{n : S_n \leq t\}$ be the total number of replacements occurred by time t. Then $\{S_n(t), n \geq 0\}$ is called a renewal process and $\{N(t), t \geq 0\}$ is called a renewal counting process.*

Exercise 5.1 *Suppose a recurrent finite-state Markov chain that starts in state i at time 0. Then the time until the first return to state i is considered a renewal occurs. Explain that the number of visits to state i is a renewal counting process.*

5.1.1 BASIC RESULTS OF RENEWAL PROCESSES

Let F be the common general distribution function of i.i.d. random variables $\{X_k, k \geq 1\}$. Denote the mean of X_k by $\mu = E[X_k] = \int_0^{\infty} x dF(x) > 0$. Furthermore, let F_n be the distribution function of $S_n, n \geq 0$. Now we present some basic results for the renewal process $\{S_n(t), n \geq 0\}$ by using the light bulb replacement example. If we assume that at time 0, we start with a brand new light bulb, then we have the cumulative distribution functions for $S_n, n \geq 0$, as follows:

$$F_0(x) = \begin{cases} 0 & x < 0, \\ 1 & x \geq 0. \end{cases} \qquad (5.1)$$

$$F_1(x) = F(x), \qquad (5.2)$$

$$F_n(x) = F^{n*}(x), \text{ for } n \geq 1, \qquad (5.3)$$

where $F^{n*}(x)$ is the n-fold convolution of F with itself. By using the strong law of large numbers (SLLN) (see Appendix A), we obtain

$$\frac{S_n}{n} = \frac{X_1 + X_2 + ... + X_n}{n} \to \mu \text{ as } n \to \infty. \qquad (5.4)$$

Next, we look at the number of events happened in a finite time period, $N(t)$, and show with probability one $N(t)$ is finite and the finiteness of the rth moments, where $r \geq 1$.

Theorem 5.1 *For a renewal counting process in a finite time period t, the followings hold: (a) $P(N(t) < \infty) = 1$; (b) There exists $s_0 > 0$ such that the moment generating function $E[e^{sN(t)}] < \infty$ for all $s < s_0$; and (c) $E[(N(t))^r] < \infty$ for all $r > 0$.*

Proof. Consider the SLLN result for a renewal process, $S_n/n \to \mu > 0$ as $n \to \infty$, which implies that $S_n \to \infty$ as $n \to \infty$. Otherwise, μ will be zero which is a contradiction to $\mu > 0$. Thus, for a finite time period t, the event $\{S_n \leq t\}$ can only be true for a finite n. It follows from the definition $N(t) = \max\{n : S_n \leq t\}$ that $N(t)$ must be finite or (a) is proved.

We assume that $F(0) = P\{X_1 = 0\} < 1$, meaning that there is some positive probability that $P\{X_1 > 0\} > 0$. Thus, there exists $x_0 > 0$ such that $P\{X_1 \geq x_0\} > 0$. Due to the freedom of scaling, without loss of generality, we assume that $x_0 = 1$. Then we deine, for $k \geq 1$,

$$\bar{X}_k = \begin{cases} 0 & X_k < 1, \\ 1 & X_k \geq 1, \end{cases} \tag{5.5}$$

$\bar{S}_n = \sum_{k=1}^{n} \bar{X}_k, n \geq 0$, and $\bar{N}(t) = \max\{n : \bar{S}_n \leq t\}$. Then, $\bar{X}_k \leq X_k, k \geq 1$, and $\bar{S}_n \leq S_n, n \geq 0$. Therefore, $\bar{N}(t) \geq N(t), t \geq 0$. Clearly, $\bar{S}_n$ follows a Binomial distribution with parameters n and $p = P\{\bar{X}_k \geq 1\}$. Then $\{\bar{N}(t) = \max\{n : \bar{S}_n \leq t\}$ is the number of renewals required to obtain n successes in time period $(0,t]$ which is actually a Negative Binomial process. Thus, the moment generating function of $\bar{N}(t)$ is given by

$$E[e^{s\bar{N}(t)}] = \left[\frac{e^s p}{1 - e^s(1 - p)} \right]^n. \tag{5.6}$$

To ensure this moment generating function to be finite, we must have $1 - e^s(1 - p) > 0$ which implies $s < \ln(1/(1 - p)) = s_0$. This also implies that the rth moment of $\bar{N}(t)$ is finite. Since $\bar{N}(t)$ is an upper bound of $N(t)$, (b) and (c) of the theorem follow. ∎

Exercise 5.2 *Derive the moment generating function of $\bar{N}(t)$ used in the proof above.*

It is also easy to show that $N(\infty) = \lim_{t \to \infty} N(t) = \infty$ with probability 1. Here is the logic. Consider event $\{N(\infty) < \infty\}$. This event has zero probability as $P\{N(\infty) < \infty\} = P\{X_n = \infty \text{ for some } n\} = P\{\cup_{n=1}^{\infty}\{X_n = \infty\}\} \leq \sum_{n=1}^{\infty} P\{X_n = \infty\} = 0$. Thus, $P\{N(\infty) = \infty\} = 1 - P\{N(\infty) < \infty\} = 1$. Due to the relation between two events $\{N(t) \geq n\} \Leftrightarrow \{S_n \leq t\}$, we can obtain the distribution of $N(t)$ in terms of the distribution of $S_n(t)$.

$$
\begin{aligned}
P(N(t) = n) &= P\{N_n(t) \geq n\} - P\{N(t) \geq n+1\} \\
&= P\{S_n \leq t\} - P\{S_{n+1} \leq t\} \\
&= F_n(t) - F_{n+1}(t).
\end{aligned}
$$

Example 5.1 *(The Geometric process) Suppose that a device is operating in a discrete mode. Each operation takes one period. With probability p it fails during each period (i.e. with probability $1 - p$, it successfully completes the operation). As soon as it fails, it is replaced with a brand-new one instantly. Let X_k be the inter-replacement time, $S_n = \sum_{k=1}^{n} X_k$, the time of nth replacement, and $N(t)$, the number of replacements occurred by time t. Note that X_k follows a geometric distribution and $P\{X_k = i\} = p(1-p)^{i-1}$. Write down the distributions of S_n and $N(t)$, respectively.*

Solution *Clearly, S_n follows a Negative Binomial distribution.*

$$
P\{S_n = k\} = \binom{k-1}{n-1} p^{n-1}(1-p)^{k-n}p, \quad \text{for } k \geq n. \tag{5.7}
$$

This is the probability that in the first $k - 1$ periods, we have made $n - 1$ replacements (or failures) and the kth period is a replacement (the nth replacement). Now we can write down the distribution of $N(t)$ as follows:

$$
P\{N(t) = n\} = F_n(t) - F_{n+1}(t) = P\{S_n \leq t\} - P\{S_{n+1} \leq t\}
$$

$$
= \sum_{k=n}^{\lfloor t \rfloor} \binom{k-1}{n-1} p^n(1-p)^{k-n} - \sum_{k=n+1}^{\lfloor t \rfloor} \binom{k-1}{n} p^{n+1}(1-p)^{k-n-1}. \tag{5.8}
$$

Next, we can obtain the mean of $N(t)$, which is called the renewal function and denoted by $M(t) = E[N(t)]$, in terms of F_n.

$$
\begin{aligned}
M(t) = E[N(t)] &= \sum_{n=1}^{\infty} P\{N(t) \geq n\} \\
&= \sum_{n=1}^{\infty} P\{S_n \leq t\} = \sum_{n=1}^{\infty} F_n(t).
\end{aligned} \tag{5.9}
$$

With this expression, we can derive the renewal equation (also called the integral equation for renewal processes) which $M(t)$ satisfies. It follows from (5.9) that

$$
\begin{aligned}
M(t) &= F_1(t) + \sum_{n=1}^{\infty} F_{n+1}(t) = F(t) + \sum_{n=1}^{\infty} (F_n * F)(t) \\
&= F(t) + (M * F)(t),
\end{aligned}
$$

where "$*$" means convolution operation. This relation can be obtained by using the conditioning argument. By conditioning on the time of the first renewal, denoted by $X_1 = x$, we have

$$M(t) = E[N(t)] = \int_0^\infty E[N(t)|X_1 = x] f(x) dx$$

$$= \int_0^t \{1 + E[N(t-x)]\} f(x) dx + \int_t^\infty \{0\} f(x) dx$$

$$= \int_0^t f(x) dx + \int_0^t M(t-x) f(x) dx = F(t) + \int_0^t M(t-x) f(x) dx,$$

$$(5.10)$$

which is equivalent to the expression obtained by using the summation of the distribution function of S_n.

Example 5.2 *Suppose a device has a finite lifetime which is uniformly distributed between $(0,1)$. When it fails, it is replaced with an identical one immediately. This process is ongoing and we want to decide the renewal function for this process.*

Solution This is an exercise of solving the renewal equation. We will solve it iteratively by starting with period $(0,1]$. Since $X_k \sim Unif(0,1)$, for $t \in (0,1]$, the renewal equation can be written as

$$M(t) = t + \int_0^t M(t-x) dx = t + \int_0^t M(x) dx.$$

Differentiating this equation yields

$$M'(t) = 1 + M(t) \Rightarrow (1 + M(t))' = 1 + M(t).$$

Thus, we have the solution $1 + M(t) = Ke^t$ or $M(t) = Ke^t - 1$. With the initial condition of $M(0) = 0$, we get $K = 1$ and obtain the solution $M(t) = e^t - 1$ for $0 \le t \le 1$. Next, we look at period $(1,2]$. For $1 \le t \le 2$ it is more convenient to write the renewal equation by conditioning on the first renewal instant (like in the derivation of (5.10)). Conditioning on the first renewal instant x, for a time point $t \in (0,2]$ (see Figure 5.1), we have

$$M(t) = \int_0^1 E[N(t)|X_1 = x] f(x) dx = \int_0^1 (1 + E[N(t-x)]) dx$$

$$= 1 + \int_0^1 M(t-x) dx = 1 + \int_{t-1}^t M(x) dx.$$

Differentiating the preceding equation yields

$$M'(t) = M(t) - M(t-1) = M(t) - (e^{t-1} - 1) = M(t) + 1 - e^{t-1}.$$

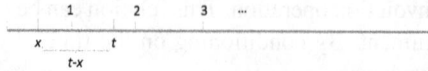

Figure 5.1   Renewal process.

Multiplying both sides of the equation above by e^{-t}, we obtain

$$e^{-t}(M'(t) - M(t)) = \frac{d}{dt}(e^{-t}M(t)) = e^{-t} - e^{-1}.$$

Integrating over t from 1 to t gives

$$e^{-t}M(t) = e^{-1}M(1) + e^{-1}\int_{1}^{t}(e^{-(s-1)} - 1)ds$$

$$= e^{-1}M(1) + e^{-1}[1 - e^{-(t-1)} - (t-1)].$$

Rearranging the terms of the above equation and using the fact $M(1) = e - 1$, we obtain

$$M(t) = e^{t} + e^{t-1} - 1 - te^{t-1} = e^{t} - 1 - e^{t-1}(t-1) \text{ for } 1 \leq t \leq 2.$$

In general for $n \leq t \leq n+1$, the renewal equation remains the same structure as

$$M(t) = 1 + \int_{t-1}^{t} M(x)dx \Rightarrow M'(t) = M(t) - M(t-1).$$

Multiplying both sides by e^{-t}, we have

$$\frac{d}{dt}(e^{-t}M(t)) = e^{-t}(M'(t) - M(t)) = -e^{-t}M(t-1).$$

Integrating both sides of the proceeding equation from t from n to t yields

$$e^{-t}M(t) = e^{-n}M(n) - \int_{n}^{t} e^{-s}M(s-1)ds.$$

Using this recursion, we can solve for $M(t)$ iteratively for any t.

It is worth noting that for a generally distributed inter-renewal time, it may not be possible to solve for the closed-form solution for $M(t)$. Uniform and exponential distributed inter-renewal times are two tractable examples. Clearly, for $X_k \sim Exp(\lambda)$, the renewal counting process is a Poisson process and all related quantities are readily available. However, we can easily get a simple expression for the Laplace Transform (LT) of $M(t)$ as shown later. More discussion on the renewal equations will be given in the next subsection.

Exercise 5.3 *Suppose that a renewal process has $P(X_i = 0) < 1$, where X_i is the ith inter-renewal interval. Show that the renewal function $M(t)$ is finite for all $t \geq 0$.*

Exercise 5.4 *A generalized renewal function can be defined as*

$$U(t) := \sum_{n=1}^{\infty} \alpha_n P(S_n \leq t),$$

where $\{\alpha_n; n \in \mathbb{N}_0\}$ is a sequence of non-negative numbers. If $\alpha_n = n^{-1}$, $U(t)$ is called harmonic renewal function; if $\alpha_n = e^{\alpha n}$, $U(t)$ is called exponential renewal function. Derive the renewal equations for these two special cases.

Exercise 5.5 *For a renewal process with the distribution function for the inter-renewal time as $P(X_1 \leq x) = 1 - (1 + \lambda x)e^{-\lambda x}$, for $x \geq 0$ and some $\lambda > 0$. Check that $M(t) = 1 + \frac{1}{2}\lambda t - \frac{1}{4}(1 - e^{-2\lambda t})$.*

5.1.2 MORE ON RENEWAL EQUATIONS

Many quantities of interest for renewal processes satisfy the integral renewal equation. Renewal equations can be developed by conditioning on the time of the first renewal as shown in analyzing $M(t)$. To further study renewal equations, we need to develop some additional concepts and tools involving measures, convolutions, and transforms. Some of the results are based on some advanced topics in measure theory and analysis for this section. You may need to review some of these topics as necessary (see Bartle (1995)). Since the quantities of interest in a renewal process often involve sum of several random variables, we need the concept of convolution of random variables in terms of their probability density functions. Here we provide a brief review.

Suppose that X and Y are independent random variables in $[0, \infty)$ with probability density functions f and g, respectively. Then by conditioning argument, we can obtain the probability density (mass) function for $X + Y$, denoted by $f * g$, in the continuous (discrete) cases, respectively,

$$(f * g)(t) = \int_0^t f(t - s)g(s)ds$$

$$(f * g)(t) = \sum_{s \in [0,t]} f(t - s)g(s).$$

In the discrete case, t is a possible value of $X + Y$, and the sum is over the countable collection of $s \in [0, t]$ with s a value of X and $t - s$ a value of Y. Moreover, the definition clearly makes sense for functions that are not necessarily probability density functions. They can be the combination of functions with some conditions (e.g., locally bounded or directly Riemann integrable) and probability density functions or distribution functions of independent random variables. Some algebraic properties regarding the convolution calculations can be established easily based on the definition. The first property is the commutativity for two distribution functions. This can be easily verified. Suppose that X and Y are two independent random variables with distribution function of F and G, respectively. Intuitively, this is true as the convolution for F and G represents the distribution function of the sum of X and Y, which is certainly commutative. Clearly, this means

$$F * G(t) = \int_0^t F(t-s)dG(s) = \int_0^t P(X+Y \le t | Y = s)f_Y(s)ds$$

$$= \int_0^t P(Y \le t - s | X = s)f_X(s)ds = \int_0^t G(t-s)dF(s) = G * F(t).$$

Recall the renewal equation $M(t) = F(t) + (M * F)(t)$. With this commutative property, we can have an alternative form for this equation, $F(t) = M(t) - (F * M)(t)$. Suppose that $f, g : [0, \infty) \to \mathbb{R}$ are locally bounded functions and that G and H are distribution functions on $[0, \infty)$. Then, the following properties exist.

1. $(f+g) * H = (f * H) + (g * H)$
2. $(cf) * H = c(f * H)$
3. $f * (G + H) = (f * G) + (f * H)$
4. $f * (cG) = c(f * G)$

The proofs of these properties are left as exercises. Using the Laplace transform of a distribution function G on $[0, \infty)$, $\tilde{G}(s) = \int_0^\infty e^{-st} dG(t)$, we can turn convolution into multiplication. Let $h = f * g$. In terms of their Laplace transforms, we claim that $\tilde{H}(s) = \tilde{F}(s)\tilde{G}(s)$ which can be shown as follows:

$$\tilde{H}(s) = \int_{t=0}^\infty e^{-st} \left(\int_{\tau=0}^t f(\tau)g(t-\tau)d\tau \right) dt$$

$$= \int_{t=0}^\infty \int_{\tau=0}^t e^{-st} f(\tau)g(t-\tau)d\tau dt$$

$$= \int_{\tau=0}^\infty \int_{t=\tau}^\infty e^{-st} f(\tau)g(t-\tau)dt d\tau$$

$$= \int_{\tau=0}^\infty \int_{t'=0}^\infty e^{-s(t'+\tau)} f(\tau)g(t')dt' d\tau, \quad \text{where } t' = t - \tau$$

$$= \left(\int_{\tau=0}^\infty e^{-s\tau} f(\tau)d\tau \right) \left(\int_{t'=0}^\infty e^{-st'} g(t')dt' \right) = \tilde{F}(s)\tilde{G}(s).$$

This rule can simplify solving the renewal equation. For example, in terms of the LTs, we can obtain the renewal equation as

$$\tilde{M}(s) = \tilde{F}(s) + \tilde{M}(s)\tilde{f}(s)$$

and by noting that $\tilde{f}(s) = s\tilde{F}(s)$ we have the solution to the renewal function for $M(t)$ as

$$\tilde{M}(s) = \frac{\tilde{F}(s)}{1 - s\tilde{F}(s)}. \tag{5.11}$$

Note that the inversion of the LT back to the distribution function may not be an easy thing to do (Abate and Whitt (1995)).

Similar to $M(t)$ we can develop the renewal equation for $S_{N(t)}$ as follows.

$$P(S_{N(t)} \leq s) = P(n\text{th renewal time is equal or smaller than } s$$

$$\text{and } (n+1)\text{st renewal is greater than } t)$$

$$= \sum_{n=0}^{\infty} P(\{S_n \leq s\} \cap \{S_{n+1} > t\})$$

$$= F^c(t) + \sum_{n=1}^{\infty} P(\{S_n \leq s\} \cap \{S_{n+1} > t\})$$

$$= F^c(t) + \sum_{n=1}^{\infty} \int_0^{\infty} P(\{S_n \leq s\} \cap \{S_{n+1} > t\} | S_n = y) dF_n(y)$$

$$= F^c + \sum_{n=1}^{\infty} \int_0^s F^c(t-y) dF_n(y)$$

$$= F^c(t) + \int_0^s F^c(t-y) d\left(\sum_{n=1}^{\infty} F_n(y)\right)$$

$$= F^c(t) + \int_0^s F^c(t-y) dM(y). \tag{5.12}$$

Here, the interchange of integral and summation is justified due to all positive terms. Note that the structure of the renewal equation above can be stated in words as that an expected value of a random renewal process (either probability or expectation) may be expressed as the sum of another expected value function and the convolution of two expected value functions. In general, for a renewal process with the inter-renewal distribution F, define $u(t) = E[U_t]$ where $\{U_t : t \geq 0\}$ is a random process associated with the renewal process. Then by conditioning on the first renewal time $S_1 = X_1$, we can develop an integral equation of form

$$u = a + u * F, \tag{5.13}$$

which is called a general renewal equation (GRE) for u, where a is a function on $[0, \infty)$. A unique solution to such a GRE can be determined.

Theorem 5.2 *(The fundamental theorem on renewal equations) A solution to the GRE exists and has the form $u = a + a * M$. If a is locally bounded, then u is locally bounded and is unique solution to the GRE.*

Proof. Suppose that $u = a + a * M$. Then $u * F = a * F + a * M * F$. But from the renewal equation for M above, $M * F = M - F$. Hence we have $u * F = a * F + a * (M - F) = a * [F + (M - F)] = a * M$. But $a * M = u - a$ by definition of u, so $u = a + u * F$ and hence u is a solution to the renewal equation. Next since a is locally bounded, so is $u = a + a * M$. Suppose now that v is another locally bounded solution of the integral equation, and let $w = u - v$. Then w is locally bounded and $w * F = (u * F) - (v * F) = [(u - a) - (v - a)] = u - v = w$. Hence $w = w * F_n$ for $n \in \mathbb{N}_+$. Suppose that $|w(s)| \leq D_t$ for $0 \leq s \leq t$. Then $|w(t)| \leq D_t F_n(t)$ for $n \in \mathbb{N}_+$. Since $M(t) = \sum_{n=1}^{\infty} F_n(t) < \infty$ it follows that $F_n(t) \to 0$ as $n \to \infty$. Hence $w(t) = 0$ for $t \in [0, \infty)$ and so $u = v$. ∎

This result can be also proved by using the Laplace transforms. Let α and θ denote the Laplace transforms of the functions a and u, respectively, and Φ the Laplace transform of the distribution F. Taking Laplace transforms through the renewal equations gives the simple algebraic equation $\theta = \alpha + \theta s \Phi$. Solving it gives $\theta = \frac{\alpha}{1 - s\Phi} = \alpha \left(1 + \frac{s\Phi}{1 - s\Phi}\right) = \alpha + \alpha \Gamma$ where $\Gamma = \frac{s\Phi}{1 - s\Phi}$ is the Laplace transform of the distribution M. Thus, θ is the transform of $a + a * M$.

From this theorem, we notice that the renewal equation for $M(t)$ is actually in the form of the solution to the GRE with $u = M$ and $a = F$. The renewal equation for $P(S_{N(t)} \leq s)$ in (5.12) is also in the form of the solution to the GRE with $u = P(S_{N(t)} \leq s)$ and $a = F^c(t)$.

Other quantities of interest for a renewal process include Residual Life and Current Age. The residual life (also called excess life) is defined as $R(t) = S_{N(t)+1} - t$, which is the time interval from the current time point t to the next renewal point, and the current age defined as $A(t) = t - S_{N(t)}$, which the time interval from the most recent renewal to the current time point. Thus, the total life cycle is $C(t) = A(t) + R(t) = S_{N(t)+1} - S_{N(t)}$. We will show that $C(t)$ and X_k have different distributions and such a phenomenon is called the "Inspection Paradox".

Exercise 5.6 *Denote by $p_n(t) = P(N(t) = n)$ and write $M(t) = \sum_{n=0}^{\infty} n p_n(t)$. Find the Laplace transform of $\tilde{M}(s)$ directly from $\tilde{M}(s) = \sum_{n=0}^{\infty} n \tilde{p}_n(s)$ with $\tilde{p}_n(s) = \int_0^{\infty} e^{-st} p_n(t) dt$.*

Exercise 5.7 *Prove the identity $E[S_{N(t)+1}] = \mu[M(t) + 1]$ by using the basic properties of a renewal process (Note that this identity can also be proved later by using the concept of the stopping time and the Wald's equation introduced later in this chapter).*

Exercise 5.8 *Show that the expected residual life of a renewal process can be computed as $E[R(t)] = \mu(1 + M(t)) - t$.*

Exercise 5.9 *Let X be a uniformly distributed random variable over $[0, 1]$. Such an X divides $[0, 1]$ into two subintervals $[0, X]$ and $(X, 1]$. Now randomly* *select one of these two subintervals by comparing the value of another uniformly distributed random variable Y over $[0,1]$ with these two intervals. Select $[0,X]$ if $Y \leq X$, otherwise, select $(X,i]$. Denote by Z the length of the subinterval selected. Then*

$$Z = \begin{cases} X & \text{if } Y \leq X, \\ 1 - X & \text{if } Y > X. \end{cases}$$

Use Z as the inter-renewal interval to construct a renewal process. Determine $E[Z]$ and write an expression for $M(t)$ for such a renewal process.

5.1.3 LIMIT THEOREMS FOR RENEWAL PROCESSES

For a renewal process, we can compute some useful limits. The first one is regarding the long-term renewal rate.

Proposition 5.1 *With probability 1,*

$$\frac{N(t)}{t} \to \frac{1}{\mu} \quad \text{as } t \to \infty.$$

Proof. It follows from $S_{N(t)} \leq t \leq S_{N(t)+1}$ that

$$\frac{S_{N(t)}}{N(t)} \leq \frac{t}{N(t)} < \frac{S_{N(t)+1}}{N(t)}.$$

For the left-hand side, by SLLN, we have $S_{N(t)}/N(t) = \sum_{i=1}^{N(t)} X_i/N(t) \to \mu$ as $N(t) \to \infty$. Since $N(t) \to \infty$ as $t \to \infty$, we get $S_{N(t)}/N(t) \to \mu$ as $t \to \infty$. For the right-hand side, we can write it as

$$\frac{S_{N(t)+1}}{N(t)} = \frac{S_{N(t)+1}}{N(t)+1} \times \frac{N(t)+1}{N(t)}.$$

Using $S_{N(t)+1}/(N(t)+1) \to \mu$ by the same SLLN and the fact of $(N(t)+1)/N(t) \to 1$ as $t \to \infty$, we have the ratio limit $S_{N(t)+1}/N(t) \to \mu$ Thus, $t/N(t) \to \mu$ which completes the proof. ∎

Another similar result regarding the limit of $M(t)/t$ is given in the following "Elementary Renewal Theorem". Although this theorem is intuitive, it needs a formal proof as for a random variable sequence, $X_n \to X$ w.p.1 does not ensure $E[X_n] \to E[X]$ as $n \to \infty$. Here is an example. Consider a sequence of random variables

$$Y_n = \begin{cases} 0 & \text{if } U > \frac{1}{n}, \\ n & \text{if } U \leq \frac{1}{n}, \end{cases} \tag{5.14}$$

where $U \sim Unif(0,1)$ and $n \geq 2$. It is easy to see that with probability 1, $\lim_{n \to \infty} Y_n \to 0$. However, $\lim_{n \to \infty} E[Y_n] = \lim_{n \to \infty} [0 \times \frac{n-1}{n} + n \times \frac{1}{n}] = 1$.

Theorem 5.3 *(The Elementary Renewal Theorem)*

$$\frac{M(t)}{t} \to \frac{1}{\mu} \text{ as } t \to \infty \ \left(where \ \frac{1}{\infty} \equiv 0 \right).$$

Before proving this theorem, we introduce the famous Wald's equation. First, we define the concept of stopping time for a sequence of independent random variables.

Definition 5.2 *Let $\{X_n : n \geq 1\}$ be a sequence of independent random variables. An integer-valued random variable $N > 0$ is said to be a stopping time with respect to $\{X_n : n \geq 1\}$ if the event $\{N = n\}$ is independent of $\{X_k : k \geq n+1\}$.*

Example 5.3 *Consider the following random variable sequences and identify which integer-valued variable N is a stoping time.*

(a) Suppose that $\{X_n : n \geq 1\}$ are i.i.d. Bernoulli(p) with possible value of 1 or -1. That is $P\{X_n = 1\} = p$ and $P\{X_n = -1\} = 1 - p$ for $n = 1,2,.....$ If we define an integer-valued random variable as

$$N = \min\{n : X_1 + \cdots + X_n = 8\},$$

then N is a stopping time. Consider this sequence of random variables represents the outcomes of playing a sequence of games. N will be the first time the cumulative winning reaches 8 dollars. This event is certainly independent of the future plays (i.e., $X_{n+1}, X_{n+2}, ...$).

(b) Consider a sequence of inter-renewal times of a renewal process $\{X_k : k \geq 1\}$. Then the renewal counting process $N(t)$ is not a stopping time but $N(t) + 1$ is a stopping time. This can be verified by checking if the event $\{N(t) = n\}$ is independent of $X_{n+1}, X_{n+2}, ...$

Note that the following two events are equivalent:

$$\{N(t) = n\} \Leftrightarrow \{X_1 + \cdots + X_n \leq t\} \cap \{X_1 + \cdots + X_{n+1} > t\},$$

which depends on future random variable X_{n+1}. This implies that $N(t)$ is not a stopping time.

On the other hand, we have the following equivalence for event $\{N(t) + 1 = n\}$

$$\{N(t)+1 = n\} \Leftrightarrow \{N(t) = n-1\} \Leftrightarrow \{X_1 + \cdots + X_{n-1} \leq t\} \cap \{X_1 + \cdots + X_n > t\},$$

which is independent of $X_{n+1}, X_{n+2}, ...$ This indicates that $N(t) + 1$ is a stopping time.

Theorem 5.4 *(Wald's Equation) If $\{X_k : k \geq 1\}$ is a sequence of i.i.d. random variables with $E[X_k] < \infty$, and if N is a stopping tume for this sequence with $E[N] < \infty$, then*

$$E\left[\sum_{i=1}^{N} X_i\right] = E[N]E[X_1].$$

Proof. First, we define the indicator variable

$$I_j = \begin{cases} 1 & \text{if } j \leq N, \\ 0 & \text{if } j > N. \end{cases}$$

Then we can write the sum of N random variables as $\sum_{j=1}^{N} X_j = \sum_{j=1}^{\infty} X_j I_j$. Taking the expectation, we have

$$E\left[\sum_{j=1}^{N} X_j\right] = E\left[\sum_{j=1}^{\infty} X_j I_j\right] = \sum_{j=1}^{\infty} E[X_j I_j]. \tag{5.15}$$

Note that I_j and X_j are independent due to the fact that N is a stopping time with respect to the sequence of random variables. This is shown by the following equivalent events:

$$\{I_j = 0\} \Leftrightarrow \{N < j\} \Leftrightarrow \{N \leq j-1\}$$

and the event $\{N \leq j-1\} = \cup_{i=1}^{j-1}\{N \leq i\}$, which depends on $X_1,...,X_{j-1}$ only, but not X_j. From (5.15), we obtain

$$E\left[\sum_{j=1}^{N} X_j\right] = \sum_{j=1}^{\infty} E[X_j I_j] = \sum_{j=1}^{\infty} E[X_j]E[I_j]$$

$$= E[X_1] \sum_{j=1}^{\infty} E[I_j] = E[X_1] \sum_{j=1}^{\infty} P(N \geq j)$$

$$= E[X_1]E[N].$$

This completes the proof. ∎

Remark: In the proof above, we interchanged expectation and summation without justification. If all random variables $\{X_k : k \geq 1\}$ are positive, this interchange is allowed. However, for a sequence of arbitrary random variables, such a justification is needed. This can be done by using the Lebesgue's dominated convergence theorem and expressing a non-negative variable as difference of two positive (absolute value) random variables.

Now we can prove the Elementary Renewal Theorem.

Proof. (Proof of ERT) For a renewal process with an inter-renewal sequence of $\{X_k : k \geq 1\}$ and $E[X_1] = \mu < \infty$, we have $S_{N(t)+1} > t$. Taking the expectation on both sides of this inequality, we have $(M(t) + 1)\mu > t$. Here we use the Wald's Equation on the left-hand side. That is

$$E[S_{N(t)+1}] = E\left[\sum_{j=1}^{N(t)+1} X_j\right] = E[N(t)+1]E[X_1] = (M(t)+1)\mu.$$

It follows from $(M(t)+1)\mu > t$ that

$$\frac{M(t)}{t} > \frac{1}{\mu} - \frac{1}{t} \Rightarrow \liminf_{t\to\infty} \frac{M(t)}{t} \geq \frac{1}{\mu}.$$

Next, we show that $\limsup_{t\to\infty} M(t)/t \leq 1/\mu$. For a fixed constant L, we define a new renewal process with truncated inter-renewal times. That is $\{\bar{X}_k = \min(X_k, L) : k \geq 1\}$ with $E[\bar{X}_k] = \mu_L$. Let $\bar{S}_n = \sum_{i=1}^{n} \bar{X}_i$, and $\bar{M}(t)$ be the renewal function of the new process. Note that $\bar{S}_{N(t)+1} \leq t + \bar{X}_{N(t)+1} \leq t + L$. Taking the expectation on both sides of this inequality and using the Wald's Equation, we obtain

$$\bar{M}(t)+1)\mu_L \leq t + L \Rightarrow \frac{1}{\mu_L} + \left(\frac{L}{\mu_L} - 1\right)\frac{1}{t},$$

which implies

$$\limsup_{t\to\infty} \frac{\bar{M}(t)}{t} \leq \frac{1}{\mu_L}.$$

It follows from $\bar{S}_n \leq S_n$ that $\bar{N}(t) \geq N(t)$ and $\bar{M}(t) \geq M(t)$. Hence, we have

$$\limsup_{t\to\infty} \frac{M(t)}{t} \leq \limsup_{t\to\infty} \frac{\bar{M}(t)}{t} \leq \frac{1}{\mu_L}.$$

Letting $L \to \infty$ gives

$$\limsup_{t\to\infty} \frac{M(t)}{t} \leq \frac{1}{\mu}.$$

This completes the proof. ∎

For the case where the inter-renewal times $X_1, X_2, \dots$ are i.i.d. continuous random variables with p.d.f. $f(x)$, the renewal function is differentiable and

$$m(t) = \frac{dM(t)}{dt} = \sum_{n=1}^{\infty} f_n(t),$$

where $f_n(t)$ is the p.d.f. for S_n.

Exercise 5.10 *Based on the elementary renewal theorem, can we conclude that $M(t)$ behaves like t/μ as t grows large? Hint: Consider a special case where the inter-renewal time is constant $X_i = 1$ for $i = 1, 2, \dots$. Then argue that*

$M(t) - t/\mu$ does not approach to 0 as a limit although the elementary renewal theorem still holds. This happens when the renewal process has the "periodic behavior" or the inter-renewal time distribution F is "lattice" with a period $d = 1$. In general, the lattice case with period d, the renewals can only occur at integral multiples of d, which will be discussed in details in Blackwell's Theorem later.

Exercise 5.11 *Show that for the non-lattice renewal process (the periodic behavior is precluded), the elementary renewal theorem implies*

$$\lim_{t \to \infty} m(t) = \lim_{t \to \infty} \frac{dM(t)}{dt} = \frac{1}{\mu}.$$

In general, $M(t)$, and the distribution of $N(t)$ for a renewal process can be quite complex and often do not have closed-form expressions for a finite time t. It is natural to explore the limiting case as $t \to \infty$. Note that $N(t)$ is upper bounded by a negative binomial process $\bar{N}(t)$ as shown earlier, which is the sum of the i.i.d. random variables (geometrically distributed). In statistics, according to the central limit theorem (CLT), the sum of i.i.d. random variables approaches to a normally distributed random variable when number of random variables is getting larger. This CLT certainly applies to $\bar{N}(t)$. Thus, we project that $N(t)$ should also follow the CLT as $t \to \infty$. Suppose that the inter-renewal time follows a general distribution with mean of μ and variance of σ^2. From the elementary renewal theorem, we know that $\lim_{t \to \infty} E[N(t)] = t/\mu$. Now we claim that $N(t) \sim N(t/\mu, Var(N(t)))$ as $t \to \infty$. This is actually so-called CLT for renewal processes (CLTRP). Next, we prove the CLTRP. The proof is a little tricky and mainly adopted from Ross (1996). We start with the limiting variance of $N(t)$ divided by t, which is $\lim_{t \to \infty} Var(N(t))/t = \sigma^2/\mu^3$. This expression has an intuitive explanation. This asymptotic result implies that as t is getting larger, the variance of $N(t)$ is proportional to the squared coefficient of variation of the inter-renewal time and the proportion coefficient is the mean of the inter-arrival time. That is

$$Var(N(t)) \sim \left(\frac{\sigma}{\mu}\right)^2 \times \left(\frac{t}{\mu}\right), \quad \text{as } t \text{ is getting large.}$$

Our claim is that $N(t)$ follows $N(t/\mu, \sigma^2 t/\mu^3)$ asymptotically which is equivalent to

$$\lim_{t \to \infty} P\left(\frac{N(t) - t/\mu}{\sqrt{t\sigma^2/\mu^3}} < x\right) = \Phi(x),$$

where $\Phi(x) = \int_{-\infty}^{x} \frac{1}{\sqrt{\pi}} e^{-z^2/2} dz$ is the CDF of the standard normal distribution.

Theorem 5.5 *(CLTRP) For a renewal process with an inter-renewal sequence* $\{X_k : k \geq 1\}$. *Assume* $E[X_k] = \mu < \infty$, *and* $Var(X_k) = \sigma^2 < \infty$. *Then*

$$P\left(\frac{N(t) - t/\mu}{\sigma\sqrt{t/\mu^3}} < x\right) \to \Phi(x) \quad as \quad t \to \infty.$$

Proof. We start with writing the equivalence of the events in terms of the probability

$$P\left(\frac{N(t) - t/\mu}{\sigma\sqrt{t/\mu^3}} < x\right)$$

$$= P\left(N(t) < \frac{t}{\mu} + \frac{\sigma}{\mu}\sqrt{\frac{t}{\mu}}x\right) = P(N(t) \leq n), \tag{5.16}$$

where $n = \lfloor \frac{t}{\mu} + \frac{\sigma}{\mu}\sqrt{\frac{t}{\mu}}x \rfloor$, which can be interpreted as x standard deviations above the mean of $N(t)$. Rewriting the probability in (5.16) based on the equivalent events yields

$$P(N(t) \leq n) = P(S_n \geq t) = P\left(\frac{S_n - n\mu}{\sqrt{n}\sigma} \geq \frac{t - n\mu}{\sqrt{n}\sigma}\right)$$

$$\to 1 - \Phi\left(\frac{t - n\mu}{\sqrt{n}\sigma}\right) \quad as \quad n \to \infty \text{ by the CLT for } S_n \tag{5.17}$$

$$= \Phi(-\frac{t - n\mu}{\sqrt{n}\sigma}).$$

Finally, we need to show

$$-\frac{t - n\mu}{\sqrt{n}\sigma} \to x \quad as \quad t \to \infty.$$

Since $n \leq \frac{t}{\mu} + \frac{\sigma}{\mu}\sqrt{\frac{t}{\mu}}$, as $t \to \infty$, we have

$$\frac{t - n\mu}{\sqrt{n}\sigma} \geq \frac{t - \left(\frac{t}{\mu} + \frac{\sigma}{\mu}\sqrt{\frac{t}{\mu}}x\right)\mu}{\sigma\sqrt{\frac{t}{\mu} + \frac{\sigma}{\mu}\sqrt{\frac{t}{\mu}}x}} = \frac{-\sigma x\sqrt{t/\mu}}{\sigma\sqrt{\frac{t}{\mu} + \frac{\sigma}{\mu}\sqrt{\frac{t}{\mu}}x}} \to -x.$$

Similarly, it follows from $n \geq \frac{t}{\mu} + \frac{\sigma}{\mu}\sqrt{\frac{t}{\mu}} - 1$, as $t \to \infty$ that

$$\frac{t - n\mu}{\sqrt{n}\sigma} \leq \frac{t - \left(\frac{t}{\mu} + \frac{\sigma}{\mu}\sqrt{\frac{t}{\mu}}x - 1\right)\mu}{\sigma\sqrt{\frac{t}{\mu} + \frac{\sigma}{\mu}\sqrt{\frac{t}{\mu}}x - 1}} = \frac{-\sigma x\sqrt{t/\mu} - \mu}{\sigma\sqrt{\frac{t}{\mu} + \frac{\sigma}{\mu}\sqrt{\frac{t}{\mu}}x - 1}} \to -x.$$

This completes the proof. ∎

Exercise 5.12 *Assume that $\mu < \infty$ and $\sigma^2 < \infty$, show that*

$$\lim_{t \to \infty} E\left[\left(\left(\frac{N(t) - t/\mu}{\sqrt{t\sigma^2/\mu^3}}\right)^+\right)^p\right] = \frac{2^{p/2-1}\Gamma(2^{-1}(p+1))}{\sqrt{\pi}}$$

for each $p > 0$, where $\Gamma(\cdot)$ is Euler's gamma function, and $x^+ = max(x,0)$.

Exercise 5.13 *Suppose that a component in a system is subject to failure. If it fails, it can be replaced with either brand A with probability 0.4 or brand B with probability 0.6. The lifetimes of brand A (or brand B) are exponentially distributed with mean of 10 days (or 7 days). However, brand A component takes exactly 2 days to install and brand B component takes exactly 1 day to install. If the component is working at the beginning of the year, what is the approximate distribution for the number of failures during the year?*

5.1.4 BLACKWELL'S THEOREM AND KEY RENEWAL THEOREM

In a machine maintenance problem, we may be interested in (i) the average number of units replaced in the interval $(t, t + h]$, where $h \geq 0$ is fixed; and (ii) of the probability of renewal at time t. As discussed above, although there exists the renewal function that $M(t)$ satisfies, it may not be solvable for the closed-form solution for $M(t)$. Thus, we hope to get the answers to the questions of our interest when t is large. To achieve this goal, we need to define the lattice random variable and distinguish between two scenarios in renewal processes.

Definition 5.3 *A non-negative random variable X is said to be lattice if there exists $d \geq 0$ such that $\sum_{n \in \mathbb{N}} P(X = nd) = 1$, where $d = \sup\{d \in \mathbb{R}^+ : \sum_{n \in \mathbb{N}} P(X = nd) = 1\}$ is called its period.*

If X is a lattice random variable, its distribution function F is also called lattice. Now, we present the Blackwell's theorem.

Theorem 5.6 *(Blackwell's Theorem) For a renewal counting process $N(t)$ with mean $M(t)$, and inter-arrival times with distribution F and mean μ. If F is not lattice, then for all $a \geq 0$,*

$$\lim_{t \to \infty} [M(t + a) - M(t)]] = \frac{a}{\mu}.$$

If F is lattice with period d, then

$$\lim_{t \to \infty} E[number\ of\ renewals\ at\ nd] = \frac{d}{\mu}.$$

The complete proof of this theorem can be found in Feller (1966). Here we only provide some intuitive justification. If F is not lattice, then the expected (or average) number of renewals in an interval of length a, when t is large, is approximately a/μ, which is quite intuitive. Denote this expected number by $g(a) \equiv \lim_{t\to\infty}[M(t+a) - M(t)]$. If the existence of this limit is proved (see Feller (1966)), then the result is the consequence of the elementary renewal theorem. To show this, we note that

$$
\begin{aligned}
g(a+b) &= \lim_{t\to\infty}[M(t+a+b) - M(t)] \\
&= \lim_{t\to\infty}[M(t+a+b) - M(t+a) + M(t+a) - M(t)] \\
&= g(b) + g(a).
\end{aligned}
$$

Then, the only increasing solution of such a g is the proportion function

$$
g(a) = ca, \forall a > 0
$$

for some constant c. Next, we determine c by defining the following sequence $\{x_n, n \in \mathbb{N}\}$ in terms of $M(t)$ as

$$
x_n = M(n) - M(n-1), \quad n \in \mathbb{N},
$$

with $M(0) = 0$. Clearly, we have $\sum_{i=1}^{n} x_i = M(n)$ and $\lim_{n\to\infty} x_n = g(1) = c$. Thus, we have

$$
\lim_{n\to\infty} \frac{\sum_{i=1}^{n} x_i}{n} = \lim_{n\to\infty} \frac{M(n)}{n} = c,
$$

where the second equality follows from the fact that a sequence $\{x_i\}$ converges to c, then the running average sequence $(\sum_{i=1}^{n} x_i)/n$ also converges to c. Applying the elementary renewal theorem, we have $c = 1/\mu$. When F is lattice with period d, then $g(a)$ does not exist. This is because the expected number of renewals in an interval far from the origin depends not only on the length of the interval but also on how many points of the form $nd, n \geq 0$, contained in the interval. However, an alternative limit $\lim_{n\to\infty} E[\text{number of renewals at } nd]$ can be considered in the lattice case. Again if it exists, then by the elementary renewal theorem, it equals d/μ. Note that at a time instant nd, the number of renewal will be either 0 or 1. Thus, the expected number of renewals at this instant is equal to the probability that a renewal occurs at this instant. Consider an interval of length kd far away from the origin (i.e., a large t). For time instants of $t + id$, $i = 1, 2, ...k$, define the indicator variable $I_{t+id} = 1$ (renewal occurred) or $I_{t+id} = 0$ (no renewal). We can compute the probability rate per time unit asymptotically as

$$
\lim_{t\to\infty} \frac{E[N(t+kd) - N(t)]}{kd} = \frac{\lim_{t\to\infty}(M(t+kd) - M(t))}{kd} = \frac{1}{\mu}, \quad k \geq 1.
$$

Thus, we have

$$\lim_{n \to \infty} P(\text{a renewal occurs at } nd) = \frac{d}{\mu} = \lim_{n \to \infty} E[\text{number of renewals at } nd],$$

which is the result for the lattice F case.

As an application of Blackwell's theorem, we answer the following interesting question: is the superposition of two renewal processes another renewal process? In a special case, the answer is yes. For example, if we superimpose two Poisson processes with parameters λ_1 and λ_2, the outcome is another Poisson process with parameter $\lambda_1 + \lambda_2$. However, in general, the answer to this question is no. Here is a counter-example.

Example 5.4 *Consider two non-Poisson renewal processes, denoted by $\{S_n, n \geq 1\}$ and $\{T_n, n \geq 1\}$, with inter-renewal distributions of F and G. Let $N_S(t)$ and $N_T(t)$ be the two corresponding renewal counting processes, respectively. Then the merged counting process can be written as*

$$N(t) := N_S(t) + N_T(t) = \sum_{n=1}^{\infty} [\mathbf{1}_{(0,t]}(S_n) + \mathbf{1}_{(0,t]}(T_n)].$$

Then the sequence of jump times of $N(t)$ is given by

$$J_n = \inf\{t : N(t) = n\}.$$

It is easy to show by a special case that $N(t)$ is not a renewal process as the inter-renewal times are no longer independent. Assume that F and G are constant distributions which are represented by $S_n = \{i, 2i, 3i, ...\}$ and $T_n = \{j, 2j, 3j, ...\}$, with $i \neq j$. Let $i = 2$ and $j = 3$, then we have

$$J_n - J_{n-1} = \begin{cases} 1 & n = 2, 3, 5, 6, 7, 9, 10, 11, ... \\ 2 & n = 1, 4, 8, 12, ... \end{cases}$$

with $J_0 = 0$. Let $M(t) = E[N(t)]$ and note that $M(n) - M(n-1) = J_n - J_{n-1}$. Then it is clear that $\lim_{t \to \infty}[M(t+1) - M(t)]$ does not exist (i.e., oscillates between 1 and 2). Thus, Blackwell's theorem does not hold and $N(t)$ is not a renewal process.

We can also show that the sum of two independent renewal processes is not a renewal process by confirming that the time intervals between two consecutive events are not i.i.d..

Example 5.5 *Suppose the inter-renewal times, X_i and Y_i with $i = 1, 2, ...,$ of the two independent renewal counting processes $\{N_S(t), t \geq 0\}$ and $\{N_T(t), t \geq 0\}$ follow uniform distributions over [0,1], denoted by $X_i \sim U[0,1]$ and*

$Y_i \sim U[0,1]$. *respectively. For the superimposed process* $N(t) = N_S(t) + N_T(t)$, *the inter-event times are denoted by* Z_j *with* $j = 1, 2, ...$ *To show that* $N(t)$ *is not a renewal counting process, we only need to show that* Z_2 *and* Z_1 *are not independent and have different distributions (i.e., they are not i.i.d..). Note that* $Z_1 = \min(X_1, Y_1)$. *Denote by* $F_{Z_j}(\cdot)$ *the distribution function of* Z_j *with* $j = 1, 2, ...$ *Then we have*

$$
\begin{aligned}
F_{Z_1}(z) &= P(Z_1 \leq z) = P(\min(X_1, Y_1) \leq z) \\
&= P(X_1 \leq z, Y_1 > z) + P(X_1 > z, Y_1 \leq z) + P(X_1 \leq z, Y \leq x) \\
&= z(1-z) + (1-z)z + z^2 \\
&= 2z - z^2
\end{aligned}
$$

for $z \in [0,1)$. *This is a triangular distribution over [0,1] with p.d.f. of* $f(z) = 2 - 2z$. *Now, we look at* Z_2. *It is easy to see that this time interval depends on value of* Z_1. *Assume that* $Z_1 = x, x \in [0,1)$. *Then,* Z_2 *will be upper bounded by* $1 - x$ *due to the residual time of one of the uniformly distributed random variables, which is denoted by* W. *Clearly* $W \sim U[0, 1-x)$. *Thus,* Z_2 *is the minimum of a uniformly distributed random variable over [0,1] (the inter-renewal time of one of* $N_S(t)$ *and* $N_T(t)$, *denoted by* U) *and a uniformly distributed random variable over* $[0, 1-x]$. *The distribution of the time from the first event to the second event of* $N(t)$ *should be denoted by* $F_{Z_2|Z_1=x}(z)$ *and is given by*

$$
\begin{aligned}
F_{Z_2|Z_1=x}(z) &= P(\min(U, W) \leq z) \\
&= P(U \leq z, W > z) + P(U > z, W \leq z) + P(U \leq z, W \leq z) \\
&= \frac{2z - z^2 - zx}{1 - x}
\end{aligned}
$$

for $z \in [0, 1-x)$. *Comparing the first and second inter-event intervals, we conclude that* $N(t)$ *is not a renewal counting process.*

Although the superimposed process $N(t)$ is not a renewal process, we may study such a process by formulating a Markov renewal process. The Markov renewal process describes the event occurrence instants by introducing a state variable for the process and the transitions among the states. The Markov renewal process is an extension of the renewal process and will be discussed in the next section. Another important result about superimposing independent renewal processes, called sources, is that the superposed process is approaching to a Poisson process when the number of sources is getting larger and larger. This is because the "local" (or micro-) behavior of the superimposed process is getting closer to that of a Poisson process as the number of individual renewal processes is getting larger. For more details about this behavior, interested readers are referred to Cox (1961).

Exercise 5.14 *Show that superimposing $k \geq 2$ independent Poisson processes with rates $\lambda_i, i = 1, 2, ...k$ (renewal processes) yields another Poisson process (renewal process) with rate $\sum_{i=1}^{k} \lambda_i$.*

To study the total effects of the renewals at a certain time point far away from the origin, we present the key renewal theorem (KRT).

Theorem 5.7 *(KRT) For a nonlattice F with $\mu = E[X] \leq \infty$, if $h(t)$ is directly Riemann integrable, then*

$$\lim_{t \to \infty} (h * M)(t) = \lim_{t \to \infty} \int_0^t h(t - s) \, dM(s) = \frac{1}{\mu} \int_0^\infty h(x) \, dx.$$

For a lattice F with period $d, \mu = E[X] \leq \infty$ and $\sum_{n=0}^{\infty} h(nd)$ exists and is finite, then

$$\lim_{n \to \infty} \sum_{k=0}^{n} h((n-k)d)[M(kd) - M((k-1)d)] = \frac{d}{\mu} \sum_{n=0}^{\infty} h(nd).$$

The complete proof of this theorem can be found in Feller (1966). Here we utilize the concept of "directly Riemann integrable" that is a more restrictive definition than the usual "Riemann integrable" in calculus. A function is said to be directly Riemann integrable if the lower and upper Riemann sums on the entire unbounded interval converge to a common number as the partition is refined. Suppose that $h : [0, \infty) \to [0, \infty)$. For $\delta \in [0, \infty)$ and $k \in \mathbb{N}$, let $m_k(h, \delta) = \inf\{h(t) : t \in [k\delta, (k+1)\delta)\}$ and $M_k(h, \delta) = \sup\{h(t) : t \in [k\delta, (k+1)\delta)$. The Riemann sums of h on $[0, \infty)$ corresponding to δ are

$$L_h(\delta) = \delta \sum_{k=0}^{\infty} m_k(h, \delta), \quad U_h(\delta) = \delta \sum_{k=0}^{\infty} M_k(h, \delta).$$

The sums exist in $[0, \infty]$ and satisfy the following properties:

- $L_h(\delta) \leq U_h(\delta)$ for $\delta > 0$
- $L_h(\delta)$ increases as δ decreases
- $U_h(\delta)$ decreases as δ decreases

We show the logic of the KRT by considering an example. Assume that a renewal counting process $N(t)$ with a non-lattice F. Each time a renewal occurs, an impulse is generated and represented by a directly Riemann integrable function $h(t)$. For example, $N(t)$ may represent the number of vehicles passing an office building and each car passing generates some traffic noise. The decibel scale measure of the noise is the value of $h(t)$. We are interested in the average noise level due to vehicle traffic at a certain time point far away from the origin. Figure 5.2 shows a possible form of $h(t)$. Denote by S_n the arrival instant of nth renewal. The effect at time $t \geq S_n$ of an impose generated

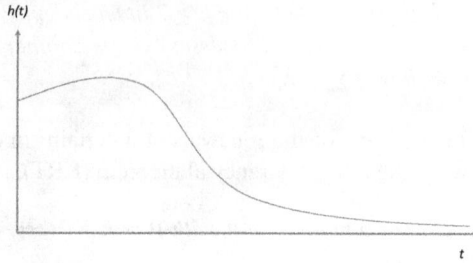

Figure 5.2 Noise function $h(t)$.

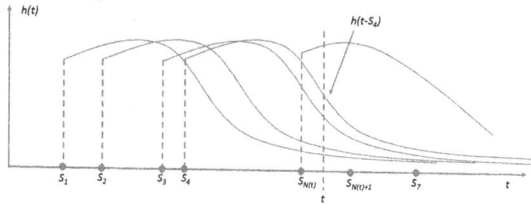

Figure 5.3 A graphic representation of $H(t)$.

at the nth renewal occurrence is given by $h(t - S_n)$. Thus, we can express the total effect of all impulses generated before time t, denoted by $H(t)$, as

$$H(t) = \sum_{n=1}^{N(t)} h(t - S_n).$$

Since $N(t)$ is a nondecreasing function, for $t \geq 0$ we can treat it as a positive random measure on $\mathbb{R}^+$ that locates a unit dirac mass function at any renewal epoch. Thus, we have

$$N(t) = \sum_{n=1}^{\infty} I_{S_n \leq t}; \quad N(dt) = dN(t) = \sum_{n=1}^{\infty} \delta_{S_n}(dt).$$

Using this representation, we can express the total impulse at time t in the alternative way

$$H(t) = \int_0^t h(t - s) dN(s). \tag{5.18}$$

Figure 5.3 shows a graphical illustration of $H(t)$.

Next, we can develop a third representation of $H(t)$. After changing the variable, the integral in equation (5.18) may be rewritten as

$$H(t) = \int_0^t h(s) dN(t - s) = \sum_{n=-N(t)}^{-1} h(S_n), \tag{5.19}$$

where we have defined S_{-n}, with $n > 0$, as the nth renewal before t. This is equivalent to taking time point t as the origin and looking backward in time, i.e., $S_{-1} = S_{N(t)}, S_{-2} = S_{N(t)-1}, ..., S_{-N(t)} = S_1$. If we let $t \to \infty$, the renewal process reaches the steady state and equation (5.19) can be written as

$$H(\infty) = \int_0^\infty h(s) dN^*(-s) = \sum_{n=-\infty}^{-1} h(S_n^*), \tag{5.20}$$

where we denoted by $S_{-n}^*, n > 0$, the nth renewal before 0 of the stationary renewal process $N^*(t), t \in (-\infty, \infty)$. From a time point t, going forward in time to the next renewal is the residual life $R(t)$ and going backward in time to the previous renewal is the age $A(t)$ of the renewal process. As shown later, these two random variables have the same distribution in limit or $\lim_{t \to \infty} P(R(t) \leq x) = \lim_{t \to \infty} P(A(t) \leq x)$. This implies that the stationary renewal process going forward in time is equal to the stationary renewal process going backward in distribution, i.e. $N^*(-t) =_d N^*(t), t \geq 0$. Thus, Equation (5.20) can be rewritten as

$$H(\infty) = \int_0^\infty h(s) dN^*(s) = \sum_{n=1}^\infty h(S_n^*). \tag{5.21}$$

Recall that from the elementary renewal theorem that the renewal function for a large $t > 0$ (i.e., stationary renewal process) is $M^*(t) = E[N*(t)] = t/\mu$. Then we have

$$E[H(\infty)] = E\left[\int_0^\infty h(s) dN^*(s)\right] = \int_0^\infty h(s) E[dN^*(s)]$$
$$= \int_0^\infty h(s) M^* d(s) = \frac{1}{\mu} \int_0^\infty h(s) ds.$$

Taking expectation in (5.18), we get

$$E[H(t)] = E\left[\int_0^t h(t-s) dN(s)\right] = \int_0^t h(t-s) E[dN(s)] = \int_0^t h(t-s) M(ds).$$

Assuming that $E[H(t)] \to E[H(\infty)]$ as $t \to \infty$, we obtain

$$\lim_{t \to \infty} \int_0^t h(t-s) M(ds) = \frac{1}{\mu} \int_0^\infty h(s) ds,$$

which is the KRT. It can be shown that the Blackwell's theorem and KRT are equivalent. Consider a special indicator function for $h(t) = \mathbf{1}(a \leq t \leq b)$ with

$0 \le a \le b < \infty$. Thus, we can apply the KRT to this $h(t)$ function as follows:

$$\int_0^t h(t-s)M(ds) = \int_0^\infty \mathbf{1}(a \le t-s \le b)\mathbf{1}(s \le t)M(ds)$$

$$= \int_0^\infty \mathbf{1}(t-b \le s \le t-a)M(ds)$$

$$= \int_{t-b}^{t-a} M(ds) = M(t-a) - M(t-b),$$

which implies the Blackwell's theorem if letting $t \to \infty$. That is

$$\lim_{t\to\infty} M(t+(b-a)) - M(t) = \lim_{t\to\infty} M(t-a) - M(t-b)$$

$$= \lim_{t\to\infty} \int_0^t h(t-s)Md(s) = \frac{1}{\mu}\int_0^\infty h(t)dt = \frac{1}{\mu}(b-a),$$

for any $b - a \ge 0$. To see the other way, we note that

$$\frac{1}{\mu}\int_0^\infty h(t)dt = \frac{1}{\mu}(b-a) = \lim_{t\to\infty} M(t+(b-a)) - M(t)$$

$$= \lim_{t\to\infty} M(t-a)) - M(t-b) = \lim_{t\to\infty} \int_0^t h(t-s)M(ds)$$

for any $0 \le a \le b < \infty$. This verification is done with the indicator functions. It is also valid for simple functions with bounded support, which can be expressed as linear combination of indicator variables. For a general directly Riemann integrable function $h(x)$, we can decompose it as the difference between two positive directly Riemann integrable functions, i.e. $h(t) = h^+(t) - h^-(t)$, where $h^+(t) = \max(h(t), 0) > 0$ and $h^-(t) = \max(-h(t), 0) > 0$. Note that any positive directly Riemann integrable function $f(t)$ can be approximated from below by a sequence of simple functions $f_n(t)$ of bounded support with the following expression:

$$f_n(t) = \sum_{k=0}^{n2^n-1} \inf\{f(t), k2^{-n} \le t \le (k+1)2^{-n}\}\mathbf{1}(k2^{-n} \le t < (k+1)2^{-n}).$$

In this way, we can verify the equivalence of Blackwell's theorem and the KRT for general directly Riemann function case.

Exercise 5.15 *Toss a coin indefinitely with probability p to get Heads H (with probability $q = 1 - p$ to get Tails T). Suppose that we are interested in a particular patten, say THTH. Let $N(t)$ be the number of times the pattern occurs by time $\lfloor t \rfloor$. For example, in a realized sequence of tossing the coin $\{H,T,H,T,H,T,H,H,H,T,H,T,H,...\}$, the pattern THTH occurs at times 5, 7, 13,... and $N(13) = 3$.*

(a) Explain that $\{N(t), t \ge 0\}$ is a delayed renewal counting process.

(b) What is the expected value of the inter-renewal time μ?

Note that the renewal process in the exercise above is a discrete-time renewal process. Also this exercise shows the potential applications to "patterns" of renewal theory. For more details about discrete-time renewal process and applications to patterns, interested readers are referred to Ross (2014). Another important application of the key renewal theorem is presented in the next section.

5.1.5 INSPECTION PARADOX

Now we consider the age process, $A(t) = t - S_{N(t)}$, and the residual life process, $R(t) = S_{N(t)+1} - t$. Again, we will focus on the cases when t is getting large (i.e., $t \to \infty$).

The key renewal theorem can be used to find the limiting distributions of age and residual life. Using the equivalent events, we have $P(R(t) > x) = P(N(t,t+x] = 0)$. If the renewal process is nonlattice, then $P(R(t) > x)$ satisfies the renewal equation that can be established by conditioning on the time of the first renewal, which will occur in one of three time intervals: $[0,t], (t,t+x]$, and $(t+x, \infty)$. Thus, we have

$$P(R(t) > x) = \int_0^\infty P(R(t) > x | X_1 = s) dF(s)$$

$$= \int_0^t P(R(t) > x | X_1 = s) dF(s) + \int_t^{t+y} P(R(t) > x | X_1 = s) dF(s)$$

$$+ \int_{t+y}^\infty P(R(t) > x | X_1 = s) dF(s)$$

$$= \int_0^t P(R(t-s) > x) dF(s) + \int_t^{t+x} 0 dF(s) + \int_{t+x}^\infty 1 dF(s)$$

$$= F^c(t+x) + \int_0^t P(R(t-s) > x) dF(s).$$

Clearly, this is a GRE of form $u = a + u * F$ with $u = P(R(t) > x)$ and $a = F^c(t+x)$. It follows from the fundamental theorem for renewal processes that $u = a + a * M$ which is

$$P(R_t > x) = F^c(t+x) + \int_0^t F^c(t+x-s) dM(s), \quad x \in [0,\infty).$$

But $F^c(t+x) \to 0$ as $t \to \infty$, and by the KRT, the integral converges to $\frac{1}{\mu} \int_0^\infty F^c(x+y) dy$. Here, $h(t-s)$ term in the KRT is $F^c(t+x-s)$. Thus, $x - s = -y$ or $x + y = s$ and $ds = dy$. Finally a change of variables in the limiting integral gives the result.

$$P(R(t) > x) \to \frac{1}{\mu} \int_x^\infty F^c(y) dy \text{ as } t \to \infty, \quad x \in [0,\infty).$$

Similarly, for the age process, using (5.12) and the equivalence of $\{A(t) \geq x\} = \{S_{N(t)} \leq t - x\}$, we have $P(A(t) \geq x)$ satisfying

$$P(A(t) \geq x) = P(S_{N(t)} \leq t - x) = F^c(t) + \int_0^{t-x} F^c(t-s)\,dM(s), \quad x \in [0,t].$$

Again, $F^c(t) \to 0$ as $t \to \infty$. The change of variables $u = t - x$ changes the integral into $\int_0^u F^c(u + x - s)\,dM(s)$. By the KRT, this integral converges $\frac{1}{\mu}\int_0^\infty F^c(y+x)\,dy = \int_x^\infty F^c(y+x)\,dy$,

$$P(A(t) \geq x) \to \frac{1}{\mu}\int_x^\infty F^c(y)\,dy \text{ as } t \to \infty, \quad x \in [0,\infty).$$

By these two results, the limiting right distribution functions of $R(t)$ and $A(t)$ can be obtained. Note that the limiting distribution function is

$$1 - \frac{1}{\mu}\int_x^\infty F^c(y)\,dy = \frac{1}{\mu}\left(\mu - \int_x^\infty F^c(y)\,dy\right).$$

But recall that $\mu = \int_0^\infty F^c(y)\,dy$ so the result follows since $\int_0^\infty F^c(y)\,dy - \int_x^\infty F^c(y)\,dy = \int_0^x F^c(y)\,dy$.

Therefore, the age and the residual life have the same limiting distribution. That is

$$\lim_{t\to\infty} P(A(t) \leq x) = \lim_{t\to\infty} P(R(t) \leq x) = \frac{1}{\mu}\int_0^x F^c(y)\,dy, x \in [0,\infty). \quad (5.22)$$

The fact that the current and remaining age processes have the same limiting distribution may seem surprising at first, but there is a simple intuitive explanation. After a long period of time, the renewal process looks just about the same backward in time as forward in time. But reversing the direction of time reverses the roles of age and residual life.

Next, we consider the cycle time at an arbitrary point t, $C(t) = A(t) + R(t) = S_{N(t)+1} - S_{N(t)}$. If $N(t) = n$, then $C(t) = X_{n+1}$. Hence, in general, $C(t) = X_{N(t)+1}$. It seems that $C(t)$ should have the same distribution as any X_k. In fact, this inter-renewal interval with the inspection point has a different distribution from F. We first find the limiting mean of age or residual life. From (5.22), we have the limiting p.d.f for the residual life $f_{R(t)} = (1/\mu)F^c(x)$. The expected value $E[R(t)]$ as $t \to \infty$ is given by

$$\lim_{t\to\infty} E[R(t)] = \int_0^\infty x\frac{1}{\mu}F^c(x)\,dx = \frac{1}{2\mu}\int_0^\infty F^c(x)\,d(x^2)$$

$$= \frac{1}{2\mu}\left(x^2 F^c(x)\big|_0^\infty - \int_0^\infty x^2\,dF^c(x)\right)$$

$$= \frac{1}{\mu}\int_0^\infty x^2\,dF(x) = \frac{E[X^2]}{2\mu}.$$

Note that $x^2 F^c(x)|_0^\infty = 0$ is due to the finiteness of the moment generating function for the inter-renewal time. Denote by $\mathcal{M}(s) = E[e^{sX}]$ for some $s > 0$ the moment generating function of inter-renewal time. It follows that $F^c(x) = P(X > x) \leq \mathcal{M}(s)e^{-sx}$ for all $x > 0$. Thus, for the finite $\mathcal{M}(s)$ at some s, $\lim_{x\to\infty} x^2 F^c(x) = 0$. Alternatively, this result, summarized in the following proposition, can be obtained by applying the KRT to the renewal equation that $E[R(t)]$ satisfies (see Ross (1996)).

Proposition 5.2 *If F is non-lattice, $E[X] = E[X_k] = \mu$, and $E[X_k^2] = E[X^2] < \infty$, then*

$$\lim_{t\to\infty} E[R(t)] = \lim_{t\to\infty} E[A(t)] = \frac{E[X^2]}{2\mu}.$$

The relation between $M(t)$ and $E[R(t)]$ can be established by taking the expectation on both sides of $S_{N(t)+1} = t + R(t)$ and using the Wald's Equation. That is

$$M(t) - \frac{t}{\mu} = \frac{E[R(t)]}{\mu} - 1,$$

which leads to the limiting result

$$\lim_{t\to\infty} \left(M(t) - \frac{t}{\mu} \right) = \frac{E[X^2]}{2\mu} - 1.$$

Now, we find that the mean of the inter-renewal interval with an inspection point t is greater than or equal to the mean of the inter-renewal interval $\mu = E[X]$ for the equilibrium renewal process (i.e., the process as $t \to \infty$).

$$\lim_{t\to\infty} E[C(t)] = \lim_{t\to\infty} E[X_{N(t)+1}] = \lim_{t\to\infty} E[A(t) + R(t)]$$

$$= \lim_{t\to\infty} E[A(t)] + \lim_{t\to\infty} E(R(t)) = \frac{E[X^2]}{2\mu} + \frac{E[X^2]}{2\mu}$$

$$= \frac{E[X^2]}{\mu} \geq \frac{\mu^2}{\mu} = \mu.$$

Such a phenomenon is called the "inspection paradox", the inspected inter-renewal interval is stochastically larger than a regular inter-renewal interval. Here, we utilize the limiting distribution to demonstrate this paradox. In fact, this paradox exists in general. We formally present this property in terms of the tail distribution. To simplifying the notations (without using the limit notation), we introduce the symbols for limiting random variables as $t \to \infty$. Define $\bar{C} = \lim_{t\to\infty} C(t)$, $\bar{A} = \lim_{t\to\infty} A(t)$, and $\bar{R} = \lim_{t\to\infty} R(t)$. It is worth noting that some important and nice formulas can be derived by using the "sample-path analysis" approach. For example, consider the age process $A(t)$ in a time period $(0, t]$. The average age can be estimated by computing the total area of the

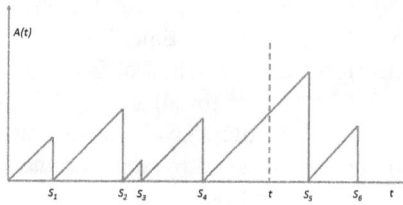

Figure 5.4 Age of a renewal process (the sloping line is 45 degrees).

sample path of $A(t)$ (i.e., the sum of the triangle areas with sides X_i) divided by t as shown in Figure 5.4. That is

$$\frac{1}{t}\int_0^t A(s)ds \approx \frac{1}{t}\sum_{i=1}^{N(t)}\frac{X_i^2}{2}.$$

In fact, we can express this approximation by bounding its value as

$$\frac{1}{t}\sum_{i=1}^{N(t)}\frac{X_i^2}{2} \le \frac{1}{t}\int_0^t A(s)ds \le \frac{1}{t}\sum_{i=1}^{N(t)+1}\frac{X_i^2}{2}.$$

Writing $1/t = (N(t)/t) \times (1/N(t))$ and using the elementary renewal theorem and SLLN, we obtain the same result $E[\bar{A}(t)] = \lim_{t\to\infty}(1/t)\int_0^t A(s)ds = E[X^2]/(2\mu)$ with probability 1. Another approximation is

$$\int_0^t \mathbf{1}\{A(s) > x\}ds \approx \sum_{i=1}^{N(t)}(X_i - x)^+,$$

where $a^+ := \max\{0,a\}$. It is easy to see from the sample path that the length of time during the ith inter-renewal interval that $A(s) > x$ is exactly $(X_i - x)^+$. Dividing both sides of the relation above and taking the limit by letting $t \to \infty$, we get

$$\lim_{t\to\infty}\frac{1}{t}\int_0^t \mathbf{1}\{A(s) > x\}ds = \lim_{t\to\infty}\frac{1}{t}\sum_{i=1}^{N(t)}(X_i - x)^+ = \frac{1}{\mu}E[(X - x)^+]$$

with probability 1. Furthermore, by taking the expectation on both sides of the above relation, we obtain

$$\lim_{t\to\infty}\frac{1}{t}\int_0^t P(A(s) > x)ds = P(\bar{A}(t) > x) = \frac{1}{\mu}E[(X - x)^+] = \frac{1}{\mu}\int_x^\infty F^c(x)ds.$$

Similar development can be done for $R(t)$ based on Figure 5.5 and is left as an exercise. For the cycle time $C(t)$, based on the sample path in Figure 5.6, we

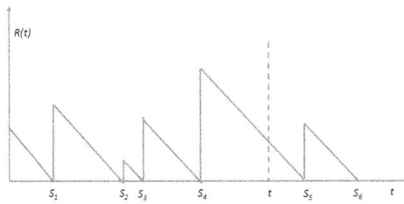

Figure 5.5 Residual life of a renewal process.

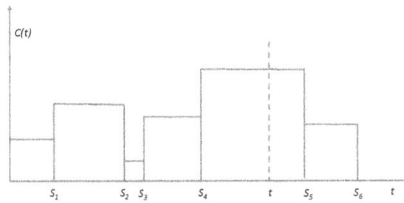

Figure 5.6 Cycle life of a renewal process.

have

$$\frac{1}{t}\int_0^t 1\{C(s) > x\}ds \approx \frac{1}{t}\sum_{i=1}^{N(t)} X_i 1\{X_i > x\}.$$

The length of time during the ith inter-renewal time $C(s) > x$ is exactly X_i if $X_i > x$ and 0 otherwise. Taking the limit of $t \to \infty$ on both sides of the relation above and using the elementary renewal theorem and SLLN we obtain

$$\lim_{t\to\infty}\frac{1}{t}\int_0^t 1\{C(s) > x\}ds = \frac{1}{\mu}E[X1\{X > x\}]$$

with probability 1. Denote by $F_{\bar{R}}(x) = \lim_{t\to\infty} P(R(t) \le x) = (1/\mu)\int_0^x F^c(y)dy$. Integrating the limiting result, we have the limiting complementary distribution function for the cycle time $\bar{C} = \lim_{t\to\infty} C(t)$ as

$$F_{\bar{C}}^c(x) = \frac{1}{\mu}E[X1\{X > x\}] = \frac{1}{\mu}\int_x^\infty xf(x)dx = -\frac{1}{\mu}\int_x^\infty xdF^c(x)$$

$$= -\left(\frac{1}{\mu}xF^c(x)\big|_x^\infty - \frac{1}{\mu}\int_x^\infty F^c(x)fx\right)$$

$$= \frac{1}{\mu}xF^c(x) + \frac{1}{\mu}\int_x^\infty F^c(x)dx$$

$$= \frac{1}{\mu}xF^c(x) + F_{\bar{R}}^c(x)$$

with density function $f_{\bar{C}}(x) = (1/\mu)xf(x)$. We have already seen that $E[\bar{C}] = \lim_{t\to\infty} E[X_{N(t)+1}] \geq E[X_n] = E[X]$. Now we present the inspection paradox in terms of tail probability.

Theorem 5.8 (*Inspection Paradox*) *For every fixed t, $C(t)$ is stochastically larger than X, i.e.*

$$P(C(t) > x) \geq P(X > x) = F^c(x), \; t \geq 0, \; x \geq 0.$$

Furthermore, $\bar{C}$ is stochastically larger than X, i.e.

$$P(\bar{C} > x) \geq P(X > x) = F^c(x), \; x \geq 0.$$

Proof. By conditioning on the number of renewals by time t and the nth renewal instant, we have

$$P(C(t) > t|N(t) = n, S_n = s) = P(X_{n+1} > x|X_{n+1} > t-s)$$
$$= \frac{P(X_{n+1} > x, X_{n+1} > t-s)}{P(X_{n+1} > t-s)}$$
$$= \frac{F^c(\max(x, t-s))}{F^c(t-s)} \geq F^c(x).$$

The last inequality follows by considering two cases: If $x > t - s$, then $\max(x, t-s) = x$ and

$$\frac{F^c(\max(x, t-s))}{F^c(t-s)} = \frac{F^c(x)}{F^c(t-s)} \geq F^c(x),$$

because of $F^c(t-s) \leq 1$. If $x \leq t-s$, then $\max(x, t-s) = t-s$ and

$$\frac{F^c(\max(x, t-s))}{F^c(t-s)} = \frac{F^c(t-s)}{F^c(t-s)} = 1 \geq F^c(x).$$

Thus, we conclude that with probability 1, $P(C(t) > x|N(t), S_{N(t)}) \geq F^c(x)$. Taking the expectation with respect to $N(t)$ and $S_{N(t)}$ on both sides yields the result, i.e. $P(C(t) > x) = E[P(C(t) > x|N(t), S_{N(t)})] \geq F^c(x)$. Next, by taking the limit, we obtain

$$P(\bar{C} > x) = \lim_{t\to\infty} \frac{1}{t} \int_0^t P(C(s) > x)ds$$
$$\geq \lim_{t\to\infty} \frac{1}{t} \int_0^t P(X > x)ds \geq P(X > x).$$

This completes the proof. ∎

Exercise 5.16 *Suppose that the inter-renewal times $X_1, X_2, \ldots$ follow the gamma distribution with p.d.f. $f(t) = te^{-t}$.*

(a) Find the derivative of the renewal function $m(t)$ and $M(t)$.

(b) Check that $\lim_{t \to \infty}[M(t) - t/\mu] = (\sigma^2 - \mu^2)/(2\mu^2)$ where μ and σ^2 are the mean and variance of X_1, respectively.

Exercise 5.17 *Show that for any non-negative function (i.e., g is 0 on $(-\infty, 0)$) and the age process $A(t)$ of a renewal process, the following relation holds*

$$E[g(A(t)] = g(t)F^c(t) + \int_0^t g(t-s)F^c(t-s)dM(s),$$

where F is the distribution function of the inter-renewal time and $M(t)$ is the renewal function of the renewal process.

Exercise 5.18 *(The insurance ruin problem) Suppose that claims to an insurance company occur according to a Poisson process with rate λ. The successive claim amounts Y_i with $i = 1, 2, \ldots$ are i.i.d. random variables with the distribution function F, p.d.f. $f(x)$, and mean μ. Let $N(t)$ be the number of claims occurred in $[0, t]$. Then $\sum_{i=1}^{N(t)} Y_i$ is the total amount paid out in claims by time t. Assume that the company starts with an initial capital x and receives revenue at a constant rate r per unit time. The probability that the company's net capital ever becomes negative (bankruptcy) is given by*

$$B(x) = P\left[\sum_{i=1}^{N(t)} Y_i > x + rt \text{ for } t \geq 0\right],$$

which is called the bankruptcy proability.

(a) Show that $B(x)$ satisfies the following differential equation

$$B'(x) = \frac{\lambda}{c}B(x) - \frac{\lambda}{c}\int_0^x B(x-y)f(y)dy - \frac{\lambda}{c}\int_x^\infty f(y)dy.$$

(b) Verify that $B(x) = E[\rho^{N(x)+1}] = \rho e^{-x(1-\rho)/\mu}$ with $\rho = \lambda\mu/c$ satisfies the differential equation in (a).

While most of the analyses above are regarding the limit behaviors of the renewal process as $t \to \infty$, for a class of so-called stationary point processes, we can obtain some results about the limiting behaviors as $t \to 0$. A counting process $\{N(t), t \geq 0\}$ that possesses stationary increments is called a stationary point process. We present a few results without proofs. Interested readers are referred to Ross (1996) for more details. The first result is regarding the limiting ratio of probability of at least an event occurring in time interval $[0,t]$ to t as $t \to 0$. It can be proved that for any stationary point process $\{N(t), t \geq 0\}$

$$\lim_{t \to 0} \frac{P(N(t) > 0)}{t} = \lambda > 0 \tag{5.23}$$

for all $t \geq 0$, where λ is a constant (including the case $\lambda = \infty$). Note that the trivial case of $P(N(t) = 0) = 1$ is excluded. λ is called the intensity and is the transition rate in the CTMCs. Clearly, a homogeneous Poisson process is a stationary point process.

Exercise 5.19 *Verify that for a Poisson process with rate λ, the intensity in (5.23) holds.*

Since for any stationary point process, we have

$$E[N(t+s)] = E[N(t+s) - N(s)] + E[N(s)] = E[N(t)] + E[N(s)],$$

which implies that for some constant c it follows $E[N(t)] = ct$. Here c can be called the rate of $E[N(t)]$.

Exercise 5.20 *Show that $c \geq \lambda$ for a stationary point process and $c = \lambda$ for the equilibrium renewal process.*

To identify a class of stationary processes with $c = \lambda$, we define the "regular stationary point process". A stationary point process is said to be regular or orderly if $P(N(t) \geq 2) = o(t)$, which means that the probability that two or more events occur simultaneously at any time point is 0. It can be shown that for a regular stationary point process, $c = \lambda$ holds (including the case $c = \lambda = \infty$). It is easy to check that a homogeneous Poisson process is a regular stationary point process.

Exercise 5.21 *Show that a Poisson process with rate λ is a regular stationary point process and explain λ is both the transition rate (i.e., "the probability rate") and the renewal function rate c (i.e., $E[N(t)]$ rate).*

5.1.6 SOME VARIANTS OF RENEWAL PROCESSES

Alternating Renewal Processes: Traditionally, an alternating renewal process is defined as a renewal process that alternates between two states, denoted by 0 and 1, over time. Such a process can model a device like a component in a machinery system that alternates between on and off states. Whenever this component fails, it will be replaced with an identical one. Here, "on" period represents the lift time of the component and "off" period represents the replacement time. Actually, this renewal process can be generalized to a process with a deterministic multiple state sequence. For example, if a device consists of two parallel parts, one is in operation and the other is a back-up one. If the operating one fails, the back-up one is turned on. When both fails, the device is replaced. Thus, the system has three states in sequence: both functioning, one functioning, and both failing. Clearly, we can always consider such a process as an alternating renewal process by grouping these states into two consolidated states. The basic assumption is that the pairs of random times successively spent in the two states form an independent and identically distributed sequence. Let $\mathbf{X} = (X_1, X_2, ...)$ denote the successive lengths of time that the system is in state 1, and let $\mathbf{Y} = (Y_1, Y_2, ...)$ denote the successive lengths of time that the system is in state 0. So to be clear, the system starts in state 1 and remains in that state for a period of time X_1, then goes to state 0 and stays in this state for a period of time Y_1, then back to state 1 for a period of time X_2, and so forth. We assume that $\mathbf{W} = ((X_1, Y_1), (X_2, Y_2), ...)$ is an independent and identically distributed sequence. It follows that $\mathbf{X}$ and $\mathbf{Y}$ each are independent, identically distributed sequences, but $\mathbf{X}$ and $\mathbf{Y}$ might well be dependent. In fact, Y_n might be a function of X_n for $n \in N_+$. Let $\mu = E(X)$ denote the mean of a generic time period X in state 1 and let $v = E(Y)$ denote the mean of a generic time period Y in state 0. Let G denote the distribution function of a time period X in state 1, and as usual, let $G^c = 1 - G$ denote the complementary distribution function of X. Let $Z_k = X_k + Y_k$ for $k \geq 1$ with distribution function F. Clearly $\{Z_k, k \geq 1\}$ forms a renewal process. The renewal process associated with $\mathbf{W} = ((X_1, Y_1), (X_2, Y_2), ...)$ as described above is called an alternating renewal process (ARP). The stochastic process can be considered as the system on a binary state space. Thus, another way of representing this ARP is to use the system state process $\mathbf{I} = \{I_t : t \in [0, \infty)\}$ on state space $\{0, 1\}$. Denote by $p(t) = P(I_t = 1)$ the probability that the system is "on" at time t. It can be shown that this probability satisfies a renewal equation.

Theorem 5.9 *For an ARP with $E[X_k + Y_k] < \infty$, the probability function $p(t)$ satisfies a renewal equation*

$$p(t) = G^c(t) + (p * F)(t).$$

If the renewal process is non-lattice, then

$$\lim_{t \to \infty} p(t) = \frac{\mu}{\mu + v}.$$

Proof. Conditioning on the first renewal Z_1, we have

$$p(t) = P(I_t = 1) = P(I_t = 1, Z_1 > t) + P(I_t = 1, Z_1 \leq t)$$

$$= P(X_1 > t) + \int_0^t P(I_t = 1 | Z_1 = s) dF(s)$$

$$= G^c(t) + \int_0^t p(t - s) dF(s), \quad \text{for } t \in [0, \infty).$$

Next, based on the fundamental theorem on renewal equation, the solution to the GRE is $p(t) = G^c(t) + (G^c * M)(t)$. Note that $G^c(t) \to 0$ as $t \to \infty$. Using the KRT, we obtain that, as $t \to \infty$,

$$p(t) \to \frac{1}{\mu + v} \int_0^\infty G^c(s)ds = \frac{\mu}{\mu + v}.$$

This completes the proof. ∎

By appropriately defining on and off periods (or states), many stochastic processes can be turned into ARPs, leading in turn to interesting limits, via the basic limit theorem above. For example, we can obtain the limiting distribution of $\bar{C} = \lim_{t \to \infty} C(t)$, $\bar{R} = \lim_{t \to \infty} R(t)$, and $\bar{A} = \lim_{t \to \infty} A(t)$. This can be left as an exercise.

Exercise 5.22 *By defining the on and off states appropriately, derive the limiting distributions for $\bar{C}, \bar{R}$, and $\bar{A}$.*

Exercise 5.23 *Consider a service station serving Poisson arriving customers with arrival rate λ. The service times are i.i.d. random variables with mean of μ_G and there is no waiting room for customers. Such as queue can be denoted by $M/G/1/1$ in Kendall notation. Let $L(t)$ be the number of customers in the system at time t (i.e., $L(t) = 1$ or 0). By appropriately defining the on and off states of an alternating renewal process, find the limiting distribution of $L(t)$, that is $\lim_{t \to \infty} P(L(t) = 1)$.*

Exercise 5.24 *Similar to the previous exercise, find the limiting distribution for $G/M/1/1$ queue.*

Delayed Renewal Processes: If the first inter-renewal time has a different distribution from that of the remaining ones, the renewal process is called a delayed renewal process (DRP). The DRP is to model the situation where the starting time $t = 0$ is not a renewal point. For example, to study a failure/replacement process, we start with observing a device that has been operating for some time. Then the distribution function for the time to the first failure (first renewal) is different from the distribution function for the life time of a brand new device. Thus, this renewal process is a DRP. Let $\{X_k, k \geq 1\}$ be a sequence of inter-renewal times with X_1 having distribution G and X_k having distribution F for $k \geq 2$. We denote the delayed renewal counting process by $N_D(t) = \max\{n : S_n \leq t\}$, where $S_0 = 0, S_n = \sum_{k=1}^n X_k, k \geq 1$. In contrast, we call the renewal process with $G = F$ the ordinary renewal process (ORP). Similar to ORP, we get $P(N_D(t) = n) = P(S_n \leq t) - P(S_{n+1} \leq t) = G * F_{n-1}(t) - G * F_n(t)$. Let $M_D(t) = E[N_D(t)]$ be the renewal function for the DRP. We can easily find the renewal function and its Laplace transform as

$$M_D(t) = \sum_{n=1}^\infty G * F_{n-1}(t), \quad \tilde{M}(s) = \frac{\tilde{G}(s)}{1 - \tilde{F}(s)}.$$

Following the development of results for the ORP in previous section, we can derive the renewal equation that $M_D(t)$ satisfies and prove the SLLN, elementary renewal theorem, Blackwell's theorem, and KRT for the DRP. These derivations are left as exercises.

5.1.7   RENEWAL REWARD PROCESSES

If we consider the economic effects of a renewal process, we need to define a renewal reward process (RRP). Assume that for a renewal process $\{N(t), t \geq 0\}$, a reward is received whenever a renewal occurs. Denote by r_k the reward earned at the kth renewal instant. Assume further that $\{r_k, k \geq 1\}$ is a sequence of i.i.d. random variables. However, it is allowed that r_k may depend on X_k. Thus, we assume that the random variable pairs $\{(X_k, r_k), k \geq 1\}$ are i.i.d. Now we can define a reward stochastic process.

Definition 5.4 *For a sequence of the inter-renewal time and reward pairs* $\{(X_k, r_k), k \geq 1\}$, *let* $R_t = \sum_{k=1}^{N(t)} r_k$ *be the accumulated reward by time t. The stochastic process* $\mathbf{R} = \{R_t, t \geq 0\}$ *is called the renewal reward process (RRP) associated with* $\{(X_k, r_k), k \geq 1\}$. *The function* $\pi(t) = E[R_t]$ *for* $t \geq 0$ *is the reward function.*

Note that an RRP can be considered as a generalization of an ordinary renewal process. Specifically, if $r_k = 1, \forall k \geq 1$, then $R_t = N(t)$, so that the reward process simply reduces to the renewal counting process, and then $\pi(t)$ reduces to the renewal function $M(t)$.

An RRP is constructed if we want to model a particular random variable that is associated with a renewal process. Here are some typical examples:

- Customers arrive at a store according to a renewal process (a special case is a Poisson process). Each customer spends a random amount of money. We want to study the total sales (i.e., total customer spending) as a function of time t.
- The random arrivals are visits to a website. Each visitor spends a random amount of time at the site. We are interested in the total amount of time spent by visitors as a function of time t.
- The arrivals are random failure times of a complex system. Each failure requires a random repair time. We want to keep track of the total repair time for this system.
- The arrivals are earthquakes at a particular location. Each earthquake has a random severity, a measure of the energy released. We want to record the total amount of energy released from a sequence of earthquakes.

The total reward at time t, R_t, is a random sum of random variables for each time $t \geq 0$. In the special case that r_k and X_k are independent, the distribution of R_t is known as a compound distribution, based on the distribution of $N(t)$ and the distribution of a generic reward $r_k = r$. Specializing further, if the renewal process is Poisson and is independent of r, the process $\{R_t, t \geq 0\}$ is a compound Poisson process. Next, we present the renewal reward theorem.

Theorem 5.10 *For an RRP with $E[r_k] = v < \infty$ and $E[X_k] = \mu < \infty$, there exist the following limits with probability 1:*

$$\lim_{t \to \infty} \frac{R_t}{t} = \frac{v}{\mu},$$

$$\lim_{t \to \infty} \frac{E[R_t]}{t} = \frac{v}{\mu}.$$

Proof. Write the reward ratio as

$$\frac{R_t}{t} = \left(\frac{R_t}{N(t)}\right)\left(\frac{N(t)}{t}\right) = \left(\frac{\sum_{k=1}^{N(t)} r_k}{N(t)}\right)\left(\frac{N(t)}{t}\right).$$

Applying the SLLN to the sequence of i.i.d. random variables r_k yields $(1/n)\sum_{k=1}^{n} r_k \to v$ as $n \to \infty$ with probability 1. Then it follows from $N(t) \to \infty$ as $t \to \infty$ that $(1/N(t))\sum_{k=1}^{N(t)} r_k \to v$ as $t \to \infty$ with probability 1. Using the SLLN for the renewal process, we know that $N(t)/t \to 1/\mu$ as $t \to \infty$ with probability 1. Combining these results proves the first claim.

Note that

$$R_t = \sum_{k=1}^{N(t)} r_k = \sum_{k=1}^{N(t)+1} r_k - r_{N(t)+1}.$$

Recall that $N(t) + 1$ is a stopping time for the sequence of inter-renewal times $\{X_k, k \geq 1\}$, and hence is also a stopping time for the $\{(X_k, r_k), k \geq 1\}$ (a stopping time for a filtration is also a stopping time for any larger filtration). By Wald's equation,

$$E\left[\sum_{k=1}^{N(t)+1} r_k\right] = E[(N(t)+1)]v = [M(t)+1]v = vM(t) + v.$$

It follows from the elementary renewal theorem that

$$\lim_{t \to \infty} \frac{vM(t) + v}{t} = \lim_{t \to \infty} \left(v\frac{M(t)}{t} + \frac{v}{t}\right) = \frac{v}{\mu}.$$

Next, we show that $E[r_{N(t)+1}]/t \to 0$ as $t \to \infty$. Let $u(t) = E[r_{N(t)+1}]$ for $t \geq 0$. Conditioning on the first renewal time X_1, we get

$$u(t) = E[r_{N(t)+1}\mathbf{1}\{X_1 > t\}] + E[r_{N(t)+1}\mathbf{1}\{X_1 \leq t\}].$$

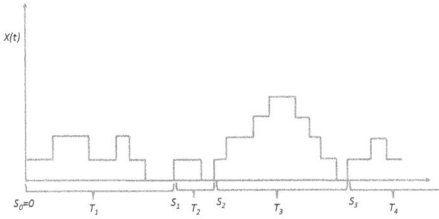

Figure 5.7 Busy cycles of a single server system.

Note that $X_1 > t$ if and only if $N(t) = 0$ so the first term, denoted by $a(t)$, is $a(t) = E[r_1 \mathbf{1}\{X_1 > t\}]$ and note that $|a(t)| \leq E[|r_1|] < \infty$. Hence, $E[r_1 \mathbf{1}\{X_1 > t\}]/t \to 0$ as $t \to \infty$. For the second term, if the first arrival occurs at time earlier than t, then the renewal process restarts, independently of the past, so

$$E[r_{N(t)+1}\mathbf{1}\{X_1 \leq t\}] = \int_0^t u(t-s)dF(s), \quad t \geq 0.$$

Thus, $u(t)$ satisfies the GRE $u = a + u * F$. Using the fundamental theorem of renewal equation, the solution is $u = a + a * M$. Now, fix $\varepsilon > 0$, there exists $T \in (0, \infty)$ such that $|a(t)| < \varepsilon$ for $t > T$. Hence, for $t > T$, we have

$$|\frac{u(t)}{t}| \leq \frac{1}{t}\left[|a(t)| + \int_0^{t-T}|a(t-s)|dM(s) + \int_{t-T}^t |a(t-s)|dM(s)\right]$$

$$\leq \frac{1}{t}[\varepsilon + \varepsilon M(t-T) + E[|r_1|][M(t) - M(t-T)].$$

Using the elementary renewal theorem again, the last expression converges to ε/μ as $t \to \infty$. Since $\varepsilon > 0$ is arbitrary, it follows that $u(t)/t \to 0$ as $t \to \infty$. ∎

5.1.8 REGENERATIVE PROCESSES

The idea of renewal cycle in a renewal process can be generalized to a stochastic process $\{X(t), t \geq 0\}$ on a discrete state space $\{0, 1, 2, ...\}$. If there exist consecutive random time intervals, denoted by $T_n, n = 1, 2, ...$, in which a process restarts itself probabilistically, this process is called a regenerative process with T_n as the nth regeneration cycle. Clearly, $\{T_n, n \geq 1\}$ constitute the inter-renewal times of a renewal process with distribution function F. Let $S_n = \sum_{k=1}^n T_n$ and $N(t) = \max\{n : S_n \leq t\}$ denote the time instant of the nth cycle and the number of cycles by time t, respectively. Thus, S_n is the nth probabilistic replica of the whole process starting at 0. An example of regenerative process is the busy cycles of *GI/G/1* queue length process as shown in Figure 5.7. By the renewal theory, we can obtain the stationary distribution of the system state.

Theorem 5.11 *For a regenerative process with $E[S_1] < \infty$, the stationary distribution for the system state is given by*

$$p_j := \lim_{t \to \infty} P\{X(t) = j\} = \frac{E[\text{time spent in state } j \text{ during } S_k]}{E[S_k]}.$$

Proof. Denote by $p(t) = P(X(t) = j)$. Conditioning on the time of occurrence of the last cycle before time t, i.e. $S_{N(t)}$ and the number of cycles occurred by time t, we have

$$p(t) = P(X(t) = j | S_{N(t)} = 0)F^c(t)$$

$$+ \sum_{k=1}^{\infty} \int_0^t P(X(t) = j | S_{N(t)} = s)F^c(t-s)dF_k(s)$$

$$= P(X(t) = j | S_{N(t)} = 0)F^c(t)$$

$$+ \int_0^t P(X(t) = j | S_{N(t)} = s)F^c(t-s)d\left(\sum_{k=1}^{\infty} F_k(s)\right)$$

$$= P(X(t) = j | S_{N(t)} = 0)F^c(t) + \int_0^t P(X(t) = j | S_{N(t)} = s)F^c(t-s)dM(s).$$

Note that $P(X(t) = j | S_{N(t)} = 0) = P(X(t) = j | S_1 > t)$ and $P(X(t) = j | S_{N(t)} = s) = P(X(t-s) = j | S_1 > t - s)$ Let $h(t) := P(X(t) = j, S_1 > t) = P(X(t) = j | S_{N(t)} = 0)P(S_{N(t)} = 0)$, which is directly Riemann integrable. It follows from the KRT that

$$\lim_{t \to \infty} p(t) = \frac{\int_0^{\infty} P(X(t) = j, S_1 > t)dt}{E[S_1]}.$$

Define

$$I(t) = \begin{cases} 1 & \text{if } X(t) = j < 1, S_1 > t, \\ 0 & \text{otherwise.} \end{cases}$$

Then, $\int_0^{\infty} I(t)dt$ is the amount of time in the first cycle that $X(t) = j$. The result follows by noting that

$$E\left[\int_0^{\infty} I(t)dt\right] = \int_0^{\infty} E[I(t)]dt = \int_0^{\infty} P(X(t) = j, S_1 > t)dt.$$

∎

Example 5.6 *(The Machine Repairman Problem) In a production line, a machine is in operation and is subject to failure. To ensure the downtime to be short, a spare machine is put as a backup and a repairman is standby for repairing any failure machine. The time to failure of an operating machine, denoted by X, follows an i.i.d. distribution with p.d.f $f(x)$. If the operating machine fails while the backup machine is functional, the failure machine will be*

*repaired and at the same time the backup machine is put in operation. The re-
pair time, denoted by R follows an i.i.d. with p.d.f. g(r). The repaired machine
will be put as a backup if the other machine is in operation when the repair is
completed. If the operating machine fails while the other machine is in repair,
the failure machine will wait for the completion of the current repair. When the
repair is completed, the repaired machine will be put in operation and at the
same time, the machine waiting for repair will get repaired. Let π_i be the long-
term proportion of time that exactly i of the machines are in working condition
(either in operation or as a functional backup) where $i = 0,1,2$. Find π_0, π_1,
and π_2.*

Solution: First, we identify a regenerative cycle which is the time interval
between two consecutive visits to a particular state. This is state $i = 1$. Denote the length of this cycle by T_c. Then $T_c = \max(X,R)$ since starting from
$i = 1$ state, the process will go through either $i = 0$ due to $X < R$ or $i = 2$ due
to $X > R$ and return to state $i = 1$ at $\max(X,R)$. The duration in state $i = 0$ or
$i = 2$ is the difference between $\max(X,R) - \min(X,R)$. Denote the duration in
state i by T_i with $i = 0,1,2$. Then we have

$$T_1 = \min(X,R), \quad T_0 = (R-X)^+, \quad T_2 = (X-R)^+,$$

where $x^+ = \max(x,0)$. It follows from the renewal reward theorem that

$$\pi_0 = \frac{E[T_0]}{E[T_c]} = \frac{E[(R-X)^+]}{E[\max(X,R)]}, \quad \pi_1 = \frac{E[T_1]}{E[T_c]} = \frac{E[\min(X,R)]}{E[\max(X,R)]},$$

$$\pi_2 = \frac{E[T_2]}{E[T_c]} = \frac{E[(X-R)^+]}{E[\max(X,R)]}.$$

Note that the following simple expression exists

$$\max(x,r) - \min(x,r) = |x-r| = \max(x-r,0) + \max(r-x,0)$$
$$= (x-r)^+ + (r-x)^+,$$

which can be written as $\max(x,r) = \min(x,r) + (x-r)^+ + (r-x)^+$. Using this
relation, we can verify $\pi_0 + \pi_1 + \pi_2 = 1$. With the p.d.f $f(x)$ and $g(r)$, we can
write the expectations for computing the stationary probabilities as follows:

$$E[\max(X,R)] = \int_0^\infty \int_0^\infty \max(x,r) f(x) g(r) dx dr$$
$$= \int_0^\infty \int_0^r r f(x) g(r) dx dr + \int_0^\infty \int_r^\infty x f(x) g(r) dx dr,$$

$$E[(R-X)^+] = \int_0^\infty \int_0^\infty (r-x)^+ f(x)g(r)dxdr$$

$$= \int_0^\infty \int_0^r (r-x)f(x)g(r)dxdr,$$

$$E[\min(X,R)] = \int_0^\infty \int_0^\infty \min(x,r)f(x)g(r)dxdr$$

$$= \int_0^\infty \int_0^r xf(x)g(r)dxdr + \int_0^\infty \int_r^\infty rf(x)g(r)dxdr,$$

$$E[(X-R)^+] = \int_0^\infty \int_0^\infty (x-r)^+ f(x)g(r)drdx$$

$$= \int_0^\infty \int_0^x (x-r)f(x)g(r)drdx.$$

These expectations involving the maximum or minimum of random variables are based on the conditioning argument. We can condition on either R or X to compute these expectations. For the first three expectations, we condition on R and for the last one we condition on X.

Exercise 5.25 *In Example 5.6, write the expectations $E[\max(X,R)]$, $E[(R - X)^+]$ and $E[\min(X,R)]$ by conditioning on X and the expectation $E[(X - R)^+]$ by conditioning on R. Then prove these expectations are equivalent to those in Example 5.6.*

Exercise 5.26 *In Example 5.6, assume that the time to failure and the repair time are exponentially distributed with the p.d.f.'s $f(x) = \lambda e^{-\lambda x}$ for X and $g(r) = \mu e^{-\mu r}$ for R, respectively.*
   (a) Compute π_0, π_1, and π_2 by using the formulas in Example 5.6.
   (b) Formulate the system as a 3 state CTMC, determine the stationary probabilities, and compare them with the results in (a).

Exercise 5.27 *In Example 5.6, assume that X and R are exponentially distributed with rate λ and μ, respectively, if an additional backup machine is added, determine the stationary probabilities as in (b) of the Exercise 5.26. How about using the renewal reward theorem to find the stationary probabilities?*

Another important application of renewal reward theorem is to solve the *system replacement problem*. Consider a system has a random life time (time to failure) X with distribution function $F(x)$ and p.d.f. $f(x)$. Under an "age replacement policy", the system can be replaced with a brand-new one either at a failure instant or at a pre-specified T. Let C_f and C_p be the replacement costs due to a failure and due to reaching age T, respectively and $C_f \geq C_p$. Now we can develop a long-term cost rate as a function of T. The replacement cycle is obviously $\min(X,T)$. By conditioning on the life time of the system, the expected value is given by

$$E[\min(X,T)] = \int_0^\infty \min(x,T)f(x)dx = \int_0^T xf(x)dx + \int_T^\infty Tf(x)dx$$

$$= xF(x)|_0^T - \int_0^T F(x)dx + T(1-F(T))$$

$$= T - \int_0^T F(x)dx = \int_0^T [(1-F(x)]dx.$$

The expected cost per cycle is $C_f F(T) + C_p[1-F(T)]$. Thus, the long-term cost rate per unit time $J(T)$ is given by

$$J(T) = \frac{C_f F(T) + C_p[1-F(T)]}{\int_0^T [(1-F(x)]dx}.$$

For some special distributions (e.g., uniform distribution), we may obtain the closed-form formula for the optimal T^* that minimizes J. Otherwise, numerical search can be used to determine the optimal T^*.

Exercise 5.28 *In the system replacement problem, assume that the life time of the system follows a Weibull distribution with parameter λ and k. That is $F(x) = 1 - e^{-(x/\lambda)^k}$ for $x \geq 0$. Develop the long-term cost rate function under the age replacement policy. If $\lambda = 10, k = 2, C_f = 50$, and $C_p = 20$, compute the optimal T^* and its associate long-term cost rate $J(T^*)$.*

5.2 MARKOV RENEWAL PROCESSES

In the previous chapters and the section above, we have studied the Markov chains and the renewal processes. Now we introduce a more general stochastic process, called Markov renewal process, by combining these two types of stochastic processes.

5.2.1 BASIC RESULTS FOR MARKOV RENEWAL PROCESSES

Let E be a finite set, $\mathcal{N}$ the set of non-negative integers, and $\mathcal{R}_+$ the non-negative real numbers (Note: these sets may be denoted by different symbols in other chapters). We define the Markov renewal process as follows:

Definition 5.5 *On a probability space $(\Omega, \mathcal{F}, P)$, two sequences of random variables are defined as*

$$X_n : \Omega \to E, \quad T_n : \Omega \to \mathcal{R}_+$$

for each $n \in \mathcal{N}$ so that $0 = T_0 \leq T_1 \leq T_2 \leq \cdots$. These random variables are said to form a Markov renewal process (X, T) with state space E if

$$P(X_{n+1} = j, T_{n+1} - T_n \leq t | X_0, ..., X_n; T_0, ..., T_n)$$
$$= P(X_{n+1} = j, T_{n+1} - T_n \leq t | X_n)$$

for all $n \in \mathcal{N}, j \in E$, and $t \in \mathcal{R}_+$.

If we assume that the process is time-homogeneous, we can denote a family of conditional probabilities by

$$Q(i, j, t) = P(X_{n+1} = j, T_{n+1} - T_n \leq t | X_n = i),$$

which is independent of n and called a semi-Markov kernel. Clearly, $Q(i, j, t)$ is right continuous non-decreasing and bounded function of t for a given pair (i, j) with $i, j \in E$. If we are only interested in the state transition probabilities regardless of the sojourn time in a state, we can obtain the transition probabilities for some Markov chain with state space E as follows:

$$P(i, j) = \lim_{t \to \infty} Q(i, j, t)$$

with the property of

$$P(i, j) \geq 0, \quad \sum_{j \in E} P(i, j) = 1, \quad i, j \in E.$$

It follows from the definition of Markov renewal process and $P(i, j)$ that

$$P(X_{n+1} = j | X_0, ..., X_n; T_0, ..., T_n) = P(X_n, j)$$

for all $n \in \mathcal{N}$ and $j \in E$. On the other hand, if we are interested in the state sojourn times regardless of the state transitions, we can define the probability distribution function of the state sojourn time as follows:

$$G(i, j, t) = Q(i, j, t) / P(i, j) = P(T_{n+1} - T_n \leq t | X_n = i, X_{n+1} = j), \quad i, j \in E, t \in \mathcal{R}_+$$

with the convention of $G(i, j, t) = 1$ by definition if $P(i, j) = 0$ (consequently $Q(i, j, t) = 0$). Thus, times between two consecutive state transitions depend on both the starting and ending states and are not i.i.d.. However, the inter-transition times can be said "conditionally i.i.d.". This term means given $X_n = i, X_{n+1} = j$, the inter-transition times are i.i.d. random variables with the distribution function $G(i, j, t)$. Furthermore, we can easily show the n consecutive inter-transition times are conditionally independent. That is for a set of numbers $t_1, ..., t_n \in \mathcal{R}_+$, we have

$$P(T_1 - T_0 \leq t_1, ..., T_n - T_{n-1} \leq t_n | X_0, X_1, ..., X_n)$$
$$= G(X_0, X_1, t_1) G(X_1, X_2, t_2) \cdots G(X_{n-1}, X_n, t_n),$$

which implies that $T_1 - T_0, T_2 - T_1, ..., T_n - T_{n-1}$ are conditionally independent given the Markov chain $X_0, X_1, ..., X_n$.

Obviously, if the state space E consists of a single state, then the inter-transition times are i.i.d. non-negative random variables and form a classical renewal process. If the state space E contains multiple states and the inter-transition times are exponentially distributed, we have a CTMC. Thus, the Markov renewal process is considered as a generalization of Markov chains and renewal processes. Another view on a Markov renewal process is to identify the (possibly delayed) renewal process by focusing on a particular state $i \in E$. Define $S_0^i, S_1^i, ...$ to be the successive T_n for which $X_n = i$. Then $S^i = \{S_n^i, n \in \mathcal{N}\}$ is (possibly delayed) renewal process. Thus, for each state i, there is a corresponding renewal process and the superposition of all these renewal processes generates instants $T_n, n \in \mathcal{N}$. The ith renewal process (i.e., for state i) contributes the point T_n (called type i) if and only if $X_n = i$. A particular type of successive points, $X_0, X_1, ...$, form a Markov chain. For a finite E, some of the states must be positive recurrent. Thus, starting from any initial state, the Markov chain will eventually reach a specific recurrent state with probability one. Then from a recurrent state i, X_n becomes i infinitely often, resulting in a nonterminating renewal process $\{S_n^i, n \in \mathcal{N}\}$ that implies that $\sup_n S_n^i = +\infty$ with probability one. Hence, we also have $\sup_n T_n = +\infty$ almost surely.

Example 5.7 *Consider the repairman problem in Example 5.6. We can formulate a Markov renewal process by specifying the semi-Markov kernel and related quantities. Suppose that we are interested in the number of state transitions in a time interval $[0,t]$. Let X_n denote the number of working condition machines at the nth state transition, where $n = 1, 2, ...$ and $X_n = i, i \in E = \{0, 1, 2\}$. Let T_n denote the time instant of the nth transition. Furthermore, we assume that $X_0 = 1$ and $T_0 = 0$. Note that the distribution functions for the operating time and repair time are $F(x)$ (with p.d.f. $f(x)$) and $G(r)$ (with p.d.f. $g(r)$), respectively. Then we have*

$$Q(1,0,t) = P(X_1 = 0, T_1 - T_0 \leq t | X_0 = 1)$$
$$= P(X < R) \int_0^\infty \frac{P(x < R < t)}{P(R > x)} f(x) dx,$$
$$Q(1,1,t) = P(X_1 = 1, T_1 - T_0 \leq t | X_0 = 1) = 0,$$
$$Q(1,2,t) = P(X_1 = 2, T_1 - T_0 \leq t | X_0 = 1)$$
$$= P(X > R) \int_0^\infty \frac{P(r < X < t)}{P(X > r)} g(r) dr,$$
$$Q(0,0,t) = P(X_2 = 0, T_2 - T_1 \leq t | X_1 = 0) = 0,$$
$$Q(0,1,t) = P(X_2 = 1, T_2 - T_1 \leq t | X_1 = 0)$$
$$= \int_0^\infty P((R-x)^+ \leq t) f(x) dx = \int_0^\infty P((r-X)^+ \leq t) g(r) dr.$$

$$Q(0,2,t) = P(X_2 = 2, T_2 - T_1 \leq t | X_1 = 0) = 0,$$
$$Q(2,0,t) = P(X_2 = 0, T_2 - T_1 \leq t | X_1 = 2) = 0,$$
$$Q(2,1,t) = P(X_2 = 1, T_2 - T_1 \leq t | X_1 = 2)$$
$$= \int_0^\infty P((X-r)^+ \leq t)g(r)dr = \int_0^\infty P((x-R)^+ \leq t)f(x)dx,$$
$$Q(2,2,t) = P(X_2 = 2, T_2 - T_1 \leq t | X_1 = 2) = 0.$$

The relevant probabilities are as follows:

$$P(X < R) = \int_0^\infty \int_0^r f(x)dx g(r)dr, \quad P(x < R < t) = \int_x^t g(r)dr,$$

$$P(R > x) = \int_x^\infty g(r)dr, \quad P(X \geq R) = 1 - P(X < R),$$

$$P(r < X < t) = \int_r^t f(x)dx, \quad P(X > r) = \int_r^\infty f(x)dx,$$

$$P((R-X)^+ \leq t) = \int_0^\infty P((r-X)^+ \leq t)g(r)dr = \int_0^\infty \int_{r-t}^r f(x)dx g(r)dr,$$

$$P((X-R)^+ \leq t) = \int_0^\infty P((x-R)^+ \leq t)f(x)dx = \int_0^\infty \int_{x-t}^x g(r)dr f(x)dx.$$

It is easy to check that the probability transition matrix $\mathbf{P} = [P(i,j)]$ *is*

$$\mathbf{P} = \begin{pmatrix} 0 & 1 & 0 \\ P(X < R) & 0 & P(X > R) \\ 0 & 1 & 0 \end{pmatrix},$$

by taking the limit $t \to \infty$.

Furthermore, since the sojourn times for the three states are $[T_{n+1} - T_n | X_n = 1] = \min(X, R), [T_{n+1} - T_n | X_n = 0] = (R-X)^+$ *and* $[T_{n+1} - T_n | X_n = 2] = (X - R)^+$, *respectively, the distribution functions for the state sojourn times are as follows:*

$$G(1,0,t) = Q(1,0,t)/P(1,0) = P(T_1 - T_0 \leq t | X_0 = 1, X_1 = 0)$$
$$= \int_0^\infty \frac{P(x < R < t)}{P(R > x)} f(x)dx,$$
$$G(1,2,t) = Q(1,2,t)/P(1,2) = P(T_1 - T_0 \leq t | X_0 = 1, X_1 = 2)$$
$$= \int_0^\infty \frac{P(r < X < t)}{P(X > r)} g(r)dr,$$
$$G(0,1,t) = Q(0,1,t)/P(0,1) = \int_0^\infty P((r-X)^+ \leq t)g(r)dr,$$
$$G(2,1,t) = Q(2,1,t)/P(2,1) = \int_0^\infty P((x-R)^+ \leq t)f(x)dx.$$

Exercise 5.29 *In Example 5.7, if X follows an exponential distribution with $f(x) = \lambda e^{-\lambda x}$ and R follows a uniform distribution over $[0, U]$ with $g(r) = 1/L$. Derive the formulas for $Q(i, j, t)$ and $G(i, j, t)$ for $i, j = 0, 1, 2$ and $t \geq 0$.*

Exercise 5.30 *In Example 5.7, Let $S_0^i, S_1^i, \ldots$ be the successive T_n at which the process visits state i. That is $X_n = i$. For the case of $i = 0$ or $i = 2$, show that $\{S_n^i, n \in \mathcal{N}\}$ is a delayed renewal process. What if $i = 1$?*

Now based on the Markov renewal process, we can define a continuous-time stochastic process, called semi-Markov process.

Definition 5.6 *A continuous-time process $Y = \{Y_t, t \in \mathcal{R}_+$ is called a semi-Markov process with state space E and transition kernel $Q = \{Q(i, j, t)\}$ if*

$$Y_t = X_n \text{ for } T_n \leq t \leq T_{n+1}.$$

A semi-Markov process can be used to model a system's state, denoted by Y_t, that deteriorates due to aging and usage or is improved due to maintenance. Suppose that the system has a finite number of states and the sojourn time in each state is random. The distribution of such a sojourn time $[T_n, T_{n+1}]$ depends on the current state X_n and the next state to be visited X_{n+1}. The successive states visited form a Markov chain. We can also define the distribution function for the sojourn time at state i as

$$H(i, t) = P(T_{n+1} - T_n \leq t | X_n = i)$$

and the transition probability from state i to state j at time t as

$$K(i, j, t) = P(X_{n+1} = j | X_n = i, T_{n+1} - T_n = t).$$

Then, we can write

$$Q(i, j, t) = \int_0^t H(i, du) K(i, j, u)$$

for all $i, j \in E$ and $t \in \mathcal{R}_+$. It is easy to see that a CTMC is a special case of the semi-Markov process. If the sojourn time is an exponentially distributed random variable with the state-dependent parameter, i.e. $H(i, t) = 1 - e^{-\lambda(i)t}$, then it is reasonable to assume that the transition probability is time-independent due to the memoryless property, i.e. $K(i, j, t) = P(i, j)$. Thus, we have

$$Q(i, j, t) = P(i, j)(1 - e^{-\lambda(i)t}),$$

which can be used to show the Markov property

$$P(Y_{t+s} = j | Y_u; u \leq t) = P(Y_{s+t} = j | Y_t)$$

for all $t, s \in \mathscr{R}_+$ and $j \in E$. Moreover, we can show that $P(Y_{s+t} = j | Y_t = i) = P(i, j, s)$ which is independent of t, indicating that Y_t is a time-homogeneous CTMC. The transition rates (i.e., the elements of the infinitesimal generator matrix, denoted by $\mathbf{A} = [A(i, j)]$) are given by

$$A(i, j) = \lambda(i)P(i, j), \quad A(i, i) = -\lambda(i)$$

for all $i, j \neq i \in E$ with $A(i, i) \neq 0$. The trivial case is for $A(i, i) = 0, \lambda(i) = 1$, and $P(i, i) = 1$ which implies $P(i, j) = 0$ for all $j \neq i$. For the conditional probability with the initial state as a condition, we denote $P(\cdot | X_0 = i)$ by $P_i(\cdot)$. Now we can define a n-step semi-Markov kernel

$$Q^n(i, j, t) = P(X_n = j, T_n \leq t | X_0 = i) = P_i(X_n = j, T_n \leq t), \quad i, j \in E, \ t \in \mathscr{R}_+$$

for all $n \in \mathscr{N}$ with $Q^0(i, j, t) = I(i, j)$ for all $t \geq 0$ where $I(i, j) = 1$ for $i = j$ and $I(i, j) = 0$ for $i \neq j$. By conditioning on the first transition time and the state to enter, we have

$$Q^{n+1}(i, k, t) = \sum_{j \in E} \int_0^t Q(i, j, du) Q^n(j, k, t - u). \tag{5.24}$$

As mentioned earlier, the successive visits to a particular state j form a (possibly delayed) renewal process. The number of renewals in $[0, t]$ is given by

$$\sum_{n=0}^{\infty} I_{\{X_n = j, T_n \leq t\}}.$$

Taking the expectation, we obtain the corresponding expected number of renewals, denoted by $M(i, j, t)$,

$$M(i, j, t) = E_i \left[\sum_{n=0}^{\infty} I_{\{X_n = j, T_n \leq t\}} \right] = \sum_{n=0}^{\infty} Q^n(i, j, t) \tag{5.25}$$

for any initial state i, which is called Markov renewal function. The set of Markov renewal functions, $M = \{M(i, j, t); i, j \in E, t \in \mathscr{R}_+\}$ is called the Markov renewal kernel corresponding to Q. For state j, let $F(j, j, t)$ be the distribution function of the time between two consecutive visits to state j. Similar to the classical renewal process, the Markov renewal function is given by

$$M(j, j, t) = \sum_{n=0}^{\infty} F^n(j, j, t), \tag{5.26}$$

where $F^n(j, j, t)$ is the n-th fold convolution of $F(j, j, t)$. By conditioning on the first time to visit state j, we obtain

$$M(i, j, t) = \int_0^t F(i, j, du) M(j, j, t - u). \tag{5.27}$$

Exercise 5.31 *In Example 5.7, if X follows an exponential distribution with* $f(x) = \lambda e^{-\lambda x}$ *and R follows a uniform distribution over* $[0,U]$ *with* $g(r) = 1/L$. *Derive the formula for* $M(i, j, t)$ *for* $i, j = 0, 1, 2$ *and* $t \geq 0$.

If $M(i, j, t)$ is determined by (5.25), the first-passage time distribution functions $F(i, j, t)$ can be solved by using (5.26) and (5.27) via Laplace transforms. Define the Laplace transforms matrices with the elements (i, j) as follows:

$$Q_{ij}(s) = \int_0^\infty e^{-st} Q(i, j, dt) \text{ for } \lambda \geq 0,$$

$$F_{ij}(s) = \int_0^\infty e^{-st} F(i, j, dt) \text{ for } \lambda \geq 0, \tag{5.28}$$

$$M_{ij}(s) = \int_0^\infty e^{-st} M(i, j, dt) \text{ for } \lambda \geq 0.$$

Next, we will use $Q(s) = [Q_{ij}(s)], F(s) = [F_{ij}(s)]$, and $M(s) = [M_{ij}(s)]$ to solve one from the others. Taking the Laplace transforms for (5.24), we have

$$Q_{ik}^{n+1}(s) = \sum_{j \in E} Q_{ij}(s) Q_{jk}^n(s), \tag{5.29}$$

for all $n \in \mathcal{N}$. In terms of matrix form, the n-step matrix $Q^n(s)$ is exactly the nth power of the one-step matrix $Q(s)$. That is $Q^n(s) = (Q(s))^n$ for all $n \geq 1$. Using the matrix form, it follows from (5.25) that

$$M(s) = I + Q(s) + Q^2(s) + \cdots$$
$$= I + Q(s) + (Q(s))^2 + \cdots. \tag{5.30}$$

Multiplying both sides of the equation above by $I - Q(s)$ yields

$$M(s)(I - Q(s)) = I,$$

which implies

$$M(s) = (I - Q(s))^{-1}. \tag{5.31}$$

Moreover, by taking the Laplace transforms, it follows from (5.27) and (5.26) that

$$M_{ij}(s) = \begin{cases} F_{ij}(s)M_{jj}(s) & \text{if } i \neq j, \\ \frac{1}{1-F_{jj}(s)} & \text{if } i = j. \end{cases} \tag{5.32}$$

These relations indicate that the matrices $Q(s), F(s)$, and $M(s)$ are determined each other uniquely. For the case with infinite state space E, all expressions except for (5.31) still hold. However, we can write an infinite system of linear equations as $M(s) = I + Q(s)M(s)$. Thus, $M(s)$ becomes the minimal solution of this equation system. Due to the Markov chain feature, the classification of states for the Markov renewal process can be done similarly as that in the DTMC and CTMC. Along similar lines of the analysis for the classical renewal process, we can establish the Markov renewal equations and obtain the limit theorems for the Markov renewal process. The Markov renewal equations is a system of integral equations that a class of functions satisfy. This class of functions are defined as a mapping

$$f : E \times \mathscr{R}_+ \to \mathscr{R}.$$

The functions in this class have the following properties. (a) For each $i \in E$, $f(i,t)$ is Borel measurable and bounded over finite intervals, and (b) $f(i,t)$ is right continuous and monotone in t. For any fixed $j \in E$, $Q^n(i,j,t)$ and $M(i,j,t)$ are special examples of $f(i,t)$. Denote this class of function by $\mathscr{X}$. A function $f(i,t) \in \mathscr{X}$ is said to satisfy a Markov renewal equation if

$$f(i,t) = g(i,t) + \sum_{j \in E} \int_0^t Q(i,j,du) f(j,t-u), \quad \text{for } i \in E, t \in \mathscr{R}_+ \qquad (5.33)$$

for some function $g \in \mathscr{X}$. The second term on the right-hand side of (5.33) can be denoted as $Q * f$, the convolution of Q and f. This Markov renewal equation has a unique solution as stated below.

Theorem 5.12 *The Markov renewal equation (5.33) has one and only one solution which is given by*

$$f(i,t) = \sum_{j \in E} \int_0^t M(i,j,ds) g(j,t-s) = M * g(i,t), \quad \text{for } i \in E, t \in \mathscr{R}_+.$$

Proof. We need to show that $f = M * g$ is the only solution to the Markov renewal equation

$$f = g + Q * f. \qquad (5.34)$$

By performing the convolution with g on both sides of the relation $M = I + Q * M$, we have

$$M * g = (I + Q * M) * g = g + Q * (M * g), \qquad (5.35)$$

which implies that $M * g$ is a solution to (5.34). To show that $M * g$ is the only solution, we assume that there exists another different solution f and then verify that the difference $d = f - M * g$ is zero. First, subtracting $M * g$ from both sides of (5.34), which is satisfied by this different solution f, gives

$$f - M * g = Q * f + g - M * g = Q * f - Q * (M * g) = Q * (f - M * g), \qquad (5.36)$$

where the second equality follows from (5.35). Thus, d must satisfy $d = Q * d$. By iterating this relation, we have $d = Q * d = Q^2 * h = \cdots = Q^n * d = \cdots$.

For fixed t, it follows from $d \in \mathscr{X}$ that there exists some constant C such that $|d(i,s)| \leq C$ for all $i \in E$ and $s \leq t$. Then, we have

$$
\begin{aligned}
|d(i,t)| = |Q^n * d(i,t)| &\leq \sum_j \int_0^t Q^n(i,j,du)|d(j,t-u)| \\
&\leq C \sum_j Q^n(i,j,t) = CP_i(T_n \leq t).
\end{aligned}
\tag{5.37}
$$

The finite E implies $\sup T_n = \infty$. Thus, as $n \to \infty$, we have

$$
P_i(T_n \leq t) \to 0,
$$

which implies that $d = 0$ by (5.37). ∎

All the results above also hold for countably infinite space E under certain conditions. We do not provide the technical details here. Interested readers are referred to Cinlar (1975).

Next, we discuss the limit theorems for the Markov renewal process. For a fixed state $j \in E$, the function $M(j,j,t)$ is an ordinary renewal function with $F(j,j,t0 = Q(j,j,t)$. We are interested in the limit

$$
M(j,j) = \lim_{t \to \infty} M(j,j,t) = M(j,j,\infty),
$$

which is the expected number of visits to j by the Makov chain starting at j. If j is transient, then $M(j,j)$ is finite and can be computed as the mean of a geometrically distributed random variable. If j is recurrent, then $M(j,j) = \infty$ and the limit theorem corresponding to the Blackwell's theorem for the ordinary renewal process is

$$
\lim_{t \to \infty}[M(j,j,t) - M(j,j,t-\tau)] = \frac{\tau}{\int_0^\infty tF(j,j,dt)}
\tag{5.38}
$$

for $\tau = a > 0$ where a is a real number if j is aperiodic and $\tau = nd$ if j is periodic with period d. Here in terms of notations, we utilize $F(j,j,dt) = [dF(j,j,t)/dt]dt = f(j,j,t)dt$. Define the frequency of visiting state j per time unit as

$$
\omega(j) = \frac{1}{\int_0^\infty tF(j,j,dt)}.
$$

Then we can get the corresponding key renewal theorem for the Markov renewal process as follows: If j is recurrent and aperiodic, for any directly Riemann integrable function h, we have

$$
\lim_{t \to \infty} \int_0^t M(j,j,ds)h(t-s) = \omega(j)\int_0^\infty h(u)du,
\tag{5.39}
$$

and a similar result holds for the periodic j case. For the recurrent irreducible class to which j belongs, we can compute $\omega(j)$ directly from $P(i,j) = \lim_{t \to \infty} Q(i,j,t)$ as stated in the following proposition.

Proposition 5.3 *Suppose X is irreducible and recurrent and let π be the stationary distribution of X, which is a solution of*

$$\sum_{i \in E} \pi(i) P(i,j) = \pi(j), \quad j \in E,$$

$$\sum_{i \in E} \pi(i) = 1.$$

Moreover, let $m(i)$ be the mean sojourn time in state i and then

$$m(i) = E[T_{n+1} - T_n | X_n = i] = \int_0^\infty [1 - \sum_k Q(i,k,t)] dt.$$

Thus,

$$\omega(j) = \frac{\pi(j)}{\sum_{i \in E} \pi(i) m(i)}, \quad j \in E.$$

Exercise 5.32 *In Example 5.7, if X follows an exponential distribution with $f(x) = \lambda e^{-\lambda x}$ and R follows a uniform distribution over $[0,U]$ with $g(r) = 1/L$. Assume that $\lambda = 0.2$ and $L = 2$. Compute $\omega(j)$ with $j = 0,1,2$, which is the limiting distribution of the semi-Markov chain.*

We can also establish the key renewal theorem for a function of time and state.

Theorem 5.13 *Suppose X is irreducible recurrent and $h(j,t)$ is directly Riemann integrable for every $j \in E$. If the states are aperiodic in (X,T), then*

$$\lim_{t \to \infty} M * h(i,t) = \frac{\sum_j \pi(j) n(j)}{\sum_j \pi(j) m(j)} \tag{5.40}$$

with

$$n(j) = \int_0^\infty h(j,s) ds, \quad j \in E.$$

If the states are periodic with period d, then the (5.40) holds for $t = x + kd, k \in \mathcal{N}, k \to \infty$, and $x \in [0,d]$, with

$$n(j) = d \sum_{k=1}^\infty h(j, x + kd - d_{ij}),$$

where $d_{ij} = \inf\{t : F(i,j,t) > 0\}$ (modulo d).

Proof. We only prove the aperiodic case and leave the periodic case as an exercise for readers. It follows from the finite E that

$$\lim_{t \to \infty} M * h(i,t) = \lim_{t \to \infty} \sum_j \int_0^t M(i,j,ds) h(j,t-s) = \sum_j \lim_{t \to \infty} \int_0^t M(i,j,ds) h(j,t-s). \tag{5.41}$$

We first find the limit of the jth term as $t \to \infty$. Since $M(i,j,\cdot)$ is the convolution of $F(i,j,\cdot)$ with $M(j,j,\cdot)$, the jth term is the convolution of $M(j,j,\cdot)$ with the function $F(i,j,\cdot) * h(j,\cdot)$ based on the commutativity and associativity for the convolutions. Note that these functions are all directly Riemann integrable. Interchanging the order of integrations, we obtain

$$\int_0^\infty \left[\int_0^t F(i,j,ds) h(j,t-s) \right] dt = \int_0^\infty F(i,j,ds) \int_s^\infty h(j,t-s) dt$$

$$= F(i,j,s)\big|_0^\infty \int_0^\infty h(j,s) ds$$

$$= F(i,j,\infty) \int_0^\infty h(j,s) ds = n(j),$$

where the last equality follows from $F(i,j,\infty) = 1$ due to the irreducibility. Using (5.39) for the aperiodic case, we have

$$\lim_{t \to \infty} \int_0^t M(i,j,ds) h(j,t-s) = \omega(j) n(j), \quad j \in E. \qquad (5.42)$$

Substituting (5.42) into (5.41) and using the expression of $\omega(j)$, we obtain the (5.40). This completes the proof of the aperiodic case. ∎

Exercise 5.33 *Show that the key renewal theorem for Markov renewal process reduces to the key renewal theorem of the renewal process as a special case.*

It is also interesting to present some additional limit relations that the functions involved in a Markov renewal process such as $M(i,j,t)$ satisfy.

Lemma 5.1 *Suppose ψ is a non-decreasing function defined on $\mathscr{R}_+$ with $\psi(t+b) - \psi(t) \le c$ for all t for some constants c and $b > 0$. Then, for any non-decreasing function g we have*

$$\lim_{t \to \infty} \frac{\psi * g(t)}{\psi(t)} = g(+\infty).$$

As an example of this proposition, we consider $\psi(t) = M(j,j,t)$ and $g(t) = F(i,j,t)$. Since these functions satisfy the conditions in Proposition 5.1. Thus, we have

$$\lim_{t \to \infty} \frac{M(i,j,t)}{M(j,j,t)} = \lim_{t \to \infty} \frac{M(j,j,t) * F(i,j,t)}{M(j,j,t)} = F(i,j). \qquad (5.43)$$

Another powerful result is as follows:

Proposition 5.4 *Let X be irreducible recurrent and let π be its stationary distribution. If $g(k,t)$ is right continuous and non-decreasing, then*

$$\lim_{t \to \infty} \frac{1}{M(h,i,t)} \int_0^t M(j,k,ds) g(k,t-s) = \frac{1}{\pi(i)} \pi(k) g(k,\infty) \qquad (5.44)$$

for any $h,i,j,k \in E$.

As a special case of this proposition, we have

$$\lim_{t \to \infty} \frac{M(j,k,t)}{M(h,i,t)} = \frac{\pi(k)}{\pi(i)}. \tag{5.45}$$

Here, we omit the proofs of these results and interested readers are referred to many references (e.g., Cinlar (1975)).

Exercise 5.34 *Provide intuitive explanations for equations (5.43) and (5.45).*

5.2.2 RESULTS FOR SEMI-MARKOV PROCESSES

Next, we present more results for the semi-Markov process defined on the Markov renewal process (X,T) with a finite E (see Definition 5.6). Define the transition probability function for the semi-Markov process as

$$P(i,j,t) = P(Y_t = j | Y_0 = X_0 = i), \ i,j \in E, t \in \mathscr{R}_+,$$

and the probability that the process is still in state j at time t if starting state is j as

$$h(j,t) = P(Y_t = j | Y_0 = j) = 1 - \sum_{k \in E} Q(j,k,t), \ j \in E, t \in \mathscr{R}_+.$$

Now we can express the transition probability function as follows.

Proposition 5.5 *For all $i,j \in E$ and $t \in \mathscr{R}_+$,*

$$P(i,j,t) = \int_0^t M(i,j,ds)h(j,t-s).$$

Proof. Let T_1 be the first transition time. If $T_1 > t$, then $Y_t = Y_0$; if $T_1 = s \leq t$, then Y_t has the same distribution as Y_{t-s} with initial value X_1. This renewal argument implies that

$$P(i,j,t) = I(i,j)h(i,t) + \sum_k \int_0^t Q(i,k,ds)P(k,j,t-s). \tag{5.46}$$

For fixed j, if we define $f(i,t) = P(i,j,t)$ and $g(i,t) = I(i,j)h(i,t)$, then (5.46) becomes the Markov renewal equation $f = g + Q * f$. Solving this equation by using Theorem 5.12, we get $f = M * g$ which is the desired result. This completes the proof. ∎

Denote by $U(i,j,t)$ the expected time spent in state j during $[0,t]$ by the semi-Markov process Y given the initial state i. That is

$$U(i,j,t) = E_i \left[\int_0^t \mathbf{1}_j(Y_s)ds \right], \ i,j \in E, t \in \mathscr{R}_+,$$

where $1_j(k) = 1$ or 0 according to $k = j$ or $k \neq j$ and the subscript i of E indicates the initial state. Using Proposition 5.5 and the fact $E_i[1_j(Y_s)] = P(Y_s = j|Y_0 = i) = P(i, j, s)$, and passing the expectation inside the integral, we have

$$U(i, j, t) = \int_0^t M(i, j, ds) \int_0^{t-s} h(j, u) du. \tag{5.47}$$

We can consider U as an operator, called the potential kernel of Y, which can be applied to a class of functions defined earlier on the state space E (i.e., $f : E \times \mathcal{R}_+ \to \mathcal{R}$). That is

$$U f(i, t) = E_i \left[\int_0^t f(Y_s, s) ds \right]. \tag{5.48}$$

For example, if $f(i, s)$ is interpreted as the rate of rewards being received at an instant s when Y is in state j (i.e., $Y_s = j$), then $U f(i, t)$ becomes the expected value of total reward received during $[0, t]$ given $Y_0 = i$. Using (5.48) and (5.47), we have

$$U f(i, t) = \sum_j \int_0^t M(i, j, du) \int_0^{t-u} h(j, s) f(j, u + s) ds. \tag{5.49}$$

By applying the limit theorems, we can obtain a number of useful results. The total amount of time spent in state j by the process Y starting at state i can be computed as $\lim_{t \to \infty} U(i, j, t) = U(i, j, +\infty)$. It follows from Proposition 5.4 that for any $i, j \in E$

$$\lim_{t \to \infty} \frac{U(i, j, t)}{M(h, j, t)} = m(j).$$

Combining this result with (5.45), we obtain

$$\lim_{t \to \infty} \frac{U(i, j, t)}{U(h, k, t)} = \frac{\pi(j)m(j)}{\pi(k)m(k)}.$$

The limiting probability distribution for the semi-Markov process can be obtained from the limiting probability distribution of the corresponding Markov chain X.

Proposition 5.6 *If X is irreducible recurrent and $m(k) < \infty$, then, for $i, j \in E$, we have*

$$v(j) = \lim_{t \to \infty} P(Y_t = j|Y_0 = i) = \frac{\pi(j)m(j)}{\sum_k \pi(k)m(k)} \tag{5.50}$$

in the aperiodic case.

For the periodic case, interested readers are referred to Cinlar (1975).

For a special case in which the state sojourn time distributions are exponential with state-dependent rate $\lambda(j)$, Y becomes a CTMC. If all states are aperiodic, then

$$m(j) = \int_0^\infty h(j,s)ds = \int_0^\infty e^{-\lambda(j)s}ds = 1/\lambda(j).$$

Exercise 5.35 *Consider a stochastic process $\{X(t), t \geq 0\}$ with N discrete states and transition probability function $P(i,j,t)$. That is $X(t) = i$ with $i = 1, 2, ..., N$. Assume that the state sojourn time in state i is uniformly distributed over $[0, 1/i]$. For $N = 3$, find $\pi(i)$ and $m(i)$ with $i = 1, 2, 3$.*

5.2.3 SEMI-REGENERATIVE PROCESSES

There is a class of non-Markovian stochastic processes that possess the strong Markov property at certain selected random time instants. Thus, if we only focus on the process embedded at these random instants, we obtain a Markov renewal process. These random time instants are called stopping times for a so-called semi-regenerative process Z based on a Markov renewal process. The process $\{Z_t, t \in \mathscr{R}_+\}$ with a topological space E is a random function of time which is right-continuous with left-hand limits (RCLL) for almost all ω. A random variable $T : \Omega \to [0, \infty]$ is called a stopping time for Z if for any $t \in \mathscr{R}_+$, the occurrence or non-occurrence of the event $\{T \leq t\}$ can be determined based on the history of the process denoted by $\mathbb{H}_t = \sigma(Z_u; u \leq t)$. For more details about stopping times and related concepts, readers are referred to Appendix B.

Definition 5.7 *A process Z is said to be semi-regenerative if there exists a Markov renewal process (X, T) with a finite state space E such that*

> *1. for each $n \in \mathscr{N}$, T_n is a stopping time for Z;*
> *2. for each $n \in \mathscr{N}$, X_n is determined by the history $\mathbb{H}_t$ for Z up to time t;*
> *3. for each $n \in \mathscr{N}$, a finite set of time points $0 \leq t_1 < \cdots < t_m$ with $m \geq 1$, and a bounded function defined on F^m, the following relation holds*

$$E[f(Z_{T_n+t_1}, ..., Z_{T_n+t_m})|\mathbb{H}_{T_n}] = E_j[f(Z_{t_1}, ..., Z_{t_m})]$$

if $X_n = j$.

Condition 3 is the most important one and implies that the stochastic process, Z, with time points $T_n, n \in \mathscr{N}$ at which, from a probabilistic point of view, the process restarts itself. Note that all Markov processes and semi-Markov processes are semi-regenerative. However, the converse may not be true. Define two conditional probabilities

$$K(i,A,t) = P(Z_t \in A, T_1 > t | Z_0 = i), \quad P(i,A,t) = P(Z_t \in A | Z_0 = i) \quad (5.51)$$

for $t \in \mathscr{R}_+, i \in E$. These conditional probabilities are transition probability functions for the semi-regenerative process and can be computed as follows.

Theorem 5.14 *For a semi-regenerative process Z with state space E, let (X,T) be the Markov renewal process embedded in Z. Further, let Q be the semi-Markovian kernel for (X,T) and M be the corresponding Marko renewal kernel. For any open set $A \subset F$, $t \in \mathscr{R}_+, i \in E$, we have*

$$P(i,A,t) = \sum_{j \in E} \int_0^t M(i,j,ds) K(j,A,t-s). \quad (5.52)$$

Proof. It follows from the total probability law that

$$\begin{aligned} P(i,A,t) &= P(Z_t \in A, T_1 > t | Z_0 = i) + P(Z_t \in A, T_1 \leq t | Z_0 = i) \\ &= K(i,A,t) + P(Z_t \in A, T_1 \leq t | Z_0 = i). \end{aligned} \quad (5.53)$$

Using the definition of semi-regenerative process, the second term of (5.53) can be written as

$$\begin{aligned} P(Z_t \in A, T_1 \leq t | Z_0 = i) &= E_i[I_{[0,t]}(T_1) P(Z_t \in A | \mathbb{H}_{T_1})] \\ &= E_i[I_{[0,t]}(T_1) P(X_1, A, t - T_1)] \\ &= \sum_j \int_0^\infty Q(i,j,ds) I_{[0,t]}(s) P(j,A,t-s). \end{aligned}$$

Substituting this expression into (5.53), we obtain

$$P(i,A,t) = K(i,A,t) + \sum_j \int_0^t Q(i,j,ds) P(j,A,t-s).$$

For a fixed A, this is a Markov renewal equation with form $f = g + Q * f$ with $f(i,t) = P(i,A,t)$ and $g(i,t) = K(i,A,t)$. The proof is completed from applying Theorem 5.12. ∎

The transient distributions of a semi-regenerative process Z with embedded Markov renewal process (X,T) and initial distribution $\mathbf{p} = (p_i, ; i \in E)$ are given by

$$P(Z_t = A) = \sum_{i \in E} p_i P(i,A,t),$$

where

$$P(i,A,t) = \sum_{n=0}^\infty \sum_{k \in E} \int_0^t K(k,A,t-u) dF^n(i,k,u)$$

for all $t \in \mathscr{R}_+$, and $i, j \in E$. The expression of $P(i,A,t)$ is obtained by conditioning upon the number n of Markov renewal points until time t and the state k which is observed at the last Markov renewal point before t.

For the limiting behavior of the semi-regenerative process, applying Theorem 5.12 to Theorem 5.14 with $h(i,t) = K(i,A,t)$, we have the following.

Proposition 5.7 *For the semi-regenerative process described in Theorem 5.14, if (X,T) is irreducible recurrent aperiodic and $m(j) < \infty$ for all j, then we have*

$$\lim_{t \to \infty} P(i,A,t) = \frac{\sum_j \pi(j) n(j,A)}{\sum_j \pi(j) m(j)} \tag{5.54}$$

with

$$n(j,A) = \int_0^\infty K(j,A,t) dt.$$

Since the limiting probability is independent of the initial state i, we can denote it by $\lim_{t \to \infty} P(Z_t = A) = \lim_{t \to \infty} P(i,A,t)$. In the next two sections, we will discuss the applications of semi-regenerative process models.

5.2.4 $M/G/1$ AND $GI/M/1$ QUEUES

In this section, we study a queueing system in which either the service time or the inter-arrival time is a non-exponential random variable. In this kind of continuous-time systems, we will not be able to formulate a CTMC as for an arbitrary time instant, most likely the time since the last customer departure (or the time since the last customer arrival) is required to predict the queue length at a future instant. In other words, the history of the process is needed for predicting the future state. However, for some selected time instants, such a history may not be needed for predicting the future state. Thus, we analyze these queueing systems by modeling the system as semi-regenerative processes.

5.2.4.1 $M/G/1$ Queue

Consider a single server queueing system with a Poisson arrival process of rate λ and general i.i.d. service times, denoted by S. Denote the number of customers in the system (i.e., the queue length) at time t by L_t. In general, the stochastic process $L = \{L_t, t \in \mathscr{R}_+\}$ with state space $E = \mathbb{N}_0$ is not a Markov process. However, we can focus on selected time instants which are customer departure instants. Since at these instants, the service is just completed, we do not need to record the service time elapsed to predict the future state of the system. Thus, the queue length at these instants forms a discrete-time Markov chain, called embedded Markov chain.

Define $T_0 := 0$ and denote by T_n the time of the nth customer departure. Let X_n be the number of customers in the system at time T_n for all $n \in \mathscr{N}$. Suppose that $0 < E[S] < \infty$ which implies $\lim_{n \to \infty} T_n = \infty$. Note that T_n's are stopping times for queue length process L. Thus, L is a semi-regenerative process with embedded Markov renewal process (X, T), where $T = \{T_n, n \in \mathcal{N}\}$ and $X = \{X_n, n \in \mathcal{N}\}$. For such a model, we have

$$Q(i, j, t) = \int_0^t e^{-\lambda s} \frac{(\lambda s)^{j-i+1}}{(j-i+1)!} dF_S(s), \quad j = i - 1 + k$$

for $i \geq 1, k \geq 0$, and $j \in \mathcal{N}$. Taking the limit of the above as $t \to \infty$, we obtain

$$P(i, j) = p_{ij} = \lim_{t \to \infty} Q(i, j, t) = \begin{cases} \int_0^\infty e^{-\lambda t} \frac{(\lambda t)^{j-i+1}}{(j-i+1)!} dF_S(t) = a_k, & j = i - 1 + k, \\ 0, & j < i - 1, \end{cases}$$

(5.55)

for $i \geq 1$ and $p_{0j} = a_j$ for all $j \in \mathcal{N}$. The transition matrix for the embedded Markov chain can be written as

$$P = \begin{pmatrix} a_0 & a_1 & a_2 & a_3 & \cdots \\ a_0 & a_1 & a_2 & a_3 & \cdots \\ 0 & a_0 & a_1 & a_2 & \cdots \\ \vdots & \ddots & \ddots & \ddots & \ddots \end{pmatrix}.$$

This matrix has a structure of skip-free to the left, which is also called an $M/G/1$ type transition matrix. It is interesting to see that the inter-transition time distribution only depends on whether the initial state is empty or not. That is

$$G(i, j, t) = P(T_{n+1} - T_n \leq t | X_n = i, X_{n+1} = j)$$
$$= \begin{cases} F_S(t), & i > 0, \\ \int_0^t \lambda e^{-\lambda u} F_S(t - u) du, & i = 0, \end{cases}$$

which is independent of $j \in \mathcal{N}$. Furthermore, we have

$$K(i, j, t) = P(L_t = j, T_1 > t | L_0 = X_0 = i)$$
$$= \begin{cases} e^{-\lambda t}, & i = j = 0, \\ \int_0^t \lambda e^{-\lambda u} e^{-\lambda(t-u)} \frac{(\lambda(t-u))^{j-1}}{(j-1)!} (1 - F_S(t - u)) du & i = 0, j > 0, \\ (1 - F_S(t)) e^{-\lambda t} \frac{(\lambda t)^{j-i}}{(j-i)!}, & 0 < i \leq j, \end{cases}$$

for all $t \in \mathcal{R}_+$. Since all a_j's are strictly positive, the matrix P implies that the embedded Markov chain is irreducible. Assume that $\sum_{k=1}^\infty k a_k < 1$, that is the expected number of arrivals during a service time period is less than one. This condition implies that on average the service rate is faster than the arrival rate so that the queueing system is stable. This stability condition can be verified by using the Foster's criterion discussed in Chapter 2. Set $\varepsilon = 1 - \sum_{k=1}^\infty k a_k$ and define the function $f(k) = k$ for all $k \in \mathcal{N}$. Then, it follows from the condition $\sum_{k=1}^{\infty} k a_k < 1$ that

$$\sum_{j=0}^{\infty} p_{ij} f(j) = \sum_{j=i-1}^{\infty} a_{j-i+1} j = \sum_{j=i-1}^{\infty} a_{j-i+1} (j-i+1) + i - 1$$

$$= \sum_{k=1}^{\infty} k a_k - 1 + f(i) \le f(i) - \varepsilon$$

for all $i \ge 1$. Thus, the function f satisfies Foster's criterion and hence the embedded Markov chain is positive recurrent. Computing

$$\sum_{k=1}^{\infty} k a_k = \sum_{k=1}^{\infty} k \int_0^{\infty} e^{-\lambda t} \frac{(\lambda t)^k}{k!} dF_S(t) = \int_0^{\infty} \sum_{k=1}^{\infty} e^{-\lambda t} \frac{(\lambda t)^{k-1}}{(k-1)!} dF_S(t),$$

$$= \lambda E[S],$$

we can write the stability condition as $\rho = \lambda E[S] < 1$. For a stable $M/G/1$ queue, we can compute the stationary distribution of the queue length via the embedded Markov chain's stationary distribution. Denote by v the stationary distribution for $X = \{X_n, n \in \mathcal{N}\}$. We need to solve $vP = v$ which can be written as

$$v_0 = v_0 a_0 + v_1 a_0$$
$$v_1 = v_0 a_1 + v_1 a_1 + v_2 a_0$$
$$v_2 = v_0 a_2 + v_1 a_2 + v_2 a_1 + v_3 a_0 \qquad (5.56)$$
$$\cdots$$

To solve this system of infinite number of equations, we need to develop the recursive relations. For each line l in (5.52), adding the first l equations and defining $r_l = 1 - \sum_{k=0}^{l} a_k = \sum_{k=l+1}^{\infty} a_k$, we obtain

$$a_0 v_1 = v_0 r_0$$
$$a_0 v_2 = v_0 r_1 + v_1 r_1$$
$$a_0 v_3 = v_0 r_2 + v_1 r_2 + v_2 r_1 \qquad (5.57)$$
$$\cdots$$

With this recursion system, we can numerically compute the stationary distribution for X as long as we can determine v_0. Define $r := \sum_{l=0}^{\infty} r_l$, which is equal to ρ. To show this fact, based on $r_l = 1 - \sum_{k=0}^{l} a_k = \sum_{k=l+1}^{\infty} a_k$ we write the following:

$$r_0 = a_1 + a_2 + a_3 + \cdots$$
$$r_1 = \quad\quad a_2 + a_3 + \cdots$$
$$r_3 = \quad\quad\quad\quad a_3 + \cdots \qquad (5.58)$$
$$\cdots$$

Adding these equations yields $r = \sum_{l=0}^{\infty} r_l = \sum_{k=1}^{\infty} ka_k = \rho < 1$. By adding all equations in (5.53) and using $r = \rho < 1, a_0 = 1 - r_0$, we have

$$(1 - r_0) \sum_{k=1}^{\infty} v_k = v_0 r + \sum_{k=1}^{\infty} v_k(r - r_0). \tag{5.59}$$

This relation can be written as

$$\sum_{k=1}^{\infty} v_k = v_0 \frac{r}{1-r} = v_0 \frac{\rho}{1-\rho},$$

which results in $v_0 = 1 - \rho$. With v_0, we can recursively compute v_k for $k \geq 1$. Next, we show that v is also the stationary distribution of Q.

Theorem 5.15 *For a stable $M/G/1$ queue, the limiting distribution of the queue length, denoted by π, is given by*

$$\pi_j = \lim_{t \to \infty} P(L_t = j) = v_j$$

for all $j \in \mathcal{N}$.

Proof. It follows from Proposition 5.7 that

$$\pi_j = \frac{1}{vm} \sum_{i=0}^{j} v_i \int_0^{\infty} K(i,j,t)dt \tag{5.60}$$

for all $j \in E$. The mean renewal interval is

$$vm = v_0 \left(\frac{1}{\lambda} + E[S] \right) + \sum_{k=1}^{\infty} v_k E[S] = \frac{1}{\lambda}.$$

Substituting $vm = \lambda^{-1}$ into (5.60) gives

$$\pi_j = \lambda \sum_{i=0}^{j} v_i \int_0^{\infty} K(i,j,t)dt = \lambda \left(v_0 \int_0^{\infty} K(0,j,t)dt + \sum_{i=1}^{j} v_i \int_0^{\infty} K(i,j,t)dt \right). \tag{5.61}$$

For $j > 0$, we first compute the integral

$$\int_0^{\infty} K(0,j,t)dt = \int_0^{\infty} \int_0^{t} \lambda e^{-\lambda u} e^{-\lambda(t-u)} \frac{(\lambda(t-u))^{j-1}}{(j-1)!} (1 - F_S(t-u))dudt$$

$$= \int_0^{\infty} e^{-\lambda s} \lambda ds \cdot \int_0^{\infty} e^{-\lambda t} \frac{(\lambda t)^{j-1}}{(j-1)!} F_S^c(t)dt$$

$$= \int_0^{\infty} K(1,j,t)dt.$$

For $0 < i \le j$, it can be shown that

$$
\int_0^\infty K(i,j,t)dt = \int_0^\infty e^{-\lambda t}\frac{(\lambda t)^{j-i}}{(j-i)!}F_S^c(t)dt
$$
$$
= \frac{1}{\lambda}\int_0^\infty \sum_{k=j-i+1}^\infty e^{-\lambda t}\frac{(\lambda t)^k}{k!}dF_S(t) = \frac{r_{j-i}}{\lambda},
$$
(5.62)

where the last equality follows from the definition of $(a_k)_{k\in\mathcal{N}}$ and r_{j-i}. Thus, for $j \ge 1$, using (5.61) and (5.57) we have

$$
\pi_j = v_0 r_{j-1} + \sum_{i=1}^{j} v_i r_{j-i} = v_j.
$$

Since π and v are probability distributions, it follows from $\pi_j = v_j$ for all $j \ge 1$ that $\pi_0 = v_0$ holds. ∎

This result can be also obtained by using the two classical results for M/G/1 queues. The first one is called Kleinrock's result, which says that for a queueing system where the system state can change at most by $+1$ or -1, the queue length distribution as seen by a departing customer will be the same as that seen by an arriving customer. The second result is called "Poisson Arrival See Time Averages" (PASTA) (see Wolff (1989)). Another important result is to obtain the moment generating function (m.g.f.), also called z-transform, of the queue length distribution for the embedded Markov chain based on (5.52). Define $v(z) = \sum_{i=0}^\infty v_i z^i$ and $a(z) = \sum_{i=0}^\infty a_i z^i = \tilde{S}(\lambda - \lambda z)$, the two m.g.f.'s for the queue length at the embedded instants and the number of arrivals during a service time, respectively, where $\tilde{S}(s)$ is the LST of the service time. By multiplying both sides of the lth equation in (5.52) by z^l with $l = 0, 1, ...$, we have

$$
v_0 z^0 = v_0 a_0 z^0 + v_1 a_0 z^0
$$
$$
v_1 z^1 = v_0 a_1 z^1 + v_1 a_1 z^1 + v_2 a_0 z^1
$$
$$
v_2 z^2 = v_0 a_2 z^2 + v_1 a_2 z^2 + v_2 a_1 z^2 + v_3 a_0 z^2
$$
$$
\cdots.
$$

Adding these equations yields

$$
v(z) = v_0 a(z) + v_1 a(z) + v_2 z a(z) + v_3 z^2 a(z) + \cdots
$$
$$
= v_0 a(z) - v_0 a(z)/z + v_0 a(z)/z + v_1 a(z) + v_2 z a(z) + v_3 z^2 a(z) + \cdots
$$
$$
= v_0 a(z) - v_0 a(z)/z + a(z)/z[v_0 + v_1 z + v_2 z^2 + v_3 z^3 + \cdots]
$$
$$
= v_0 a(z) - v_0 a(z)/z + (a(z)/z)v(z)
$$
$$
= v_0 a(z) + (v(z) - v_0)a(z)/z,
$$

which leads to

$$v(z) = \frac{(z-1)v_0 a(z)}{z - a(z)}. \tag{5.63}$$

To determine v_0, we let $z \to 1$ and use the fact $\lim_{z \to 1} a(z) = \sum_{i=0}^{\infty} a_i = 1$ to have

$$\lim_{z \to 1} v(z) = v_0 \lim_{z \to 1} \frac{z-1}{z - a(z)} = v_0 (1 - a'(1))^{-1},$$

where the last equality follows from L'hospital rule. Note that $a'(1) = \sum_{i=0}^{\infty} i a_i = \rho$. Using $\lim_{z \to 1} v(z) = v_0/(1-\rho) = 1$, we obtain $v_0 = 1 - \rho$ if $\rho < 1$. Since $v = \pi$, we have

$$v(z) = \pi(z) = \frac{(1-\rho)(z-1)a(z)}{z - a(z)}$$

from which the moments of the queue length can be computed by taking the first and higher derivatives of $v(z)$ with respect to z and setting $z = 1$. For example, the mean of the queue length, denoted by $E[L]$, is given by

$$E[L] = \rho + \frac{\lambda^2 \sigma^2 + \rho^2}{2(1-\rho)},$$

where σ^2 is the variance of the service time.

Exercise 5.36 *Assume exponential service times for the M/G/1. Show that the results for the asymptotic distribution coincide with the results obtained for the M/M/1 queue.*

Exercise 5.37 *Consider an M/G/1 queue with batch arrivals. That is the arrival process is a compound Poisson process with arrival rate λ. The batch size B distribution is $P(B = i) = b_i$ with $i = 1, 2, \dots$.*

(a) Let $B(z) := \sum_{n=1}^{\infty} b_n z^n$ be the z-transform of the batch size. Show that the z-transform of the number of customers arriving in $[0,t]$ is $N(t,z) = e^{-\lambda t(1-B(z))}$.

(b) Denote by $S^*(s)$ the LST of the service time distribution. Show that z-transform of the number of customers arriving during a service period is given by $S^*(\lambda - \lambda B(z))$.

(c) Write down the transition matrix for the embedded Markov chain for such a queueing system.

5.2.4.2 GI/M/1 **Queue**

Now we consider a single server queue where the arrivals occur according to a renewal process and the service times are i.i.d. exponential random variables. Denote the number of customers in the system at time t by L_t. Due to the generally distributed i.i.d. inter-arrival times, denoted by A, the stochastic process $L = \{L_t, t \in \mathcal{R}_+\}$ with state space $E = \mathbb{N}_0$ is not a Markov process. Again we select the customer arrival instants as the embedded points and formulate a semi-regenerative process. Define T_n as the time of the nth arrival and $X_n := L_{T_n} - 1$ as the number of customers in the system immediately before the nth arrival. It is easy to verify that L is semi-regenerative with embedded Markov renewal process (X, T) where $X = \{X_n, n \in \mathcal{N}\}$ and $T = \{T_n, n \in \mathcal{N}\}$ with $T_0 := 0$. The semi-Markov kernel can be obtained as

$$Q(i,j,t) = \begin{cases} \int_0^t e^{-\mu t} \frac{(\mu t)^{i+1-j}}{(i+1-j)!} dF_A(t), & 1 \le j \le i+1, \\ 1 - \sum_{k=0}^i Q(i,k,t), & j = 0, \\ 0, & j > i+1. \end{cases}$$

The elements of the transition probability matrix $P = [p_{ij}]$ of X are given by taking the limit of $Q(i,j,t)$ as $t \to \infty$ as

$$p_{ij} = \lim_{t \to \infty} Q(i,j,t) = \begin{cases} \int_0^t e^{-\mu t} \frac{(\mu t)^{i+1-j}}{(i+1-j)!} dF_A(t) = a_{i+1-j}, & 1 \le j \le i+1, \\ 1 - \sum_{k=0}^i a_k = \sum_{k=i+1}^\infty a_k = b_i, & j = 0, \\ 0, & j > i+1, \end{cases}$$

for all $i, j \in \mathcal{N}$. The transition matrix for the embedded Markov chain can be written as

$$P = \begin{pmatrix} b_0 & a_0 & 0 & 0 & 0 & \cdots \\ b_1 & a_1 & a_0 & 0 & 0 & \cdots \\ b_2 & a_2 & a_1 & a_0 & 0 & \cdots \\ \vdots & \vdots & \ddots & \ddots & \ddots & \ddots \end{pmatrix}.$$

This matrix has a structure of skip-free to the right, which is also called an $GI/M/1$ type transition matrix. Using the independence between the arrival process and the service process, we can write

$$K(i,j,t) = \begin{cases} (1 - F_A(t))e^{-\mu t} \frac{(\mu t)^{i+1-j}}{(i+1-j)!}, & 1 \le j \le i+1, \\ (1 - F_A(t))e^{-\mu t} \sum_{k=i+1}^\infty \frac{(\mu t)^k}{k!}, & j = 0, \\ 0, & j > i+1, \end{cases}$$

for all $t \in \mathcal{R}_+$ and $i, j \in \mathcal{N}$. The mean sojourn time $m(j) = E[T_1 | X_0 = j] = E[A]$ is the mean of inter-arrival time. Thus vector m is constant. Next, assuming that the queue is stable, we determine the stationary distribution ν of the embedded Markov chain X by solving $\nu P = \nu$. The set of equations can be written as

$$v_0 = \sum_{n=0}^\infty v_n b_n$$

$$v_k = \sum_{n=k-1}^\infty v_n a_{n-k+1} \text{ for } k \ge 1.$$

(5.64)

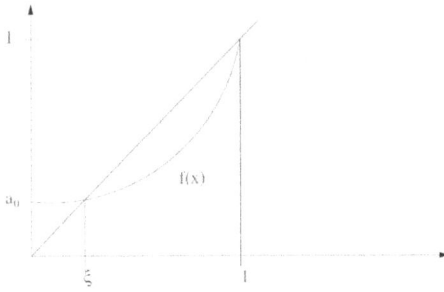

Figure 5.8 A fixed point ξ satisfying $\xi = \sum_{n=0}^{\infty} \xi^n a_n$.

For a $GI/M/1$ type transition matrix, we can assume the stationary probability distribution has the geometric distribution form. That is $v_n = (1-\xi)\xi^n$ for all $n \in \mathcal{N}$ and some $0 < \xi < 1$. Substituting such a solution into (5.60) yields

$$v_0 = 1-\xi \quad \text{and} \quad \xi = \sum_{n=0}^{\infty} \xi^n a_n. \tag{5.65}$$

Now the key is to determine ξ via the second equation. Let $g(x) = \sum_{n=0}^{\infty} a_n x^n$ which is well-defined over the interval $[0,1]$. It is easy to verify that $g(x)$ is an increasing and convex function of x by taking the first- and second-order derivatives with respect to x. That is $g'(x) = \sum_{n=0}^{\infty} a_n n x^{n-1} > 0$ and $g''(x) = \sum_{n=0}^{\infty} a_n n(n-1)x^{n-2} > 0$. With the two boundary values of $g(0) = a_0 > 0$ and $g(1) = 1$, a fixed point ξ satisfying $\xi = \sum_{n=0}^{\infty} \xi^n a_n$ can be determined by the intersection of a 45 degree straight line and function $g(x)$ in Figure 5.8. Further, it follows from the properties of g and the mean value theorem that such a fix point ξ exists if and only if the condition $g'(1) > 1$ holds. To interpret this condition, we compute

$$g'(1) = \sum_{n=1}^{\infty} na_n = \int_0^{\infty} \sum_{n=1}^{\infty} ne^{-\mu t} \frac{(\mu t)^n}{n!} dF_A(t)$$

$$= \int_0^{\infty} e^{-\mu t} \sum_{n=1}^{\infty} \frac{(\mu t)^{n-1}}{(n-1)!} (\mu t) dF_A(t) = \mu \int_0^{\infty} t dF_A(t).$$

Thus, the condition turns out to be $\mu E[A] > 1$ which is equivalent to $\rho = (1/E[A])(1/\mu)) = \lambda E[S] < 1$, the stability condition for a single server queue. Note that ξ can be numerically computed by using the iteration approach. We can start with $\xi_0 := 0$ and iterate by $\xi_{n+1} := g(\xi_n)$ for all $n \in \mathcal{N}$. Then $\xi = \lim_{n\to\infty} \xi_n$. To validate this approach, we first show that the sequence $(\xi_n : n \in \mathcal{N})$ generated by the iterations is strictly increasing. This property follows by induction. Note that $\xi_1 = a_0 > 0 = \xi_0$. Assume $\xi_n > \xi_{n-1}$ holds.

Then we have

$$\xi_{n+1} = \sum_{k=0}^{\infty} a_k \xi_n^k > \sum_{k=0}^{\infty} a_k \xi_{n-1}^k = \xi_n$$

by the induction assumption and the fact that all a_n are strictly positive. Next, we show that this sequence of ξ_n is upper bounded (i.e., $\xi_n < 1$ for all $n \in \mathcal{N}$). First, $\xi_0 = 0 < 1$. Then assume $\xi_n < 1$. Now we have

$$\xi_{n+1} = \sum_{k=0}^{\infty} a_k \xi_n^k < \sum_{k=0}^{\infty} a_k = 1.$$

Thus, $\xi_n < 1$ for all $n \in \mathcal{N}$ by induction. Since the sequence $(\xi_n : n \in \mathcal{N})$ is increasing and bounded, it converges to ξ_∞ which is the fixed point of g. This can be shown as follows:

$$\xi_\infty = \lim_{n \to \infty} \xi_n = \lim_{n \to \infty} \sum_{k=0}^{\infty} a_k \xi_n^k = \sum_{k=0}^{\infty} a_k \left(\lim_{n \to \infty} \xi_n \right)^k$$

$$= \sum_{k=0}^{\infty} a_k \xi_\infty^k.$$

Note that the sequence $(\xi_n : n \in \mathcal{N})$ starts with $\xi_0 = 0$ and is strictly increasing and upper bounded by 1. Hence, $\xi = \xi_\infty$ is the only fix point smaller than 1. Now we can apply Proposition 5.7 with v, m, and K to obtain the stationary distribution of the $GI/M/1$ queue as follows:

$$
\begin{aligned}
\pi_j := \lim_{t \to \infty} P(L_t = j) &= \frac{1}{vm} \sum_{i \in E} v_i \int_0^\infty K(i,j,t) dt \\
&= \frac{1}{vm} \sum_{i=j-1}^{\infty} v_i \int_0^\infty e^{-\mu t} \frac{(\mu t)^{i+1-j}}{(i+1-j)!} (1 - F_A(t)) dt \\
&= \frac{(1-\xi)\xi^{j-1}}{E[A]} \int_0^\infty e^{-\mu t} \sum_{i=j-1}^{\infty} \xi^{i+1-j} \frac{(\mu t)^{i+1-j}}{(i+1-j)!} F_A^c(t) dt \\
&= \frac{(1-\xi)\xi^{j-1}}{E[A]} \int_0^\infty e^{-\mu t(1-\xi)} F_A^c(t) dt
\end{aligned}
\tag{5.66}
$$

for all $j \geq 1$. Also we have

$$\pi_0 := \lim_{t \to \infty} P(L_t = 0) = 1 - \frac{1}{E[A]} \int_0^\infty e^{-\mu t(1-\xi)} F_A^c(t) dt. \tag{5.67}$$

To simplify these expressions, we find

$$\xi = \sum_{n=0}^{\infty} a_n \xi^n = \int_0^\infty e^{-\mu t} \sum_{n=0}^{\infty} \frac{(\mu t \xi)^n}{n!} dF_A(t) = \int_0^\infty e^{-\mu t(1-\xi)} dF_A(t)$$

and use it to obtain

$$\int_0^\infty F_A(t)e^{-\mu t(1-\xi)}dt = \frac{-1}{\mu(1-\xi)}\left[F_A(t)e^{-\mu t(1-\xi)}\right]_0^\infty$$
$$- \frac{-1}{\mu(1-\xi)}\int_0^\infty e^{-\mu t(1-\xi)}dF_A(t)$$
$$= \frac{\xi}{\mu(1-\xi)}.$$

With this integral result, (5.66) and (5.67) reduce to

$$\pi_j = (1-\xi)\xi^{j-1}\rho \quad \text{for } j \geq 1 \text{ and } \pi_0 = 1-\rho.$$

Using the concept of subinvariant measure, we can rigorously show that the stationary distribution does not exist if $\rho > 1$.

Exercise 5.38 *Assume a Poisson arrival process for the $GI/M/1$ queue. Show that the results for the asymptotic distribution coincide with the results obtained for the $M/M/1$ queue.*

Exercise 5.39 *Consider a $GI/M/1$ queue with an N threshold policy. In this system, the server stops attending the queue when the system becomes empty and resumes serving the queue when the number of customers reaches a threshold value N. Outline the procedures of deriving the stationary distribution of the queue length.*

Exercise 5.40 *Extend the analysis of $GI/M/1$ queue to the case of multiserver case. That is $GI/M/c$ queue with $c \geq 2$.*

5.2.5 AN INVENTORY MODEL

In this section, we use the semi-regenerative process to study a (s,Q) inventory system where demands occur once every T units and there are two suppliers. This model was proposed by Haughton and Isotupa (2018) motivated by a US manufacturer with one international supplier and one domestic supplier. The international supplier, called a regular supplier, offers a low price but has a random lead time due to the uncertainty of the border-crossing times. The domestic supplier, called an emergent supplier, offers a high price but has a deterministic lead time. The demand is deterministic and occurs every T time units. At the demand occurrence instant, only one unit is demanded. The reorder policy with the regular supplier is the (s,Q) type, meaning whenever the inventory level reduces to s, an order of size Q is placed with the regular supplier. Assume that the lead time follows an exponential distribution. This implies that the maximum inventory level is upper bounded by $s+Q$. When the inventory level reaches zero, the demand is lost (i.e., lost sales). Thus, the manufacturer places an emergency order with the emergent supplier when the stock-out occurs. It is also assumed that the lead time of the emergent supplier is zero. However, the placed order with the emergent supplier is not filled with 100% probability. In other words, with the emergency order placed, the inventory level gets replenished with probability p instantaneously.

Let $I(t)$ denote the inventory level at time t. Then, the process $\{I(t), t \in \mathscr{R}_+\}$ with state space $E = \{0, 1, ..., Q+s\}$ is a semi-regenerative process with the deterministic regeneration instants, which are a special case of a sequence of stopping times. Let $T = \{T_0, T_1, T_2, ..., \} = \{0, T, 2T, ...\}$ be the successive time instants at which demands occur. If $I_n = I(T_n^-)$, then $(I, T) = \{I_n, T_n^-; n \in \mathscr{N}_0\}$ is the Markov renewal process.

We try to determine the stationary distribution of $I(t)$ with which the system performance measure such as long-term cost rate can be computed for a given (s, Q) policy. To achieve this goal, we write two conditional probabilities as

$$P(i,j,t) = P(I(t) = j|I_0 = i), \quad K(i,j,t) = P(I(t) = j, T_1 > t|I_0 = i), \quad (5.68)$$

and denote the derivative of the semi-Markov kernel Q by q which is defined as

$$q(i,j,t) = Q(i,j,dt) = \lim_{\Delta \to 0} P(I_1 = j, t \le T_1 \le t+\Delta|I_0 = i)/\Delta. \quad (5.69)$$

Then, it is clear that $P(i,j,t)$ satisfies the Markov renewal equation

$$P(i,j,t) = K(i,j,t) + \sum_{l=0}^{s} \int_0^t q(i,l,u)P(l,j,t-u)du. \quad (5.70)$$

We first find $P(i,j) = \lim_{t \to \infty} Q(i,j,t) = \int_0^\infty q(i,j,t)dt$ as follows:

$$P(i,j) = \begin{cases} e^{-\mu T} & i = j = 0; 2 \le i \le s+1, j = i-1, \\ 1 - e^{-\mu T} & i = 0, j = Q; 2 \le i \le s+1, j = i+Q-1, \\ (1-p)e^{-\mu T} & i = 1, j = 0, \\ (1-p)(1-e^{-\mu T}) & i = 1, j = Q, \\ pe^{-\mu T} & i = 1, j = s, \\ p(1-e^{-\mu T}) & i = 1, j = Q+s, \\ 1 & s+2 \le i \le Q+s, j = i-1, \\ 0 & \text{otherwise.} \end{cases}$$

$$(5.71)$$

The stationary distribution of the embedded Markov chain can be obtained by solving the equation $\pi(j) = \sum_i \pi(i)P(i,j)$ and $\sum_j \pi(j) = 1$ due to the finite E.

The results are as follows:

$$\pi(j) = \frac{(e^{\mu T} - 1)e^{(j-1)\mu T}}{1-p}\,\pi(0), \qquad\qquad\qquad 1 \le j \le s,$$

$$\pi(j) = \frac{(e^{\mu T} - 1)e^{s\mu T}}{1-p}\,\pi(0) - \frac{p(e^{\mu T} - 1)}{1-p}\,\pi(0), \qquad s+1 \le j \le Q,$$

$$\pi(j) = \frac{(e^{\mu T} - 1)e^{s\mu T}}{1-p}\,\pi(0) - \frac{p(e^{\mu T} - 1)}{1-p}\,e^{(j-Q-1)\mu T}\,\pi(0), \quad Q+1 \le j \le Q+s.$$

$$(5.72)$$

Using the normalization condition, we have

$$\pi(0) = \left[1 + \frac{Q(e^{\mu T} - 1)}{1-p}e^{s\mu T} + \frac{(Q-s)p(e^{\mu T} - 1)}{1-p}\right]^{-1}.$$

Next, we need to write down $K(i,j,t)$ functions. According to the system description and the inventory policy, these functions are given by

$$K(i,j,t) = \begin{cases} e^{-\mu t} & i=j=0; 2\le i\le s+1, j=i-1, \\ 1-e^{-\mu t} & i=0, j=Q; 2\le i\le s+1, j=i+Q-1, \\ (1-p)e^{-\mu t} & i=1, j=0, \\ (1-p)(1-e^{-\mu t}) & i=1, j=Q, \\ pe^{-\mu t} & i=1, j=s, \\ p(1-e^{-\mu t}) & i=1, j=Q+s, \\ 1 & s+2\le i\le Q+s, j=i-1, \\ 0 & \text{otherwise.} \end{cases}$$

$$(5.73)$$

Then, due to the deterministic $T_1 = T$, we can get

$$\int_0^\infty K(i,j,t) = \begin{cases} \frac{1-e^{-\mu T}}{\mu} & i=j=0; 2\le i\le s+1, j=i-1, \\ T - \frac{1-e^{-\mu T}}{\mu} & \begin{array}{l} i=0, j=Q, \\ 2\le i\le s+1, j=i+Q-1, \end{array} \\ \frac{(1-p)(1-e^{-\mu T})}{\mu} & i=1, j=0, \\ (1-p)T - \frac{(1-p)(1-e^{-\mu T})}{\mu} & i=1, j=Q, \\ \frac{p(1-e^{-\mu T})}{\mu} & i=1, j=s, \\ pT - \frac{p(1-e^{-\mu T})}{\mu} & i=1, j=Q+s, \\ T & s+2\le i\le Q+s, j=i-1, \\ 0 & \text{otherwise.} \end{cases}$$

$$(5.74)$$

Now we can obtain the stationary distribution of the inventory process, denoted by $P(j) = \lim_{t\to\infty} P(I_t = j)$, by applying Proposition 5.7 and the fact

$\sum_{i=0}^{s} \pi(i)m(i) = T$ as follows:

$$P(j) = \frac{\sum_{i=0}^{s} \pi(i) \int_0^\infty K(i,j,t)}{\sum_{i=0}^{s} \pi(i)m(i)}$$

$$= \begin{cases} \frac{e^{-\mu T}-1}{\mu T}\pi(0) & j=0, \\ \frac{(e^{\mu T}-1)^2 e^{(j-1)\mu T}}{\mu T(1-p)}\pi(0) & 1 \le j \le s, \\ \frac{(e^{\mu T}-1)e^{s\mu T}}{1-p}\pi(0) - \frac{p(e^{\mu T}-1)}{1-p}\pi(0) & \\ \quad \text{for} \quad s+1 \le j \le Q-1, \\ \frac{(e^{\mu T}-1)e^{s\mu T}}{1-p}\pi(0) - \frac{p(e^{\mu T}-1)}{1-p}\pi(0) + \pi(0) - \frac{e^{\mu T}-1}{\mu T}\pi(0) & j=Q, \\ \frac{(e^{\mu T}-1)e^{s\mu T}}{1-p}\pi(0) - \frac{p(e^{\mu T}-1)^2 e^{(j-Q-1)\mu T}}{\mu T(1-p)}\pi(0) & \\ \quad \text{for} \quad Q+1 \le j \le Q+s. \end{cases}$$

$$(5.75)$$

With this stationary distribution, we can compute the various performance measures for this inventory system such as the long-term cost rate of operations.

Exercise 5.41 *For the inventory system described above, consider the following cost parameters (symbols and values):*

- $K_1 = 150$: *the set-up cost per order for the regular order.*
- $K_2 = 80$: *the set-up cost per order for the emergency order. This cost is incurred whether or not the order is filled.*
- $c_1 = 20$: *the cost per item for the regular order.*
- $c_2 = 25$: *the cost per item for the emergency order. This cost is only incurred if the order is filled.*
- $g = 200$: *the shortage cost/unit short.*
- $h = 2$: *the inventory carrying cost/unit/unit time.*
- $T = 0.05, \mu = 2$

Develop the long-term average cost rate function and use numerical approach to determine the optimal (s,Q) policy that minimizes the long-term average cost rate (Using Matlab is recommended).

5.2.6 SUPPLEMENTARY VARIABLE METHOD

It is clear that having a generally distributed (i.e., non-exponentially distributed) random variable in a stochastic system makes the model non-Makovian and hence more difficult to be analyzed. This is because that the non-exponential random variable lacks the "memoryless property". While the semi-regenerative process is an effective approach to overcome this challenge, there is another approach called supplementary variable method that can build a Markovian model by introducing an additional continuous state variable. We show this method by applying it to the $M/G/1$ queue. Let $S_e(t)$ be the elapsed service time for customer currently in service at time t and $L(t)$ be the number of customers in system at time t. By definition, $S_e(t) = 0$ if $L(t) = 0$. Then, the two-dimensional process $\{(L(t), S_e(t)), t \geq 0\}$ forms a continuous-time Markov process although the process $\{L(t), t \geq 0\}$ is not a Markov chain. The continuous state variable $S_e(t)$ is called the supplementary variable. We can obtain the joint stationary distribution of $(L(t), S_e(t))$ in terms of the transform from which the transform for the stationary distribution of $L(t)$ can be derived. First, we introduce the joint probability density at time t, $f_k(t, x)$, which is defined as

$$f_j(t,x)dx = P(L(t) = j, x < S_e(t) \leq x + dx)$$

for $j = 1, 2, \dots$ Denote by $P_j(t) = P(L(t) = j) = \int_0^\infty f_j(t,x)dx$ the queue length distribution at time t. Under the stability condition, the $M/G/1$ queue reaches steady state. Thus, we have

$$f_j(x)dx = \lim_{t \to \infty} f_j(t,x)dx = \lim_{t \to \infty} P(L(t) = j, x < S_e(t) \leq x + dx)$$
$$= P(L = j, x < S_e \leq x + dx)$$

with $f_0(x) = 0$, and

$$\pi_j = \lim_{t \to \infty} P_j(t) = \int_0^\infty f_j(x)dx.$$

Let $f_{S_c}(x)$ be the conditional p.d.f. of the service time S given $S > x$. Thus, we have

$$f_{S_c}(x)dx = P(x < S < x + dx | S > x) = \frac{f_S(x)}{1 - F_S(x)}dx,$$

where $F_S(x)$ and $f_S(x)$ are the cdf and pdf of the service time S, respectively. Considering each state as a node and using the node-flow balance equations, we have

$$\lambda \pi_0 = \int_0^\infty f_1(x) f_{S_c}(x)dx \quad \text{for } j = 0,$$
$$f_j(x + \Delta x)dx = \lambda \Delta x[1 - f_{S_c}(x)\Delta x]f_{j-1}(x)dx + (1 - \lambda \Delta x)[1 - f_{S_c}(x)\Delta x]f_j(x)dx$$

for $j = 1, \dots$, where the LHS represents the flow out of the node with j customers in the system and the RHS represents the flow into the node. In the second equation (i.e., for $j \geq 1$), the first term on the RHS represents only an arrival to the state with $j - 1$ customers occurs during the time interval Δx and the second term on the RHS represents that nothing (no arrival or departure) happens except for the progress of the service in the state with j customers during the time interval Δx. Note that it is not possible for an "service completion" (customer departure) during Δx that brings state $j+1$ to the state $(j, x+\Delta x)$ as a departure would lead to the elapsed service time to 0 rather than $x+\Delta x$. The second flow balance equation can be written as

$$f_j(x+\Delta x) = \lambda \Delta x f_{j-1}(x) + [1 - \Delta x(\lambda + f_{S_c}(x))]f_j(x),$$

which by taking limit $\Delta x \to 0$ leads to

$$\frac{df_j(x)}{dx} + [\lambda + f_{S_c}(x)]f_j(x) = \lambda f_{j-1}(x). \tag{5.76}$$

Such a differential equation has the following boundary conditions:

$$f_1(0) = \lambda \pi_0 + \int_0^\infty f_2(x) f_{S_c}(x) dx \text{ for } j = 1$$

$$f_j(0) = \int_0^\infty f_{j+1}(x) f_{S_c}(x) f_{S_c}(x) dx \text{ for } j = 2, 3, \ldots$$

The boundary conditions are due to the fact that $f_j(0)$ is the flow rate at the instant when the service just started (i.e., the elapsed service time $x = 0$). The normalization condition is given by

$$\sum_{j=0}^\infty \pi_j = \pi_0 + \sum_{j=1}^\infty \infty \int_0^\infty f_j(x) dx = 1.$$

To solve the differential equation (5.76), we utilize the transform approach. Define

$$F(z, x) = \sum_{j=1}^\infty f_j(x) z^k.$$

Multiplying the jth equation of (5.76) by z^j and summing them up yields

$$\frac{\partial F(z, x)}{\partial x} = [\lambda z - \lambda - f_{S_c}(x)]F(z, x), \tag{5.77}$$

with initial condition in terms of transforms as

$$F(z, 0) = \lambda z \pi_0 + \sum_{j=1}^\infty z^j \int_0^\infty f_{j+1}(x) f_{S_c}(x) dx$$

$$zF(z, 0) = \lambda z(z-1)\pi_0 + \int_0^\infty f_{S_c}(x)F(z, x)dx. \tag{5.78}$$

To facilitate the solution, we introduce the transform for the function defined by

$$g_j(x) = \frac{f_j(x)}{1 - f_S(x)} \quad \text{for } j = 1, 2, \ldots$$

with $g_0(x) = 0$. That is

$$G(z,x) = \sum_{j=1}^{\infty} g_j(x)z^j = \frac{F(z,x)}{1 - f_S(x)}. \tag{5.79}$$

Using (5.79) in (5.77) yields

$$\frac{\partial G(z,x)}{\partial x} + \lambda(1-z)G(z,x) = 0. \tag{5.80}$$

Clearly, the solution to (5.80) is given by

$$G(z,x) = G(z,0)e^{-\lambda(1-z)x}.$$

Using the initial conditions (5.78), $F(z,0) = G(z,0)$, and $f_j(0) = g_j(0)$ for $j = 0,1,2,...$, we have

$$zG(z,0) = \lambda z(z-1)\pi_0 + \int_0^{\infty} f_S(x)G(z,0)e^{-\lambda(1-z)x}dx$$
$$= \lambda z(z-1)\pi_0 + G(z,0)\tilde{S}(\lambda - \lambda z).$$

Solving this equation, we obtain

$$G(z,0) = \frac{\lambda z(1-z)\pi_0}{\tilde{S}(\lambda - \lambda z) - z}$$

and

$$G(z,x) = \frac{\lambda z(1-z)\pi_0}{\tilde{S}(\lambda - \lambda z) - z}e^{-\lambda(1-z)x}. \tag{5.81}$$

It follows from the definition of $G(z,x)$ that

$$F(z,0) = \frac{\lambda z(1-z)\pi_0}{\tilde{S}(\lambda - \lambda z) - z}$$

$$\tag{5.82}$$

$$F(z,x) = \frac{\lambda z(1-z)\pi_0}{\tilde{S}(\lambda - \lambda z) - z}[1 - F_S(x)]e^{-\lambda(1-z)x}.$$

Define $F(z) := \int_{x=0}^{\infty} F(z,x)dx$. Then it holds that

$$\sum_{j=1}^{\infty} \pi_j z^j = \sum_{j=1}^{\infty} z^j \int_{x=0}^{\infty} f_j(x)dx = \int_0^{\infty} F(z,x)dx = F(z).$$

Note that $\pi(z) = \sum_{j=0}^{\infty} \pi_j z^j = \pi_0 + F(z)$. It follows from (5.82) that

$$F(z) = \int_0^{\infty} F(z,x)dx = \int_0^{\infty} \left(\frac{\lambda z(1-z)\pi_0}{\tilde{S}(\lambda - \lambda z) - z}[1 - F_S(x)]e^{-\lambda(1-z)x} \right) dx$$

$$= \left(\frac{\lambda z(1-z)\pi_0}{\tilde{S}(\lambda - \lambda z) - z} \right) \left(\frac{1 - \tilde{S}(\lambda - \lambda z)}{\lambda(1-z)} \right)$$

$$= \frac{z\pi_0(1 - \tilde{S}(\lambda - \lambda z))}{\tilde{S}(\lambda - \lambda z) - z}.$$

Using the condition $F(1) = 1$ and applying L'Hôpital's rule, we obtain $\pi_0 = 1 - \lambda E[S] = 1 - \rho$. Thus, we finally get

$$\pi(z) = \pi_0 + F(z) = \pi_0 \left\{ 1 + \frac{z[1 - \tilde{S}(\lambda - \lambda z)]}{\tilde{S}(\lambda - \lambda z) - z} \right\}$$

$$= \frac{(z-1)(1-\rho)\tilde{S}(\lambda - \lambda z)}{z - \tilde{S}(\lambda - \lambda z)}$$

$$= \frac{(1-\rho)(z-1)a(z)}{z - a(z)},$$

where the last equality follows by noting that $\tilde{S}(\lambda - \lambda z) = a(z)$ used in the previous subsection. Supplementary variable method offers a possible approach to analyzing stochastic models with generally distributed random variables.

Exercise 5.42 *Use the supplementary variable method to derive the stationary distribution of the queue length in a $GI/M/1$ queue. (Hint: Using the residual inter-arrival time in the $GI/M/1$ queue as supplementary variables.)*

REFERENCE NOTES

This chapter studies the renewal processes and embedded Markov chains to overcome the difficulty caused by the non-exponential random variables in stochastic modeling. The basis of building an embedded Markov chain is the structure of the Markov renewal process that generalizes both the renewal process and the continuous-time Markov chain. The Markov renewal theory presented in this chapter is mainly adopted from Cinlar [1]. Some results for the renewal processes, such as the CLT for renewal processes, are based on Ross [5] and [6]. Some more theoretical developments, such as the rigorous proof of the Blackwell's theorem, are referred to the classical book by Feller [3]. The inventory model based on the Markov renewal process is based on the work by Haughton and Isotupa [4].

REFERENCES

1. E. Cinlar, Markov renewal theory: A survey. *Management Science*, 21 (7), 727–752, 1975.
2. D.R. Cox, "Renewal Theory", Methuen Co.Lid Science Paperbacks, 1961.
3. W. Feller, "An introduction to probability theory and its applications", Vol. II. John Wiley & Sons Inc., New York, 2nd ed., 1971.
4. M. Haughton and K.P.S. Isotupa, A continuous review inventory system with lost sales and emergency orders. *American Journal of Operations Research*, 8, 343–359, 2018.
5. S. Ross, "Stochastic Processes", John Wiley & Sons Inc., New York, 2nd ed., 1996.

6. S. Ross, "Introduction to Probability Models", Academic Press, 11th ed., 2014.

7. R. W. Wolff, "Stochastic Modeling and the Theory of Queues", Prentice – Hall, 1989.

6 Random Walks and Brownian Motions

The stochastic processes to be studied in this chapter are the random walk and Brownian motion that are the Markov chain and Markov process, respectively, with infinite state space. These two models are closely related as the Brownian motion can be viewed as a limiting version of a random walk. These stochastic processes have wide applications in many areas such as finance, economics, engineering, biology, and others. In addition, they play an important role in developing more complex stochastic models. We will provide both the transient and steady-state analysis on these models. The first part of this chapter is on the random walk process. A Binomial securities market model is presented to illustrate the application of the random walk process. The second part focuses on the Brownian motion that is a building block of formulating the stochastic models for large systems based on the stochastic process limits, a topic in the next two chapters.

6.1 RANDOM WALK PROCESSES

Most random walk processes belong to the class of the DTMC's. We start with the simple random walk.

6.1.1 SIMPLE RANDOM WALK – BASICS

Consider a simple game by flipping a coin. If the outcome is a head, the player wins \$ 1; otherwise, he loses \$ 1. Suppose that the probability of getting heads is p. Denote by $\{X_k, k \geq 1\}$ the sequence of outcomes of playing the game. Clearly, this is a sequence of Bernoulli random variables. For a player, the cumulative winning is a quantity of interest. Define

$$S_0 = a, \quad S_n = \sum_{i=1}^{n} X_i, \ n \geq 1,$$

where $P(X_i = 1) = p, P(X_i = -1) = 1 - p = q$, $0 < p < 1$ and $a > 0$ is the initial wealth. Then $S_n = \sum_{i=1}^{n} X_i, \ n \geq 1$, representing the total wealth of the player at step n, is called a simple random walk. A special case of the simple random walk is to toss a fair coin with $p = 1/2$, called symmetric random walk. Furthermore, we can define a state variable, Z_n, to indicate the status of

DOI: 10.1201/9781003150060-6

the player as follows:

$$Z_n = \begin{cases} 0, & \text{if } S_n = a, \\ 1, & \text{if } S_n > a, \\ -1, & \text{if } S_n < a, \end{cases}$$

which represents break even, winning, and losing, respectively, st step n. Then $\{Z_n, n \geq 1\}$ is a regenerative process with three discrete states. To understand this regenerative process, we first note some properties of simple random walk. A simple random walk is

1. time-homogeneous if $P(S_{n+k} = j|S_k = a) = P(S_n = j|S_0 = a)$
2. spatial homogeneous if $P(S_{n+k} = j+b|S_k = a+b) = P(S_n = j|S_0 = a)$
3. Markovian if $P(S_{n+k} = j|S_0 = j_0, S_1 = j_1, \cdots, S_k = j_k) = P(S_{n+k} = j|S_k = j_k)$.

The simple random walk can be graphed as a two-dimensional chart with the horizontal axis for the time n and vertical axis as the state S_n, called "time-state" chart. Denote the time-state point on the chart by (n, S_n). We can count the number of sample paths from the starting point $(0, a)$ to a future point (n, b) after n steps, denoted by $N_n(a, b)$. It is easy to see that a sample path exists if it has $(n+b-a)/2$ steps of moving up (winning) in n steps. Thus, the number of paths is equal to $N_n(a, b) = \binom{n}{(n+b-a)/2}$ if $(n+b-a)/2$ is a positive integer, or 0 otherwise. Since all these sample paths have the same probability, we have the n-step transition probability as

$$P(S_n = b|S_0 = a) = \binom{n}{\frac{n+b-a}{2}} p^{(n+b-a)/2} q^{(n+a-b)/2}. \tag{6.1}$$

As an example, we obtain the probability that a symmetric random walk starting with $a = 0$ returns to 0 at step $2n$, denoted by u_{2n}, as

$$u_{2n} = P(S_{2n} = 0|S_0 = 0) = \binom{2n}{n} \left(\frac{1}{2}\right)^{2n}.$$

It is easy to show the recursive relation

$$u_{2n} = \frac{2n-1}{2n} u_{2n-2}.$$

Similarly, we can write recursive relations for the transition probabilities in a non-symmetric random walk (left as an exercise). Using Stirling's approximation $n! \approx \sqrt{2\pi n}(n/e)^n$ for large n, we obtain

$$u_{2n} = \frac{(2n)!}{n!(2n-n)!} \left(\frac{1}{2}\right)^{2n} \approx \frac{\sqrt{4\pi n} \left(\frac{2n}{e}\right)^{2n}}{\left(\sqrt{2\pi n} \left(\frac{n}{e}\right)^n\right)^2} \left(\frac{1}{2}\right)^{2n} = \frac{1}{\sqrt{\pi n}}.$$

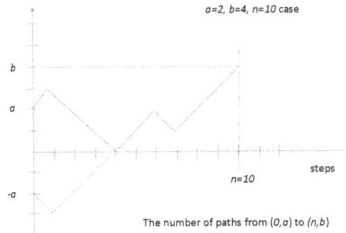

The number of paths from $(0,a)$ to (n,b)

Figure 6.1 Reflection principle in random walk.

As mentioned in Chapter 2, for a symmetric random walk which is a DTMC, the fact $\sum_n u_{2n} = \infty$ indicates that $\{S_n = 0\}$ is a recurrent event when $p = q = 1/2$ but the DTMC is null recurrent.

Exercise 6.1 *Consider a non-symmetric random walk starting with $a = 0$. Write the recursive relation for the transition probabilities u_{2n} from 0 to 0 in even number of steps.*

An important property about simple random walk is called "reflection principle".

Proposition 6.1 *(Reflection Principle) Let $N_n^0(a,b)$ be the number of sample paths from $(0,a)$ to (n,b) that touch or cross the x-axis. If $a, b > 0$, then $N_n^0(a,b) = N_n(-a,b)$. That is, there is a one-to-one correspondence between any path in $N_n^0(a,b)$ and a unique one in $N_n(-a,b)$.*

This can be seen by establishing, as shown in Figure 6.1, a one-to-one correspondence between paths from $(0,a)$ to (n,b) and paths from $(0,-a)$ to (n,b) that touch the time axis at some $0 < t < n$.

With this principle, we can verify the so-called Ballot theorem.

Suppose that two candidates, Adam and Bob, are in competition for some title. In the end, Adam wins by b votes in n casts. If the votes are counted in a random order, and we are interested in the event A that Adam leads throughout the count. Denote by $P(A)$ the probability of this event. Note that a sample path from $(0,0)$ to (n,b) that does not revisit the axis must move from $(0,0)$ to $(1,1)$. The number of paths from $(1,1)$ to (n,b), such that $b > 0$, which do not touch x-axis is

$$N_{n-1}(1,b) - N_{n-1}^0(1,b) = N_{n-1}(1,b) - N_{n-1}(-1,b)$$

$$= \binom{n-1}{\frac{n+b-2}{2}} - \binom{n-1}{\frac{n-b-2}{2}} = \frac{(n-1)!}{\left(\frac{n+b-2}{2}\right)!\left(\frac{n-b}{2}\right)!} - \frac{(n-1)!}{\left(\frac{n-b-2}{2}\right)!\left(\frac{n+b}{2}\right)!}$$

$$= \frac{(n-1)!}{\left(\frac{n+b}{2}\right)!} \left[\frac{\left(\frac{n+b}{2}\right)!}{\left(\frac{n-b}{2}\right)! \left(\frac{n+b}{2}-1\right)!} - \frac{1}{\left(\frac{n-b}{2}-1\right)!} \right]$$

$$= \frac{(n-1)!}{\left(\frac{n+b}{2}\right)!} \left[\frac{\frac{n+b}{2}}{\left(\frac{n-b}{2}\right)!} - \frac{1}{\left(\frac{n-b}{2}-1\right)!} \right]$$

$$= \frac{(n-1)!}{\left(\frac{n+b}{2}\right)! \left(\frac{n-b}{2}-1\right)!} \left[\frac{\frac{n+b}{2}}{\frac{n-b}{2}} - 1 \right] = \frac{(n-1)!}{\left(\frac{n+b}{2}\right)! \left(\frac{n-b}{2}-1\right)!} \frac{b}{\frac{n-b}{2}}$$

$$= \frac{(n-1)!}{\left(\frac{n+b}{2}\right)! \left(\frac{n-b}{2}\right)!} \frac{nb}{n} = \frac{n!}{\left(\frac{n+b}{2}\right)! \left(\frac{n-b}{2}\right)! n} \frac{b}{n} = \frac{b}{n} N_n(0,b).$$

The absolute value of b can accommodate both $b > 0$ or $b < 0$ case. Clearly, we have

$$P(A) = \frac{\frac{b}{n} N_n(0,b)}{N_n(0,b)} = \frac{b}{n}.$$

This interesting result is called the Ballot theorem. With this result, we can find more probabilities of interest. For example, we may want to compute the probability that the process does not visit 0 over the first n steps and reaches b in step n given the starting state is 0. The probability of this event is

$$P(\{S_1 \neq 0\} \cap \{S_2 \neq 0\} \cap \cdots \cap \{S_n \neq 0\} \cap \{S_n = b\} | S_0 = 0)$$
$$= P(A \cap \{S_n = b\} | S_0 = 0)$$

$$= P(A | S_n = b, S_0 = 0) P(S_n = b | S_0 = 0) = \frac{|b|}{n} P(S_n = b | S_0 = 0).$$

Next, we can get the probability that the random walk never visits 0 over the first n steps.

$$P(\{S_1 \neq 0\} \cap \{S_2 \neq 0\} \cap \cdots \cap \{S_n \neq 0\}) = \sum_{b=-\infty}^{+\infty} \frac{|b|}{n} P(S_n = b | S_0 = 0) = \frac{E[|S_n|]}{n}.$$

Another interesting probability is the probability of the first passage through b at step n, denoted by $f_n(b)$, also called hitting time probability, given the starting at $S_0 = 0$. That is

$$f_b(n) = P(\{S_1 \neq b\} \cap \{S_2 \neq b\} \cap \cdots \cap \{S_{n-1} \neq b\} \cap \{S_n = b\} | S_0 = 0)$$
$$= \frac{|b|}{n} P(S_n = b). \tag{6.2}$$

This result follows as if we reverse the time, the sample paths become those that reach b after n-steps without touching the x-axis. Later, we can simplify the notation by writing joint events $(\{S_1 \neq b\} \cap \cdots \cap \{S_{n-1} \neq b\} \cap \{S_n = b\})$ as $(S_1 \neq b, \cdots, S_{n-1} \neq b, S_n = b)$. Another problem of interest is to find the

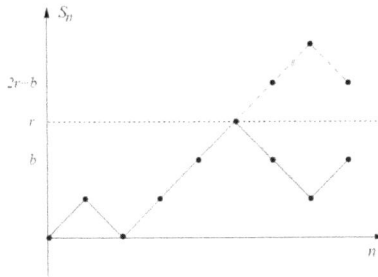

Figure 6.2 Mirroring for maximum.

maximum value that the simple random walk can reach before it settles at $S_n = b$ with $b > 0$. This is so-called the extreme value problem of a stochastic process. Define $M_n = \max\{S_i, i = 1, ..., n\}$. The complementary distribution function (i.e., tail probability) can be obtained as follows:

Theorem 6.1 *For a simple random walk with $S_0 = 0$, given $r \geq 1$,*

$$(a) \quad P(M_n \geq r, S_n = b) = \begin{cases} P(S_n = b), & \text{if } b \geq r, \\ \left(\dfrac{q}{p}\right)^{r-b} P(S_n = 2r - b), & \text{if } b < r \end{cases}$$

and

$$(b) \quad P(M_n \geq r) = P(S_n \geq r) + \sum_{b=-\infty}^{r-1} \left(\frac{q}{p}\right)^{r-b} P(S_n = 2r - b)$$

$$= P(S_n = r) + \sum_{c=r+1}^{\infty} \left[1 + \left(\frac{q}{p}\right)^{c-r}\right] P(S_n = c).$$

For $p = q = 1/2$,

$$P(M_n \geq r) = 2P(S_n \geq r + 1) + P(S_n = r).$$

Proof. If $b \geq r$, the event $\{S_n = b\}$ is a subset of the event $\{M_n \geq r\}$. Thus, $P(M_n \geq r, S_n = b) = P(S_n = b)$. If $b < r$, we can draw a horizontal line $y = r$. For each sample path belongs to $\{M_n \geq r, S_n = n\}$, we get a partial mirror image: it retains the part until the sample path touches the line of $y = r$, and completes the remaining part with the mirror image from there (see Figure 6.2). Thus the number of sample paths is the same as the number of sample paths from 0 to $2r - b$. The corresponding probability for this mirror sample paths is $P(S_n = 2r - b)$. To get the probability of the original sample paths, we need to exchange the roles of the $r - b$ pairs of p and q. This means that $(r-b)$ p's are replaces with q's which are equivalent to increasing the power of q by $r-b$ and decreasing the power of p's by $r-b$. Thus $P(M_n \geq r, S_n = b) = (q/p)^{r-b} P(S_n = 2r-b)$ for $b < r$. This gives the second part of (a). For (b), we simply sum over all possible values of b as

$$P(M_n \geq r) = \sum_{b=-\infty}^{\infty} P(M_n \geq r, S_n = b)$$

$$= \sum_{b=r}^{\infty} P(S_n = b) + \sum_{b=-\infty}^{r-1} \left(\frac{q}{p}\right)^{r-b} P(S_n = 2r-b)$$

$$= P(S_n \geq r) + \sum_{c=r+1}^{\infty} \left(\frac{q}{p}\right)^{c-r} P(S_n = c)$$

$$= P(S_n \geq r) + \sum_{c=r+1}^{\infty} P(S_n = c) + \sum_{c=r+1}^{\infty} \left(\frac{q}{p}\right)^{c-r} P(S_n = c)$$

$$= P(S_n = r) + \sum_{c=r+1}^{\infty} \left[1 + \left(\frac{q}{p}\right)^{c-r}\right] P(S_n = c).$$

Here we utilize the change of variable $c = 2r - b$. It is easy to check the result for the $p = q = 1/2$ case. ∎

For a simple random walk, let Z_b be the number of times that the process visits a particular state b before returning the starting state $S_0 = 0$. Then Z_b can be expressed as the infinite sum of indicator random variables. Define $Y_n = I(S_1 S_2 \cdots S_n \neq 0, S_n = b)$, representing state b is visited in the nth step and the process has never visited 0 yet. Therefore, $Z_b = \sum_{n=1}^{\infty} Y_n$. The sum of infinite number of Y_n's indicate that as long as the walk has not returned to 0, its visit to b is recorded (i.e., accumulated). Note that $E[Y_n] = f_b(n)$. We can get the mean of Z_b, as

$$E[Z_b] = \sum_{n=1}^{\infty} E[Y_n] = \sum_{n=1}^{\infty} f_b(n).$$

Consider a game of tossing a fair coin. The rule of the game is based on the difference between the number of heads and the number of tails obtained in a sequence of coin tosses. Whenever the number of heads exceeds the number of tails by b, the player received one dollar. The game stops when first time the number of heads equals the number of tails (i.e., the first equalization of the accumulated numbers of heads and tails). Since the expected payoff of such a game is $1 regardless of b, the fair entrance fee for this game should be $ 1 and independent of b. This is a highly counter-intuitive result. To show $E[Z_b] = 1$ for a symmetric random walk, we introduce the concept of "dual walk".

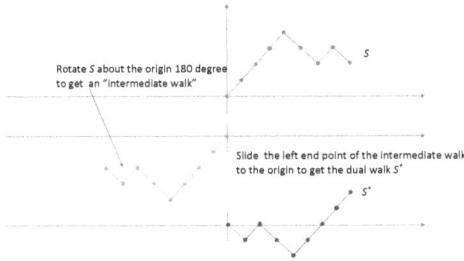

Figure 6.3 Transforming from the original random walk to its dual.

Definition 6.1 *For a simple random walk $S_n = X_1 + \cdots + X_n$ with $S_0 = 0$, define a new simple random walk S_n^* of length n by reversing the order of the X_k's as*

$$X_1^* = X_n, ..., X_n^* = X_1,$$

and for $t = 1, ..., n$

$$S_t^* = X_1^* + \cdots + X_t^* = X_n + \cdots + X_{n-t+1} = S_n - S_{n-t}.$$

The simple random walk S^* is called the dual of S.

Since both walks are sums of the same X_k's, they have the same terminal values, i.e. $S_n = S_n^*$. Graphically, we can obtain the dual walk S^* by first counter-clock-wise rotating a sample path of S 180 degree around the origin (time reversal) and then sliding the left endpoint to the origin (see Figure 6.3). Note that every event related to S has a corresponding (dual) event related to S^*. For example, the probability of one sample path for S is equal to the probability of the corresponding sample path for S^*. Thus, we can get an important relation in terms of some interesting probabilities:

$$P(S_1 > 0, ..., S_{n-1} > 0, S_n = b) = P(S_n^* > S_1^*, ..., S_n^* > S_{n-1}^*, S_n^* = b). \quad (6.3)$$

In this relation, the left-hand side is the probability that the original simple random walk S attains b in step n without visiting state 0 in the first $n-1$ steps; and the right-hand side is the probability that the dual walk S^* visits state b first time in step n. This relation can be clearly seen in Figure 6.3. By using the dual walk, we can easily show that $E[Z_b] = 1$ in the game discussed above. In fact, we have used this "dual walk" idea in developing the formula for $f_b(n)$ earlier. That is $f_b(n) = P(S_1 > 0, ..., S_{n-1} > 0, S_n = b) = P(S_n^* > S_1^*, ..., S_n^* > S_{n-1}^*, S_n^* = b)$, which is the probability that the dual walk S^* first visit to b occurs at step n. Note that $\sum_{n=1}^{\infty} f_b(n)$ is equal to the probability that S^* will ever reach b, which is denoted by f_b. We show that f_b for a simple random walk will be no more than 1 (i.e., a legitimate probability). By independence and Markov property, we have $f_b = f_1^b$ (the bth power of f_1). We first find $f_1 = P$(a simple random walk will ever reach $b = 1$). Conditioning on the first step, we get

$$f_1 = p \cdot 1 + q \cdot f_2 = p + q f_1^2,$$

which leads to

$$q f_1^2 - f_1 + p = 0.$$

The solutions to this quadratic equation are given by

$$f_1^* = \frac{1 \pm \sqrt{1 - 4pq}}{2q} = \frac{1 \pm (p - q)}{2q} = \begin{cases} 1 & \text{if } p \geq q, \\ \frac{p}{q} & \text{if } p < q. \end{cases}$$

The second equality is due to the following relation

$$1 = p + q = (p + q)^2 = p^2 + 2pq + q^2,$$

which can be written as

$$1 - 4pq = p^2 - 2pq + q^2 = (p - q)^2.$$

This implies

$$\sqrt{1 - 4pq} = |p - q|.$$

Thus, we obtain

Proposition 6.2 *For a simple random walk,*

$$f_b = \begin{cases} 1, & \text{if } p \geq q, \\ \left(\frac{q}{p}\right)^b, & \text{if } p < q. \end{cases}$$

Now for a symmetric simple random walk with $p = q = 1/2$, we have

$$E[Z_b] = \sum_{n=1}^{\infty} f_b(n) = f_b = 1,$$

and for non-symmetric simple random walk with $p < q$, this mean is less than 1.

Exercise 6.2 *Suppose that $\{X_k, k \geq 1\}$ is a sequence of i.i.d. random variables with $E[X_1] > 0$. Then, $S_n = \sum_{i=1}^{n} X_i$ with $n \geq 1$ is a general random walk. Define the first time that random walk becomes positive, denoted by T. That is $T := \min\{n : S_n > 0\}$. Given that $S_0 < 0$, show that $E[T] < \infty$.*

Exercise 6.3 *The duality relationship holds under a weaker condition than the one that the random variables $X_1, \ldots, X_n$ are i.i.d.. Such a condition is called "exchangeable" and defined as follows: A sequence of random variables is called exchangeable if the sequence $X_{i_1}, \ldots, X_{i_n}$ has the same joint distribution for all permutations $(i_1, \ldots, i_n)$ of $(1, 2, \ldots, n)$. Consider an urn containing M balls of which W are white. If we define*

$$X_k = \begin{cases} 1 & \text{if the } k\text{th selection is white,} \\ 0 & \text{otherwise,} \end{cases}$$

argue that the sequence $X_1, \ldots, X_n$ will be exchangeable but not independent.

Exercise 6.4 *Define the range of a random walk, denoted by R_n, as the number of distinct values of set $(S_0, S_1, \ldots, S_n)$, which is the cardinality of the set. It is the number of distinct points visited by the random walk up to time n. Show*

$$\lim_{n \to \infty} \frac{E[R_n]}{n} = P(\text{random walk never returns } 0).$$

For a symmetric simple random walk, we can also study the time-dependent transition probability and the first passage time when state 0 is absorbing. Denote by $m_n := \min_{0 \le k \le n} S_k, n = 0, 1, \ldots$ the minimum ever reached over n periods. Since the state 0 is absorbing, we define the n-step transition probability $a_{ij}(n)$ as

$$a_{ij}(n) = P(S_n = j, m_n > 0 | S_0 = i), \quad n = 0, 1, \ldots,$$

where $i, j > 0$. Note that the restriction due to the absorbing state is represented by the condition $\{m_n > 0\}$. From (6.1), we denote a special n-step transition probability in this symmetric random walk by

$$v_b(n) = P(S_n = b | S_0 = 0) = \binom{n}{\frac{n+b}{2}} \left(\frac{1}{2}\right)^n.$$

Using the reflection principle, we can obtain the following result.

Proposition 6.3 *For a symmetric random walk $\{S_n\}$ with absorbing state 0, the n-step transition probabilities are given by*

$$a_{ij}(n) = v_{j-i}(n) - v_{i+j}(n), \quad n = 0, 1, 2 \ldots, \tag{6.4}$$

where $i, j > 0$.

Proof. For a symmetric random walk $\{S_n\}$ without restriction on the movement, we have

$$P(S_n = j | S_0 = i) = P(S_n = j, m_n > 0 | S_0 = i) + P(S_n = j, m_n \le 0 | S_0 = i)$$
$$= a_{ij}(n) + P(S_n = j, m_n \le 0 | S_0 = i),$$

which leads to

$$a_{ij}(n) = P(S_n = j|S_0 = i) - P(S_n = j, m_n \leq 0|S_0 = i).$$

Note that $P(S_n = j, m_n \leq 0|S_0 = i)$ is the probability that the symmetric random walk touches the x-axis and then reaches state $j > 0$ at time n given the intial state i. It follows from the reflection principle that this probability equals the probability that this random walk crosses the x-axis and then reaches $-j$ at time n. That is

$$P(S_n = j, m_n \leq 0|S_0 = i) = P(S_n = -j|S_0 = i) = v_{-i-j}(n).$$

Using the symmetry of the transition probabilities, we have $v_{-i-j} = v_{i+j}$. This completes the proof. ∎

Let T_j be the first passage time of $\{S_n\}$ to state j, which is an example of stopping time. We are interested in the probability of event $\{T_0 = n\}$. It follows from the skip-free property of the random walk that

$$P(T_0 = n|S_0 = i) = P(X_n = -1, S_{n-1} = 1, m_{n-1} > 0|S_0 = i), \quad i > 0.$$

From the independence of X_n and S_{n-1}, we can compute the conditional probability above as

$$P(T_0 = n|S_0 = i) = \frac{1}{2}\{v_{i-1}(n-1) - v_{i+1}(n-1)\}, n = 0, 1, 2....$$

For a special case $i = 1$, we have

$$P(T_0 = 2n + 1|S_0 = 1) = \frac{(2n)!}{n!(n+1)!}\left(\frac{1}{2}\right)^{2n+1}$$

and $P(T_0 = 2n|S_0 = 1) = 0$ for $n = 0, 1, 2, ...$ It is a good exercise to check for a symmetric random walk that

$$P(T_0 \leq 2n + 1|S_0 = 1) = 1 - v_1(2n + 1), \quad n = 0, 1, 2, ...$$

It follows from $v_1(n) \rightarrow 0$ as $n \rightarrow \infty$ that

$$P(T_0 < \infty|S_0 = 1) = 1,$$

which is consistent with what we discussed in Chapter 2 on the null recurrence for the symmetric random walk. Another reason for presenting more results on the symmetric random walk is because a more general simple random walk can be transformed to the symmetric random walk by change of measures.

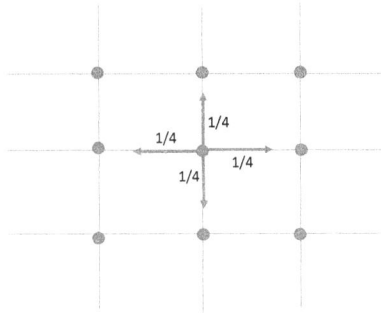

Figure 6.4 Two-dimensional random walk.

Exercise 6.5 *Consider a game that is being played by two players, A and B. In such a game, gambler A, flips a coin, and gains 1 dollar if it lands on heads (i.e., B loses 1 dollar), and loses 1 dollar if it lands on tails (i.e., B wins 1 dollar). Furthermore, suppose that the gambler wins the game if he reaches n dollars, and loses the game if he reaches 0 dollars. This game is represented by a random walk with the fortune of the gambler at time t given by S_t. we know the game must terminate, since either point n or 0 is eventually reached. Assume that A, and B have a total of N dollars. Treat this game as a DTMC, write the transition probability matrix, and compute the probability that one of the gambler is ruined for the case of $N = 5$ (more general result for an arbitrary N will be discussed in later section as the gambler's ruin problem).*

Exercise 6.6 *(Higher dimensions random walk) A simple symmetric random walk can be extended from one-dimension to two dimensions random walk as shown in Figure 6.4 with transition probabilities*

$$P_{(ij)(i'j')} = \begin{cases} \frac{1}{4} & if\,|i - i'| = 1\ or\ |j - j'| = 1, \\ 0 & otherwise, \end{cases}$$

where i and j are the row and column numbers of cells, respectively.

Suppose that a mouse is moving inside the maze with 9 cells as shown in Figure 6.5, from one cell to another, in search of food. When at a cell, the mouse will move to one of the adjoining cells randomly with equal probability of $1/k$ if the cell has k doors to other cells. For $n \geq 0$, let X_n be the cell number the mouse will visit after having changed cells n times. Then $\{X_n : n = 0, 1, ...\}$ is a Markov chain with state space $\{1, 2, ..., 9\}$, which can be considered as a two-dimensional random walk with a finite state space. Write down the transition probability matrix.

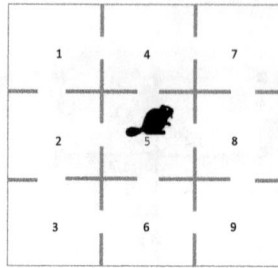

Figure 6.5 The maze problem for a mouse.

6.1.2 SPITZER'S IDENTITY LINKING RANDOM WALKS TO QUEUE-ING SYSTEMS

The relation between position S_n and the maximum M_n at time n can be established and is called Spitzer's Identity. There are different ways of proving such a relation. We present an easy and intuitive one.

Theorem 6.2 *For a right-continuous random walk and $|s|, |t| < 1$,*

$$\ln\left(\sum_{n=0}^{\infty} t^n E\left[s^{M_n}\right]\right) = \sum_{n=1}^{\infty} \frac{1}{n} t^n E\left[s^{S_n^+}\right], \qquad (6.5)$$

where $S_n^+ = \max\{0, S_n\}$ (i.e., the non-negative part of S_n).

Proof. Note that $f_b(n) = P(T_b = n)$ where T_b is the first time when $S_n = b$ for some $b > 0$ given that $S_0 = 0$. We also define the z-transform of T_b as $F_b(z) = E[z^{T_b}]$. To have $M_n = b$, the random walk must reach b at some time $1 \leq j \leq n$, and at the same time, does not reach $b+1$ in the next $n-j$ steps. Hence,

$$P(M_n = b) = \sum_{j=0}^{n} P(T_b = j)P(T_1 > n - j). \qquad (6.6)$$

Multiply both sides of (6.6) by $s^b t^n$ and sum over $b, n \geq 0$. Then, the left-hand side becomes

$$\sum_{n=0}^{\infty} t^n \sum_{b=0}^{\infty} s^b P(M_n = b) = \sum_{n=0}^{\infty} t^n E\left[s^{M_n}\right], \qquad (6.7)$$

which is the argument of the logarithmic function on the left-hand side of (6.5). To work out the right-hand side of (6.5), we need to evaluate $\sum_{n=0}^{\infty} t^n P(T_1 > n)$.

First note

$$\sum_{n=0}^{\infty} t^n P(T_1 > n) = \sum_{n=0}^{\infty} t^n \left[1 - P(T_1 \le n)\right] = \sum_{n=0}^{\infty} t^n - \sum_{n=0}^{\infty} t^n P(T_1 \le n)$$

$$= \frac{1}{1-t} - \sum_{n=0}^{\infty} t^n P(T_1 \le n). \tag{6.8}$$

Then we work out the summation term on the right-hand side of the equation above

$$\sum_{n=0}^{\infty} t^n P(T_1 \le n) = t \left[P(T_1 = 1)\right]$$

$$+ t^2 \left[P(T_1 = 1) + P(T_1 = 2)\right]$$
$$+ t^3 \left[P(T_1 = 1) + P(T_1 = 2) + P(T_1 = 3)\right]$$
$$+ t^4 \left[P(T_1 = 1) + P(T_1 = 2) + P(T_1 = 3) + P(T_1 = 4)\right]$$

$$\vdots$$

$$= \frac{t}{1-t} P(T_1 = 1) + \frac{t^2}{1-t} P(T_1 = 2) + \frac{t^3}{1-t} P(T_1 = 3) + \cdots$$

$$= \frac{1}{1-t} \left\{ t P(T_1 = 1) + t^2 P(T_1 = 2) + t^3 P(T_1 = 3) + \cdots \right\}$$

$$= \frac{1}{1-t} F_1(t). \tag{6.9}$$

Substituting (6.9) into (6.8) yields the relation $\sum_{n=0}^{\infty} t^n P(T_1 > n) = (1 - F_1(t))/(1-t)$ where $F_1(t) = E[t^{T_1}]$. Now using this relation, we can work out the right-hand side of (6.5) after multiplying both sides of (6.6) by $s^b t^n$ and summing over $b, n \ge 0$ as follows:

$$\sum_{b=0}^{\infty} \sum_{n=0}^{\infty} s^b t^n \sum_{j=0}^{n} P(T_b = j) P(T_1 > n - j)$$

$$= \sum_{b=0}^{\infty} s^b \sum_{j=0}^{\infty} \left[\sum_{n=j}^{\infty} t^{n-j} P(T_1 > n - j) \right] t^j P(T_b = j)$$

$$= \sum_{b=0}^{\infty} s^b \sum_{j=0}^{\infty} \frac{1 - F_1(t)}{1-t} t^j P(T_b = j) \tag{6.10}$$

$$= \frac{1 - F_1(t)}{1-t} \sum_{b=0}^{\infty} s^b F_b(t) = \frac{1 - F_1(t)}{1-t} \sum_{b=0}^{\infty} s^b [F_1(t)]^b$$

$$= \frac{1 - F_1(t)}{(1-t)(1 - s F_1(t))}$$

$$= D(s,t).$$

It follows from the hitting time result (6.2)

$$nP(T_1 = n) = P(S_n = 1) = \sum_{j=0}^{n} P(T_1 = j)P(S_{n-j} = 0) \qquad (6.11)$$

that the following relation in terms of the generating functions holds

$$tF_1'(t) = F_1(t)U(t), \qquad (6.12)$$

where $U(s)$ is the probability generating function for the time returning to 0. Using this relation, we develop another expression as follows:

$$
\begin{aligned}
\frac{\partial}{\partial t} \ln[1 - sF_1(t)] &= \frac{-sF_1'(t)}{1 - sF_1(t)} \\
&= -\frac{s}{t} F_1(t)U(t) \sum_{k=0}^{\infty} s^k [F_1(t)]^k \\
&= -\sum_{k=0}^{\infty} \frac{s^{k+1}}{t} [F_1(t)]^{k+1} U(t) \\
&= -\sum_{k=1}^{\infty} \frac{s^k}{t} [F_1(t)]^k U(t).
\end{aligned}
\qquad (6.13)
$$

It follows from

$$P(S_n = k) = \sum_{j=0}^{n} P(T_k = j)P(S_{n-j} = 0)$$

that the term $[F_1(t)]^k U(t)$ is the generating function of $P(S_n = k)$ in n. That is

$$[F_1(t)]^k U(t) = \sum_{n=0}^{\infty} t^n P(S_n = k).$$

Thus, using this expression, (6.13) can be written as

$$\frac{\partial}{\partial t} \ln[1 - sF_1(t)] = -\sum_{n=1}^{\infty} t^{n-1} \sum_{k=1}^{\infty} s^k P(S_n = k). \qquad (6.14)$$

Now, we have

$$
\begin{aligned}
\frac{\partial}{\partial t} \ln D(s,t) &= -\frac{\partial}{\partial t} \ln(1-t) + \frac{\partial}{\partial t} \ln[1 - F_1(t)] - \frac{\partial}{\partial t} \ln[1 - sF_1(t)] \\
&= \sum_{n=1}^{\infty} t^{n-1} \left(1 - \sum_{k=1}^{\infty} P(S_n = k) + \sum_{k=1}^{\infty} s^k P(S_n = k) \right) \\
&= \sum_{n=1}^{\infty} t^{n-1} \left(P(S_n \le 0) + \sum_{k=1}^{\infty} s^k P(S_n = k) \right) \\
&= \sum_{n=1}^{\infty} t^{n-1} E[s^{S_n^+}].
\end{aligned}
\qquad (6.15)
$$

Finally, integrating both sides of the equation above with respect to t yields the final result.   ∎

Exercise 6.7 *Prove the relation (6.12).*

The Spitzer's identity can be also expressed in terms of the moment generating functions or characteristic functions. Thus, (6.5) can be written as

$$\ln\left(\sum_{n=0}^{\infty} t^n E[\exp\{\gamma M_n\}]\right) = \sum_{n=1}^{\infty} \frac{1}{n} t^n E[\exp\{\gamma S_n^+\}] \qquad (6.16)$$

or

$$\ln\left(\sum_{n=0}^{\infty} t^n E[\exp\{i\lambda M_n\}]\right) = \sum_{n=1}^{\infty} \frac{1}{n} t^n E[\exp\{i\lambda S_n^+\}] \qquad (6.17)$$

by replacing the z-transforms (also called probability generating function or p.g.f.) of M_n and S_n^+ with their corresponding moment generating functions or with their corresponding characteristic functions.

We can develop another version of the Spitzter identity that allows us to use the random walk to study a queueing system.

First, a key relation between the waiting time for a queueing system and the maximum of a random walk process can be established. Consider an $G/G/1$ queue. Suppose customers, labelled as $0, 1, 2, \cdots$, arrive at a single server facility at times $0, X_1, X_1 + X_2, \cdots$. This means that customer 0 arrives at an empty system at time 0. Denote by X_n the inter-arrival time between the nth arrival and $(n-1)$st arrival. Let $A_n = X_1 + X_2 + \cdots + X_n$ be the instant of the nth arrival after time 0. Denote by Y_n the service time of the nth customer, where $n = 0, 1, \ldots$. Inter-arrival times are i.i.d. random variables with distribution F_X and service times are i.i.d. random variables with distribution F_Y. We also assume that X_n and Y_n are mutually independent. Let W_n be the waiting time (queueing delay) for the nth customer, $n \geq 1$. The system sojourn time (waiting plus service) for the nth customer is defined as $W_n + Y_n$. Clearly, if $X_n < W_{n-1} + Y_{n-1}$ (i.e., if customer n arrives before the departure of customer $n-1$), then customer n must wait in the queue until the service of customer $n-1$ is finished. Otherwise, customer n's waiting time is zero. Thus, we have

$$W_n = \max[W_{n-1} + Y_{n-1} - X_n, 0], \qquad (6.18)$$

for $n \geq 1$ and $W_0 = 0$. Equation (6.18) is the well-known Lindley recursion which is developed based on the sample path of the $G/G/1$ system. Since Y_{n-1} and X_n are coupled together for each customer n, we define $Z_n = Y_{n-1} - X_n$, which is a sequence of i.i.d. random variables. Thus, (6.18) can be written as

$W_n = \max[W_{n-1} + Z_n, 0]$. Using this relation iteratively yields the following

$$
\begin{aligned}
W_n &= \max[\max[W_{n-2} + Z_{n-1}, 0] + Z_n, 0] \\
&= \max[(W_{n-2} + Z_{n-1} + Z_n), Z_n, 0] \\
&= \max[(W_{n-3} + Z_{n-2} + Z_{n-1} + Z_n), (Z_{n-1} + Z_n), Z_n, 0] \\
&= \cdots \cdots \\
&= \max[(Z_1 + Z_2 + \cdots + Z_n), (Z_2 + Z_3 + \cdots + Z_n), ..., (Z_{n-1} + Z_n), Z_n, 0].
\end{aligned}
$$

$$(6.19)$$

Note that if the maximization is achieved at $Z_i + Z_{i+1} + \cdots + Z_n$, then we have

$$
\begin{aligned}
W_n &= \max[(W_{i-1} + Z_i + Z_{i+1} + \cdots + Z_n), (Z_{i+1} + Z_{i+2} + \cdots + Z_n), \\
&\quad ..., (Z_{n-1} + Z_n), Z_n, 0] \\
&= Z_i + Z_{i+1} + \cdots + Z_n,
\end{aligned}
$$

which implies that $W_{i-1} = 0$. This result indicates that a busy period must start with the arrival of customer $i - 1$ and continue at least through the service of customer n. Now we define $U_1^n = Z_n, U_2^n = Z_n + Z_{n-1}$ and for $i \le n$ $U_i^n = Z_n + Z_{n-1} + \cdots + Z_{n-i+1}$. Using this defined sequence, (6.19) can be written as

$$W_n = \max[0, U_1^n, U_2^n, ..., U_n^n].$$
$$(6.20)$$

It is clear that $\{U_i^n, 1 \le i \le n\}$ are the first n terms of a random walk that goes backward and starts with Z_n. Also note that W_{n+1} is the maximum of a different set of random variables (i.e., a random walk going backward from Z_{n+1}). However, the ordered random variables $(Z_n, Z_{n-1}, ..., Z_1)$ are statistically identical to $(Z_1, Z_2, ..., Z_n)$ since they are i.i.d. sequences. Now the probability that the waiting time of customer n is at least w is

$$P(W_n \ge w) = P(\max[0, U_1^n, U_2^n, ..., U_n^n] \ge w).$$

Another interpretation of this relation is that $P(W_n \ge w)$ is equal to the probability that the random walk $\{U_i^n, 1 \le i \le n\}$ crosses a threshold at w by the nth trial. Since we assume that customer 0 arrives at time 0, W_n is the waiting time of the nth arrival after the beginning of a busy period (though the nth arrival might belong to a later busy period than the initial busy period). Note that $P(W_n \ge w)$ is the same as the probability that the first n terms of a random walk based on $\{Z_i, i \ge 1\}$ crosses a threshold w. Since the first $n + 1$ terms of this random walk provide one more opportunity to cross w than the first n terms, we have

$$\cdots \le P(W_n \ge w) \le P(W_{n+1} \ge w) \le \cdots \le 1.$$

Since this sequence of probabilities is non-decreasing and upper bounded, it must have a limit as $n \to \infty$. Such a limit can be denoted by $P(W \ge w)$ and

$P(W \geq w) = \lim_{n \to \infty} P(W_n \geq w)$. The implication of the existence of this limit is that the queueing system is stable and can reach the steady state. As mentioned earlier, the analysis above is based on the sample path of a stochastic process, reflected by (6.18) and (6.19), and can be further developed into a powerful technique, called system point level crossing (SPLC) method, for deriving the probability distributions in stochastic models. This method was originated by Brill in 1975. We will show it briefly later by analyzing a queueing system (or an inventory system). For more details about the level crossing method, interested readers are referred to the book by Brill (2008).

Now we return to the Spitzer identity with $S_n = U_n^n$, $W_n \stackrel{d}{=} M_n$ and $W \stackrel{d}{=} M$, where $\stackrel{d}{=}$ denotes "equal in distribution". As $n \to \infty$, $M_n \uparrow M$ if and only if the random walk $\{S_n\}$ drifts to $-\infty$. That is, $E[Z_n] = E[Y_{n-1}] - E[X_n] < 0$ where $E[Y_{n-1}] = E[Y]$ and $E[X_n] = E[X]$ are the mean service time and mean inter-arrival time of a stable $G/G/1$ queue (i.e., $\rho := E[Y]/E[X] < 1$). First, re-write (6.17) as

$$\sum_{n=0}^{\infty} t^n E[\exp\{i\lambda M_n\}] = \exp\left\{ \sum_{n=1}^{\infty} \frac{1}{n} t^n E[\exp\{i\lambda S_n^+\}] \right\}. \qquad (6.21)$$

Note that for the stationary M, the left-hand side of (6.21) can be written as

$$\sum_{n=0}^{\infty} t^n E[\exp\{i\lambda M\}] = \frac{1}{1-t} E[\exp\{i\lambda M\}]. \qquad (6.22)$$

Using (6.22), (6.21) can be written as

$$
\begin{aligned}
E[\exp\{i\lambda M\}] &= \lim_{t \to 1}(1-t)\exp\left\{ \sum_{n=1}^{\infty} \frac{1}{n} t^n E[\exp\{i\lambda S_n^+\}] \right\} \\
&= \lim_{t \to 1}\exp\left\{ -\sum_{n=1}^{\infty} \frac{t^n}{n} \right\}\exp\left\{ \sum_{n=1}^{\infty} \frac{1}{n} t^n E[\exp\{i\lambda S_n^+\}] \right\} \\
&= \lim_{t \to 1}\exp\left\{ \sum_{n=1}^{\infty} \frac{1}{n} t^n E[\exp\{i\lambda S_n^+\}] - \sum_{n=1}^{\infty} \frac{t^n}{n} \right\} \\
&= \exp\left\{ \sum_{n=1}^{\infty} \frac{1}{n} (E[\exp\{i\lambda S_n^+\}] - 1) \right\},
\end{aligned}
\qquad (6.23)
$$

where the second equality follows from the fact $\ln(1-t) = -\sum_{n=1}^{\infty} t^n/n$. This version of Spitzer identity is useful in analyzing a queueing system as the steady-state waiting time has the same distribution as M. In terms of the first moments, the Spitzer identity can be expressed as

$$E[M_n] = \sum_{j=1}^{n} \frac{1}{j} E[S_j^+]. \qquad (6.24)$$

Note that this identity can also be proved as shown in Ross (1996) without using the Spitzer identity in terms of the moment generating function or the characteristic function.

Exercise 6.8 *Verify the expression (6.24).*

Next, we provide an example of utilizing the Spitzer identity to analyze a queueing system.

Example 6.1 *(Discrete $D/G/1$ Queue). A discrete $D/G/1$ queue is a single server system where customers arrive with discrete and deterministic inter-arrival times, denoted by s. Customers are served according to a FCFS discipline and their service times are i.i.d. discrete random variables, denoted by A. Thus, the waiting time of the nth customer W_n satisfies*

$$W_n = \max[W_{n-1} + A_{n-1} - s, 0]$$

for $n = 0, 1, \ldots$ If $E[A] < s$, the steady state can be reached and the stationary waiting time $W = \lim_{n \to \infty} W_n$ exists. The probability generating function (z-transform) $W(z)$ can be derived (see Servi (1986)) as follows

$$W(z) = \frac{s - E[A]}{z^s - A(z)} (z - 1) \prod_{k=1}^{s-1} \frac{z - z_k}{1 - z_k}, \tag{6.25}$$

where $A(z)$ is the p.g.f. of A and $z_0 = 1, z_1, \ldots, z_{s-1}$ are the s roots of $z^s = A(z)$ in $|z| \leq 1$. Here we assume that $A(z)$ is an analytic function in disk $|z| \leq 1 + \varepsilon$ with $\varepsilon > 0$ and $A(0) = P(A = 0) > 0$. Note that it is common to utilize the explicit factorization which requires the s roots of a characteristic equation on and inside the unit circle (see Servi (1986)). We show that the Spitzer's identity can be used to develop approximate expressions for both transient and stationary waiting time in a continuous $G/G/1$ queue by using its discrete $G/G/1$ counterpart.

Now we focus on the discrete $G/G/1$ queue in which the inter-arrival time between customer n and $n+1$ is C_n and the service time of customer n is B_n. Assume that B_n and C_n are i.i.d. discrete random variables and can be denoted by B and C, respectively. Introducing a constant value $s \geq C$, we have W_n in this discrete $G/G/1$ queue satisfying

$$W_n = \max[W_{n-1} + B_{n-1} - C_{n-1}, 0] = \max[W_{n-1} + A_{n-1} - s, 0], \quad n = 0, 1, \ldots, \tag{6.26}$$

where A_n are i.i.d as $A = B - C + s$. Thus, we can utilize (6.25) and the root finding approach to obtain the information for the stationary waiting time but no information about the transient waiting time. If we assume that customer

0 arrives at time 0 to an empty system (i.e., $W_0 = 0$), the joint probability generating function of W_n is given by Spitzer identity. That is for $0 \le t < 1, |z| \le 1$,

$$\sum_{n=0}^{\infty} t^n E[z^{W_n}] = \exp\left\{\sum_{l=1}^{\infty} \frac{t^l}{l} E[z^{S_l^+}]\right\}, \tag{6.27}$$

where S_l is the position of a random walk at instant l with the step $A - s$ (i.i.d. discrete random variables). Thus, the mean of W_n is given by

$$\begin{aligned}
E[W_n] &= \sum_{l=1}^{n} \frac{1}{l} E[S_l^+] \\
&= \sum_{l=1}^{n} \frac{1}{l} \sum_{j=ls}^{\infty} (j - ls) P(A^{*l} = j),
\end{aligned} \tag{6.28}$$

where A^{*l} denotes the l-fold convolution of A (i.e., $A^{*l} = \sum_{i=1}^{l} A_i$ with A_i being i.i.d. as A). Note that (6.27) can be rewritten as

$$(1-t) \sum_{n=0}^{\infty} t^n E[z^{W_n}] = \exp\left\{\sum_{l=1}^{\infty} \frac{t^l}{l} (E[z^{S_l^+}] - 1)\right\}. \tag{6.29}$$

It follows from Abel's theorem and (6.29) that $W(z)$ can be obtained as

$$\begin{aligned}
W(z) &= \lim_{t \uparrow 1}(1-t) \sum_{n=0}^{\infty} t^n E[z^{W_n}] = \lim_{t \uparrow 1} \exp\left\{\sum_{l=1}^{\infty} \frac{t^l}{l} (E[z^{S_l^+}] - 1)\right\} \\
&= \exp\left\{-\sum_{l=1}^{\infty} \frac{1}{l} P(S_l > 0)\right\} \exp\left\{\sum_{l=1}^{\infty} \frac{1}{l} E[z^{S_l} \mathbf{1}\{S_l > 0\}]\right\},
\end{aligned}$$

where $\mathbf{1}\{x\}$ equals 1 if x is true and 0 otherwise. For a p.g.f. $g(z)$ (i.e., z-transform), denote by $C_{z^j}[g(z)]$ the coefficient of z^j in $g(z)$. Let $w_j = P(W = j)$. Then, the stationary distribution for the waiting time of the discrete $G/G/1$ queue can be written as

$$w_j = w_0 C_{z^j} \left[\exp\left\{\sum_{l=1}^{\infty} \frac{1}{l} \sum_{i=ls}^{\infty} P(A^{*l} = i) z^{i-ls}\right\}\right], \quad j = 0, 1, \dots, \tag{6.30}$$

where

$$w_0 = \exp\left\{-\sum_{l=1}^{\infty} \frac{1}{l} \sum_{i=ls}^{\infty} P(A^{*l} = i)\right\}. \tag{6.31}$$

Expressions (6.28), (6.30), and (6.31) provide explicit representations of waiting time characteristics solely in terms of infinite series of convolutions of

A. A feasible approach is to determine the distribution of A^{*l} from the distribution of $A^{*(l-1)}$. A fast Fourier transform algorithm was suggested by Ackroyd (1980). For the details about the fast Fourier transform approach to invert a p.g.f., interested readers are referred to Abate and Whitt (1992).

The relation between the Brownian motion (the limiting version of random walk) and $G/G/1$ queue will be discussed in the next chapter.

Exercise 6.9 *Denote by $G(\alpha, \beta)$ the gamma distribution with shape parameter α and rate parameter β. Consider a single server queue with gamma distributed inter-arrival times, with the distribution denoted by $G(s, \lambda)$ and gamma distributed service times, with the distribution denoted by $G(r, \mu)$. Using Spitzer's identity, show that the delay for the $(n+1)$st customer is given by*

$$E[W_{n+1}] = n\left(\frac{r}{\mu} - \frac{s}{\lambda}\right) + \sum_{k=1}^{n}\frac{1}{k}\sum_{i=0}^{ks=1}\frac{ks-i}{\lambda}\binom{kr+i-1}{i}\left(\frac{\mu}{\lambda+\mu}\right)^{kr}\left(\frac{\lambda}{\lambda+\mu}\right)^{i}.$$

The symmetric simple random walk can be extended to the case where X_i's are finite integer-valued random variables with $E[X_i] = 0$. Such a random walk $S_n = \sum_{i=1}^{n}X_i$ with $n \geq 1$ can be shown to be a martingale. For a review on martingales, readers are referred to Chapter B. Hence, a random walk can be analyzed by using martingales. For example, for a random walk $\{S_n, n \geq 0\}$ with X_i which can only take on one of the values $0, \pm 1, ..., \pm M$ for some $M < \infty$, it can be shown that $\{S_n, n \geq 0\}$ is a recurrent DTMC if and only if $E[X_i] = 0$. Interested readers are referred to Ross (1996) for using martingales to prove this claim.

6.1.3 SYSTEM POINT LEVEL CROSSING METHOD

Continuing the sample-path approach, in this section, we briefly introduce the level crossing method for deriving the probability distributions for the performance random variables in a stochastic model. We re-write the Lindley recursion (6.18) in an $M/G/1$ queue as

$$W_{n+1} = \max[W_n + Y_n - X_{n+1}, 0], \quad n = 1, 2, ...,$$

which is fundamental in constructing the sample path for the virtual waiting time process. Denote by $F_Z(\cdot)$ and $f_Z(\cdot)$ the distribution function and density function of random variable Z, respectively. Further, define $P_{W_n}(0) = F_{W_n}(0)$ as the probability mass at 0 for W_n for the nth customer. Then, we can write

$$F_{W_n}(\infty) = P_{W_n}(0) + \int_{x=0}^{\infty} f_{W_n}(x)dx = 1$$

for $n = 1, 2, ...$. We also assume that the stability condition for the system holds and the steady state is reached with the notation without subscript as the

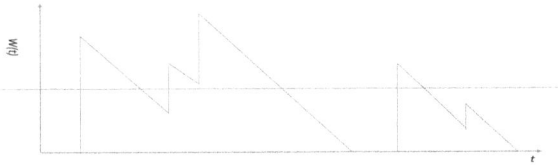

Figure 6.6   Sample path of $W(t), t \geq 0$ with a level x.

stationary distribution. For example $\lim_{n \to \infty} F_{W_n}(x) = F(x)$. Let $W(t)$ be the virtual waiting time process in a single server queueing system (the total workload at time t). A sample path is a single realization of the process over time and can be denoted by $X(t) = W(t, \omega)$. $X(t)$ is real valued and right continuous function on the reals. This implies that $X(t)$ has jump or removable discontinuities on a sequence of strictly increasing time instants A_n with $n = 0, 1, \ldots$, where $A_0 = 0$ without loss of generality. Clearly, A_n represent arrival instants of customers in queues or demand or replenishment instants for inventories. Figure 6.6 shows a sample path for an $M/G/1$ queue based on the Lindley recursion. It is easy to check that the two consecutive upward jumps are related by the Lindley recursion in terms of the customer waiting times. Note that a sample path decreases continuously on time segments between jump points as shown in Figure 6.6. The decreasing rate is denoted by $dX(t)/dt = -r(X(t))$ for $A_n \leq t < A_{n+1}, n = 0, 1, \ldots$ and $r(x) \geq 0$ for all $x \in (-\infty, \infty)$. In a standard $M/G/1$ queue, the state space for $W(t)$ is $[0, \infty)$, and $r(x) = 1$ for $x > 0$ and $r(0) = 0$. In Figure 6.6, for a virtual wait level x, there are two types of level crossings. A continuous down-crossing of level x occurs at time instant $t_0 > 0$ if $\lim_{t \to t_0^-} X(t) = x$ and $X(t) > x$ for $t < t_0$. A jump up-crossing of level x occurs at a time instant $t_1 > 0$ if $\lim_{t \to t_1^-} X(t) < x$ and $X(t) \geq x$ for $t > t_1$. Similarly, the continuous up-crossing and jump down-crossing can be defined similarly for other types of stochastic processes. For example, in an inventory-production system with constant production rate and randomly arrival customer order of random amount, when the production is on and the inventory level is built up gradually from below level x, then a continuous up-crossing of inventory level x will occur; when a customer order arrives at an instant with current inventory level y and the amount (called downward jump magnitude) is greater than $y - x$, then a jump down-crossing of a level x occurs. Let $g(x)$ be the p.d.f. for the process of the interest in steady state (e.g., $\lim_{t \to \infty} W(t) = W$). Then an integral equation for $g(x)$ can be set up based on the following theorems. To state these theorems, we introduce more notations. Assume that the upward (downward) jumps occur at Poisson rate λ_u (λ_u).

These upward and downward jumps are independent of each other and of the state of the system. Let the corresponding upward and downward jump magnitudes have the c.d.f.'s denoted by B_u and B_d, respectively. Denote by $D_t^c(x)$ $(U_t^c(x))$ the total number of continuous down-crossings (up-crossing) of level x and by $D_t^j(x)$ $(U_t^j(x))$ the number of jump down-crossing (up-crossing) of level x over $(0,t)$ due to the external Poisson rate λ_d (λ_u). Then, the following result holds.

Theorem 6.3 *(Brill 1975, for $r(x) = 1$). With probability 1*

$$\lim_{t \to \infty} D_t^c(x)/t = r(x)g(x)$$

$$\lim_{t \to \infty} D_t^j(x)/t = \lambda_d \int_{y=x}^{\infty} (1 - B_d(y-x))g(y)dy = \lambda_d \int_{y=x}^{\infty} B_d^c(y-x)g(y)dy$$

$$(6.32)$$

for all x.

The proof of this theorem regarding down-crossings can be found in Brill (2008). Intuitively, both sides of the first equation of (6.32) represent the long-run rate of continuous decays by a typical sample path into level x from above and both sides of the second equation of (6.32) represent long-run rate of downward jumps occurring at rate λ_d from state space set (x, ∞) into $(-\infty, x]$. Now we present the theorem regarding the up-crossings.

Theorem 6.4 *(Brill 1975). With probability 1*

$$\lim_{t \to \infty} U_t^j(x)/t = \lambda_u \int_{y=-\infty}^{x} (1 - B_u(x-y))g(y)dy = \lambda_u \int_{y=-\infty}^{x} B_u^c(x-y)g(y)dy$$

$$(6.33)$$

for all x.

Again, both sides of (6.33) represent the long-run rate of upward jumps by a sample path occurring at rate λ_u from state space set $(-\infty, x]$ into (x, ∞). Next we can state the conservation law regarding the crossing rates. That is, for every state-space level x and every sample path, in the long run, we have

Total down-crossing rate = Total up-crossing rate.

For a stable process with two types of down-crossings and only jump up-crossings, this conservation law can be written as

$$\lim_{t \to \infty} D_t^c(x)/t + \lim_{t \to \infty} D_t^j(x)/t = \lim_{t \to \infty} U_t^j(x)/t. \qquad (6.34)$$

Using Theorems 6.3 and 6.4 in (6.34) yields

$$r(x)g(x) + \lambda_d \int_{y=x}^{\infty} B_d^c(y-x)g(y)dy = \lambda_u \int_{y=-\infty}^{x} B_u^c(x-y)g(y)dy, \quad (6.35)$$

which is the integral equation for $g(x)$. In practice, we start with a typical sample path, set up the integral equation for $g(x)$, and solve it for the steady-state p.d.f. for the process. To demonstrate the level crossing approach to solving for the steady-state distribution for a stochastic process, we first consider the $M/G/1$ queue with arrival rate λ and i.i.d. service times denoted by S. Denote by $B_S(x)$ and $b_S(x)$ the c.d.f. and p.d.f. for S. Let $F_W(x)$ and $f_W(x)$ be the c.d.f. and p.d.f. for the steady-state virtual wait process with $P_W(0) = F_W(0)$. Under the stability condition $\lambda E[S] < 1$, $f_W(x)$ satisfies the following integral equation according to (6.34) with $r(x) = 1$ and no jump down-crossings

$$f_W(x) = \lambda B_S^c(x) P_W(0) + \lambda \int_0^x B_S^c(x-y) f_W(y) dy \quad (6.36)$$

for $x > 0$, where $f_W(x)$ is the down-crossing rate of level x, the first term on the RHS is the jump up-crossing rate of level x from level 0, and the second term on the RHS is the jump up-crossing rate of level x from levels in $(0, x)$. Clearly, we have $f_W(0) = \lambda P_W(0)$ from the integral equation by setting $x = 0$ and $P_W(0) + \int_0^{\infty} f_W(y) dy = 1$ from the definition of the distribution of W. For a specific service time distribution, we may obtain the explicit solution to the p.d.f. of the stationary waiting time. Let's consider a $M/E_k/1$ queue where the service times are Erlang-k distributed i.i.d. random variables with parameter (k, μ). That is

$$b_S(x) = \frac{\mu^k x^{k-1} e^{-\mu x}}{(k-1)!}$$

$$B_S(x) = \int_{y=0}^x \frac{\mu^k y^{k-1} e^{-\mu y}}{(k-1)!} dy = 1 - \sum_{n=0}^{k-1} \frac{(\mu x)^n}{n!} e^{-\mu x}$$

with mean $E[S] = k/\mu$. Using these expressions in (6.36), we have

$$f_W(x) = \lambda P_W(0) e^{-\mu x} \left(\sum_{n=0}^{k-1} \frac{(\mu x)^n}{n!} \right)$$

$$+ \lambda \int_{y=0}^x e^{-(x-y)} \left(\sum_{n=0}^{k-1} \frac{[\mu(x-\mu)]^n}{n!} \right) f_W(y) dy \quad (6.37)$$

for $x > 0$, where $P_W(0) = 1 - \lambda E[S] = 1 - k\lambda/\mu$. As a special case of $k = 1$, we can get the stationary waiting time distribution quickly. We present two special cases of $M/E_k/1$ queue.

Example 6.2 *Find the stationary waiting time distribution for $M/M/1$ queue and $M/E_2/1$ queue.*

By setting $k = 1$ in (6.37) which is the $M/M/1$ case, we obtain

$$f_W(x) = \lambda P_W(0)e^{-\mu x} + \lambda \int_{y=0}^{x} e^{-\mu(x-y)} f_W(y)dy.$$

Differentiating both sides of the equation above with respect to x yields

$$f_W'(x) = -\lambda \mu P_W(0)e^{-\mu x} + \lambda f_W(x) + \lambda \int_{y=0}^{x} -\mu e^{-\mu(x-y)} f_W(y)dy.$$

It follows from these two equations that

$$f_W'(x) + (\mu - \lambda)f_W(x) = 0,$$

which is a first-order homogeneous differential equation. The general solution to this equation is $f_W(x) = Ce^{-(\mu-\lambda)x}$. Using the initial condition, we can determine the constant $C = \lambda P_W(0) = \lambda(1 - \lambda/\mu)$. Thus the p.d.f. of the stationary waiting time for the $M/M/1$ queue is given by

$$f_W(x) = \lambda \left(1 - \frac{\lambda}{\mu}\right) e^{-(\mu-\lambda)x}.$$

Next, we consider the case of $M/E_2/1$ queue. By setting $k = 2$ in (6.37), we obtain

$$f_W(x) = \lambda P_W(0)e^{-\mu x}(1+\mu x) + \lambda \int_{y=0}^{x} e^{-\mu(x-y)}(1+\mu(x-y))f_W(y)dy.$$

Differentiating the equation above with respect to x twice yields two differential equations

$$f_W'(x) = -\lambda \mu P_W(0)e^{-\mu x}\mu x + \lambda f_W(x)$$
$$+ \lambda \int_{y=0}^{x} -\mu^2(x-y)e^{-\mu(x-y)}(1+\mu(x-y))f_W(y)dy$$

$$f_W''(x) = -\lambda \mu^2 P_W(0)e^{-\mu x} + \lambda \mu^3 x P_W(0)e^{-\mu x} + \lambda f_W'(x)$$
$$+ \lambda \int_{y=0}^{x} [-\mu^2(x-y)e^{-\mu(x-y)} + \mu^3(x-y))e^{-\mu(x-y)}]f_W(y)dy.$$

Multiplying $f_W(x)$ and $f_W'(x)$ by μ^2 and $-\mu$, respectively, yields

$$\mu^2 f_W(x) = \lambda \mu^2 P_W(0)e^{-\mu x}(1+\mu x)$$
$$+ \lambda \int_{y=0}^{x} \mu^2 e^{-\mu(x-y)}(1+\mu(x-y))f_W(y)dy$$

$$-\mu f_W''(x) = \lambda \mu^3 x P_W(0)e^{-\mu x} - \lambda \mu f_W(x)$$
$$+ \lambda \int_{y=0}^{x} \mu^3(x-y)e^{-\mu(x-y)} f_W(y)dy,$$

Using these two equations and $f_W''(x)$ expression, we can obtain a second-order homogeneous differential equation as

$$f_W''(x) + (2\mu - \lambda)f'(x) + (\mu^2 - \lambda)f_W'(x) + (\mu^2 - 2\lambda\mu)f_W(x) = 0.$$

To solve this equation, we try $f_W(x) = e^{rx}$ so that $f_W'(x) = re^{rx}$ and $f_W''(x) = r^2 e^{rx}$. Substituting these expressions into the second-order differential equation, we have

$$r^2 + (2\mu - \lambda)r + (\mu^2 - 2\lambda\mu) = 0,$$

which has two roots as

$$r = \frac{(\lambda - 2\mu) \pm \sqrt{\lambda^2 + 4\lambda\mu}}{2}.$$

Thus, the general solution is given by

$$f_W(x) = C_1 e^{r_1 x} + C_2 e^{r_2 x}. \tag{6.38}$$

Using the conditions $f_W(0) = \lambda P_W(0)$ and $P_W(0) + \int_0^\infty f_W(y)dy = 1$, we can determine constants C_1 and C_2 as

$$C_1 = \frac{r_1 r_2 \left[1 - \left(1 - \frac{\lambda}{r_2}\right)\right]}{r_1 - r_2}, \quad C_2 = \lambda P_W(0) - C_1,$$

with $P_W(0) = 1 - \frac{2\lambda}{\mu}$. Now $f_W(x)$ can be computed completely by (6.38).

For more discussions about using the level crossing methods, interested readers are referred to the book by Brill (2008).

Exercise 6.10 *Using the level crossing method to obtain the stationary distribution for the waiting time in $E/M_3/1$ queue.*

Exercise 6.11 *Using the level crossing method to derive the stationary distribution for the waiting time in $M/M/2$ queue. Extend this analysis to the $M/M/c$ queue.*

Exercise 6.12 *Consider a continuous review inventory system with a (s, S) policy where $s \geq 0$ is the re-order point and S is the order-up-to level. Suppose that customer orders arrive at the system according to a Poisson process with rate λ. Each order has an exponentially i.i.d. demand size with mean $1/\mu$. Furthermore, it is assumed that the stock decays at constant rate $d \geq 0$ when the stock level is in $(s, S]$. The inventory is replenished instantly whenever the stock level either decays continuously to s or jumps downward to s or below. Derive the steady-state p.d.f. of the stock on hand $g(x)$ using the level crossing method.*

Exercise 6.13 *Based on our discussion on the level crossing method, summarize the advantages and limitations of using this method in analyzing stochastic systems such as a queueing or an inventory control system.*

6.1.4 CHANGE OF MEASURES IN RANDOM WALKS

Now we turn back to random walks and discuss the change of measures. To find the probability of an event regarding a random variable or a stochastic process, we can define a new probability measure which possesses some desirable properties. This technique, called the change of probability measures, is presented with some details in the Chapter E. The basic idea is to define a new probability measure $\mathbb{P}^*$ based on the original probability measure $\mathbb{P}$ for a random variable or a stochastic process defined on $(\Omega, \mathscr{F}, \mathbb{P})$. The relation between these two measures for an event $A \in \mathscr{F}$ is given by

$$\mathbb{P}^*(A) = E^{\mathbb{P}}[1_A z],$$

where z is a positive and integrable random variable defined on $(\Omega, \mathscr{F}, \mathbb{P})$ and called the Radon–Nikodym derivative due to $z = d\mathbb{P}^*/d\mathbb{P}$ and 1_A denotes the indicator function. Now we can show how to transform a simple random walk to a symmetric simple random walk via change of measures. Call a simple random walk with probability of going up p a p-random walk. The moment generating function of each step $m(\theta)$ is given by

$$m(\theta) := E[e^{\theta X_n}] = pe^{\theta} + qe^{-\theta}, \quad n = 1, 2, \dots,$$

where $q = 1 - p$. We consider the random walk starting with $S_0 = i$ over T periods. Define

$$Y_T := m^{-T}(\theta) e^{\theta(S_T - i)} = \prod_{i=1}^{T} m^{-1}(\theta) e^{\theta X_i}, \quad n = 1, 2, \dots,$$

where the second equality follows from $S_n - i = \sum_{i=1}^{n} X_i$ and the independence of X_n's. Clearly, $E[Y_T] = 1$ and it is easy to check that $\{Y_n\}$ is a positive martingale. For $T < \infty$, we can define the new probability measure as

$$\mathbb{P}^*(A) = E^{\mathbb{P}}[1_A Y_T], \quad A \in \mathscr{F}_T. \tag{6.39}$$

With this "change of measures" formula, we can transform a random walk to another random walk with different probability parameters.

Proposition 6.4 *For some* $\theta \in \mathscr{R}$, *let*

$$p^* = \frac{pe^{\theta}}{pe^{\theta} + qe^{-\theta}}.$$

Using (6.39), the p-random walk $\{S_n\}$ *is transformed to a* p^*-*random walk under* $\mathbb{P}^*$.

Proof. The proof can be done by verifying the probability $\mathbb{P}^*(X_n = 1)$ as follows:

$$\mathbb{P}^*(X_n = 1) = E^{\mathbb{P}^*}[1_{\{X_n=1\}}] = E^{\mathbb{P}}[1_{\{X_n=1\}}Y_T] = E^{\mathbb{P}}\left[1_{\{X_n=1\}}\prod_{i=1}^{T}m^{-1}(\theta)e^{\theta X_i}\right]$$

$$= E^{\mathbb{P}}\left[m^{-1}(\theta)1_{\{X_n=1\}}e^{\theta X_n}\right]\prod_{i\neq n}E\left[m^{-1}(\theta)e^{\theta X_i}\right]$$

$$= E^{\mathbb{P}}\left[m^{-1}(\theta)1_{\{X_n=1\}}e^{\theta X_n}\right] = m^{-1}(\theta)E^{\mathbb{P}}[1_{\{X_n=1\}}e^{\theta X_n}]$$

$$= \frac{pe^{\theta}}{pe^{\theta}+qe^{-\theta}} = p^*.$$

Similarly, we have $\mathbb{P}^*(X_n = -1) = q^* = 1 - p^*$. Furthermore, $X_1,...,X_T$ are also independent under $\mathbb{P}^*$. Thus, $\{S_n\}$ is a p^*-random walk under $\mathbb{P}^*$. ∎

By choosing appropriate θ, we can transform a non-symmetric random walk into a symmetric random walk and vice versa. Let $\theta = \ln \sqrt{q/p}$. For a p-random walk, the probability of going up under the new probability measure is given by

$$p^* = \frac{pe^{\ln \sqrt{q/p}}}{pe^{\ln \sqrt{q/p}}+qe^{-\ln \sqrt{q/p}}} = \frac{p\sqrt{q/p}}{p\sqrt{q/p}+q\sqrt{p/q}}$$

$$= \frac{\sqrt{pq}}{\sqrt{pq}+\sqrt{pq}} = \frac{1}{2}.$$

This indicates that a p-random walk has been transformed into a symmetric random walk. We can also do the reverse by selecting $\theta = \ln \sqrt{p/q}$. That is

$$p^* = \frac{\frac{1}{2}e^{\ln \sqrt{p/q}}}{\frac{1}{2}e^{\ln \sqrt{p/q}}+\frac{1}{2}e^{-\ln \sqrt{p/q}}} = \frac{\sqrt{p/q}}{\sqrt{p/q}+\sqrt{q/p}} = \frac{p}{p+q} = p.$$

The advantage of being able to transform a random walk via change of measures is that the results obtained by analyzing a symmetric random walk can be transferred to the asymmetric random walk under new probability measure. As an example, we show how to obtain the n-step transition probability $a_{ij}(n)$ of the p-random walk when state 0 is absorbing. That is the generalization of the (6.4). Let $a_{ij}(n)$ be the transition probabilities of the p-random walk when state 0 is absorbing for $n \leq T$ under a probability measure $\mathbb{P}^*$. That is

$$a_{ij}(n) = \mathbb{P}^*(S_n = j, m_n > 0|S_0 = i), n = 0,1,2,...,$$

where $i, j > 0$. By the change of measures, we have

$$a_{ij}(n) = E^{\mathbb{P}}[1_{\{S_n=j,m_n>0\}}Y_T|S_0 = i].$$

Due to $S_T = S_n + \sum_{i=n+1}^{T} X_i$ and the fact that $X_{n+1}, \ldots, X_T$ are independent of S_n under measure $\mathbb{P}$, we get

$$
\begin{aligned}
a_{ij}(n) &= E^{\mathbb{P}}[1_{\{S_n = j, m_n > 0\}} Y_n | S_0 = i] E^{\mathbb{P}}\left[\prod_{i=n+1}^{T} m^{-1}(\theta) e^{\theta X_i}\right] \\
&= m^{-n}(\theta) E^{\mathbb{P}}[1_{\{S_n = j, m_n > 0\}} e^{\theta(S_n - i)} | S_0 = i] \\
&= m^{-n}(\theta) e^{\theta(j-i)} \mathbb{P}(S_n = j, m_n > 0 | S_0 = i).
\end{aligned}
\tag{6.40}
$$

Note that for the symmetric random walk under $\mathbb{P}$

$$
\mathbb{P}(S_n = j, m_n > 0 | S_0 = i) = v_{j-i}(n) - v_{j+i}(n),
\tag{6.41}
$$

where the symmetry $v_j(n) = v_{-j}(n)$ is used in the symmetric random walk. By selecting $\theta = \ln \sqrt{p/q}$, we have

$$
e^{\theta j} = \left(\frac{p}{q}\right)^{j/2}
$$

and

$$
m(\theta) = \frac{e^\theta + e^{-\theta}}{2} = \frac{1}{2}\left(\sqrt{\frac{p}{q}} + \sqrt{\frac{q}{p}}\right) = (4pq)^{-1/2}.
$$

Using these expressions and substituting (6.41) into (6.40) yields

$$
a_{ij}(n) = 2^n p^{(n+j-i)/2} q^{(n-j+i)/2}\{v_{j-i}(n) - v_{j+i}(n)\}, \quad i, j > 0.
$$

Using the similar approach, we can also obtain the distribution of $m_n = \lim_{1 \le k \le n} S_k$ of a p-random walk $\{S_n\}$. This is left as an exercise.

Exercise 6.14 *Using the change of measures, derive the distribution of $m_n = \lim_{1 \le k \le n} S_k$ of a p-random walk $\{S_n\}$.*

6.1.5   THE BINOMIAL SECURITIES MARKET MODEL

In this subsection, we present a Binomial security price model in terms of a random walk and use it to demonstrate the application of change of measures in pricing a security derivative such as a call option. Let $S(t)$ be the price of a risky security such as a stock at time t with no dividends paid and $S(0) = S$. It is assumed that the financial market is frictionless, that is, there are no transaction costs or taxes. Let $\{W_n\}$ be a random walk with i.i.d. steps $X_i = 1$ with probability p and $X_i = 0$ with probability $q = 1 - p$. At period t, we have $W_t = \sum_{i=1}^{t} X_i$. In a Binomial price model, the stock price either goes up by u percent (called up factor) or down by d percent (called down factor) in each period based on the associated random walk. $X_i = 1$ corresponds a price up movement and $X_i = 0$ corresponds a price down movement. Therefore, for a finite T-period horizon, we have

$$S(t) = u^{W_t} d^{t-W_t} S, \quad t = 0, 1, 2, ..., T. \tag{6.42}$$

Using $W_{t+1} = W_t + X_{t+1}$, we can write

$$S(t+1) = u^{X_{t+1}} d^{1-X_{t+1}} S(t), \quad t = 0, 1, ..., T-1,$$

which leads to

$$\frac{\Delta S(t)}{S(t)} = u^{X_{t+1}} d^{1-X_{t+1}} - 1,$$

where $\Delta S(t) = S(t+1) - S(t)$. Clearly, the probability of the stock price at time t resulted from k steps of price up and $t - k$ steps of price down is

$$\mathbb{P}(S(t) = u^k d^{t-k} S) = \binom{t}{k} p^k q^{t-k}, \quad k = 0, 1, 2, ..., t. \tag{6.43}$$

We consider a very simple investment portfolio which consists of a risk-free security and the risky security described above. Denote by $B(t)$ the value of the risk-free account at time t when \$ 1 is deposited at time 0. Then we have

$$B(t) = R^t, \quad t = 0, 1, 2, ..., T,$$

where $R = 1 + r$ and r is the interest rate. Let $\{(f(t), g(t)); t = 1, 2, ..., T\}$ denote the portfolio process. Then the value process is given by

$$V(t) = f(t)B(t) + g(t)S(t), \quad t = 1, 2, ..., T. \tag{6.44}$$

Due to self-financing, we have

$$\Delta V(t) = f(t+1)\Delta B(t) + g(t+1)\Delta S(t) \quad t = 0, 1, ..., T-1.$$

Taking the risk-free account as the numeraire, the denominated value process satisfies

$$\Delta V^*(t) = g(t+1)\Delta S^*(t) \quad t = 0, 1, ..., T-1,$$

where $S^*(t) = S(t)/B(t)$ (see Kijima (2013)).

Next, we show how to calculate the expected value of a payoff function of a derivative security written on the stock. Let $h(x)$ be a real value function. Then, the payoff function $h(S(T))$ may represent the amount received by an investor at the maturity T of the claim. The expected discounted value of $h(S(T))$ can be written as

$$E\left[\frac{h(S(T))}{B(T)}\right] = \frac{1}{R^T} \sum_{k=0}^{T} h(u^k d^{T-k} S) \binom{T}{k} p^k q^{T-k}.$$

For a special case $h(x) = x$, the equation above reduces to

$$E[S^*(T)] = \frac{S}{R^T} \sum_{k=0}^{T} \binom{T}{k} (up)^k (dq)^{T-k} = S\left(\frac{up+dq}{R}\right)^T, \qquad (6.45)$$

where $S^*(T) = S(T)/B(T)$.

Now we consider the case where the claim is a call option with strike price K written on the stock. In this case, $h(x) = \{x - K\}^+$. It can be shown that the expected present value of the payoff is given by

$$E\left[\frac{\{S(T) - K\}^+}{B(T)}\right] = E[S^*(T)]\bar{B}_k(T, \tilde{p}) - KR^{-T}\bar{B}_k(T, p), \qquad (6.46)$$

where $\tilde{p} = up/(up + dq)$,

$$k = \min\{i : i > \ln[K/Sd^T]/\ln[u/d]\},$$

and $\bar{B}_k(n, p)$ denotes the tail probability of the binomial distribution with parameter (n, p). This formula has some flavor of the Black Scholes formula for pricing options in continuous-time model. Let A be the event that the call option becomes in-the-money at the maturity T, i.e. $A = \{S(T) \geq K\}$. Then, we can write

$$E[\{S(T) - K\}^+] = E[S(T)1_A] - KE[1_A]. \qquad (6.47)$$

The second term on the right-hand side of (6.46) follows from (6.47) and $E[1_A] = \mathbb{P}(S(T) \geq K) = \bar{B}_k(T, p)$. To verify the first term on the right-hand side of (6.46), which is the expectation of the product of two non-independent random variables ($S(T)$ and 1_A), we need to utilize the change of measure formula. By choosing $\theta = \ln[u/d]$ in $m(\theta)$, we have

$$m(\theta) = E[e^{\theta X_n}] = pe^{\ln[u/d]\cdot 1} + qe^{\ln[u/d]\cdot 0} = p\frac{u}{d} + q = \frac{pu+qd}{d}.$$

Define $Y_T = m^{-T}(\theta)e^{\theta W_T}$. Thus, we have

$$\mathbb{P}^*(A) = E[Y_T 1_A] = \frac{E[e^{\theta W_T} 1_A]}{m^T(\theta)}. \qquad (6.48)$$

Using $\theta = \ln[u/d]$ and (6.42), we can write $S(T) = d^T e^{\theta W_T} S$. Now the first term on the right-hand side of (6.47) is given by

$$\begin{aligned}
E[S(T)1_A] &= Sd^T E[e^{\theta W_T} 1_A] = Sd^T m^T(\theta)E[Y_T 1_A] \\
&= Sd^T m^T(\theta)\mathbb{P}^*(A) = S(up+dq)^T \mathbb{P}^*(S(T) \geq K),
\end{aligned} \qquad (6.49)$$

where the second and third equalities follow from (6.48). Using (6.45),(6.49), and (6.47) verify the first term on the right-hand side (6.46). Here, $\bar{B}_k(T, \tilde{p})$

is the probability that the call option becomes in-the-money at the maturity T under the probability measure $\mathbb{P}^*$. The p-random walk has been transformed to the $\tilde{p}$-random walk under the new probability measure for some θ. The relation is

$$\tilde{p} = \frac{pe^\theta}{pe^\theta + q} = \frac{up}{up + dq}.$$

Exercise 6.15 *While European options can only exercise the right at the maturity, the payoff of an American option can occur at any time on or before the maturity. Consider an American call option written on the stock that pays no dividends. Show that it is optimal to wait until maturity.*

6.1.6 THE ARC SINCE LAW

Consider a symmetric simple random walk with $S_0 = 0$ and $X_k = \{-1, 1\}$, $k \geq 1$. If we want to find the probability that this walk revisits state 0, the number of periods must be an even number. Now we find the probability that the last visit to 0 up to period $2n$ occurred at time $2k$. Denote this probability by $\alpha_{2n}(2k)$.

Theorem 6.5 *(Arc Sine Law for last visit to the origin) For a simple random walk with $p = q = 1/2$ and $S_0 = 0$,*

$$\alpha_{2n}(2k) = u_{2k}u_{2n-2k}.$$

Proof. First note that

$$
\begin{aligned}
\alpha_{2n}(2k) &= P(S_{2k} = 0, S_{2k+1} \neq 0, ..., S_{2n} \neq 0) \\
&= P(S_{2k+1} \neq 0, ..., S_{2n} \neq 0 | S_{2k} = 0)P(S_{2k} = 0) \\
&= P(S_1 \neq 0, ..., S_{2n-2k} \neq 0 | S_0 = 0)P(S_{2k} = 0).
\end{aligned}
$$

The third equality follows as time point $2k$ is a regeneration point. Next, we need to show that $P(S_1 \neq 0, ..., S_{2n-2k} \neq 0 | S_0 = 0) = u_{2n-2k}$. To simplify the notation, the conditional part is omitted and $2n - 2k$ is denoted by $2m$. We have

$$
\begin{aligned}
P(S_1 \neq 0, ..., S_{2m} \neq 0 | S_0 = 0) &= \sum_{b \neq 0} P(S_1 \neq 0, ..., S_{2m} \neq 0, S_{2m} = b | S_0 = 0) \\
&= \sum_{b \neq 0} \frac{|b|}{m} P(S_{2m} = b) = 2\sum_{b=1}^{m} \frac{2b}{2m} P(S_{2m} = b) \\
&= 2\sum_{b=1}^{m} \frac{N_{2m-1}(1, 2b) - N_{2m-1}^0(1, 2b)}{2^{2m}} \\
&= 2\sum_{b=1}^{m} \frac{N_{2m-1}(1, 2b) - N_{2m-1}(-1, 2b)}{2^{2m}}
\end{aligned}
$$

$$= 2 \sum_{b=1}^{m} \frac{\binom{2m-1}{m-b} - \binom{2m-1}{m-b-1}}{2^{2m}} = 2 \binom{2m-1}{m-1} \left(\frac{1}{2}\right)^{2m}$$

$$= 2 \frac{(2m-1)!}{m!(m-1)!} \left(\frac{1}{2}\right)^{2m}$$

$$= 2 \left(\frac{m}{m}\right) \frac{(2m-1)!}{m!(m-1)!} \left(\frac{1}{2}\right)^{2m}$$

$$= \frac{(2m)!}{m!m!} \left(\frac{1}{2}\right)^{2m} = \binom{2m}{m} \left(\frac{1}{2}\right)^{2m} = u_{2m}.$$

This completes the proof. ∎

Using the Stirling approximation, we know $u_{2k} \approx (\pi k)^{-1/2}$ and then $\alpha_{2n}(2k) \approx \pi^{-1}[k(n-k)]^{-1/2}$. Let t_{2n} be the time of the last visit to 0 up to time $2n$. Introducing $0 < x < 1$, we can obtain the approximate distribution function of t_{2n} as

$$P(t_{2n} \leq 2xn) \approx \sum_{k=0}^{xn} \frac{1}{\pi \sqrt{k(n-k)}}$$

$$\approx \frac{1}{\pi} \int_0^{xn} \frac{1}{\sqrt{y(n-y)}} dy$$

$$= \frac{1}{\pi} \int_0^x \frac{1}{\sqrt{u(1-u)}} du = \frac{2}{\pi} \arcsin(\sqrt{x}).$$

This result implies that the limiting t_{2n} has a probability density function described by $\arcsin(\sqrt{x})$. There are some interesting implications. For a symmetric random walk ($p = q = 1/2$), first, one may think that the equalization of heads and rails should occur very often. Actually, such an intuition is not true. In fact, after $2n$ tosses with n large (i.e., $x = 0.5$), the probability that the walk never visits 0 after n tosses is still 50%. Second, the probability that the last time before $2n$ that the walk touches 0 first decreases and then increases with $2k$. In other words, $\alpha_{2n}(2k)$ is symmetric about the midpoint n as shown in Figure 6.7. This means that the last visit to 0 before the ending time occurs more likely at the beginning or near the ending point. Such a behavior can be explained as follows: Since the process starts at 0, it is more likely that the walk visits 0 again soon. After it wonders away from 0, it becomes less likely to touch 0. However, if by chance the walk touches 0 at some point, it is more likely that it touches 0 again in near future. Thus it pushes the most recent visit to 0 closer and closer to the end. Next, we present another result for the event that a symmetric random walk remains positive for a certain number of periods.

Theorem 6.6 (*Arc Sine law for sojourn times of positive values*) *For a simple random walk with $p = q = 1/2$ and $S_0 = 0$, the probability that the walk*

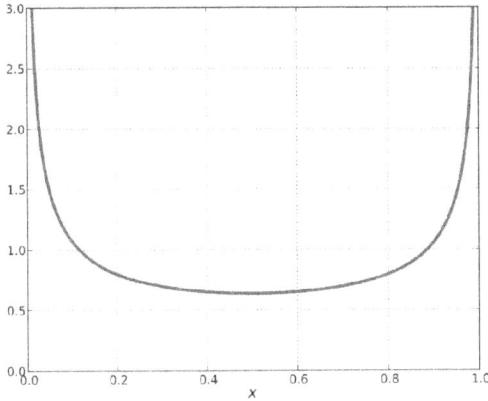

Figure 6.7 Arcsine law.

remains positive value for exactly 2k periods of time up to time 2n, denote by $\beta_{2n}(2k)$ *is the same as* $\alpha_{2n}(2k)$.

Proof. We prove this result by using mathematical induction. First, we consider the case of $k = n = m$. It has been proved that for a walk with $p = q = 1/2$ and $S_0 = 0$

$$u_{2m} = P(S_1 \neq 0, S_2 \neq 0, ..., S_{2m} \neq 0).$$

Since these sample paths are either always above or always below 0, we have

$$u_{2m} = 2P(S_1 > 0, S_2 > 0, ..., S_{2m} > 0).$$

Note that

$$P(S_1 > 0, S_2 > 0, ..., S_{2m} > 0) = P(S_1 = 1, S_2 \geq 1, ..., S_{2m} \geq 1)$$
$$= P(S_1 - 1 = 0, S_2 - 1 \geq 0, ..., S_{2m} - 1 \geq 0)$$
$$= P(S_1 = 1 | S_0 = 0)P(S_2 - 1 \geq 0, S_2 - 1 \geq 0, ..., S_{2m} - 1 \geq 0 | S_1 - 1 = 0)$$
$$= P(S_1 = 1 | S_0 = 0)P(S_2 - 1 \geq 0, S_2 - 1 \geq 0, ..., S_{2m} - 1 \geq 0 | S_1 - 1 = 0)$$
$$= (1/2)P(S_2 \geq 0, S_3 \geq 0, ..., S_{2m} \geq 0 | S_1 = 0)$$
$$\quad \text{(moving the } x\text{-axis up by one unit)}$$
$$= (1/2)P(S_1 \geq 0, S_2 \geq 0, ..., S_{2m-1} \geq 0 | S_0 = 0)$$
$$= (1/2)P(S_1 \geq 0, S_2 \geq 0, ..., S_{2m-1} \geq 0, S_{2m} \geq 0 | S_0 = 0)$$
$$\quad \text{due to } \{S_{2m-1} \geq 0\} \Leftrightarrow \{S_{2m} \geq 0\}$$
$$= (1/2)\beta_{2m}(2m).$$

Thus, it follows from $u_0 = 1$ that

$$\alpha_{2m}(2m) = u_{2m} = \beta_{2m}(2m).$$

We introduce a symbol Z_n representing the number of periods that the walk has positive values between 0 and n. For $1 \le k \le n-1$, if $Z_{2n} = 2k$, then we must have $S_{2r} = 0$ for some $1 \le r \le n-1$ (otherwise, it is contradiction to $k \le n-10$. Let T_0 be the time interval from origin to the time point that the walk returns to 0 and $f_{2r} = P(T_0 = 2r)$ its probability mass function.

$$\beta_{2n}(2k) = \sum_{r=1}^{n-1} P(T_0 = 2r)P(Z_{2n} = 2k | T_0 = 2r)$$

$$= \sum_{r=1}^{n-1} f_{2r} \frac{1}{2} P(Z_{2n-2r} = 2k) + \sum_{r=1}^{n-1} f_{2r} \frac{1}{2} P(Z_{2n-2r} = 2k - 2r)$$

$$= \frac{1}{2} \sum_{r=1}^{n-1} f_{2r}\beta_{2n-2r}(2k) + \frac{1}{2} \sum_{r=1}^{n-1} f_{2r}\beta_{2n-2r}(2k - 2r).$$

Since $\beta_{2n-2r}(2k) = 0$ if $k > n - r$ (or $r > n - k$) and $\beta_{2n-2r}(2k - 2r) = 0$ if $k < r$, we can write the above as

$$\beta_{2n}(2k) = \frac{1}{2} \sum_{r=1}^{n-k} f_{2r}\beta_{2n-2r}(2k) + \frac{1}{2} \sum_{r=1}^{k} f_{2r}\beta_{2n-2r}(2k - 2r).$$

Note that this expression proves the special cases of $k = 1$ and $k = n = m$. To complete the mathematical induction, we assume that the claim ($\beta_{2n}(2k) = \alpha_{2n}(2k) = u_{2k}u_{2n-2k}$) is true for $n < m$. Then we have

$$\beta_{2m}(2k) = \frac{1}{2} \sum_{r=1}^{m-k} f_{2r}\beta_{2m-2r}(2k) + \frac{1}{2} \sum_{r=1}^{k} f_{2r}\beta_{2m-2r}(2k - 2r)$$

$$= \frac{1}{2} \sum_{r=1}^{m-k} f_{2r}u_{2k}u_{2m-2r-2k} + \frac{1}{2} \sum_{r=1}^{k} f_{2r}u_{2k-2r}u_{2m-2k}$$

$$= \frac{1}{2} u_{2k} \sum_{r=1}^{m-k} f_{2r}u_{2m-2r-2k} + \frac{1}{2} u_{2m-2k} \sum_{r=1}^{k} f_{2r}u_{2k-2r}$$

$$= \frac{1}{2} u_{2k}u_{2m-2k} + \frac{1}{2} u_{2m-2k}u_{2k} = u_{2k}u_{2m-2k} = \alpha_{2m}(2k).$$

The equality in the last line is due to the renewal equation for the u_{2j}. That is $u_{2j} = \sum_{r=1}^{j} f_{2r}u_{2j-2r}$. ∎

Exercise 6.16 *For a simple symmetric random walk, show*

$$P(S_1 > 0, ..., S_{2n-1} > 0, S_{2n} = 0) = \frac{1}{n2^{2n}} \binom{2n-2}{n-1}.$$

Exercise 6.17 *Suppose a fair coin is tossed N times and the number of "heads", and "tails", recorded after each toss are n_h and n_t, respectively (i.e.,*

$n_h + n_t = N$). *Over the course of the trials, how many times is the number of heads recorded greater than the number of tails? Naive intuition might lead to the expectation that as N increases the number of times "heads" is in the lead increases in proportion. Use the arcsine law for the symmetric simple random walk to explain that such an intuition is not true.*

6.1.7 THE GAMBLER'S RUIN PROBLEM

Suppose that two players A and B play a game for an indefinite number of rounds until one player becomes bankruptcy. For each round, the winner takes $ 1 from the loser. Player A starting with a dollars wins with probability p and player B starting with b dollars wins with probability $q = 1 - p$ each round. The game is over when either of the players becomes bankrupt (or called ruined). This problem can be modeled as a simple random walk from the player A's perspective with two absorbing states 0 (i.e., A is ruined) and $a + b$ (i.e., B is ruined) and starting state a. This is equivalent to the simple random walk starting at 0 with two absorbing states b and $-a$. Some quantities of interest include the absorbing probabilities and absorbing times. To find the absorbing probabilities, we define $A_k = P$(player A ruins B when he has k dollars). Thus, $A_0 = 0$ and $A_{a+b} = 1$ and we want to know A_a. Conditioning on the outcome of the first round of the game with player A having k dollars, we have

$$A_k = pA_{k+1} + qA_{k-1}, \qquad (6.50)$$

which is a homogeneous difference equation. This equation can be solved by determining the roots of the characteristic polynomial, $z = pz^2 + q$, which can be re-written as $z^2 - (1/p)z + q/p = 0$ for $p \neq q$. There are two roots $z_1 = 1$ and $z_2 = q/p$. Therefore, the general solution to (6.50) is

$$A_k = C_1 \cdot 1^k + C_2 \cdot \left(\frac{q}{p}\right)^k,$$

where C_1 and C_2 are determined by boundary conditions

$$A_0 = 0 \Rightarrow C_1 + C_2 = 0,$$

$$A_{a+b} = 1 \Rightarrow C_1 + C_2 \left(\frac{q}{p}\right)^{a+b} = 1,$$

which gives

$$C_1 = -\frac{1}{\left(\frac{q}{p}\right)^{a+b} - 1}, \quad C_2 = \frac{1}{\left(\frac{q}{p}\right)^{a+b} - 1}.$$

Thus, we obtain

$$A_k = -\frac{1}{\left(\frac{q}{p}\right)^{a+b} - 1} + \frac{1}{\left(\frac{q}{p}\right)^{a+b} - 1}\left(\frac{q}{p}\right)^k = \frac{\left(\frac{q}{p}\right)^k - 1}{\left(\frac{q}{p}\right)^{a+b} - 1}.$$

For the case of $p = q$, we can obtain the result by letting the ratio $r = q/p \to 1$ and using the L'hospital rule to get $A_k = k/(a+b)$. These results can be summarized as follows:

$$
A_k = \begin{cases} \dfrac{\left(\frac{q}{p}\right)^k - 1}{\left(\frac{q}{p}\right)^{a+b} - 1}, & \text{if } p \neq q, \\[2ex] \dfrac{k}{a+b}, & \text{if } p = q. \end{cases}
$$

For more details about solving difference equations that occur in probability and stochastic processes, see the Chapter H.

Exercise 6.18 *For the gambler's ruin problem, let S_n denote the winning of player A at time n. Define $T = \min\{n : S_n = a + b, \text{ or } S_n = 0\}$, the time to ruin. Find $E[T]$.*

6.1.8 GENERAL RANDOM WALK

A general random walk is a discrete-time stochastic process with the initial condition of $X_0 = x$ and the update rule of

$$
X_{n+1} = X_n + \xi_n, \quad n = 0, 1, 2, ..., \tag{6.51}
$$

where $\xi = (\xi_n)_{n \in \mathbb{Z}_+}$ are some i.i.d. continuous random variables. Such a process can represent the stock price process if an investor only observes the price one time per period (i.e., the price at the end of each day), or some stochastic signal observed at selected instants in time. For example, the stock price on day n is $X_n = \sum_{k=1}^n \xi_k$.

Note that to be consistent with the notations used in the literature, we have changed the symbols of variables from those used in the simple random walk. That is S_n, the position of the walk at time n, is replaced with X_n; and X_k, each step change, is replaced with ξ_k. It is clear that the simple random walk presented in previous section is a special case when ξ follows a Bernoulli distribution with two values of $+1$ and -1. Relation (6.51) can be realized in the form $X_{t+1} = f(X_t, \xi_t)$ with $f(x, u) = x + u$. Thus, this random walk is a discrete-time Markov process (DTMP). Due to the continuous state space, we cannot utilize the transition probability matrix to describe the dynamics of the system as in a DTMC. Thus, we need to introduce some new terminology and generalize the DTMC (a simple random walk treated as a Markov chain as shown in Chapter 2) to DTMP.

6.1.8.1 A Brief Introduction to DTMP

For a DTMP, we first introduce the transition kernel $P(x, A)$ for all $x \in \mathscr{S}$ and $A \subset \mathscr{S}$, which is defined as the probability of reaching the measurable set A

from state x. Then, we define the DTMP. Likewise, the concepts of irreducibility, recurrence, and positive recurrence, have to be re-defined on a continuous state space. We briefly describe some definitions for the DTMP that are paralleled to these for the DTMC. To evaluate the transition kernel, we need to define the transition kernel density $p(x, \cdot)$.

Definition 6.2 *For all $x \in \mathscr{S}$, $p(x, y)$ is defined as a non-negative function such that*

$$P(x, A) = \int_{y \in A} p(x, y) dy.$$

Note that $p(x, \cdot)$ is a pdf for a given x. Thus, we have

$$P(x, S) = \int_{y \in S} p(x, y) dy = 1.$$

Definition 6.3 *Given a transition kernel P, a sequence $\{X_n\}_{n \geq 0}$ is a discrete-time Markov process, if for any n and any $x_0, \ldots, x_n$, the conditional distribution of X_{n+1} given $X_n = x_n, X_{n-1} = x_{n-1}, \ldots, X_0 = x_0$ is*

$$P(X_{n+1} \in A | X_n = x_n, X_{n-1} = x_{n-1}, \ldots, X_0 = x_0) = P(X_{n+1} \in A | X_n = x_n)$$

$$= P(x_n, A) = \int_A p(x_n, x) dx.$$

A DTMP is also called a "Markov chain with continuous state space". It is time-homogeneous, if the transition density does not depend on n. For each measurable set A with positive measure function $\phi(A) > 0$, for each x there exists n such that $P^n(x, A) > 0$, the non-negative kernel $P(x, A)$ is called ϕ-irreducible. For every measurable set B with $\phi(B) > 0$, if

$$P(X_1, X_2, \ldots \in B \text{ i.o.} | X_0) > 0$$

for all X_0, where *i.o.* means "infinitely often", and

$$P(X_1, X_2, \ldots \in B \text{ i.o.} | X_0) = 1$$

for ϕ- almost-everywhere X_0, then the DTMP is recurrent. Furthermore, if

$$P(X_1, X_2, \ldots \in B \text{ i.o.} | X_0) = 1$$

for all X_0, the DTMP has the property of Harris recurrence. A probability distribution π is called the stationary distribution for the DTMP with transition density p (or transition kernel P) if

$$\pi(y) = \int p(x, y) \pi(x) dx \qquad (6.52)$$

for all $y \in \mathscr{S}$, or

$$\pi(A) = \int P(x,A)\pi(x)dx$$

for all measurable sets $A \subset \mathscr{S}$, in which case we may have $\pi P = \pi$. Periodicity can be also defined similarly in the DTMP as in DTMC. A DTMP with period n can be defined as follows: There exists $n \geq 2$ and a sequence of nonempty, disjoint measurable sets $A_1, A_2, ..., A_n$ such that for all $j < n$, $P(x, A_{j+1}) = 1, \forall x \in A_j$, and $P(x, A_1) = 1 \forall x \in A_n$. If the transition kernel has density $p(x, (x - \varepsilon, x + \varepsilon))$, a sufficient condition for aperiodicity is that this kernel density is positive in a neighborhood of x, since the chain can remain in this neighborhood for an arbitrary number of times before visiting other set A. We present the following theorem without proof which is parallel to the DTMC case.

Theorem 6.7 *Let P be a Markov transition kernel, with invariant distribution π, and suppose that P is π-irreducible. Then π is unique and is called invariant distribution of P. If P is aperiodic, then for π almost-every x*

$$||P^n(x,A) - \pi(A)||_{TV} \to 0,$$

where the distance between two probability measures π_1 and π_2 is measured using the Total Variation Norm: $||\pi_1 - \pi_2||_{TV} = \sup_A |\pi_1(A) - \pi_2(A)|$. If P is Harris recurrent, then the convergence occurs for all $x \in \mathscr{S}$.

The stationary analysis of the DTMP can be done similarly. Clearly, the continuous state analog of the one-step transition probability p_{ij} for DTMC is the one-step transition kernel density. This is not the probability that the DTMP makes a move from state x to state y. Instead, it is a probability density function in y which describes a curve under which the area represents probability. Hence, x can be thought of as a parameter of this density. For example, given a DTMP is currently in state x, the next value y might be drawn from a normal distribution centered at x. In this case, we would have the transition density

$$p(x,y) = \frac{1}{\sqrt{2\pi}}e^{-\frac{1}{2\sigma^2}(y-x)^2}.$$

The analogue of the n-step transition probability $p_{ij}^{(n)}$ for the DTMC is the n-step transition density denoted by $p^{(n)}(x;y)$. Thus, we must have

$$\int_{\mathscr{S}} p(x,y)dy = 1 \quad \text{and} \quad \int_{\mathscr{S}} p^{(n)}(x,y)dy = 1$$

for all $n \geq 1$. Next, we can develop the C-K equations for the DTMP. Suppose that the chain is currently at state x. The location of the chain 2 time steps later could be described by conditioning on the intermediate state visited at time 1 as follows.

$$p^{(2)}(x,y) = \int_{\mathscr{S}} p(x,z) \cdot p(z,y)dz.$$

Due to the independence of transitions, once the DTMP is at the intermediate state z, the next move to y is independent of the previous move from x to z. In general, for any $0 \leq m \leq n$, we have

$$p^{(n)}(x,y) = \int_{\mathscr{S}} p^{(m)}(x,z) \cdot p^{(n-m)}(z,y)dz,$$

which is the C-K equation in terms of transition densities. Here, we define $p^{(0)}(x,y)$ as the Dirac delta function. The C-K equation can be also expressed in terms of the transition kernel as follows: For all $n,m \in N \times N, x \in \mathscr{S}, A \in \mathscr{B}(\mathscr{S})$,

$$P(X_{n+m} \in A | X_0 = x) = P^{(m+n)}(x,A) = \int_{\mathscr{S}} p^{(n)}(y,A)p^{(m)}(x,y)dy.$$

Similar to DTMC, if the DTMP is Harris recurrent, we can obtain the stationary distribution density equation (6.52).

Example 6.3 *(General Random Walk)* Note that

$$X_{n+1} = X_n + \xi_n.$$

Thus, we have

$$P(X_{n+1} \leq y | X_n = x) = P(x + \xi_n \leq y | X_n = x)$$
$$= P(\xi_n \leq y - x) = \int_{-\infty}^{y-x} g(z)dz,$$

where $g(z)$ is the normal distribution density with mean of 0 and variance of σ^2, and the transition density is given by

$$p(x,y) = \frac{\partial}{\partial y} P(X_{n+1} \leq y | X_n = x) = g(y - x).$$

Another example is the Gaussian Autoregression Process of order 1 denoted by AR(1), which can be considered as a variant of GRW.

Example 6.4 *Consider a DTMP with the following recursive relation*

$$X_{n+1} = \theta X_n + \xi_n.$$

where $\xi_n \sim N(0, \sigma^2)$, is i.i.d. random variables, and independent of X_0 and $\theta \in \mathscr{R}$. Then X_{n+1} is Gaussian and conditionally independent of $X_{n-1}, \dots, X_0$ given X_n. The transition density is obtained as

$$p(x, y) = \frac{1}{\sqrt{2\pi}} e^{-\frac{1}{2\sigma^2}(y - \theta x)^2}.$$

For AR(1), if $|\theta| < 1$, the stationary distribution exists and can be determined. The stationary distribution density equation (6.52) becomes

$$\pi(y) = \int_{-\infty}^{\infty} \frac{1}{\sqrt{2\pi}\sigma} e^{-\frac{(y - \theta x)^2}{2\sigma^2}} \pi(x) dx.$$

It can be checked that the stationary distribution density, that satisfies the equation above, is $N(0, \sigma^2/(1 - \theta^2))$.

Exercise 6.19 For a GRW, if ξ_n with $n \geq 1$ are i.i.d. random variables uniformly distributed over $[-L, L]$. Derive the transition density function and write the stationary distribution density equation.

6.1.9 BASIC PROPERTIES OF GRW

Often, we characterize the system change of a DTMP by developing some distributions such as the marginal distribution at a particular time point, joint distribution of system state at different points in time, and some descriptive statistics for the process. We start with developing some properties for the general random walk. An important property of this process is the independent increments.

Theorem 6.8 A general random walk has independent increments.

Proof. Consider two points in time $0 \leq s < t$ and write

$$X_t = X_t - X_0 = (X_t - X_s) + (X_s - X_0).$$

Iterating (6.51) yields

$$X_t = X_t - X_0 = \sum_{i=0}^{t-1} \xi_i$$

for every t. Thus,

$$X_t - X_s = (X_t - X_0) - (X_s - X_0) = \sum_{i=0}^{t-1} \xi_i - \sum_{i=0}^{s-1} \xi_i = \sum_{i=s}^{t-1} \xi_i.$$

This implies that $X_s - X_0$ is a function of $\xi_0, \dots, \xi_{s-1}$, while $X_t - X_s$ is a function of $\xi_s, \dots, \xi_{t-1}$. Because ξ is an i.i.d. random variable, and independence is preserved by grouping, we see that the increments $X_t - X_s$ and $X_s - X_0$ are independent. ∎

Next, we can compute the mean and variance of the general random walk (GRW) process at any time t. Denote by $\mu = E(\xi)$ and $\sigma^2 = Var(\xi)$, the common mean and variance of the ξ_i's. Then, it follows from the linearity of expectation that

$$E(X_t) = E\left(\sum_{i=0}^{t-1} \xi_i\right) = \sum_{i=0}^{t-1} E(\xi_i) = t\mu,$$

$$\sigma_X^2(t) = Var(X_t) = Var\left(\sum_{i=0}^{t-1} \xi_i\right) = \sum_{i=0}^{t-1} Var(\xi_i) = t\sigma^2.$$

While the variance of X_t always increases linearly with t, the mean of X_t can increase linearly ($\mu > 0$ case) or decrease linearly ($\mu < 0$ case) or unchange ($\mu = 0$ case), respectively. Similarly, we can compute the mean and variance for the increments of X_t. For $s < t$, we have

$$E(X_t - X_s) = E(X_t) - E(X_s) = (t - s)\mu,$$

$$Var(X_t - X_s) = Var\left(\sum_{i=s}^{t-1} \xi_i\right) = \sum_{i=s}^{t-1} Var(\xi_i) = (t - s)\sigma^2.$$

An interesting fact is that although the GRW has independent increments, the states at two different time points are not independent due to the Markovian property. Thus, we can compute the autocorrelation function by using again the property of independent increments. Without loss of generality, we assume that $X_0 = 0$. Fix $s < t$, the autocorrelation function, denoted by $R_X(s,t)$, can be written as

$$
\begin{aligned}
R_X(s,t) &= E(X_s X_t) = E[(X_s - X_0)(X_t - X_0)] \\
&= E[(X_s - X_0)(X_t - X_s + X_s - X_0)] \\
&= E[(X_s - X_0)(X_t - X_s)] + E[(X_s - X_0)^2].
\end{aligned}
\tag{6.53}
$$

Due to the independent increments, we have

$$E[(X_s - X_0)(X_t - X_s)] = E[(X_s - X_0)]E[(X_t - X_s)] = s\mu \cdot (t - s)\mu = s(t - s)\mu^2.$$

and

$$E[(X_s - X_0)^2] = E(X_s^2) = s\sigma^2 + s^2\mu^2.$$

Substituting these two expressions in (6.53) yields

$$R_X(s,t) = s(t - s)\mu^2 + s\sigma^2 + s^2\mu^2 = st\mu^2 + s\sigma^2.$$

On the other hand, if $t \leq s$, we obtain

$$R_X(s,t) = st\mu^2 + t\sigma^2.$$

In general, for any two time points, we can write

$$R_X(s,t) = st\mu^2 + \sigma^2 \min(s,t). \tag{6.54}$$

The autocovariance, denoted by $C_X(s,t)$, is then given by

$$C_X(s,t) = R_X(s,t) - E(X_s)E(X_t) = \sigma^2 \min(s,t). \tag{6.55}$$

The autocorrelation (or autocovariance) of the GRW implies that even we know the distribution of a DTMP X at every period (the marginal distribution), we cannot completely describe the probability behavior of the DTMP. Using two simple discrete-time stochastic processes with discrete state space can demonstrate this fact easily. Consider two stochastic processes as follows:

- $X = (X_t)_{t \in \mathbb{Z}_+}$ is an i.i.d. Bernoulli (1/2) process. That is each X_t equals +1 with probability of 1/2 and 0 with probability of 1/2, respectively.
- $Y = (Y_t)_{t \in \mathbb{Z}_+}$ is a Markov chain with binary state space $S = \{0,1\}$ with initial state Y_0 following a Bernoulli (1/2) and one-step transition probabilities $p_{00} = p_{11} = 0.7$ and $p_{10} = p_{01} = 0.3$.

It is easy to check that these two processes have the same one-time marginal distributions, i.e., $P(X_t = 0) = P(Y_t = 0)$ for all t (since the state space of both processes has only two states). Note that $P(Y_1 = 0) = P(Y_1 = 0|Y_0 = 0)P(Y_0 = 0) + P(Y_1 = 0|Y_0 = 1)P(Y_0 = 1) = 0.7 \times 0.5 + 0.3 \times 0.5 = 0.5$. Inductively, we can show that $P(Y_t = 0) = 0.5$ for all t. Furthermore, we can easily determine that the $\pi = (0.5, 0.5)$ is an equilibrium distribution of Markov chain Y. Since X_0 is a Bernoulli(1/2), we know that X_t follows Bernoulli(1/2) for all t. This shows that the two processes have the identical marginal distribution for all t. However, the two processes are very different. For example, if $X_{100} = 0$, then the probability that $X_{101} = 0$ is 0.5. However, for the Y process, if $Y_{100} = 0$, then the probability that $Y_{101} = 0$ will be 0.7! This is because that the Y_t's are not independent, i.e., Markov processes have memory, albeit a very short one. This simple example reveals that to fully describe a DTMP, we must be able to specify all probabilities of the form (joint probability distribution)

$$P(a_1 \leq X_{t_1} \leq b_1, a_1 \leq X_{t_2} \leq b_1, ..., a_n \leq X_{t_n} \leq b_n)$$

for all n, all $t_1 < t_2 < ... < t_n$ and all finite or infinite intervals $[a_1, b_1], ... [a_n, b_n]$ of state space S. Of course, it may not be easy to obtain closed-form expressions for these probabilities even for $n = 2$ (two period duration) — but, as long as we can do this in principle, we are done. In many situations, however, we do not need such a fine-grained description of X. For example, we may want to know what value X_t will take "on average" and how much we can expect it to fluctuate around this average value or the degree of dependence between two time points. This information is encoded in the numbers introduced above. i.e., $E(X_t), Var(X_t)$, and $R_X(s,t)$. Due to the independent increment property, we can obtain the Laplace transform of the GRW at a certain time point t_n. If we assume that the Laplace transform for ξ_i is $\tilde{F}_\xi(s)$, then the Laplace transform for the distribution of X_{t_n} is $\tilde{F}_{X_{t_n}}(s) = (\tilde{F}_\xi(s))^{t_n}$. To obtain the distribution, numerical inversion may be applied (see book by Abate and Whitt (1992)).

Exercise 6.20 *In a GRW, assume that ξ_i follows an exponential distribution with rate μ. Use the Laplace transform approach to develop the distribution of X_{t_n}.*

6.1.9.1   Central Limit Theorem for GRW

Assume that the i.i.d. random variable $\xi_n = \xi$ has the mean 0 and variance of σ_ξ^2. Denote by $M_\xi(s) = E[e^{s\xi}]$ the moment generating function (MGF) of ξ which is assumed to exist (for s in certain range) in most practical situations (not true for all random variables, see the exercise following). Define the associated GRW with variance 1, $Z_n = X_n / \sqrt{n\sigma_\xi^2} = \sum_{i=1}^{n}(\xi_i / \sqrt{n\sigma_\xi^2})$. Then we have

$$M_{X_n}(s) = (M_\xi(s))^n \quad M_{Z_n}(s) = \left(M_\xi \left(\frac{s}{\sigma_\xi \sqrt{n}} \right) \right)^n.$$

It follows from the Taylor's theorem that the MGF can be written as

$$M_\xi(s) = M_\xi(0) + sM_\xi'(0) + \frac{1}{2}s^2 M_\xi''(0) + o(s^2),$$

where $o(s^2) \to 0$ as $s \to 0$. Note that $M_\xi(0) = 1$, by definition, $M_\xi'(0) = E[\xi] = 0$, and $M_\xi''(0) = Var(\xi) = \sigma_\xi^2$. Thus, we obtain

$$M_\xi(s) = 1 + \frac{\sigma_\xi^2}{2}s^2 + o(s^2).$$

Letting $s = t/(\sigma_\xi \sqrt{n})$, we have

$$M_{Z_n(t)}(t) = \left(1 + \frac{\sigma_\xi^2}{2}\left(\frac{t}{\sigma_\xi \sqrt{n}}\right)^2 + o(t^2/(\sigma_\xi^2 n)) \right)^n = \left(1 + \frac{t^2}{2n} + o(n) \right)^n,$$

where $o(n) \to 0$ as $n \to \infty$. Taking limits of the above relation as $n \to \infty$ yields

$$\lim_{n \to \infty} M_{Z_n(t)} = \lim_{n \to \infty} \left(1 + \frac{t^2/2}{n} + o(n) \right)^n = e^{t^2/2},$$

which is the MGF of a standard normal distribution. Here, we use $\lim_{n\to\infty}(1 + a/n)^n = e^a$, a well-known limit in calculus This implies that Z_n converges in distribution to $N(0,1)$. In general, for a GRW with the i.i.d. ξ_k's having mean of $\mu > 0$ and finite variance of σ^2, X_n approaches to a normally distributed stationary random variable. That is

$$\frac{X_n - n\mu}{\sqrt{n\sigma^2}} = \frac{\sum_{k=1}^{n} \xi_k - n\mu}{\sqrt{n\sigma^2}} = \sum_{k=1}^{n} \frac{\xi_k - \mu}{\sqrt{n\sigma^2}} \to N(0,1) \text{ as } n \to \infty,$$

by the same reasoning. This implies that a GRW, X_n, approaches a stationary normally distributed random variable with mean of $n\mu$ and standard deviation $\sqrt{n}\sigma$. That is $X_n \sim N(n\mu, n\sigma^2)$ for large n, which is the central limit theorem (CLT) for a GRW. This result is also called De Moivre–Laplace theorem that can also be proved without using the MGF. The derivation is via using the limits of binomial distribution and Stirling's formula by considering a simple random walk. This result can be generalized to the functional space. That is a random walk process will approach to a continuous-time and continuous state Brownian motion. Such a generalization is called the the Donsker's theorem, an important functional Limit Theorem. The Functional Central Limit Theorem (FCLT) will be discussed in more details in the next chapter.

Exercise 6.21 *Clearly, the CLT for a GRW requires that the first two moments of ξ_i be finite. This is not true for all cases. For example, a heavy tailed random walk, called the Cauchy random walk, has ξ_i following the Cauchy distribution with the p.d.f. given by*

$$f_{\xi_i}(x) = \frac{1}{\pi} \frac{1}{1+x^2},$$

which has undefined mean and variance.

(a) Show that X_n/n for each n is also Gauchy, thus the CLT does not hold.

(b)Use Matlab or other software package to simulate this Gauchy random walk and compare it with the symmetric simple random walk. What can you observe?

6.2 BROWNIAN MOTION

Now we consider continuous-time Markov processes (CTMPs) which have a continuous state space. A fundamental CTMP is the Brownian motion that can be introduced in different ways. For example, it can be obtained by taking the limiting process of a simple random walk process (a DTMC) or a general random walk (a DTMP), or as a solution to a differential equation that describes some stochastic processes (see Chapter H).

6.2.1 BROWNIAN MOTION AS A LIMIT OF RANDOM WALKS

The main idea is to re-scale the time and space to convert a DTMC to a CTMP. Such a result is the most important functional limit theorem, called Donsker's theorem. Consider a simple symmetric random walk, a special case of GRW, with

$$X_0 = 0, \quad X_n = \xi_1 + \xi_2 + \cdots + \xi_n, \quad n \geq 1,$$

where $\xi_1, \xi_2, \ldots$ is a sequence of Bernoulli random variables with parameter 1/2 (we use the notations for the GRW). That is

$$\xi_i = \begin{cases} -1 & \text{with probability of } 1/2, \\ 1 & \text{with probability of } 1/2. \end{cases}$$

Suppose now that we re-scale the random walk in the following way: the time interval between steps is decreased from 1 to some small number Δt, and accordingly the step size is decreased from 1 to another small number Δx. Δt and Δx are chosen so that after 1 unit of time, i.e. after $n = 1/\Delta t$ steps, the standard deviation of the resulting random variable X_n is normalized to 1. Due to the independence of steps (i.e., increments) of the random walk and the fact that the random variable which takes values Δx and $-\Delta x$ with equal probabilities, the variance is then $(\Delta x)^2$. Clearly, we have

$$Var(X_n) = Var(\xi_1) + Var(\xi_2) + \cdots + Var(\xi_n) = n(\Delta x)^2 = \frac{(\Delta x)^2}{\Delta t}.$$

Thus, we should choose $\Delta x = \sqrt{\Delta t}$ to avoid trivial results (0 or ∞). For each Δt, we can define a continuous-time process $(X_n)_{t \in [0,\infty)}$ by the following procedure:

1. Multiply all steps of a simple random walk, $(X_n)_{n \in \mathbb{N}_0}$, by $\Delta x = \sqrt{\Delta t}$ to obtain a new process $(X_n^{\Delta x})_{n \in \mathbb{N}_0}$ as

$$X_0^{\Delta x} = 0, \quad X_n^{\Delta x} = \sqrt{\Delta t}\xi_1 + \sqrt{\Delta t}\xi_2 + \cdots + \sqrt{\Delta t}\xi_n = \sqrt{\Delta t}X_n, \quad n \geq 1.$$

2. Define the continuous time process $(B^{\Delta t})_{t \in [0,\infty)}$ by

$$B_t^{\Delta t} = \begin{cases} X_n^{\Delta x}, & \text{if } t \text{ is at } n\Delta t, \\ \text{interpolate linearly}, & \text{otherwise}. \end{cases}$$

It follows from the CLT for the GRW that as $\Delta t \to 0$ the processes $(B^{\Delta t})_{t \in [0,\infty)}$ converge in a mathematically precise sense (in distribution) to a continuous time stochastic process, B_t, which is normally distributed with mean of 0 and variance of s at time $s > 0$ given that $B_0 = 0$. Such a process is called a Brownian motion (BM). For more details about the convergence of random variables and stochastic processes, readers are referred to Chapter F and Chapter G. Before presenting the formal definition, we note that the existence of BM can be shown by constructing it. The construction is based on the intuitive "connect-the-dots" approach that is presented in Chapter H.

Definition 6.4 *Brownian motion is a continuous-time, infinite-horizon stochastic process* $(B_t)_{t \in [0,\infty)}$ *such that*

1. $B_0 = 0$ *(Brownian motion starts at zero),*
2. *for* $t > s \geq 0$,
3. *the increment* $B_t - B_s$ *is normally distributed with mean* $\mu = 0$ *and variance* $\sigma^2 = t - s$ *(the increments are normally distributed).*
4. *the random variables* $B_{s_m} - B_{s_{m-1}}, B_{s_{m-1}} - B_{s_{m-2}}, ..., B_{s_1} - B_0$, *are independent for any* $m \in \mathcal{N}$ *and any* $s_1 < s_2 < \cdots < s_m$. *(the increments of the Brownian motion are independent).*
5. *the trajectories of a Brownian motion are continuous functions.*

As a CTMP, we need to define the finite-dimensional distribution for a Brownian motion, which is useful in determining its full probabilistic structure.

Definition 6.5 *The finite-dimensional distributions of a CTMP* $(X_t)_{t \in \mathcal{I}}$ *are all distribution functions of the form*

$$F_{(X_{t_1}, X_{t_2}, ..., X_{t_n})}(x_1, x_2, ..., x_n) \triangleq P(X_{t_1} \leq x_1, X_{t_2} \leq x_2, ..., X_{t_n} \leq x_n)$$

for all $n \in \mathbb{N}$ *and all n-tuples* $(t_1, t_2, ..., t_n)$ *of indices in* $\mathcal{I}$.

The finite-dimensional distribution of a Brownian motion is a multivariate normal distribution.

Theorem 6.9 *For any n-tuple of indices* $0 \leq t_1 < t_2 < ... < t_n$, *the random vector* $(B_{t_1}, B_{t_2}, ..., B_{t_n})$ *has the multivariate normal distribution with mean* $\mu = (0, 0, ..., 0)$ *and the variance-covariance matrix*

$$\Sigma = \begin{pmatrix} t_1 & t_1 & \cdots & t_1 \\ t_1 & t_2 & \cdots & t_2 \\ \vdots & \vdots & \ddots & \vdots \\ t_1 & t_2 & \cdots & t_n \end{pmatrix},$$

where $\Sigma_{ij} = \min(t_i, t_j) = t_{\min(i,j)}$.

Proof. To prove $(B_{t_1}, B_{t_2}, ..., B_{t_n})$ having the multivariate normal distribution, it is enough to prove that $X \triangleq a_1 B_{t_1} + a_2 B_{t_2} + ... + a_n B_{t_n}$ is a normally distributed random variable for each vector $a = (a_1, a_2, ..., a_n)$. If we rewrite X as

$$X = a_n(B_{t_n} - B_{t_{n-1}}) + (a_n + a_{n-1})(B_{t_{n-1}} - B_{t_{n-2}})$$
$$+ (a_n + a_{n-1} + a_{n-2})(B_{t_{n-2}} - B_{t_{n-3}}) + ...$$

Thus, X is a linear combination of the increments $X_k \triangleq B_{t_k} - B_{t_{k-1}}$ which are mutually independent. Note that the last increment X_n is independent of $B_{t_1}, B_{t_2}, ..., B_{t_{n-1}}$ by definition, and so it is also independent of $B_{t_2} - B_{t_1}, B_{t_3} - B_{t_2}, ...$ Similarly the increment $B_{t_{n-1}} - B_{t_{n-2}}$ is independent of 'everybody' before it by the same argument. We have therefore proven that X is a sum of independent normally distributed random variables, and therefore normally distributed itself. This is true for each vector $a = (a_1, a_2, ..., a_n)$, and therefore the random vector $(B_{t_1}, B_{t_2}, ..., B_{t_n})$ is multivariate normal. Next, we determine the mean and variance-covariance matrix of $(B_{t_1}, B_{t_2}, ..., B_{t_n})$. For $m = 1, ..., n$, we write the telescoping sum

$$E(B_{t_m}) = E(X_m) + E(X_{m-1}) + ... + E(X_1),$$

where $X_k \triangleq B_{t_k} - B_{t_{k-1}}$. It follows from $E(B_{t_m}) = 0$ that $\mu = (0, 0, ...0)$. For $i < j$, we have

$$\Sigma_{ij} = E(B_{t_i} B_{t_j}) = E(B_{t_i}^2) - E(B_{t_i}(B_{t_j} - B_{t_i})) = E(B_{t_i}^2) = Var(B_{t_i}) = t_i,$$

due to the fact that B_{t_i} is normally distributed with mean 0 and variance t_i. The cases of $i = j$ and $i \geq j$ can be treated similarly. This completes the proof. ∎

An important characteristics of the Brownian motion, denoted by BM, is that its path is not differentiable but continuous at any point (see Chapter H for the more theoretical results about BM). This is in contrast to another stochastic process with the same type of finite-dimensional distribution. That is the "independent process" $(X_t)_{t \in [0, \infty)}$ for which X_t and X_s are independent for $t \neq s$, and normally distributed. The independent process, denoted by IP, is not continuous at any point and can be called a complete random function $[0, \infty) \to \mathbb{R}$.

Exercise 6.22 *Suppose that B_t is a Brownian motion and U is an independent random variable, which is uniformly distributed on $[0, 1]$. Then the process $\{\tilde{B}_t : t \geq 0\}$ is defined as*

$$\tilde{B}_t = \begin{cases} B_t & \text{if } t \neq U, \\ 0 & \text{if } t = U. \end{cases}$$

Argue that $\{\tilde{B}_t : t \geq 0\}$ is not a Brownian motion.

Exercise 6.23 *Consider a standard BM $\{B_t, t \geq 0\}$ and two time points s and t such that $s \leq t$. Find the distribution of $aB_s + bB_t$, where a and b are two constants.*

6.2.2 GAUSSIAN PROCESSES

Although BM and IP are completely different stochastic processes, they share a common characteristic that their finite-dimensional distributions are multivariate normal. A class of stochastic processes with this feature are called Gaussian Processes.

Definition 6.6 *A continuous-time stochastic process* $(X_t)_{t\in[0,\infty)}$ *is called a Gaussian process if its finite-dimensional distributions are multivariate normal.*

Besides BM and IP, there are other examples of Gaussian processes.

Example 6.5 *(BM with drift) Consider a stochastic process defined as*

$$X_t = B_t + at.$$

This process is still a Gaussian process with the expectation and covariance functions as

$$\mu_X(t) = at, \quad c_X(t,s) = \min(t,s).$$

Example 6.6 *(Brownian Bridge) This process is obtained from a BM on the finite interval [0,1], $(B_t)_{t\in[0,1]}$, by requiring $B_1 = 0$. That is $X_t = B_t - tB_1$. It can be shown that X_t is a Gaussian process with the expectation and covariance functions. For an arbitrary n-tuple of indices $(t_1,t_2,...,t_n)$ in [0,1], it can be shown that the finite-dimensional distribution of $(X_{t_1},X_{t_2},...,X_{t_n})$ is multivariate normal. For a set of constants $a_1,a_2,...,a_n$, we write*

$$Y = a_1 X_{t_1} + a_2 X_{t_2} + ... + a_n X_{t_n}$$
$$= -(a_1 + a_2 + \cdots + a_n)tB_1 + a_1 B_{t_1} + a_2 B_{t_2} + \cdots + a_n B_{t_n},$$

which is normal because $(B_1, B_{t_1}, B_{t_2}, ..., B_{t_n})$ is a multivariate normal random vector. Thus, the Brownian Bridge is a Gaussian process. Using the fact that $E(B_t B_s) = \min(t,s)$, for $t,s \le 1$, we have

$$\mu_X(t) = E(X_t) = E(B_t - B_1 t) = 0$$
$$c_X(t,s) = Cov(X_t,X_s) = E((B_t - tB_1)(B_s - sB_1))$$
$$= E(B_t B_s) - sE(B_t B_1) - tE(B_s B_1) + tsE(B_1 B_1)$$
$$= \min(t,s) - ts \ (= s(1-t), \ if \ s < t).$$

BM has some nice scaling invariance property. It follows from $B_t \sim N(0,t)$ that

$$B_{\alpha t} \sim N(0, \alpha t) \sim \sqrt{\alpha} N(0,t) \sim \sqrt{\alpha} B_t, \quad for \ \alpha > 0,$$

which implies

$$\frac{1}{\sqrt{\alpha}} B_{\alpha t} \sim B_t.$$

Such a scaling is often called Brownian scaling.

Another important result is the "Law of Large Numbers" for the BM, which is stated as a lemma.

Lemma 6.1 *Almost surely,* $\lim_{t \to \infty} B_t / t = 0$.

Proof. First, we show that for positive integer n $\lim_{n \to \infty} \frac{B_n}{n} = 0$ holds. For an $\varepsilon > 0$, using the fact that $B_n / \sqrt{(n)}$ has a standard normal distribution, we have

$$\sum_{n=1}^{\infty} P\left(\frac{B_n}{n} > \varepsilon\right) = \sum_{n=1}^{\infty} P\left(\frac{B_n}{\sqrt{n}} > \varepsilon \sqrt{n}\right) \leq \sum_{n=1}^{\infty} \frac{1}{\varepsilon \sqrt{n}} \frac{1}{\sqrt{2\pi}} e^{-(n\varepsilon^2)/2}. \quad (6.56)$$

Here, the second inequality follows due to the following bound for the tail probability of standard normal distribution

$$P(Z > z) = \frac{1}{\sqrt{2\pi}} \int_z^{\infty} 1 \cdot e^{-t^2/2} dt \leq \frac{1}{\sqrt{2\pi}} \int_z^{\infty} \frac{t}{z} e^{-t^2/2} dt = \frac{e^{-z^2/2}}{z\sqrt{2\pi}},$$

where the inequality follows from $t \geq z > 0$ which implies $1 \leq t/z$. Since the last sum of (6.56) is finite, the Bore–Cantelli Lemma implies that there exists N such that, for all $n \geq N$, $B(n)/n < \varepsilon$ almost surely. Using the property of symmetry, $P(B_n/n < -\varepsilon) = P(B_n/n > \varepsilon)$, we have $B_n/n > -\varepsilon$ almost surely for sufficiently large n. Thus, $\lim_{n \to \infty} B_n/n = 0$. Next, we extend this result to all non-negative times t. It is sufficient to show

$$\limsup_{n \to \infty} \frac{1}{n} \left(\max_{n \leq t \leq n+1} B_t - B_n\right) = 0.$$

For $\varepsilon > 0$, using the reflection principle (to be discussed in the next section) and the fact that B_1 has a standard normal distribution, we have

$$\sum_{n=1}^{\infty} P\left(\max_{0 \leq t \leq 1} B_t > \varepsilon n\right) = \sum_{n=1}^{\infty} 2P(B_1 \leq \varepsilon n) \leq 2 \sum_{n=1}^{\infty} \frac{1}{\varepsilon n} \frac{1}{\sqrt{2\pi}} e^{-(\varepsilon^2 n^2)/2}.$$

Since the last sum is finite, the Borel–Cantelli Lemma implies that

$$P(\limsup_{n \to \infty} (\max_{0 \leq t \leq 1} B_t) > \varepsilon n) = 0.$$

Equivalently, $\max_{0 \leq t \leq 1} B_t < \varepsilon n$ or $\frac{1}{n} \max_{0 \leq t \leq 1} B_t < \varepsilon$ almost surely. Note that the distributions of $\max_{0 \leq t \leq 1} B_t$ and $\max_{n \leq t \leq n+1} B_t - B_n$ are the same by the time-shift invariance. Thus, for sufficiently large n, $\frac{1}{n} \max_{n \leq t \leq n+1} B_t - B_n < \varepsilon$ almost surely. This implies that $\limsup_{n \to \infty} \frac{1}{n} (\max_{n \leq t \leq n+1} B_t - B_n) = 0$. Using the symmetry of B_t, we have $\lim_{n \to \infty} \frac{1}{n} (\max_{n \leq t \leq n+1} |B_t - B_n|) = 0$. Since B_t/t is bounded by $\frac{1}{n} (B_n + \max_{n \leq s \leq n+1} |B_s - B_n|)$ for $n \leq t \leq n+1$, it follows from the squeeze theorem that $\lim_{t \to \infty} B_t / t = 0$. ∎

Exercise 6.24 *For a range* $a < 0 < b$, *define* $T(a,b) = \inf\{t \geq 0 : B_t = a \text{ or } B_t = b\}$, *the first exit time of this range for the standard BM. Show (a)* $E[T(-b,b)]$ *is a constant multiple of* b^2; *and (b)* $P(\{B_t : t \geq 0\} \text{ exits } [a,b] \text{ at } a)$ *is only a function of the ratio* b/a.

Exercise 6.25 *Define the process* $\{X(t) : t \geq 0\}$ *as*

$$X(t) = \begin{cases} 0 & for\ t = 0, \\ tB_{1/t} & for\ t > 0. \end{cases}$$

Show that $\{X(t) : t \geq 0\}$ *is also a standard BM (time inversion). (Hint: show that the finite dimensional marginals are Gaussian random vectors and continuity at zero by using the law of large numbers for BM).*

We will discuss the re-scaling properties in details for the BM in the next section.

6.2.3 SAMPLE PATH PROPERTIES

6.2.3.1 Infinite Zeros and Non-Differentiability

Standard Broanian Motion (SBM) has some interesting and counter-intuitive properties. To discuss these properties, we focus on one-dimensional SBM although higher-dimensional BM can be studied. Each sample path is a continuous function of time. The first property is that a SBM B_t has infinitely many zeros in any finite interval with probability 1 (almost all sample paths). Instead of presenting a rigorous proof, we intuitively argue this property holds. First, it is believable that a SBM will visit 0 infinitely many times in the time interval (c, ∞) where c is an arbitrary large number. This is analogous to the simple random walk case which is positive recurrent. With the scaling property of SBM, we know that the process $Y_t = tB_{1/t}$ is also a SBM (Exercise 6.22). Since we believe that $Y_t = 0$ for infinitely many times in (c, ∞) which implies that $B_t = 0$ for infinitely many times in the interval $(0, 1/c)$. Since we can make c arbitrarily large, this infinite zero property holds for any finite interval no matter how small it is. The next property is that with probability 1, a sample path does not have a derivative at any time. This non-differentiability of the SBM can be justified by the property of independent increments. For any $\delta > 0$, the increment $B_{t+\delta} - B_t$ is independent of the increment $B_t - B_{t-\delta}$. Thus, it is almost impossible that these increments of both sides of t "matched up" well enough to be differentiable at time t. We can also understand why the SBM has these properties by re-scaling the time of the process. If we are looking at the process over a large time interval, such as 1 second, and are not concerned about the fluctuations over short time intervals (i.e., sub-intervals of 1 second interval), then the sample path observed at these integer instants is easy to understand, since $B_0, B_1, B_2, \ldots$ constitutes a GRW with iid standard normal increments. Note that the increments in this sample path have the variance and the standard deviation equal to 1 which is the same as the length of time interval. The ratio of the standard deviation to the length of interval, called variation ratio, indicates the level of variability of the process. Now we re-scale the process to get a more detailed process dynamics over shorter time intervals. Let's zoom in on the first tenth of a second and sample the process in time intervals of length 0.01 instead of length 1 second. Then we might get a sample path as shown in Figure 6.8. The increments now have variance of 0.01 and standard deviation is 0.1 and the variation ratio is 10, meaning the level variability is 10 times of that in the case with larger time intervals. We can continue to zoom in on smaller

Figure 6.8 Brownian motion.

and smaller time scales until we are satisfied with the details of the process. If we sampled 10,000 times every second instead of 100 times. That is the sampling interval is 0.0001 second. The variation ratio is 100. The phenomenon that the variability of the process increases as the time interval decreases implies the two properties discussed above. Hopefully, it is convinced that these two properties hold for the SBM by these arguments although rigorous proofs are available in more advanced textbooks on Brownian motions. Here, we only scale the time not the state variable above. If we appropriately scale both time and state variable, a SBM is resulted.

6.2.3.2 Re-Scaling a Process

As shown earlier, a SBM, a continuous-time process, can be considered as a limit process of GRW, a discrete-time process by re-scaling time and space. Since this technique will be used in the chapters on macroscopic stochastic models, we provide more comments on re-scaling a random function. This process is analogous to the sampling process in an infinite population case. In general, re-scaling time (or state space) axis can be done by defining the basic time interval Δt (or Δy) and appropriately changing it. A fixed time period T can be divided into n sub-intervals Δt's. For example, we can zoom out on a process in an XY plot by reducing Δt (the physical size in the plot) while keeping the same time value it represents (on x-axis, such as 1 min) and also appropriately re-scaling the state variable (on y-axis). This implies that a fixed time period $T = n\Delta t$ (i.e., with fixed n and the time value of Δt) becomes smaller and smaller interval in physical size on the horizontal axis. In other words, a finite interval of the fixed length on the horizontal axis will represent a longer and longer time period as Δt becomes smaller and smaller and n becomes larger and larger. In fact, the time horizon will increase to the n times of the original time horizon. Note the equivalence of $\{\Delta t \to 0\} \Leftrightarrow \{n \to \infty\}$. Selecting $T = 1$ (i.e., $n\delta t = 1$), then we can re-scale the time either using Δt or n. Then a time period $(0, t]$ will become $(0, nt]$ after re-scaling (i.e., each unit on x-axis after re-scaling is n times of each unit on x-axis before re-scaling). Re-scaling the y-axis can be done similarly. For each unit on y-axis, we have $n_y \Delta y = 1$. The relation between the two axis re-scaling sub-intervals can be represented by $\Delta y = (\Delta t)^\alpha$ where $0 \le \alpha \le 1$, which implies $n_y = n^\alpha$. This means each unit on y-axis after re-scaling is n^α times of each unit on the y-axis before re-scaling. Re-scaling both axis of a time function (either deterministic or random) $f(t)$ to produce a function $g(t)$ can be done as follows:

$$n^\alpha g(t) = f(nt) \Rightarrow g(t) = \frac{1}{n^\alpha} f(nt).$$

Such a re-scaling is equivalent to zooming out on a process n times from $f(t)$ to $g(t)$. In the discussion on the symmetric simple random walk limit, we know that at time t, the variance of the process is $(\Delta y)^2 (t/\Delta t)$. It is easy to see that at $\alpha = 0$ (i.e., no re-scaling for y-axis, $\Delta y = 1$ or $n_y = 1$), the variance of the process at any finite t goes to infinity as $\Delta t \to 0$; and at $\alpha = 1$ (i.e., the same re-scaling for both y-axis and t-axis or $\Delta y = \Delta t$), the variance at any time t goes to 0 as $\Delta t \to 0$. Thus, as α increases from 0 to 1, the variability of the process decreases from infinity to 0. Different α values correspond to different scalings. The two most useful scalings are $\alpha = 1$ and $\alpha = 1/2$ that yield the fluid limit and diffusion limit of the stochastic models (e.g., Brownian motion as shown above), which will be the topics in later chapters.

Example 6.7 *Find the mean and variance of a BM with drift by re-scaling a non-symmetric simple random walk. We use the notations above. Consider a simple random walk with*

$$X_k = \begin{cases} 1, & \text{with probability } p, \\ -1, & \text{with probability } 1-p. \end{cases}$$

Hence, $E[X_k] = 2p - 1$ and $Var(X_k) = 1 - (2p-1)^2$. Then the discrete-time process (DTMC) is

$$f(t) = S_t = X_1 + \cdots + X_t = \sum_{k=1}^{[t]} X_k.$$

Now we divide the time unit into n sub-intervals such that $n\Delta t = 1$. Then the corresponding sub-interval on y axis is $\Delta y = (\Delta t)^\alpha$. Consequently, $n_y = n^\alpha$. Denote the re-scaled process by $g(t)$. Then we have $n^\alpha g(t) = f(nt) = \sum_{k=1}^{[nt]} X_k$ which gives

$$g(t) = \frac{1}{n^\alpha} f(nt) = \frac{1}{n^\alpha} \sum_{k=1}^{[nt]} X_k = \Delta y (X_1 + \cdots + X_{[t/\Delta t]}) = (\Delta t)^\alpha (X_1 + \cdots + X_{[t/\Delta t]}).$$

Here, we choose $\alpha = 1/2$. *Thus, the mean and variance of* $g(t)$ *can be obtained as*

$$E[g(t)] = \Delta y[t/\Delta t]E[X_k] = \sqrt{\Delta t}[t/\Delta t](2p-1)$$

$$= \frac{1}{\sqrt{[\Delta t]}}(2p-1)t = [n_y(2p-1)]t,$$

$$Var(g(t)) = (\Delta y)^2[t/\Delta t]Var(X_k) = (\Delta t)[t/\Delta t][1-(2p-1)^2]$$

$$= [1-(2p-1)^2]t.$$

Noting that $n_y(2p-1) \to \mu$ *and* $\sigma^2 = 1-(2p-1)^2 \to 1$, *as* $\Delta t \to 0$, *we have*

$$E[g(t)] \to \mu t, \quad Var(g(t)) \to t.$$

6.2.3.3 The Reflection Principle – Hitting Times and Maximum of BM

Similar to a simple random walk, we define the first passage time to a certain state, called hitting time, and find its probability. Denote by

$$T_b = \inf\{t : B_t \geq b\} = \inf\{t : B_t = b\},$$

where the second equality holds due to the path continuity. We want to find the probability $P(T_b \leq t)$, the distribution function of T_b. Conditioning on B_t, we have

$$P(T_b \leq t) = P(T_b \leq t, B_t < b) + P(T_b \leq t, B_t > b)$$
$$= P(B_t < b|T_b \leq t)P(T_b \leq t) + P(B_t > b). \tag{6.57}$$

Here, we utilize $\{T_b \leq t, B_t > b\} = \{B_t > b\}$. Clearly, $P(B_t > b) = 1 - \Phi(b/\sqrt{t})$ due to the fact of $B_t \sim N(0,t)$. Next, we show that $P(B_t < b|T_b \leq t) = 1/2$. Given that $T_b \leq t$, the process is equally likely to continue to be above b or below b at time t as shown in Figure 6.9. This property is called the reflection principle. This implies that $P(B_t < b|T_b \leq t) = 1/2$. Using this conditional probability in (6.57) yields

$$P(T_b \leq t) = 2P(B_t > b) = 2\left[1 - \Phi\left(\frac{b}{\sqrt{t}}\right)\right].$$

Clearly, $P(T_b < \infty) = \lim_{t \to \infty} P(T_b \leq t) = 1$, i.e., T_b is finite with probability 1. However, $E[T_b] = \int_0^\infty P(T_b > t)dt = \infty$, i.e., its expected value is infinite (left as an exercise). Such a result is consistent with that of the discrete counterpart, the symmetric random walk.

Using this reflection principle, we can also obtain the joint distribution of BM and its maximum. Let $M_t = \sup_{0 \leq s \leq t} B_s$, the maximum level reached by

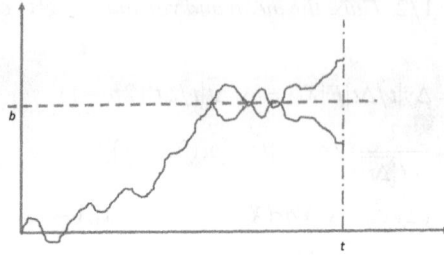

Figure 6.9 Mirroring the maximum.

BM in the time interval $(0,t]$. Note that $M_t \geq 0$ and is non-decreasing in t and $\{M_t \geq b\} = \{T_b \leq t\}$. Thus, $P(M_t \geq b) = P(T_b \leq t)$. Taking $T = T_b$ in the reflection principle, for $b \geq 0, b \geq x$, and $t \geq 0$, we have

$$
\begin{aligned}
P(M_t \geq b, B_t \leq x) &= P(T_b \leq t, B_t \leq x) \\
&= P(T_b \leq t, B_t \geq 2b - x) \\
&= P(B_t \geq 2b - x) \quad\quad\quad\quad (6.58) \\
&= 1 - \Phi\left(\frac{2b - x}{\sqrt{t}}\right) = F(b,x).
\end{aligned}
$$

An application of the hitting time distribution is to derive the Arc sine law for the standard BM. Denote by $Z_{(t_1,t_2)}$ the event that the standard BM takes on the value 0 at least once in (t_1,t_2). Conditioning on B_{t_1}, we have

$$
\begin{aligned}
P(Z_{(t_1,t_2)}) &= \frac{1}{\sqrt{2\pi t_1}} \int_{-\infty}^{\infty} P(Z_{(t_1,t_2)}|B_{t_1} = x) e^{-x^2/2t_1} dx \\
&= \frac{1}{\sqrt{2\pi t_1}} \int_{-\infty}^{\infty} P(T_{|x|} \leq t_2 - t_1) e^{-x^2/2t_1} dx \\
&= \frac{1}{\pi \sqrt{t_1(t_2 - t_1)}} \int_0^{\infty} \int_x^{\infty} e^{-y^2/2(t_2-t_1)} dy e^{-x^2/2t_1} dx \\
&= 1 - \frac{2}{\pi} \arcsin \sqrt{t_1/t_2},
\end{aligned}
$$

where the second equality follows from the symmetry of BM about the origin and its path continuity. Thus, we have the arcsine law for the BM as follows:

For $0 < x < 1$,

$$
P(Z^c_{(xt,t)}) = \frac{2}{\pi} \arcsin \sqrt{x}. \quad\quad\quad\quad (6.59)
$$

Next, we can find the probability that the first passage time to a certain state for a BM with drift, a more general case. Here we need to utilize a technique called "change of drift for a BM as change of probability measure" (see Chapter E). Consider a Brownian motion with drift, $X_t = B_t + \mu t$ for on a probability space $(\Omega, \mathscr{A}, \mathbb{P})$. Then, by the change of probability measure, X_t is a SBM with respect to the probability measure Q which is

$$dQ := dQ_T := \exp\left(-\mu B_T - \frac{1}{2}\mu^2 T\right) d\mathbb{P}.$$

In other words, X_t becomes a SBM on probability space $(\Omega, \mathscr{A}, Q)$. From (6.58), the joint distribution density can be expressed as

$$dQ = Q(X_t \in dx, M_t \in dy) = \frac{\partial^2 F(y,x)}{\partial x \partial y} dx dy$$

$$= \frac{2(2y-x)}{\sqrt{2\pi t^3}} \exp\left(-\frac{(2y-x)^2}{2t}\right) \mathbf{1}_{(-\infty,y]}(x) dx dy. \tag{6.60}$$

Now it follows from the Girsanov Theorem (Change of Measure) that

$$P(T_b \le t) = P(M_t \ge b) = \int \mathbf{1}_{[a,\infty)} M_t d\mathbb{P}$$

$$= \int \mathbf{1}_{[b,\infty)} M_t \exp\left(\mu B_t + \frac{1}{2}\mu^2 t\right) dQ$$

$$= \int \mathbf{1}_{[b,\infty)} M_t \exp\left(\mu X_t - \frac{1}{2}\mu^2 t\right) dQ.$$

Note that X_t is a SBM with respect to Q. Using (6.60), we have

$$P(T_b \le t) = \exp\left(-\frac{1}{2}\mu^2 t\right) \int_{y \ge b} \int_{x \le y} e^{\mu x} \frac{2(2y-x)}{\sqrt{2\pi t^3}} \exp\left(-\frac{(2y-x)^2}{2t}\right) dx dy.$$

Computing the integral expression by using Fubini's theorem as

$$P(T_b \le t) = \frac{1}{\sqrt{2\pi t}} \exp\left(-\frac{1}{2}\mu^2 t\right) \left(\int_{x \ge b} e^{\mu x} I_1(x) dx + \int_{x \le b} e^{\mu x} I_2(x) dx\right)$$

$$:= J_1 + J_2, \tag{6.61}$$

where

$$I_1(x) := \int_{y \ge x} \frac{2(2y-x)}{t} \exp\left(-\frac{(2y-x)^2}{2t}\right) dy$$

$$= \left[-\exp\left(-\frac{(2y-x)^2}{2t}\right)\right]_{y=x}^{\infty} = \exp\left(-\frac{x^2}{2t}\right),$$

$$I_2(x) := \int_{y \ge b} \frac{2(2y-x)}{t} \exp\left(-\frac{(2y-x)^2}{2t}\right) dy = \exp\left(-\frac{(2b-x)^2}{2t}\right).$$

Thus, we get

$$J_1 = \frac{1}{\sqrt{2\pi t}} \exp\left(-\frac{1}{2}\mu^2 t\right) \int_{x \geq b} e^{\mu x} I_1(x) dx = \frac{1}{\sqrt{2\pi t}} \int_{x \geq b} \exp\left(-\frac{(x - \mu t)^2}{2t}\right) dx$$

$$= \frac{1}{\sqrt{\pi}} \int_{z \geq \frac{b - \mu t}{\sqrt{2t}}} e^{-z^2} dz = \frac{1}{\sqrt{2\pi}} \int_{z \geq \frac{b - \mu t}{\sqrt{t}}} e^{-\frac{z^2}{2}} dz,$$

$$(6.62)$$

and

$$J_2 = \frac{1}{\sqrt{2\pi t}} \exp\left(-\frac{1}{2}\mu^2 t\right) \int_{x \leq b} e^{\mu x} I_2(x) dx$$

$$\overset{u := 2b - x}{=} \frac{1}{\sqrt{2\pi t}} \exp\left(-\frac{1}{2}\mu^2 t\right) \int_{u \geq b} e^{\mu(2b - u)} \exp\left(-\frac{u^2}{2t}\right) du \qquad (6.63)$$

$$= \cdots = \frac{e^{2b\mu}}{\sqrt{\pi}} \int_{z \geq \frac{b + \mu t}{\sqrt{2t}}} e^{-z^2} dz = \frac{e^{2b\mu}}{\sqrt{2\pi}} \int_{z \geq \frac{b + \mu t}{\sqrt{t}}} e^{-\frac{z^2}{2}} dz.$$

Summarizing the above, we obtain

$$P(T_b \leq t) = 1 - \Phi\left(\frac{b - \mu t}{\sqrt{t}}\right) + e^{2b\mu} \Phi\left(-\frac{b + \mu t}{\sqrt{t}}\right).$$

Next, we derive the limiting (stationary) distribution of the maximum of a BM with drift μ, denoted by M_∞^μ, by utilizing the "strong independent increments property of SMB. For the notational convenience, let $B_t = B(t)$ for the SMB. We first state this property without proof.

Proposition 6.5 (*Strong Independent Increments*) *Let T be a stopping time of $B(t)$. Suppose that both T and $B(t)$ are defined on the probability space $(\Omega, \mathscr{F}, P)$.*

(a) If T is finite almost surely, then

$$\{\tilde{B}(t), t \geq 0\} := \{B(T + t) - B(T), t \geq 0\} \qquad (6.64)$$

is a SBM process that is independent of events determined by B up to time T.

(b) If T has $P(T = \infty) > 0$, then on the trace probability space

$$(\Omega', \mathscr{F}', P') := (\Omega \cap [T < \infty], \{B \cap [T < \infty] : B \in \mathscr{F}\}, P\{\cdot | [T < \infty]\}),$$

then $\tilde{B}$ defined in (6.64) is a SBM independent of events in $\mathscr{F}'$ determined by B up to time T.

The proof of this proposition can be found, for example, in Resnick (1992). Now we use this strong independent increments property to derive a functional equation for the tail of the distribution of M_∞^μ. Again, denote the BM with a drift by $X_t = X(t) = B(t) + \mu t$ for notational convenience. This time we do not use the change of measures. Instead, we define the hitting time for $X(t)$ to state x as

$$\tau_x = \inf\{t > 0 : X(t) = x\} = \inf\{t > 0 : B(t) + \mu t = x\},$$

which is a stopping time with respect to the SBM B. We continue using the notation for the SBM hitting time $T_x = \inf\{s > 0 : B(s) = x\}$. To develop an equation that M_∞^μ satisfies, we consider $x_1 > 0$ and $x_2 > 0$ and note that

$$P(M_\infty^\mu \geq x_1 + x_2) = P(X(t) = x_1 + x_2, \text{ for some } t > 0) = *,$$

where $*$ is the intermediate result. Since $X(t)$ cannot hit $x_1 + x_2$ without first hitting x_1, we have

$$* = P(B(t) + \mu t = x_1 + x_2, \text{ for some } t > 0, \tau_{x_1} < \infty)$$
$$= P(B(s + \tau_{x_1}) + \mu(s + \tau_{x_1}) = x_1 + x_2, \text{ for some } s > 0, \tau_{x_1} < \infty) = *.$$

Using $B(\tau_{x_1}) + \mu \tau_{x_1} = x_1$ yields

$$* = P(B(s + \tau_{x_1}) - B(\tau_{x_1}) + \mu s = x_2, \text{ for some } s > 0, \tau_{x_1} < \infty) = *.$$

It follows from Proposition 6.5 that the post-τ_{x_1} process is a SBM independent of the process up to τ_{x_1}. Thus, we get

$$* = P'(B(s + \tau_{x_1}) - B(\tau_{x_1}) + \mu s = x_2, \text{ for some } s > 0)P(\tau_{x_1} < \infty)$$
$$= P(M_\infty^\mu \geq x_2)P(M_\infty^\mu \geq x_1) = *.$$

This means that $P(M_\infty^\mu \geq x)$ satisfies the functional equation

$$P(M_\infty^\mu \geq x_1 + x_2) = P(M_\infty^\mu \geq x_1)P(M_\infty^\mu \geq x_2).$$

Note that the only solutions to this functional equation are the exponential functions. Thus, for some $c = c(\mu)$, we must have

$$P(M_\infty^\mu \geq x) = e^{-cx}, \quad x > 0. \tag{6.65}$$

We determine c by writing

$$P(M_\infty^\mu \geq x) = P(B(s) + \mu s = x, \text{ for some } s > 0) = *.$$

Since $X(t)$ crosses x after $B(t)$ due to the negative drift, we have

$$* = P(B(s + T_x) + \mu(s + T_x) = x \text{ for some } s > 0) = *.$$

Using $B(T_x) = x$ gives

$$* = P(B(s + T_x) - B(T_x) + \mu(s + T_x) = 0 \text{ for some } s > 0) = *.$$

Note that $\{\tilde{B}(s) = B(s+T_x) - B(T_x), s \geq 0\}$ is a SBM independent of the path of B up to time T_x and hence $\tilde{B}$ is independent of T_x. Conditioning on T_x, we have

$$* = \int_0^\infty P(B(s) + \mu(s+y) = 0 \text{ for some } s > 0) P(T_x \in dy)$$

$$= \int_0^\infty P(B(s) + \mu s = |\mu|y \text{ for some } s > 0) P(T_x \in dy)$$

$$= \int_0^\infty P(M_\infty^\mu \geq |\mu|y) P(T_x \in dy)$$

$$= \int_0^\infty e^{-c|\mu|y} P(T_x \in dy)$$

$$= E[e^{-c|\mu|T_x}],$$

where the forth equality follows from (6.65). Note that the last expression is the Laplace transform of T_x at $c|\mu|$ which is $e^{-\sqrt{2c|\mu|}x}$ (left as an exercise). It follows from $e^{-\sqrt{2c|\mu|}x} = e^{-cx}$ for $x > 0$ that $c = 2|\mu|$. For a more general case of $X(t) = \sigma B(t) + \mu t$, it can be shown that $c = 2|\mu|/\sigma^2$ (left as an exercise). We summarize these results.

Theorem 6.10 *For a BM $X(t) = \sigma B(t) + \mu t$ with parameters $\mu < 0$ and σ, the limiting tail probability of the maximum of $X(t)$, denoted by $M_\infty^{(\mu,\sigma)}$ is given by*

$$P(M_\infty^{(\mu,\sigma)} > x) = e^{-(2|\mu|/\sigma^2)x}, \quad x \geq 0. \tag{6.66}$$

Exercise 6.26 *Recall that the distribution function of T_b (the hitting time to b of a SBM) is given by $P(T_b \leq t) = 2[1 - \Phi(b/\sqrt{t})]$. Prove its Laplace tranform is*

$$E[e^{-sT_b}] = e^{-\sqrt{2s}b}, \quad s > 0.$$

Exercise 6.27 *Prove Theorem 6.10.*

Exercise 6.28 *For a BM $\{X_t, t \geq 0\}$ with drift coefficient μ and variance parameter σ^2, show that*

$$P(M_t \geq y | X(t) = x) = e^{-2y(y-x)/t\sigma^2}, \quad y \geq x,$$

where $M_t = \sup_{0 \leq s < t} X_s$.

6.2.3.4 Conditional Distribution of BM

For two time points $0 < s < t$, we determine the conditional probability of B_t given B_s or B_s given B_t. Denote $Y = B(t)$ given $X = B(s)$. Then the conditional density of $Y|X$ is given by

$$f_{Y|X}(y|x) = \frac{1}{\sqrt{2\pi(t-s)}} \exp\left(-\frac{(y-x)^2}{2(t-s)}\right).$$

This indicates that the conditional $Y|X = B_t|B_s$ follows a normal distribution with mean of x and variance of $t-s$. Next, we compute the conditional distribution of $X|Y = B_s|B_t$. First, we write the joint density of (X,Y) as

$$f_{X,Y}(x,y) = \frac{1}{2\pi\sqrt{s(t-s)}} \exp\left(-\frac{x^2}{2s} - \frac{(y-x)^2}{2(t-s)}\right),$$

and the marginal density of Y as

$$f_Y(y) = \frac{1}{\sqrt{2\pi t}} \exp\left(-\frac{y^2}{2t}\right).$$

The conditional density of X given $Y = y$ is

$$
\begin{aligned}
f_{X|Y}(x|y) &= \frac{\frac{1}{2\pi\sqrt{s(t-s)}}\exp\left(-\frac{x^2}{2s} - \frac{(y-x)^2}{2(t-s)}\right)}{\frac{1}{\sqrt{2\pi t}}\exp\left(-\frac{y^2}{2t}\right)} \\
&= \frac{1}{\sqrt{2\pi\frac{s(t-s)}{t}}} \exp\left(\frac{y^2}{2t} - \frac{x^2}{2s} - \frac{(y-x)^2}{2(t-s)}\right) \\
&= \frac{1}{\sqrt{2\pi\frac{s(t-s)}{t}}} \exp\left(\frac{y^2}{2t} - \frac{x^2}{2s} - \frac{y^2}{2(t-s)} + \frac{xy}{t-s} - \frac{x^2}{2(t-s)}\right) \\
&= \frac{1}{\sqrt{2\pi\frac{s(t-s)}{t}}} \exp\left(-\frac{tx^2}{2s(t-s)} - \frac{sy^2}{2t(t-s)} + \frac{xy}{t-s}\right) \\
&= \frac{1}{\sqrt{2\pi\frac{s(t-s)}{t}}} \exp\left(-\frac{\left(x - \frac{sy}{t}\right)^2}{2\frac{s(t-s)}{t}}\right),
\end{aligned}
$$

which indicates that $X|Y = B_s|B_t$ follows a normal distribution with mean of $\frac{sy}{t}$ and variance of $\frac{s(t-s)}{t}$. It is worth noting that the important results for the simple random walk can be established for the BM due to the close relationship between these two models. For example, we can derive the Arcsine law for the BM, shown earlier by using the hitting time distribution, in an alternative way as follows. For a time interval $[t_0, t_1]$, conditioning on $B_{t_0} = u > 0$, we have

$$P(B_t = 0 \text{ for some } t_0 < t < t_1 | B_{t_0} = u)$$
$$= P(B_t > u \text{ for some } t_0 < t < t_1 | B_{t_0} = 0)$$

$$= P(\sup_{t_0 < t < t_1} B_t > u | B_{t_0} = 0) = P(\sup_{0 < t < t_1 - t_0} B_t > u)$$

$$= P(|B_{t_1 - t_0}| > u) = 2 \int_u^\infty \frac{1}{\sqrt{2\pi(t_1 - t_0)}} \exp\left(-\frac{x^2}{2(t_1 - t_0)}\right) dx.$$

For $B_{t_0} = u < 0$, we can get similar expression. Thus,

$$P(B_t = 0 \text{ for some } t_0 < t < t_1 | B_{t_0} = u)$$

$$= 2 \int_{|u|}^\infty \frac{1}{\sqrt{2\pi(t_1 - t_0)}} \exp\left(-\frac{z^2}{2(t_1 - t_0)}\right) dz.$$

Un-conditioning the above expression, we obtain

$$P(B_t = 0 \text{ for some } t_0 < t < t_1)$$

$$= 2 \int_0^\infty P(B_t = 0 \text{ for some } t_0 < t < t_1 | B_{t_0} = u) \frac{1}{\sqrt{2\pi t_0}} e^{-\frac{u^2}{2t_0}} du$$

$$= 4 \int_0^\infty \int_u^\infty \frac{1}{\sqrt{2\pi(t_1 - t_0)}} \exp\left(-\frac{z^2}{2(t_1 - t_0)} - \frac{u^2}{2t_0}\right) dz du$$

$$= 4 \int_0^\infty \int_{\sqrt{\frac{t_0}{t_1 - t_0}} y}^\infty \frac{1}{2\pi} \exp\left(-\frac{x^2 + y^2}{2}\right) dx dy = \frac{2}{\pi} \arccos\left(\sqrt{\frac{t_0}{t_1}}\right)$$

$$= 1 - \frac{2}{\pi} \arcsin\left(\sqrt{\frac{t_0}{t_1}}\right),$$

where the third equality is due to the change of variables $x = z/\sqrt{t_1 - t_0}$ and $y = u/\sqrt{t_0}$.

Exercise 6.29 *In a 800 meter race, two frontrunners compete in the final lap. Let $X(t)$ denote the amount of time in seconds by which the runner started in the inside position is ahead when $100t$ percent of the race has been completed. Assume that $\{X(t), 0 \le t \le 1\}$ can be effectively modeled as a BM with variance parameter σ^2.*

(a) If the insider runner is leading by $\sigma/2$ seconds at the midpoint of the race, what is the probability that he is the winner?

(b) If the insider runner wins the race by $\sigma/4$ seconds, what is the probability that he was behind his competitor at the midpoint?

6.2.3.5 BM as a Martingale

To make the notations easier in this subsection, we use the equivalent expressions $B_t = B(t)$. It follows from the independent increments property for $t > s$ that $B(s + t) = B(s) + (B(s + t) - B(s))$, in which $B(s)$ and $(B(s + t) - B(s))$ are independent. Furthermore, $B(s + t) = B(s) + (B(s + t) - B(s))$ is also independent of the past history before s, $\{B(u) : 0 \le u < s\}$. Hence, we conclude that the future $B(s+t)$, given the present state, $B(s)$, only depends on the random variable, $(B(s+t)-B(s))$, that is independent of the past. This indicates that BM is a Markov process. Since the increments are also stationary, a BM is a time-homogeneous Markov process. We can show that Standard BM satisfies the definition of a Martingale as follows:

$$
\begin{aligned}
E(B(t+s)|B(u):0\leq u\leq s) &= E(B(t+s)|B(s))\\
&= E(B(s)+(B(s+t)-B(s))|B(s))\\
&= B(s)+E(B(s+t)-B(s)|B(s))\\
&= B(s)+E(B(s+t)-B(s))\\
&= B(s)+0\\
&= B(s).
\end{aligned}
$$

Here, the first equality is due to the Markov property and the forth equality is due to the independent increments. A martingale captures the notion of a fair game, in that regardless of one's current and past fortunes, his or her expected fortune at any time in the future is the same as the current fortune: on average, he or she neither wins nor loses any money. For the details about martingale, readers are referred to the Chapter B. Another important martingale is the exponential martingale that is motivated by modeling the stock price via an exponential function of a BM. Using a Binomial model, we can justify the exponential function of the BM for modeling the stock price process. Consider a time period of 1 which is divided into n periods. We start with an asset of S_0 whose value fluctuates with time. This variability is measured by σ called volatility. Assume that the interest rate is zero. A simple model, as illustrated in Figure 6.10, is to have a change either "going up" or "going down" for each period with the following ratios, respectively:

$$
u_n = 1+\frac{\sigma}{\sqrt{n}},\quad d_n = 1-\frac{\sigma}{\sqrt{n}}.
$$

For time period t, we have an nt-period model. The risk-neutral probabilities, denoted by p for going up and q for going down, for one-period satisfy

$$
pu_n + qd_n = 1
$$
$$
p+q = 1,
$$

which yields

$$
p = \frac{1-d_n}{u_n-d_n} = \frac{1-\left(1-\frac{\sigma}{\sqrt{n}}\right)}{\left(1+\frac{\sigma}{\sqrt{n}}\right)-\left(1-\frac{\sigma}{\sqrt{n}}\right)} = \frac{1}{2}
$$

$$
q = \frac{u_n-1}{u_n-d_n} = \frac{\left(1+\frac{\sigma}{\sqrt{n}}\right)-1}{\left(1+\frac{\sigma}{\sqrt{n}}\right)-\left(1-\frac{\sigma}{\sqrt{n}}\right)} = \frac{1}{2}.
$$

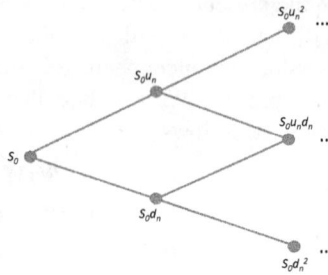

Figure 6.10 A simple binomial model.

Thus, we can treat the nt-period model as tossing a fair coin nt times. Let h be the number of steps of price moving up and l be the number of steps of price moving down. Denote by m_{nt} the position of the random walk. We have $h + l = nt$ and $h - l = m_{nt}$ which give $h = (nt + m_{nt})/2$ and $l = (nt - m_{nt})/2$. This implies that the stock price at time t is

$$S_t^{(n)} = S_0 u_n^h d_n^l = S_0 \left(1 + \frac{\sigma}{\sqrt{n}} \right)^{\frac{nt + m_{nt}}{2}} \left(1 - \frac{\sigma}{\sqrt{n}} \right)^{\frac{nt - m_{nt}}{2}}.$$

Theorem 6.11 *As $n \to \infty$, the distribution of $S_t^{(n)}$ converges to the distribution of*

$$S_t = S_0 \exp \left(\sigma B(t) - \frac{\sigma^2 t}{2} \right),$$

where $B(t) \sim N(0, t)$.

Proof. We show that the distribution of $\log S_t^{(n)}$ converges to the distribution of $\log S_t$. Note that

$$\log S_t^{(n)} = \log S_0 + \frac{nt + m_{nt}}{2} \log \left(1 + \frac{\sigma}{\sqrt{n}} \right) + \frac{nt - m_{nt}}{2} \log \left(1 - \frac{\sigma}{\sqrt{n}} \right)$$

$$= \log S_0 + \frac{nt + m_{nt}}{2} \left(\frac{\sigma}{\sqrt{n}} - \frac{\sigma^2}{2n} + O(n^{-\frac{3}{2}}) \right)$$

$$+ \frac{nt - m_{nt}}{2} \left(-\frac{\sigma}{\sqrt{n}} - \frac{\sigma^2}{2n} + O(n^{-\frac{3}{2}}) \right)$$

$$= \log S_0 + nt \left(-\frac{\sigma^2}{2n} + O(n^{-\frac{3}{2}}) \right) + m_{nt} \left(\frac{\sigma}{\sqrt{n}} + O(n^{-\frac{3}{2}}) \right)$$

$$= \log S_0 - \frac{\sigma^2 t}{2} + O(n^{-\frac{1}{2}}) + \sigma B_t^{(n)} + O(n^{-1}) B_t^{(n)},$$

where the last equality follows from $m_{nt}/\sqrt{n} = B_t^{(n)}$, which converges in distribution as $n \to \infty$ to

$$\log S_0 + \sigma B(t) - \frac{\sigma^2 t}{2} = \log S_t.$$

This completes the proof. ∎

The exponential martingale with a parameter $\sigma \in \mathcal{R}$ is defined by

$$Z_t = \exp\left(\sigma B(t) - \frac{\sigma^2}{2} t\right),$$

where $B(t)$ is a standard BM with filtration $(\mathcal{F}_t)$. The martingale property can be easily demonstrated as follows: for $0 \le s \le t$,

$$E[Z_t|\mathcal{F}_s] = E\left[\exp\left(\sigma B(t) - \frac{\sigma^2 t}{2}\right)|\mathcal{F}_s\right]$$

$$= E[\exp(\sigma B(t))|\mathcal{F}_s] \cdot \exp\left(-\frac{\sigma^2 t}{2}\right)$$

$$= E[\exp(\sigma(B(t) - B(s)) + \sigma B(s))|\mathcal{F}_s] \cdot \exp\left(-\frac{\sigma^2 t}{2}\right)$$

$$= E[\exp(\sigma(B(t) - B(s)))|\mathcal{F}_s] \cdot \exp(\sigma B(s)) \cdot \exp\left(-\frac{\sigma^2 t}{2}\right).$$

Using the fact

$$E[\exp(\sigma(B(t) - B(s)))|\mathcal{F}_s] = E[\exp(\sigma(B(t) - B(s)))] = \exp\left(\frac{1}{2}\sigma^2(t - s)\right),$$

we obtain

$$E[Z_t|\mathcal{F}_s] = \exp\left(\frac{1}{2}\sigma^2(t - s)\right) \cdot \exp(\sigma B(s)) \cdot \exp\left(-\frac{\sigma^2 t}{2}\right)$$

$$= \exp\left(\sigma B(s) - \frac{1}{2}\sigma^2 s\right) = Z_s.$$

For a non-standard BM, $X(t) = \mu t + \sigma B(t)$, we can utilize the result above to get the exponential martingale corresponding to $X(t)$ as follows:

$$\exp\left(X(t) - \left(\mu + \frac{1}{2}\sigma^2\right)t\right),$$

by noting that $(X(t) - \mu t)/\sigma$ is a standard BM. Furthermore, for any parameter θ, the more general exponential martingale for $\theta X(t)$ (the scaled BM) is

$$\exp\left(\theta X(t) - \left(\theta \mu + \frac{1}{2}\theta^2 \sigma^2\right)t\right).$$

Note that here we utilize the moment generating function (m.g.f) of BM at a certain time point, which is a normal random variable $X \sim N(\mu, \sigma^2)$. To obtain this m.g.f., we start with computing the m.g.f. of standard normal random variable, denoted by Z,

$$E(e^{sZ}) = \frac{1}{\sqrt{2\pi}} \int_{-\infty}^{\infty} e^{sx} e^{-\frac{x^2}{2}} dx = \frac{1}{\sqrt{2\pi}} \int_{-\infty}^{\infty} e^{-\frac{x^2-2sx}{2}} dx$$

$$= e^{s^2/2} \frac{1}{\sqrt{2\pi}} \int_{-\infty}^{\infty} e^{-\frac{(x-s)^2}{2}} dx = e^{s^2/2}.$$

For a non-standard normal random variable $X = \sigma Z + \mu$, we have

$$E(e^{sX}) = e^{s\mu} E(e^{\sigma sZ}) = e^{s\mu} e^{(\sigma s)^2/2} = e^{s\mu + s^2\sigma^2/2}.$$

A powerful result, called the optional stopping theorem (also called optional sampling theorem), says that under regularity conditions, when Y is a martingale with respect to Z and T is a stopping time relative to Z, that $E[Y(T)] = E[Y(0)]$. For more details about optional stopping theorem, we refer to Chapter B.

Exercise 6.30 *Show that the $\{B(t)^2 - t, t \geq 0\}$ is a martingale with respect to BM. Such a process is called the quadratic martingale.*

Exercise 6.31 *(Gambler's ruin problem) Let $X(t) = \sigma B(t) + \mu t$ for $t > 0$, a Brownian motion with drift μ and variance parameter σ^2. Suppose that we start with an initial amount of money w and gamble according to the $X(t)$. Thus, our wealth at time t is given by*

$$W(t) = w + X(t) = w + \sigma B_t + \mu t, \quad t \geq 0.$$

We are interested in the probability that we win x (reach wealth $w + x$ before we lose our initial wealth w (reach wealth 0)). Denote by T the first time that we either win x or lose w, that is

$$T := \inf\{t \geq 0 : W(t) = 0 \text{ or } w + x\} = \inf\{t \geq 0 : X(t) = -w \text{ or } +x\}.$$

The probability of reaching wealth $w + x$ before being ruined is

$$p := P(W(T) = w + x) = P(X(T) = x).$$

Use the optional stopping theorem to show the followings:
(a) For the case without drift (i.e., $\mu = 0$), use the linear martingale (i.e., $B(t)$ itself) and quadratic martingale to show

$$p = \frac{w}{x+w}, \quad E[T] = \frac{xw}{\sigma^2}.$$

(b) For the case with drift (i.e., $\mu \neq 0$), use the exponential martingale and linear martingale to show

$$p = \frac{e^{+2\mu w/\sigma^2} - 1}{e^{+2\mu w/\sigma^2} - e^{-2\mu w/\sigma^2}} \quad for \ \mu > 0,$$

$$p = \frac{1 - e^{+2\mu w/\sigma^2}}{e^{-2\mu w/\sigma^2} - e^{+2\mu w/\sigma^2}} \quad for \ \mu < 0,$$

$$E[T] = \frac{px - (1-p)w}{\mu}.$$

6.2.4 TRANSITION PROBABILITY FUNCTION OF BM

Since BM is a CTMP, its dynamics can be studied in terms of transition probability functions in a similar way used in a CTMC or DTMC. In this subsection, we present some basics. Define the transition probability function for the BM as

$$F(y,s|x,t) = P(B(s) \leq y|B(t) = x), \quad t < s, \tag{6.67}$$

where $x, y \in \mathcal{R}$. If follows from Definition 6.4 that

$$F(y,s|x,t) = P(B(s) - B(t) \leq y - x) = \Phi\left(\frac{y-x}{\sqrt{s-t}}\right), \quad t < s. \tag{6.68}$$

Since a standard BM is homogeneous in time as well as in space, we have

$$F(y,s|x,t) = F(y-x,s-t|0,0), \quad t < s$$

with the density represented by $P(y,s|x,t) = dF(y,s|x,t)/dy$. Thus, the transition probability function for a BM depends only on the differences $s - t$ and $y - x$. With understanding that $B(0) = 0$, we can denote the transition probability function by

$$F(y,s) = F(y,s|0,0) = \Phi\left(\frac{y}{\sqrt{s}}\right), \quad s > 0, \ y \in \mathcal{R} \tag{6.69}$$

with the density denoted by $P(y,s)$. Using the Markov property of the BM, we can write the Chapman–Kolmogorov (C-K) equation as follows:

$$F(x,s+t) = \int_{-\infty}^{\infty} P(y,s)F(x-y,t)dy, \tag{6.70}$$

which is satisfied by $F(y,s)$ in (6.69). Next, we develop the partial differential equation (PDE) that the transition probability density satisfies. Instead of presenting this equation rigorously (which can be found in many references such as Schilling (2021)), we establish such a PDE by using intuitive arguments.

Another advantage of using this approach is to demonstrate the connection between the BM and the random walk. A more general approach is presented in the Chapter H. First consider a standard BM. Using the symmetric simple random walk approximation, we have

$$P(B(t+\Delta t) = B(t) + \Delta x) = P(B(t+\Delta t) = B(t) - \Delta x) \approx \frac{1}{2},$$

where $\Delta x = \sqrt{\Delta t}$. This approximation implies

$$P(x, t+\Delta t) \approx \frac{1}{2} P(x - \Delta x, t) + \frac{1}{2} P(x + \Delta x, t). \tag{6.71}$$

This relation indicates that a forward C-K equation is to be developed (i.e., the time increment Δt is the last step to the destination state time). Now consider

$$\frac{P(x, t+\Delta t) - P(x, t)}{\Delta t} \approx \frac{P(x + \Delta x, t) + P(x - \Delta x, t) - 2P(x, t)}{2(\Delta x)^2}$$

$$= \frac{1}{2\Delta x} \left[\frac{P(x + \Delta x, t) - P(x, t)}{\Delta x} - \frac{P(x, t) - P(x - \Delta x, t)}{\Delta x} \right], \tag{6.72}$$

where the approximation follows from (6.71). Taking the limit of $\Delta t \to 0$ ($\Delta x \to 0$) on both sides of (6.72) yields

$$\frac{\partial}{\partial t} P(x, t) = \lim_{\Delta t \to 0} \frac{P(x, t+\Delta t) - P(x, t)}{\Delta t}$$

$$= \lim_{\Delta x \to 0} \frac{1}{2\Delta x} \left[\frac{P(x + \Delta x, t) - P(x, t)}{\Delta x} - \frac{P(x, t) - P(x - \Delta x, t)}{\Delta x} \right]$$

$$= \frac{1}{2} \lim_{\Delta x \to 0} \frac{\frac{\partial}{\partial x} P(x, t) - \frac{\partial}{\partial x} P(x - \Delta x, t)}{\Delta x}$$

$$= \frac{1}{2} \frac{\partial^2}{\partial x^2} P(x, t), \tag{6.73}$$

which is also called the heat equation in physics. Second, for the BM with variance, also called diffusion coefficient, σ^2, we can use the same argument by noting that

$$P(B(t+\Delta t) = B(t) + \sigma\Delta x) = P(B(t+\Delta t) = B(t) - \sigma\Delta x) \approx \frac{1}{2},$$

and obtain

$$\frac{\partial}{\partial t} P(x, t) = \frac{\sigma^2}{2} \frac{\partial^2}{\partial x^2} P(x, t). \tag{6.74}$$

Third, we consider the BM with a linear drift μ. It follows from $P(x,t+\Delta t) = P(x-\mu\Delta t,t)$ that

$$\frac{P(x,t+\Delta t)-P(x,t)}{\Delta t} = \frac{P(x-\mu\Delta t,t)-P(x,t)}{\Delta t}$$
$$= -\mu\frac{P(x,t)-P(x-\mu\Delta t,t)}{\mu\Delta t}. \qquad (6.75)$$

Taking the limit of $\Delta t \to 0$ on both sides of the equation above, we have

$$\frac{\partial}{\partial t}P(x,t) = -\mu\lim_{\Delta t\to 0}\frac{P(x,t)-P(x-\mu\Delta t,t)}{\mu\Delta t} = -\mu\frac{\partial}{\partial x}P(x,t). \qquad (6.76)$$

Combining (6.74) and (6.76) yields

$$\frac{\partial}{\partial t}P(x,t) = -\mu\frac{\partial}{\partial x}P(x,t) + \frac{\sigma^2}{2}\frac{\partial^2}{\partial x^2}P(x,t). \qquad (6.77)$$

This equation is called forward C-K equation for a Brownian Motion. It is easy to check that transition probability function for the BM

$$P(y,t|x,0) = \frac{1}{\sigma\sqrt{2\pi t}}\exp\left\{-\frac{(y-x-\mu t)^2}{2t\sigma^2}\right\}, \quad t>0 \qquad (6.78)$$

satisfies (6.77). Without loss of generality, we can set $x = 0$ (i.e., $B(0) = 0$) and write the transition probability density as $P(y,t)$. BM can be considered as a special case of diffusion processes (see Chapter H). For a diffusion process $\{X(t)\}$, we define the infinitesimal mean and infinitesimal variance of the increment $\Delta X(t) = X(t+\Delta t) - X(t)$ conditional on $X(t) = x$ as follows:

$$\mu(x,t) = \lim_{\Delta t\to 0}\frac{1}{\Delta t}E[\Delta X(t)|X(t) = x]$$
$$\sigma^2(x,t) = \lim_{\Delta t\to 0}\frac{1}{\Delta t}E[\{\Delta X(t)\}^2|X(t) = x]. \qquad (6.79)$$

For the BM, we have $\mu(x,t) = \mu, \sigma^2(x,t) = \sigma^2$.

Similar to the CTMC, we can also establish the backward C-K equation for the BM. Again, based on the random walk assumption, we can apply Taylor's expansion for $P(y,t+\Delta t|x,0)$ that is assumed to be sufficiently smooth. Recall that a BM with drift μ and diffusion coefficient σ can be approximated by a random walk $S_t = \sum_{n=0}^t X_n$ with X_n taking values of either $\pm\Delta x$ and time step of Δt, where

$$\sigma^2\Delta t = (\Delta x)^2,$$

since the random walk $\{S_n\}$ converges to the BM as $\Delta t \to 0$. Furthermore, we can specify the $p = P(X_n = \Delta x)$ and $q = P(X_n = -\Delta x) = 1-p$. Based on

$E[X_n] = p\Delta x + q(-\Delta x) = \mu\Delta t$ and $\Delta t = (\Delta x)^2/\sigma^2$, we have

$$p - q = \frac{\mu\Delta x}{\sigma^2}$$
$$p + q = 1,$$

which can be solved to generate

$$p = \frac{1}{2} + \frac{\mu}{2\sigma^2}\Delta x,$$
$$q = \frac{1}{2} - \frac{\mu}{2\sigma^2}\Delta x. \tag{6.80}$$

From the backward C-K equation for the random walk, we have

$$P(y, t - \Delta t | x, 0) = \left(\frac{1}{2} + \frac{\mu}{2\sigma^2}\Delta x\right) P(y, t | x + \Delta x, 0)$$
$$+ \left(\frac{1}{2} - \frac{\mu}{2\sigma^2}\Delta x\right) P(y, t | x - \Delta x, 0). \tag{6.81}$$

Using Taylor's expansion, we obtain

$$P(y, t - \Delta t | x, 0) = P(y, t | x, 0) - \Delta t \frac{\partial}{\partial t} P(y, t | x, 0) + \cdots$$

$$P(y, t | x \pm \Delta x, 0) = P(y, t | x, 0) \pm \Delta x \frac{\partial}{\partial y} P(y, t | x, 0) + \frac{(\Delta x)^2}{2} \frac{\partial^2}{\partial y^2} P(y, t | x, 0) + \cdots. \tag{6.82}$$

Without loss of generality, setting $x = 0$ and substituting (6.82) into (6.81) yields

$$-\Delta t \frac{\partial}{\partial t} P(y, t) = \frac{\mu}{\sigma^2} (\Delta x)^2 \frac{\partial}{\partial y} P(y, t) + \frac{(\Delta x)^2}{2} \frac{\partial^2}{\partial y^2} P(y, t) + o(\Delta t). \tag{6.83}$$

Substituting $(\Delta x)^2 = \sigma^2 \Delta t$ into (6.83), dividing the resulting equation by Δt, and then taking the limit $\Delta t \to 0$, we obtain

$$-\frac{\partial}{\partial t} P(y, t) = \mu \frac{\partial}{\partial y} P(y, t) + \frac{\sigma^2}{2} \frac{\partial^2}{\partial y^2} P(y, t), \tag{6.84}$$

which is the backward C-K equation for the BM.

Exercise 6.32 *Derive the probability density function for the BM with drift coefficient μ and variance coefficient σ^2 by solving the forward C-K equation. (Hint: One option is to use the Fourier transform to solve the partial differential equation.)*

Exercise 6.33 *Verify that the solution from the previous exercise (the p.d.f. of the BM) satisfies the backward C-K equation.*

6.2.5 THE BLACK-SCHOLES FORMULA

In this subsection, we extend the discrete-time option pricing model (Binomial pricing formula) to the continuous-time model. Note that for pricing a derivative security, only the distribution under the risk-neutral probability measure is relevant. We consider an European call option written on a stock with strike price K at maturity T. The stock price process, $S(t)$, follows a geometric BM given by

$$S(t) = S(0)e^{X(t)}, \quad t \geq 0,$$

where $X(t) = \ln[S(t)/S(0)]$ is a BM with drift μ and diffusion coefficient σ. That is

$$\ln \frac{S(t)}{S} = \mu t + \sigma B(t), \quad 0 \leq t \leq T. \tag{6.85}$$

where the right-hand side can be approximated by a random walk $X_t = \sum_{i=1}^{t} Y_i, t = 1, 2, ..., n$. if the period $[0, T]$ has been divided into n equally spaced sub-intervals with width $h = T/n$. Here Y_i's are i.i.d. Bernoulli random variables. That is

$$Y_i = \begin{cases} x, & \text{with probability } p, \\ y, & \text{with probability } 1 - p = q. \end{cases}$$

To ensure the Binomial model to converge to the Black–Scholes model, we need to compute the required values of x and y by solving

$$E[X_n] = E\left[\sum_{i=1}^{n} Y_i\right] = n(px + qy) = \mu T,$$

$$Var(X_n) = Var\left(\sum_{i=1}^{n} Y_i\right) = npq(x - y)^2 = \sigma^2 T,$$

for x and y. The results are

$$x = \frac{\mu T}{n} + \sigma \sqrt{\frac{1-p}{p}} \sqrt{\frac{T}{n}},$$

$$y = \frac{\mu T}{n} - \sigma \sqrt{\frac{1-p}{p}} \sqrt{\frac{T}{n}}. \tag{6.86}$$

Now recall the Binomial model defined in (6.42) for the stock price, which is the discrete-time version of geometric BM. Clearly, the relation between (u, d) and (x, y) is $x = \ln u$ and $y = \ln d$ (or $u = e^x$ and $d = e^y$). Also it follows from $R^n = e^{rT}$ that $R = e^{rT/n}$. Denote by p_n^* the risk-neutral upward probability for the Binomial model of (6.42) with n periods. It follows from $p_n^* u + (1 - p_n^*)d =$

R that

$$p_n^* = \frac{R-d}{u-d} = \frac{e^{rT/n} - \exp\left\{\frac{\mu T}{n} - \sigma\sqrt{\frac{1-p}{p}}\sqrt{\frac{T}{n}}\right\}}{\exp\left\{\frac{\mu T}{n} + \sigma\sqrt{\frac{1-p}{p}}\sqrt{\frac{T}{n}}\right\} - \exp\left\{\frac{\mu T}{n} - \sigma\sqrt{\frac{p}{1-p}}\sqrt{\frac{T}{n}}\right\}}$$

$$= \frac{\exp\left\{(r-\mu)\frac{T}{n}\right\} - \exp\left\{-\sigma\sqrt{\frac{p}{1-p}\frac{T}{n}}\right\}}{\exp\left\{\sigma\sqrt{\frac{1-p}{p}\frac{T}{n}}\right\} - \exp\left\{-\sigma\sqrt{\frac{p}{1-p}\frac{T}{n}}\right\}}.$$

$$(6.87)$$

It is easy to check that $\lim_{n\to\infty} p_n^* = p$ (left as an exercise for readers). Similar to the discrete-time Binomial model, the call option premium is determined through risk-neutral probability p_n^*. We can denote the risk-neutral probability measure by $\mathbb{Q}$ which is

$$\mathbb{Q}(Y_i = x) = 1 - \mathbb{Q}(Y_i = y) = p_n^*, \quad i = 1, 2, ..., n.$$

Under $\mathbb{Q}$, we can obtain the following limits. Denoting the mean by $\mu_n = E^{\mathbb{Q}}[Y_1]$ and the variance by $\sigma_n^2 = Var(Y_1)$, we have

$$\mu_n = p_n^* x + q_n^* y, \quad \sigma_n^2 = p_n^* q_n^* (x-y)^2. \tag{6.88}$$

Furthermore, we can prove

$$\lim_{n\to\infty} E^{\mathbb{Q}}[X_n] = \lim_{n\to\infty} n\mu_n = \left(r - \frac{\sigma^2}{2}\right)T,$$

$$\lim_{n\to\infty} Var^{\mathbb{Q}}[X_n] = \lim_{n\to\infty} n\sigma_n^2 = \sigma^2 T. \tag{6.89}$$

The proof can be done based on $\exp(\pm x) = 1 \pm x + x^2/2 + o(x^2)$. That is

$$e^{r/n} = 1 + r/n + o(1/n)$$

$$e^{\pm\sigma/\sqrt{n}} = 1 + \sigma/\sqrt{n} + \sigma^2/(2n) + o(1/n).$$

Substituting these expressions into (6.87), using (6.88), and applying L'Hospital's rule will produce these limits (The details are left as an exercise for readers). It follows from the CLT that

$$\lim_{n\to\infty} \mathbb{Q}\left\{\frac{X_n - n\mu_n}{\sigma_n\sqrt{n}} \le x\right\}$$

$$= \mathbb{Q}\left\{\frac{\ln[S(T)/S] - (r-\sigma^2/2)T}{\sigma\sqrt{T}} \le x\right\} = \Phi(x).$$

Thus, X_n converges to the normally distributed random variable $\ln[S(T)/S]$ with mean $(r - \sigma^2/2)T$ and variance $\sigma^2 T$ under the risk-neutral measure $\mathbb{Q}$.

For an European call option of the discrete-time version, we can write the premium as

$$C_n = R^{-n} E^Q \left[\{Se^{X_n} - K\}^+ \right].$$

Clearly, X_n converges in distribution to $X = \ln[S(T)/S]$ if and only if $E[e^{tX_n}]$ converges to $E[e^{tX}]$ for which the moment generating functions exist. Moreover, we have

$$\lim_{n \to \infty} E[f(X_n)] = E[f(X)]$$

for any bounded and continuous function $f(x)$, for which the expectations exist. Therefore, we obtain

$$\lim_{n \to \infty} C_n = C = e^{-rT} E^Q \left[\{Se^{\ln[S(T)/S]} - K\}^+ \right], \qquad (6.90)$$

where $\ln[S(T)/S] \sim N((r - \sigma^2/2)T, \sigma^2 T)$ under Q.

Exercise 6.34 (a) Verify $p_n^* \to p$ as $n \to \infty$, where p_n^* is given by (6.87).
(b) Show (6.89).

Define $A = \{S(T) \geq k\}$. Similar to the change of measures in the random walk case, we can also transform the BM with drift 0 to the BM with drift μ and vice versa via the change of measures. Corresponding to (6.39), we have

$$\mathbb{P}^*(A) = E^{\mathbb{P}}[1_A Y(T)], \quad Y(t) = e^{\mu B(t) - \mu^2 t/2}. \qquad (6.91)$$

It can be verified that a BM with drift 0 under the probability measure $\mathbb{P}$ can be transformed into a BM with drift μ under the new probability measure $\mathbb{P}^*$. This is similar to transforming from a symmetric random walk into a non-symmetric random walk. To show this, consider a random walk $X_t = \sum_{n=1}^{t} Y_n$ and let Y_n have the following distribution

$$\mathbb{P}^*(Y_n = \Delta x) = 1 - \mathbb{P}^*(Y_n = -\Delta x) = p.$$

It follows from

$$p\Delta x + (1 - p)(-\Delta x) = \mu \Delta t = \mu(\Delta x)^2,$$

that $p = (1 + \mu \Delta x)/2$. Letting $\theta \Delta x = \ln \sqrt{p/(1 - p)}$, we have

$$\theta = \frac{1}{2\Delta x} \ln \left(\frac{1 + \mu \Delta x}{1 - \mu \Delta x} \right). \qquad (6.92)$$

With such a selection of θ, a symmetric random walk under the original probability measure $\mathbb{P}$ is transferred to a p-random walk under $\mathbb{P}^*$ For the symmetric X_n, we have

$$m(\theta) = \frac{1}{2} \left(e^{\theta \Delta x} + e^{-\theta \Delta x} \right) = (1 - \mu^2 (\Delta x)^2)^{-1/2} = (1 - \mu^2 \Delta t)^{-1/2}.$$

Using $n = t/\Delta t$, we have

$$\lim_{\Delta t \to 0} m^{-n}(\theta) = \lim_{\Delta t \to 0} (1 - \mu^2 \Delta t)^{t/2\Delta t} = e^{-\mu^2 t/2}. \tag{6.93}$$

Taking the limit of $\Delta x \to 0$ on (6.92) yields

$$\lim_{\Delta x \to 0} \theta = \frac{1}{2} \lim_{\Delta x \to 0} \left\{ \frac{\mu}{1 + \mu \Delta x} + \frac{\mu}{1 - \mu \Delta x} \right\} = \mu. \tag{6.94}$$

According to the CLT, as $\Delta t \to 0$ (consequently $n \to \infty$), the random variable defined by

$$Y_n = m^{-n}(\theta) e^{\theta(X_n - X_0)}, \quad n = 1, 2, ...,$$

converges in distribution to

$$Y(t) = \exp \left\{ -\frac{\mu^2 t}{2} + \mu B(t) \right\}$$

under the original measure $\mathbb{P}$ which proves (6.91). Now we evaluate the expectation term in (6.90). Letting $A = \{S(T) \geq K\}$, we can write

$$E^Q[\{S(T) - K\}^+] = E^Q[S(T) 1_A] - K E^Q[1_A]. \tag{6.95}$$

Due to $\ln[S(T)/S] \sim N((r - \sigma^2/2)T, \sigma^2 T)$ under the risk-neutral probability measure Q, the second term on the right-hand side of (6.95) is given by

$$KE^Q[1_A] = KQ(S(T) \geq K) = KQ(\ln[S(T)/S] \geq \ln[K/S])$$

$$= KQ \left(z \geq \frac{\ln[K/S] - \left(r - \frac{\sigma^2}{2}\right) T}{\sigma \sqrt{T}} \right) = K\Phi(d'), \tag{6.96}$$

where

$$d' = \frac{\ln(S/K) + \left(r - \frac{1}{2}\sigma^2\right) T}{\sigma \sqrt{T}}.$$

To find the first term on the right-hand side of (6.95), we note that a new probability measure Q^* can be induced by defining

$$Y(T) = \exp \left\{ \ln[S(T)/S] - rT \right\}. \tag{6.97}$$

Since $\ln[S(T)/S] = (r - \sigma^2/2)T + \sigma B(T)$ under Q, the exponent of $Y(T)$ is actually $-\sigma^2 T/2 + \sigma B(T)$. That is

$$Y(T) = \exp \left\{ -\frac{\sigma^2 T}{2} + \sigma B(T) \right\}, \tag{6.98}$$

which is the same function as (6.91) by replacing μ with σ at time T. Using the formula of change of measures, we have

$$Q^*(A) = E^Q \left[e^{\ln[S(T)/S] - rT} 1_A \right] = e^{-rT} S^{-1} E^Q[S(T) 1_A],$$

which can be re-written as

$$E^Q[S(T) 1_A] = Se^{rT} Q^*(A) = Se^{rT} Q^*(S(T) \geq K). \tag{6.99}$$

Here, the expectation of the product of the two non-independent random variables is expressed in terms of the tail probability under the new probability measure. It follows from $Y(T)$ of (6.98) that a BM with a drift of $-\sigma$ is transferred to a BM with a drift 0. That is a BM $B^*(t) = B(t) - \sigma t$ is a standard BM under the probability measure Q^*. That is $dQ^*/dQ = Y(T)$. This implies that while

$$\ln[S(T)/S] = (r - \sigma^2/2)T + \sigma B(T)$$

holds under Q,

$$\ln[S(T)/S] = (r - \sigma^2/2)T + \sigma B(T) = (r - \sigma^2/2)T + \sigma(\sigma T + B^*(T))$$
$$= (r + \sigma^2/2)T + \sigma B^*(T)$$

holds under Q^*. Now we can compute $Q^*(S(T) \geq K)$ in (6.99) as

$$Q^*(S(T) \geq K) = Q^*(\ln[S(T)/S] \geq \ln[K/S])$$
$$= Q^* \left(z \geq \frac{\ln[K/S] - \left(r + \frac{\sigma^2}{2} \right) T}{\sigma \sqrt{T}} \right) = \Phi(d), \tag{6.100}$$

where

$$d = \frac{\ln(S/K) + \left(r + \frac{1}{2}\sigma^2 \right) T}{\sigma \sqrt{T}} = d' + \sigma \sqrt{T}.$$

Substituting (6.100) into (6.99) and using (6.96), (6.95), we have

$$E^Q[\{S(T) - K\}^+] = Se^{rT} \Phi(d) - K\Phi(d - \sigma \sqrt{T}). \tag{6.101}$$

Finally, it follows from (6.90) and (6.101) that

$$C = S\Phi(d) - Ke^{-rT} \Phi(d - \sigma \sqrt{T}), \tag{6.102}$$

which is the famous Black–Scholes formula for the valuation of an European call option.

Exercise 6.35 *The Black–Scholes formula for an European call option can also be obtained by solving the backward C-K equation. In this approach, the value of the payoff at time t for the call option is first defined and denoted by* $p(S,t)$. *Then, a partial differential equation can be established for the present value of* $p(S,t)$ *(i.e.,* $V(S,t) = e^{-r(T-t)}p(S,t)$*). Finally, solve this partial differential equation by transforming it to the heat equation. Write down the partial differential equation and solve it for the Black–Scholes formula by using this approach.*

REFERENCE NOTES

This chapter studies the random walks and Brownian motions, the two important stochastic processes in stochastic modeling. There are many good books on these two stochastic processes such as [5] and [8]. The level crossing method as a special tool is taken from [3]. The example of option pricing is mainly based on [4]. More applications of Brownian motions in queueing systems can be found in [6].

REFERENCES

1. J. Abate and W. Whitt, The Fourier-series method for inverting transforms of probability distributions. *Queueing Systems*, 10, 5–87, 1992.
2. M.H. Ackroyd, Computing the waiting time distribution for the G/G/1 queue by signal processing methods. *IEEE Transactions on Communications*, 28, 52–58, 1980.
3. P.H. Brill, "Level Crossing Methods in Stochastic Models", Springer Science, 2008.
4. M. Kijima, "Stochastic Processes with Applications to Finance", CPC Press, 2nd ed., 2013.
5. G. Lawler and V. Limic, "Random Walk: A Modern Introduction", Cambridge University Press, 2010.
6. S. Resnick, "Adventures in Stochastic Processes", Birkhuser, Boston, 1992.
7. S. Ross, "Stochastic Processes", John Wiley & Sons Inc., New York, 2nd ed., 1996.
8. R. L. Schilling, "Brownian Motion: A Guide to Random Processes and Stochastic Calculus", Berlin, Boston: De Gruyter, 2021.
9. L. D. Servi, D/G/1 queues with vacations. *Operations research*, 34 (4), 619–629, 1986.

7 Reflected Brownian Motion Approximations to Simple Stochastic Systems

In this chapter, we focus on a variant of Brownian motion, called reflected Brownian motion (Reflected BM or RBM), which is a BM with a reflecting boundary at zero. Such a process is restricted to the non-negative real numbers (i.e., $\mathbb{R}^d_+$). Since in many practical stochastic systems, the state variables do not take on negative values such as the queue lengths in a service system or the inventory levels in a production system without back-orders, the reflected BM is a natural CTMP model for analyzing these practical systems. In fact, the reflected BMs can be basic elements for more complex and larger stochastic systems. This chapter presents the theoretical foundation for generating the reflected BM and demonstrates the applications of using the reflected BM approximation in queueing systems and production-inventory systems.

7.1 APPROXIMATIONS TO $G/G/1$ QUEUE

The building block of large-scale waiting line systems is a single server queue with arrivals as a renewal or non-renewal process and general i.i.d. service times. Such a queue, denoted by $G/G/1$, has been discussed in the previous chapter. Since we will utilize the asymptotic analysis to study large-scale stochastic systems, a different approach is introduced in this chapter by analyzing the $G/G/1$ queue. The main idea is to utilize the reflected Brownian motion to approximate the stochastic process of our interest under certain realistic conditions. Then, we present a simple tandem queue and a simple production-control system as examples for applying this approach. The causes for the queue phenomenon or congestion are the arrival process and service process which along with the initial queue length are called primitive input data. We construct the queue length process from these input data. To facilitate the asymptotic analysis, we first define some cumulative processes. Let u_i be the inter arrival time between the $(i\text{-}1)$st and the ith arrivals, $i = 1, 2, \dots$. Let v_i be the required service time of the job that is the ith to be served, including the $Q(0)$ jobs that are present in the system at time zero. Let $u = \{u_i, i = 1, 2, \dots\}$

and $v = \{v_i, i = 1, 2, \dots\}$. Assume these inter-arrival times and service times are non-negative and have finite means. Define the cumulative nth event times of arrival and service processes, respectively as

$$U(0) := 0, \ U(n) := \sum_{i=1}^{n} u_i, \ n \geq 1,$$

$$V(0) := 0, \ V(n) := \sum_{i=1}^{n} v_i, \ n \geq 1 \tag{7.1}$$

and the number of jobs that have arrived during the time interval $(0,t]$, denoted by $A(t)$, and the number of jobs the server can potentially complete during the first t units of time (i.e., provided that the server is busy all the time), denoted by $S(t)$, can be defined, respectively, as

$$A(t) := \sup\{n : U(n) \leq t\}, \ S(t) := \sup\{n : V(n) \leq t\}. \tag{7.2}$$

$A(t)$ and $S(t)$ are called arrival and service processes, respectively. These two processes interact to produce the queue length and workload processes that are the key performance measures of the queue. Let $Q(t)$ be the number of jobs in the system at time t, and $Z(t)$ be the workload at time t, which is the amount of time required to complete all jobs present at time t. Thus, the queue length and workload process can be denoted by $Q = \{Q(t), t \geq 0\}$ and $Z = \{Z(t), t \geq 0\}$, respectively. Using the flow balance principle and work-conserving assumption (the server is busy if and only if the system is nonempty), we have

$$Q(t) = Q(0) + A(t) - S(B(t)),$$

$$B(t) = \int_0^t 1\{Q(t) > 0\}ds, \tag{7.3}$$

where $B(t)$ denotes the cumulative time when the server is busy over time interval $[0,t]$, and thus $S(B(t))$ is the number of service completions (i.e., the number of jobs that have departed from the system in the same time interval. $B = \{B(t), t \geq 0\}$ is called the busy process. Based on the definition of $V(n)$, we can express the workload process at time t as

$$Z(t) = V(Q(0) + A(t)) - B(t). \tag{7.4}$$

Another performance measure is the idle time of the system. The idle process, denoted by $I = \{I(t), t \geq 0\}$, can be defined as

$$I(t) = t - B(t) = \int_0^t 1\{Q(s) = 0\}ds. \tag{7.5}$$

These input and performance processes become the basis for developing stochastic process limits.

Exercise 7.1 *Graph the sample paths of $A(t), S(B(t))$, and $Q(t)$ and explain the relations of these processes in (7.3) graphically.*

Exercise 7.2 *Explain the advantage of introducing the counting processes $A(t)$ and $S(t)$ to study the queueing process $Q(t)$.*

7.2 QUEUE LENGTH AS REFLECTION MAPPING

Denote the finite mean interarrival time and mean service time by $1/\lambda$ and $m \equiv 1/\mu$, respectively. Under certain conditions, it follows from the limit theorems (the strong law of large numbers) that $A(t)$ and $S(t)$ are asymptotically close to λt and μt, respectively, as $t \to \infty$. The queue length process $Q(t)$ of (7.3) can be rewritten as the sum of two terms. One, denoted by $X(t)$, is written as some constant terms plus the difference of two centered cumulative processes and the other, denoted by $Y(t)$, represents the potential cumulative processed workload which is not realized due to the idle period.

$$Q(t) = X(t) + Y(t), \tag{7.6}$$

where

$$X(t) := Q(0) + (\lambda - \mu)t + [A(t) - \lambda t] - [S(B(t)) - \mu B(t)], \\ Y(t) := \mu I(t). \tag{7.7}$$

Furthermore, for all $t \geq 0$, we have

$$Q(t) \geq 0, \tag{7.8}$$
$$dY(t) \geq 0, \ Y(0) = 0, \tag{7.9}$$
$$Q(t)dY(t) = 0. \tag{7.10}$$

The relations among $Q(t), X(t)$, and $Y(t)$ can be clearly illustrated graphically. These relations indicate (i) the queue length should be non-negative; Y is non-decreasing from the initial value zero; and the non-idling condition. The idle time accumulates only when the queue length is zero (empty system). An alternative expression to $Q(t)dY(t) = 0$ is

$$\int_0^\infty Q(t)dY(t) = 0. \tag{7.11}$$

Note that $X(t)$ can be asymptotically determined by the primitive data (input processes - arrival process $A(t)$, service process $S(t)$) and the busy period $B(t)$. This is because $B(t)$ approaches to $(\lambda/\mu)t$ as $t \to \infty$. Our main interest is to obtain the queue length process $Q(t)$ (performance measure) from the input processes. Fortunately, both $Q(t)$ and $Y(t)$ can be uniquely determined by $X(t)$

if the above relations hold. Thus, we can consider these relations as the mapping from X to (Q,Y). Such a mapping is called the one-dimensional reflection mapping as stated in the following theorem, called reflection mapping theorem (RMT). Denote the J-dimensional real-valued functional space for the right-continuous with left limit (RCLL) functions on time interval $[0,T]$ by D^J with convention of $D^1 = D$. The subspace for the functions with condition $x(0) \geq 0$ is denoted by D_0.

Theorem 7.1 *(Reflection Mapping Theorem) For any $x \in D_0$, there exists a unique pair (y,z) in D^2 satisfying*

$$z = x + y \geq 0, \tag{7.12}$$

$$dy \geq 0, \ y(0) = 0, \tag{7.13}$$

$$zdy = 0. \tag{7.14}$$

In fact, the unique y and z can be expressed as

$$y(t) = \sup_{0 \leq s \leq t} [-x(s)]^+, \tag{7.15}$$

$$z(t) = x(t) + \sup_{0 \leq s \leq t} [-x(s)]^+. \tag{7.16}$$

z is called the reflected process of x and y is called the regulator of x. Denote the unique y and z by $y = \Psi(x)$ and $z = \Phi(x)$. Then the mappings Ψ and Φ are Lipschitz continuous on $D_0[0,t]$ under the uniform topology for any fixed t (including $t = \infty$).

For the proof of this theorem we refer to Chen and Yao (2001). Since the proofs of this chapter can be found in Chen and Yao (2001), we omit these proofs to make this chapter more concise. Interested readers are referred to Chen and Yao (2001) for more technical details. Here, we focus on intuitive explanations of these results. For example, to intuitively understand this theorem, we can provide a graphical presentation of this reflection mapping by using the workload process in a fluid queue model, a special queueing setting.

Consider a fluid queue with $A(t) = \lambda t$ as the cumulative input of fluid over $[0,t]$ and $S(t) = \mu t$ as the cumulative available processing over $[0,t]$, where λ is the constant input rate and μ is the constant output (i.e., processing) rate. It is assumed that $\lambda < \mu$ and the service facility's processing or "on" period is random. Let $Q(t)$ be the workload in the system at time t (note: for a classical queue with discrete entities such as $G/G/1$ queue, we distinguish the queue length and workload by using notations $Q(t)$ and $Z(t)$, respectively). If we assume that the system starts with $Q(0)$ workload, then we have the potential workload process

$$X(t) = Q(0) + A(t) - S(t),$$

Legends:

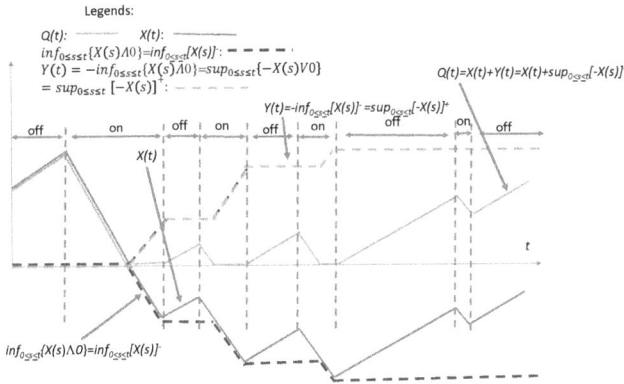

Figure 7.1   One-dimensional reflection mapping of a fluid queue.

which represents what the workload would be if we ignore the emptiness condition. Thus, the actual workload process for $t \geq 0$ is given by

$$Q(t) = X(t) - \inf_{0 \leq s \leq t} \{X(s) \wedge 0\} = X(t) + Y(t),$$

where $a \wedge b = \min\{a, b\}$ and $Y(t) := -\inf_{0 \leq s \leq t}\{X(s) \wedge 0\}$, which is called the regulator of $X(t)$. Such a relation is clearly shown in Figure 7.1. We can see that the "on" period length is random. When the service facility is off, the fluid level in the system is built up and when it is "on", the fluid level is dropping or stays at zero. As the "on" and "off" periods alternate, the system fluid level fluctuates. It is also observed in Figure 7.1 that $Y(t)$ accumulates the idle period of the system.

Exercise 7.3 *Consider a fluid queue with a finite buffer of size K. Assume all parameters are the same as the fluid queue discussed above except for the finite buffer. Write down a set of equations for the reflection mapping of X to (Q, Y) and create the sample path graph like Figure 7.1 to explain these equations.*

Using Theorem 7.1, for an $G/G/1$ queue, the queue length process and idle time process can be written as $Q(t) = \Phi(X)$ and $I = m\Psi(X)$. Due to the Lipschitz continuity of the mappings Φ and Ψ, if X can be approximated by another (usually simpler) process, say B (Brownian motion), then the queue length process can be approximated with the same accuracy by $Q(t) = \Phi(B)$. Similar to $Q(t)$, the workload process $Z(t)$ can be also considered as the mapping from an input process (from the primitive data) to (Z, Y) as follows:

$$Z(t) = X(t) + Y(t), \tag{7.17}$$
$$X(t) = V(Q(0) + A(t)) - t, \tag{7.18}$$
$$Y(t) = I(t) = t - B(t). \tag{7.19}$$

Reflection (Skorohod) Mapping Theorem (RMT) is extremely useful in developing various approximations to queue length and workload processes based on limit theorems. This is because the input process can be approximated by the Brownian motion which is a fundamental continuous-time Markov process via fluid scaling and diffusion scaling. For the $G/G/1$ system, it can be shown that X can be approximated by a one-dimensional Brownian motion that starts at $x = X(0) \geq 0$, with drift θ and standard deviation σ. In this case, the process $Z = \{Z(t), t \geq 0\} := \Phi(X)$, which is called a one-dimensional reflected Brownian motion as shown in Theorem 7.1.

Exercise 7.4 *Graph sample paths for $Z(t), X(t),$ and $Y(t)$ and explain the relations in (7.17)–(7.19) graphically.*

We can denote the Z by $Z = RBM_x(\theta, \sigma^2)$. Such an approximation is appropriate for anlyzing the single server queue in heavy traffic (to be discussed in details later). To fully describe the system dynamics of $Z = RBM_x(\theta, \sigma^2)$, we need to determine the limiting or stationary distribution of the process. A subtle issue is to verify that the limiting distribution is the stationary distribution. A formal definition of the stationary distribution is as follows:

Definition 7.1 *A probability distribution of Z, denoted by π and defined on $(\mathscr{R}_+, \mathscr{B}(\mathscr{R}_+))$, is stationary if*

$$E_\pi[f(Z(t))] = \int_{\mathscr{R}_+} E_z[f(Z(t)]d\pi(z) = E_\pi[f(Z(0))]$$

for all bounded continuous function f on $\mathscr{R}_+$.

Theorem 7.2 *Reflected Brownian Motion $Z(t) = RBM_x(\theta, \sigma^2)$ converges in distribution to some limiting random variable $Z(\infty)$ if and only if $\theta < 0$. In this case, the stationary distribution of $Z(\infty)$ is exponential with parameter $-2\theta/\sigma^2$.*

The rigorous proof of the above theorem can be found in any standard textbook on Brownian motion (e.g., Chen and Yao 2001). Below we give some justification of this theorem.

Let $X(t)$ be the Brownian motion with the initial position $X(0) = 0$. We can write

$$P(Z(t) \geq z) = P(X(t) + \sup_{0 \leq s \leq t} [-X(s)]^+ \geq z)$$

$$= P(\sup_{0 \leq u \leq t} \tilde{X}_t(u) \geq z),$$

where

$$\tilde{X}_t(u) = \begin{cases} X(t) - X(t-u), & 0 \leq u \leq t, \\ X(u), & t < u < \infty. \end{cases}$$

It is easy to see that for any fix t, $\tilde{X}_t = \{\tilde{X}_t(u), u \geq 0\}$ is a Gaussian process with drift θ and covariance $Cov(\tilde{X}_t(u), \tilde{X}_t(v)) = \sigma^2(u \wedge v)$. Thus, $\tilde{X}_t$ is a Brownian motion with drift θ and standard deviation σ. Let $T_{\tilde{z}}^t := \inf\{u \geq 0 : \tilde{X}_t(u) \geq z\}$. We have

$$P(Z(t) \geq z) = P(T_{\tilde{z}}^t \leq t) = P(T_{\tilde{z}}^0 \leq t) = P\left(\sup_{0 \leq u \leq t} X(u) \geq z\right),$$

where the last equality holds since $\tilde{X}_t = \{\tilde{X}_t(u), u \geq 0\}$ and X are equal in distribution. It is well known that the supremum of a Brownian motion, $\sup_{0 \leq u \leq t} X(s)$, follows an exponential distribution if $\theta < 0$ and drifts to $+\infty$ if $\theta \geq 0$ as $t \to \infty$ (see Theorem 6.10). That is

$$\lim_{t \to \infty} \left(\sup_{0 \leq u \leq t} X(u) \geq z\right) = \begin{cases} 1, & \text{if } \theta \geq 0, \\ e^{-\frac{2|\theta|}{\sigma^2} z}, & \text{if } \theta < 0. \end{cases}$$

Using the stopping time $T := \inf\{t \geq 0, X(t) = 0\}$, we can show that starting at $X(0) = x > 0$, the reflected Brownian motion also follows the same exponential distribution for the case of $\theta < 0$ and drifts to positive infinity with probability 1 for the case of $\theta \geq 0$. Based on the Markov property and the Feller continuity of RBM, we can show that the limiting distribution is the stationary distribution. That is

$$\int_{\mathcal{R}_+} f(z) d\pi(z) = \lim_{s \to \infty} E_0[f(Z(t+s))] = \lim_{s \to \infty} E_0[E_{Z(t)}[f(Z(t+s))]]$$

$$= \int_{\mathcal{R}_+} E_x[f(Z(t))].$$

A negative θ indicates a stable $G/G/1$ queue. A pathwise bound on the RBM with a negative drift (the maximum queue length) can be established as follows:

$$\sup_{0 \leq t \leq T} |Z(t)| \overset{a.s.}{=} O(\log T). \tag{7.20}$$

The derivation of such a bound can be found in Chen and Yao (2001).

We can also establish that the convergence of a sequence of input processes x_n will lead to the convergence of the sequences of mapped processes z_n and y_n.

Let $\{\theta_n, n \geq 1\}$ be a sequence of real numbers and let $\{x_n, n \geq 1\}$ be a sequence in $D[0, \infty)$ with $x_n(0) \geq 0$. Consider a sequence of reflected mapped process with a drift θ_n:

$$z_n(t) = x_n(t) + \theta_n t + y_n(t),$$
$$y_n(t) = \sup_{0 \leq u \leq t}[-x_n(u) - \theta_n u]^+. \tag{7.21}$$

Suppose that as $n \to \infty$,

$$x_n \to x \quad u.o.c. \tag{7.22}$$

with $x \in C[0, \infty)$. Here, u.o.c. represents the convergence "uniformly on all compact sets" which is equivalent to "almost surely convergence" for a sequence of stochastic processes in the space of J-dimensional real-valued functions on $[0, T]$ that are right-continuous and with left limits (RCLL), denoted by $\mathscr{D}^J[0, T]$, endowed with the uniform topology. Then, we have the following results:

1. If $\theta_n \to \theta$ as $n \to \infty$, then

$$(y_n, z_n) \to (y, z), \tag{7.23}$$

   where
$$z(t) = x(t) + \theta t + y(t),$$
$$y(t) = \sup_{0 \leq u \leq t}[-x(u) - \theta u]^+. \tag{7.24}$$

2. If $\theta_n \to -\infty$ as $n \to \infty$ and $x(0) = 0$ then

$$(y_n(t) + \theta_n t, z_n(t)) \to (-x(t), 0). \tag{7.25}$$

3. If $\theta_n \to \infty$ as $n \to \infty$, then

$$(y_n(t), z_n(t) - \theta_n t) \to (0, x(t)). \tag{7.26}$$

These results are useful in establishing the weak convergence results to be discussed next. The proof of these results can be found in Chen and Yao (2001).

Exercise 7.5 *Explain the difference between the Brownian motion with a drift and the reflected Brownian motion with a negative drift in terms of the limiting process behaviors. (Hint: Argue such a difference from the ergodicity of the Markov chains for the random walk and reflected random walk.)*

7.3 FUNCTIONAL STRONG LAW OF LARGE NUMBERS (FSLLN) – FLUID LIMIT

This is an extension of the strong law of large numbers (SLLN) in basic statistics and probability theory. While the classical SLLN is the limit theorem for a random variable, the FSLLN is the limit theorem for a stochastic process (a collection of random variables indexed by a time set). Thus, at any point in time, the FSLLN is reduced to the SLLN. Recall that in the SLLN, we have to keep increasing the sample size (or the number of observations in an experiment) for a particular quantitative random variable to make the average of the observations to approach to the true mean value of the random variable. Similarly, we need to increase the number of observations of the process. This can be done by scaling up the time and space. If we scale the time and space appropriately, we can obtain some meaningful "average limits". In this section, we scale up the time and space at the same rate. This scaling process works as follows. Consider a sequence of $G/G/1$ queues indexed by n. For the nth system, we tend to increase the time by n times and at the same time increase the space by n times. For the arrival and service processes (primitive data), graphically, this means that the horizontal axis is increased by n times. For the same size of the graph, it means now each subinterval on the time axis represents more basic time units. For example, a subinterval of one minute now becomes n minutes or a 10 subinterval horizontal axis now has changed from 10 minute time horizon has been scaled up to $10n$ minute time horizon. Consequently, the cumulative number of events (arrivals or service completions) will increase proportionally. Thus, the vertical axis will represent the "$n-$average value" for the nth system. Thus each subinterval on the vertical axis (space) will represent n times of the actual cumulative value that occurs (non-average or $n = 1$ case) so that the same size of the graph can accommodate this scaling up. This is in some sense that the vertical axis is also scaled up by n times. Such a scaling is called fluid scaling and the limits obtained are the "fluid limits". Below are the main fluid scaled processes in $G/G/1$ queue (arrival process, service process, queue length, and busy period)

$$\overline{A}^n(t) := \frac{1}{n}A(nt), \ \overline{S}^n(t) := \frac{1}{n}S(nt), \ \overline{Q}^n(0) := \frac{1}{n}Q^n(0), \qquad (7.27)$$

and

$$\overline{Q}^n(t) = \frac{1}{n}Q^n(nt), \ \overline{Z}^n(t) = \frac{1}{n}Z^n(nt), \overline{B}^n(t) = \frac{1}{n}B^n(nt). \qquad (7.28)$$

It is worth mentioning the notation convention for the scaling process. First, for the fluid scaling (or the diffusion scaling discussed later), the scaled process's time argument may be omitted if it is obvious. That is, either $\overline{A}^n(t)$ or $\overline{A}^n$ can be used. Second, for simplicity, the superscript for the nth system may be omitted if it is obvious. That is, either $Q^n(nt)$ or $Q(nt)$ can be used. The starting point is to recognize that the input processes approach to the deterministic mean process as $n \to \infty$ according to FSLLN

$$\overline{A}^n(t) \to \lambda t, \quad \overline{S}^n(t) \to \mu t, \quad u.o.c. \text{ as } n \to \infty. \tag{7.29}$$

Moreover, we assume that the initial queue length depends on n and almost surely as $n \to \infty$, we have

$$\overline{Q}^n(0) := \frac{1}{n} Q^n(0) \to \overline{Q}(0) \geq 0. \tag{7.30}$$

These results are the direct extension of the SLLN to the stochastic arrival and service processes. Fortunately, the fluid scaled performance processes $\overline{Q}^n(t), \overline{Z}^n(t)$, and $\overline{B}^n(t)$ will approach to the fluid process limits via the reflection mapping theorem (RMT).

Theorem 7.3 *Suppose that (7.29) and (7.30) hold. Then as $n \to \infty$,*

$$(\overline{Q}^n, \overline{Z}^n, \overline{B}^n) \to (\overline{Q}, \overline{Z}, \overline{B}), \qquad u.o.c. \tag{7.31}$$

where

$$\begin{aligned} \overline{Q} &= \overline{X} + \overline{Y} \\ \overline{X}(t) &= \overline{Q}(0) + (\lambda - \mu)t, \\ \overline{Y} &= \Psi(\overline{X}), \overline{Z} = m\overline{Q}, \overline{B}(t) = t - m\overline{Y}(t), \end{aligned} \tag{7.32}$$

and $m = 1/\mu$. In addition,

$$\overline{Q}(t) = [\overline{Q}(0) + (\lambda - \mu)t]^+, \tag{7.33}$$

$$\overline{Y}(t) = [-\overline{Q}(0) - (\lambda - \mu)t]^+. \tag{7.34}$$

The proof of this theorem can be found in Chen and Yao (2001).

The fluid limiting process $\overline{Q}(t)$ can be interpreted as the water level of a container at time t with a constant inflow rate λ, a constant outflow rate μ, and an initial water level $\overline{Q}(0)$. $\overline{Y}(t)$ represents the cumulative amount of lost outflow capacity due to the emptiness of the container during the time interval $[0,t]$. Under the fluid scaling, the single server queue becomes a fluid model, where the queue length approaches a deterministic and continuous flow, the fluid limiting process. Now consider three cases with different traffic intensities. If $\rho = \lambda/\mu < 1$ (i.e., $\lambda < \mu$), then after a finite time period $T \geq 0$, the fluid queue process $\overline{Q}(t)$ will reduce to zero at a rate $\mu - \lambda$ and stay zero afterward. That is $\overline{Q}(t) = 0$ for all $t > T$ where $T = \overline{Q}(0)/(\mu - \lambda)$. If $\rho > 1$, $\overline{Q}(t)$ will increase linearly at a rate $\lambda - \mu$ with time t without a bound. If $\rho = 1$, $\overline{Q}(t)$ remains at the initial level. These three cases are the fluid limit processes corresponding to the positive recurrent, transient, and null recurrent cases in a *M/M/1* queue. The fluid model has wide applications in congestion systems with time-varying arrival rate and/or time-varying service rate.

Exercise 7.6 *The fluid limit of a stochastic system can be used to study its stability conditions. Use a GI/G/1 queue show that the stability condition for the fluid limit queue is a necessary condition for the original stochastic queue. (Note: Such a condition is not sufficient in a queueing network setting (see Gautam (2012)).*

7.4 FUNCTIONAL CENTRAL LIMIT THEOREM (FCLT) – DIFFUSION LIMIT

Now we generalize the classical Central Limit Theorem (CLT) for random variables to the Functional Central Limit Theorem (FLCT) for stochastic processes. Recall the CLT for a random variable $N(t)$, representing the number of arrivals during time interval $[0,t]$ for a renewal process. The time between arrivals is a generally distributed i.i.d. random variable with mean of μ and standard deviation of σ. Denote the arrival rate by $\lambda = 1/\mu$. It follows from the CTL that $N(t)$ approaches to a normal distribution as $t \to \infty$ as shown in Theorem 5.5 in Chapter 5. Such a convergence is said to be convergence in distribution or "weak convergence". For the convergence modes of random variables, readers are referred to Chapter F.

Theorem 7.4 *(CLT for renewal process) For a renewal process, as $t \to \infty$,*

$$\frac{N(t) - \lambda t}{\sqrt{\lambda t \lambda} \sigma} \to Z \tag{7.35}$$

where Z is the standard normal random variable with the distribution denoted by $N(0,1)$ (or Φ).

The weak convergence of the CLT can be extended to the stochastic processes. For a review on the convergence of a stochastic process to a stochastic process limit, we refer to Chapter G. Let us scale the time by n times in the renewal process. Then according to the CLT, the centered process $N(nt) - \lambda nt$ will approach to the normal distribution with mean 0 and standard deviation of $\sqrt{\lambda nt} \lambda \sigma$ as $n \to \infty$. Therefore, $(N(nt) - \lambda nt)/\sqrt{n}$ will approach to the normal distribution with mean 0 and standard deviation of $\sqrt{\lambda t} \lambda \sigma$. Such a scaling is equivalent to centering the fluid scaled process around its fluid limit and scaling it up by a factor of $\sqrt{n}$. We can apply this type of scaling to the input processes of the $G/G/1$ queue:

$$\widehat{A}^n(t) := \sqrt{n}[\bar{A}^n(t) - \lambda t] = \frac{A(nt) - n\lambda t}{\sqrt{n}}, \tag{7.36}$$

$$\widehat{S}^n(t) := \sqrt{n}[\bar{S}^n(t) - \mu t] = \frac{S(nt) - n\mu t}{\sqrt{n}}, \tag{7.37}$$

$$\widehat{V}^n(t) := \sqrt{n}[\bar{V}^n(t) - mt] = \frac{V(\lfloor nt \rfloor) - n\lambda t}{\sqrt{n}}. \tag{7.38}$$

Such a scaling is called diffusion scaling. Since we magnify the centered fluid process by a factor of $\sqrt{n}$, the random fluctuations of the process around the fluid limit can be observed. This is similar to the case where the variability around the mean of a random variable can be observed by the CLT. Similarly, by using the fluid limit results, we can apply the diffusion scaling to the performance processes as follows:

$$\widehat{Q}^n(t) := \sqrt{n}[\overline{Q}^n(t) - (\lambda - \mu)^+ t] = \frac{Q(nt) - (\lambda - \mu)^+ nt}{\sqrt{n}}, \tag{7.39}$$

$$\widehat{Z}^n(t) := \sqrt{n}[\overline{Z}^n(t) - (\rho - 1)^+ t] = \frac{Z(nt) - (\rho - 1)^+ nt}{\sqrt{n}}, \tag{7.40}$$

$$\widehat{B}^n(t) := \sqrt{n}[(\rho \wedge 1)t - \overline{B}^n(t)] = \frac{(\rho \wedge 1)nt - B(nt)}{\sqrt{n}}. \tag{7.41}$$

Basically, the diffusion scaling involves scaling up the time (horizontal axis) by a factor of n (replacing t by nt) and scaling up the space (vertical axis) by a factor of $\sqrt{n}$ (dividing the process by $\sqrt{n}$). This can be done by either centering or not centering the process by its fluid limit as seen below.

Similar to fluid scaling, we establish the diffusion limits for the performance processes from the diffusion limits of the input processes via one-dimensional RMT. Based on the FCLT (see Chapter G), we have the following joint input process weak convergence if the sequence of interarrival times (u) and the sequence of service times (v) are independent and each is an i.i.d. sequence with a finite second moment:

$$(\widehat{A}^n, \widehat{S}^n, \widehat{V}^n) \xrightarrow{d} (\widehat{A}, \widehat{S}, \widehat{V}) \text{ as } n \to \infty. \tag{7.42}$$

Here, $\widehat{A}$ and $\widehat{S}$ two independent driftless Brownian motions with variances λc_a^2 and μc_s^2, respectively. Such a result is called the Donsker's Theorem (see Chapter G). The total amount of work contributed by all jobs arrived by time t, $\widehat{V}(t) = -m\widehat{S}(mt)$. This relation can be shown using the scaling argument. The weak convergence of the major performance measure processes of the G/G/1 queue can be summarized in the following theorem (see Chen and Yao (2001) for a detailed proof).

Theorem 7.5 *If the weak convergence of input processes, (7.42) holds, then*

$$(\widehat{Q}^n, \widehat{Z}^n, \widehat{I}^n) \xrightarrow{d} (\widehat{Q}, \widehat{Z}, \widehat{I}) \text{ as } n \to \infty, \tag{7.43}$$

with the diffusion limit $(\widehat{Q}, \widehat{Z}, \widehat{I})$ taking the following forms:
(i) if $\rho < 1$, then

$$\widehat{Q} = 0, \quad \widehat{Z} = 0, \quad \widehat{I} = m[S(\rho t) - A(t)], \tag{7.44}$$

(ii) if $\rho = 1$, then

$$\widehat{Q} = \widehat{X} + \widehat{Y} \geq 0, \quad \widehat{X} = \widehat{A} - \widehat{S}, \quad \widehat{Y} = \Psi(\widehat{X}), \tag{7.45}$$

$$\widehat{Z} = m\widehat{Q}, \quad \widehat{I} = m\widehat{Y}, \tag{7.46}$$

(iii) if $\rho > 1$, then

$$\widehat{Q} = \widehat{A} - \widehat{S}, \quad \widehat{Z} = m\widehat{Q}, \quad \widehat{I} = 0. \tag{7.47}$$

This theorem is valid if the weak convergence of primitives $(\widehat{A}^n, \widehat{S}^n, \widehat{V}^n)$ holds.

The first case, $\rho < 1$, is a stable queue, which can be also called a "non-heavy traffic case". While the fluid limit process becomes zero after a finite length of time, the diffusion limit $\widehat{Q}^n = \sqrt{n}\, \overline{Q}^n$ converges weakly to zero. Since it is scaled up by a factor of $\sqrt{n}$, it provides a lower bound on the rate at which the fluid scaled process approaches to zero. However, in this case, the limit of the centered idle period process is a Brownian motion (as the difference between two Brownian motions).

The case of $\rho = 1$ is also called a critical loaded queue. In this case, $\widehat{X} = \widehat{A} - \widehat{S}$ is a Brownian motion, with a zero drift and a variance equal to $\lambda c_a^2 + \mu c_s^2 = \lambda(c_a^2 + c_s^2)$. The limiting process $\widehat{Q}$ is a reflected Brownian motion. The results of this case are more interesting and can be used to analyze the heavy traffic service systems.

The case of $\rho > 1$ is for a overloaded queue. The diffusion scaled process $\widehat{Q}^n(t) = \sqrt{n}[\overline{Q}^n(t) - (\lambda - \mu)t]$ may not be always non-negative due to centering. The implication of $\widehat{I} = 0$ is that the cumulative idle time converges to zero at a rate higher than the one for the convergence of fluid limit.

Exercise 7.7 *For the total amount of work contributed by all jobs arrived by time t, show the relation $\widehat{V}(t) = -m\widehat{S}(mt)$ where $m = 1/\mu$.*

7.5 HEAVY TRAFFIC APPROXIMATION TO $G/G/1$ QUEUE

The diffusion limit theorem for the $G/G/1$ queue indicates that more information can be obtained for the critically loaded queue case ($\rho = 1$). To apply the results of this case, we consider a sequence of queues index by n under a heavy traffic condition. The queue length process is diffusion-scaled without centering as

$$\widehat{Q}^n(t) = \frac{1}{\sqrt{n}} Q^n(nt) \tag{7.48}$$

with the heavy traffic condition on the nth queue with $\sqrt{n}[\lambda_n - \mu_n] \to \theta$, as $n \to \infty$. Here, we try to adjust the arrival rate and service rate to make them closer as n increases so that the heavy traffic condition is approached. Accordingly, the input processes can be diffusion-scaled up as

$$\widehat{A}^n(t) = \frac{A^n(nt) - n\lambda_n t}{\sqrt{n}}, \ \widehat{S}^n(t) = \frac{S^n(nt) - n\mu_n t}{\sqrt{n}}. \tag{7.49}$$

Applying Theorem 7.5 to the diffusions scaling here, under the weak convergence of the input processes, we know that the limiting process for a sequence of queue lengths $\widehat{Q}^n(t)$ approaches to the case of $\rho = 1$, as a reflected Brownian motion, except that a drift θ should be added to the Brownian motion $\widehat{X}$. Here a little notation abuse happens. We denote the limiting process for the queue length by $\widehat{Q}$ (which is uncentered while the same notation is used in Theorem 7.5 for the centered limiting process). Also note that it is not necessary to assume that the arrival process A^n and the service process S^n are renewal processes. This is because we are concerned with the cumulative number of events in these processes only regardless of properties of the inter-event times. The purpose of considering the sequence of queues under heavy traffic condition is to avoid Case 1 in Theorem 7.5, which gives the diffusion limit of zero. However, Case 1 is the stable queue situation ($\rho < 1$) where the stationary distribution exists. This is certainly a practical case in which the performance evaluation is needed. To develop the approximations for such a case based on Case 2, we assume that the stable queue of our interest is among a sequence of queues whose traffic intensities approach to 1 from below. This process is equivalent to satisfying the heavy traffic condition and obtaining the process limit. Clearly, the drift of the Brownian motion $\widehat{X}$ is $\theta = \sqrt{n}(\lambda - \mu) < 0$. Since the limit is obtained by diffusion-scaling, we need to invert the scaling to get the approximation to the queue length process by a reflected Brownian motion. The weak convergence for the arrival process suggests the following approximation for large n:

$$\widehat{A}^n(t) \equiv \sqrt{n}\left[\frac{1}{n}A(nt) - \lambda t\right] \overset{d}{\approx} \widehat{A}(t), \tag{7.50}$$

where $\overset{d}{\approx}$ means approximately equal in distribution. To invert scaling, we replace nt by t and t by t/n (i.e., scaling time down by a factor of n). The approximation above can be written as

$$\sqrt{n}\left[\frac{1}{n}A(t) - \lambda\frac{t}{n}\right] \overset{d}{\approx} \widehat{A}(t/n), \tag{7.51}$$

which is equivalent to

$$[A(t) - \lambda t] \overset{d}{\approx} \sqrt{n}\widehat{A}(t/n) \overset{d}{=} \widehat{A}(t), \tag{7.52}$$

where the last equality in distribution is due to the scaling property of a driftless Brownian motion (see Chapter H). Similarly, for the service process, we have

$$[S(t) - \mu t] \overset{d}{\approx} \widehat{S}(t). \tag{7.53}$$

Replacing t with $B(t)$ in the above yields

$$[S(B(t)) - \mu B(t)] \overset{d}{\approx} \widehat{S}(B(t)). \tag{7.54}$$

Since the cumulative busy time should be proportional to $(\rho \wedge 1)t$, we can replace $B(t)$ in the limit $\widehat{S}$ by $(\rho \wedge 1)t$ and have

$$[S(B(t)) - \mu B(t)] \overset{d}{\approx} \widehat{S}((\rho \wedge 1)t). \tag{7.55}$$

Using these approximations and noting that

$$Q(t) = Q(0) + A(t) - S(B(t)) = X(t) + Y(t),$$

where

$$X(t) = Q(0) + [A(t) - \lambda t] - [S(B(t)) - \mu B(t)] + (\lambda - \mu)t,$$
$$Y(t) = \mu[t - B(t)],$$

we obtain an approximation to X

$$\widehat{X}(t) = Q(0) + \widehat{A}(t) - \widehat{S}((\rho \wedge 1)t) - (\lambda - \mu)t. \tag{7.56}$$

Since $\widehat{A}$ and $\widehat{S}$ are two independent driftless Brownian motions, $\widehat{X}$ is also a Brownian motion with an initial position of $Q(0)$, drift of $(\lambda - \mu)$, and variance of

$$\sigma^2 = \lambda c_a^2 + (\lambda \wedge \mu)c_s^2. \tag{7.57}$$

Therefore, the reflected Brownian motion can be used as an approximation for the queue length process from the RMT. That is

$$Q \approx \widehat{Q} = \Phi(\widehat{X}) = RBM_{Q(0)}(\lambda - \mu, \lambda c_a^2 + (\lambda \wedge \mu)c_s^2). \tag{7.58}$$

For an overloaded queue or $\rho > 1$, we can use the $\widehat{X}$, the Brownian motion, directly as an approximation to the queue length process based on Theorem 7.5.

$$Q \approx \widehat{X} = BM_{Q(0)}(\lambda - \mu, \lambda c_a^2 + \mu c_s^2). \tag{7.59}$$

The disadvantage of this Brownian motion approximation is that there is chance that the $\widehat{X}$ can be negative while the queue-length process is non-negative. However, if the initial queue length is positive, the probability of

$\widehat{X} < 0$ for large t is very small or negligible due to the positive drift. The other performance processes such as the workload and busy period can be approximated in a similar way. For a stable queue (i.e., $\lambda < \mu$), the RMBs which approximate the queue length and workload, respectively, have stationary exponential distributions with means

$$Q \approx \frac{\rho(c_a^2 + c_s^2)}{2(1-\rho)}, \quad Z \approx \frac{\lambda(c_a^2 + c_s^2)}{2(1-\rho)\mu^2}, \qquad (7.60)$$

respectively. Simulations show that these approximations are reasonably good for the systems with the mediate to heavy traffic loads.

Exercise 7.8 *Derive the limiting cumulative distribution function of the reflected Brownian motion by solving the C-K forward equation.*

Exercise 7.9 *Use a software package (Matlab or Excel) to simulate $M/G/1$ queue with the arrival rate λ and service rate μ and to evaluate the accuracy of the approximations based on the diffusion limits. The analytical formula for the mean queue length for such a queue L, as shown in Chapter 5, is given by*

$$L = \rho + \frac{\lambda^2 \sigma^2 + \rho^2}{2(1-\rho)},$$

where $\rho = \lambda/\mu < 1$ (stable queue) is the traffic intensity and σ^2 is the variance of the service time. Simulate the queueing system with different types of distributions for service times: exponential, normal, and Pareto distributions under different traffic intensities from low to high. Based on the simulated results, make comments on this type of heavy-traffic approximations.

7.6   BOUNDS FOR FLUID AND DIFFUSION LIMIT APPROXIMATIONS

From the discussion above, we know that the fluid limit and diffusion limit approximations for the input and performance processes in a $G/G/1$ queue are obtained from the FSLLN and FCLT combined with RMT. Next, we would like to quantify how far the actual processes will make excursions from these process limits. In statistics, the law of the iterated logarithm (LIL) tells very precisely how far the number of successes in a coin-tossing game (n tosses) will be from the average value (np, where p is the probability of getting a success in each toss).

Now we extend the LIL to functional law of the iterated logarithm (FLIL). If both the arrival sequence u and the service sequence v having finite second moments are assumed, then the input processes jointly follow the FLIL as follows:

$$\sup_{0 \le t \le T} |A(t) - \lambda t| \stackrel{a.s.}{=} O\left(\sqrt{T \log \log T}\right),\qquad(7.61)$$

$$\sup_{0 \le t \le T} |S(t) - \mu t| \stackrel{a.s.}{=} O\left(\sqrt{T \log \log T}\right),\qquad(7.62)$$

$$\sup_{0 \le t \le T} |V(t) - mt| \stackrel{a.s.}{=} O\left(\sqrt{T \log \log T}\right),\qquad(7.63)$$

as $T \to \infty$.

We do not provide the detailed proofs of these results (interested readers are referred to some stochastic process limits books such as Whitt (2002)). However, to demonstrate the basic idea of proving these upper bounds on the supremum of the absolute deviation between the stochastic process and its fluid limit, we provide a detailed proof of the LIL for the symmetric random walk process in Chapter D, a simple version of the FLIL.

If the input processes satisfy the FLIL, then, as $T \to \infty$,

$$\sup_{0 \le t \le T} |Q(t) - \overline{Q}(t)| \stackrel{a.s.}{=} O\left(\sqrt{T \log \log T}\right),\qquad(7.64)$$

$$\sup_{0 \le t \le T} |Z(t) - \overline{Z}(t)| \stackrel{a.s.}{=} O\left(\sqrt{T \log \log T}\right),\qquad(7.65)$$

$$\sup_{0 \le t \le T} |B(t) - (\rho \wedge e)t| \stackrel{a.s.}{=} O\left(\sqrt{T \log \log T}\right),\qquad(7.66)$$

where Q, Z, and B are the queue length, the workload, and the busy period processes, respectively, with the fluid limit processes satisfying

$$Q = X + Y,\qquad(7.67)$$
$$X(t) = Q(0) + (\lambda - \mu)t,\qquad(7.68)$$
$$Y = \Psi(X),\qquad(7.69)$$
$$Z = mQ,\qquad(7.70)$$

where $Q = \Phi(X)$ is the reflection mapping as defined in Theorem 7.1. The proof can be found in Chen and Yao (2001). These results reveal the bounds on the absolute deviation between the major performance measures of the queueing system and their fluid limits.

These bounds provide some information about the quality of the approximations based on the fluid limits in asymptotic analysis. Also they are strong approximations due to the almost surely convergence.

If the rth moments ($r > 2$) for the arrival sequence u and service sequence v are finite, the deviation of the input processes from their diffusion limit processes can be also bounded when time horizon increases. These bounds are established based on the functional strong approximation theorem (FSAT) (See Chapter G). We use the symbol $\tilde{X}$ to represent the strong approximation of the process X. It follows from FSAT that

$$\sup_{0 \le t \le T} |A(t) - \tilde{A}(t)| \stackrel{a.s.}{=} o\left(T^{1/r}\right), \tag{7.71}$$

$$\sup_{0 \le t \le T} |S(t) - \tilde{S}(t)| \stackrel{a.s.}{=} o\left(T^{1/r}\right), \tag{7.72}$$

$$\sup_{0 \le t \le T} |V(t) - \tilde{V}(t)| \stackrel{a.s.}{=} o\left(T^{1/r'}\right), \tag{7.73}$$

where

$$\tilde{A}(t) = \lambda t + \lambda^{1/2} c_a W^0(t), \tag{7.74}$$

$$\tilde{S}(t) = \mu t + \mu^{1/2} c_s W^1(t), \tag{7.75}$$

$$\tilde{V}(t) = mt - m^{1/2} c_s W^0(mt). \tag{7.76}$$

Here, W^0 and W^1 are two independent standard Brownian motions. In (7.73), $r' = r$ if $r < 4$ and any $r' < 4$ if $r \ge 4$. Again, using the RMT, we have the following result.

Theorem 7.6 *If the FAST (7.74)–(7.76) hold with $r > 2$, then as $T \to \infty$,*

$$\sup_{0 \le t \le T} |Q(t) - \tilde{Q}(t)| \stackrel{a.s.}{=} o\left(T^{1/r'}\right), \tag{7.77}$$

$$\sup_{0 \le t \le T} |Z(t) - \tilde{Z}(t)| \stackrel{a.s.}{=} o\left(T^{1/r'}\right), \tag{7.78}$$

$$\sup_{0 \le t \le T} |B(t) - \tilde{B}(t)| \stackrel{a.s.}{=} o\left(T^{1/r'}\right), \tag{7.79}$$

where

$$\tilde{Q} = \tilde{X} + \tilde{Y},$$
$$\tilde{X}(t) = Q(0) + (\lambda - \mu)t + \lambda^{1/2} c_a W^0(t) - \mu^{1/2} c_s W^1((\rho \wedge 1)t),$$
$$\tilde{Y} = \Psi(\tilde{X}),$$
$$\tilde{Z}(t) = m\tilde{Q}(t) + m^{1/2} c_s [W^1((\rho \wedge 1)t) - W^1(\rho t)],$$
$$\tilde{B}(t) = t - \mu \tilde{Y}(t),$$

Ψ is the reflection mapping, $r' = r$ if $r < 4$, and any $r' < 4$ if $r \ge 4$.

Note that the little "o" notation indicates some loose upper bounds of these approximations. However, the growth rates of these bounds are in the order of

$T^{1/r}$ with $r \geq 2$. This theorem also implies that if higher moments for the arrival sequence and the service sequence exist, we can obtain more information about the deviations between the actual process and its approximation process. For example, if the fourth or higher moments exist (i.e., $r \geq 4$), then these bounds can be improved as

$$\sup_{0 \leq t \leq T} |Q(t) - \widetilde{Q}(t)| \stackrel{a.s.}{=} o\left(T^{1/r'}\right), \tag{7.80}$$

$$\sup_{0 \leq t \leq T} |Z(t) - \widetilde{Z}(t)| \stackrel{a.s.}{=} o\left(T^{1/r'}\right), \tag{7.81}$$

$$\sup_{0 \leq t \leq T} |B(t) - \widetilde{B}(t)| \stackrel{a.s.}{=} o\left(T^{1/r'}\right). \tag{7.82}$$

We do not present the proofs of these results. To get rigorous details of developing these results, interested readers are referred to the two papers by Komlos, Major, and Tusnfidy (1975, 1976) or the books by Csörgö and L. Horvath (1993) and Csörgö and Revesz (1981) although we also provide some basic ideas of establishing these results in Chapter G. For more discussion on the bounds of fluid and diffusion limit approximations, also see Chen and Yao (2001). To further refine the FSLLN, they also present the rate of convergence for fluid limits. The characterization of the rate of convergence is demonstrated by the probability that the deviation of the process of the nth system, denoted by $\bar{X}^n$, from its corresponding fluid limit, denoted by $\bar{X}$. Such a probability is bounded by an exponential function in the following form.

$$P(\|\bar{X}^n - \bar{X}\|_T \geq \varepsilon) \leq Ce^{-\phi(\varepsilon)n},$$

where C is a constant, and ϕ is a positive function on $(0, \infty)$.

The basic idea of characterizing the rate of convergence for fluid limits is based on the large deviations estimate that is presented in Chapter C. For rigorous proofs of these results, interested readers are referred to Csörgö and Revesz(1981).

Exercise 7.10 *For the $M/G/1$ queue considered in Exercise 7.9, use simulations to verify the input process FLIL (7.61)–(7.63) and the performance process FLIL (7.64)–(7.66) for the $M/G/1$ queue and extend the simulation analysis to the $G/G/1$ queue. Comment on the relation between T and the distribution parameters of service times and inter-arrival times in these FLIL.*

7.7 APPLICATIONS OF RBM APPROACH

7.7.1 A TWO-STATION TANDEM QUEUE

Consider a simple tandem queue with two servers in series. Customers arrive at the system according to a Poisson process with rate λ. Service times at stage

i are i.i.d. random variables with rate μ_i where $i = 1, 2$. Let $Q_i(t)$ be the queue process, $S_i(t)$ the service process, and $B_i(t)$ the busy period at stage i with $i = 1, 2$. Further, let $A(t)$ denote the arrival process. Now we can write

$$Q_1(t) = Q_1(0) + A(t) - S_1(B_1(t)),$$
$$Q_2(t) = Q_2(0) + S_1(B_1(t)) - S_2(B_2(t)),$$

where $B_i(t) = \int_0^t 1\{Q_i(s) > 0\} ds$ with $i = 1, 2$. Under heavy traffic condition (to be discussed in the next chapter):

$$(\lambda^n, \mu_1^n, \mu_2^n) \to (\lambda, \mu_1, \mu_2),$$
$$\theta_1^n := \sqrt{n}(\lambda^n - \mu_1^n) \to \theta_1,$$
$$\theta_2^n := \sqrt{n}(\mu_1^n - \mu_2^n) \to \theta_2.$$

The queue length processes for the nth system are scaled as

$$\hat{Q}_1^n = \frac{Q_1^n(nt)}{\sqrt{n}} = \hat{Q}_1^n(0) + \hat{A}^n(t) + \theta_1^n t - \hat{S}_1^n\left(\int_0^t 1\{\hat{Q}_1^n > 0\} ds\right)$$
$$+ \sqrt{n}\mu_1^n \int_0^t 1\{\hat{Q}_1^n(s) = 0\} ds,$$

$$\hat{Q}_2^n = \frac{Q_2^n(nt)}{\sqrt{n}} = \hat{Q}_2^n(0) + \hat{S}_1^n\left(\int_0^t 1\{\hat{Q}_1^n > 0\} ds\right) - \hat{S}_2^n\left(\int_0^t 1\{\hat{Q}_2^n > 0\} ds\right)$$
$$+ \theta_2^n t - \sqrt{n}\mu_1^n \int_0^t 1\{\hat{Q}_1^n(s) = 0\} ds + \sqrt{n}\mu_2^n \int_0^t 1\{\hat{Q}_2^n(s) = 0\} ds,$$

where $\hat{A}^n(t) = \frac{A^n(nt) - n\lambda^n t}{\sqrt{n}}$ and $\hat{S}_i^n(t) = \frac{S_i^n(nt) - n\mu_i^n t}{\sqrt{n}}$ with $i = 1, 2$. We can obtain a diffusion limit for this model. Specifically, $\hat{Q}_1^n$ and $\hat{Q}_2^n$ converge in distribution to $\hat{Q}_1$ and $\hat{Q}_2$, respectively, which are the solutions of

$$\hat{Q}_1(t) = \hat{Q}_1(0) + \hat{A}(t) - \hat{S}_1(B_1(t)) + \theta_1 t + Y_1(t),$$
$$\hat{Q}_2(t) = \hat{Q}_2(0) + \hat{S}_1(B_1(t)) - \hat{S}_2(B_2(t)) + \theta_2 t - Y_1(t) + Y_2(t),$$

which can be written in vector form as

$$\hat{\mathbf{Q}}(t) = \hat{\mathbf{Q}}(0) + \begin{pmatrix} \hat{A}(t) - \hat{S}_1(B_1(t)) \\ \hat{S}_1(B_1(t)) - \hat{S}_2(B_2(t)) \end{pmatrix} + \begin{pmatrix} \theta_1 \\ \theta_2 \end{pmatrix} t$$
$$+ \begin{pmatrix} 1 \\ -1 \end{pmatrix} Y_1(t) + \begin{pmatrix} 0 \\ 1 \end{pmatrix} Y_2(t)$$
$$= \hat{\mathbf{Q}}(0) + \begin{pmatrix} \hat{A}(t) - \hat{S}_1(B_1(t)) \\ \hat{S}_1(B_1(t)) - \hat{S}_2(B_2(t)) \end{pmatrix} + \begin{pmatrix} \theta_1 \\ \theta_2 \end{pmatrix} t$$
$$+ \begin{pmatrix} 1 & 0 \\ -1 & 1 \end{pmatrix} \begin{pmatrix} Y_1(t) \\ Y_2(t) \end{pmatrix}$$

$$= \hat{Q}(0) + \left(\begin{array}{c} \hat{A}(t) - \hat{S}_1(B_1(t)) \\ \hat{S}_1(B_1(t)) - \hat{S}_2(B_2(t)) \end{array} \right) + \left(\begin{array}{c} \theta_1 \\ \theta_2 \end{array} \right) t$$

$$+ (\mathbf{I} - \mathbf{P}') \left(\begin{array}{c} Y_1(t) \\ Y_2(t) \end{array} \right), \qquad (7.83)$$

where

$$\mathbf{I} = \left(\begin{array}{cc} 1 & 0 \\ 0 & 1 \end{array} \right), \quad \mathbf{P} = \left(\begin{array}{cc} 0 & 1 \\ 0 & 0 \end{array} \right).$$

Note that the approximations $B_1(t) \approx (\rho_1 \wedge 1)t = ((\lambda/\mu_1) \wedge 1)t$ and $B_2(t) \approx (\rho_2 \wedge 1)t = ((\lambda/\mu_2) \wedge 1)t$. Equation (7.83) is a special case of the diffusion limit approximation for a queueing network to be discussed in the next chapter. There are two purposes of presenting this example. First, it shows that the reflected BM approximation developed for a $G/G/1$ queue can be extended to a more complex queueing system. Secondly, it provides a good numerical example to demonstrate the general results for a complex queueing network to be presented in the next chapter. Using (7.83), we can derive the joint stationary distribution of the two queues using the reflected BM approximation approach. Readers may delay reading the following until after studying the next chapter. In other words, the following illustration can be served as a simple numerical example of the BM network presented in Section 8.1. To apply the formulas in Section 8.1, we need to write the following matrices for this example:

$$\mathbf{R} = \mathbf{I} - \mathbf{P}' = \left(\begin{array}{cc} 1 & 0 \\ -1 & 1 \end{array} \right),$$

$$\theta = \alpha - (\mathbf{I} - \mathbf{P}')\mu = \left(\begin{array}{c} \lambda \\ 0 \end{array} \right) - \left(\begin{array}{cc} 1 & 0 \\ -1 & 1 \end{array} \right) \left(\begin{array}{c} \mu_1 \\ \mu_2 \end{array} \right) = \left(\begin{array}{c} \lambda - \mu_1 \\ \mu_1 - \mu_2 \end{array} \right).$$

It is easy to check the condition

$$\mathbf{R}^{-1}\theta = \left(\begin{array}{cc} 1 & 0 \\ 1 & 1 \end{array} \right) \left(\begin{array}{c} \lambda - \mu_1 \\ \lambda - \mu_2 \end{array} \right) = \left(\begin{array}{c} \lambda - \mu_1 \\ \lambda - \mu_2 \end{array} \right) < 0.$$

To obtain the stationary distribution, three more matrices (with the same notations as those in Chapter 8) required for the case with the exponential service and inter-arrival times are given by

$$\mathbf{D} = \left(\begin{array}{cc} 1 & 0 \\ 0 & 1 \end{array} \right), \quad \Gamma = \left(\begin{array}{cc} 2\lambda & -\lambda \\ -\lambda & \lambda \end{array} \right), \quad \Lambda = \left(\begin{array}{cc} 2\lambda & 0 \\ 0 & \lambda \end{array} \right).$$

Now, we can check the condition for a product-form joint distribution for a queueing network (to be presented in Chapter 8)

$$\mathbf{R}\mathbf{D}^{-1}\Lambda + \Lambda\mathbf{D}^{-1}\mathbf{R}' = \left(\begin{array}{cc} 1 & 0 \\ -1 & 1 \end{array} \right) \left(\begin{array}{cc} 1 & 0 \\ 0 & 1 \end{array} \right) \left(\begin{array}{cc} 2\lambda & 0 \\ 0 & \lambda \end{array} \right)$$

$$+ \left(\begin{array}{cc} 2\lambda & 0 \\ 0 & \lambda \end{array} \right) \left(\begin{array}{cc} 1 & 0 \\ 0 & 1 \end{array} \right) \left(\begin{array}{cc} 1 & -1 \\ 0 & 1 \end{array} \right)$$

$$= \begin{pmatrix} 2\lambda & 0 \\ -2\lambda & \lambda \end{pmatrix} + \begin{pmatrix} 2\lambda & -2\lambda \\ 0 & \lambda \end{pmatrix}$$

$$= \begin{pmatrix} 4\lambda & -2\lambda \\ -2\lambda & 2\lambda \end{pmatrix} = 2\Gamma.$$

Define the stationary distribution of the diffusion limit model $F(z_1,z_2) = \lim_{t\to\infty} P(\hat{Q}_1 \le z_1, \hat{Q}_2 \le z_2)$ with the density function of $f(z_1,z_2)$. With the condition above, we know that this distribution has a product form as expected with $f(z_1,z_2) = \prod_{j=1}^{2} f_j(z_j)$ with

$$f_j(z) = \begin{cases} \eta_j e^{-\eta_j z}, & z \ge 0, \\ 0, & z < 0, \end{cases} \quad j = 1,2,$$

where

$$\eta = -2\Lambda^{-1} \mathbf{D} \mathbf{R}^{-1} \theta$$

$$= -2 \begin{pmatrix} \frac{1}{2\lambda} & 0 \\ 0 & \frac{1}{\lambda} \end{pmatrix} \begin{pmatrix} 1 & 0 \\ 0 & 1 \end{pmatrix} \begin{pmatrix} \lambda - \mu_1 \\ \lambda - \mu_2 \end{pmatrix} = \begin{pmatrix} \frac{\mu_1 - \lambda}{\lambda} \\ \frac{2(\mu_2 - \lambda)}{\lambda} \end{pmatrix}.$$

Next, we show that with general i.i.d. inter-arrival times and service times, the stationary distribution does not necessarily have the product form. This is done by checking that the condition $2\Gamma = \mathbf{R} \mathbf{D}^{-1} \Lambda + \Lambda \mathbf{D}^{-1} \mathbf{R}'$ does not hold in such a general case. The Γ matrix can be determined by the following formula presented in the next chapter

$$\Gamma_{kl} = \sum_{j=1}^{N} (\lambda_j \wedge \mu_j)[p_{jk}(\delta_{kl} - p_{jl}) + c_{s_j}^2 (p_{jk} - \delta_{jk})(p_{jl} - \delta_{jl})]$$

$$+ \lambda_k c_{A_j}^2 \delta_{kl}, \quad k,l = 1,\ldots,N,$$

where $c_{s_j}^2$ and $c_{A_j}^2$ are the squared coefficients of variation for the service time of server j and the inter-arrival times of external arrivals to server j. With $N=2$, Γ and Λ are given by

$$\Gamma = \begin{pmatrix} \lambda(c_{A_1}^2 + c_{s_1}^2) & -\lambda c_{s_1}^2 \\ -\lambda c_{s_1}^2 & \lambda c_{s_2}^2 \end{pmatrix}, \quad \Lambda = \begin{pmatrix} \lambda(c_{A_1}^2 + c_{s_1}^2) & 0 \\ 0 & \lambda c_{s_2}^2 \end{pmatrix}.$$

In this general case, the condition for a product-form joint distribution does not hold. That is

$$\mathbf{R} \mathbf{D}^{-1} \Lambda + \Lambda \mathbf{D}^{-1} \mathbf{R}' = \begin{pmatrix} 1 & 0 \\ -1 & 1 \end{pmatrix} \begin{pmatrix} 1 & 0 \\ 0 & 1 \end{pmatrix} \begin{pmatrix} \lambda(c_{A_1}^2 + c_{s_1}^2) & 0 \\ 0 & \lambda c_{s_2}^2 \end{pmatrix}$$

$$+ \begin{pmatrix} \lambda(c_{A_1}^2 + c_{s_1}^2) & 0 \\ 0 & \lambda c_{s_2}^2 \end{pmatrix} \begin{pmatrix} 1 & 0 \\ 0 & 1 \end{pmatrix} \begin{pmatrix} 1 & -1 \\ 0 & 1 \end{pmatrix}$$

$$= \begin{pmatrix} \lambda(c_{A_1}^2 + c_{s_1}^2) & 0 \\ -\lambda(c_{A_1}^2 + c_{s_1}^2) & \lambda c_{s_2}^2 \end{pmatrix}$$

$$+ \begin{pmatrix} \lambda(c_{A_1}^2 + c_{s_1}^2) & -\lambda(c_{A_1}^2 + c_{s_1}^2) \\ 0 & \lambda c_{s_2}^2 \end{pmatrix}$$

$$= \begin{pmatrix} 2\lambda(c_{A_1}^2 + c_{s_1}^2) & -\lambda(c_{A_1}^2 + c_{s_1}^2) \\ -\lambda(c_{A_1}^2 + c_{s_1}^2) & 2\lambda c_{s_2}^2 \end{pmatrix} \neq 2\Gamma.$$

Therefore, we do not expect to have the simple product-form stationary distribution in general for a queueing network. It is clear that when the coefficients of variation of the service times and inter-arrival times are 1, then we can obtain the product-form stationary distribution.

Exercise 7.11 *Extend the two-station tandem queueing model to the n-station tandem queueing model by using the reflected BM approximation approach.*

7.7.2 A PRODUCTION-INVENTORY MODEL

Now we consider a production-inventory system with partial backlogging in which the inventory level can be modeled as a reflected BM and the production rate is controlled by a two-threshold policy. It is assumed that this model is resulted from appropriate diffusion scaling of the production-inventory system with discrete items. This example is taken from Lin (2017). Specifically, the mean demand rate is denoted by μ and the production rate can be set at one of two levels. Denote by r_i where $i = 1, 2$ and $0 \leq r_1 < \mu < r_2$. The production control policy, denoted by (l, L) with $0 \leq l < L \leq \infty$, prescribes that the rate is switched from r_1 (low rate) to r_2 (high rate) whenever the inventory level reduces to the low threshold l and the rate is switched from r_2 to r_1 whenever the inventory level reaches the high threshold L. Moreover, it is assumed that (a) the system is a single production facility with infinite capacity; (b) the shortages are allowed and the total amount of stock-out is treated as a mixture of back-ordered and lost sales; and (c) the inventory process is driven by two reflected BMs under the two production rates, respectively. With these assumptions, we can model the inventory process as the reflected BMs. When the production starts with rate r_1 at the inventory level L, the inventory process from L to l is driven by a BM

$$X^{(1)}(t) = X^{(1)}(0) + (r_1 - \mu)t + \sigma B(t) \tag{7.84}$$

with $r_1 - \mu < 0$ and $B(t)$ as the standard BM. When the production rate changes from r_1 to r_2, the inventory process from l to L is driven by the reflected BM

$$\hat{Z}(t) = X^{(2)}(t) + (1 - \beta)Y(t), \tag{7.85}$$

where

$$X^{(2)}(t) = X^{(2)}(0) + (r_2 - \mu)t + \sigma B(t). \tag{7.86}$$

Here $X^{(2)}(0) = l$, $r_2 - \mu > 0$, and $0 \leq \beta \leq 1$ is the fraction of demand during the stock-out period that will be back-ordered.

Note that when $\beta = 0$ (i.e., no back-ordering), we have the case with completely lost sales. Then $\hat{Z}(t) = X^{(2)}(t) + Y(t) \geq 0$ where $X^{(2)}(t) = l + (r_2 - \mu)t + \sigma B(t)$, and $Y(t) = \sup_{0 \leq s \leq t}[-X^{(2)}(s)]^+$ which satisfies (1) $Y(t)$ is continuous and increasing with $Y(0) = 0$, $X^{(2)}(0) = l$; (2) $\hat{Z}(t) \geq 0$ for all $t \geq 0$, and (3) $Y(t)$ increases only when $\hat{Z}(t) = 0$ as specified earlier. $Y(t)$ is also called the local time of $X^{(2)}(t)$ at level 0 in the literature on inventory models.

Clearly, $\hat{Z}(t)$ is a reflected BM which arises naturally as a diffusion approximation in inventory theory (also called regulated BM with a lower control barrier at zero). We consider a cost structure consisting of holding cost, switch-over cost, and stock-out cost. Usually, the holding cost function denoted by $h(x)$ is assumed to be a linear or convexly increasing function of the inventory level x. Here we consider the simplest case of $h(x) = x$. An inventory cycle, denoted by C, can be defined as the time interval between two consecutive r_1 to r_2 switching instants. The switch-over cost is a fixed value for two switchings (i.e., from r_1 to r_2 and from r_2 to r_1 per inventory cycle and is denoted by S). The stock-out cost is assumed to have two components. For every unit short, there is a base-cost c, which may represent the future price discount. Further, there is also an additional cost c_0 per unit if the demand is completely lost, which represents the opportunity cost of the lost sales. Denote by $T_x^X(y)$ the first time at which the process X, starting at state x, hits state y. That is

$$T_x^X(y) = \inf\{t \geq 0 | X(0) = x \text{ and } X(t) = y\}.$$

With this definition, the expected cycle time, denoted by $E[C]$ can be written as

$$E[C] = E_L[T_L^{X^{(1)}}(l)] + E_l[T_l^{\hat{Z}}(L)], \qquad (7.87)$$

where $E_x[\cdot]$ is the expectation operator associated with the law of the underlying process when the process starts at x. The expected holing cost during an inventory cycle, denoted by $EHC(l, L)$ can be written as

$$EHC(l, L) = E_L\left[\int_0^{T_L^{X^{(1)}}(l)} X^{(1)}(s)ds\right] + E_l\left[\int_0^{T_l^{\hat{Z}(L)}} \hat{Z}(s)ds\right]. \qquad (7.88)$$

The other two types of costs during an inventory cycle, switch-over cost and stock-out cost, are simply S and $[c + c_0(1 - \beta)]E_l[Y(T_l^{\hat{Z}}(L))]$ as the expected shortage during the cycle is given by $E_l[Y(T_l^{\hat{Z}}(L))]$. Thus, the total expected cost during the cycle, denoted by $TEC(l, L)$ is given by

$$TEC(l, L) = S + EHC(l, L) + [c + c_0(1 - \beta)]E_l[Y(T_l^{\hat{Z}}(L))]. \qquad (7.89)$$

According to the renewal reward theorem, we can find the long-term expected cost rate, denoted by $J(l,L)$, as $J(l,L) = TEC(l,L)/E[C]$. To determine the optimal two-threshold policy, we can formulate the following problem:

$$\min_{(l,L)} J(l,L) = \frac{TEC}{E[C]}$$

$$= \frac{1}{E_L[T_L^{X^{(1)}}(l)] + E_l[T_l^{\hat{Z}}(L)]} \left\{ S + E_L \left[\int_0^{T_L^{X^{(1)}}(l)} X^{(1)}(s) ds \right] \right.$$

$$\left. + E_l \left[\int_0^{T_l^{\hat{Z}}(L)} \hat{Z}(s) ds \right] + c E_l[Y(T_l^{\hat{Z}}(L))] + c_0(1-\beta) E_l[Y(T_l^{\hat{Z}}(L))] \right\}.$$

$$(7.90)$$

Next, we derive the explicit expressions of the key terms in $J(l,L)$. We introduce $\mu_1 = r_1 - \mu < 0$ and $\mu_2 = r_2 - \mu > 0$ used below. All results are based on the following lemma:

Lemma 7.1 *Let $f : \mathscr{R} \to \mathscr{R}$ be a twice continuously differentiable function, then*

$$E_x[f(\hat{Z}(t \wedge T_x(y)))] = f(x) + E_x \left[\int_0^{t \wedge T_x(y)} \Gamma f(\hat{Z}(s)) ds \right] + f'(0) E_x[Y(t \wedge T_x(y))],$$

$$(7.91)$$

where $\Gamma f = \frac{1}{2}\sigma^2 f'' + \mu_2 f'$.

This result follows from the Itô's formula and the property that Y is increasing and continuous with $Y(0) = 0$ and Y increases only when $\hat{Z} = 0$. For a brief introduction to stochastic calculus, readers are referred to Chapter H. Using this lemma, we can obtain the following results:

1. The total expected holding cost incurred in the period from the instant when the inventory level is at l to the instant when the inventory level reaches L is given by

$$E_l \left[\int_0^{T_l^{\hat{Z}}(L)} \hat{Z}(s) ds \right] = \frac{L^2 - l^2}{2\mu_2} - \frac{\sigma^2(L^2 - l^2)}{2\mu_2^2}$$

$$+ \frac{\sigma^2(e^{-2\mu_2 l/\sigma^2} - e^{-2\mu_2 L/\sigma^2})}{4\mu_2^3}.$$

$$(7.92)$$

2. The total expected holding cost incurred in the period from the instant when the inventory level is at L to the instant when the inventory level

reaches l is given by

$$E_L \left[\int_0^{T_L^{X^{(1)}}(l)} X^{(1)}(s)ds \right] = -\frac{L^2 - l^2}{2\mu_1} + \frac{\sigma^2(L^2 - l^2)}{2\mu_1^2}$$
$$- \frac{\sigma^4(e^{2\mu_1(L-l)/\sigma^2} - 1)}{4\mu_1^3}.$$

(7.93)

3. The expected shortage per inventory cycle is given by

$$E_l[Y(T_l^{\hat{Z}}(L))] = \frac{\sigma^2(e^{-2\mu_2 l/\sigma^2} - e^{-2\mu_2 L/\sigma^2})}{2\mu_2}.$$

(7.94)

4. The expected time interval from the instant at which the inventory level is at l to the instant at which the inventory level reaches L is given by

$$E_l[T_l^{\hat{Z}}(L)] = \frac{L-l}{\mu_2} + \frac{\sigma^2 e^{-2\mu_2 l/\sigma^2}(e^{-2\mu_2(L-l)/\sigma^2} - 1)}{2\mu_2^2}.$$

(7.95)

5. The expected time interval from the instant at which the inventory level is at L to the instant at which the inventory level reaches l is given by

$$E_L[T_L^{X^{(1)}}(l)] = -\frac{L-l}{\mu_1}.$$

(7.96)

We provide the derivation of these results and start with the expected shortage per inventory cycle of (7.94).

Choosing $f(z) = e^{\lambda z}$ with $\lambda > 0$ in Lemma 7.1, we have $\Gamma f(z) = g(\lambda)f(z)$ where $g(\lambda) = (1/2)\sigma^2 \lambda^2 + \mu_2 \lambda > 0$. It follows from $\hat{Z}(t) \geq 0$ for all $t \geq 0$ and $f(\hat{Z}(t)) = e^{\lambda \hat{Z}(t)} \geq 1$ that

$$E_x \left[\int_0^{t \wedge T_x(y)} \Gamma f(\hat{Z}(s))ds \right] = E_x \left[\int_0^{t \wedge T_x(y)} g(\lambda)f(\hat{Z}(s))ds \right] \geq g(\lambda)E_x[t \wedge T_x(y)].$$

(7.97)

Based on Lemma 7.1, we have

$$E_x[f(\hat{Z}(t \wedge T_x(y)))] \geq f(x) + g(\lambda)E_x[t \wedge T_x(y)] + \lambda E_x[Y((t \wedge T_x(y)))]. \quad (7.98)$$

Note that $f(\hat{Z}(t \wedge T_x(y))) \leq f(y)$ as f is an increasing function and $Z(t \wedge T_x(y)) \leq y$ and $Y(\cdot) \geq 0$. Thus, Using these inequalities and (7.98), we get

$$f(y) \geq E_x[f(\hat{Z}(t \wedge T_x(y)))]$$
$$\geq f(x) + g(\lambda)E_x[t \wedge T_x(y)] + \lambda E_x[Y((t \wedge T_x(y)))]$$
$$\geq f(x) + g(\lambda)E_x[t \wedge T_x(y)],$$

which results in

$$E_x[t \wedge T_x(y)] \leq \frac{1}{g(\lambda)}[f(y) - f(x)]. \tag{7.99}$$

In addition, $(t \wedge T_x(y) \uparrow T_x(y)$ as $t \to \infty$ which implies that $E_x[T_x(y)]$ is upper bounded by a finite number. Thus, it follows from the monotone convergence theorem that $E_x[T_x(y)] < \infty, 0 \leq x \leq y$, leads to $T_x(y) < \infty$ a.s. as $t \to \infty$. Since f and $\hat{Z}$ are continuous, we have $f(\hat{Z}(t \wedge T_x(y))) \to f(\hat{Z}(T_x(y)))$ as $t \to \infty$. Furthermore, by the dominated convergence, we have

$$E_x[f(\hat{Z}(t \wedge T_x(y)))] \to E_x[f(\hat{Z}(T_x(y)))] = E_x[f(y)] = f(y), \tag{7.100}$$

which will be used in (7.91) in the limiting case. To treat the second term on the right-hand side of (7.91), we introduce a measure on Borel subsets on the state space of the reflected RM process. For any Borel set B of $[0, y]$, we define the measure v on B by

$$v(B) = E_x\left[\int_0^{T_x(y)} I_B(\hat{Z}(s))ds\right],$$

where $I_B(x)$ is the indicator of set B. This measure can be interpreted as the expected duration that $\hat{Z}(t)$ stays in the subset B in the time interval $[0, T_x(y)]$. For a special case of $B = [0, y]$, $v([0, y]) = E_x[T_x(y)]$ holds. Then, we get

$$E_x\left[\int_0^{t \wedge T_x(y)} \Gamma f(\hat{Z}(s))ds\right] = \int_{[0,y]} \Gamma f(z)v(dz).$$

Such a relation can be justified by the following: it holds by definition if Γf is the indicator of a set; then it holds by linearity if Γf is a simple function; finally, it holds in general by the monotone convergence. Since $f(z) = e^{\lambda z}$ for some real $\lambda > 0$, then $\Gamma f(z) = (\frac{1}{2}\sigma^2\lambda^2 + \mu_2\lambda)e^{\lambda z}$ and $f'(0) = \lambda$. To evaluate the last term on the right-hand side of (7.89), we set $\lambda = -2\mu_2/\sigma^2$ so that $\Gamma f(z) = 0$ to make the second term on the right-hand side vanished. By taking the limit of $t \to \infty$ on both sides of (7.91) and using (7.100), we obtain

$$f(y) - f(x) = \int_{[0,y]} \Gamma f(z)v(dz) + f'(0)E_x[Y(T_x(y))]$$
$$= \lambda E_x[Y(T_x(y))], \tag{7.101}$$

which results in the expression of the expected shortage per inventory, $E_l[Y(T_l^{\hat{Z}}(L))]$ in (7.94), if $x = l$ and $y = L$, which is utilized in evaluating the last term of (7.89).

Second, letting $f(z) = z$ in

$$f(y) - f(x) = \int_{[0,y]} \Gamma f(z)v(dz) + f'(0)E_x[Y(T_x(y))] \tag{7.102}$$

yields

$$
\begin{aligned}
y - x &= \int_{[0,y]} \mu_2 v(dz) + E_x[Y(T_x(y))] \\
&= v([0,y]) + E_x[Y(T_x(y))] \\
&= E_x[T_x(y)] + E_x[Y(T_x(y))].
\end{aligned}
\tag{7.103}
$$

Re-writing (7.103) with $x = l, y = L$ and (7.94), we obtain the expression of $E_l[T_l^{\hat{Z}}(L)]$ in (7.95).

Third, we consider the process $X(t) = x + \mu_1 t + \sigma B(t)$ and define $\tau_1 = \inf\{t \geq 0 | X(t) = l\}$ and $\tau_2 = \inf\{t \geq 0 | X(t) = L\}$. Setting $x = 0$ and using itô formula with $f(x) = x$ we obtain

$$
X(t) = 0 + \int_0^t \mu_1 ds + \int_0^t \sigma dB(t).
$$

Replace t with $t \wedge \tau_i, i = 1, 2$ and take the expectation E_0 on both sides of the equation above. Then it follows from the martingale stopping theorem that

$$
E_0[X(t \wedge \tau_i)] = \mu_1 E_0[t \wedge \tau_i].
\tag{7.104}
$$

Due to $X(t \wedge \tau_i) \leq M, i = 1, 2$, we can use the bounded and monotone theorems to get

$$
E_0[X(t \wedge \tau_i)] \to E_0[X(\tau_i)] = \begin{cases} l & \text{if } i = 1, \\ L & \text{if } i = 2, \end{cases} \quad \text{as } t \to \infty,
$$

and

$$
E_0[t \wedge \tau_i] \to E_0[\tau_i] \text{ as } t \to \infty.
$$

Thus, $l = \mu_1 E_0[\tau_1]$ and $L = \mu_2 E_0[\tau_2]$. Since $\tau_1 - \tau_2 \overset{d}{=} \tau_{l-L}$, where $\tau_{l-L} = \inf\{t \geq 0 | X(t) = l - L\}$, we have

$$
E_L[T_L^{X^{(1)}}(l)] = E_0[\tau_1] - E_0[\tau_2],
$$

which is the result (7.96).

Finally, we derive the expected holding costs (7.92) and (7.93). Let $T_x^{\hat{Z}}(y) = \inf\{t \geq 0 | \hat{Z}(0) = x \text{ and } \hat{Z} = y\}$ for $0 \leq x \leq y$. Applying Itô's formula to $\hat{Z}(t)$, we have

$$
f(\hat{Z}(t)) = f(x) + \sigma \int_0^t f'(\hat{Z}(s)) dB(s) + \int_0^t \phi(\hat{Z}(s)) ds + (1 - \beta) f'(0) \hat{Y}(t),
$$

where $\phi(z) \equiv \frac{1}{2}(z) f''(z) + \mu_2 f'(z)$. Replacing t with $T_x^{\hat{Z}}(y)$, taking the expectation E_x, specializing to $f(x) = 0$ and $f'(0) = 0$, and recognizing that $\int_0^t f'(\hat{Z}(s)) dB(s)$ is a martingale, we get

$$
E_x\left[f\left(\hat{Z}\left(T_x^{\hat{Z}}(y)\right)\right)\right] = E_x\left[\int_0^{T_x^{\hat{Z}}(y)} \phi(\hat{Z}(s)) ds\right],
$$

which is equivalent to

$$f(y) = E_x \left[\int_0^{T_x^{\hat{Z}}(y)} \phi(\hat{Z}(s)) ds \right].$$

(7.105)

To utilize this relation to find the expected holding cost incurred for the half of the cycle (i.e., the time period for the inventory level changing from l to L), we obtain f by solving the differential equation

$$\frac{1}{2}\sigma^2 f''(z) + \mu_2 f'(z) = \phi(z)$$

with conditions $f(x) = 0$ and $f'(0) = 0$. Note that this is a second-order non-homogeneous differential equation with constant coefficients and f term missing. It follows from the routine solution that

$$f(z) = \int_x^z \int_0^\eta \frac{2\phi(\xi)}{\sigma^2} e^{-2\mu_2\eta/\sigma^2 + 2\mu_2\xi/\sigma^2} d\xi d\eta.$$

(7.106)

Using (7.105) and (7.106) and choosing $\phi(z) = z$ with $x = l, z = L$, we have

$$E_l \left[\int_0^{T_l^{\hat{Z}}(L)} \hat{Z}(s) ds \right] = \int_l^L \int_0^\eta \frac{2\xi}{\sigma^2} e^{-2\mu_2\eta/\sigma^2 + 2\mu_2\xi/\sigma^2} d\xi d\eta,$$

which results in (7.92). Similarly, we consider that the sample path of $L - X^{(1)}(t)$ goes from 0 to $L - l$ with positive drift $-\mu_1$ and variance σ^2. Then we can get

$$E_L \left[\int_0^{T_L^{X^{(1)}}(l)} X^{(1)}(s) ds \right] = \frac{L(l - L)}{\mu_1} - \int_0^{L-l} \int_0^\eta \frac{2\xi}{\sigma^2} e^{2\mu_1\eta/\sigma^2 - 2\mu_1\xi/\sigma^2} d\xi d\eta,$$

which, after some calculations, leads to (7.93).

Using (7.92)–(7.96) in (7.90), we can determine the optimal production control policy (l^*, L^*) that minimizes the long-term average cost rate for the system. Due to the complexity of the expression of $J(l, L)$, we need the mathematical software package such as Mathematica or Matlab to find the optimal policy. For example, using Mathematica, we can iteratively solve simultaneous equations that are obtained by differentiating $J(l, L)$ with respect to l and L, respectively, and setting the derivatives to zero. The results for minimizing $J(l, L)$ from the necessary conditions for the optimality are checked by numerical experiment and graphical illustrations. Here is an example taken from Lin (2017).

Example 7.1 *Suppose that a production facility makes items with two production rates, $r_1 = 0.55$ units/day and $r_2 = 3.83$ units/day, to meet the random market demand with mean $\mu = 2.19$ units/day and standard deviation
$\sigma = \sqrt{20}$ units/day. The set-up cost is $S = \$2000$. The stock-out cost param-
eters are $c = \$50$/unit, $c_0 = \$150$/unit and the proportion of lost sales is $\beta =
0.8$. For such a system, by setting the first-order conditions $(\partial J(l, L)/\partial l = 0$
and $\partial J(l, L)/\partial L = 0)$ and using numerical search, we obtain $l = 2.89$ and
$L = 66.25$. The best of all four possible rounded integer solutions is $l^* = 3$
and $L^* = 66$ with $J^*(3, 66) = \$66.26$ and $E_L\left[T_L^{X^{(1)}}(l^*)\right] = 38.41$ days and
$E_l\left[T_L^2(L^*)\right] = 36.14$ days.*

For more discussions on the numerical examples for this production-inventory
model, interested readers are referred to Lin (2017).

Exercise 7.12 *Use a software package such as Matlab or Mathematica to develop a computational program based on the production-inventory model discussed. Based on the same data as Example 7.1, examine the effects of changes
in the values of the parameters β and c_0 on the optimal policy (l^*, L^*) and its
associated long-term expected cost rate $J(l^*, L^*)$. Choose the range for β to be
$[0.0, 1.0]$ with increment of 0.2 and the range for S to be $[2000, 3000]$ with increment of 150. Make comments on the relations between the back-order policy
and the optimal production policy.*

Exercise 7.13 *An advantage of using the reflected BM approximation-based
model to study a production-inventory system is the robustness of the results.
Explain this claim.*

REFERENCE NOTES

This relatively short chapter only presents an introduction of using the reflected BM to analyze simple stochastic systems such as a simple queueing network or production-inventory system. This basic approach starts with properly
scaling the input and performance processes, applying FSLLN or FCLT combined with RMT to obtain the fluid or diffusion process limits, characterizing
the deviation between the original processes and their corresponding process
limits via FLIL, and further refining the approximations by FSAT and exponential rate of convergence. For more details of this approach, there are many
excellent books such as [7] and[1]. The inventory-production model is based
on the work by [6].

REFERENCES

1. H. Chen and D.D. Yao, "Fundamentals of Queueing Networks: Performance, Asymptotics, and Optimization", Springer, New York, 2001.
2. M. Csorgo and P. Revesz, "Strong Approximations in Probability and Statistics", Academic Press, 1981.
3. M. Csörgö and L. Horvath, "Weighted Approximations in Probability and Statistics", Wiley and Sons, 1993.
4. J. Komlos, P. Major, and G. Tusnady, An approximation of partial sums of independent rv's and the sample df. I. *Wahrsch verw Gebiete/Probability Theory and Related Fields,* 32, 111–131, 1975.
5. J. Komlos, P. Major, and G. Tusnady, An approximation of partial sums of independent rv's and the sample df. I. *Wahrsch verw Gebiete/Probability Theory and Related Fields*, 34, 33–58, 1976.
6. H. Lin, Two-critical-number control policy for a stochastic production inventory system with partial backlogging. *International Journal of Production Research*, 55, 14, 4123–4135, 2017.
7. W. Whitt, "Stochastic-Process Limits: An Introduction to Stochastic-Process Limits and Their Application to Queues", Springer 2002.

8 Large Queueing Systems

In this chapter, we discuss some large stochastic systems such as general queueing networks in which the arrival processes, routing processes, and service processes are all general stochastic processes. Due to the complexity and scale of such systems, using the exact analysis is difficult. Thus, we focus on the approximation methods. We utilize an open network model to demonstrate the approximation approach based on stochastic process limits. Then we show the decomposition approach which is based on the $G/G/1$ approximation formulas presented in the previous chapter. To make this chapter concise, we refer the proofs of theorems and propositions to the books and research papers where the details are available.

8.1 MULTI-DIMENSIONAL REFLECTED BROWNIAN MOTION APPROXIMATION TO QUEUEING NETWORKS

Consider a queueing network with N single server stations, indexed by $i = 1,...,N$. Customers arrive to the network externally according to a renewal process and service times at each station (node) follow general distributions. Specifically, customers arrive at node i externally according to a renewal process with rate λ_i and squared coefficient of variation (SCOV) $C_{a_i}^2$ for the inter-arrival time (between the $(i-1)$st and ith arrivals), denoted by a_i. The service times at node i are i.i.d. random variables, denoted by s_i, with mean $1/\mu_i$ and SCOV $C_{s_i}^2$. After a customer's service at node i, the customer exits the network with probability q_i, or joins the queue at node j with probability p_{ij}. Clearly, $q_i = 1 - \sum_{j=1}^{N} p_{ij}$. The $N \times N$ matrix $P = [p_{ij}]$ is called a routing matrix. If P has a spectral radius (or equivalently, its largest real eigenvalue) less than one, we call the network an open network. (A number of equivalent conditions for such a property are given in Chen and Yao (2001)). Similar to the treatment of $G/G/1$ queue in the previous chapter, we start with the primitive data (input processes). Denote by $E = \{i : \lambda_i > 0\}$ the set of indices for those nodes that have external arrivals. For $i \in E$, denote by $A_i = \{A_i(t), t \geq 0\}$ the number of customers arrived at node i from outside of the network during $(0,t]$. That is $A_i(t) := \sup\{n : \sum_{j=1}^{n} a_i(j) \leq t\}$ where $a_i(j)$ is the inter-arrival time between $(j-1)$st and jth arrivals at node i. Clearly, $A_i \equiv 0$ for $i \notin E$. Similarly, denote by $S_i = \{S_i(t), t \geq 0\}$ the number of customers served at node i during $(0,t]$. That is $S_i(t) := \sup\{n : \sum_{j=1}^{n} s_i(j) \leq t\}$, where $s_i(j)$ is the service time of the jth job at node i. While these arrival and service renewal processes are similar to those defined in a $G/G/1$ queue, we need to introduce the routing sequence to model the queueing network. Define a sequence of N-dimensional i.i.d. random vectors, denote by $r^k = \{r^k(j), j \geq 1\}$, $k = 1, ..., N$, taking values in set $\{e^1, ..., e^N, \mathbf{0}\}$, where e^i is the ith unit vector in $\mathscr{R}^N$. The event $r^k(j) = e^i$ represents that the jth customer completing service at node k will join node i. Note that $r^k(j) = \mathbf{0}$ indicates that this customer will leave the network after the service at node k. Now we can define the following N N-dimensional vectors:

$$R^k(0) = \mathbf{0} \quad \text{and} \quad R^k(n) = \sum_{j=1}^{n} r^k(j), \quad n \geq 1,$$

where the ith component, denoted by $R_i^k(n)$, $k, i = 1, ..., N$, which represents the number of customers completed service at node k joining node i when the total number of customers departed from node k is n. The routing vector sequence then can be written as $R^k = \{R^k(n), n \geq 0\}$ when n customers have been served by node k. The initial queue length at node i is denoted by $Q_i(0)$. The arrival process, service process, routing sequence, and initial queue length form the primitive data that can also be called input processes. Based on these input processes, we can construct some performance measure processes that can be also called output processes. These processes include the queue-length process $Q = \{Q(t), t \geq 0\}$, the workload process $Z = \{Z(t), t \geq 0\}$, and busy period process $B = \{B(t), t \geq 0\}$. Note that these performance measure processes are N-dimensional processes with the ith component representing the performance measures at node i. Based on the flow balance principle, we can write

$$Q_i(t) = Q_i(0) + A_i(t) + \sum_{k=1}^{N} R_i^k(S_k(B_k(t))) - S_i(B_i(t)), \tag{8.1}$$

for all $t \geq 0$ and $i = 1, ..., N$. It follows from the work-conserving rule that

$$B_i(t) = \int_0^t 1\{Q_i(s) > 0\} ds. \tag{8.2}$$

for all $t \geq 0$ and $i = 1, ..., N$. Again, we can show that these two relations uniquely determine the queue length process and the busy period process. The ith component of the workload process and the idle period process, respectively, can be written as

$$Z_i(t) = V_i \left(Q_i(0) + A_i(t) + \sum_{k=1}^{N} R_i^k(S_k(B_k(t))) \right) - B_i(t)$$
$$= V_i(Q_i(t) + S_i(B_i(t))) - B_i(t),$$

and

$$I_i(t) = t - B_i(t) = \int_0^t 1\{Q_i(s) = 0\} ds$$

for all $t \geq 0$ and $k = 1, ..., N$. For the ith component of each input process (i.e., node i), using the FSLLN, we have

$$\frac{1}{t}A_i(t) \to \lambda_i, \quad \frac{1}{t}S_i(t) \to \mu_i, \quad \frac{1}{t}R_j^i(t) \to p_{ij}, \tag{8.3}$$

as $t \to \infty$. In terms of matrix form with centering, we can re-write (8.1) as

$$\mathbf{Q}(t) = \mathbf{X}(t) + (\mathbf{I} - \mathbf{P}')\mathbf{Y}(t), \tag{8.4}$$

where the ith component of $\mathbf{X}(t)$ and $\mathbf{Y}(t)$ are given by

$$X_i(t) = Q_i(0) + \left(\lambda_i + \sum_{k=1}^{N} \mu_k p_{ki} - \mu_i\right)t + [A_i(t) - \lambda_i t]$$

$$+ \sum_{k=1}^{N} p_{ki}[S_k(B_k(t)) - \mu_k B_k(t)] - [S_i(B_i(t)) - \mu_i B_i(t)]$$

$$+ \sum_{k=1}^{N} [R_i^k(S_k(B_k(t))) - p_{ki}S_i(B_i(t))], \tag{8.5}$$

$$Y_i(t) = \mu_i[t - B_i(t)] = \mu_i I_i(t) = \mu_i \int_0^t 1\{Q_i(s) = 0\}ds,$$

for $k = 1, ..., N$. Similar to the $G/G/1$ queue case, for all $t \geq 0$,

$$\mathbf{Q}(t) \geq 0,$$
$$d\mathbf{Y}(t) \geq 0 \text{ and } \mathbf{Y}(0) = 0,$$
$$Q_i(t)dY_i(t) = 0, \quad i = 1, ..., N.$$

These relations uniquely determine a pair of mappings from $\mathbf{X}$ to $\mathbf{Q} := \Phi(\mathbf{X})$ and to $\mathbf{Y} := \Psi(\mathbf{X})$. Using the same logic, if process $\mathbf{X}$ can be approximated by a simple process η, then we can approximate the queue-length process $\mathbf{Q} = \Phi(\mathbf{X})$ by $\Phi(\eta)$ via the Lipschitz continuity of the reflection mapping Φ. Due to the routing process occurred in the network, the mapping becomes a more general oblique reflection mapping than the simple reflection mapping for the $G/G/1$ case (i.e., $\mathbf{P}' = 0$). Such a mapping can be as described below.

Exercise 8.1 *Verify the ith component of* $\mathbf{X}(t)$ *in (8.5) by using (8.1).*

8.1.1 OBLIQUE REFLECTION MAPPING

Let $\mathbf{R}$ be a $N \times N$ matrix and $\mathscr{D}_0^N = \{\mathbf{x} \in \mathscr{D}^N : \mathbf{x}(0) \geq 0\}$ be the space of N-dimensional real-valued functions on $[0, \infty)$ that are right-continuous and with left limits (RCLL). We can generalize the one-dimensional reflection mapping into the multi-dimensional reflection mapping as follows. For an $\mathbf{x} \in \mathscr{D}_0^N$, find a pair $(\mathbf{y}, \mathbf{z})$ in $\mathscr{D}^{2N}$ such that

$$\mathbf{z} = \mathbf{x} + \mathbf{R}\mathbf{y} \geq \mathbf{0},$$
$$d\mathbf{y} \geq \mathbf{0} \text{ and } \mathbf{y}(0) = \mathbf{0},$$
$$z_i dy_i = 0, \quad i = 1, ..., N.$$

This is known as the Skorohod problem. If the pair exists and is unique for the given $\mathbf{x}$, then we can write $\mathbf{y} = \Psi(\mathbf{x})$ and $\mathbf{z} = \Phi(\mathbf{x})$. (Ψ, Φ) is called an oblique reflection mapping (ORM) and $\mathbf{R}$ is called the reflection matrix. In the Skorohod problem for a queueing network, matrix $\mathbf{R}$ belongs to a class of matrices, called M-matrices. An M-matrix is a square matrix with positive diagonal elements, non-positive off-diagonal elements, and a non-negative inverse. For a review on the M matrix, readers are referred to Chapter J. For an $N \times N$ M-matrix $\mathbf{R} = [r_{ij}]$, let $\mathbf{D}$ be an $N \times N$ diagonal matrix whose kth diagonal element is $r_{kk}, k = 1, ..., N$. Then, we can write $\mathbf{R} = (\mathbf{I} - \mathbf{G})\mathbf{D}$, where $\mathbf{I}$ is the identity matrix and $\mathbf{G} := \mathbf{1} - \mathbf{R}\mathbf{D}^{-1}$ is a non-negative matrix. It is clear that $(\mathbf{I} - \mathbf{G})$ is an M-matrix if and only if $\mathbf{R}$ is an M-matrix. We summarize some equivalent properties of a non-negative matrix as follows. For a non-negative matrix $\mathbf{G}$, the followings are equivalent.

- matrix $(\mathbf{I} - \mathbf{G})$ is an M-matrix;
- matrix $\mathbf{G}$ is convergent, i.e., $\lim_{m \to \infty} \mathbf{G}^n = 0$;
- $(\mathbf{I} - \mathbf{G})^{-1}$ exists and is non-negative;
- matrix $\mathbf{G}$ has a special radius less than unity;
- all principal minors of $(\mathbf{I} - \mathbf{G})$ are M-matrices and have non-negative inverses;
- $(\mathbf{I} - \mathbf{G}')^{-1}$ exists.

When applied to the open network model, the reflection matrix in (8.4) takes the form $\mathbf{R} = \mathbf{I} - \mathbf{P}'$, where $\mathbf{P} = [P_{ij}]$ is a substochastic matrix whose (i, j)th component represents the probability that a customer after service completion at station i goes next to node j. The assumption that matrix $\mathbf{P}$ is convergent signifies that the model is an open network. Next, we present the ORM theorem which is a generalization of the RMT in the $G/G/1$ queue case.

Theorem 8.1 *(Oblique Reflection Mapping Theorem) Assume that $\mathbf{R}$ is an $N \times N$ M-matrix. For every $\mathbf{x} \in \mathscr{D}_0^N$, there exists a unique pair $(\mathbf{y}, \mathbf{z})$ in $\mathscr{D}^{2N}$ satisfying*

$$\mathbf{z} = \mathbf{x} + \mathbf{R}\mathbf{y} \geq \mathbf{0},$$
$$d\mathbf{y} \geq \mathbf{0} \text{ and } \mathbf{y}(0) = \mathbf{0},$$
$$z_i dy_i = 0, \quad i = 1, ..., N.$$

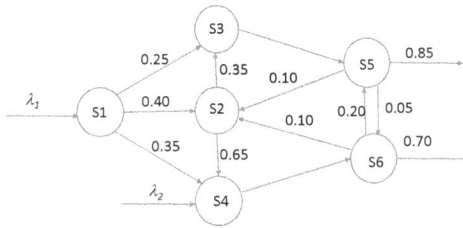

Figure 8.1   A six-node network with two external arrivals.

The mappings Φ and Ψ are Lipschiz continuous mappings from $\mathcal{D}_0^N$ to $\mathcal{D}^N$ under the uniform norm.

Again, the proofs of this theorem and other theoretical results can be found in Chen and Yao (2001).

Exercise 8.2 *Consider a queueing network with six single-server stations shown in Figure 8.1. There are two external arrival steams (to S1 and S4) with rates $\lambda_1 = 15$ customers per hour and SCOV 2 and $\lambda_2 = 9$ customers per hour and SCOV 1. The service time at each station is generally distributed with mean 2.3, 2.5, 2.5, 2, and 2 (in minutes) and SCOV of 1.5,2, 0.25,0.36,1, and 1.44, respectively at stations 1 to 6. Write matrices $\mathbf{P}, \mathbf{R}, \mathbf{D}$, and $\mathbf{G}$. Check all conditions for the open network numerically with this system.*

8.1.2   A FLUID NETWORK

Consider an open network consisting of N nodes within which fluid is circulating. Let λ and μ be the external arrival rate vector and service rate vector, respectively; and $\mathbf{P}$ be the flow-routing matrix. Furthermore, we define $\mathbf{z}(t)$ and $\mathbf{y}(t)$ as the buffer content vector at time t and the cumulative amount of potential outflow lost due to empty buffer during $(0,t]$, respectively. Such a fluid network can be denoted by $(\lambda, \mu, \mathbf{P})$ with initial fluid level $\mathbf{z}(0) = \mathbf{x}(0)$. It is easy to verify that this fluid network satisfies the ORM theorem. Using the flow balance, we can write

$$z_i(t) = z_i(0) + \lambda_i t + \sum_{k=1}^{N} (\mu_i t - y_i(t)) p_{ki} - (\mu_i t - y_i(t))$$

for node i, which can be written in matrix form as

$$\mathbf{z}(t) = \mathbf{x}(t) + (\mathbf{I} - \mathbf{P}')\mathbf{y}(t), \quad t \geq 0, \tag{8.6}$$

where

$$\mathbf{x}(t) = \mathbf{x}(0) + (\lambda - (\mathbf{I} - \mathbf{P}')\mu)t, \quad t \geq 0.$$

Since the fluid level cannot be negative and the cumulative loss process is non-decreasing, we have

$$\mathbf{z}(t) \geq 0, \ t \geq 0,$$
$$d\mathbf{y} \geq 0 \ \text{and} \ \mathbf{y}(0) = 0. \tag{8.7}$$

In addition, it follows from the work-conserving condition that

$$z_i(t)dy_i(t) = 0 \quad \text{for all } t \geq 0 \text{ and } i = 1, ..., N. \tag{8.8}$$

Clearly, the fluid level process $\mathbf{z} = \{\mathbf{z}(t), t \geq 0\}$ and the cumulative loss process $\mathbf{y} = \{\mathbf{y}(t), t \geq 0\}$ satisfy the conditions in ORM theorem. Therefore, we have $\mathbf{z} = \Phi(\mathbf{x})$ and $\mathbf{y} = \Psi(\mathbf{x})$. To characterize the fluid network's system behavior, we define an effective inflow rate vector a, whose i component a_i denotes the sum of exogenous and endogenous inflow rates to node i. The effective outflow rate from a node is the minimum of its effective inflow rate and its potential service rate. Note that if every node is a stable system, then the effective outflow rate should be the same as the effective inflow rate. Thus, the effective inflow rate for node i satisfies the following so-called traffic equation

$$a_i = \lambda_i + \sum_{k=1}^{N} (a_k \wedge \mu_k) p_{ki}, \ i = 1, ..., N,$$

which can be written in vector form as

$$\mathbf{a} = \lambda + \mathbf{P}'(\mathbf{a} \wedge \mu). \tag{8.9}$$

It can be proved that the equation (8.6) has a unique solution. Denote by $\rho_i = a_i/\mu_i$ the traffic intensity of node i, $i = 1, ..., N$. If $\rho \leq \mathbf{e}$, the traffic equation reduces to $\mathbf{a} = \lambda + \mathbf{P}'\mathbf{a}$ which has an explicit solution: $\mathbf{a} = (\mathbf{I} - \mathbf{P}')^{-1}\lambda$. Hence, the condition $\rho \leq \mathbf{e}$ is equivalent to $\mathbf{a} = (\mathbf{I} - \mathbf{P}')^{-1}\lambda \leq \mu$. In a queueing network, nodes can be classified into two or three categories. Node i is called a non-bottleneck i, if $\rho_i < 1$; and called a bottleneck if $\rho_i \geq 1$. For a bottleneck node, if $\rho_i = 1$, it is called a balanced bottleneck; and if $\rho > 1$, it is called a strict bottleneck. The fluid network model is characterized by the following theorem.

Theorem 8.2 *For a fluid network* $(\lambda, \mu, \mathbf{P})$ *with initial fluid level* $\mathbf{z}(0)$*, the fluid level process and the cumulative loss process can be determined as follows: there exists a finite time* τ *such that*
(1) if $\rho_i < 1$ *(a non-bottleneck node), then*

$$z_i(t) = 0, \ \text{for } t \geq \tau,$$
$$y_i(t) = y_i(\tau) + (\mu_i - a_i)(t - \tau), \ \text{for } t \geq \tau,$$

(2) if $\rho_i \geq 1$ (a bottleneck node), then

$$z_i(t) = z_i(\tau) + (a_i - \mu_i)(t - \tau), \; for \; t \geq \tau,$$
$$y_i(t) = 0, \;\; for \; t \geq 0.$$

Note that if $\mathbf{z}(0) = 0$, then we can choose $\tau = 0$ and have $\mathbf{z}(t) = (\mathbf{a} - \mu)^+ t$ and $\mathbf{y}(t) = (\mu - \mathbf{a})^+ t$. For a non-bottleneck node, $\tau_i = \tau_i(0)/(\mu_i - a_i)$. Although this theorem is intuitive, its proof is quite technical and can be found in Chen and Yao (2001). Another important question about a queueing network is how to identify the bottleneck and non-bottleneck nodes. This can be done by applying a computational algorithm. This algorithm starts with assuming all nodes are bottleneck. Denote by f and b the set of non-bottleneck (flow through) nodes and the set of bottleneck nodes, respectively. Here is the algorithm:

- Step 0 Set $f := \emptyset$ and $b := \{1,...,N\}$;
- Step 1 Calculate

$$\hat{\lambda}_b := \lambda_b + \mathbf{P}'_{fb}[\mathbf{I} - \mathbf{P}'_f]^{-1}\lambda_f,$$
$$\hat{\mathbf{P}}_b := \mathbf{P}_b + \mathbf{P}_{bf}[\mathbf{I} - \mathbf{P}_f]^{-1}\mathbf{P}_{fb},$$
$$\hat{\theta}_b := \hat{\lambda}_b - [\mathbf{I} - \hat{\mathbf{P}}_b]\mu_b;$$

- Step 2 If $\hat{\theta}_b \geq 0$, return f as the set of non-bottlenecks, b as the set of bottlenecks, then STOP; otherwise, set $f := f \cup \{k : \hat{\theta}_k < 0\}$ and $b := f^c$, then GOTO Step 1.

The basic idea of this algorithm is to transform the original fluid network $(\lambda, \mu, \mathbf{P})$ with N nodes to a subnetwork $(\hat{\lambda}_b, \mu_b, \hat{\mathbf{P}}_b)$ with nodes b (bottleneck set). Intuitively, the flows (both exogenous and endogenous) of the subnetwork are modified from the original network as though the passage flow through buffers in f were instantaneous. Specifically, the first equation says that the equivalent external arrival rate for set b equals the actual external arrival rate plus transfer rate from set f to set b. The second equation computes the equivalent flow-transfer matrix for set b, which includes both the actual inter-bottleneck transfer probability and the transfer probability from set b to set f and then back to set b. Finally, the third equation computes the difference between the overall arrival rate and the potential service rate for nodes in set b. For more discussion and rigorous justification for this algorithm, see Chen and Yao (2001).

Exercise 8.3 *In the six station network of Exercise 8.2, compute the effective inflow rate vector* **a**. *Show that there are no bottleneck stations in this network. Keep increasing λ_1 until one of the stations becomes bottleneck. What is the minimum value for λ_1 for this event to happen?*

Exercise 8.4 *Use a mathematical software package such as Matlab or Mathematica to build a simulation model for the network in Exercise 8.2. Simulate the flow level processes of the network and verify the behaviors stated in Theorem 8.2. (Hint: you need to assume the initial fluid level $\mathbf{z}(0)$ and adjust other system parameters to see both the behaviors of the systems with and without bottleneck stations, respectively).*

8.1.3   A BROWNIAN MOTION NETWORK

While the fluid network model discussed above is the result of fluid limit of the queueing network, now we consider the diffusion limit the queueing network, which leads to a reflected Brownian motion model. We assume that $\mathbf{X} = \{\mathbf{X}(t), t \geq 0\}$ is a N-dimensional Brownian motion with initial position $\mathbf{x} = \mathbf{X}(0) \geq 0$, drift vector θ, and covariance matrix Γ. Then, $\mathbf{Z} = \Phi(\mathbf{X})$ is known as a reflected Brownian motion on the non-negative orthant, and the process $\mathbf{Y} = \Psi(\mathbf{X})$ is known as the regulator of the reflected Brownian motion. The queue length process $\mathbf{Z}$ can be denoted by $\mathbf{Z} = RBM_x(\theta, \Gamma; \mathbf{R})$, where $\mathbf{R}$ is the reflection matrix. This model is very similar to the one for approximating the $G/G/1$ queue except now we use the multi-dimensional reflected Brownian motion. That is $\mathbf{Z}$ evolves like some N-dimensional Brownian motion in the interior of the orthant $\mathscr{R}_+^N$. When it reaches the ith face, $F_i := \{x = (x_k) \in \mathscr{R}_+^N : x_i = 0\}$, of the non-negative orthant, the process is reflected instantaneously toward the interior of the non-negative orthant in the direction determined by the ith column of matrix $\mathbf{R}$, and the ith component of the regulator $Y_i(t)$ represents the cumulative amount of such effort in pushing it back. As expected, the following results can be established (see Chen and Yao (2001) for the detailed proofs).

Theorem 8.3 *The reflected Brownian motion $\mathbf{Z} = RBM_x(\theta, \Gamma; \mathbf{R})$ is a diffusion process, which is a strong Markov process with continuous sample paths.*

The stationary distribution of such a process exists under certain condition.

Theorem 8.4 *The stationary distribution of $\mathbf{Z} = RBM_x(\theta, \Gamma; \mathbf{R})$ exists if and only if $\mathbf{R}^{-1}\theta < 0$. When the stationary distribution exists, it is unique.*

It can be shown that $\mathbf{R} = \mathbf{I} - \mathbf{P}'$ and $\theta = \lambda - (\mathbf{I} - \mathbf{P}')\mu$ (a good exercise). Thus, the condition $\mathbf{R}^{-1}\theta < 0$ is equivalent to $\mathbf{a} := (\mathbf{I} - \mathbf{P}')^{-1}\lambda < \mu$ (i.e., the traffic intensity for each node is less than 1). Clearly, this is consistent with the stability condition for the generalized Jackson network (arrivals and services follow renewal processes). Such a result is natural as $\mathbf{Z}$ can be used to approximate the queue length process of the network via stochastic process limits. Note that the covariance matrix Γ does not play a role in the stability condition. Even the stationary distribution for $\mathbf{Z}$ exists, determining this distribution is not an easy task except for some special cases. The stationary distribution of a queueing network satisfies a so-called basic adjoint relation (BAR) which is established based on Ito's formula. To present the result, we define the following operators:

$$\mathscr{L} = \frac{1}{2} \sum_{i,j=1}^{N} \Gamma_{ij} \frac{\partial^2}{\partial z_i \partial z_j} + \sum_{j=1}^{N} \theta_j \frac{\partial}{\partial z_j},$$

$$\mathscr{D}_j = \sum_{i=1}^{N} r_{ij} \frac{\partial}{\partial z_i} \equiv r_j' \nabla,$$

where r_j is the jth column of $\mathbf{R}$. We can write

$$\mathbf{Z}(t) = \mathbf{X}(0) + \theta t + \Gamma^{1/2} \mathbf{W}(t) + \mathbf{R} \mathbf{Y}(t), \tag{8.10}$$

where $\Gamma^{1/2} \mathbf{W}(t) = \mathbf{X}(t) - \mathbf{X}(0) - \theta t$ is an N-dimensional standard Brownian motion. Based on Ito's formula, for any function f defined on $\mathscr{R}_+^N$ with continuous and bounded first- and second-order derivatives, we have

$$f(\mathbf{Z}(t)) - f(\mathbf{Z}(0)) = \int_0^t \Delta f(\mathbf{Z}(s)) d\mathbf{W}(s) + \int_0^t \mathscr{L} f(\mathbf{Z}(s)) ds$$
$$+ \sum_{j=1}^{N} \int_0^t \mathscr{D}_j f(\mathbf{Z}(s)) dY_j(s), \tag{8.11}$$

where $\Delta f(z) = (\partial f/\partial z_1, ..., \partial f/\partial z_N)$. Now taking the conditional expectation with respect to the stationary distribution, denoted by π on both sides of the above equation and noting that the first integral on the right-hand side is a martingale, we have

$$0 = E_\pi \int_0^t \mathscr{L} f(\mathbf{Z}(s)) ds + \sum_{j=1}^{N} E_\pi \int_0^t \mathscr{D}_j f(\mathbf{Z}(s)) dY_j(s)$$

$$= t E_\pi [\mathscr{L} f(\mathbf{Z}(0))] + \sum_{j=1}^{N} E_\pi \int_0^t \mathscr{D}_j f(\mathbf{Z}(s)) dY_j(s) \tag{8.12}$$

$$= t \int_{\mathscr{R}_+^J} \mathscr{L} f(z) d\pi(z) + \sum_{j=1}^{N} E_\pi \int_0^t \mathscr{D}_j f(\mathbf{Z}(s)) dY_j(s).$$

Using the fact that Y_j, as a continuous additive function of Z with support on F_j, increases at time t only when $Z_j(t) = 0$, we obtain

$$E_\pi \int_0^t \mathscr{D}_j f(\mathbf{Z}(s)) dY_j(s) = t \int_{F_j} \mathscr{D}_j f(z) dv_j(z),$$

where v_j is a finite Borel measure. Summarizing these results, we can show that a stationary distribution of $\mathbf{Z} = RBM_x(\theta, \Gamma; \mathbf{R})$, π, satisfies a basic adjoint relation (BAR).

Theorem 8.5 *Suppose that a stationary distribution of* $\mathbf{Z} = RBM_x(\theta, \Gamma; \mathbf{R})$, π, *exists. Then for each* $j = 1, ..., N$, *there exists a finite Borel measure* v_j *on* F_j *such that a BAR*

$$\int_{\mathscr{R}_+^N} \mathscr{L} f(z) d\pi(z) + \sum_{j=1}^{N} \int_{F_j} \mathscr{D}_j f(z) dv_j(z) = 0, \tag{8.13}$$

holds for all continuous and bounded functions f *defined on* $\mathscr{R}_+^N$ *whose derivatives up to order 2 are also continuous and bounded.*

Using the BAR, we can derive the stationary distribution in a special case where this distribution has a product form (like the classical Jackson networks). Denote by $\mathbf{D}$ the N-dimensional diagonal matrix whose jth diagonal element is r_{jj}, the jth diagonal element of matrix $\mathbf{R}$, $j = 1, ..., N$, and denote by Λ the N-dimensional matrix whose jth diagonal element is Γ_{jj}, the jth diagonal element of matrix Γ, $j = 1, ..., N$.

Theorem 8.6 *If* $\mathbf{R}^{-1}\theta < 0$ *and* $2\Gamma = \mathbf{R}\mathbf{D}^{-1}\Gamma + \Gamma\mathbf{D}^{-1}\mathbf{R}'$, *then the density function of* π *can be given by*

$$f(z_1, ..., z_N) = \prod_{j=1}^{N} f_j(z_j), \tag{8.14}$$

and

$$f_j(z) = \begin{cases} \eta_j e^{-\eta_j z}, & z \geq 0, \\ 0, & z < 0, \end{cases} \tag{8.15}$$

where

$$\eta = -2\Lambda^{-1}\mathbf{D}\mathbf{R}^{-1}\theta,$$

and $j = 1, ..., N$.

Recall that the product-form stationary density function occurs in a classical Jackson network. Here, for the stationary distribution of $\mathbf{Z}$ to have the product form, some conditions must be satisfied. Conversely, if we can show that the stationary density function has the product form, then these conditions must hold with exponential form f_j. Note that the two-station tandem queue discussed in Section 7.7 is a good example.

While we refer interested readers to Chen and Yao (2001) for the proofs of Theorems 8.5 and 8.6, we show some simple examples to gain the intuition of these results. The two-station tandem queue shown in Section 7.7.1 in the previous chapter can be served as an example of Theorem 8.6. Readers are encouraged to refer to Section 7.7.1. To show a simple example of Theorem 8.5, we consider the $G/G/1$ queue case, which is the special case of the network with a single node (i.e., $N = 1$). In this case, (8.10) is reduced to

$$Z(t) = Z(0) + \theta t + \sigma W(t) + Y(t),$$

where W is a SBM (denoted by B in Chapter 7) and Y is the regulator of the RBM. Accordingly, (8.11) becomes

$$f(Z(t)) - f(Z(0)) = \sigma \int_0^t f'(Z(s))dW(s)$$
$$+ \int_0^t \left[\frac{1}{2}\sigma^2 f''(Z(s)) + \theta f'(Z(s)) \right] ds.$$
$$+ \int_0^t f'(Z(s))dY(s).$$

Note that $\int_0^t f'(Z(s))dW(s)$ is a martingale (see Chapter H) and $\int_0^t f'(Z(s)) dY(s) = f'(0)Y(t)$ owing to the fact that Y increases at t only when $Z(t) = 0$. Taking the conditional expectation with respect to the stationary distribution π on both sides of the equation above, we obtain the BAR for this single queue case, the special case of (8.12), as follows:

$$0 = E_\pi \int_0^t \left[\frac{1}{2}\sigma^2 f''(Z(s)) + \theta f'(Z(s)) \right] ds + f'(0)E_\pi[Y(t)]$$
$$= t \int_0^\infty \left[\frac{1}{2}\sigma^2 f''(z) + \theta f'(z) \right] d\pi(z) + f'(0)E_\pi[Y(t)],$$

which can be written as

$$0 = \int_0^\infty \left[\frac{1}{2}\sigma^2 f''(z) + \theta f'(z) \right] d\pi(z) - \theta f'(0),$$

by letting $f(z) = z$ and noting that $E_\pi[Y(t)] = -\theta t = \lambda - \mu$.

Exercise 8.5 *Use the BAR for the $G/G/1$, derive the stationary distribution from the RBM approximation.*

Using the multi-dimensional reflected Brownian motion to approximate the queue length process of the network is an extension of the heavy traffic approximation method for the single server queue discussed in Chapter 7. This more general approach can be applied to multi-class queueing networks and the reflected Brownian motion in such cases is called a semimartingale reflected Brownian motion (SRBM). SRBM is defined as a real-valued stochastic process that can be decomposed as the sum of a local martingale and an adapted finite-variation process. More discussion on the general Brownian approximations to networks and the related concepts can be found in Chen and Yao (2001).

8.1.4   FLUID AND DIFFUSION APPROXIMATIONS TO QUEUEING NETWORKS

With these results, we can develop the Brownian approximations to the performance measures, such as queue-length process, of the general queueing networks via oblique reflection theorem. Note that it follows $\mathbf{Q} = \Phi(\mathbf{X})$ that if $\mathbf{X}$ can be approximated by a Brownian motion, the queue-length process Q can be approximated by a reflected Brownian motion. Such a development is parallel to the fluid and diffusion approximations developed for the $G/G/1$ queue in the previous chapter. We start with the fluid limit approximation for the queueing network. Based on the FSLLN, we have for arrival process, service process, and routing sequence, respectively

$$\mathbf{A}(t) \approx \lambda t,$$
$$\mathbf{S}(t) \approx \mu t,$$
$$\mathbf{R}(t) \approx \mathbf{P}'t,$$

where $\mathbf{R}(t) := (R^1(\lfloor t \rfloor), ..., R^N(\lfloor t \rfloor))$. Applying these fluid approximations to the expression (8.5), we obtain in matrix form

$$\mathbf{X}(t) \approx \bar{\mathbf{X}}(t) = \mathbf{Q}(0) + \theta t,$$

where $\theta = \lambda - (\mathbf{I} - \mathbf{P}')\mu$. It follows from the Lipschitz continuity of the oblique reflection mapping that the queue length $\mathbf{Q} = \Phi(\mathbf{X})$ and $\mathbf{Y} = \Psi(\mathbf{X})$ can be approximated by $\bar{\mathbf{Q}} = \Phi(\bar{\mathbf{X}})$ and $\bar{\mathbf{Y}} = \Psi(\bar{\mathbf{X}})$. Similarly, we can develop the following approximations:

$$\mathbf{Y}(t) \approx \bar{\mathbf{Y}}(t) \approx (\mu - a)^+ t,$$
$$B_j(t) \approx (1 \wedge \rho_j)t,$$
$$S_j(B_j(t)) \approx \mu_j(1 \wedge \rho_j)t = (a_j \wedge \mu_j)t.$$

Note that these fluid limit based approximations are the first-moment approximations (i.e., means) which do not offer the information about the probability distribution of the processes. To obtain such information, we need to develop more refined diffusion approximations (i.e., the second-moment-based approximations).

Suppose that the inter-arrival times of the external arrival process $A_j (j \in E)$ and the service times $s_j, j = 1, ..., N)$ have finite second moments. Denote by c_{A_j} the coefficient of variation (COV) of the inter-arrival times of A_j for $j \in E$, and denote by c_{s_j} the COV of service times s_j for $j = 1, ..., N$. It follows then from the FCLT that

$$\mathbf{A}(t) - \lambda t \overset{d}{\approx} \hat{\mathbf{A}}(t),$$
$$\mathbf{S}(t) - \mu t \overset{d}{\approx} \hat{\mathbf{S}}(t), \tag{8.16}$$
$$\mathbf{R}(t) - \mathbf{P}'t \overset{d}{\approx} \hat{\mathbf{R}}(t),$$

where $\hat{\mathbf{A}}, \hat{\mathbf{S}}$, and $\hat{\mathbf{R}}^j$ (the jth coordinate of $\hat{\mathbf{R}}$), $j = 1, ..., N$, are $N + 2$ independent N-dimensional driftless Brownian motions. They have the following covariance matrices (the covariance matrices for $\hat{\mathbf{A}}$ and $\hat{\mathbf{S}}$ are indexed by 0 and $N+1$, respectively):

$$\Gamma^0 = (\Gamma^0_{jk}) \quad \text{where} \quad \Gamma^0_{jk} = \lambda_j c^2_{A_j} \delta_{jk},$$

$$\Gamma^{N+1} = (\Gamma^{N+1}_{jk}) \quad \text{where} \quad \Gamma^{N+1}_{jk} = \mu_j c^2_{s_j} \delta_{jk},$$

$$\Gamma^l = (\Gamma^l_{jk}) \quad \text{where} \quad \Gamma^l_{jk} = p_{lj}(\delta_{jk} - p_{lk}) \ \text{for} \ l = 1, ... N.$$

Based on the first- and second-order approximations, we obtain

$$S_j(B_j(t)) - \mu_j B_j(t) \overset{d}{\approx} \hat{S}_j((1 \wedge \rho_j)t),$$

$$R^i_j(S_i(B_i(t))) - p_{ij} S_i(B_i(t)) \overset{d}{\approx} \hat{R}^i_j((a_i \wedge \mu_i)t),$$

(8.17)

$i,j = 1, ..., N$. Using (8.13) and (8.14), we can approximate $X_j(t)$ by a deterministic linear function plus a linear combination of some independent N-dimensional driftless Brownian motions as follows:

$$\tilde{X}_j(t) = Q_j(0) + \theta_j t + \hat{A}_j(t) - \hat{S}_j((1 \wedge \rho_j)t)$$

$$+ \sum_{i=1}^{N} \left[\hat{R}^i_j((a_i \wedge \mu_i)t) + p_{ij} \hat{S}_i((1 \wedge \rho_i)t) \right].$$

In vector form, we have

$$\tilde{\mathbf{X}}(t) = \mathbf{Q}(0) + \theta t + \hat{\mathbf{A}}(t) + \sum_{i=1}^{N} \hat{\mathbf{R}}^i((a_i \wedge \mu_i)t) - (\mathbf{I} - \mathbf{P}')\hat{\mathbf{S}}((e \wedge \rho)t),$$

which is a N-dimensional Brownian motion starting at $\mathbf{Q}(0)$ with drift vector θ and covariance matrix $\Gamma = (\Gamma_{kl})$:

$$\Gamma_{kl} = \sum_{j=1}^{N} (a_j \wedge \mu_j)[p_{jk}(\delta_{kl} - p_{jl}) + c^2_{s_j}(p_{jk} - \delta_{jk})(p_{jl} - \delta_{jl})]$$

$$+ \lambda_k c^2_{A_j} \delta_{kl}, \quad k,l = 1, ..., N.$$

Thus, the queue-length process $\mathbf{Q}$ can be approximated by the reflected Brownian motion $\Phi(\tilde{\mathbf{X}}) = RBM_{Q(0)}(\theta, \Gamma; \mathbf{I} - \mathbf{P}')$.

If the queueing network does not have any bottleneck node (i.e., $\mathbf{a} = (\mathbf{I} - \mathbf{P}')^{-1}\lambda < \mu$), the RBM has a stationary distribution, which is characterized by the BAR. The stationary distribution can be determined numerically based on the BAR. The details of developing the numerical solutions to the stationary distribution of queueing network can be found in Chapter 10 of Chen and Yao (2001). For a queueing network with the stationary distribution, the product-form condition reduces to

$$2\Gamma_{jk} = -(p_{jk}\Gamma_{jj} + p_{kj}\Gamma_{kk})$$

(8.18)

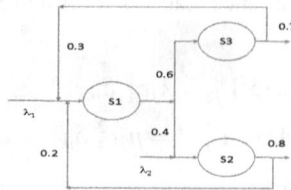

Figure 8.2 A three-node network with two external arrivals.

for all $j \neq k$. This implies that if $\Gamma = (\Gamma_{jk})$ satisfies the above condition, the density function of the stationary distribution takes a product form $f(x_1,...,x_N) = \prod_{j=1}^{N} f_j(x_j)$ where $f_j(x) = \eta_j e^{-\eta_j x}$ for $x \geq 0$ and $f_j(x) = 0$ for $x < 0$ with $\eta_j = 2(\mu_j - a_j)/\Gamma_{jj}, j = 1,...,N$. It can be verified that a queueing network with parameters $c_{A_j} = 1$ and $c_{s_j} = 1$ for all $j = 1,...,N$ satisfies the product-form condition. A Jackson network with constant service rates at all nodes is one such special case.

It is worth noting that the fluid and diffusion limit processes are the stochastic processes on continuous state spaces while the queue length process for a queueing network is a stochastic process on a discrete state space. To get around this inconsistency, we can focus on the workload process, denoted by $\mathbf{W}(t)$, which is the amount of time it would take to complete service of every customer in the system at time t including any remaining service at the server. The workload and the queue length are related by the simple approximate relation $\mathbf{W}(t) = \mathbf{Q}(t)/\mu$. Hence, the workload process $\{\mathbf{W}(t), t \geq 0\}$ can also be approximated by a multidimensional reflected Brownian motion. Similar to the single server queue case discussed in the previous chapter, the diffusion limits (FCLT), FLIL, and strong approximations can be discussed for the general Jackson networks (including the closed networks). For the details on this topic, interested readers are referred to Chen and Yao (2001).

Exercise 8.6 *Consider a three-station network shown in Figure 8.2. Assume that there are several possible parameter vectors:*

$$(\lambda_1, \lambda_2) = \{(0.5, 0.5), (0.75, 0.75)\},$$
$$(m_1, m_2, m_3) = \{(0.60, 0.60, 0.80), (0.20, 0.20, 0.50), (0.40, 0.50, 0.80)\},$$
$$(SCOV_1, SCOV_2, SCOV_3) = \{(0.25, 0.25, 0.25), (1, 1, 1), (2, 2, 2)\}.$$

Use simulations to evaluate the accuracy of the diffusion approximations to the performance measures such as the average sojourn time in each station of such a network. (Note that for $SCOV_1 = SCOV_2 = SCOV_3 = 1$ is a special case of Jackson network and the exact formulas exist).

8.2 DECOMPOSITION APPROACH

Our focus in the previous section is on the joint distribution of the queueing network. Only under some specific conditions, the product-form distribution exists. In general, the stationary distribution is very complex. Thus, this general approach may not be practical for real situations. This motivates us to develop a simpler approximation approach based on decomposition. We demonstrate this approach by considering a queueing network with single server nodes. The main idea is to treat each node as an isolated $G/G/1$ queue that can be approximated by the formulas developed in the previous chapter. Thus, we mainly focus on the first moment performance measures such as the mean queue length or mean waiting time for each node. Recall that the first moment performance measures for $G/G/1$ queue can be computed approximately by using the first two moments of the inter-arrival times and service times. We utilize the equivalent quantities of the average arrival rate, the squared coefficient of variation (SCOV) of inter-arrival times, the average service rate, and the SCOV of the service times to compute the mean queue length of each node in the network. To figure out these quantities for each node, we need to consider three basic flow operations in a feed-forward type queueing network, which are superposition of flows, splitting of a flow, and flowing through a queue. We first develop the formulas to compute average rates and SCOV's of the outflows after each of these operations by using average rates and SCOV's of the inflows.

8.2.1 SUPERPOSITION OF FLOWS

Suppose that there are m flows that are superimposed into a single flow. Let θ_i and c_i^2 be the average rate and the SCOV of the inter-arrival times on flow i, respectively, where $i = 1, ..., m$. Let θ and c^2 be the effective arrival rate and the effective SCOV of the inter-arrival times for the aggregate flow as a result of superposition of the m flows. By considering the superposition as the sum of independent renewal processes, we can obtain

$$\theta = \theta_1 + \cdots + \theta_m,$$

$$c^2 = \sum_{i=1}^{m} \frac{\theta_i}{\theta} c_i^2.$$

It is worth noting that the actual inter-arrival times of the aggregated superimposed process are not i.i.d. and hence the process is not truly a renewal process. Treating the superimposed process as a renewal process can be considered as an approximation.

8.2.2   FLOWING THROUGH A QUEUE

Consider a $G/G/1$ queue with arrival rate θ_A and SCOV c_A^2 of the inter-arrival times. The service times follow a general distribution with rate θ_s (1/mean) and SCOV c_s^2. Now we find the relation between the arrival rate and SCOV of the inter-arrival times and the departure rate and SCOV of the inter-departure times for the $G/G/1$ queue, denoted by θ_D and c_D^2, respectively. For a steady-state system, clearly we have

$$\theta_A = \theta_D.$$

Using the mean value analysis, we can derive a relation between the SCOVs:

$$c_D^2 = c_A^2 + 2\rho^2 c_s^2 + 2\rho(1-\rho) - 2\theta_A W(1-\rho), \qquad (8.19)$$

where $\rho = \theta_A/\theta_D$ and

$$W \approx \frac{1}{\theta_s} + \frac{\rho^2(c_A^2 + c_s^2)}{2\theta_A(1-\rho)}.$$

With these expressions, we can develop an approximate relation

$$c_D^2 \approx (1-\rho^2)c_A^2 + \rho^2 c_s^2.$$

Insert: Mean Value Analysis The less intuitive relation (8.19) can be obtained by the mean value analysis (MVA) which is a powerful approach to obtaining the first moment of the performance measure for a stochastic system. We briefly introduce this approach here. MAV is based on the two important properties on multiple random variables. For a set of random variables, $X_1, X_2, ..., X_n$ with their respective finite means, $E[X_1], E[X_2], ..., E[X_n]$, the followings hold:

1. Expected value of a linear combination of random variables is equal to the linear combination of the expected values. That is for a set of constant numbers $a_1, a_2, ..., a_n$, we have

$$E[a_1 X_1 + a_2 X_2 + \cdots + a_n X_n] = a_1 E[X_1] + a_2 E[X_2] + \cdots + a_n E[X_n].$$

2. Expected value of a product of independent random variables $X_1, X_2, ..., X_n$ is equal to the product of the expected values of those random variables. That is

$$E[X_1 X_2 \cdots X_n] = E[X_1] E[X_2] \cdots E[X_n].$$

Exercise 8.7 *Prove the two important properties regarding the expectation of the linear combination of multiple random variables and product of on multiple independent random variables, which the MAV is based.*

The MVA consists of three steps: (a) Develop a recursive relation for some basic quantity of interest such as the queue length or waiting time of a customer. (b) Take expectations on the relation developed in (a) and the squared recursive relation. (c) Manipulate these expected value relations to obtain the mean quantity of interest.

As an example of using MVA, we consider a $G/G/1$ queue and develop a bound on the expected waiting time and show the relation (8.19). We first introduce some notations:

A_n	Time of the nth arrival
$T_{n+1} = A_{n+1} - A_n$	The $(n+1)$st inter-arrival time
S_n	Service time of the nth customer
W_n	Time spent (sojourn) in the system by the nth customer
$D_n = A_n + W_n$	Time of the nth departure
$I_{n+1} = (A_{n+1} - A_n - W_n)^+$	Idle time between the nth and $(n+1)$st service
$V_{n+1} = D_{n+1} - D_n$	Time between the nth and $(n+1)$st departures

With the assumptions of i.i.d. for inter-arrival times and service times, we can express the system parameters and the stationary performance measures in terms of the expected values of the quantities introduced as follows.

$\lambda = 1/E[T_n]$	Mean arrival rate
$\mu = 1/E[S_n]$	Mean service rate
$c_A^2 = Var[T_n]/(E[T_n])^2 = \lambda^2 Var[T_n]$	Squared coefficient of variation for inter-arrival time
$c_S^2 = Var[S_n]/(E[S_n])^2 = \mu^2 Var[S_n]$	Squared coefficient of variation for service time
$W = \lim_{n \to \infty} E[W_n]$	Stationary expected waiting time
$I_d = \lim_{n \to \infty} E[I_n]$	Stationary expected idle time
$I^{(2)} = \lim_{n \to \infty} E[I_n^2]$	Stationary second moment of idle time

Write the recursive relation as follows:

$$
\begin{aligned}
W_{n+1} &= S_{n+1} + \max(D_n - A_{n+1}, 0) = S_{n+1} + (D_n - A_{n+1})^+ \\
&= S_{n+1} + (W_n + A_n - A_{n+1})^+ \\
&= S_{n+1} + W_n + A_n - A_{n+1} + (A_{n+1} - A_n - W_n)^+ \\
&= S_{n+1} + W_n - T_{n+1} + I_{n+1}.
\end{aligned}
\tag{8.20}
$$

Taking the expectation on both sides of the equation above, we have

$$
E[W_{n+1}] = E[S_{n+1}] + E[W_n] - E[T_{n+1}] + E[I_n],
$$

which leads to the limit

$$W = \frac{1}{\mu} + W - \frac{1}{\lambda} + I_d,$$

as $n \to \infty$. Thus, for a stable system (i.e., $W < \infty$), we have the relation $I_d = 1/\lambda - 1/\mu = (1-\rho)/\lambda > 0$, where $\rho = \lambda/\mu$, which is $\rho < 1$. Re-arranging (8.20) yields

$$W_{n+1} - S_{n+1} - I_{n+1} = W_n - T_{n+1}.$$

Now squaring both sides of the equation above, we have

$$
\begin{aligned}
&(W_{n+1} - S_{n+1})^2 - 2(W_{n+1} - S_{n+1})I_{n+1} + I_{n+1}^2 = (W_n - T_{n+1})^2 \\
\Rightarrow\ & W_{n+1}^2 - 2W_{n+1}S_{n+1} + S_{n+1}^2 + I_{n+1}^2 - 2(W_{n+1} - S_{n+1})I_{n+1} \\
&= W_n^2 - 2W_n T_{n+1} + T_{n+1}^2 \\
\Rightarrow\ & W_{n+1}^2 - 2(W_{n+1} - S_{n+1})S_{n+1} - S_{n+1}^2 + I_{n+1}^2 - 2(W_{n+1} - S_{n+1})I_{n+1} \\
&= W_n^2 - 2W_n T_{n+1} + T_{n+1}^2.
\end{aligned}
\tag{8.21}
$$

Now the product terms in the equation above are either zero or multiplication of two independent random variables, which make the expectations of these terms to be computable. Note that (1) the second term on the LHS is the product of the two independent random variables. That is $W_{n+1} - S_{n+1}$ is independent of S_{n+1}; (2) the fifth term on the LHS is zero. That is $(W_{n+1} - S_{n+1})I_{n+1} = 0$, which is due to the fact that any waiting implies zero idle time and no waiting implies positive idle time; and (3) the second term on the RHS is the product of the two independent random variables. That is W_n is independent of T_{n+1}. Taking the expectation on both sides of the last line in (8.21), we obtain

$$
\begin{aligned}
&E[W_{n+1}^2] - 2E[W_{n+1} - S_{n+1}]E[S_{n+1}] - E[S_{n+1}^2] + E[I_{n+1}^2] \\
&= E[W_n^2] - 2E[W_n]E[T_{n+1}] + E[T_{n+1}^2].
\end{aligned}
\tag{8.22}
$$

By taking the limit $n \to \infty$ of the equation above and noting $E[S_{n+1}] = 1/\mu, E[S_{n+1}^2] = (c_s^2 + 1)/\mu^2, E[T_{n+1}] = 1/\lambda, E[T_{n+1}^2] = (c_A^2 + 1)/\lambda^2, \rho = \lambda/\mu$, we have

$$W = \frac{1}{\mu} + \frac{\rho^2 c_s^2 + c_A^2 + (1-\rho)^2 - \lambda^2 I^{(2)}}{2\lambda(1-\rho)}.
\tag{8.23}$$

The unknown term in this expression is the second moment of the idle period $I^{(2)}$. However, a simple bound can be found as $I^{(2)} \geq I_d^2 = (1-\rho)^2/\lambda^2$. Using such a bound in (8.23), we get an upper bound for the expected waiting time in a $G/G/1$ queue as

$$W \leq \frac{1}{\mu} + \frac{\rho^2 c_s^2 + c_A^2}{2\lambda(1-\rho)},$$

which reduces to the exact expected waiting time in $M/G/1$ queue.

To derive (8.19), we consider the following relation

$$V_{n+1} = D_{n+1} - D_n = \max(A_{n+1} - D_n, 0) + S_{n+1} = (A_{n+1} - D_n)^+ + S_{n+1}$$
$$= I_{n+1} + S_{n+1}.$$

(8.24)

By taking the limit as $n \to \infty$ and the expectation on both sides of the equation above yields

$$E[V] = \lim_{n \to \infty} E[V_{n+1}] = I_d + \frac{1}{\mu} = \frac{1}{\lambda} - \frac{1}{\mu} + \frac{1}{\mu} = \frac{1}{\lambda}. \tag{8.25}$$

It follows from (8.24) and the independence of I_{n+1} and S_{n+1} that

$$Var[V_{n+1}] = Var[I_{n+1}] + Var[S_{n+1}],$$

which leads to

$$Var[V] = I^{(2)} - I_d^2 + \frac{c_s^2}{\mu^2}. \tag{8.26}$$

Using (8.25) and (8.26), we have the squared coefficient of variation for the inter-departure times as

$$c_D^2 = \frac{Var[V]}{E(V)^2} = \lambda^2 \left(I^{(2)} - I_d^2 + \frac{c_s^2}{\mu^2} \right). \tag{8.27}$$

Using (8.23) to obtain an expression for $I^{(2)}$ in terms of W, substituting such an expression into (8.27), and noting $I_d = (1-\rho)/\lambda$, we obtain (8.19).

Although both MVA and fluid limit analysis focus on first moments of performance measures, there is a big difference between these two approaches. In fluid limit model, the system dynamics approaches to a deterministic process with no random fluctuations. Thus, it is usually under the heavy traffic assumption. On the other hand, in the MVA, the system dynamics is stochastic and we are only interested in the means of certain random variables at a certain time point or when the system reaches the steady state.

Exercise 8.8 *Use the MVA to derive the mean queue length of the $M/G/1$ queue. (Hint: Define X_n as the number of customers in the system immediately after the nth departure and develop the recursive relation for it).*

8.2.3 SPLITTING A FLOW

Consider a flow as the renewal process $\{N(t), t \geq 0\}$ with i.i.d. inter-arrival times with mean $1/\theta$ and SCOV c^2. This flow is split into n sub-flows according to a Bernoulli splitting process. That is with probability p_i, each arrival is classified into an arrival of sub-flow i, where $i = 1,...,n$. Denote by $1/\theta_i$ and c_i^2 the mean and SCOV of the inter-arrival times of sub-flow i. Let $X_1, X_2,...$ be the inter-arrival times of the original flow. Similarly, let $Y_1^i, Y_2^i,...$ be the inter-arrival times of the sub-flow i. Based on the Bernoulli splitting process, the inter-arrival time Y_j^i is the sum of geometrically distributed number of X_k values for $j = 1,...$ and appropriate chosen k. For example, $Y_1^i = X_1 + X_2 + \cdots + X_N$ where N is a geometric random variable with success probability p_i. Note that $E[Y_1^i] = 1/\theta_i$ and $Var[Y_1^i] = c_i^2/\theta_i^2$. Using $Y_1^i = X_1 + X_2 + \cdots + X_N$ and conditioning on N, we have

$$E[Y_1^i] = E[E[Y_1^i|N]] = E\left[\frac{N}{\theta}\right] = \frac{1}{p_i\theta},$$

$$Var[Y_1^i] = E[Var[Y_1^i|N]] + Var[E[Y_1^i|N]]$$

$$= E\left[\frac{Nc^2}{\theta^2}\right] + Var\left[\frac{N}{\theta}\right] = \frac{c^2}{p_i\theta^2} + \frac{1-p_i}{p_i^2\theta^2}.$$

Therefore, we obtain the following relation

$$\theta_i = p_i\theta,$$

$$c_i^2 = c^2 p_i + 1 - p_i$$

for $i = 1,...,n$.

8.2.4   DECOMPOSITION OF A QUEUEING NETWORK

If we are only interested in finding the first moment performance measure for each node of a queueing network and each node has only one server, we can treat each node as a $G/G/1$ queue and apply the mean queue length formula. This requires us to figure out the effective arrival rate, denoted by a_i, and the SCOV of the inter-arrival times, denoted by $c_{A_i}^2$, for node i. The effective arrival rates can be obtained by solving

$$a_j = \sum_{i=1}^{N} p_{ij}a_i + \lambda_j, \tag{8.28}$$

for $j = 1,...,N$. This set of equations can be written in the matrix form as

$$\mathbf{a} = \lambda[\mathbf{I} - \mathbf{P}]^{-1}. \tag{8.29}$$

For an open queueing network without bottleneck nodes (i.e., $\rho_i = a_i/\mu_i < 1$ for $i = 1,...,N$.), we now develop a formula to approximate the SCOV of the inter-arrival times for node i. Here, we need to apply the three rules developed for a feed-forward network: flowing a queue, splitting a flow, and superimposing multiple flows. It is worth noting that these rules are utilized here in a general queueing network as if the network is "feed-forward" type. Thus, the formula developed for the SCOV is an approximation. For each node i, if the inter-arrival time has the mean of $1/a_i$ and SCOV $c_{A_i}^2$, then the inter-departure time will have the mean of $1/a_i$ and SCOV $(1-\rho_i^2)c_{a_i}^2 + \rho_i^2 c_{S_i}^2$ based on the "flowing a queue" rule. Since the probability that a departing customer from node i will join node j is p_{ij}, the inter-arrival times of customers from node i to node j has the mean arrival rate $a_i p_{ij}$ and SCOV $1 - p_{ij} + p_{ij}[(1-\rho_i^2)c_{a_i}^2 + \rho_i^2 c_{S_i}^2]$ based on the "splitting a flow" rule. Finally, since the aggregate arrivals to node j from all other nodes in the network plus the external arrivals, the effective inter-arrival times to node j has the SCOV as

$$c_{a_j}^2 = \frac{\lambda_j}{a_j}c_{A_j}^2 + \sum_{i=1}^{N}\frac{a_i p_{ij}}{a_j}\{1 - p_{ij} + p_{ij}[(1-\rho_i^2)c_{a_i}^2 + \rho_i^2 c_{S_i}^2]\} \qquad (8.30)$$

by using the superposition rule for $j = 1, ..., N$. Thus, we have N equations with N unknowns $c_{a_1}^2, ..., c_{a_N}^2$. These unknowns can be solved by using these equations. To solve a large queueing network, we write these equations in matrix forms. First, we define the following vectors and matrices:

$$\mathbf{c_a}^2 = \begin{bmatrix} c_{a_1}^2 \\ c_{a_2}^2 \\ \vdots \\ c_{a_N}^2 \end{bmatrix}, \lambda = \begin{bmatrix} \lambda_1 \\ \lambda_2 \\ \vdots \\ \lambda_N \end{bmatrix}, \frac{1}{\mathbf{a}} = \begin{bmatrix} \frac{1}{a_1} \\ \frac{1}{a_2} \\ \vdots \\ \frac{1}{a_N} \end{bmatrix}, \mathbf{c_A}^2 = \begin{bmatrix} c_{A_1}^2 \\ c_{A_2}^2 \\ \vdots \\ c_{A_N}^2 \end{bmatrix}, \mathbf{a} = \begin{bmatrix} a_1 \\ a_2 \\ \vdots \\ a_N \end{bmatrix},$$

$$\mathbf{P} = \begin{bmatrix} p_{11} & p_{12} & \cdots & p_{1N} \\ p_{21} & p_{22} & \cdots & p_{2N} \\ \vdots & \vdots & \cdots & \vdots \\ p_{N1} & p_{N2} & \cdots & p_{NN} \end{bmatrix}, \rho^2 = \begin{bmatrix} \rho_1^2 \\ \rho_2^2 \\ \vdots \\ \rho_N^2 \end{bmatrix}, \mathbf{c_s}^2 = \begin{bmatrix} c_{S_1}^2 \\ c_{S_2}^2 \\ \vdots \\ c_{S_N}^2 \end{bmatrix}.$$

We also introduce the Hadamard product of matrices with identical dimensions, denoted by $\mathbf{A} \odot \mathbf{B}$, as follows:

$$(\mathbf{A} \odot \mathbf{B})_{ij} = (A_{ij})(B_{ij}).$$

It is easy to show that the matrix form of the set of N equations for the SCOVs is given by

$$\mathbf{c_a}^2 = \lambda \odot \frac{1}{\mathbf{a}} \odot \mathbf{c_A}^2 + \left(\frac{1}{\mathbf{a}}\mathbf{a}^T\right) \odot [\mathbf{P} \odot (\mathbf{J_N} - \mathbf{P})]^T \mathbf{e}^T$$

$$+ \left(\frac{1}{\mathbf{a}}\mathbf{a}^T\right) \odot [\mathbf{P} \odot \mathbf{P}]^T [(\mathbf{e} - \rho^2) \odot \mathbf{c_a}^2] + \left(\frac{1}{\mathbf{a}}\mathbf{a}^T\right) (\rho^2 \odot \mathbf{c_s}^2), \qquad (8.31)$$

where $\mathbf{J_N}$ is an all-ones matrix of $N \times N$ and $\mathbf{e}$ is all-ones column vector of size N. As long as all nodes are stable (i.e., no bottleneck nodes in the queueing network), this set of equations has a unique solution that gives the SCOV of the effective inter-arrival times for each node. Such a solution can be obtained by using some standard mathematical software package such as Matlab. The procedure of finding the mean queue length for each node is as follows:

1. Determine the effective arrival rate a_j of node j for $j = 1, ..., N$ by solving the set of equations (8.29).
2. Compute the traffic intensity for each node $\rho_j = a_j / \mu_j$ and verify the stability condition $\rho_i < 1$ for $i = 1, ..., N$.
3. Determine the SCOV $c_{a_j}^2$ of node j for $j = 1, ..., N$ by solving the set of equation (8.31).
4. Use the formula for $G/G/1$ queue to compute the mean queue length of each node with $\rho_i, c_{a_i}^2$, and $c_{s_i}^2$ for $i = 1, ..., N$.

With this algorithm, we can develop approximations to the first-moment performance measures of a large queueing network (i.e., networks with large N).

Exercise 8.9 *Suppose that an emergency department of a hospital can be approximately modeled as a six service station (nodes) network as shown in Figure 8.3. Customers arrive externally only into node S0 (station 0) and depart the system only from node S5. Assume that inter-arrival time has a mean of 2 min and standard deviation 0.5 min. The mean and standard deviation of service times (in minutes) at each node is given by*

Node i		0	1	2	3	4	5
Mean $1/\mu_i$		1	1.20	1.95	1.2	1.0	0.8
Standard deviation $\sqrt{c_{s_i}^2/\mu_i^2}$	1	1	1	1	1	1

The table below gives the routing probabilities from Nodes on the Left to Nodes on the Top:

	0	1	2	3	4	5
0	0	0.6	0.4	0	0	0
1	0	0	0	1	0	0
2	0	0.3	0	0.2	0.5	0
3	0	0	0	0	0	1
4	0	0	0	0.3	0	0.7
5	0	0	0	0	0.3	0

Use the decomposition approach to compute the steady-state average number of customers at each node of the network and the mean sojourn time spent by customers in the network.

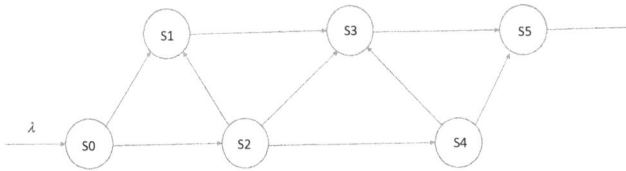

Figure 8.3 A six-node network with one external arrival stream for an ER in a hospital.

8.3 ONE-STAGE QUEUEING SYSTEM WITH MANY SERVERS

While the previous section is focused on a large queueing network with many nodes, this section treats a single queue system (single node) with many servers. To analyze such a large stochastic service system, we again utilize the diffusion approximation approach which is more practical than the exact approach. This is because the exact approach is either cumbersome (e.g., for an $MAP/PH/s$ queue discussed in Chapter 4) or intractable (e.g., for a $G/G/s$ queue). As shown earlier, the main idea of developing this diffusion approximation is to scale the time by a factor of n and to scale the space by a factor of $\sqrt{n}$. If we consider the stochastic process for a major performance measure, denoted by $Y(t)$, the corresponding diffusion-scaled process, denoted by $\hat{Y}_n(t)$, can be written as

$$\hat{Y}_n(t) = \frac{Y(nt) - \bar{Y}(nt)}{\sqrt{n}},$$

where $\bar{Y}(nt)$ is the deterministic fluid limit of the original process $Y(t)$ (i.e., $\bar{Y}(nt) = \bar{Y}_l(nt) = Y(lnt)/l$ as $l \to \infty$). Usually, we choose $\bar{Y}(nt) = E[Y(t)]$ or a heuristic approximation for it. Based on the FCLT, as $n \to \infty$, the scaled process $\hat{Y}_n(t)$ converges to a diffusion process. A diffusion process is a CTMP with almost surely continuous sample paths. Examples of diffusion processes are Brownian motion, Reflected Brownian motion, Ornstein-Uhlenbeck process, Brownian bridge process, branching process, etc. For the rigorous proof for the convergence of the stochastic process $\{\hat{Y}_n(t), t \geq 0\}$ as $n \to \infty$ to a diffusion process $\{\hat{Y}_\infty(t), t \geq 0\}$, interested readers are referred to Whitt (2002). In many applications, we are interested in the stationary distribution of a stochastic process. That is the distribution of $Y(\infty)$. To utilize the converged process to develop the approximation, we consider the stationary distribution of the scaled process, $\hat{Y}_\infty(\infty)$, which can be approximated by the distribution of $\hat{Y}_n(\infty)$ for the nth system when n is large. It follows from $\hat{Y}_\infty(\infty) \approx \hat{Y}_n(\infty)$ for a large n that

$$Y(\infty) \approx \bar{Y}(\infty) + \sqrt{n}\hat{Y}_n(\infty)$$

in distribution. The key issue of using this kind of approximation is how to select a large n to ensure the approximation to be reasonable. For a multi-server queue with arrival rate λ, service rate μ, and s servers, there are two choices: (i) $n = 1/(1-\rho)^2$ with $\rho = \lambda/s\mu$; and (ii) $n = s$. The first choice works (i.e., n becomes large) when $\rho \to 1$ from below and is appropriate for a heavy traffic system. The second choice works for a queueing system with a large number of servers. In practice, such a large-scale system is often under heavy traffic as well (i.e., a large call center). We demonstrate several different scalings for n in developing diffusion approximations for an $M/M/s$ queue. For more general $G/G/s$ queues, interested readers are referred to Whitt (2002). An advantage of considering the $M/M/s$ queue is that the queue-length process is Markovian and naturally converges to a diffusion process via an appropriate scalings for time and space. However, in the $G/G/s$ queue, the queue-length process may not converge to a diffusion process although the marginal distribution at any time in steady state converges to Gaussian.

8.3.1 MULTI-SERVER QUEUES WITHOUT CUSTOMER ABANDONMENTS

Consider an $M/M/s$ queue with arrival rate λ and service rate μ. Let $X(t)$ be the number of customers in the system at time t. Applying diffusion scaling to the queue-length process $X(t)$, we have

$$\hat{X}_n(t) = \frac{X(nt) - \bar{X}(nt)}{\sqrt{n}}$$

for any $n > 0$ and $t \geq 0$. As a heuristic approximation for the deterministic fluid limit $\bar{X}(nt)$ under a stability condition $\rho = \lambda/(s\mu) < 1$, we utilize the expected queue length for a steady-state system

$$L = \frac{\lambda}{\mu} + \frac{\rho(\lambda/\mu)^s}{s!(1-\rho)^2} p_0, \tag{8.32}$$

where

$$p_0 = \left[\sum_{i=0}^{s-1} \frac{(\lambda/\mu)^i}{i!} + \frac{(\lambda/\mu)^s}{s!} \frac{1}{1-\rho} \right]^{-1}.$$

We present three different scalings for n to show how to develop diffusion approximations for the queueing systems. While the first one is for the system with small number of servers, the other two are for the system with a large number of servers. Since we do not consider customer balking and reneging (abandonments) and are interested in stable systems, we assume $\rho < 1$ for all three types of scalings.

8.3.1.1 Increasing ρ with fixed s and μ

This scaling is achieved by increasing λ to make ρ approach to 1 and taking $n = 1/(1-\rho)^2$. As ρ increases, n increases so that the scaled process

$$\hat{X}_n(t/n) = \frac{X(t) - L}{\sqrt{n}}$$

converges to a diffusion process. Then the queue-length process $X(t)$ can be approximated by $\hat{X}_n(t/n)$ in distribution. Note that the diffusion scaling used here is a little different from the one used in Chapter 7. We fix the time horizon t and divide it by n. Such a scaling is actually equivalent to the standard diffusion scaling (i.e., the space is scaled by $\sqrt{n}$ and the time is scaled by n) and is appropriate for analyzing a queueing system with high traffic intensity but not many servers. The $G/G/1$ queue treated in Chapter 7 belongs to this type of scaling.

Exercise 8.10 *Provide a practical queueing system that can be studied by the scaling scheme of increasing ρ with fixed s and μ. Make comments on the efficiency and customer service for this type of systems.*

8.3.1.2 Increasing λ and s with fixed ρ

Now we consider a sequence of $M/M/s$ queues with a fixed service rate μ where λ and s are increased in such a way that ρ remains constant. These systems are indexed by n and n is set to be s (i.e., $n = s$). As n (or s) increases, the scaled process

$$\hat{X}_n(t/n) = \frac{X(t) - L}{\sqrt{n}}$$

converges to a diffusion process. Such a scaling is appropriate for developing approximations for the performance measures of a queueing system with a large number of servers but not very high traffic intensity.

Exercise 8.11 *Provide a practical queueing system that can be studied by the scaling scheme of increasing λ and s with fixed ρ. Make comments on the efficiency and customer service for this type of systems.*

8.3.1.3 Increasing λ and s with an increasing ρ

To study a queueing system with a large number of servers and heavy traffic intensity, we propose a special scaling. Again, we increase λ and s. However, their increases are related by a relation

$$\beta = (1 - \rho)\sqrt{s},$$

where β is a constant number. Since in this scaling, $s \to \infty$ and $\rho \to 1$, we can take the scale $n = s$ or $n = 1/(1-\rho)^2$. Again, as n increases, the scaled process

$$\hat{X}_n(t/n) = \frac{X(t) - L}{\sqrt{n}}$$

converges to a diffusion process. This scaling is appropriate for studying a queueing system with a large number of servers and heavy traffic intensity. This is a very practical regime. For example, a large inbound call center fits such a regime. Halfin and Whitt (1981) first introduced this scaling that has been applied to solve the staffing problems in large-scale call centers. They show that the scaled process

$$\hat{X}_n(t) = \frac{X(nt) - s}{\sqrt{n}}$$

with $n = s$ converges to a reflected Brownian motion with negative drift if $\hat{X}_n(t) \geq 0$. This is because for $X(nt) \geq s$, the queue-length process is a birth-and-death process with constant parameters λ and $s\mu$ which is a CTMC. This is a random walk that converges to a Brownian motion upon diffusion scaling. However, for $\hat{X}_n(t) < 0$, this process converges to an Ornstein–Uhlenbeck process upon diffusion scaling. Here, we intuitively explain why the deterministic fluid limit is chosen to be s in diffusion scaling. Note that the second term of L in (8.32) is approaching to zero as $n = s$ increases (scaling up the system), thus $L \to \lambda/\mu$ which converges to s. This is because under the Halfin–Whitt regime, as $s = n$ increases, $\rho = \lambda/(s\mu) \to 1$ which implies $L \to \lambda/\mu \to s$. The probability of delay converges to a constant

$$\lim_{n \to \infty} P\{\hat{X}_n(\infty) \geq 0\} = \theta$$

with

$$\theta = \frac{1}{1 + \sqrt{2\pi}\beta\Phi(\beta)e^{(\beta^2/2)}},$$

where $\Phi(x)$ is the CDF of a standard normal random variable and $\beta = (1 - \rho_n)\sqrt{s_n}$ with $\lambda_n \to \infty$ as $n \to \infty$ and $\mu_n = \mu$, $s_n = n$, and $\rho_n = \lambda_n/(n\mu) < 1$ for all n. Note that θ is the probability that a customer will experience any delay before the service. A larger β will decrease the probability of service delay. For the scaling with an increasing ρ and fixed s and μ, β will be zero and the probability of service delay is 1. On the other hand, for the scaling with an increasing s (and λ) and fixed ρ, β will be a large value which makes θ to be zero. This means that the probability of no delay will be 1. These results can be extended to the $G/M/s$ queue case (see Halfin and Whitt (1981)).

Exercise 8.12 *Besides the call center example, provide another practical queueing system that can be studied by the scaling scheme of increasing λ and s with an increasing ρ.*

8.3.2 MULTI-SERVER QUEUES WITH CUSTOMER ABANDONMENTS

Now we consider a queueing system where not every customer is patient enough to get service. There are two types of customer abandonments in the queueing system: balking and reneging. Balking means an arriving customer leaves immediately without joining the queue when he or she sees that the queue is too long. It is a customer behavior in an observable queue case. Reneging means that a customer in the queue decides to leave because service has not began after a certain amount of waiting. Reneging behavior is more common in an unobservable queue such as a call center. There are three classical Markovian multi-server queueing models: $M/M/s$ with infinite buffer, which is also called the Erlang-C model; $M/M/s/K$ with a finite buffer K, which is called the Erlang-A model; and $M/M/s/s$ with no waiting space, which is called the Erlang-B model. To model a call center, we assume that customers have random patient times. This means that if a customer's wait time is more than his or her patience time, he or she will abandon the queue. These customer patient times are modeled as i.i.d. random variables that are exponentially distributed with mean patience time $1/\alpha$. In this section, we focus on Erlang-A model with customer abandonments as the other models with customer abandonments can be treated similarly. The queue-length process (the number of customers in the system at time t), $X(t)$, can be modeled as a birth and death process with birth rate λ and death rate $\min(s,x)\mu + (x-s)^+\alpha$ when $X(t) = x$. Consequently, under diffusion scaling, this CTMC will converge to a diffusion process which can be used to develop diffusion approximations. Note that with customer abandonments, the queueing system is always stable. This means that there can be three service regimes: (i) $\rho < 1$ is called quality-driven (QD) regime because it strives for quality of service (typically $\lambda \to \infty$ and $s \to \infty$ and $\lambda/(s\mu) \to \rho < 1$); (ii) $\rho > 1$ is called efficiency-driven (ED) regime because it strives for efficient use of service (typically $\lambda \to \infty$ and $s \to \infty$ and $\lambda/(s\mu) \to \rho > 1$); (iii) $\rho \approx 1$ is called quality-and-efficiency-driven (QED) regime as it strives for both (especially as $\lambda \to \infty, s \to \infty$ such that $(1-\rho)\sqrt{s} \to \beta$). In fact, β in QED can be negative. To show how to develop the diffusion approximations for a queueing system with a large number of servers under heavy traffic, we pick up the Erlang-A model under the ED regime. The other combinations of models and service regimes can be studied similarly and interested readers are referred to Whitt (2004). First, we index a sequence of $M/M/s/K+M$ queues by $s = n$, where the sth (or nth) system has s servers and a buffer size of K_s and customer patient times follow an exponential distribution. We will scale the arrival rate λ_s and the buffer size K_s while the service rate and abandonment rate are kept constant (not scaled). Furthermore, the traffic intensity $\rho > 1$ is held constant to ensure that the entire sequence of queues are in ED regime. Since this sequence of scaled queues are all overloaded with increasing arrival rate, we do not have to scale the time to invoke the FCLT. Define q as the average number of customers waiting in the queue for each server. Then for each server (each service channel) in the sth system, the flow balance equation (input rate = output rate) is given by

$$\frac{\lambda_s}{s} = \mu + \alpha q,$$

which can be re-written as

$$q = \frac{\mu(\rho - 1)}{\alpha}. \tag{8.33}$$

As s increases, we scale up the arrival rate and the buffer size as follows:

$$\lambda_s = \rho s \mu, \\ K_s = s(\eta + 1) \tag{8.34}$$

for some $\eta > q$. Note that choosing $\eta > q$ can ensure that asymptotically no arriving customers are rejected due to a full system. The diffusion-scaled queue-length process is given by

$$\hat{X}_s(t) = \frac{X_s(t) - \bar{X}_s(t)}{\sqrt{s}},$$

where $\bar{X}_s(t)$ is the fluid limit of the queue-length process. Clearly, under the flow balance condition, we have

$$\bar{X}_s(t) = s + sq = s(1 + q). \tag{8.35}$$

Thus, the diffusion-scaled process can be written as

$$\hat{X}_s(t) = \frac{X_s(t) - s(1 + q)}{\sqrt{s}} \tag{8.36}$$

for all $t \geq 0$. It can be shown that such a scaled process $\{\hat{X}_s(t), t \geq 0\}$ converges to an Ornstein–Uhlenbeck process (also called O-U process) as $s \to \infty$. In state x, the two parameters of the O-U process are the infinitesimal mean $-\alpha x$ and infinitesimal variance $2\mu \rho$. Moreover, the stationary distribution of $\hat{s}(\infty)$ converges to a normal distribution with mean 0 and variance $\rho \mu / \alpha$. For the rigorous proofs of the convergence results, interested readers are referred to Whitt (2004). Here, we provide the explanations of obtaining these results in an intuitive way. First, if the original queue-length process in a queueing model is a birth-and-death process, it can be shown that the appropriately diffusion scaled process converges to a diffusion process such as a reflected Brownian motion or an O-U process. Second, the O-U process has a property of "mean-reverting". For the system considered here, if state $x > s(1 + q)$, then the death rate exceeds the birth rate, hence, the queue-length process gets pulled back to $s(1 + q)$. Similarly, if state $x \leq s(1 + q)$, then the birth rate exceeds the death rate, hence, the queue-length process gets pushed up to $s(1+q)$. With this mean-reverting property, it is expected that the scaled process will converge to an O-U process. Note that $X_s(t)$ takes on any non-negative integer k (i.e., the number of customers in the system), accordingly, the scaled process $\hat{X}_s(t)$ (scaled queue-length) should also take on integer value $[(k-s(1+q))/\sqrt{s}]$. However, in the limit, the scaled process will approach to a continuous state process. That means the scaled process state x_s will converge to a real value x, i.e., $x_s \to x$ as $s \to \infty$. Clearly, it follows from the diffusion-scaling formula that

$$x_s = \frac{[s(1+q)+x\sqrt{s}]-s(1+q)}{\sqrt{s}}$$

is the appropriate choice. Now we can compute the infinitesimal mean (also called mean drift rate), denoted by $m_s(x_s)$, as follows:

$$
\begin{aligned}
m_s(x_s) &= \lim_{h\to 0} E[(\hat{X}_s(t+h) - \hat{X}_s(t))/h|\hat{X}_s(t) = x_s] \\
&= \lim_{h\to 0} E[(X_s(t+h) - X_s(t))/(h\sqrt{s})|X_s(t) = \sqrt{s}x_s + s(1+q)] \quad (8.37) \\
&= \lim_{h\to 0} \frac{(\lambda_s - s\mu - \alpha(sq+x\sqrt{s}))h + o(h)}{h\sqrt{s}}
\end{aligned}
$$

for any $x_s \geq -q\sqrt{s}$ where $o(h)$ is a collection of terms of order less than h such that $o(h)/h \to 0$ as $h \to 0$. The first equality of (8.37) follows from the definition of the infinitesimal mean; the second equality follows from the diffusion scaling definition; and the third equality is based on the Taylor expansion of the expected queue-length process. Note that the first term of the numerator of the last line in (8.37) is the expected queue-length change in a small time interval h. The rate of change is the birth rate λ_s minus the total death rate $s\mu + \alpha(sq + x\sqrt{(s)})$ where $sq + x\sqrt{s}$ is the expected number of customers waiting in the line (not including s customers in service). Thus, we require $sq + x\sqrt{s} \geq 0$ which leads to $x_s \geq -q\sqrt{s}$.

By using $\lambda_s = \rho s\mu$, $q = \mu(\rho-1)/\alpha$, and taking the limit $h \to 0$ in (8.37), we have

$$m_s(x_s) = \mu\rho\sqrt{s} - \mu\sqrt{s} - \alpha x_s - \alpha q\sqrt{s} = -\alpha x_s.$$

Next, taking the limit $s \to \infty$, we obtain $m_s(x_s) \to m(x)$ such that

$$m(x) = -\alpha x.$$

Similarly, we can compute the infinitesimal variance as follows:

$$
\begin{aligned}
v_s(x_s) &= \lim_{h\to 0} E[(\hat{X}_s(t+h) - \hat{X}_s(t))^2/h|\hat{X}_s(t) = x_s] \\
&= \lim_{h\to 0} E[(X_s(t+h) - X_s(t))^2/(hs)|X_s(t) = \sqrt{s}x_s + s(1+q)] \quad (8.38) \\
&= \lim_{h\to 0} \frac{\lambda_s h + s\mu h + \alpha h(sq + x\sqrt{s}) + o(h)}{hs}
\end{aligned}
$$

for any $x_s \geq -q\sqrt{s}$. The third equality in (8.38) follows from the fact that the arrival process, service process, and reneging process are all Poisson processes. By using the expressions of λ_s and q and taking the limit $h \to 0$, we have

$$v_s(x_s) = \mu\rho + \mu + \alpha x_s/\sqrt{s} + \alpha q = 2\rho\mu + \alpha x_s/\sqrt{s}.$$

Taking the limit $s \to \infty$ leads to $v_s(x_s) \to v(x)$ such that

$$v(x) = 2\mu\rho.$$

It is worth noting that a diffusion process can be defined by a stochastic differential equation (SDE). We provide an brief introduction to the SDEs in the Chapter H and interested readers are referred to some excellent books such as Kurtz (1978). For the two popular diffusion processes, Brownian motion and O-U process, the following SDEs

$$dX(t) = \mu dt + \sigma dW(t),$$
$$dX(t) = -\kappa X(t)dt + \sigma dW(t)$$

are satisfied, respectively. We have obtained $\kappa = \alpha$ and $\sigma^2 = 2\mu\rho$ for the O-U process. Therefore, the stationary distribution of the O-U process, $\hat{X}_s(\infty)$ as $s \to \infty$, converges to a normal distribution with mean 0 and standard variance of $\sigma^2/(2\kappa) = \rho\mu/\alpha$. For a large s, the steady-state $X_s(\infty)$ can be approximated by a normally distributed random variable with mean $s(1+q)$ and variance $s\rho\mu/\alpha$. Thus, some simple approximate performance measures for a queue in the steady state (i.e., stationary performance measure) can be obtained easily. For example, for a not very large α, $X_s(\infty)$ is expected to exceed s with a very high probability (approximately 1). Then it follows from $L \approx s(1+q)$ and $L_q \approx L - s$ for a large s that $L_q \approx sq = s\mu(\rho - 1)/\alpha$. Let $P(ab)$ be the probability that an arriving customer in steady state will abandon without service. Now we use the flow balance equation

$$\alpha L_q = \lambda_s P(ab).$$

Substituting $L_q \approx sq = s\mu(\rho - 1)/\alpha$ and $\lambda_s = s\rho\mu$ into the equation above yields

$$P(ab) \approx \frac{\rho - 1}{\rho}.$$

So far our analysis is focused on developing approximations for the steady-state performance measures of a large queueing system. In the next section, we show that using the fluid and diffusion limits, we can also develop approximations for the transient (time-dependent) performance measures for a large queueing system.

Exercise 8.13 *Consider a large call center with 1000 operators each with service rate* $\mu = 1$ *customer per minute. The arrival rate of calls is* $\lambda = 1100$ *customers per minute and the average abandonment time is 5 minutes for each customer. Compute approximately the major performance measures of this system such as the mean and the standard deviation for the number of customers waiting for service to begin and the probability of abandonment. Make comments on the accuracy of the approximations to the performance measures of this call center.*

8.4   QUEUES WITH TIME-VARYING PARAMETERS

In this section, we show that the approach of developing approximations based on the fluid and diffusion limit processes can be applied to analyzing the transient behavior of a queueing system. Consider a generic single queue system that is modulated by Poisson processes with time-varying parameters. These Poisson processes represent arrival, service, reneging, and balking processes and can be deterministically or stochastically non-homogeneous over time. Here we focus on a multi-server Markovian queue with a non-homogeneous Poisson arrivals and time-varying number of exponential servers, denoted by $M_t/M/s_t$. The service rate μ remains a constant. Our interest is to develop reasonable approximations for the time-dependent performance measures such as the queue-length at time t, $X(t)$, which is the number of customers in the system. First, we introduce some notations for the non-homogeneous Poisson process. Let $\Lambda(t)$ be the expected number of events in a non-homogeneous Poisson process over time interval $(0,t]$. Then the non-homogeneous Poisson process, as a counting process, can be denoted by $N(\Lambda(t))$, which represents the number of events occurred over $(0,t]$. For the $M_t/M/s_t$ queue, the arrival process with a time-varying arrival rate can be written as

$$N_a\left(\int_0^t \lambda_u du\right).$$

Here we use the subscript to indicate the type of the counting process (i.e., "a" means "arrival"). The other counting process that affects the dynamics of the queueing system is the departure process which is not only non-homogeneous but also state dependent. The departure process can be written as

$$N_d\left(\int_0^t \min\{X(u),s_u\}\mu du\right).$$

If we do not consider other customer behaviors such as reneging or balking, the queue-length process is modulated by these two non-homogeneous Poisson processes and expressed as follows:

$$X(t) = X(0) + N_a\left(\int_0^t \lambda_u du\right) - N_d\left(\int_0^t \min\{X(u),s_u\}\mu du\right),$$

where $X(0)$ is the initial state of the system. In general, a single queue can be modulated by more non-homogeneous Poisson processes if additional factors affecting the queue-length are considered. For example, let us assume that (i) customer reneging behavior exists and is modeled by an exponentially distributed patient time with mean $1/\alpha$; and (ii) a new stream of customers arriving according to a homogeneous Poisson process with rate β will join the queue with probability of $p(X(t))$ upon seeing the queue length $X(t)$ at the arrival instant t. Then the queue-length process can be written as

$$X(t) = X(0) + N_a \left(\int_0^t \lambda_u du \right) - N_d \left(\int_0^t \min\{X(u), s_u\} \mu du \right)$$

$$- N_r \left(\alpha \int_0^t \max\{X(u) - s_u, 0\} du \right) + N_b \left(\beta \int_0^t p(X(u)) du \right).$$

Note that the expected number of events occurred over $(0,t]$ of the non-homogeneous Poisson process in the examples above depends on time t (via λ_t and s_t) and/or state $X(t)$. Thus, the integrand for the expected number of events can be written as a function of time and state $f(t, X(t))$. With this notation, we can write down the general queue-length process modulated by k Poisson processes as follows:

$$X(t) = X(0) + \sum_{j=1}^k I_j N_j \left(\int_0^t f_j(u, X(u)) du \right), \tag{8.39}$$

where $N_j(\cdot)$ is an independent non-homogeneous Poisson process, $I_j = \pm 1$, and $f_j(\cdot, \cdot)$ is a continuous function for all $j \in \{1, 2, ...k\}$. We can perform fluid scaling or diffusion scaling on Equation (8.39) to develop first-moment-based or second moment-based approximations for the queue-length of the system.

8.4.1 FLUID APPROXIMATION

We utilize a different fluid scaling to develop a way of computing the approximate expected queue length $E[X(t)]$. Instead of scaling up the time and space by n, the scaled process is obtained by taking n times faster rates of events in a fixed time interval $(0,t]$ and scaling space up by n times. Such a fluid-scaling, called "uniform acceleration" (see Massey and Whitt (1998)), is performed on (8.39) and the scaled queue-length process is given by

$$\bar{X}_n^{ua}(t) = \frac{nX_n(0) + \sum_{j=1}^k I_j N_j \left(\int_0^t n f_j(u, \bar{X}_n^{ua}(u)) du \right)}{n} \tag{8.40}$$

with $nX_n(0) = X(0)$ so that the initial state is also scaled. An intuitive reason of using the uniform acceleration scaling is because the time horizon for transient analysis should not be scaled up. With such a fluid scaling, as $n \to \infty$, the scaled process $\{\bar{X}_n^{ua}(t), t \geq 0\}$ converges to a deterministic process almost surely by invoking the FSLLN. That is $\lim_{n\to\infty} \bar{X}_n^{ua}(t) = \bar{X}(t)$ almost surely, where $\bar{X}(t)$ is a deterministic fluid process. It is worth noting that this type of scalings (both fluid and diffusion scalings) require some assumptions that are fairly mild and satisfied by most queueing systems. These assumptions are (i) the number of modulating Poisson processes k is finite; (ii) for any j, t, and x, $|f_j(t,x)| \leq C_j(1+x)$ for some $C_j < \infty$ and $T < \infty$ such that $t \leq T$; and (iii) $f_j(t,x)$ is Lipschitz continuous. That is $|f_j(t,x) - f_j(t,y)| \leq M|x-y|$ for all j, x, and $t \leq T$ and some $T < \infty$.

The main purpose of making these assumptions is to justify the interchange of expectation and limit and to ensure the existence of some integrals occurred in the following analysis. It is easy to check that in the previous examples these assumptions are satisfied. Note that when these assumptions are satisfied, we expect $\bar{X}(t) = \lim_{n\to\infty} E[\bar{X}_n^{ua}(t)]$. Now by taking expectations on both sides of (8.40), we have

$$
\begin{aligned}
E[\bar{X}_n^{ua}(t)] &= \frac{nX_n(0) + \sum_{j=1}^{k} I_j E\left[N_j\left(\int_0^t nf_j(u, \bar{X}_n^{ua}(u))du\right)\right]}{n} \\
&= \frac{nX_n(0) + \sum_{j=1}^{k} I_j E\left[\int_0^t nf_j(u, \bar{X}_n^{ua}(u))du\right]}{n} \qquad (8.41) \\
&= X_n(0) + \sum_{j=1}^{k} I_j E\left[\int_0^t f_j(u, \bar{X}_n^{ua}(u))du\right],
\end{aligned}
$$

where the second equality follows from the mean property of Poisson process. Since we are interested in the limit of the expected value of the scaled process, we use a nonparametric approach. First, we develop a limiting relation. It follows from the Lipschitz continuous property of $f_j(t,x)$ that

$$
|f_j(u, \bar{X}_n^{ua}(u)) - f_j(u, E[\bar{X}_n^{ua}(u)])| \leq M|\bar{X}_n^{ua}(u) - E[\bar{X}_n^{ua}(s)]|,
$$

which results in

$$
|E[f_j(u, \bar{X}_n^{ua}(u))] - f_j(u, E[\bar{X}_n^{ua}(u)])| \leq ME[|\bar{X}_n^{ua}(u) - E[\bar{X}_n^{ua}(s)]|], \quad (8.42)
$$

after taking the expectation. Taking the limit $n \to \infty$ on both sides of (8.42) leads to RHS being 0 which implies

$$
\lim_{n\to\infty} E[f_j(u, \bar{X}_n^{ua}(u))] = \lim_{n\to\infty} f_j(u, E[\bar{X}_n^{ua}(u)]). \qquad (8.43)
$$

Taking the limit on both sides (8.41) and using (8.43) yields

$$
\lim_{n\to\infty} E[\bar{X}_n^{ua}(t)] = X_n(0) + \sum_{j=1}^{k} I_j \int_0^t \lim_{n\to\infty} f_j(u, E[\bar{X}_n^{ua}(u)])du.
$$

Denoting $\bar{X}(t) = \lim_{n \to \infty} E[\bar{X}_n^{ua}(t)]$, we can write the equation above

$$\bar{X}(t) = \bar{X}(0) + \sum_{j=1}^{k} I_j \int_0^t f_j(u, \bar{X}(u)) du, \tag{8.44}$$

which can be used for solving $\bar{X}(t)$ for any t numerically. By choosing a reasonably large n, we can use the deterministic $\bar{X}(t)$ to approximate $\bar{X}_n^{ua}(t)$. Furthermore, with $X(t) = n\bar{X}_n^{ua}(t)$, an approximation for $X(t)$ can be developed.

Exercise 8.14 *Consider an international border-crossing station between the United States and Canada. The hourly arrival rate varies between 200 and 500 vehicles. The border waiting line can be approximately modeled as a $M_t/M/s_t$ queue where the time-varying arrival process is accommodated by the time-varying staffing level. The mean service time for each server is 1.2 minutes. The estimated hourly arrival rates and the number of servers (inspection booths) s_i for a 8-hour period (discretized into 8 hourly intervals) in a summer season are shown in the Table below.*

t	(0,1]	(1,2]	(2,3]	(3,4]	(4,5]	(5, 6]	(6, 7]	(7, 8]
λ_t	200	250	310	400	460	500	430	360
s_t	4	6	7	8	10	11	9	8

Develop a fluid scaling for this system by numerically describing the deterministic process $\{\bar{X}(t), 0 \le t \le 8\}$. Compare this deterministic process against a simulated path of the number in the system process $\{X(t), 0 \le t \le 8\}$ (one replication). Suppose that there are 90 vehicles in the system at time zero. That is $X(0) = 90$. Find an approximation for $E[X(t)]$ and compare against the average of 100 replications from a simulation experiment.

8.4.2 DIFFUSION APPROXIMATION

To obtain the second moment-based approximations, we need to consider the diffusion-scaled process $\{\hat{X}_n(t) \ge 0\}$ where $\hat{X}_n(t)$ is given by

$$\hat{X}_n(t) = \sqrt{n}(\bar{X}_n^{ua}(t) - \bar{X}(t)).$$

To determine the limit $\lim_{n \to \infty} \hat{X}_n(t)$, define a new function $F(x,t)$ as

$$F(t,x) = \sum_{j=1}^{k} I_j f_j(t,x). \tag{8.45}$$

Based on the assumptions made earlier, the function $F(t,x)$ has a property of

$$\left| \frac{d}{dx} F(t,x) \right| \le M,$$

for some finite M and $0 \le t \le T$. Shreve (2004) shows that under those conditions the limit $\lim_{n \to \infty} \hat{X}_n(t) = \hat{X}(t)$ exists and $\{\hat{X}(t), t \ge 0\}$ is a diffusion process. It can be shown that this diffusion process $\hat{X}(t)$ is the solution to

$$\hat{X}(t) = \sum_{j=1}^{k} I_j \int_0^t \sqrt{f_j(u, \bar{X}(u))} dW_j(u) + \int_0^t F'(u, \bar{X}(u)) \hat{X}(u) du, \qquad (8.46)$$

where $W_j(\cdot)$'s are independent standard Brownian motions, and $F'(t,x) = dF(t,x)/dx$. It is worth noting that Equation (8.46) is an SED that a diffusion process satisfies. The first term of the SED is the diffusion (volatility) part and the second term is the mean drift part. For more technical details, interested readers are referred to Shreve (2004).

For a large n, we have

$$\bar{X}_n^{ua}(t) \approx \bar{X}(t) + \frac{\hat{X}(t)}{\sqrt{n}} \qquad (8.47)$$

in distribution. For a given t, $\bar{X}_n^{ua}(t)$ can be approximated by a Gaussian random variable when n is large. The expected value and variance of $\bar{X}_n^{ua}(t)$ are given by

$$E[\bar{X}_n^{ua}(t)] \approx \bar{X}(t) + \frac{E[\hat{X}(t)]}{\sqrt{n}}, \qquad (8.48)$$

$$Var[\bar{X}_n^{ua}(t)] \approx \frac{Var[\hat{X}(t)]}{\sqrt{n}}, \qquad (8.49)$$

It can be shown that $Var[\hat{X}(t)]$ satisfies the differential equation:

$$\frac{dVar[\hat{X}(t)]}{dt} = \sum_{j=1}^{k} f_j(t, \bar{X}(t)) + 2F'(t, \bar{X}(t)) Var[\hat{X}(t)], \qquad (8.50)$$

with initial condition $Var[\hat{X}(0)] = 0$. The derivation of this ordinary differential equation (ODE) can be found in Arnold (1992). Again, this ODE can be solved numerically in practice.

Exercise 8.15 *Consider the $M_t/M/s_t$ queue that models the land border-crossing station in Exercise 8.14. Using the results obtained in Exercise 8.14, develop a diffusion model for that system. Then find an approximation for $Var[X(t)]$ and compare against simulations by making 100 replications and obtaining sample variances.*

8.5 MEAN FIELD METHOD FOR A LARGE SYSTEM WITH MANY IDENTICAL INTERACTING PARTS

To efficiently analyze Markovian models consisting of a large finite number of identical interacting entities, the mean field method is often utilized to understand behaviors of this kind of models when the number of entities tends to infinity and this method suggests an approximation when the number of entities is sufficiently large. We demonstrate this methodology by considering a large queueing system consisting of N identical Markovian queues, call entities. Note that this queueing system serves as an example only. The mean field method can be applied to other stochastic systems with a large number of interacting entities. Thus, if the word "queue" in this section is replaced with the word "entity", we can apply this method to other stochastic systems satisfying the assumptions made. The following example and analysis is mainly taken from Bobbio et al. (2008). Since all queues are Markovian, the state transitions of the CTMCs depend on the current state only, thus, the entire system remains Markovian. The state of queue i $(i = 1, 2, ..., N)$ at time t is denoted by $X_i(t)$. It is assumed that all these queues are identical and indistinguishable. Therefore, the behavior of queue i (any one of these queues) does not depend directly on a particular queue j, but it may depend on the global number of queues in each state. Since all queues are identical, the state of a randomly selected (tagged) queue is denoted by $X(t)$ with a finite state space S composed of $M = |S|$ states. Let $N_i(t)$ denote the number of queues that are in state i for $i \in S$ at time t. Assume that each queue has a buffer size of $M - 1$. Then, we can use the M-dimensional vector $\mathbf{N}(t) = (N_1(t), N_2(t), ..., N_M(t))$ with $\sum_{i=1}^{M} N_i(t) = N$ to describe the queueing system. The global behavior of the set of N queues form a CTMC over the state space of size M^N. However, since all queues are identical and indistinguishable, the state space can be lumped into the aggregate state space S' with $|S'| = \binom{N+M-1}{M-1}$, where a state of the overall CTMC is identified by the number of queues staying in each state of S. That is $\mathbf{N}(t) = (N_1(t), N_2(t), ..., N_M(t))$. Here we refer the particular queue of interest (i.e., queue i) as a local CTMC and the whole system with the lumped states (i.e., the environment of queue i) as the overall CTMC. The evolution of the local CTMC is such that there are no synchronous transitions in different queues and the transition rates of a given queue may depend on the global behavior of the overall CTMC, the actual value of $\mathbf{N}(t)$. Thus, the transition rates, denoted by Q_{ij}, for the local CTMC or queue i are given by

$$Q_{ij}(\mathbf{N}(t)) = \begin{cases} \lim_{\Delta \to 0} \frac{1}{\Delta} P(X(t+\Delta) = j | X(t) = i, \mathbf{N}(t)) & \text{if } N_i(t) > 0, \\ 0, & \text{if } N_i(t) = 0, \end{cases}$$

$$Q_{ii}(\mathbf{N}(t)) = - \sum_{j \in S, j \neq i} Q_{ij}(\mathbf{N}(t)).$$

$$(8.51)$$

The first condition in (8.51) $N_i(t) \geq 1$ implies that at least the tagged queue is in state i. For the overall CTMC, to accommodate large N values, we replace $\mathbf{N}(t)$ with the normalized vector $\mathbf{n}(t) = \mathbf{N}(t)/N$, where the element i $0 \leq n_i(t) \leq 1$ representing the proportion of queues in state i at time t and $\sum_{i \in S} n_i(t) = 1$. For a given N, the transition rate function based on the normalized vector can be written as

$$q_{ij}^{(N)}(\mathbf{n}(t)) = Q_{ij}(N \cdot \mathbf{n}(t)). \tag{8.52}$$

Thus, (8.51) and (8.52) describe the same transition rate matrix. To investigate the limiting behavior as N tends to infinity, we need a $q_{ij}(\mathbf{n}(t))$ function which is defined for all feasible $\mathbf{n}(t)$ vectors and satisfies $q_{ij}^{(N)}(\mathbf{n}(t)) = q_{ij}(\mathbf{n}(t))$ for all $N \geq 1$. The existence of such $q_{ij}(\cdot)$ functions with the desired properties plays an important role in the applicability of the mean field method. In fact, there are some examples in which $q_{ij}(\cdot)$ exists but may not be bounded and/or continuous. As long as $q_{ij}(\cdot)$ is independent of N for a system, the mean field is most likely to be valid. We present a queueing system to demonstrate such a phenomenon. This observation suggests that the practical application of the mean field approach for CTMCs requires more relaxed conditions than the ones in the studies by Le Boudec et al. (2007) and Benaim and Le Boudec (2008). The main theorem for applying the mean field is given as follows.

Theorem 8.7 *The normalized state vector of the lumped process, $\mathbf{n}(t)$, tends to be deterministic, in distribution, as N tends to infinity and satisfies the differential equation*

$$\frac{d}{dt}\mathbf{n}(t) = \mathbf{n}(t)q(\mathbf{n}(t)), \tag{8.53}$$

where $\mathbf{q}(n(t)) = \{q_{ij}(\mathbf{n}(t))\}$.

The proof of this theorem is omitted here and requires further investigation on the conditions that $\mathbf{q}(\mathbf{n}(t))$ must satisfy. For example, in Le Boudec et al. (2007), with some strong conditions for $\mathbf{q}(\mathbf{n}(t))$, this theorem can be proved. Theorem 8.7 holds in the limiting sense. In practice, N is large but not to infinity. Thus, the following proposition provides an approximation method for the case when N is finite but sufficiently large.

Proposition 8.1 *When N is sufficiently large, the normalized state vector of the lumped process $\mathbf{n}(t)$ is a random vector whose mean can be approximated by the following differential equations*

$$\frac{d}{dt}E[n_i(t)] \approx \sum_{j \in S} E[n_j(t)]q_{ji}(E[n(t)]). \tag{8.54}$$

The proof can be found in a technical report by Gribaudo et al. (2008). While Theorem 8.7 is exact as $N \to \infty$, Proposition 8.1 is approximate for a sufficiently large N. The accuracy depends on the distribution of $\mathbf{n}(t)$. The closer

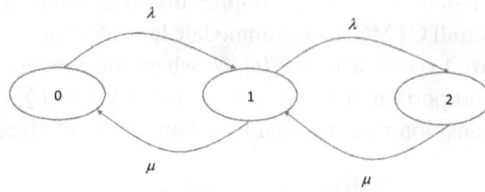

Figure 8.4   State transition in an $M/M/1/1$ queue.

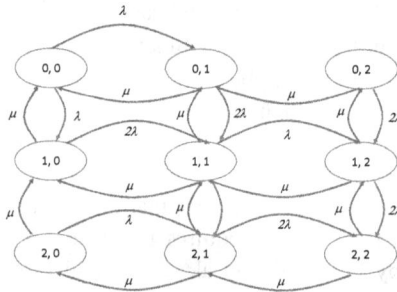

Figure 8.5   State transition in a system with two $M/M/1/1$ queues.

$\mathbf{n}(t)$ is to deterministic, the better is the approximation provided by the proposition.

To illustrate the mean field methodology, we consider a queueing system consisting of N identical subsystems. We start with $N = 1$ case where a single server queue has a buffer of size 1. Assume that the inter-arrival times and service times are two independent sequences of exponentially distributed i.i.d. random variables with mean of $1/\lambda$ and $1/\mu$, respectively. Such a system can be denoted by $M/M/1/1$ and depicted in Figure 8.4. If the system state is the number of customers in the system, then the state space is $S = \{0,1,2\}$ with $M = 3$. Now we consider $N > 1$ cases which consists of N identical $M/M/1/1$ subsystems with a Poisson process with rate $N\lambda$. Furthermore, we assume that an arriving customer chooses the shortest queue to join if the buffer is not full. This customer's joining policy makes the different queues inter-dependent. For $N = 2$, a two-dimensional CTMC can be formed with the first (second) state variable representing the number of customers in the first (second) queue. The state-transition diagram is shown in Figure 8.5. Clearly, the transition rates of the arrivals to queue 1 depend on the state of queue 2 as shown in Figure 8.6.

Since the $N = 2$ queues are identical with $M = 3$, we can lump the states to get the lumped state (environmental) process $\mathbf{N}(t) = (N_1(t), N_2(t), N_3(t))$, where $N_1(t)$ is the number of queues in state 0, $N_2(t)$ is the number of queues in

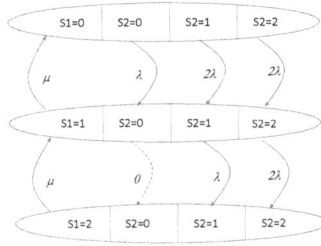

Figure 8.6 Transition rates in the first $M/M/1/1$ queue depending on the second $M/M/1/1$ queue.

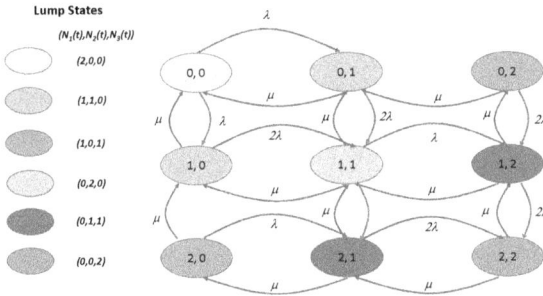

Figure 8.7 Markov chain with lumping states from the two $M/M/1/1$ queues.

state 1, and $N_3(t)$ is the number of queues in state 2. To determine the number of states in this lumped state process for an N queue system with M states for each queue, we consider an equivalent problem of allocating N identical objects into M bins. Graphically, we can represent the problem by finding the number of ways of using $M-1$ bars to separate N stars putting in a line. This problem can be considered as follows: there are $N+M-1$ things that need to be placed, and $M-1$ of those placements are chosen for the bars. Thus, there are $\binom{N+M-1}{N-1}$ possible allocations of N identical objects among M distinct bins. In our example with $N=2$ and $M=3$, there are $\binom{2+3-1}{3-1}=6$ states. That is

$$N(t) \in \{(2,0,0),(1,1,0),(0,2,0),(1,0,1),(0,0,2),(0,1,1)\}.$$

In Figure 8.7, each lumped state consists of the states that are not distinguishable due to identical queues.

There are two ways of viewing the dynamics of the lumped state process. The first one shown in Figure 8.8 is to regard $N(t)$ as a CTMC.

The other is from the viewpoint of a tagged queue. In this case, the arrival rates (i.e., birth rates) depend on the states of the lumped CTMC while the

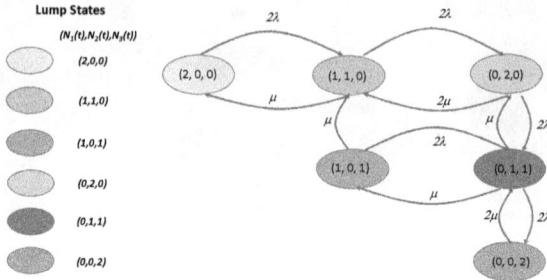

Figure 8.8 Overall view of the Markov chain.

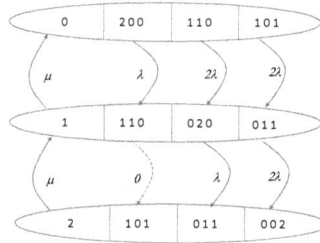

Figure 8.9 Transition rates in the tagged queue depending on the Markov chain with lumped states.

service rate (i.e., death rate) remains the same. This view is shown in Figure 8.9, where the left cell represents the state of this tagged queue and the three states in the right cell are the possible lumped states of $N(t)$ to which this tagged queue state belongs. Thus, the process of the tagged queue can be described as a birth-and-death process with state dependent birth rates, which is shown in Figure 8.10. These birth rates can be considered as functions of the lumped states and are given by

$$\Lambda_0(N(t)) = \begin{cases} \lambda & \text{if } N(t) = (2,0,0), \\ 2\lambda & \text{if } N(t) = (1,1,0), \\ 2\lambda & \text{if } N(t) = (1,0,1). \end{cases}$$

$$\Lambda_1(N(t)) = \begin{cases} 0 & \text{if } N(t) = (1,1,0), \\ \lambda & \text{if } N(t) = (0,2,0), \\ 2\lambda & \text{if } N(t) = (0,1,1). \end{cases}$$

Using the normalized vector $\mathbf{n}(t) = \mathbf{N}(t)/N$, we obtain

$$\mathbf{n}(t) \in \{(1,0,0),(0.5,0.5,0),(0,1,0),(0.5,0,0.5),(0,0,1),(0,0.5,0.5)\}.$$

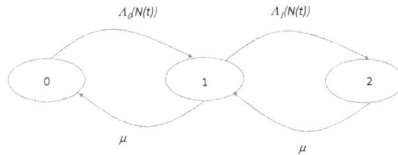

Figure 8.10   Markov chain with identical entities.

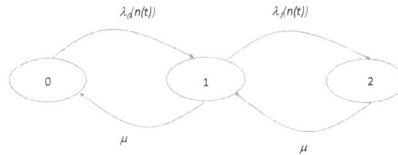

Figure 8.11   Markov chain of one of identical entities.

Consequently, the birth rates can be written as

$$\lambda_0(\mathbf{n}(t)) = \begin{cases} \lambda & \text{if } n(t) = (1,0,0), \\ 2\lambda & \text{if } n(t) = (0.5, 0.5, 0), \\ 2\lambda & \text{if } n(t) = (0.5, 0, 0.5). \end{cases}$$

$$\lambda_1(\mathbf{n}(t)) = \begin{cases} 0, & \text{if } n(t) = (0.5, 0.5, 0), \\ \lambda & \text{if } n(t) = (0, 1, 0), \\ 2\lambda & \text{if } n(t) = (0, 0.5, 0.5). \end{cases}$$

To make these transition rates independent of N, we can use the expressions in terms of non-full states. That is

$$\lambda_0(\mathbf{n}(t)) = \begin{cases} \frac{\lambda}{n_0(t)} & \text{if } n_0(t) \neq 0, \\ 0 & \text{if } n_0(t) = 0, \end{cases}$$

$$\lambda_1(\mathbf{n}(t)) = \begin{cases} \frac{\lambda}{n_1(t)} & \text{if } n_0(t) = 0 \text{ and } n_1(t) > 0, \\ 0 & \text{if } n_0(t) > 0 \text{ or } n_1(t) = 0. \end{cases} \qquad (8.55)$$

The transition rate diagram is shown in Figure 8.11. For this birth-and-death process, the death rates are fixed and independent of the state of other queue and the birth rates depend on the lumped states. Such a dependency of the other queues is the action of the mean field. The transition rates, depending on the states of the lumped process (i.e., other queues) but independent of N, is the key to evaluate the asymptotic behavior when $N \to \infty$. According Theorem 8.7,

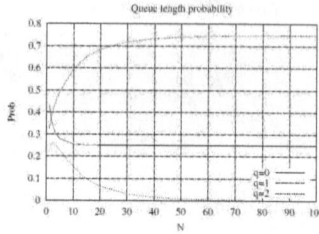

Figure 8.12 Probability of queue length as function of N. Source: Adapted from A. Bobbio, M. Gribaudo, M. Telek, Analysis of large scale interacting systems by mean field method. *Quantitative Evaluation of Systems, IEEE Computer Society*, 215-224, 2008.

we have the deterministic $(n(t))$ satisfying

$$\frac{d}{dt}(n_0(t), n_1(t), n_2(t)) = (n_0(t), n_1(t), n_2(t))$$

$$\times \begin{pmatrix} -\lambda_0(n(t)) & \lambda_0(n(t)) & 0 \\ \mu & -\mu - \lambda_1(n(t)) & \lambda_1(n(t)) \\ 0 & \mu & -\mu \end{pmatrix}.$$

$$(8.56)$$

With the initial state $\mathbf{n}(0) = (1,0,0)$, the transient behavior of $\mathbf{n}(t)$ can be computed by using the numerical method. The limiting behavior as $t \to \infty$ can be obtained as the limit of the transient results as follows:

$$\lim_{t \to \infty} \mathbf{n}(t) = \begin{cases} (1 - \frac{\lambda}{\mu}, \frac{\lambda}{\mu}, 0) & \text{if } \frac{\lambda}{\mu} < 1, \\ (0,0,1) & \text{if } \frac{\lambda}{\mu} \geq 1, \end{cases} \qquad (8.57)$$

which agree with our intuition on the system behavior. From (8.55) and the limiting behavior of $n_0(t)$, we can see that $\lambda_0(\mathbf{n}(t))$ can be unbounded as λ approaches to μ. However, since $\lambda_0(\mathbf{n}(t))$ is always multiplied by $n_0(t)$ in the right-hand side of the ODE. Thus, the unboundedness of $\lambda_0(\mathbf{n}(t))$ is not an issue in the numerical analysis based on (8.56). Note that the expression of (8.55) also indicates the discontinuity of the product $n_0(t)\lambda_0(n(t))$ at $n_0(t) = 0$ as it equals λ when $n_0(t) > 0$ and 0 otherwise. The same situation occurs with $n_1(t)\lambda_1(n(t))$. Figure 8.12, adapted from Bobbio et al. (2008), shows the limit proposed in equation (8.57) by plotting the probability of each state as a function of N in a case with $\lambda = 1.5$ and $\mu = 2.0$.

Now we demonstrate the convergence as $N \to \infty$ by performing numerical analysis on a simpler example consisting of N identical $M/M/1/0$ queues. That is each queue has only two states 0 and 1. In this case, the normalized vector for the stationary lumped process (n_0, n_1) can be determined by n_0 (i.e.,

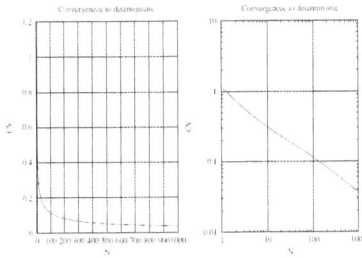

Figure 8.13 The coefficient of variation of queue length 0 and 1 as functions of N. Source: Adapted from A. Bobbio, M. Gribaudo, and M. Telek, Analysis of large scale interacting systems by mean field method. *Quantitative Evaluation of Systems, IEEE Computer Society*, 215-224, 2008.

$\lim_{t \to \infty} n_0(t) = n_0$ and $\lim_{t \to \infty} n_1(t) = n_1$). Again, with $\lambda = 1.5$ and $\mu = 2$, Figure 8.13, adapted from Bobbio et al. (2008), shows how the COV of n_0 (the proportion of the queues in the first state) approaches to zero as $N \to \infty$. For more results and analysis of this type of systems, interested readers are referred to Bobbio et al. (2008).

Exercise 8.16 *Provide a practical system that can be effectively studied by using the mean field method. Discuss the pros and cons of this method.*

REFERENCE NOTES

This chapter extends the stochastic limit approximation method to large stochastic system analysis. The aim of this chapter is to provide readers with an introduction to this approach. The multi-dimensional BM approximations are basically based on [4]. Thus, we omit all proofs and refer to [4] for more details and further discussions. For the decomposition approach and queues with time-varying parameters, we refer to [6] for more development. The one-stage queueing systems with many servers is based on [9] and [10]. The introduction to mean-field method is taken from [3].

REFERENCES

1. L. Arnold, "Stochastic Differential Equations: Theory and Applications", Krieger Publishing Company, Melbourne,FL, 1992.
2. M. Benaim and J.Y. Le Boudec, A class of mean field interaction models for computer and communication systems, *Performance Evaluation*, doi:10.1016/j.peva.2008.03.005, 2008.

3. A. Bobbio, M. Gribaudo, M. Telek, Analysis of large scale interacting systems by mean field method, *Quantitative Evaluation of Systems, IEEE Computer Society*, 215–224, 2008.

4. H. Chen and D.D. Yao, "Fundamentals of Queueing Networks: Performance, Asymptotics, and Optimization", Springer, New York, 2001.

5. M. Gribaudo, M. Telek, and A. Bobbio, Mean field methods in performance analysis. Technical report, Dip Informatica - Universit'a Piemonte Orientale; www.di.unipmn.it/Tecnical-R/index.htm, 2008.

6. N. Gautam, " Analysis of Queues: Methods and Applications", CRC Press, London, 2012.

7. J.Y. Le Boudec, D. McDonald, and J. Mundinger, A generic mean field convergence result for systems of interacting objects. In 4th Int Conf on Quantitative Evaluation of Systems - QEST2007, 2007.

8. S. Shreve, "Stochastic Calculus for Finance I: The Binomial Asset Pricing Model", Springer Finance, 2004.

9. W. Whitt, "Stochastic-Process Limits: An Introduction to Stochastic-Process Limits and Their Application to Queues", Springer 2002.

10. W. Whitt, Efficiency-Driven Heavy-Traffic Approximations for Many-Server Queues with Abandonments, *Management Science*, 50, (10), 1449–1461, 2004.

9 Static Optimization in Stochastic Models

In this and the next chapters, we consider the optimization problems in stochastic models. According to the nature of the problems, we classify them into static and dynamic optimization types. This chapter focuses on the static optimization problems. We first present the optimization models based on renewal cycles. A maintenance model and a vacation queueing model are utilized to demonstrate this approach. Next, we discuss more optimization models based on stationary performance measures and address the optimization issues in the stochastic service models with customer choices. Then, we present an example of static optimal policy, called $c\mu$ rule, for a multi-class queue, which can be considered as a degeneracy of the dynamic policy. As a prelude to the next chapter, we also study a customer assignment problem in a queueing system with multi-class servers and characterize the optimal policy as a c/μ rule. Finally, we discuss the performance measures that are often used in optimizing stochastic systems.

9.1 OPTIMIZATION BASED ON REGENERATIVE CYCLES

9.1.1 OPTIMAL AGE REPLACEMENT POLICY FOR A MULTI-STATE SYSTEM

Consider a system with a finite number of states of different operating efficiencies. Denote the state of the system by $\phi(t)$ at time t. Assume that the system has $M + 1$ states with the perfect working state M and failure state 0. At time 0, the system is brand new or $\phi(0) = M$. As time goes on, the system deteriorates by transferring from more efficient to less efficient states. That is from the initial state $\phi(0) = M$ to $\phi(t) \in E = \{M, M - 1, M - 2, ..., 1, 0\}$ with $t \geq 0$. Such a state transition structure fits both a single component system and a multiple component system. For simplicity, here we focus on a single component system in the following analysis. The main issue for the system manager is to decide when such a multi-state system should be replaced based on its age under a given cost structure consisting of planned and unplanned replacement costs. This model is general enough to be applicable in many practical situations. For example, if a medical procedure such as knee replacement is not performed until the condition reaches a complete joint failure state (i.e., $\phi(t) = 0$), the chances of recovery can be reduced significantly. Thus, we consider a preventive age replacement policy (ARP) for such a multi-state system.

DOI: 10.1201/9781003150060-9

The analysis of this section is mainly based on the work by (Sheu and Zhang (2013)). The multi-state system evolves as a non-homogeneous CTMC. That is

$$P(\phi(t_{n+1}) = k_{n+1} | \phi(t_1) = k_1, \phi(t_2) = k_2, \cdots, \phi(t_1) = k_n,)$$
$$= P(\phi(t_{n+1}) = k_{n+1} | \phi(t_n) = k_n)$$

for all $0 \leq t_1 < t_2 < \cdots < t_n < t_{n+1}$ where $k_i \in E$ is the actual system state at time instant t_i. Denote for any $t \geq 0, s > 0$ the transition probability by

$$P_{ij}(t, t+s) = P(\phi(t+s) = j | \phi(t) = i). \tag{9.1}$$

Due to the Markovian property, for any $t \geq 0, s \geq 0, u \geq 0$ and $i, j, k \in E$, we can write down the finite number of C-K equations as

$$P_{ij}(t, t+s+u) = \sum_{k=0}^{M} P_{ik}(t, t+s) P_{kj}(t+s, t+s+u) \tag{9.2}$$

with

$$\lim_{s \to 0} P_{ij}(t, t+s) = \begin{cases} 1 & \text{if } i = j, \\ 0 & \text{if } i \neq j. \end{cases}$$

In matrix form, the C-K equations (9.2) can be written as

$$\mathbf{P}(t, t+s+u) = \mathbf{P}(t, t+s) \mathbf{P}(t+s, t+s+u), \tag{9.3}$$

where $\mathbf{P}(t, t+s) = [P_{ij}(t, t+s)]_{i,j=0}^{M}$ and $\mathbf{P}(t, t) = \mathbf{I}$ (an identity matrix). In this model, we assume that the initial state is a brand new system, i.e., $\phi(0) = M$. Then the most interesting transition probability function is the probability that the system is in state j at time t, denoted by $P_j(t) = P_{Mj}(0, t)$. It is easy to write the differential equations that these probabilities satisfy as

$$P_j'(t) = P_j(t) \left(- \sum_{k=0, k \neq j}^{M} \lambda_{jk}(t) \right) + \sum_{k=0, k \neq j}^{M} P_k(t) \lambda_{kj}(t), \quad j \in E \tag{9.4}$$

with $P_M(0) = 1$ where $\lambda_{ij}(t)$ is the transition rate from state i to state j at time t. To simplify the notation, we let $\lambda_{jj}(t) = \left(- \sum_{k=0, k \neq j}^{M} \lambda_{jk}(t) \right)$. In matrix form, we can write (9.4) as follows:

$$\mathbf{P}'(t) = \mathbf{P}(t) \mathbf{A}(t), \tag{9.5}$$

where

$$\mathbf{P}(t) = [P_M(t), P_{M-1}(t), \cdots, P_1(t), P_0(t)],$$
$$\mathbf{P}'(t) = [P_M'(t), P_{M-1}'(t), \cdots P_1'(t), P_0'(t)],$$

$$\mathbf{A}(t) = \begin{bmatrix} \lambda_{MM}(t) & \lambda_{MM-1}(t) & \lambda_{MM-2} & \cdots & \lambda_{M1}(t) & \lambda_{M0}(t) \\ 0 & \lambda_{M-1M-1}(t) & \lambda_{M-1M-2}(t) & \cdots & \lambda_{M-11}(t) & \lambda_{M-10}(t) \\ 0 & 0 & \lambda_{M-2M-2}(t) & \cdots & \lambda_{M-21}(t) & \lambda_{M-20}(t) \\ \vdots & \vdots & \vdots & \ddots & \vdots & \vdots \\ 0 & 0 & 0 & \cdots & 0 & 0 \end{bmatrix}.$$

We can solve (9.5) by using a recursive approach that takes advantage of the structure of the infinitesimal matrix. This approach becomes more efficient for large M cases and is summarized in the following theorem.

Theorem 9.1 *For a multi-state system with $E = \{M, M-1, ..., 0\}$ and the degradation process modeled as nonhomogeneous CTMC $\{\phi(t), t \geq 0\}$, the transition probability function $P_i(t), i = M, M-1, ..., 1, 0$ can be recursively solved as follows:*

$$P_M(t) = P(\phi(t) = M) = \exp\left(\int_0^t \lambda_{MM}(s)ds\right)$$
$$= \exp\left(-\int_0^t \sum_{k=0}^{M-1} \lambda_{Mk}(s)ds\right),$$

(9.6)

and

$$P_i(t) = P(\phi(t) = i)$$
$$= \sum_{k=i+1}^{M} \int_0^t P_k(\tau_{M+1-k})\lambda_{ki}(\tau_{M+1-k}) \cdot \exp\left(\int_{\tau_{M+1-k}}^t \lambda_{ii}(s)ds\right) d\tau_{M+1-k}$$
$$= \sum_{k=i+1}^{M} \int_0^t P_k(\tau_{M+1-k})\lambda_{ki}(\tau_{M+1-k}) \cdot \exp\left(-\int_{\tau_{M+1-k}}^t \sum_{j=0}^{i-1} \lambda_{ij}(s)ds\right) d\tau_{M+1-k}$$

(9.7)

for $i = M-1, M-2, ..., 1, 0$ with the initial conditions of $P_M(0) = P(\phi(0) = M)$ and $P_k(0) = P(\phi(0) = k) = 0$ for $k = M-1, M-2, ..., 1, 0$.

The proof is left as an exercise for readers. In practice, a system may be composed of several multi-state elements. To analyze the aging behavior of such a system, one has to know the degradation process of each element. Element $l, l = 1, ..., n$, has different $M_l + 1$ states with a performance vector denoted by $g_l = \{g_{l0}, g_{l1}, \cdots, g_{lM_l}\}$, where g_{lj} is the performance rate of element l in the state $j \in \{0, 1, ..., M_l\}$. Let $G_l(t)$ denote the performance rate of element l at time $t \geq 0$. Then $G_l(t)$ follows the nonhomogeneous CTMC with state space g_l. The probability distribution of performance rate for element l at time t can be written as

$$P_l(t) = \{P_{l0}(t), P_{l1}(t), \cdots, P_{lM_l}(t)\},$$

(9.8)

where $P_{lj}(t) = P(G_l(t) = g_{lj}), j = 0, 1, \cdots, M_l$. The n element multi-state system's structure function, defined as $G(t) = \psi(G_1(t), G_2(t), \cdots, G_n(t))$, also follows a nonhomogeneous CTMC by using the Lz-transform method (see Lisnianski (2012) Chapter 6, Proposition 1, and Property 2). The Lz-transform exists only when all transition intensities are continuous functions of time t, which is our case here. Hence, we can obtain the performance distribution of the entire multi-state system via the Lz-transform method. Using $P_{lj}(t), l = 1, 2, ..., n; j = 0, 1, 2, ..., M_l$, we can define the Lz-transform for element l as

$$L_z[G_l(t)] = \sum_{j=0}^{M_l} P_{lj}(t) z^{g_{lj}}. \tag{9.9}$$

The overall performance rate of this multi-state system can be determined by the structure function $\psi(G_1(t), G_2(t), \cdots, G_n(t))$. Let Ω denote the connection operator. For example, $\Omega_{\psi p}$ represents elements connected in parallel structure and $\Omega_{\psi s}$ represents elements connected in series structure. These operators are applied over the Lz-transforms of individual elements and their combinations to obtain the Lz-transform for the entire multi-state system

$$L_z[G(t)] = \Omega\{L_z[G_1(t)], L_z[G_2(t)], \cdots, L_z[G_n(t)]\}$$

$$= \Omega\left\{ \sum_{j_1=0}^{M_1} P_{1j_a}(t) z^{g_{1j_1}}, \cdots, \sum_{j_n=0}^{M_n} P_{nj_n}(t) z^{g_{njn}} \right\}$$

$$= \sum_{j_1=0}^{M_1} \cdots \sum_{j_n=0}^{M_n} \left(\prod_{l=1}^{M_n} P_{lj_l}(t) z^{\psi(g_{1j_1}, \cdots, g_{njn})} \right) = \sum_{j=0}^{N} P_j(t) z^{g_j},$$

where $N+1$ is the total number of the system states (the total number of feasible combinations of elements' states), and g_j is the entire system performance rate in the corresponding state $j, j = 0, 1, 2, \cdots, N$. The system status depends on the relation between the actual performance level and the desired performance level or user demand level. We assume that the user demand level is a constant w. The actual performance level (rate) $G(t)$ for the multi-state system is a non-homogeneous CTMC with the state space $g = \{g_0, g_1, \cdots, g_N\}$ where $g_{j+1} \geq g_j$. Thus, the user demand threshold is defined as $k(w) = \inf\{j : g_j \geq w\}$. Then, the set of acceptance states is $S_A = \{k(w), k(w) + 1, \cdots, N\}$ and its complement, $S_A^c = \{0, 1, \cdots, k(w) - 1\}$, is the set of unacceptable states. Assume that at time 0 a brand new system is in operation. Whenever the system's performance rate $G(t)$ enters S_A^c, the system is considered as "failed" and is replaced with a brand new one. Under an age replacement policy (ARP), the system is replaced with a brand new one either at a failure instant or at age T, whichever comes first.

To determine the optimal ARP, we consider a reward and cost structure as follows: Let r_i be the reward rate (revenue rate – operating cost rate) for a working state i, where $i = k(w), k(w) + 1, \cdots, N$ and $r_N \geq r_{N-1} \geq \cdots, \geq r_{k(w)+1} \geq r_{k(w)}$. Let c_u and c_p denote the cost of unplanned replacement due to a system failure and the cost of planned replacement due to reaching age T, respectively. We assume that $c_u > c_p > 0$, which signifies that an unplanned replacement is more costly than a planned replacement.

Denote the time interval from $t = 0$ to a system failure by Y. Then the length of a replacement cycle, denoted by Y^*, is given by $Y^* = Y \wedge T = \min(Y, T)$. To derive the long-term cost or net benefit rate, we define the survival probability function for a given user demand level w as $R(t, w) = \sum_{j=k(w)}^{N} P_j(t)$. Consequently, let $F(t, w) = 1 - R(t, w)$ be the failure probability function and denote by $f(t, w)$ its probability density. Then the expected replacement cycle is given by

$$E[Y^*] = E[Y \wedge T] = \int_0^T t f(t, w) dt + T R(T, w)$$

$$= \int_0^T R(t, w) dt = \int_0^T \sum_{j=k(w)}^{N} P_j(t) dt = \sum_{j=k(w)}^{N} \int_0^T P_j(t) dt.$$

Based on the renewal reward theorem, the expected net benefit rate is given by

$$J_{ARP}(T, w) = \frac{\sum_{j=k(w)}^{N} r_j v_j - c_u F(T, w) - c_p R(T, w)}{\int_0^T R(t, w) dt},$$

where $v_j = \int_0^T P_j(t) dt$ is the expected sojourn time in acceptance state $j \in \{k(w), \cdots, N\}$. Similarly, we can get the expected cost rate as

$$C_{ARP}(T, w) = \frac{c_u F(T, w) + c_p R(T, w)}{\int_0^T R(t, w) dt}. \tag{9.10}$$

Under certain conditions, we can prove that there exists a unique optimal replacement age.

Theorem 9.2 *Suppose that $F(t, w)$ has an increasing failure rate (IFR) property. Denote by $\lambda(t, w)$ the failure rate function. If $\lambda(\infty, w) > (1/v)[c_u/(c_u - c_p)]$ where $v = \int_0^\infty R(t, w) dt$, then there exists a finite and unique $T^*(w)$ that minimizes $C_{ARP}(T, w)$ and satisfies*

$$\lambda(T^*(w), w) \int_0^{T^*(w)} R(t, w) dt - F(T^*(w), w) = \frac{c_p}{c_u - c_p}, \tag{9.11}$$

and

$$C_{ARP}(T^*(w), w) = (c_u - c_p) \lambda(T^*(w), w). \tag{9.12}$$

Proof. Taking the first-order derivative of $C_{ARP}(T, w)$ with respect to T and setting it to be zero, we have

$$\lambda(T, w) \int_0^T R(t, w) dt - F(T, w) = \frac{c_p}{c_u - c_p}. \tag{9.13}$$

Differentiating the left-hand side of the above with respect to T, we have

$$\lambda'(T,w)\int_0^T R(t,w)dt + \lambda(T,w)R(T,w) - f(T,w) = \lambda'(T,w)\int_0^T R(t,w)dt \geq 0$$
(9.14)

due to the fact that $\lambda(T,w)$ is increasing in T. Thus, the left-hand side of (9.13) is an increasing function of T with the limit as maximum. This maximum is $\lambda(\infty,w)v - 1$ as $T \to \infty$. This result implies that there exists a unique solution to (9.13) if $\lambda(\infty,w) - 1 > c_p/(c_u - c_p)$, which is the condition $\lambda(\infty,w) > (1/v)[c_u/(c_u - c_p)]$. Using the condition (9.11) in (9.10) yields (9.12). ∎

We present an example of a single element system. For a multi-element example, we refer readers to Sheu and Zhang (2013) in which a two element system is numerically analyzed.

Example 9.1 *Consider a single element system which degrades through 6 different states $\phi(t) = 5,4,3,2,1,0$ with 5 as the brand new system and 0 as completely failed system. Let T_{ij} be the sojourn time in state i before entering state j. Assume that T_{ij} follows a Weibull distribution with a scale parameter $\alpha_{ij} = \delta(i - \theta j)$ with $\delta > 0, 0 < \theta \leq 1$ and a shape parameter $\beta > 1$. Thus, the density function is*

$$f_{ij}(t) = \beta \left(\frac{1}{\alpha_{ij}}\right)^\beta t^{\beta-1} e^{-\left(\frac{t}{\alpha_{ij}}\right)^\beta}.$$

Suppose that we have a set of parameters as $\delta = 1, \theta = 0.5, \beta = 3, i = 5$, and $j = 0,1,2,3,4$. Transition rate functions (verifying them is a good exercise) become

$$\lambda_{ij}(t) = 3t^2/(i - 0.5j)^3, \quad i = 5,4,3,2,1, 0 \leq j \leq i - 1,$$

$$\lambda_{ii}(t) = -\sum_{k=0, k\neq i}^5 3t^2/(i - 0.5k)^3.$$

Solution: We first write down the infinitesimal generator matrix as follows:

$$\mathbf{A} = \begin{bmatrix} -0.28487t^2 & 0.1111t^2 & 0.06997t^2 & 0.046875t^2 & 0.03292t^2 & 0.024t^2 \\ 0 & -0.41994t^2 & 0.192t^2 & 0.1111t^2 & 0.06997t^2 & 0.046875t^2 \\ 0 & 0 & -0.6781t^2 & 0.375t^2 & 0.192t^2 & 0.1111t^2 \\ 0 & 0 & 0 & -1.2639t^2 & 0.8889t^2 & 0.375t^2 \\ 0 & 0 & 0 & 0 & -3t^2 & 3t^2 \\ 0 & 0 & 0 & 0 & 0 & 0 \end{bmatrix}$$

The probability distribution functions $P_i(t)$ are solved and illustrated in Figure 9.1. Assume that $k(w) = 3$. Thus, this one-element system is considered to be in a working state if $\phi(t) = 3,4,5$, and in a failure state if $\phi(t) = 0,1,2$. Next, we can determine the optimal T^* and its associated minimum $C_{ARP}(T^*,w)$ or

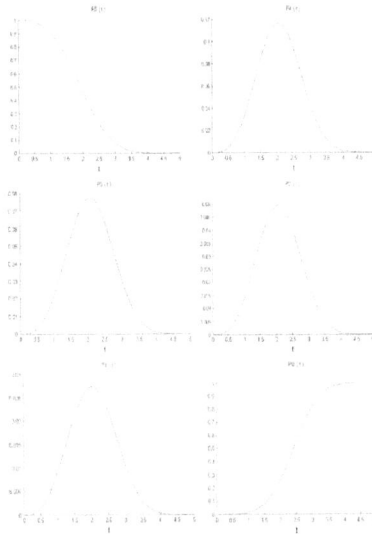

Figure 9.1 Probabilities that the one element system is in state $i = 5,4,3,2,1,0$ at time t. Source: Adapted from S.H. Sheu and Z.G. Zhang, An optimal age replacement policy for multi-state systems, *IEEE Transactions on Reliability*, 62, (3), 722-723, 2013.

TABLE 9.1

Impact of θ on T^*, $C_{ARP}(T^*, w)$, and $J_{ARP}(T^*, w)$ for a single element multiple-state system with $r_3 = 1000, r_4 = 1100, r_5 = 1200, c_u = 1500, c_p = 1000$

θ	T^*	$C_{ARP}(T^*, w)$	T^*	$J_{ARP}(T^*, w)$
1.0	1.18	1096.20	1.16	71.16
0.9	1.48	907.96	1.45	267.04
0.8	1.74	793.42	1.70	384.82
0.7	1.97	717.52	1.93	463.69
0.6	2.18	662.94	2.14	520.57
0.5	2.37	621.57	2.32	563.82
0.4	2.54	588.97	2.49	597.99
0.3	2.70	562.49	2.64	625.80
0.2	2.84	540.50	2.78	648.95
0.1	2.97	521.89	2.91	668.75
0.0	3.09	505.90	3.03	685.45

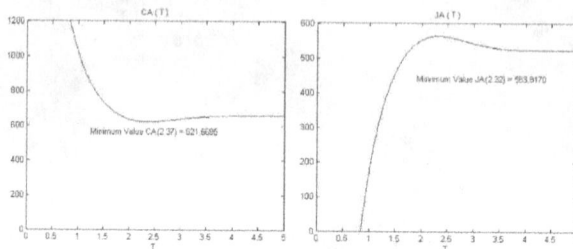

Figure 9.2 $C_{ARP}(T,w)$ and $J_{ARP}(T,w)$ functions for $\theta = 0.5$. Source: Adapted from S.H. Sheu and Z.G. Zhang, An optimal age replacement policy for multi-state systems, *IEEE Transactions on Reliability*, 62, (3), 722-723, 2013.

maximum $J_{ARP}(T^*,w)$. The reward/cost parameters and results are summarized in a Table 9.1 for various θ values.

For a particular $\theta = 0.5$ case, we graph the average cost rate and average profit rate functions in Figure 9.2.

Exercise 9.1 *Prove Theorem 9.1.*

Exercise 9.2 *Use Matlab or Mathematica to compute the optimal solution in Example 9.1. Create more graphs by changing the parameters and perform the sensitivity analysis.*

9.1.2 OPTIMAL THRESHOLD POLICY FOR A $M/G/1$ QUEUE

A popular control policy in a stochastic service or production system is the threshold policy. We use a single server queueing model to demonstrate the approach to determining the optimal policy of this type. Consider an $M/G/1$ queue with server vacations. The server vacation is a period of time that the server is not attending the queue. In other words, the server may take a break or work on other type of non-queueing jobs. This class of queueing models have been well studied in the literature. For a comprehensive discussion on these models, interested readers are referred to the excellent books by Takagi (1996) and Tian and Zhang (2001). In this section, we present a special $M/G/1$ queue with vacations and threshold policy and show how to develop a long-term cost rate function based on the regenerative cycles. With such a cost (or profit) function, we can find the optimal threshold policy. Assume that customers arrive at system according to a Poisson process with rate λ and their service times are general i.i.d. random variables denoted by S. The single server keeps serving customers until the system becomes empty. Then the server will take a vacation. After finishing the vacation, the server will check if there are any waiting customers. If the number of waiting customers is below a

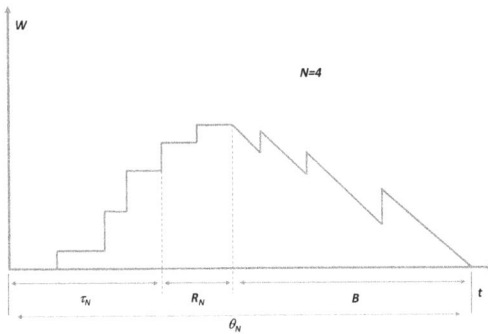

Figure 9.3   Work process over a service cycle in an $M/G/1$ queue with server vacations.

threshold N at the vacation completion instant, the server will take another vacation. The server will keep taking vacations until there are at least N waiting customers at a vacation completion instant and then the server resumes serving the queue. For the vacations, we assume that the first vacation is stochastically larger than the successive vacations if any. This assumption is reasonable if there is set-up or switch-over time when the server is changing from "serving the queue" mode to "taking vacation" mode. Thus, there are two types of vacations that are exponentially distributed and denoted by $V_i, i = 1, 2$. Such a model is also called the $M/G/1$ queue with a threshold policy and an exceptional first vacation. The analysis here is mainly taken from the work by Zhang and Love (1998). Under an N threshold policy, we define a service cycle as the time interval between two consecutive service completion (i.e., customer departure) instants when the system becomes empty. The work process over such a cycle is illustrated in Figure 9.3. Clearly, the starting (or the ending) instant of the cycle is a regenerative point. This service cycle can be decomposed into three parts which are (i) the duration of accumulating N customers, denoted by τ_N; (ii) the forward recurrence time (FRT) of the last vacation before resuming service, denoted by R_N; and (iii) the server's serving customer time to reach an empty system, denoted by B. First, the means of these random periods can be determined easily. Clearly, $E[\tau_N] = N/\lambda$ due to the Poisson arrival process. Since R_N can be the FRT of either type 1 or type 2 vacation and the FRT has the same distribution as that of full vacation due to the memoryless property of exponential random variable, the distribution function of R_N can be written as

$$F_{R_N}(t) = P_N^1 F_{V_1}(t) + (1 - P_N^1)F_{V_2}(t),$$

where P_N^1 is the probability that R_N is the FRT of type 1 vacation and is given by

$$P_N^1 = \left(\frac{\lambda E[V_1]}{1 + \lambda E[V_1]} \right)^N.$$

With this distribution, we obtain

$$E[R_N] = P_N^1 E[V_1] + (1 - P_N^1) E[V_2].$$

To find $E[B]$, we note that the expected number of customers arriving during B is $\lambda E[B]$, which implies

$$E[B] = (N + \lambda E[R_N] + \lambda E[B]) E[S].$$

By solving the equation above for $E[B]$, we have

$$E[B] = \frac{E[S]}{1 - \rho} (N + \lambda E[R_N]),$$

where $\rho = \lambda E[S]$. Now we get the expected value of the service cycle as

$$E[\theta_N] = E[\tau_N] + E[R_N] + E[B]$$
$$= \frac{N}{\lambda} + E[R_N] + \frac{E[S]}{1 - \rho} (N + \lambda E[R_N]) = \frac{1}{1 - \rho} \left(\frac{N}{\lambda} + E[R_N] \right)$$
$$= \frac{1}{1 - \rho} \left(\frac{N}{\lambda} + P_N^1 E[V_1] + (1 - P_N^1) E[V_2] \right).$$

Next, we will determine the total expected cost occurred in the service cycle, denoted by $E[TC_N]$. We assume that a cost/revenue structure is imposed. This structure consists of a set-up (or switch-over) cost, denoted by c_0, a linear holding cost with a unit cost parameter h, and two revenue rates for two types of vacations, denoted by r_1 and r_2, respectively. Again, $E[TC_N]$ can be decomposed into three parts which are the expected costs occurred in τ_N, R_N, and B, denoted by $E[C_{\tau_N}], E[C_{R_N}]$, and $E[C_B]$, respectively. Due to the Poisson arrivals, we have

$$E[C_{\tau_N}] = \frac{h}{\lambda} + \frac{2h}{\lambda} + \cdots + \frac{(N - 1)h}{\lambda}$$
$$= \frac{N(N - 1)h}{2\lambda}.$$

By conditioning on R_N and the number of arrivals k during R_N, and noting that the inter-arrival times have mean $t/(k + 1)$ given k Poisson arrivals occurred in $(0, t]$ (see Ross), we obtain the expected cost during R_N as

$$E[C_{R_N}] = hNE[R_N] + \sum_{k=0}^{\infty} \int_0^{\infty} e^{-\lambda t} \frac{(\lambda t)^k}{k!} \left[\sum_{l=0}^{k} \frac{hlt}{k + 1} \right] dF_{R_N}(t)$$
$$= hNE[R_N] + \frac{1}{2} \lambda h R_N^{(2)},$$

where $R_N^{(2)}$ is the second moment of R_N. To determine $E[C_B]$, we define an $(l-1)$-busy period of the $M/G/1$ queue, denoted by θ_l^1, as the time interval from the instant at which the server starts serving with l customers in the system to the instant at which a departure customer leaves $l-1$ customers in the system. Here, we need some results about the busy period of a classical $M/G/1$ queue. Let θ denote such a busy period. Then the distribution of the busy period length can be studied via the Laplace transform. Denoting by A the number of arrivals during the first service time, denoted by S, of the busy period, we have

$$\theta = S + \theta_1 + \theta_2 + \cdots + \theta_A, \quad A = 0, 1, 2, \ldots,$$

where θ_i is the busy period initiated by the ith arriving customer during the first customer service time. Note that $\theta_1, \theta_2, \ldots, \theta_A$ follow the same distribution as θ. Based on this observation, we can establish a functional equation in terms of the Laplace transform of the busy period, denoted by $\theta^*(s)$ as follows:

$$\theta^*(s) = E[e^{-s\theta}] = E[E[E[e^{-s\theta}|A, S]S]] = E[E[E\left[e^{-s(S+\theta_1+\cdots+\theta_A)}|A, S\right]S]]$$

$$= E[E\left[e^{-sS}E\left[e^{-s(\theta_1+\cdots+\theta_A)}|A, S\right]S\right]] = E[E\left[e^{-sS}E[e^{-s\theta}]^A|S\right]]$$

$$= E[e^{-sS}E\left[\theta^*(s)^A|S\right]] = E\left[e^{-sS}\sum_{k=0}^{\infty}e^{-\lambda S}\frac{(\lambda S)^k}{k!}\theta^*(s)^k\right]$$

$$= E\left[e^{-sS}e^{-\lambda S}e^{\lambda S\theta^*(s)}\right] = E\left[e^{-(s+\lambda-\lambda\theta^*(s))S}\right]$$

$$= S^*(s + \lambda - \lambda\theta^*(s)).$$

Using this equation, we can get the mean of busy period easily by taking the first-order derivative of both sides of this functional equation and noting $E[\theta] = -\theta^{*'}(0)$. The result is $E[\theta] = E[S]/(1-\rho)$. Then we can develop the recursive relation to compute the expected cost for a $(l-1)$ busy period, denoted by C_l^1. First, using the mean of the busy period of $M/G/1$ queue, we can find C_1^1 as follows

$$C_1^1 = \frac{h}{1-\rho}\left(\frac{\lambda S^{(2)}}{2(1-\rho)} + E[S]\right).$$

Then, C_l^1 is given by

$$C_l^1 = h(l-1)E[\theta] + C_1^1, \quad l > 1.$$

Now, we can compute $E[C_B]$. Let D be the number of customers present in the system when B begins and let $d_k = P(D=k)$. Then, $E[C_B]$ is obtained as

$$E[C_B] = \sum_{k=N}^{\infty}\left(\sum_{l=1}^{k}C_l^1\right)d_k = \frac{hE[\theta]}{2}D^{(2)} + \left(C_1^1 - \frac{hE[\theta]}{2}\right)E[D].$$

An easy way to compute the first two moments of D is through its z-transform, $D(z)$, which is given by

$$D(z) = z^N R_N^*(\lambda - \lambda z).$$

The first two moments are then computed as

$$E[D] = N + \lambda E[R_N],$$
$$D^{(2)} = N^2 + (2N+1)\lambda E[R_N] + \lambda^2 R_N^{(2)}.$$

The revenue rate earned during the vacation period of a service cycle can be written as

$$\pi_N = r_1 E[V_1] + r_2 \left(\frac{N}{\lambda} + E[R_N] - E[V_1] \right).$$

It follows from the renewal reward theorem that the long-term net cost rate is given by

$$
g_N = \frac{c_0 + E[C_{\tau_N}] + E[C_{R_N}] + E[C_B] - \pi_N}{E[\theta_N]}
$$

$$
= \frac{c_0 + \frac{N(N-1)h}{2\lambda} + hNE[R_N] + \frac{1}{2}R_N^{(2)} + \frac{hE[\theta]}{2}D^{(2)} + \left(C_1^1 - \frac{hE[\theta]}{2} \right)E[D] - \pi_N}{\frac{1}{1-\rho}\left(\frac{N}{\lambda} + E[R_N] \right)},
$$

where $R_N^{(2)} = 2P_N^1 E[V_1]^2 + 2(1 - P_N^1)E[V_2]^2$. Since we are unable to prove the convexity of g_N or to obtain an explicit expression for N from the first-order condition, the optimal threshold N can be found by a numerical search. A finite upper bound for searching optimal N has been established in Zhang and Love (1998). A more general two threshold policy in an $M/G/1$ queue with vacations has been studied by Zhang et al. (1997).

Exercise 9.3 *A similar case to the queue with vacations and threshold policy is a queue with threshold policy. That is when the system becomes empty, the server stops serving customers and resumes service at the customer arrival instant that makes the queue length to reach the threshold N. With the same $M/G/1$ queue discussed above, derive the long-term cost rate.*

Exercise 9.4 *Use Matlab or Mathematica to develop a search algorithm for determining the optimal two-threshold policy of the vacation queueing model discussed above.*

9.2 OPTIMIZATION BASED ON STATIONARY PERFOR-MANCE MEASURES – ECONOMIC ANALYSIS OF STA-BLE QUEUEING SYSTEMS

In many stochastic models, we focus on the steady-state analysis, which is usually much simpler than the transient (time-dependent) analysis. The system performance can be improved from different perspectives by formulating and solving an optimization problem with an appropriate objective function based on the stationary performance measures. To demonstrate this class of problems, we consider a stochastic service system with customer choice modeled as a stable $M/G/1$ queue over an infinite time horizon. Customers arrive to the system according to a Poisson process with rate λ and join a queue if getting the service after a possible waiting generates a positive economic benefit. The service times are i.i.d. random variables which are denoted by S and characterized by the mean $E[S] = 1/\mu$ and variance $Var(S) = \sigma^2$. Each customer will receive a reward or utility from the service. The average total customer utility per unit time is a function of arrival rate and denoted by $U(\lambda)$. Assume that the system has reached steady state, we denote the average stationary delay (or waiting) cost per customer, as a function of system parameters, by $D(\lambda,\mu,\sigma)$. Each joining customer is charged with a toll denoted by p. The average service cost per unit time is denoted by $c(\mu)$. If the system described is an abstraction of a real service system, the system can be optimized from different perspectives by two decision variables λ and μ. We consider decision-making problems from three perspectives. These are perspectives from the customers, server (service facility), and social planner, respectively and correspond to three optimization problems.

9.2.1 INDIVIDUAL OPTIMIZATION

Assume that every customer is self-interested. The decision facing every arriving customer is whether or not to join the system based on his own net utility. Since a customer may wait in the line before getting the service, we define the full price of the customer, denoted by $\pi(\lambda,\mu)$, as the toll plus the delay cost, i.e., $\pi(\lambda,\mu) = p + D(\lambda,\mu,\sigma)$. The customer makes the joining or balking decision based on the comparison between his full price and service utility. Note that his service utility is determined by the marginal utility $U'(\lambda)$. For the concept of marginal utility, readers are referred to Stidham (2009). This customer's individual optimization prescribes that the customer joins the system if $U'(\lambda)$ is greater than $\pi(\lambda,\mu)$, and becomes indifferent between joining and balking if $U'(\lambda)$ equals $\pi(\lambda,\mu)$. To ensure that a customer must join an empty system, we must assume that $U'(0) \geq \pi(\lambda,\mu)$ (Note that $\lambda = 0$ implies nobody joins the system, hence, the system is empty). The individual optimal arrival rate, also called equilibrium arrival rate, is reached when $U'(\lambda) = \pi$. It follows form the boundary condition of $U'(0) \geq \pi$ and the fact that $U'(\lambda)$ is continuous and non-increasing in λ that the individual optimal condition must be satisfied. We can write the individual optimality condition as

$$U'(\lambda) \leq p + D(\lambda,\mu,\sigma), \quad \lambda \geq 0, \text{ and}$$
$$U'(\lambda) = p + D(\lambda,\mu,\sigma), \quad \text{if } \lambda > 0. \tag{9.15}$$

It is worth noting that the "optimization" in this individual choice problem is not the traditional optimization by mathematical programming. It is the "optimization" from the perspective of maximizing the individual's utility through his or her choice, which is very different from the next two optimization problems to be discussed.

Exercise 9.5 *Explain the individual self-interest equilibrium condition (9.15).*

9.2.2 SOCIAL OPTIMIZATION

From the social planner's viewpoint, the total net utility, denoted by $\mathscr{S}$, should be maximized by optimizing λ and μ. Then given a fixed toll p, the optimization problem can be written as the

$$\max_{\{\lambda,\mu\}} \mathscr{S}(\lambda,\mu) = U(\lambda) - \lambda D(\lambda,\mu,\sigma) - c(\mu)$$

$$\begin{aligned} s.t. \quad &p + D(\lambda,\mu,\sigma) - U'(\lambda) \geq 0, \\ &p + D(\lambda,\mu,\sigma) - U'(\lambda) = 0, \text{ if } \lambda > 0, \\ &\lambda \geq 0. \end{aligned} \tag{9.16}$$

Here, the constraints are the equilibrium conditions (9.15). This non-linear programming problem with two decision variables can be solved by converting it to a single variable problem. This is done by using the constraints to solve for λ in terms of μ. Since only λ and μ are treated as the decision variables and σ (standard deviation of service time) is assumed to be a constant and omitted in the delay cost notation (i.e., $D(\lambda,\mu)$). Define

$$\mu_0 := \inf\{\mu > 0 : U'(0) - D(0,\mu) > p\},$$

which is the minimum service rate that ensures at least some customers will join the system. Then, we consider two cases:

Case 1: $\mu > \mu_0$. It follows from the definition of μ_0 that $U'(0) - D(0,\mu) > p$. Since $U'(\lambda) - D(\lambda,\mu)$ is strictly decreasing in λ and approaches to $-\infty$ as $\lambda \uparrow \mu$, there exists a unique solution, $\lambda \in (0,\mu)$ to

$$U'(\lambda) - D(\lambda,\mu) = p.$$

Denote such a solution by $\lambda(\mu)$.

Case 2: $0 \leq \mu \leq \mu_0$. In this case, we have $U'(0) - D(0,\mu) \leq p$. Again, due to the decreasing property of $U'(\lambda) - D(\lambda,\mu) < p$ for all $\lambda > 0$. Thus, this is the case where no customer joins the queue. That is $\lambda(\mu) = 0$.

Since the second case is not interesting, we only focus on the first case. Substituting $\lambda(\mu)$ in the objective function in (9.16) converts the original problem into the single decision variable optimization problem as follows:

$$\max_{\mu > \mu_0} \phi(\mu) = \mathscr{S}(\lambda(\mu),\mu) = U(\lambda(\mu)) - \lambda(\mu)D(\lambda(\mu),\mu) - c(\mu)$$

for $\mu \in (\mu_0, \infty)$. Note that $\lambda(\mu)$ is the unique solution to $U'(\lambda(\mu)) - D(\lambda(\mu), \mu) = p$, satisfying $0 < \lambda(\mu) < \mu$. Next, we study the property of the socially optimal solution, denoted by (λ_s, μ_s). Taking the first-order derivative of $\phi(\mu)$ with respect to μ and setting it to zero results in the following necessary condition for μ_s

$$
\begin{aligned}
\frac{d\phi(\mu)}{d\mu} &= U'(\lambda(\mu))\lambda'(\mu) - \lambda'(\mu)D(\lambda(\mu), \mu) \\
&\quad - \lambda(\mu)\left[\frac{\partial}{\partial\lambda}D(\lambda(\mu), \mu)\lambda'(\mu) + \frac{\partial}{\partial\mu}D(\lambda(\mu), \mu)\right] - c'(\mu) = 0.
\end{aligned}
\tag{9.17}
$$

Re-arranging the terms, this condition can be written as

$$
\begin{aligned}
&\left[U'(\lambda(\mu)) - D(\lambda(\mu), \mu) - \lambda(\mu)\frac{\partial}{\partial\lambda}D(\lambda(\mu), \mu)\right]\lambda'(\mu) \\
&\quad - \lambda(\mu)\frac{\partial}{\partial\mu}D(\lambda(\mu), \mu) - c'(\mu) = 0.
\end{aligned}
$$

Using $U'(\lambda(\mu)) - D(\lambda(\mu)) = p$, we obtain an equivalent necessary condition for the social optimality as follows:

$$
\left[p - \lambda(\mu)\frac{\partial}{\partial\lambda}D(\lambda(\mu), \mu)\right]\lambda'(\mu) = \lambda(\mu)\frac{\partial}{\partial\mu}D(\lambda(\mu), \mu) + c'(\mu). \tag{9.18}
$$

To derive an expression for $\lambda'(\mu)$, we differentiate with respect to μ both sides of the equation, $U'(\lambda(\mu)) - D(\lambda(\mu), \mu) = p$, and obtain

$$
\left[U''(\lambda(\mu)) - \frac{\partial}{\partial\lambda}D(\lambda(\mu), \mu)\right]\lambda'(\mu) - \frac{\partial}{\partial\mu}D(\lambda(\mu), \mu) = 0.
$$

From this equation, we have

$$
\lambda'(\mu) = \frac{-\frac{\partial}{\partial\mu}D(\lambda, \mu)}{\frac{\partial}{\partial\lambda}D(\lambda, \mu) - U''(\lambda)} := f(\lambda(\mu), \mu). \tag{9.19}
$$

It is easy to confirm that $\lambda'(\mu) = f(\lambda, \mu) > 0$ due to the fact that $D(\lambda, \mu)$ is increasing in λ and decreasing in μ and $U(\lambda)$ is concave. It is clear that the socially optimal (λ_s, μ_s) with $\mu_s > \mu_0$ satisfies the following necessary conditions:

$$
\begin{aligned}
\left[p - \lambda\frac{\partial}{\partial\lambda}D(\lambda, \mu)\right]f(\lambda, \mu) &= \lambda\frac{\partial}{\partial\mu}D(\lambda, \mu) + c'(\mu), \\
U'(\lambda) - D(\lambda, \mu) &= p.
\end{aligned}
\tag{9.20}
$$

Note that the term $\lambda \frac{\partial}{\partial \lambda} D(\lambda, \mu)$ represents the external effect incurred when the arrival rate is λ (i.e., when a customer joins the queue). The optimal arrival rate for a given μ satisfies $p = \lambda \frac{\partial}{\partial \lambda} D(\lambda, \mu)$, which is not only justifiable but also shows that such a toll can induce customer's socially optimal behavior. The term $p - \lambda \frac{\partial}{\partial \lambda} D(\lambda, \mu)$ can be interpreted as the discrepancy between the fixed toll, p, and the (negative) external effect a customer brings to the system at (λ, μ). For a system with customer choice where the customer equilibrium joining behavior must be taken into account, the necessary and sufficient condition for socially optimal μ_s at a given socially optimal λ_s is

$$\lambda_s \frac{\partial}{\partial \mu} D(\lambda_s, \mu_s) + c'(\mu_s) = 0. \tag{9.21}$$

Denote by $\Delta = \lambda \frac{\partial}{\partial \mu} D(\lambda, \mu) + c'(\mu)$. Since $D(\lambda, \mu)$ and $c(\mu)$ are both convex in μ, Δ increases in μ. For any fixed λ, there is a corresponding socially optimal service rate, denoted by μ', such that the pair (λ, μ') satisfies the necessary condition (9.20). This μ' can be smaller, equal to, or greater than μ_s which is the socially optimal service rate associated with the socially optimal arrival rate λ_s, depending on the sign of $\Delta < 0, = 0$, or > 0. It follows from (9.21) these three cases correspond to three pricing scenarios: the fixed toll less than the external effect, the fixed toll equal to the external effect, and the fixed toll greater than the external effect, which in turn implies the service system is under-utilized ($\mu' < \mu_s$), properly-utilized ($\mu' = \mu_s$), and over-utilized ($\mu' > \mu_s$) from the socially optimal viewpoint. It is worth noting that the condition (9.20) is only the necessary condition. Thus, there may be multiple solutions to the necessary conditions for optimality, only one of which is globally optimal. In fact, the optimization problem for an $M/G/1$ queue with (λ, μ) as decision variables and no customer equilibrium constraints can be complicated due to the lack of joint convexity (or joint concavity). We will discuss this issue after giving an example of the current model.

Example 9.2 *Find the socially optimal solution* (λ_s, μ_s) *for a stable* $M/M/1$ *queue with a linear utility function* $U(\lambda) = r \cdot \lambda$ *and* $r > p$.

Solution: With $U(\lambda) = r\lambda$ and the assumption of $r > p$, it follows from $U'(0) = r$ that μ_0 (the minimum μ) is the unique solution to $D(0, \mu) = r - p$. Using $U''(\lambda) = 0$ in (9.19), we can simplify the socially optimal necessary conditions (9.20) to

$$p\lambda'(\mu) = c'(\mu)$$
$$D(\lambda(\mu), \mu) = r - p. \tag{9.22}$$

Note that $D(\lambda(\mu), \mu)$ is independent of μ under the customer equilibrium condition in this linear utility function case. Usually, these conditions are only necessary ones. If they are also sufficient, we can solve the second equation in (9.22) for $\lambda(\mu)$ and then differentiate and substitute $\lambda'(\mu)$ into the first equation of (9.22) and solve for μ_s. Finally, we can get the associated λ_s by substituting μ_s into the expression for $\lambda(\mu)$. Now we consider a stable $M/M/1$ queue with a linear waiting cost. Thus, we have

$$D(\lambda, \mu) = \frac{h}{\mu - \lambda},$$

where h is the unit waiting cost parameter. Using this expression in (9.22), we can obtain

$$\lambda(\mu) = \mu - \frac{h}{r - p}. \tag{9.23}$$

With this expression, the first condition of (9.22) is simplified to $p = c'(\mu)$, which can be used for solving for the interior solution $\mu_s > \mu_0$. To discuss the solution to this equation, we first use $\lambda'(\mu) = 1$ and $D(\lambda(\mu), \mu) = r - p$ to obtain the following expressions for $\phi(\mu)$ and $\phi'(\mu)$ for this specific queueing model

$$\phi(\mu) = \left(\mu - \frac{h}{r - p} \right) p - c(\mu)$$

$$\phi'(\mu) = p - c'(\mu).$$

If we assume that the service cost is a convex function of μ, then $c'(\mu)$ is non-decreasing. Hence, we have three cases. Case 1: $c'(\mu_0) \geq p$. This means $\phi'(\mu) \leq 0$ for all $\mu > \mu_0$. Thus, $\phi(\mu)$ for social optimality is non-increasing for all $\mu > \mu_0$. In this case, we conclude $\mu_s = 0$ and $\lambda_s = 0$. Case 2: $\lim_{\mu \to \infty} c'(\mu) \leq p$. This is the case that $\phi'(\mu) \geq 0$ for all $\mu > \mu_0$. Thus, $\phi(\mu)$ for social optimality is non-decreasing. This is not a realistic case as the implication is that socially optimal solution is to increase the service rate to infinity and obtain the infinite total net utility. Case 3: $c'(\mu_0) < p < \lim_{\mu \to \infty} c'(\mu)$. This is the case where a unique finite $\mu > \mu_0$ exists that satisfies $c'(\mu) = p$. The solution to this equation (i.e., $\phi'(\mu) = 0$) can be the socially optimal μ_s.

Since only Case 3 is interesting, we can use a numerical example to demonstrate the socially optimal solution for the $M/M/1$ queue.

Exercise 9.6 *Consider an $M/M/1$ queue with the following parameters: $c(\mu) = 0.01\mu^2, p = 1, h/(r - p) = 20$. Make a graph showing the relation between the social welfare (total net utility) and the service rate $\mu \in [0, 100]$. Indicate μ_0 and μ_s on the graph.*

9.2.3   SERVICE PROVIDER OPTIMIZATION

Now we consider the optimization problem from the service provider's (SP's) (or server's) perspective. The goal of the SP is to maximize its profit, denoted by $\Pi(\lambda, \mu)$, by serving a stable queue. The profit-maximization problem can be written as

$$\max_{\{\lambda, \mu\}} \Pi(\lambda, \mu) = \lambda p - c(\mu)$$

$$\text{s.t.} \quad p + D(\lambda, \mu) - U'(\lambda) \geq 0,$$
$$p + D(\lambda, \mu) - U'(\lambda) = 0, \quad \text{if } \lambda > 0,$$
$$0 \leq \lambda \leq \mu.$$

Again, we define $\mu_0 = \inf\{\mu > 0 : U'(0) - D(0, \mu) > p\}$. The only practically meaningful solution to the SP problem is the one over $\mu > \mu_0$ range. Therefore, we focus on the following equivalent problem:

$$\max_{\{\lambda, \mu : \mu > \mu_0\}} \Pi(\lambda, \mu) = \lambda p - c(\mu)$$

$$\text{s.t.} \quad U'(\lambda) - D(\lambda, \mu) = p, \tag{9.24}$$
$$0 \leq \lambda < \mu.$$

We use the same approach as the one for finding the socially optimal solution. Let $\lambda(\mu)$ denote the solution of the equilibrium constraint, $U'(\lambda) - D(\lambda, \mu) = p$, associated with service rate $\mu > \mu_0$. Then, this profit-maximization problem can be written as

$$\max_{\{\mu > \mu_0\}} \lambda(\mu) p - c(\mu).$$

The first-order necessary condition for the SP-optimal service rate, denoted by μ_p, is $\lambda'(\mu)p = c'(\mu)$. Note that this necessary condition is the same as that for the socially optimal solution in the linear utility function case. This is due to the two problems have identical objective functions and constraints in this case. When the utility function is not linear, the SP-optimal problem becomes a non-linear program (NLP) with constraints. Since $U'(\lambda)$ is non-increasing and $D(\lambda, \mu)$ is non-decreasing in λ and non-increasing in μ, we can get a relaxation of (9.24) by replacing the equality constraint with an inequality constraint without loss of optimality. This NLP is as follows:

$$\max_{\{\lambda, \mu : \mu > \mu_0\}} \Pi(\lambda, \mu) = \lambda p - c(\mu)$$

$$\text{s.t.} \quad U'(\lambda) - D(\lambda, \mu) \geq p, \tag{9.25}$$
$$0 \leq \lambda < \mu.$$

The solution to (9.25) is also the solution to the SP-optimal problem (9.24). From the NLP theory, we can get

Theorem 9.3 *If $U'(\lambda)$ is concave in λ, $D(\lambda,\mu)$ is jointly convex in (λ,μ), and $c(\mu)$ is convex in μ, then the NLP (9.25) has a unique global optimal solution (λ,μ), which is the unique solution to the KKT conditions,*

$$p+\gamma\left(U''(\lambda)-\frac{\partial}{\partial\lambda}D(\lambda,\mu)\right)=0,$$

$$-c'(\mu)-\gamma\frac{\partial}{\partial\mu}D(\lambda,\mu)=0,$$

$$U'(\lambda)-D(\lambda,\mu)-p=0.$$

The proof of this theorem is straightforward and left as an exercise. Next, we can establish the relation between the socially optimal solution (λ_s,μ_s) and the SP-optimal solution (λ_p,μ_p).

Theorem 9.4 *Over the range $(\mu > \mu_0)$, we have $\lambda_s > \lambda_p$ and $\mu_s > \mu_p$.*

Proof. First, we prove that $\lambda(\mu)$ is strictly increasing in μ over the range (μ_0,∞). Suppose $\mu_0 < \mu_1 < \mu_2$. We know that

$$U'(\lambda(\mu_1))-D(\lambda(\mu_1),\mu_1)=p.$$

Since $D(\lambda,\mu)$ is strictly decreasing in μ, we have

$$U'(\lambda(\mu_1))-D(\lambda(\mu_1),\mu_2)>p.$$

To make this inequality again, we must make the right-hand side smaller by replacing $\lambda(\mu_1)$ with $\lambda(\mu_2)$. However, we know that $U(\lambda)$ is concave and $D(\lambda,\mu)$ is strictly increasing in λ. Thus, $U'(\lambda)-D(\lambda,\mu_2)$ is strictly decreasing in λ. It follows from

$$U'(\lambda(\mu_2))-D(\lambda(\mu_2),\mu_2)=p$$

that $\lambda(\mu_2)>\lambda(\mu_1)$, which implies that $\lambda(\mu)$ is strictly increasing in μ.

With this property, it suffices to prove $\mu_s > \mu_p$. Writing the socially optimal and SP-optimal problems in the same format by substituting $\lambda(\mu)$ for λ in the objective functions, we have

$$\max_{\mu}\mathscr{S}(\lambda(\mu),\mu):=U(\lambda(\mu))-\lambda(\mu)D(\lambda(\mu),\mu)-c(\mu),$$

$$\max_{\mu}\Pi(\lambda(\mu),\mu):=\lambda(\mu)U'(\lambda(\mu))-\lambda(\mu)D(\lambda(\mu),\mu)-c(\mu),$$

respectively. Subtracting one objective function from the other yields

$$\mathscr{S}(\lambda(\mu),\mu)-\Pi(\lambda(\mu),\mu)=U(\lambda(\mu))-\lambda(\mu)U'(\lambda(\mu)).$$

Note that $\lambda(\mu)$ is non-decreasing in μ as shown above and the concavity of $U(\lambda)$ implies that $U(\lambda) - \lambda U'(\lambda)$ is non-decreasing in λ (this can be shown easily by taking the first-order derivative and showing its non-negativity). Thus, $\mathscr{S}(\lambda(\mu), \mu) - \Pi(\lambda(\mu), \mu)$ is non-decreasing in μ. Using this property, it follows from

$$\mathscr{S}(\lambda(\mu_s), \mu_s) - \Pi(\lambda(\mu_s), \mu_s) \geq \mathscr{S}(\lambda(\mu_p), \mu_p) - \Pi(\lambda(\mu_p), \mu_p)$$

that $\mu_s \geq \mu_p$. This completes the proof. ∎

To further explain the difference between the two optimization scenarios, we note that the constraint for both social and SP optimality, which defines $\lambda(\mu)$, is the customer self-interest equilibrium (i.e., individual optimality) condition for a given toll and service rate μ:

$$U'(\lambda(\mu)) - D(\lambda(\mu), \mu) = p. \tag{9.26}$$

Differentiating both sides of the equation above with respect to μ, we obtain

$$U''(\lambda(\mu))\lambda'(\mu) = \frac{\partial}{\partial \lambda}D(\lambda(\mu), \mu)\lambda'(\mu) + \frac{\partial}{\partial \mu}D(\lambda(\mu), \mu). \tag{9.27}$$

Recall the expression of the first-order condition for socially optimal μ:

$$\frac{d}{d\mu}[U(\lambda(\mu)) - \lambda(\mu)D(\lambda(\mu), \mu) - c(\mu)] = U'(\lambda(\mu))\lambda'(\mu) - \lambda'(\mu)D(\lambda(\mu), \mu)$$

$$- \lambda(\mu)\left[\frac{\partial}{\partial \lambda}D(\lambda(\mu), \mu)\lambda'(\mu) + \frac{\partial}{\partial \mu}D(\lambda(\mu), \mu)\right] - c'(\mu) = 0.$$

Substituting (9.26) and (9.27) into the condition above yields the equivalent condition for the social optimal μ:

$$\lambda'(\mu)p - c'(\mu) = \lambda(\mu)U''(\lambda(\mu))\lambda'(\mu). \tag{9.28}$$

Note that the right-hand side is non-positive due to the fact that $U(\lambda)$ is concave, $\lambda(\mu)$ is non-negative and non-decreasing in μ. On the other hand, the first-order condition for the SP-optimal μ is

$$\lambda'(\mu)p - c'(\mu) = 0. \tag{9.29}$$

The left-hand side of both (9.28) and (9.29) can be interpreted as the marginal net revenue with respect to μ. These two conditions imply that the SP sets the marginal net revenue equal to zero, while the social planner sets it equal to a non-positive number. For the linear utility function case, we have $U''(\lambda) = 0$ that makes the two optimal solutions coincide. Further, note that

$$\frac{d}{d\mu}D(\lambda(\mu), \mu) = \frac{\partial}{\partial \lambda}D(\lambda(\mu), \mu)\lambda'(\mu) + \frac{\partial}{\partial \mu}D(\lambda(\mu), \mu)$$

$$= U''(\lambda(\mu))\lambda'(\mu) \leq 0$$

due to the concavity of the utility function and the non-decreasing property of $\lambda(\mu)$. This implies that the delay cost per customer does not increase as the service rate increases. For the linear utility function case, the above inequality is replaced with equality that indicates that the delay cost becomes constant.

Total net utility is the most general objective function in this class of optimization problems. It can be considered as the sum of the net utility of customers and the profit of the SP as follows:

$$\mathscr{S}(\lambda,\mu) = [U(\lambda) - \lambda(p + D(\lambda,\mu))] + [\lambda p - c(\mu)]$$
$$= U(\lambda) - \lambda D(\lambda,\mu) - c(\mu).$$

Note that the term λp does not occur in the expression as it represents an internal transfer from one party to another within the system.

Before concluding this section, we would like to point out that the optimization problem with respect to (λ,μ) in a basic queueing model such as $M/M/1$ queue can be more complex in the case without customer equilibrium constraint due to the lack of the joint concavity of the two decision variables. For example, if we use $M/M/1$ queue to model a production system, then customers are jobs to be processed and the server is the production facility. In this case, we can formulate an optimization problem with the net benefit of the production facility as the objective function as follows:

$$\max_{\{\lambda,\mu\}} B(\lambda,\mu) = \lambda r - hL(\lambda,\mu) - c\mu, \quad 0 \le \lambda < \mu,$$

where r is the revenue per job, h is the unit waiting cost, $L = \lambda/(\mu - \lambda)$ is the average queue length, and c is the unit processing cost. Note that this objective function has the same structure of the total net utility in the social optimization problem. The first-order conditions are given by

$$\frac{\partial}{\partial \lambda} B(\lambda,\mu) = r - h\frac{\partial}{\partial \lambda} L(\lambda,\mu) = 0,$$

$$\frac{\partial}{\partial \mu} B(\lambda,\mu) = -h\frac{\partial}{\partial \mu} L(\lambda,\mu) - c = 0.$$

It follows from

$$\frac{\partial}{\partial \lambda} L(\lambda,\mu) = \frac{\mu}{(\mu - \lambda)^2}, \quad \frac{\partial}{\partial \mu} L(\lambda,\mu) = -\frac{\lambda}{(\mu - \lambda)^2},$$

that the unique solution to the first-order conditions is given by

$$\lambda = \frac{hc}{(r - c)^2}, \quad \mu = \frac{hr}{(r - c)^2}. \tag{9.30}$$

Although such a solution is feasible as $\lambda < \mu$ due to $c < r$, it is not necessarily globally optimal. It is easy to check that $B(\lambda,\mu)$ is both concave in λ

and concave in μ. However, $B(\lambda,\mu)$ is not jointly concave in (λ,μ) because $L(\lambda,\mu) = \lambda/(\mu-\lambda)$ is not jointly convex due to the form of the $M/M/1$ queue length formula.

Exercise 9.7 *Verify $B(\lambda,\mu)$ is not jointly concave in (λ,μ). Discuss why the solution (9.30) may give a negative profit.*

The discussion in this section only provides a sample of an active research area in queueing literature, the economic analysis of equilibrium queueing behaviors. There are many variants and extensions of the models presented. For example, customers can be heterogeneous (or multi-classes) in terms of delay sensitivity and can be served by multiple servers. The system considered can be more complex with multiple queues for customer to choose or a network that customers have to go through. For a more comprehensive treatment of this class of problems, interested readers are referred to the books by Stidham (2009), Hassin and Haviv (2003), and Hassin (2016) can explore the extensive studies in this area.

Exercise 9.8 *Prove Theorem 9.3.*

9.3 OPTIMAL SERVICE-ORDER POLICY FOR A MULTI-CLASS QUEUE

9.3.1 PRELIMINARY RESULTS FOR A MULTI-CLASS M/G/1 QUEUE

As another example of static optimization in stochastic models is the scheduling problem for a queueing system with multi-class customers. We first consider an $M/G/1$ queue with M classes of customers where the arrival process for class i is Poisson process with rate λ_i, $i = 1, 2, ..., M$. The service times are general i.i.d. random variables with mean $E[S_i] = 1/\mu_i$, second moment $E[S_i^2]$, c.d.f. $F_i(\cdot)$, and $\rho_i = \lambda_i/\mu_i$ for class i. We first start with a multi-class $M/G/1$ queue with the FCFS order by aggregating all M classes into a single class. This is the case where the server does not differentiate between classes and serves in the order of customer arrivals. Thus, the aggregated arrival process is Poisson with arrival rate $\lambda = \lambda_1 + \lambda_2 + \cdots + \lambda_M$. Denote by S the effective service time for an arbitrary customer. Then the distribution and the first two moments of this random variable are given by

$$F_S(t) = P(S \le t) = \frac{1}{\lambda}\sum_{i=1}^{M}\lambda_i F_i(t),$$

$$E[S] = \frac{1}{\mu} = \frac{1}{\lambda}\sum_{i=1}^{M}\lambda_i E[S_i],$$

$$E[S^2] = \sigma^2 + \frac{1}{\mu^2} = \frac{1}{\lambda}\sum_{i=1}^{M}\lambda_i E[S_i^2],$$

$$\rho = \lambda E[S],$$

where μ and σ^2 are the mean and the variance of the aggregated service times, respectively, and $\rho = \rho_1 + \cdots + \rho_M$. To obtain the performance measures for a particular class i customers, we first write down the aggregated performance measures as follows:

$$L = \rho + \frac{\lambda^2 E[S^2]}{2(1-\rho)}, \quad W = \frac{L}{\lambda},$$

$$W_q = W - \frac{1}{\mu}, \quad L_q = \lambda W_q = \frac{\lambda^2 E[S^2]}{2(1-\rho)}.$$

Clearly, these formulas follow from the fact that the system can be treated as a standard $M/G/1$ queue with the aggregated arrival process and service times. Since the aggregated arrival process is still a Poisson process, the PASTA holds. This implies that the average workload in the system in steady state is as seen by the aggregate arrivals (i.e., all classes). Thus, class i customers also see the time average. That is $W_{iq} = W_q$ for $i = 1, ..., M$. By using the relation between the total sojourn time in system and the waiting time in queue combined with the Little's law, we have the performance measures for class i customers, expected waiting time, queue length, and system size, as follows:

$$W_{iq} = W_q = \frac{\lambda^2 E[S^2]}{2(1-\rho)},$$

$$L_{iq} = \lambda_i W_{iq}, \quad W_i = W_{iq} + \frac{1}{\mu_i},$$

$$L_i = \lambda_i W_i = \rho_i + L_{iq}.$$

It is easy to check that $L = L_1 + L_2 + \cdots + L_M$ holds.

Next, we analyze the $M/G/1$ queue where customers are classified and served according to a priority rule. We number the classes from the highest to the lowest by $1, 2, ..., M$ and assume that the server always starts serving a customer of the highest class in the system upon a service completion. There are two types of priority rules – nonpreemptive and preemptive ones. The nonpreemtive priority is that a customer in service does not get preempted (or interrupted) by another customer arrival of higher priority. Otherwise, the priority rule is preemptive and can be further classified into "preemptive resume" and "preemptive repeat" cases. Here we consider the nonpreemptive priority case. Further, we assume that the service discipline within a class is FCFS. The same notations are utilized in the following analysis. Consider a class-i customer arriving into the steady-state priority $M/G/1$ system that serves M classes of customers at time 0. Let S_R denote the remaining service time of the customer in service at time 0 (the arrival instant of the class i customer), T_j^0 denote the time to serve all customers of type j who are waiting in the queue at time 0 for $1 \leq j \leq i$, and T_j denote the time to serve all customers of type j who arrive during the wait time of this class i customer arriving at time 0 for $1 \leq j \leq i$. Clearly, the waiting time for this class i customer, denoted by W_i^q, can be written as

$$W_i^q = S_R + \sum_{j=1}^{i} T_j^0 + \sum_{j=1}^{i-1} T_j.$$

Taking the expectation, we have

$$E[W_i^q] = W_{iq}' = E[S_R] + \sum_{j=1}^{i} E[T_j^0] + \sum_{j=1}^{i-1} E[T_j]. \tag{9.31}$$

Note that we use the notation W_{iq}' to differentiate it from the expected waiting time for the FCFS case. We now develop the expressions for the three terms on the right-hand side of this equation. First, we give the expression of $E[S_R] = (\lambda/2)E[S^2]$, which is the result of using PASTA and i.i.d. service times (left as an exercise for readers). Again, using PASTA, this arriving customer at time 0 will see L_{jq} customers of class j waiting in line. Let X_j be the actual number of class-j customers in steady-state waiting for service to begin at time 0. Then we have $E[T_j^0] = E[E[T_j^0|X_j]] = E[X/\mu_j] = L_{jq}/\mu_j = \lambda_j W_{jq}'/\mu_j = \rho_j W_{jq}'$. Since $E[T_j]$ is the expected time of serving all type $j < i$ customers who arrive during this class i customer's waiting time (i.e., $[0, W_i^q]$), by conditioning on W_i^q, we have $E[T_j] = E[E[T_j|W_i^q]] = E[\lambda_j W_i^q/\mu_j] = \rho_j E[W_i^q] = \rho_j W_{iq}'$. Substituting these expressions of $E[S_R], E[T_j^0]$, and $E[T_j]$ into (9.31), we obtain

$$W_{iq}' = \frac{\lambda}{2} E[S^2] + \sum_{j=1}^{i} \rho_j W_{jq}' + W_{iq}' \sum_{j=1}^{i-1} \rho_j \tag{9.32}$$

for $i = 1, ..., M$. Thus, we have M equations with M unknowns. Starting from $i = 1$, we can compute W_{iq}' recursively by using (9.32). Fortunately, we can get an explicit solution to the expected waiting time for class i customers. After these developments, to simplify the notation, we drop the "prime" in the notation for the expected waiting time, i.e., $W_{iq} = W_{iq}'$ below.

Theorem 9.5 *In an $M/G/1$ queue with M classes of customers, if class i customers have the ith priority where $i = 1, 2, \cdots, M$, then the expected waiting time for class i customers is given by*

$$W_{iq} = \frac{\frac{1}{2} \sum_{j=1}^{M} \lambda_j E[S_j^2]}{(1 - \beta_i)(1 - \beta_{i-1})}, \tag{9.33}$$

where $\beta_i := \rho_1 + \rho_2 + \cdots + \rho_i$ with $\beta_0 := 0$.

Proof. We use mathematical induction to prove this theorem. It is easy to check that (9.33) holds for $i = 1$ by solving (9.32). Assume that formula (9.33) holds for $i > 1$. Then, we write equation (9.32) for class $i+1$ customers as

$$W_{i+1,q} = \frac{\lambda}{2}E[S^2] + \sum_{j=1}^{i+1} \rho_j W_{jq} + W_{i+1,q} \sum_{j=1}^{i} \rho_j$$

$$= \frac{\lambda}{2}E[S^2] + \sum_{j=1}^{i} \rho_j W_{jq} + \rho_{i+1} W_{i+1,q} + W_{iq} \sum_{j=1}^{i-1} \rho_j - W_{iq} \sum_{j=1}^{i-1} \rho_j$$

$$+ W_{i+1,q} \sum_{j=1}^{i} \rho_j$$

$$= \left(\frac{\lambda}{2}E[S^2] + \sum_{j=1}^{i} \rho_j W_{jq} + W_{iq} \sum_{j=1}^{i-1} \rho_j \right) - W_{iq} \sum_{j=1}^{i-1} \rho_j + \rho_{i+1} W_{i+1,q}$$

$$+ W_{i+1,q} \sum_{j=1}^{i} \rho_j$$

$$= W_{iq} - W_{iq} \sum_{j=1}^{i-1} \rho_j + W_{i+1,q} \sum_{j=1}^{i+1} \rho_j = W_{iq} \left(1 - \sum_{j=1}^{i-1} \rho_j \right) + W_{i+1,q} \sum_{j=1}^{i+1} \rho_j,$$

which leads to

$$W_{i+1,q} \left(1 - \sum_{j=1}^{i+1} \rho_j \right) = W_{iq} \left(1 - \sum_{j=1}^{i-1} \rho_j \right). \tag{9.34}$$

It follows from the definition of β_i and (9.34) that

$$W_{i+1,q} = W_{iq} \left(\frac{1 - \beta_{i-1}}{1 - \beta_{i+1}} \right)$$

$$= \frac{\frac{1}{2} \sum_{j=1}^{M} \lambda_j E[S_j^2]}{(1 - \beta_i)(1 - \beta_{i-1})} \left(\frac{1 - \beta_{i-1}}{1 - \beta_{i+1}} \right)$$

$$= \frac{\frac{1}{2} \sum_{j=1}^{M} \lambda_j E[S_j^2]}{(1 - \beta_{i+1})(1 - \beta_i)},$$

where the second equality follows from the induction hypothesis. This completes the proof. ∎

Based on W_{iq}, we can compute other performance measures for class i customers as follows:

$$L_{iq} = \lambda_i W_{iq}, \quad W_i = W_{iq} + E[S_i],$$
$$L_i = \lambda_i W_i = L_{iq} + \rho_i.$$

The aggregate performance measures can be obtained by starting with $L = L_1 + L_2 + \cdots + L_M$. Then, we have $W = L/\lambda$, $W_q = W - 1/\mu$, and $L_q = \lambda W_q$.

Exercise 9.9 *Consider a multi-class $M/G/1$ queue with preemptive resume priority. For such a system, all assumptions are the same as the nonpreemptive priority case discussed above except for one modification. That is during the service of a customer, if another customer with higher priority arrives, then the customer in service is preempted and service starts for this new higher priority customer. When the preempted customer returns to service, service resumes from where it was interrupted. This is also called a work-conserving discipline. (Note that, if the resumed service has to start from scratch, called preemptive repeat, it is not work conserving as the server wasted some time serving; and of the resumed service is sampled again from the same distribution, called preemptive identical, it is not work conserving either). For the preemptive resume case, develop all major performance measures as those in the nonpreemptive case. Let W_i be the expected sojourn time in the system for class i. Then we need to show*

$$W_i = \frac{1}{\mu_i(1-\beta_{i-1})} + \frac{\sum_{j=1}^{i} \lambda_j E[S_j^2]}{2(1-\beta_i)(1-\beta_{i-1})} \quad i = 1, 2, ..., M,$$

where $\beta_i = \rho_1 + \rho_2 + \cdots + \rho_i$ and $\beta_0 = 0$.

Exercise 9.10 *Consider an $M/G/1$ queue with M classes. Using the expressions for W_{iq}, the expected queueing time for class i, for both FCFS and nonpreemptive priority service disciplines, show that $\sum_{i=1}^{M} \rho_i W_{iq}$ has the same expression. Furthermore, if service times for all M classes are exponentially distributed, show that the preemptive resume policy also yields the same expression for $\sum_{i=1}^{M} \rho_i W_{iq}$ as that in a nonpreemptive priority service case. Explain why this happens in the case with exponentially distributed service times.*

9.3.2 OPTIMAL SERVICE-ORDER POLICY FOR A MULTI-CLASS QUEUE WITH NONPREEMTIVE PRIORITY – $c\mu$ RULE

Now we can consider the problem of determining the optimal service order in a multi-class $M/G/1$ queue. Such a model may fit a manufacturing setting. The production facility is considered as the server and jobs to be processed are considered as customers. Assume that there are M types of jobs to be processed which arrive according to Poisson processes. There is a waiting cost for each type i job with a unit cost parameter c_i for $i = 1, 2, ..., M$. This means that it costs a type i customer $c_i W_i$ if this customer's sojourn time in the system is W_i. We need to determine the optimal priority service policy to minimize the total expected waiting cost in this multi-class queueing system. The optimal policy turns out to be a simple $c\mu$ rule if there is no switch-over time and cost. Under such a rule, the higher priority is always given to the class with higher $c\mu$ value.

Let T_C be the average cost incurred per unit time if the priorities of M classes are numbered as $1, 2, ..., M$ from the highest to lowest for the system. Such a policy can be called a "base policy". That is a class i customer has the ith priority. Then the average cost per unit time is as follows:

$$T_C = \sum_{n=1}^{M} \lambda_n c_n W_n.$$

To show that the $c\mu$ rule is optimal, we utilize an exchange argument. Basically, we evaluate the total cost change if the priority assignment is changed from a base policy by swapping the priorities of two classes. These two classes, denoted by i and j, can be arbitrarily chosen among M classes. To make it simple, we can select two neighboring classes. That is class i and class $j = i + 1$. Denote by T_C' the average cost per unit time for the new policy after swapping two neighboring classes in a base policy. For example, in a system with $M = 6, i = 3$, and $j = 4$. Then T_C is the average cost per unit time when the priority order is 1-2-3-4-5-6 and T_C' is the average cost per unit time when the priority order is 1-2-4-3-5-6. Now we compute the cost difference $T_C - T_C'$ with a generic i and $j = i + 1$ in an M class $M/G/1$ queue. If this cost difference is positive, it means there is cost reduction by swapping the two classes, and hence the new policy is an improved priority assignment compared with the base policy. Otherwise, stick with the base policy. To simplify the calculations, we first develop a convenient expression for the expected cost per unit time for class n customers. Note that

$$W_n = W_{nq} + E[S_n] = \frac{\frac{1}{2}\sum_{k=1}^{M} \lambda_k E[S_k^2]}{(1 - \beta_n)(1 - \beta_{n-1})} + \frac{1}{\mu_n},$$

and the identity

$$\frac{\lambda_n}{(1 - \beta_n)(1 - \beta_{n-1})} = \frac{\mu_n}{1 - \beta_n} - \frac{\mu_n}{1 - \beta_{n-1}}.$$

Using these two expressions to compute the expected cost per unit time for class n customers yields

$$\lambda_n c_n W_n = \frac{1}{2} c_n \sum_{k=1}^{M} \lambda_k E[S_k^2] \left(\frac{\mu_n}{1 - \beta_n} - \frac{\mu_n}{1 - \beta_{n-1}} \right) + \frac{\lambda_n c_n}{\mu_n}.$$

Since we only swap two neighboring classes i and $j = i + 1$, in computing $T_C - T_C'$, all terms except for the ith and jth terms in $\sum_n \lambda_n c_n W_n$ would be identical and cancelled out. In terms of symbols, this means that $c_n = c_n', \mu_n = \mu_n'$, and $\beta_n = \beta_n'$ for $n \leq i - 1$ and $n \geq i + 2$. For the two terms i and $j = i + 1$, we have $c_i' = c_j, \beta_i' = \beta_{i-1} + \rho_j, \mu_i' = \mu_j, c_j' = c_i, \beta_j' = \beta_{i-1} + \rho_j + \rho_i = \beta_i + \rho_j,$

and $\mu'_j = \mu_i$. Further, we note that there is a common factor, denoted by Δ, for every class's expected cost per unit time which is $\Delta = (1/2)\sum_{k=1}^{M} \lambda_k E[S_k^2]$. Based on these observations, we can compute

$$
\begin{aligned}
\frac{T_C - T_C'}{\Delta} &= \sum_{n=1}^{M} c_n \left(\frac{\mu_n}{1-\beta_n} - \frac{\mu_n}{1-\beta_{n-1}} \right) - \sum_{r=1}^{M} c_r' \left(\frac{\mu_r'}{1-\beta_r'} - \frac{\mu_r'}{1-\beta_{r-1}'} \right) \\
&= \left\{ c_i \left(\frac{\mu_i}{1-\beta_i} - \frac{\mu_i}{1-\beta_{i-1}} \right) + c_j \left(\frac{\mu_j}{1-\beta_j} - \frac{\mu_j}{1-\beta_{j-1}} \right) \right\} \\
&\quad - \left\{ c_i' \left(\frac{\mu_i'}{1-\beta_i'} - \frac{\mu_i'}{1-\beta_{i-1}'} \right) + c_j' \left(\frac{\mu_j'}{1-\beta_j'} - \frac{\mu_j'}{1-\beta_{j-1}'} \right) \right\} \\
&= \left\{ c_i \left(\frac{\mu_i}{1-\beta_i} - \frac{\mu_i}{1-\beta_{i-1}} \right) + c_j \left(\frac{\mu_j}{1-\beta_j} - \frac{\mu_j}{1-\beta_{j-1}} \right) \right\} \\
&\quad - \left\{ c_j \left(\frac{\mu_j}{1-\beta_{i-1}-\rho_j} - \frac{\mu_j}{1-\beta_{i-1}} \right) \right. \\
&\quad \left. + c_i \left(\frac{\mu_i}{1-\beta_{i-1}-\rho_j-\rho_i} - \frac{\mu_i}{1-\beta_{i-1}-\rho_j} \right) \right\} \\
&= \frac{c_i \mu_i}{1-\beta_i} + \frac{c_j \mu_j}{1-\beta_j} - \frac{c_i \mu_i}{1-\beta_{i-1}} - \frac{c_j \mu_j}{1-\beta_{j-1}} - \frac{c_j \mu_j}{1-\beta_{i-1}-\rho_j} \\
&\quad - \frac{c_i \mu_i}{1-\beta_{i-1}-\rho_j-\rho_i} + \frac{c_j \mu_j}{1-\beta_{i-1}} + \frac{c_i \mu_i}{1-\beta_{i-1}-\rho_j} \\
&= (c_i \mu_i - c_j \mu_j) \left[\frac{1}{1-\beta_i} - \frac{1}{1-\beta_{i-1}} + \frac{1}{1-\beta_{i-1}-\rho_j} - \frac{1}{1-\beta_i-\rho_j} \right] \\
&= \frac{(c_i \mu_i - c_j \mu_j)\rho_i \rho_j (\beta_j - 2)}{(1-\beta_i)(1-\beta_{i-1})(1-\beta_{i-1}-\rho_j)(1-\beta_i-\rho_j)}.
\end{aligned}
$$

Since $\beta_j - 2 < 0$, we can make the following conclusion: If $c_i \mu_i - c_j \mu_j > 0$, then $T_C - T_C' < 0$. This implies that if $c_i \mu_i < c_j \mu_j$, then we should switch the priorities of i and j since $T_C - T_C' > 0$. In this manner, if we compare $c_i \mu_i$ and $c_j \mu_j$ for all pairs of neighbors, the final priority rule would converge to one that is in decreasing order of $c_n \mu_n$. This means that the optimal priority assignment is to give class i higher priority than class j if $c_i \mu_i > c_j \mu_j$ for $i \neq j$. Thus, we should sort the value of $c_i \mu_i$ and call the highest $c_i \mu_i$ as class-1 and the lowest as class M. This optimal service order policy is called $c\mu$-rule that minimizes the total expected cost per unit time of the system. Since such a priority assignment does not depend on the system state (i.e., the number of customers of each type in the system), it is a static policy.

Exercise 9.11 *Consider an $M/G/1$ queue with four classes, called $C1, C2, C3$, and $C4$. The arrival rates, service rates, standard deviations of service times, and holding cost rates for these four classes are listed below*

Class	C1	C2	C3	C4
λ	0.1	0.05	0.5	0.01
μ	0.5	0.25	1	0.20
σ_s	0.5	1	0.3	2
c	3	5	2	5

(a) Determine the optimal service order for these four classes to minimize the average cost per unit time if the nonpreemptive priority is implemented.

(b) Compare the optimal service order in (a) with the FCFS service order in terms of the average cost per unit time and the expected customer waiting time.

9.4 CUSTOMER ASSIGNMENT PROBLEM IN A QUEUE ATTENDED BY HETEROGENEOUS SERVERS

In this section, we consider a customer (or job) assignment problem in a queueing setting with heterogeneous servers. The content in this section is mainly taken from Xia, Zhang, and Li (2021). Note that this problem is a dynamic optimization problem in nature that is the topic of the next chapter. There are two reasons of presenting this problem here. First, we will apply the method called the "sensitivity base optimization" (SBO) (Cao 2007) to characterize the optimal policy structure and this method is different from the traditional dynamic programming approach to be discussed in the next chapter. In fact, as we show below, this approach has a similar argument of comparing "two neighboring policies" in terms of performance change as that in proving the $c\mu$ rule and can be considered as an extension of the approach used in the previous section to a dynamic policy case. Second, under certain conditions, we show that the optimal dynamic (i.e., state-dependent) policy can be reduced to a static (i.e., state-independent) policy, called c/μ rule, similar to the $c\mu$ rule presented in the previous section. Thus, the analysis of this problem forms a prelude to the dynamic optimization in stochastic models of the next chapter.

9.4.1 PROBLEM DESCRIPTION

Homogeneous customers arrive to the queueing system with heterogeneous servers according to a Poisson arrival process with rate λ. The waiting room is infinite and the service discipline is first come first serve (FCFS). The service times at servers are assumed to be independent and exponentially distributed. The heterogeneous servers are classified into K groups (also called pools). Group k has M_k servers and customers are assigned to idle servers dynamically, $k = 1, 2, \ldots, K$. Such a queue is also called a group-server queue. When a server in group k is busy with an assigned customer, it will work at service rate μ_k and consume an operating cost c_k per unit of time. Idle servers have no operating costs. A customer can be served by a server from any group. By "heterogeneous server", we mean that servers are different in terms of service rate and operating cost rate. Servers in the same group are homogeneous, i.e., they have the same service rate μ_k and operating cost rate c_k, $k = 1, 2, \ldots, K$. Servers in different groups are heterogeneous in μ_k and c_k. Without loss of generality, we assume $\mu_1 \geq \mu_2 \geq \cdots \geq \mu_K$. When all servers are busy, arriving customers have to form a queue and a holding cost (wait cost) is incurred. Reassignment scheme (customer migration) is allowed. That is a customer being served at a busy server can be reassigned to the waiting room or another idle server. Due to the memoryless property of exponential service time, such reassignment has no effect on customer's remaining service time from the viewpoint of statistics. We define the system state as the number of total customers in the system (including those in service), denoted by n. Note that the busy/idle status of each server is not needed in the definition of the system state because free customer migrations among servers are allowed in the model. When a scheduling policy is appropriately given, the distribution of n customers among servers and the queue is determined. Thus, the state space is the non-negative integer set $\mathbb{N}$, which is infinite. At each state $n \in \mathbb{N}$, we determine the number of working servers in each group, which can be represented by a K-dimensional row vector as

$$\mathbf{m} := (m_1, m_2, \cdots, m_K),$$

where m_k is the number of working servers in group k, i.e., $m_k \in \mathbb{Z}_{[0,M_k]}$, $k = 1, 2, \ldots, K$. We call $\mathbf{m}$ the action (or decision) at state n, a term used in Markov decision process to be discussed in Chapter 10). Obviously, to save costs, the number of working servers should be not greater than the number of customers, i.e., $\mathbf{m1} \leq n$, where $\mathbf{1}$ is a column vector with appropriate dimensions and all its elements are 1's. Thus, the action space at state n is defined

$$\mathscr{A}(n) := \{\text{all } \mathbf{m} : \mathbf{m} \in \mathbb{Z}_{[0,M_1]} \times \mathbb{Z}_{[0,M_2]} \times \cdots \times \mathbb{Z}_{[0,M_K]} \text{ and } \mathbf{m1} \leq n\}, \quad n \in \mathbb{N},$$

where $\times$ is the Cartesian product. Here, we focus on deterministic stationary policy d which is a mapping from the infinite state space $\mathbb{N}$ to the finite action space $\mathscr{A}$, i.e., $d : \mathbb{N} \to \mathscr{A}$. If d is determined, we will adopt action $d(n)$ at state n and $d(n,k)$ is the number of working servers of group k, where $n \in \mathbb{N}$ and $k = 1, 2, \ldots, K$. All the possible d's form the policy space $\mathscr{D}$.

When the system state is n and the scheduling action $\mathbf{m} = d(n)$ is adopted, a *holding cost* $h(n)$ and an *operating cost* $o(\mathbf{m})$ will be incurred per unit of time. In the literature, it is commonly assumed that the operating cost is increasing with respect to (w.r.t.) the number of working servers. Here, we define the operating cost function $o(\mathbf{m})$ as follows.

$$o(\mathbf{m}) := \sum_{k=1}^{K} m_k c_k = \mathbf{mc},$$

where $\mathbf{c} := (c_1, c_2, \cdots, c_K)^T$ is a K-dimensional column vector and c_k represents the operating cost rate per server in group k. We can view c_k as a function of μ_k and will further discuss a concave function for guaranteeing the effect of scale economies to be defined. Idle servers have no operating cost. Therefore, the total cost rate function of the whole system is defined as

$$f(n, \mathbf{m}) := h(n) + \mathbf{mc}. \tag{9.35}$$

We make the following assumption about the customer's holding cost (waiting cost).

Assumption 9.1 *Holding cost $h(n)$ is an increasing convex function in n and satisfies $\sum_{n=0}^{\infty} h(n)\rho^n < \infty$, for any $0 < \rho < 1$.*

Similarly to the convexity in continuous domain, we define the convexity of $h(n)$ in discrete domain

$$h(n+2) - 2h(n+1) + h(n) \geq 0, \quad \forall n \in \mathbb{N}.$$

The assumption of increasing convex holding cost fits the situation where the delay cost grows more rapidly as the system becomes more congested. The second condition $\sum_{n=0}^{\infty} h(n)\rho^n < \infty$ means that the increase of $h(n)$ is controlled by the order of ρ^n, such that the long-run average holding cost of the queueing system is bounded. Examples of $h(n)$ can be $h(n) = n$, $h(n) = n^2$, or other polynomials with positive parameters.

Denote by n_t the number of customers in the system at time $t \geq 0$. The long-run average cost of the group-server queue under policy d can be written as

$$\eta^d := \liminf_{T \to \infty} \mathbb{E}\left\{ \frac{1}{T} \int_0^T f(n_t, d(n_t)) dt \right\}, \tag{9.36}$$

which is independent of the initial state n_0 for ergodic systems. Here we focus on the ergodic policy set

$$\mathscr{D} := \{\text{all } d : 0 < d(n)\mathbf{1} \leq n, \forall n \geq 1\}.$$

Our objective is to find the optimal policy d^* such that the associated long-run average cost is minimized. That is,

$$d^* = \arg\min_{d \in \mathscr{D}} \{\eta^d\}. \tag{9.37}$$

Since the buffer is infinite, this optimization problem (9.37) belongs to a class of continuous-time Markov decision processes with countably infinite state spaces. The existence of deterministic stationary policies in such a problem is mathematically sophisticated, which is beyond the scope of this book. Interested readers can refer to Xia et al. (2020), which can be directly applied to our problem (9.37) and guarantee the existence of optimal stationary policies.

9.4.2 CHARACTERIZATION OF OPTIMAL POLICY

The key idea of the SBO theory is to utilize performance sensitivity information, such as performance difference or derivatives, to optimize Markov processes. For problem (9.37), we define the *performance potential* as

$$g(n) := \lim_{T \to \infty} \mathbb{E} \left\{ \int_0^T [f(n_t, d(n_t)) - \eta] dt \,\Big|\, n_0 = n \right\}, \quad n \in \mathbb{N}, \qquad (9.38)$$

where η is defined in (9.36) and we omit the superscript 'd' for simplicity. The
definition (9.38) indicates that $g(n)$ quantifies the long-run accumulated effect
of the initial state n on the average performance η. In the traditional Markov
decision process theory, $g(n)$ can also be viewed as the *relative value function*
or *bias*.

Using the ergodicity and strong Markov property, we decompose the right-
hand side of (9.38) into two parts

$$
\begin{aligned}
g(n) \;=\;& \mathbb{E}\{\tau\}[f(n, d(n)) - \eta] + \mathbb{E}\left\{ \int_\tau^\infty [f(n_t, d(n_t)) - \eta] dt \,\Big|\, n_0 = n \right\} \\
\;=\;& \frac{1}{\lambda + d(n)\mu}[f(n, d(n)) - \eta] \\
&+ \frac{\lambda}{\lambda + d(n)\mu} \mathbb{E}\left\{ \int_\tau^\infty [f(n_t, d(n_t)) - \eta] dt \,\Big|\, n_\tau = n+1 \right\} \\
&+ \frac{d(n)\mu}{\lambda + d(n)\mu} \mathbb{E}\left\{ \int_\tau^\infty [f(n_t, d(n_t)) - \eta] dt \,\Big|\, n_\tau = n-1 \right\}, \qquad (9.39)
\end{aligned}
$$

where τ is the sojourn time at the current state n and $\mathbb{E}\{\tau\} = \frac{1}{\lambda + d(n)\mu}$, μ is a K-
dimensional column vector of service rates defined as $\mu := (\mu_1, \mu_2, \ldots, \mu_K)^T$.

Combining (9.38) and (9.39), we have the recursion

$$
\begin{array}{ll}
[\lambda + d(n)\mu] g(n) = f(n, d(n)) - \eta + \lambda g(n+1) + d(n)\mu g(n-1), & n \geq 1, \\
\lambda g(n) = f(n, d(n)) - \eta + \lambda g(n+1), & n = 0.
\end{array}
$$
$$(9.40)$$

We denote $\mathbf{B}$ as the infinitesimal generator of the MDP under policy d,
which is written as

$$
\mathbf{B} =
\begin{bmatrix}
-\lambda & \lambda & 0 & 0 & 0 & \cdots \\
d(1)\mu & -\lambda - d(1)\mu & \lambda & 0 & 0 & \cdots \\
0 & d(2)\mu & -\lambda - d(2)\mu & \lambda & 0 & \cdots \\
0 & 0 & d(3)\mu & -\lambda - d(3)\mu & \lambda & \cdots \\
\vdots & \vdots & \vdots & & \ddots & \ddots
\end{bmatrix}.
$$
$$(9.41)$$

Hence, we can rewrite (9.40) as follows:

$$-\mathbf{B}(n,n)g(n) = f(n,d(n)) - \eta + \mathbf{B}(n,n+1)g(n+1) + \mathbf{B}(n,n-1)g(n-1),$$
$$\text{for } n \geq 1,$$
$$-\mathbf{B}(n,n)g(n) = f(n,d(n)) - \eta + \mathbf{B}(n,n+1)g(n+1),$$
$$\text{for } n = 0.$$

$$(9.42)$$

We further denote $\mathbf{g}$ and $\mathbf{f}$ as the column vectors whose elements are $g(n)$'s and $f(n,d(n))$'s, respectively. We can rewrite (9.42) in a matrix form as below:

$$\mathbf{f} - \eta \mathbf{1} + \mathbf{B}\mathbf{g} = \mathbf{0}. \qquad (9.43)$$

The above equation is also called *the Poisson equation* for continuous-time Markov decision process with the long-run average criterion. As $\mathbf{g}$ is called performance potential or relative value function, we can set $g(0) = \zeta$ and recursively solve $g(n)$ based on (9.42), where ζ is any real number. Using matrix operations, we can also obtain g by solving the Poisson equation (9.43) through numerical computation techniques, such as RG-factorizations Li (2004).

Since we focus on the stationary analysis, the stability condition of the queueing system must be satisfied for any feasible policy d. That is,

$$\lambda < \sum_{k=1}^{K} m_k \mu_k.$$

Thus, we consider the policy set whose element can guarantee the queueing system to be stable. A stabilizing policy in such a set can be characterized by the following proposition.

Proposition 9.1 *If* $\sum_{n=1}^{\infty} \left(\frac{\lambda}{d(n)\mu} \right)^n < \infty$, *then the group-server queue under the policy d is stable and its steady-state distribution π exists.*

This proposition can be verified by studying the balance equation of steady states of the birth-death process with infinitesimal generator (9.41). The condition of stabilizing policies in Proposition 9.1 is also equivalent to the following necessary and sufficient condition: If and only if there exists a constant $\tilde{n}$ and for any $n \geq \tilde{n}$, we always have $d(n)\mu > \lambda$, then d is a stabilizing policy and the associated π exists. Thus, we have $\pi \mathbf{B} = 0$ and $\pi \mathbf{1} = 1$. The long-run average cost of the system can be written as $\eta = \pi \mathbf{f}$.

Below, we quantify the effect of different policies on the system average cost by using the SBO theory. Similar to the previous section, here we utilize an "exchange argument" to compare policies. Suppose That the scheduling policy is changed from d to d', where $d, d' \in \mathscr{D}$. All the associated quantities under the new policy d' are denoted by $\mathbf{B'}$, $\mathbf{f'}$, π', η', etc. Obviously, we have $\pi'\mathbf{B'} = \mathbf{0}$, $\pi'\mathbf{1} = 1$, and $\eta' = \pi'\mathbf{f'}$. Left-multiplying π' on both sides of (9.43), we have

$$\pi'\mathbf{f} - \eta\pi'\mathbf{1} + \pi'\mathbf{Bg} = 0.$$

Using $\pi'\mathbf{B'} = \mathbf{0}$, $\pi'\mathbf{1} = 1$, and $\eta' = \pi'\mathbf{f'}$, we can rewrite the above equation as

$$\eta' - \pi'\mathbf{f'} + \pi'\mathbf{f} - \eta + \pi'\mathbf{Bg} - \pi'\mathbf{B'g} = \mathbf{0},$$

which leads to the *performance difference formula* as follows

$$\eta' - \eta = \pi'[(\mathbf{B'} - \mathbf{B})\mathbf{g} + (\mathbf{f'} - \mathbf{f})]. \tag{9.44}$$

The performance difference formula (9.44) is one of the key results of the SBO method as it provides the sensitivity information for performance optimization. It clearly quantifies the performance change due to the policy change from d to d'. Although the exact value of π' is unknown for every new policy d', all its entries are always positive for ergodic states. Therefore, if we choose a proper new policy (with associated $\mathbf{B'}$ and $\mathbf{f'}$) such that the elements of the column vector represented by the square brackets in (9.44) are always nonpositive, then we have $\eta' - \eta \leq 0$ and the long-run average cost of the system will be reduced. If there is at least one negative element in the square brackets, then we have $\eta' - \eta < 0$ and the system average cost will be reduced strictly. This is the main idea of the policy improvement based on the performance difference formula (9.44).

Using (9.44), we examine the performance sensitivity of scheduling policies on the long-run average cost of the group-server queue. Suppose that we choose a new policy d' that is the same as the current policy d except for the action at a particular state n. For this state n, policy d selects action $\mathbf{m}$ and policy d' selects action $\mathbf{m'}$, where $\mathbf{m}, \mathbf{m'} \in \mathscr{A}(n)$. Substituting (9.35) and (9.41) into (9.44), we have

$$
\begin{aligned}
\eta' - \eta &= \pi'[(\mathbf{B'} - \mathbf{B})\mathbf{g} + (\mathbf{f'} - \mathbf{f})] \\
&= \pi'(n)[(\mathbf{B'}(n,:) - \mathbf{B}(n,:))\mathbf{g} + (f'(n,d(n)) - f(n,d(n)))] \\
&= \pi'(n)\left[\sum_{k=1}^{K}(m'_k - m_k)\mu_k(g(n-1) - g(n)) + (\mathbf{m'c} - \mathbf{mc})\right] \\
&= \pi'(n)\sum_{k=1}^{K}(m'_k - m_k)[c_k - \mu_k(g(n) - g(n-1))], \tag{9.45}
\end{aligned}
$$

where $g(n)$ is the performance potential of the system under the current policy d. The value of $g(n)$ can be numerically computed based on (9.43) or online estimated based on (9.38). More detail can be referred to Chapter 3 of the book by Cao (2007).

For the purpose of analysis, we define a new quantity $G(n)$ as below

$$G(n) := g(n) - g(n-1), \quad n = 1, 2, \ldots. \tag{9.46}$$

Note that $G(n)$ quantifies the performance potential difference between neighboring states n and $n-1$. According to the theory of perturbation analysis (PA) $G(n)$ is called the *perturbation realization factor* (PRF) which measures the effect on the average performance when the initial state is perturbed from $n-1$ to n. For our job assignment problem (9.37), $G(n)$ can be understood as the benefit of reducing the long-run average cost due to a service completion. In the following analysis, we can see that $G(n)$ plays a fundamental role for directly determining the optimal scheduling policy.

Based on the recursive relation of g in (9.40), we derive the following recursions for computing $G(n)$'s

Lemma 9.1 *The PRF $G(n)$ can be computed by the following recursive equations*

$$G(n+1) = \frac{d(n)\mu}{\lambda} G(n) + \frac{\eta - f(n, d(n))}{\lambda}, \quad n \geq 1,$$
$$G(1) = \frac{\eta - f(0, d(0))}{\lambda}. \tag{9.47}$$

Proof. From the second equation in (9.40), we have

$$G(1) = g(1) - g(0) = \frac{\eta - f(0, d(0))}{\lambda}.$$

Using the first equation in (9.40), we have

$$\lambda(g(n+1) - g(n)) = d(n)\mu(g(n) - g(n-1)) + \eta - f(n, d(n)), \quad n \geq 1.$$

Substituting (9.46) into the above equation, we directly have

$$G(n+1) = \frac{d(n)\mu}{\lambda} G(n) + \frac{\eta - f(n, d(n))}{\lambda}, \quad n \geq 1.$$

Thus, the recursion for $G(n)$ is proved. ∎

Substituting (9.46) into (9.45), we obtain the following performance difference formula in terms of $G(n)$ when the scheduling action at a single state n is changed from m to m':

$$\eta' - \eta = \pi'(n) \sum_{k=1}^{K} (m'_k - m_k)(c_k - \mu_k G(n)). \tag{9.48}$$

This difference formula can be extended to a general case when d is changed to d', i.e., $d(n)$ is changed to $d'(n)$ for all $n \in \mathbb{N}$. Substituting the associated $(\mathbf{B}, \mathbf{f})$ and $(\mathbf{B}', \mathbf{f}')$ into (9.44) yields

$$\eta' - \eta = \sum_{n \in \mathbb{N}} \pi'(n) \sum_{k=1}^{K} (d'(n, k) - d(n, k))(c_k - \mu_k G(n)). \tag{9.49}$$

Based on (9.49), we can directly obtain a condition for generating an improved policy as follows.

Theorem 9.6 *If a new policy $d' \in \mathcal{D}$ satisfies*

$$(d'(n,k) - d(n,k))(c_k - \mu_k G(n)) \le 0 \qquad (9.50)$$

for all $k = 1, 2, \ldots, K$ and $n \in \mathbb{N}$, then $\eta' \le \eta$. Furthermore, if for at least one state-group pair (n,k), the inequality in (9.50) strictly holds, then $\eta' < \eta$.

Proof. Since (9.50) holds for every n and k and $\pi'(n)$ is always positive for ergodic processes, it directly follows from (9.49) that $\eta' - \eta \le 0$. Thus, the first part of the theorem is proved. The second part can be proved using a similar argument. ∎

Theorem 9.6 provides a way to generate improved policies based on the current policy. For the system under the current policy d, we compute or estimate $G(n)$'s based on its definition. The term $c_k - \mu_k G(n)$ can be understood as the *marginal cost rate* if we assign an extra customer to an idle server of group k, which can directly determine the action selection as follows. For every state n and server group k, if $c_k - \mu_k G(n) > 0$, it indicates that utilizing an additional server in group k is not economic and we should choose a smaller $d'(n,k)$; if $c_k - \mu_k G(n) < 0$, it indicates that utilizing more servers in group k is beneficial for the average cost and we should choose a larger $d'(n,k)$ satisfying the condition $d'(n)1 \le n$. Therefore, according to Theorem 9.6, the new policy d' obtained from this procedure will perform better than the current policy d. This procedure can be repeated to continually reduce the system average cost.

Note that the condition above is only a sufficient one to generate improved policies. Now, we establish a *necessary and sufficient condition* for the optimal policy as follows.

Theorem 9.7 *A policy d^* is optimal if and only if its element $d^*(n)$, i.e., $(d^*(n,1), \ldots, d^*(n,K))$, is the solution to the following integer linear programs*

ILP(n) Problem:
$$\begin{cases} \min\limits_{d(n,k)} \left\{ \sum_{k=1}^K d(n,k)(c_k - \mu_k G^*(n)) \right\} \\ s.t. \quad 0 \le d(n,k) \le M_k, \\ \quad\quad \sum_{k=1}^K d(n,k) \le n, \end{cases} \qquad (9.51)$$

for every state $n \in \mathbb{N}$, where $G^*(n)$ is the PRF defined in (9.46) under policy d^*.

The proof is left as an exercise. Theorem 9.7 indicates that the optimal policy is the solution to a series of ILP problems (9.51) indexed by states $n \in \mathbb{N}$. However, (9.51) is just a condition for verifying optimal policies and cannot be conveniently used for solving the optimal policy due to the infinite state space. To determine the optimal policy, we further investigate the structure of the solution to these ILPs.

Exercise 9.12 *Prove Theorem 9.7.*

By analyzing (9.51), we find that the value of the marginal cost rate $c_k - \mu_k G^*(n)$ directly determines the solution to the ILP problem, which has the following structure:

- For those groups with $c_k - \mu_k G^*(n) > 0$, we have $d^*(n,k) = 0$;
- For those groups with $c_k - \mu_k G^*(n) < 0$, we let $d^*(n,k) = M_k$ or as large as possible in an ascending order of $c_k - \mu_k G^*(n)$, subject to the constraint $\sum_{k=1}^{K} d^*(n,k) \leq n$.

Therefore, we can further specify the above necessary and sufficient condition of the optimal policy and derive a theorem as follows. The proof is straightforward and left as an exercise.

Theorem 9.8 *A policy d^* is optimal if and only if its element $d^*(n)$ satisfies the condition: If $G^*(n) > \frac{c_k}{\mu_k}$, then $d^*(n,k) = M_k \wedge (n - \sum_{l=1}^{k-1} d^*(n,l))$; If $G^*(n) \leq \frac{c_k}{\mu_k}$, then $d^*(n,k) = 0$, for $k = 1, 2, \ldots, K$, where the index of server groups should be renumbered in an ascending order of $c_k - \mu_k G^*(n)$ at every state n, $n \in \mathbb{N}$.*

Exercise 9.13 *Prove Theorem 9.8.*

With Theorem 9.8, we can see that the optimal policy can be completely determined by the marginal cost rate $c_k - \mu_k G^*(n)$. Such form of policy can be called an *index policy* and the marginal cost rate $c_k - \mu_k G^*(n)$ can be viewed as an index. The index policy has a simple form and is studied widely in the literature, such as the *Gittins' index* or *Whittle's index* in solving multi-armed bandit problems.

Theorem 9.8 also reveals the *quasi bang-bang control* structure of the optimal policy d^*. That is, the optimal number of working servers in group k is either 0 or M_k, except for the group that violates the condition **m1** $\leq n$. For each state n, if the group index is properly renumbered according to Theorem 9.8, the optimal policy has the following form:

$$d^*(n) = (M_1, M_2, \ldots, M_{\hat{k}-1}, M_{\hat{k}} \wedge (n - \sum_{l=1}^{\hat{k}-1} M_l), 0, 0, \ldots, 0), \qquad (9.52)$$

where $\hat{k}$ is the first group index violating the constraint $\sum_{l=1}^{\hat{k}} M_l \leq n$ or $c_{\hat{k}+1} - \mu_{\hat{k}+1} G^*(n) < 0$, i.e.,

$$\hat{k} := \min \left\{ k : \sum_{l=1}^{k} M_l > n, \text{ or } \frac{c_{k+1}}{\mu_{k+1}} \geq G^*(n) \right\}. \qquad (9.53)$$

Therefore, $\hat{k}$ can also be viewed as a *threshold* and we have $\hat{k} \in \{0, 1, \ldots, K\}$. Such a policy can be called a *quasi threshold policy* with threshold $\hat{k}$. Under this policy, the number of working servers in each group is as follows:

$$
\begin{cases}
d^*(n, l) = M_l, & \text{if } l < \hat{k}, \\
d^*(n, l) = 0, & \text{if } l > \hat{k}, \\
d^*(n, l) = M_{\hat{k}} \wedge (n - \sum_{l=1}^{\hat{k}-1} M_l), & \text{if } l = \hat{k}.
\end{cases}
\tag{9.54}
$$

If threshold $\hat{k}$ is determined, $d^*(n)$ is also determined. Thus, finding $d^*(n)$ becomes finding the associated threshold $\hat{k}$, which simplifies the search for the optimal policy. However, we note that the index order of groups is renumbered according to the ascending values of the marginal cost rate $c_k - \mu_k G^*(n)$, which is dynamically varied at different states n or different value of $G^*(n)$. The threshold $\hat{k}$ also dynamically depends on the system state n. Therefore, the index order of groups and the threshold $\hat{k}$ will dynamically vary at different states n, which makes the policy state-dependent (i.e., dynamic) and difficult to implement in practice. To further characterize the optimal policy, we explore its other structural properties.

Difference formula (9.49) and Theorem 9.8 indicate that the marginal cost rate $c_k - \mu_k G(n)$ is an important quantity to differentiate the server groups. If $c_k - \mu_k G(n) < 0$, assigning jobs to servers in group k reduces the system average cost. We call group k an *economic group* for the system under the current policy associated with G. Therefore, we define $\mathbb{K}_n$ as the economic group set at the current state n:

$$
\mathbb{K}_n := \left\{ k : G(n) > \frac{c_k}{\mu_k} \right\}.
\tag{9.55}
$$

We should turn on as many servers in the economic groups $\in \mathbb{K}_n$ as possible, subject to $d(n)\mathbf{1} \leq n$. Note that $G(n)$ reflects the reduction of the holding cost due to operating a server and completing a service, from a long-run average perspective. The condition $G(n) > \frac{c_k}{\mu_k}$ indicates that the benefit of utilizing more servers is larger than the extra cost rate $\frac{c_k}{\mu_k}$, thus the servers in group k are economic and should be utilized.

With Theorems 9.7 and 9.8, the optimization problem (9.37) can be solved by finding the solution to each subproblem in (9.51), using the structure of quasi bang-bang control or quasi threshold form as (9.52). However, it is difficult to directly solve (9.51) because $G^*(n)$ recursively depends on solution d^*. Below, we establish the monotone property of PRF $G(n)$ which can convert the infinite state space search to a finite state space search for the optimal policy. To achieve this goal, we first establish the convexity of performance potential $g^*(n)$, as stated in Theorem 9.9. Its lengthy proof is technical and provided in Xia et al. (2021).

Theorem 9.9 *The performance potential $g^*(n)$ under optimal policy d^* is increasing and convex in n.*

Since $G(n) = g(n) - g(n-1)$, from Theorem 9.9, we can directly derive the following corollary about the monotone property of $G^*(n)$.

Corollary 9.1 *The PRF $G^*(n)$ under optimal policy d^* is always non-negative and increasing in n.*

Note that $G(n)$ plays a fundamental role in (9.48) and (9.49). Thus, the monotonically increasing property of $G^*(n)$ enables us to establish the monotone structure of optimal policy d^* as follows.

Theorem 9.10 *The optimal number of total working servers is increasing in n. In other words, we have $||d^*(n+1)||_1 \geq ||d^*(n)||_1$, $\forall n \in \mathbb{N}$.*

The proof of this theorem is left as an exercise. Theorem 9.10 rigorously confirms an intuitive result that when the queue length increases, more servers should be utilized to alleviate the system congestion, which is also the essence of the congestion-based staffing policy Zhang (2009). It is worth noting that Theorem 9.10 does not imply that the number of working servers in a particular group is increasing in n. However, as shown in the next subsection, under some reasonable conditions, this number does increase in n.

Exercise 9.14 *Prove Theorem 9.10.*

Based on Theorem 9.10, we can further obtain the following corollary.

Corollary 9.2 *For a given $\bar{n}$, if $d^*(\bar{n}, k) = M_k$ for all k, then for any $n \geq \bar{n}$, we always have $d^*(n, k) = M_k$ for all k.*

Corollary 9.2 again confirms an intuitive result that once the optimal action is utilizing all servers at certain state $\bar{n}$, then the same action is optimal for all states larger than $\bar{n}$. Therefore, the search for the optimal policy can be limited to the states $n < \bar{n}$ and the infinite state space search is turned to a finite state space search. This is an advantage of characterizing the optimal policy structure.

Based on Theorems 9.6, 9.8 and Corollary 9.2, we can develop an computational algorithm to determine the optimal scheduling policy.

Exercise 9.15 *Develop an iterative algorithm for computing the optimal customer assignment policy and argue that such an algorithm converges with probability one.*

Since the server groups are ranked by the index based on $c_k - \mu_k G^*(n)$, the index sequence varies with state n. Moreover, although the number of total working servers $\|d^*(n)\|_1$ is increasing in n, $d^*(n,k)$ is not necessarily monotonically increasing in n for a particular group k. This means that it is possible that for some n and k, we have $d^*(n,k) > d^*(n+1,k)$. Thus, the optimal policy is dynamic in nature. However, if the ratio of the cost rate to service rate satisfies a reasonable condition as shown in the next subsection, we can develop a much simpler and static optimal policy obeying a so-called c/μ-rule, which is easier to implement in practice.

Exercise 9.16 *Consider a Markovian queueing system (all random variables are exponentially distributed) with 3 groups of servers with the system parameters as follows:*

- *Holding cost rate function: $h(n) = n$;*
- *Arrival rate: $\lambda = 10$;*
- *Number of groups: $K = 3$;*
- *Number of servers in groups: $\mathbf{M} = (M_1, M_2, M_3) = (3, 4, 3)$;*
- *Service rates of groups: $\mu = (6, 4, 2)$;*
- *Operating cost rates of groups: $\mathbf{c} = (7, 4, 3)$.*

Determine the optimal policy.

9.4.3 OPTIMAL MULTI-THRESHOLD POLICY – c/μ RULE

In this section, we further study the optimal scheduling policy for the group-server queue when the condition of scale economies in terms of c/μ ratios exists. In addition, a more efficient algorithm for finding the optimal policy is developed.

Assumption 9.2 *(Economies of Scale) If the server groups are sorted in the order of $\mu_1 \geq \mu_2 \geq \cdots \geq \mu_K$, then they also satisfy $\frac{c_1}{\mu_1} \leq \frac{c_2}{\mu_2} \leq \cdots \leq \frac{c_K}{\mu_K}$.*

This assumption is reasonable in practice as it means that a faster server has a smaller operating cost rate per unit of service rate. This can be explained by *the effect of the scale economies.* In fact, some reasonable service/operating cost function will lead to such an assumption. Let the operating cost rate $c_k = c(\mu_k)$ be a function of service rate μ_k. The total operating cost rate can be written as $o(\mu) = \sum_{k=1}^{K} m_k c(\mu_k)$. If the operating cost rate is an increasingly concave function in μ, for example, $o(\mu) \propto \mu^q$ with $0 \leq q \leq 1$, we can easily verify that the condition in Assumption 9.2 holds. Therefore, any *increasingly concave operating cost* in μ can guarantee the economies of scale. This kind of operating cost function form is reasonable in some practical group-server queues. For example, in a data center, a faster computer usually has a lower energy cost per unit of computing capacity. This means that the marginal cost of increasing the computing power is diminishing due to the advanced new technology. Therefore, this effect of scale economies is more obvious when the system has both new and old generations of computers. In service systems with human servers, such as a call center, different groups of operators may have different service rates and costs. For example, the call center may have two operator groups: expert group and trainee group. It is reasonable to expect that the experienced experts have a high service rate and a low cost per service rate, although their cost (salary) per person is high (this is why they are retained). The trainees have a low service rate and a high cost per service rate, although their cost (salary) per person is low (this is why they need to be trained). Therefore, for human servers, although the salary is increasing with the server's experience (productivity), the service/operating cost modeled as an increasingly concave function in μ is reasonable due to the capped salary for the job. For the non-human servers, such a service/operating cost function is also reasonable due to the high efficiency brought by the advanced technology.

With Assumption 9.2, we can verify that the group index according to the ascending order of $c_k - \mu_k G^*(n)$ remains the same for all states n as long as $G^*(n) > \frac{c_k}{\mu_k}$ holds. Thus, the ascending order of $c_k - \mu_k G^*(n)$ is always the same as the ascending order of c_k/μ_k, which is *static* and *independent* of $G^*(n)$ or system state. Therefore, the optimal policy structure in Theorem 9.8 can be further characterized.

Theorem 9.11 (The c/μ-rule) *With Assumption 9.2, a policy d^* is optimal if and only if it satisfies the condition: If $G^*(n) > \frac{c_k}{\mu_k}$, then $d^*(n,k) = M_k \wedge (n - \sum_{l=1}^{k-1} d^*(n,l))$; If $G^*(n) \leq \frac{c_k}{\mu_k}$, then $d^*(n,k) = 0$, for $k = 1, 2, \ldots, K, n \in \mathbb{N}$.*

Proof. As indicated by Theorem 9.7, the optimal policy d^* is equivalent to the solution of ILP problems (9.51). First, we let $y_k := c_k - \mu_k G^*(n)$. As stated in Assumption 9.2, we have $\mu_1 \geq \mu_2 \geq \cdots \geq \mu_K$ and $\frac{c_1}{\mu_1} \leq \frac{c_2}{\mu_2} \leq \cdots \leq \frac{c_K}{\mu_K}$. If $G^*(n) > \frac{c_k}{\mu_k}$ for any group k, i.e., if $y_k < 0$, we can easily derive $y_1 \leq y_2 \leq \cdots \leq y_k < 0$. Therefore, to minimize the objective in (9.51), we should let $d^*(n,1) = M_1$, and then $d^*(n,2) = M_2$, until $d^*(n,k) = M_k$, at the condition that the constraint $\sum_{k=1}^{M} d^*(n,k) \leq n$ in (9.51) is satisfied. Writing this decision rule in symbol representation, we have $d^*(n,k) = M_k \wedge (n - \sum_{l=1}^{k-1} d^*(n,l))$. Thus, the first half of the theorem is proved.

Second, if $G^*(n) \leq \frac{c_k}{\mu_k}$, then $c_k - \mu_k G^*(n) \geq 0$ and the corresponding $d^*(n,k)$ in (9.51) should be 0 obviously. Thus, the last half of the theorem is also proved. ∎

Theorem 9.11 implies that the optimal policy d^* follows a simple rule called the c/μ-rule: *Servers in the group with smaller c/μ ratio should be utilized* *with higher priority*. This rule is very easy to implement as the group index renumbering at each state in Theorem 9.8 is not needed anymore. As mentioned earlier, the c/μ-rule can be viewed as a counterpart of the $c\mu$-*rule* for the scheduling of polling queues in the previous section, where queues with greater $c\mu$ will be given higher priority to be served by the single service facility.

Using the monotonically increasing property of $G^*(n)$ in Corollary 9.1 and Assumption 9.2, we can further characterize the monotone structure of the optimal policy d^* as follows, where the proof is left as an exercise.

Theorem 9.12 *The optimal scheduling action $d^*(n,k)$ is increasing in n, for any group $k = 1, 2, \ldots, K$.*

Theorem 9.12 implies that $d^*(n)$ is increasing in n in vector sense. That is $d^*(n+1) \geq d^*(n)$ in vector comparison. Therefore, we certainly also have $||d^*(n+1)||_1 \geq ||d^*(n)||_1$, as indicated in Theorem 9.10.

Exercise 9.17 *Prove Theorem 9.12.*

Using the monotone property of $G^*(n)$ in Corollary 9.1 and the c/μ-rule in Theorem 9.11, we can obtain the optimality of multi-threshold policy.

Theorem 9.13 *The optimal policy d^* has a multi-threshold form with thresholds θ_k: If $n \geq \theta_k$, the maximum number of servers in group k should be utilized, for any $n \in \mathbb{N}$, $k = 1, 2, \ldots, K$.*

Proof. We know that $G^*(n)$ increases in n from Corollary 9.1. For any particular group k, we can define a threshold as

$$\theta_k := \min\left\{n : G^*(n) > \frac{c_k}{\mu_k}\right\}, \quad k = 1, 2, \ldots, K.$$

Therefore, for any $n \geq \theta_k$, we always have $G^*(n) > \frac{c_k}{\mu_k}$ and the servers in group k should be utilized as many as possible, according to the c/μ-rule in Theorem 9.11. Thus, the optimal scheduling for servers in group k has a form of threshold θ_k and the theorem is proved. ∎

Note that "the maximum number of servers in group k should be utilized" in Theorem 9.13 means that the optimal action $d^*(n,k)$ should obey the constraint $\sum_{k=1}^{K} d^*(n,k) \leq n$, i.e., $d^*(n,k) = M_k \wedge (n - \sum_{l=1}^{k-1} d^*(n,l))$. Theorem 9.13 implies that the policy of the original problem (9.37) can be represented by a K-dimensional threshold vector θ as below, without loss of optimality:

$$\theta := (\theta_1, \theta_2, \ldots, \theta_K),$$

where $\theta_k \in \mathbb{N}$. With the monotone property of $G^*(n)$ and Theorem 9.11, we can directly derive that θ_k is monotone in k, i.e.,

$$\theta_1 \leq \theta_2 \leq \cdots \leq \theta_K.$$

Moreover, we can further obtain the following corollary about the optimal threshold of group 1.

Corollary 9.3 *The optimal threshold of group 1 is always $\theta_1^* = 1$, that is, we should always utilize the most efficient server group whenever any customer presents in the system.*

The proof is left as an exercise. Theorem 9.13 indicates that the optimization problem (9.37) over an infinite state space is converted to the problem of finding the optimal thresholds θ_k^*, where $k = 1, 2, \ldots, K$. Denoting by $\mathbb{N}_{\uparrow}^K$ a K-dimensional positive integer space with its elements satisfying $\theta_1 \leq \theta_2 \leq \cdots \leq \theta_K$, the original problem (9.37) is converted to

$$\theta^* = \arg \min_{\theta \in \mathbb{N}_{\uparrow}^K} \{\eta^\theta\}. \tag{9.56}$$

Therefore, the state-action mapping policy $(d : \mathbb{N} \to \mathscr{A})$ is replaced by a parameterized policy with thresholds θ. The original policy space is reduced from an infinite dimensional space $\mathscr{D}$ to a K-dimensional integer space $\mathbb{N}_{\uparrow}^K$. This simplification also implies that a dynamic optimal policy is reduced to a static optimal policy in this customer assignment problem. It is worth noting that the SBO is a powerful approach to determining the optimal policy in a stochastic system. Interested readers are referred to Cao (2007) for more details on this method and related applications.

Exercise 9.18 *Explain that the policy space is significantly reduced by converting problem (9.37) to problem (9.56).*

Exercise 9.19 *Consider the system in Exercise 9.16 with all system parameters the same except for the operating cost rate vector. Now assume that the operating cost rate vector is given by $\mathbf{c} = (7, 8, 5)$. Thus, the scale economies assumption is satisfied. Compute the optimal threshold policy.*

Exercise 9.20 *Compare the $c\mu$-rule for a queue with a single server and multi-class customers with the c/μ-rule for a queue with multi-class servers and single class customers. (a) Find the differences between these two optimal rules. (b) Explain why these rules are optimal intuitively.*

9.5 PERFORMANCE MEASURES IN OPTIMIZATION OF STOCHASTIC MODELS

In most of the optimization problems for a stochastic production or service system, the objective function is in terms of the expected value of a random variable. For example, in a queueing system, the objective of the optimization problem can be either to minimize the expected waiting cost of customers or to maximize the expected social welfare or expected server's profit. However, in many practical situations, the first-moment-based objective function may not be sufficient to represent the goals of system managers. In this section, we introduce some more general performance measures for stochastic models that can tailor to different optimization goals. Compared with deterministic models, the main feature of stochastic models is the uncertainty or randomness in system performance. Thus, the risk measure is critical in designing or optimizing a stochastic system. Since the first moment of a random variable does not capture the variability information of the random variable, it cannot be used as a risk measure for evaluating the performance of a stochastic system. The variance, as a measure of variability, can be used to measure the risk. Thus, the mean and variance, or the moment-based measures, are often jointly used for evaluating the performance of a stochastic system from the return-risk trade-off perspective. We now present a class of quantile-based risk measures and then demonstrate the optimization of a simple stochastic model, the newsvendor problem, under this class of risk measures.

9.5.1 VALUE AT RISK AND CONDITIONAL VALUE AT RISK

Let X be the random variable, $X : \Omega \to \mathbb{R}$, that can represent the loss or gain of operating a stochastic system (i.e., X is a performance variable). A risk measure is a function that maps from $\mathbb{X}$, the set of all risks (all real functions on Ω) to $\mathbb{R}$. We first define the value at risk (VaR).

Definition 9.1 *Value at Risk (VaR$_\alpha$): For a given time horizon T and at* $(\alpha) * 100\%$ *confidence level, VaR$_\alpha$ is the loss that cannot be exceeded with probability of* $(\alpha) * 100\%$ *(or can only be exceeded with probability of at most* $(1 - \alpha) * 100\%$*).*

This definition is made in terms of loss or cost. VaR is the maximum dollar amount expected to be lost over a given time horizon, at a pre-defined confidence level. For example, if the 95% ($\alpha = 0.95$) one-month VaR for the loss of an investment portfolio is $1 million, it means that there is 95% confidence that over the next month the portfolio will not lose more than $1 million. In other words, there is 5% chance that the portfolio will lose more than $1 million over the next month. To express this definition in mathematical notation, we first need to decide whether we use the same random variable to represent

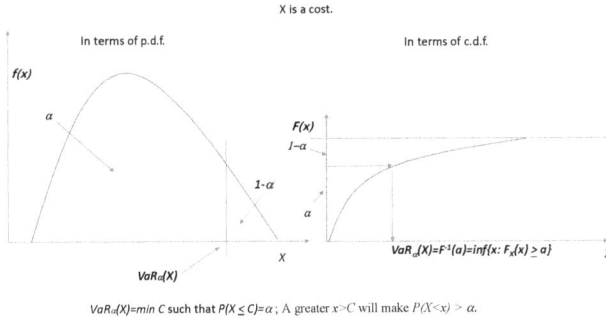

X is a cost.

In terms of p.d.f.

$f(x)$

α

In terms of c.d.f.

$F(x)$

$1-\alpha$

$1-\alpha$

α

$VaR_\alpha(X)=F^{-1}(\alpha)=\inf\{x: F_x(x) \geq \alpha\}$

X

$VaR_\alpha(X)$

$VaR_\alpha(X)=min\ C$ such that $P(X \leq C)=\alpha$; A greater $x>C$ will make $P(X<x) > \alpha$.

Figure 9.4 VaR definition.

both gain (profit) and loss (cost). If yes, a positive value indicates a profit and negative value indicates a cost. If a random variable only represents the cost (like in the example above), then we can use the positive value to represent the cost. Such a random variable is called a loss distribution. Then the definition can be stated as

$$
\begin{aligned}
VaR_\alpha(X) &= \sup\{x \in \mathbb{R} | P(X \geq x) > 1 - \alpha\} \\
&= \inf\{x \in \mathbb{R} | P(X \leq x) > \alpha\} \\
&= \inf\{x \in \mathbb{R} | F_X(x) > \alpha\} \\
&= F_X^{-1}(\alpha),
\end{aligned}
\tag{9.57}
$$

where $F_X(\cdot)$ is the distribution function of X. This definition is shown graphically in Figure 9.4. Note that the subscript α indicates the confidence level for the VaR and is usually a number close to 100%, a high probability. In some other books or research papers, α is defined as (1-confidence level) and is usually a number close to zero, a low probability (similar to the significance level in testing hypothesis in statistics). The inconsistent definitions for α can be confusing when reading the literature.

Similarly, we can define the VaR in terms of gain or profit. In such a case, the VaR is the minimum dollar amount expected to be gain over a given time horizon, at a pre-defined confidence level. For example, if the 95% one-month VaR for the profit of a company is $1 million, then there is a 95% confidence that over the next month the company will make at least $1 million (or 5% chance of making less than $1 million).

The sign of the random variable, the definition of α, and whether the gain or loss is our concern can make the definition of VaR confusing. A simple guideline for defining or interpreting the VaR is that the value of VaR is a threshold value that over a finite time interval, the probability of "good thing" or "desirable event", which is represented by the random variable value greater than or smaller than the threshold, is the confidence level or close to 100%. On the other hand, the probability of "bad thing" or " undesirable event" is (1-confidence level) or close to zero. These two types of events are clearly complement to each other. In the previous examples, the "undesirable event" would be "the loss over a month will be more than $ 1 million" in the cost example and the "desirable event" would be " the gain over a month is at least $1 million" in the profit example.

Artzner et al. (1997) consider how risk measures should behave. They suggest a set of properties that a coherent risk measure should satisfy.

Definition 9.2 *A risk measure ρ is called coherent if and only if it satisfies the following axioms:*

- *Sub-additivity: for any $X, Y \in \mathbb{X}$, then $\rho(X+Y) \leq \rho(X) + \rho(Y)$.*
- *Positive homogeneity: for any $X \in \mathbb{X}$ and $\lambda \geq 0$, then $\rho(\lambda X) = \lambda \rho(X)$.*
- *Translation invariance: for a fixed $X \in \mathbb{X}$ and $a \in \mathbb{R}$, then $\rho(X+a) = \rho(X) + a$.*
- *Monotonicity: Let $X, Y \in \mathbb{X}$ be such that $X \leq Y$ almost surely, then $\rho(X) \leq \rho(Y)$.*

It can be verified that the VaR is not a coherent risk measure (left as an exercise). However, we can define another risk measure which is coherent. That is the conditional value at risk (CVaR).

Definition 9.3 *Conditional Value at Risk ($CVaR_\alpha$): For a given time horizon T and at $(\alpha) * 100\%$ confidence level, $CVaR_\alpha$ is the conditional expectation of X subject to $X \geq VaR_\alpha(X)$. That is*

$$CVaR_\alpha(X) = E[X|X > VaR_\alpha(X)].$$

Since the minimum of X in the CVaR definition is the VaR, we have an alternative expression as

$$CVaR_\alpha(X) = \min_C \left\{ C + \frac{1}{1-\alpha} E[X - C]^+ \right\}. \tag{9.58}$$

The sub-Figure on the left in Figure 9.5 shows this definition graphically. Furthermore, we can show another expression for CVaR called Acerbi's Integral Formula. Suppose that X is the random loss distribution with distribution function $F_X(x) = P(X \leq x)$. Denote by $X^\alpha = (X|X > VaR_\alpha(X))$ the conditional random variable with the distribution function given by

$$F_X^\alpha(x) := \begin{cases} 0 & \text{if } x < VaR_\alpha(X), \\ \frac{F_X(x) - \alpha}{1 - \alpha} & \text{if } x \geq VaR_\alpha(X), \end{cases}$$

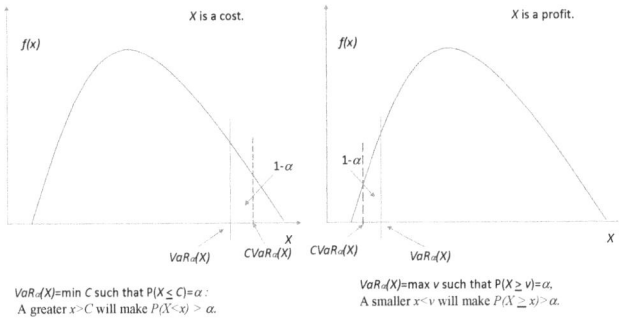

Figure 9.5 CVaR definition.

where the second expression on the RHS is due to the fact

$$P(X \leq x | X \geq VaR_\alpha(X)) = \frac{P(VaR_\alpha(X) \leq X \leq x)}{P(X \geq VaR_\alpha)} = \frac{P(X \leq x) - P(X \leq VaR_\alpha)}{P(X \geq VaR_\alpha)}$$

$$= \frac{F_X(x) - \alpha}{1 - \alpha}.$$

By taking the derivative of the distribution function, we have the density function

$$f_X^\alpha(x) = \frac{dF_X^\alpha(x)}{dx} = \begin{cases} 0 & \text{if } x < VaR_\alpha(X), \\ \frac{f_X(x)}{1-\alpha} & \text{if } x \geq VaR_\alpha(X). \end{cases}$$

Therefore, we can compute CVaR as follows:

$$CVaR_\alpha(X) = E[X | X > VaR_\alpha(X)] = E[X^\alpha] = \int_{-\infty}^{\infty} z f_X^\alpha(z) dz$$

$$= \int_{-\infty}^{VaR_\alpha(X)} z f_X^\alpha(z) dz + \int_{VaR_\alpha(X)}^{\infty} z f_X^\alpha(z) dz \qquad (9.59)$$

$$= \int_{VaR_\alpha(X)}^{\infty} z f_X^\alpha(z) dz = \int_{VaR_\alpha(X)}^{\infty} z \frac{f_X(z)}{1-\alpha} dz.$$

Now we define a new variable $\beta = F_X(z)$. Taking the derivative of β with respect to z yields

$$\frac{d\beta}{dz} = f_X(z) \Rightarrow f_X(z) dz = d\beta.$$

Using $f_X(z)dz = d\beta$, $VaR_\alpha(X) = F_X^{-1}(\alpha)$, and $z = F_X^{-1}(\beta) = VaR_\beta(X)$ in the last term on the RHS of (9.59), we obtain

$$CVaR_\alpha(X) = \frac{1}{1-\alpha} \int_\alpha^1 VaR_\beta(X) d\beta$$

by change of variable from z to β for the integral. Note that the integral limit change is due to the fact that if $z = VaR_\alpha(X)$, then $\beta = F_X(z) = F_X(VaR_\alpha(X)) = F_X(F_X^{-1}(\alpha)) = \alpha$, and if $z = \infty$, then $\beta = F_X(z) = F_X(\infty) = 1$.

Exercise 9.21 *Verify that VaR is not a coherent risk measure but CVaR is.*

If the performance random variable is the gain such as profit or revenue, then (9.58) can be given by

$$
\begin{aligned}
CVaR_\alpha(X) &= \max_v \left\{ v - \frac{1}{1-\alpha} E[v - X]^+ \right\} \\
&= \max_v \left\{ v + \frac{1}{1-\alpha} E[X - v]^- \right\}.
\end{aligned}
\tag{9.60}
$$

The sub-Figure on the right in Figure 9.5 shows this expression graphically.

9.5.2 A NEWSVENDOR PROBLEM

In this section, we utilize a risk-averse newsvendor model to show the optimization of a simple stochastic system with CVaR-based performance measures. The content of this section is mainly taken from Chen et al. (2009). We consider a risk-averse newsvendor problem with random and price-dependent demand. At the beginning of a selling season, a decision and selling price is made for the season. Since this problem has only one-period and one random factor (i.e., demand), the random profit function can be considered as a simple stochastic model. Let c be the unit purchase cost. During the selling season, the demand is met through the newsvendor's inventory; and shortage cost beyond the loss of profit is zero. The leftover inventory at the end of the selling season is salvaged at value $0 \le s < c$. The demand is random and is denoted by $D(p) = d(p,X)$, where p is the selling price, and $X \in [L,U]$ is the random price-independent component, which represents the demand risk. It is assumed that $E[d(p,X)] = d(p)$ is a continuous. strictly decreasing, non-negative, twice-differentiable function and defined on $[c,\bar{p}]$, where $\bar{p}$ is the maximum admissible price. In addition, $d(p)$ is assumed to have an increasing price elasticity (see Chen et al. (2009)). It is also assumed that X has the increasing failure rate (IFR). Denote by $F(p,x)$ and $f(p,x)$ the cumulative distribution function (c.d.f.) and the probability density function (p.d.f.) for any given price p, respectively, and $\Phi(x)$ and $\phi(x)$ the c.d.f. and p.d.f. of X, respectively. Let y be the order quantity. Then the newsvendor's profit, denoted by $\pi(y,p)$, can be written as

$$
\begin{aligned}
\pi(y,p) &= p\min(y,d(p,X)) + s(y - d(p,X))^+ - cy \\
&= (p-c)y - (p-s)(y - d(p,X))^+.
\end{aligned}
\tag{9.61}
$$

Letting $\eta = 1 - \alpha$ and using (9.60), we have the CVaR for the profit

$$CVaR_\alpha(\pi(y,p)) = \max_{v\in\mathbb{R}}\left\{v+\frac{1}{\eta}E[\pi(y,p)-v]^-\right\}$$

$$= \max_{v\in\mathbb{R}}\left\{v-\frac{1}{\eta}E[v-\pi(y,p)]^+\right\}$$

$$= \max_{v\in\mathbb{R}}\left\{v-\frac{1}{\eta}\int_0^\infty[v-(p-c)y+(p-s)(y-t)^+]^+dF(p,t)\right\}$$

$$= \max_{v\in\mathbb{R}}\left\{v-\frac{1}{\eta}\int_0^y[v+(c-s)y-(p-s)t]^+dF(p,t)\right.$$
$$\left.-\frac{1}{\eta}\int_y^\infty[v-(p-c)y]^+dF(p,t)\right\}.$$

$$(9.62)$$

Under the CVaR criterion, we can obtain the optimal joint quantity-price decision for a specific demand function form (either additive or multiplicative). For
the details, we refer to Chen et al. (2009). To provide readers with some basic
idea of the optimization analysis under the CVaR, we just show the optimal
order quantity for a given selling price p as an example. For a given price p,
let

$$h(y,p,v) = \left\{v-\frac{1}{\eta}\int_0^y[v+(c-s)y-(p-s)t]^+dF(p,t)\right.$$
$$\left.-\frac{1}{\eta}\int_y^\infty[v-(p-c)y]^+dF(p,t)\right\}.$$

$$(9.63)$$

Then, the newsvendor problem is now stated as

$$\max_y CVaR_\alpha(\pi(y,p)) = \max_y(\max_v h(y,p,v)).$$

The optimal solution to this problem is given in the following proposition and
the proof highlights the flavor of the analysis.

Proposition 9.2 *Suppose that the selling price* $p(>c)$ *is fixed. If the newsvendor is risk-averse with the* $CVaR_\alpha$ *criterion, then the optimal order quantity
is*

$$y^*(p) = F^{-1}\left(p,\eta\frac{p-c}{p-s}\right) = d\left(p,\Phi^{-1}\left(\eta\frac{p-c}{p-s}\right)\right).$$

$$(9.64)$$

Proof. The proof has two parts: (a) Finding v that maximizes the LHS of (9.62)
to determine the CVaR of $\pi(y,p)$ (i.e., finding v that maximizes $h(y,p,v)$); (b)
Determining the optimal y that maximizes $CVaR_\alpha(\pi(y,p))$ for a given p.
    We consider v values in three intervals:

Case 1: $v \leq (s-c)y$. In this case, $h(y,p,v) = v$ and thus $\frac{\partial h(y,p,v)}{\partial v} = 1 > 0$. This is the interval that $h(y,p,v)$ is increasing in v.

Case 2: $(s-c)y \leq v \leq (p-c)y$. In this case, we have

$$h(y,p,v) = v - \frac{1}{\eta} \int_0^{\frac{v+(c-s)y}{p-s}} [v + (c-s)y - (p-s)t] dF(p,t),$$

and then

$$\frac{\partial h(y,p,v)}{\partial v} = 1 - \frac{1}{\eta} F\left(p, \frac{v+(c-s)y}{(p-s)}\right).$$

At the two boundary points, we get

$$\frac{\partial h(y,p,v)}{\partial v}\Big|_{v=(s-c)y} = 1 > 0, \quad \frac{\partial h(y,p,v)}{\partial v}\Big|_{v=(p-c)y} = 1 - \frac{1}{\eta} F(p,y).$$

This is the interval that $h(y,p,v)$ is increasing in v at the lower boundary point and its monotonicity at the higher boundary point depends on system parameters.

Case 3: $v > (p-c)y$. In this case, we have

$$h(y,p,v) = v - \frac{1}{\eta} \int_0^y [v + (c-s)y - (p-s)t] dF(p,t)$$
$$- \frac{1}{\eta} [v - (p-c)y](1 - F(p,y)).$$

Taking the derivative with respect to v yields

$$\frac{\partial h(y,p,v)}{\partial v} = 1 - \frac{1}{\eta}[F(p,y) + 1 - F(p,y)] = 1 - \frac{1}{\eta} < 0.$$

This is the interval that $h(y,p,v)$ is decreasing in v.

Denote by $v^*(p)$ the v value that maximizes $h(y,p,v)$ for fixed y and p. Based on the behaviors of $h(y,p,v)$ in the three intervals, we conclude that $v^*(p) \in [(s-c)y, (p-c)y]$. Now for a given p, we find the optimal y, denoted by $y^*(p)$, under $v^*(p)$.

The key is to consider two situations for the first-order derivative $\partial h(y,p,v)/\partial v$ at the boundary point $v = (p-c)y$. The first situation is

$$\frac{\partial h(y,p,v)}{\partial v} = 1 - \frac{1}{\eta} F(p,y) < 0 \Rightarrow F(p,y) > \eta \Rightarrow y \geq F^{-1}(p,\eta),$$

which implies that the maximum of $h(y,p,v)$ can be reached at the point that satisfied the first-order condition. That is

$$1 - \frac{1}{\eta} F\left(p, \frac{v+(c-s)y}{(p-s)}\right) = 0,$$

which results in $v^*(p) = (p-s)F^{-1}(p,\eta) - (c-s)y$. Under this value, we have

$$h(y,p,v^*(p)) = -(c-s)y + \frac{1}{\eta}\int_0^{F^{-1}(p,\eta)}[(p-s)x]dF(p,x),$$

from which it follows

$$\frac{dh(y,p,v^*(p))}{dy} = s - c < 0.$$

This implies that optimal y cannot be reached.

For the second situation of $y < F^{-1}(p,\eta)$, the maximizer of $h(y,p,v)$ occurs at the upper boundary. That is $v^*(p) = (p-c)y$. Thus, we have

$$h(y,p,v^*(p)) = (p-c)y - \frac{1}{\eta}\int_0^y[(p-s)y-(p-s)x]dF(p,x),$$

then

$$\frac{dh(y,p,v^*(p))}{dy} = (p-c) - \frac{1}{\eta}(p-s)F(p,y),$$

which leads to the optimal order quantity $y^*(p) = F^{-1}\left(p, \eta\frac{p-c}{p-s}\right)$. ∎

For the analysis of optimal pricing and comparative statics of parmater changes, interested readers are referred to Chen et al. (2009) for details. In this section, we introduced two popular risk measures, VaR and CvaR, for optimization problems in stochastic models. It is worth noting that a more general class of risk measures called distortion risk measures (DRM) can be utilized to capture the decision maker's risk attitudes. For more details about DRM, interested readers are referred to Shao and Zhang (2021).

While the newsvendor problem discussed above has the CVaR-based objective function, we can also formulate the optimization problem subject to CVaR-based constraints.

Exercise 9.22 *(The Newsvendor Model with CVaR Constraints.) Consider the optimization problem of a newsvendor subject to the CVaR-based constraint. Assume that the selling price p is given or determined by the market and the demand is random with the distribution function of $F_D(\cdot)$. Other assumptions are the same as the ones made in the newsvendor problem discussed above. Thus, the newsvendor profit is a function of the order quantity denoted by $\pi(y)$. If the newsvendor wants to maximize the expected profit of the selling season and ensure that the CVaR of the profit exceeds a threshold, the problem can be written as*

$$\max_y E[\pi(y)]$$

$$s.t. \ CVaR_\alpha(\pi(y)) \geq x_0.$$

(a) *Show that this constrained problem is equivalent to the following optimization problem*

$$\max_{y} E[\pi(y)] + \lambda CVaR_\alpha(\pi(y)),$$

where λ is a parameter that quantifies the trade-off between risk and return.
(b) *Solve the equivalent optimization problem in (a).*
(Note: this problem is taken from Shao and Zhang (2021).)

REFERENCE NOTES

For the optimization problems with the infinite time horizon, the objective functions are usually in terms of long-term average cost or profit rates. Such functions can be obtained by using the renewal reward theorem based on the regenerative cycles. The two examples for this class of problems are taken from [11] and [17]. The economic analysis of stationary queueing systems with customer choices is mainly based on [12]. More references in this area can be found in [5] and [6]. The $c\mu$ rule for queueing systems with multi-class customers is based on [4]. The c/μ rule for the customer assignment problem in a queue with heterogeneous servers is mainly taken from [15]. For more details about the sensitivity-based optimization, interested readers are referred to [2].

REFERENCES

1. P. Artzner, F. Delbaen, and J. Eber, Coherent measures of risk. *Mathematical Finance*, 9(3), 203–208, 1999.
2. X.R. Cao, "Stochastic Learning and Optimization – A Sensitivity-Based Approach", Springer, New York, 2007.
3. Y. Chen, M. Xu, and Z.G. Zhang, A risk-averse newsvendor model under the CVaR criterion, *Operations Research*, 57, (4), 1040–1044, 2009.
4. N. Gautam, "Analysis of Queues: Methods and Applications", CRC Press, London, 2012.
5. R. Hassin and M. Haviv, "To Queue Or Not to Queue Equilibrium Behavior in Queueing Systems", Springer Science, 2003.
6. R. Hassin, "Rational queueing", CRC Press, 2016.
7. Q. Li, "Constructive Computation in Stochastic Models with Applications: The RG-Factorizations", Tsinghua University Press, 2010.
8. A. Lisnianski, Lz-Transform for a Discrete-State Continuous-Time Markov Process and Its Applications to Multi-State System Reliability, "Recent Advances in System Reliability, Signatures, Multi-state Systems and Statistical Inference.",Springer-Verlag, New York, 2012.
9. S. Ross, "Stochastic Processes", John Wiley and Sons, 2nd ed., 1996.
10. H. Shao and Z.G. Zhang, Extreme-case Distortion Risk Measures: A Uni cation and Generalization of Closed-form Solutions, Working paper, CORMS-Series001, 2021.

11. S.H. Sheu, Z.G. Zhang, An optimal age replacement policy for multi-state systems, *IEEE Transactions on Reliability*, 62, (3), 722–723, 2013.

12. S. Stidham, "Optimal Design of Queueing Systems", CRC Press, 2009.

13. H. Takagi, "Queueing Analysis, Vol 1, Vacation and Priority Systems", Elsevier, North-Holand, 1991.

14. N. Tian and Z.G. Zhang, "Vacation Queueing Models: Theory and Applications", Springer Science, 2006.

15. L. Xia, Z.G. Zhang, and Q. Li, A c/μ-Rule for job assignment in heterogeneous group-server queues, *Production and Operations Management*, Forthcoming, 2021.

16. Z.G. Zhang, R.G. Vickson, and M. Eenige, Optimal two-threshold policies in an $M/G/1$ queue with two vacation types. *Performance Evaluation*, 29, 63–80, 1997.

17. Z.G. Zhang and C.E. Love, The threeshold policy in $M/G/1$ queue with an exceptional first vacation. *INFOR*, 36, 193–204, 1998.

18. Z.G. Zhang, Performance analysis of a queue with congestion-based staffing policy. *Management Science*, 55(2), 240–251, 2009.

10 Dynamic Optimization in Stochastic Models

A dynamic optimization problem under uncertainty can be modeled as a Markov decision process (MDP). Such a problem involves a sequence of decisions made by an agent at different points in time over either a finite or an infinite time horizon. The MDP is a Markov process in which some decisions (also called actions) can be made for each state at an instant in time (called a decision epoch). The time interval between two consecutive decision epochs can be either deterministic or random. The formal is called a discrete-time MDP, denoted by DTMDP, and the latter is called a continuous-time MDP, denoted by CTMDP. We give two examples of these MDPs.

The first one, a DTMDP, is an inventory system with a random demand for each period. The planning horizon, N, for this system can be either finite or infinite. The demand for each period is an i.i.d. random variable. The decision for this system is how much product to order based on the inventory level at a certain inventory inspection point (i.e., periodic review system). The inter-decision time interval, called a period, is deterministic such as a day or a week. Assume that the decision epoch is at the beginning of each period and an order placed is received immediately. The demand is met immediately if the inventory is sufficient and is backlogged otherwise. We consider a cost structure imposed on the system that consists of costs of different types. These are ordering cost for placing order, holding cost for carrying inventory, backlogged cost for back-orders, and terminal cost for trashing leftover or penalty of not fulfilling back-orders (if the planning horizon is finite). The objective of the system is to minimize the expected cost (discounted expected cost or the expected cost per time unit) for the finite (infinite) planning horizon by finding the optimal inventory policy. There are four components for the MDP: (a) a state space; (b) an action space; (c) a transition probability kernel, and (d) a reward/cost function which is also called one-step function. In this inventory system, a state is the inventory level before ordering at the beginning of each period and a set of feasible values becomes the sample space. Since the back-order is allowed, the state space can be the whole real line. For each state, the order quantity or the inventory level after ordering can be considered as the action (or decision) to be taken at the decision epoch. The set of actions is the action space. The transition probability kernel is a transition probability matrix that assigns each state-action pair a probability over the next state. The one-step function is the expected (possibly state-dependent) reward/cost over

DOI: 10.1201/9781003150060-10

the next period. In the example, the transition probability is the probability that the inventory level will change from the current level (before ordering) to the level at the beginning of the next period after the ordering decision is made at the current period. The reward/cost function in the system can be derived with or without using the transition matrix after the cost structure is specified.

The second example, a CTMDP, is a queueing system with work conservation (i.e., the server is never idle when the system is not empty). Suppose that there are two types of customers arriving at a service station with only one server. The server can serve either type of customers. However, there is a random switch-over time when the server switches from serving one type to serving another type of customers. At any service completion instant, the server can decide which type of customer to serve based on the number of customers waiting in each queue. We assume that customers of each type arrive at the system according to Poisson process and service times and switch-over times are all i.i.d. and exponentially distributed. The cost structure consists of the customer waiting costs, switch-over cost, and possibly server operating cost. The objective of the system is to determine the server assignment policy based on the queue lengths of both queues. Since the decision epochs are service completion instants, the inter-decision period is random. Thus, we model the system as a CTMDP. The state is a vector with three variables: the number of customers in queue 1, the number of customers in queue 2, and the server status. For each state, assigning the server to serve one of the two queues becomes the action or decision. The transition probability is the probability that after a decision, the system changes from the current state to another state at the next decision epoch (next service completion instant). The one-step function can be determined based on the cost structure and the transition probability matrix. In this chapter, we will start with the finite DTMDP and then discuss CTMDP and the extension to stochastic games which are MDPs with multiple decision makers.

10.1 DISCRETE-TIME FINITE MARKOV DECISION PROCESS

In this section, we present the formal formulation of a DTMDP. The terminology, notations, and definitions are introduced by using a simple finite state and finite time horizon DTMDP. However, this model formulation process can be extended to the MDP with infinite state space and/or infinite time horizon. We first introduce the general notations. An MDP is defined by:

- A set of states $\mathscr{S}$, called a sample space, can be either discrete or continuous;
- A set of actions $\mathscr{A}$, called an action space that is state-dependent, can be either discrete or continuous;

- A transition probability matrix P, which defines the probability distribution over next states given the current state and current action, $P(S_{t+1}|S_t,A_t)$ where $S_{t+1} = s' \in \mathscr{S}$, $S_t = s \in \mathscr{S}$, and $A_t = a \in \mathscr{A}(s)$;
- A one-step reward/cost function $R(S_{t+1}|S_t,A_t)$;
- A starting state;
- A terminal state if the MDP is finite or episodic.

The one-step reward/cost function can be considered as a mapping $R : \mathscr{S} \times \mathscr{S} \times \mathscr{A} \to \mathscr{R}$ where $\mathscr{R} \subset \mathbb{R}$ is bounded by some constant. The transition probability matrix can be generalized to the transition probability kernel that assigns each state-action pair $(s,a) \in \mathscr{S} \times \mathscr{A}$ a probability measure over the next state-reward/cost pair $(s',r) \in \mathscr{S} \times \mathscr{R}$. This is for the case where the one-step reward/cost function depends on the next state in addition to the current state-action pair. If this one-step function is independent of the next state, then the transition probability kernel reduces to the transition probability matrix. The "Markov" property of the MDP means the future state only depends on the current state and action pair (not the history). That is

$$P(S_{t+1} = s'|S_t = s_t, A_t = a_t, S_{t-1} = s_{t-1}, A_{t-1} = a_{t-1}, ..., S_0 = s_0)$$
$$= P(S_{t+1} = s'|S_t = s_t, A_t = a_t),$$

which governs the system dynamics. Next, we define the policy and value function.

Definition 10.1 *A policy is a mapping from states to actions* $\pi : \mathscr{S} \to \mathscr{A}$. *A policy can be either deterministic or stochastic.*

Thus, for a given policy π, we can write $\pi(s_t) = a$, where $s_t \in \mathscr{S}$ and $a \in \mathscr{A}(s_t)$. Now we turn to the performance measure of the MDP. We define "return" as a specific function of the reward/cost sequence under a policy π. Suppose that the current time is t. Then the return discounted by discount rate $\gamma, (0 \le \gamma \le 1)$, given a starting state s_t and under a policy π, can be written as

$$R_t^{\pi}(s_t) = r_{t+1} + \gamma \cdot r_{t+2} + \gamma^2 \cdot r_{t+3} + \cdots = \sum_{k=0}^{\infty} \gamma^k \cdot r_{t+k+1}$$

for an infinite planning horizon, where r_{t+1} is the reward over period t. Such a return is for an infinite horizon MDP. For an MDP with a finite time horizon, the return can be written as

$$R_t^{\pi}(s_t) = r_{t+1} + \gamma \cdot r_{t+2} + \gamma^2 \cdot r_{t+3} + \cdots + \gamma^{N-t-1} r_N = \sum_{k=0}^{N-t-1} \gamma^k \cdot r_{t+k+1},$$

where N can be either a deterministic terminal time, called a finite horizon MDP, or a stopping time based on an random event, called an episodic MDP.

As shown above, r_{t+1} is a random variable, depending on the state-action pair $(s_t, \pi(s_t))$ and the next state to be reached s_{t+1}. i.e., $r_{t+1} = R(s_{t+1}|s_t, \pi(s_t))$ (from the one-step reward/cost function $R(S_{t+1}|S_t, A_t)$). Now we can define the value function for a given starting state and a policy π.

Definition 10.2 *A value function for a state at time t is defined as the expected total discounted return if the process starts at that state and under a given policy π from t onward. It is denoted by $V^{\pi}(s_t)$.*

Thus, we have for an infinite-horizon DTMDP

$$V^{\pi}(s_t) = E(R_t^{\pi}(s_t)) = E(\sum_{k=0}^{\infty} \gamma^k \cdot r_{t+k+1}) = \sum_{k=0}^{\infty} \gamma^k \cdot E(r_{t+k+1})$$

$$= \sum_{s_{t+1} \in \mathscr{S}} P(s_{t+1}|s_t, \pi(s_t))(R(s_{t+1}|s_t, \pi(s_t)) + \gamma V^{\pi}(s_{t+1})) \qquad (10.1)$$

$$= R(s_t, \pi(s_t)) + \gamma \sum_{s_{t+1} \in \mathscr{S}} P(s_{t+1}|s_t, \pi(s_t)) V^{\pi}(s_{t+1}),$$

where $R(s_t, \pi(s_t)) = \sum_{s_{t+1} \in \mathscr{S}} P(s_{t+1}|s_t, \pi(s_t)) R(s_{t+1}|s_t, \pi(s_t))$. To facilitate the analysis, since the state space is finite, the recursive relation (10.1) can be expressed in terms of matrix and vector notations. Denote by $\mathbf{V}_t^{\pi}$, $\mathbf{R}(\pi_t)$, and $\mathbf{P}(\pi_t)$ are the vector of value functions, the vector of expected returns in period t, and the transition probability matrix under a given policy π. Thus, the equation (10.1) can be expressed in vector/matrix form as

$$\mathbf{V}_t^{\pi} = \mathbf{R}(\pi_t) + \gamma \mathbf{P}(\pi_t) \mathbf{V}_{t+1}^{\pi}. \qquad (10.2)$$

Another level of abstraction is to introduce the operator notation. An operator, denoted by T, can be considered as a mapping or a function which takes an element of a set U and converts it into an element of a set V. We use the notation $T : U \rightarrow V$, which means if $u \in U$, then $Tu \in V$. For a finite DTMDP problem, we consider $U = V := \mathbf{R}^{|\mathscr{S}|}$, the set of $|\mathscr{S}|$-vectors, called the set of admission vectors or the set of admissible value functions. In this section, the operator for a DTMDP under a given policy π, denoted by $T(\pi_t)$ (also written as T_{π_t}), is defined as follows:

$$T(\pi_t)\mathbf{V}_{t+1}^{\pi} := \mathbf{R}(\pi_t) + \gamma \mathbf{P}(\pi_t)\mathbf{V}_{t+1}^{\pi}. \qquad (10.3)$$

With this operator, we can express (10.2) in a more compact way as

$$\mathbf{V}_t^{\pi} = T(\pi_t)\mathbf{V}_{t+1}^{\pi}. \qquad (10.4)$$

It is not difficult to show that the value of using policy π when starting at t in state s depends at most on $\pi_t(s)$, the decision prescribed at time point t, and the decision rules for the remaining periods, π_{τ} for $\tau = t+1, t+2, ..., N$. This fact also implies that the process under a policy is Markovian.

Exercise 10.1 *Prove that the value of an MDP under a policy π only depends on the current decision and the decision rules in the future periods.*

So far we have described the system dynamics of a DTMDP under a given policy from a starting state by the recursive relation in terms of the value function. Next we formulate the optimization problem for determining the optimal policy for the DTMDP. First, we define the optimal value function.

Definition 10.3 *The optimal value function for each state at time t, $s_t \in \mathscr{S}$, is defined as the maximum expected total discounted return if the process starts at that state and under the optimal policy π^* from t onward. It is denoted by $V^*(s_t)$.*

Based on this definition, we can write the optimality equation as

$$V^*(s_t) = V^{\pi^*}(s_t) = \max_{\pi} V^{\pi}(s_t)$$

$$= \max_{\pi} \sum_{s_{t+1} \in \mathscr{S}} P(s_{t+1}|s_t, \pi(s_t)) \left(R(s_{t+1}|s_t, \pi(s_t)) + \gamma V^{\pi}(s_{t+1}) \right) \quad (10.5)$$

$$= \max_{\pi(s_t) \in A(s_t)} \left\{ R(s_t, \pi(s_t)) + \gamma \sum_{s_{t+1} \in \mathscr{S}} P(s_{t+1}|s_t, \pi(s_t)) V^*(s_{t+1}) \right\}.$$

If the transition probability is time invariant[1], then a more convenient notational rule is to move the time index from the state subscript to the value function and action subscript. Thus, we can write the optimality equation (10.5) by noting $\pi_t(s) = a$ as

$$V_t^*(s) = \max_{a \in A(s)} \left\{ R(s, a) + \gamma \sum_{j \in \mathscr{S}} P(j|s, a) V_{t+1}^*(j) \right\}. \quad (10.6)$$

For a finite horizon MDP, the time index, $t \, (\in \{1, 2, ..., N\})$ is essential and $V_{N+1}^*(j) := f_E(j)$, a terminal function, for all $j \in \mathscr{S}$. The optimal policy for a MDP can be extracted from the optimal value function. For a finite horizon MDP, this means

$$\pi_t^*(s) = \arg\max_a \left\{ R(s, a) + \gamma \sum_{j \in \mathscr{S}} P(j|s, a) V_{t+1}^*(j) \right\}.$$

Using the vector and matrix form, we can define the optimal value operator M as

$$MV := \max_{\delta \in \Delta} T(\delta) V,$$

where δ is a decision rule for a one-period problem and Δ is the set of all admissible decision rules with V as arriving value function vector. For each

---

[1]time-variant transition probability may be needed if a learning process exists in an MDP.

state s and an action $a \in \mathscr{A}(s)$, there exits a $\delta \in \Delta$ such that $\delta(s) = a$. For an MDP with a finite horizon N, a strategy is a collection of decision rules for N periods, denoted by $\pi = (\pi_1, \pi_2, ..., \pi_N) = (\delta_1, \delta_2, ..., \delta_N)$. The following notations are utilized. The action in a state s at a time instant t is denoted by $\delta_t(s)$ (the time subscript may be omitted), the decision rule at a time instant t by δ_t, and the strategy or a policy by π. The optimality equations can be written as follows in operator notation:

$$\mathbf{V}_t^* = M\mathbf{V}_{t+1}^*, \quad \text{for every } t. \tag{10.7}$$

Definition 10.4 *An operator T is said to be isotone on V if $\mathbf{u}, \mathbf{v} \in V$ and $\mathbf{u} \leq \mathbf{v}$ imply that $T\mathbf{u} \leq T\mathbf{v}$.*

Such a property is a generalization of an increasing function defined on the real line. Both $T(\pi_t)$ and M are isotone on V. It is a good exercise to show that $T(\pi_t)$ is isotone by using its definition in (10.3). Then, we can use it to show that M is also isotone as follows. Suppose that $\mathbf{u} \leq \mathbf{v}$ with $\mathbf{u}, \mathbf{v} \in V$. Since $T(\delta)$ is isotone, $T(\delta)\mathbf{u} \leq T(\delta)\mathbf{v}$. Also note that $T(\delta)\mathbf{v} \leq M\mathbf{v}$ by definition of M. It follows from these two inequalities that $T(\delta)\mathbf{u} \leq M\mathbf{v}$ for every δ (including δ^* which is the optimal decision rule prescribed by M). Thus, we have $M\mathbf{u} \leq M\mathbf{v}$.

Exercise 10.2 *Prove the isotone property of $T(\pi_t)$ based on (10.3).*

    The isotone property is the key to solve the MDP via recursive relations. Then the multi-period problem can be decomposed into a series of one-period problem as shown below. With the operator notation along with the vector and matrix form expressions, it is not difficult to show the following properties for a finite-state DTMDP with finite time horizon (the proofs for some results are left as exercises). Note that in a more general setting, some of the results presented below may not hold. For example, if the state space is continuous and admissible decision rules must be measurable, some additional conditions are required for some properties to hold. In this chapter, we provide the detailed analysis on the finite-state and action space case and only mention if the results hold in a more general case without proofs. This approach gives readers some basic ideas of the proofs and refers to references for more technical treatments in general cases (see Puterman 1994).

Proposition 10.1 *For any vector value function $\mathbf{v} \in V$, there exists a decision rule, denoted by $\delta^* \in \Delta$, which is the optimal value operator for the one-period problem with $\mathbf{v}$ as the arrival value function. That is $T(\delta^*)\mathbf{v} = M\mathbf{v}$, which means that one-step optimum is attained by δ^* for a given $\mathbf{v}$ and the optimal value function exists.*

This proposition can be proved by using the assumption that all decision rules are admissible. However, δ^*, called the optimal decision rule, at a time point, for the finite MDP may not be unique. It can be shown that there exists an optimal policy (also called a strategy) which can be determined via optimal vector value function recursively for a finite horizon MDP.

Theorem 10.1 *For an N-period DTMDP with finite state space and action space, the followings are true.*

> *1. The optimal vector value functions, $\mathbf{V}_1^*, \mathbf{V}_2^*, ..., \mathbf{V}_N^*$, exist.*
> *2. There exists an optimal decision rule π_t^* for every t.*
> *3. The optimality equation $\mathbf{V}_t^* = M\mathbf{V}_{t+1}^*$ holds for every t.*
> *4. A strategy $\pi = (\pi_1, \pi_2, ..., \pi_N)$ is optimal for every period t if and only if $\pi_t(s)$ attains the optimum in the optimality equations for every period t and state s.*

Proof. First, (1)–(3) are proved by backward induction. Since the initial step is a specialization of these statements, we can simply show the inductive step. Assume that (1)–(3) are true for $i \geq t + 1$, then we demonstrate (1)–(3) also hold for $i \geq t$. We introduce the notation $\mathbf{V}_t(\pi)$ as a value function vector under strategy π at time t. We need to show now a maximum value function vector exists at time t. That is we can (1) find $\mathbf{V} \geq \mathbf{V}_t(\pi)$ for every π, and (2) there exists π^* such that $\mathbf{V} = \mathbf{V}_t(\pi^*)$, which implies that optimal value function $\mathbf{V}_t^*$ exists, equals $\mathbf{V}$, and attained by the optimal strategy. Thus, π^* is optimal for t and $M\mathbf{V}_{t+1}^*$ will play the role of $\mathbf{V}$, so that (1)–(3) hold for $i \geq t$.

Take an arbitrary π. Then by definition $\mathbf{V}_{t+1}^* \geq \mathbf{V}_{t+1}(\pi)$. Thus, there exists a $\mathbf{V} = M\mathbf{V}_{t+1}^*$, such that

$$MV_{t+1}^* \geq MV_{t+1}(\pi)$$
$$\geq T(\pi_t)V_{t+1}(\pi)$$
$$= V_t(\pi),$$

where the first inequality follows from the fact that M is isotone; the second inequality follows from the definition of M; and the last equality holds due to the definition of the operator T.

It follows from the inductive assumption that there exists a strategy π^* that is optimal for period $t + 1$. Note that such a strategy is independent of the decision rules for periods 1 through t (the past periods). So to show that this claim also holds for period t, we only need to show the decision rule π_t^* such that (a) $T(\pi_t^*)V_{t+1}^* = MV_{t+1}^*$; and (b) $MV_{t+1}^* = V_t(\pi^*)$. While (a) follows from Proposition 10.1, (b) is true due to the followings:

$$MV_{t+1}^* = T(\pi_t^*)V_{t+1}^*$$
$$= T(\pi_t^*)V_{t+1}(\pi^*)$$
$$= V_t(\pi^*),$$

where the first equality follows from (a), the second equality is due to the inductive assumption (i.e., $V_{t+1}(\pi^*) = V^*_{t+1}$), and the third equality holds by the definition of the operator. This completes the proof of (1)–(3).

Next, we prove (4). First, prove $\Leftarrow$. Suppose that π^* attains the optimum in the OE for each period, that is

$$T(\pi^*_t)V^*_{t+1} = MV^*_{t+1}$$

for every t. Suppose inductively that π^* is optimal for $t+1$. That is $V_{t+1}(\pi^*) = V^*_{t+1}$. Select an arbitrary strategy π. Then

$$\begin{aligned}
V_t(\pi^*) &= T(\pi^*_t)V_{t+1}(\pi^*) \\
&= T(\pi^*_t)V^*_{t+1} \\
&= MV^*_{t+1} \\
&\geq MV_{t+1}(\pi) \\
&\geq T(\pi_t)V_{t+1}(\pi) \\
&= V_t(\pi).
\end{aligned}$$

Thus, π^* is optimal for t as well.

Now, prove $\Rightarrow$. Suppose that π^* is optimal for every t. That is $V_t(\pi^*) = V^*_t$ for every t. Then

$$\begin{aligned}
T(\pi^*_t)V_{t+1}(\pi^*) &= V_t(\pi^*) \\
&= V^*_t \\
&= MV^*_{t+1}.
\end{aligned}$$

That is, π^* attains the optimum in the OE for each period t. ∎

The proof of the above theorem is relatively easy due to the finite state space and finite action space. However, the optimality equations and the operator's isotone property also hold in more general dynamic optimization problems in which the value functions can be expressed by recursions. Thus, the results in Theorem 10.1 still hold for the cases where the state space and action space are infinite and the transition probabilities are not stationary. Of course, the proof for such a general setting will be more complex (see, Puterman 1994). We summarize more general results as follows:

Theorem 10.2 *For a finite time horizon DTMDP, the optimal value functions and an optimal strategy exist, and optimality equations hold under such an optimal policy if the following conditions are satisfied:*

- *Every admissible decision (also called an action) is represented by at least one admissible decision rule at a time period.*

- *Any collection of admissible decision rules over N periods (a decision rule for each period) form an admissible strategy. That is, if $\pi_t \in \Delta$ for each t, then $\pi = (\pi_1, \pi_2, ..., \pi_N)$ is admissible.*
- *There exist terminal value functions $V_{N+1}^{\pi} = V_E \in V$.*
- *There exists an operator $T_t(\delta)$ for every t and $\delta \in \Delta$ such that under a strategy π the recursion $V_t^{\pi} = T_t(\pi_t)V_{t+1}^{\pi}$ holds for every t.*
- *$T_t(\delta) : V \to V$ for every t and $\delta \in \Delta$.*
- *$T_t(\delta)$ is isotone on V for every t and $\delta \in \Delta$.*
- *The optimal operator M_t, defined for every t by*

$$M_t V := \sup_{\delta \in \Delta} T_t(\delta)V,$$

  for $V \in V$ satisfies $M_t : V \to V$.
- *For every t and $V \in V$, there exists $\delta \in \Delta$ such that $T_t(\delta)V = M_t V$.*

Exercise 10.3 *Prove Proposition 10.1.*

Exercise 10.4 *Consider an MDP with continuous state and action spaces. That is $\mathscr{S} = [0,1]$ and $A(s) = [0,1]$ for every $s \in \mathscr{S}$. Assume that the MDP starts period t in state $s_t = s$ and decision $\delta_t(s) = a \in A(s)$ is made. Let $R(s,a)$ be the immediate expected return and $\Psi(x|s,a) = P(s_{t+1} \leq x|s_t = s, \delta_t(s) = a)$ be the conditional probability distribution function of the state of the MDP at the beginning of the next period.*
  (a) Define Δ, V and $T_t(\delta)$ for each t and $\delta \in \Delta$.
  (b) What conditions are needed for $R(\cdot, \cdot)$ so that conditions 5,7, and 8 in Theorem 10.2 hold?

For a DTMDP, it is ideal that we can characterize the structure of the optimal policies (strategies). Whether or not we can accomplish this goal depends on the properties of optimal value functions. We briefly describe the general approach for the finite time horizon MDPs. The key to characterize the optimal policy is to identify a nonempty optimal value function set V^*. These optimal value functions have certain properties and can be called structured value functions. For example, some structured value functions are convex and/or increasing. Denote by Δ_t^* the set of structured decision rules for period t. Structured decision rules are much simpler than general admissible decision rules. For example, for an inventory system, a structured decision rule for a period t can be a based-stock policy. Denote by Π^* the set of structured strategies, which is the Cartesian product of sets of structured decision rules. The following theorem provides the conditions for the existence of the structured value functions and structured optimal strategies.

Theorem 10.3 *For a finite horizon $(N + 1$ periods) DTMDP with the optimality equations satisfied, if*

1. $M_t : V^* \to V^*$, called preservation under optimal value operator,
2. if $\mathbf{V} \in V^*$, then there exists $\delta^* \in \Delta_t^*$ such that $T_t(\delta^*)\mathbf{V} = M_t\mathbf{V}$,
called attainment of one-step optimum on a structured value function
vector $\mathbf{V}$,
   3. $\mathbf{V}_T \in V^*$,

then

- *the optimal value function $\mathbf{V}_t^*$ is structured (i.e., $\mathbf{V}_t^* \in V^*$) for every t,*
  and
- *there exists a structured strategy $\pi^* \in \Pi^*$ that is optimal for every t.*

Proof. Note that $\mathbf{V}_{N+1}^* = \mathbf{V}_T \in V^*$ which by Condition 3 is in V^*. Suppose $\mathbf{V}_{t+1}^* \in V^*$. It follows from Condition 2 that a structured decision rule for period t is selected that attains the optimum in the OE. Thus, we have a decision rule $\pi_t \in \Delta_t^*$ such that $T_t(\pi_t)\mathbf{V}_{t+1}^* = M_t\mathbf{V}_{t+1}^*$. This means that the optimal and structured decision at time t is constructed. By the OE, we have $\mathbf{V}_t^* = M_t\mathbf{V}_{t+1}^*$. It follows from Condition 1 that $\mathbf{V}_t^*$ is structured ($\mathbf{V}_t^* \in V^*$). Then, the optimal and structured decision at $t - 1$ can be constructed. Then we can recursively construct a structured policy that, by the optimality criterion, is optimal for every period and verify that the optimal value functions are structured in the process. ■

To show the structured value function leading to the structured optimal policy, we present an stochastic dynamic inventory control system, which is mainly taken from Porteus (2002). Consider an inventory system that is operated for N periods. At the beginning of each period, the manager decides how much to order depending on the current stock level. The unit purchase cost is c. The demand for each period is a continuous, strictly positive, generally i.i.d. random variable, denoted by D, with distribution function $F_D(x)$, density function $f_D(x)$, and mean μ. At the end of each period, any leftover stock can be carried over to the next period for sales with a holding cost of $c_h > 0$ per period; and any unmet demand is backlogged rather than lost. A unit penalty cost of $c_b > 0$ is incurred for each unit backlogged per period. The one-period discount factor is γ. Clearly, the state variable can be defined as the stock level x before ordering at the beginning of each period. The state space then is the real line due to the continuous demand and the backlogging. Positive x indicates the leftover inventory and negative x indicates the number of units backlogged. The backlog must be met first before any future demands can be satisfied. The decision epoch is at the beginning of each period and the decision space is the positive real number (order quantity). Note that while the state and action spaces are continuous, the state space is one dimensional. This feature combined with the simple state transition mechanism makes the vector or matrix form expressions (OE or recursive relations) unnecessary. If the initial stock level at the end of period N is x, then the terminal cost $V_T(x)$ is incurred. It is assumed that V_T is convex. A simple case is

$$V_T(x) = \begin{cases} -cx & \text{if } x \leq 0, \\ 0 & \text{otherwise.} \end{cases}$$

Such a terminal value is a backlog cost if there is any unsold stock or zero otherwise. To avoid triviality, we assume $c_b > (1 - \alpha)c$. The decision variable at the beginning of period t is chosen to be the inventory level after ordering, denoted by y_t. If the order quantity is Q_t and the state is x_t, then we have $y_t = x_t + Q_t$ (or $Q_t = y_t - x_t$). Denote by D_t the demand for period t. Then the state in the next period is $x_{t+1} = y_t - D_t$. Since the main idea of stochastic dynamic optimization (i.e., MDP) is to decompose the multi-period problem into multiple single-period problems that are recursively related and focus on a single-period problem, the specific period t is immaterial, so index t for the state and action is suppressed. For a period starting with state $s_t = x$, by taking action $\pi(s_t) = y$, the expected one-period cost can be written as

$$R(x,y) = c(y - x) + H(y)$$
$$H(y) := E\mathscr{H}(y - D) = c_h E[(y - D)^+] + c_b E[(D - y)^+] \tag{10.8}$$
$$= c_h \int_{v=0}^{y} (y - v) f_D(v) dv + c_b \int_{v=y}^{\infty} (v - y) f_D(v) dv,$$

where $\mathscr{H}(z) = c_h z^+ + c_b(-z)^+$. Denote by $V_t^*(x)$ the optimal value function. Then the optimality equation is given by

$$V_t^*(x) = \min_{y \geq x} \left\{ c(y - x) + H(y) + \gamma \int_0^{\infty} V_{t+1}^*(y - v) f_D(v) dv \right\},$$

for $1 \leq t \leq N$, where $V_{N+1}^*(x) := V_T(x)$, for each x. Letting

$$G_t(y) = cy + H(y) + \gamma \int_0^{\infty} V_{t+1}^*(y - v) f_D(v) dv, \tag{10.9}$$

we can write the optimality equation as

$$V_t^*(x) = \min_{y \geq x} \{ G_t(y) - cx \}, \tag{10.10}$$

which facilitates the analysis. Now we can show that the property of optimal value function results in the simple structured optimal inventory policy. It follows from (10.10) that the optimal decision at state x in period t (i.e., the inventory level at the beginning at period t) is determined by minimizing $G_t(y)$ over $\{y | y \geq x\}$.

This can be illustrated graphically by looking for the minimum to the right of x (see Figures 10.1–10.2). If G_t is as shown in Figure 10.1, then if we start

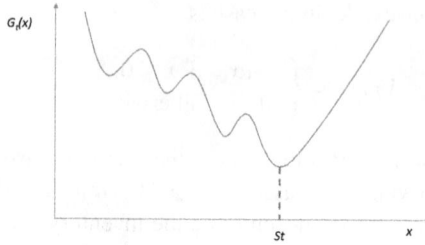

Figure 10.1　Base-stock policy is optimal.

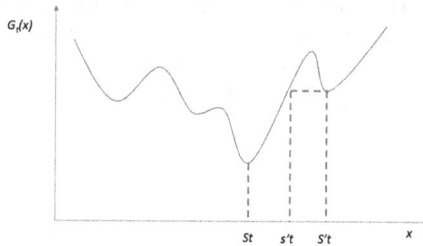

Figure 10.2　Base-stock policy is not optimal.

below (to the left) of S_t, then we should move up (to the right) to it. If we start above (to the right) of S_t, we should stay where we are. Such a decision rule is called a base-stock policy and S_t is called the base-stock level. Thus, the decision rule at state x can be written as

$$\delta(x) = \begin{cases} S_t & \text{if } x < S_t, \\ x & \text{otherwise}. \end{cases}$$

Clearly, such a simple "structured" decision rule depends on the property of $G_t(x)$ and consequently the property of the optimal value function $V_t^*(x)$. For example, if $G_t(x)$ is as shown in Figure 10.2, then the base-stock policy is not optimal. This is because the optimal decision rule depends on the state x in a more complex way. For example, if $x \leq s_t'$, then the base-stock policy is optimal. However, if $s_t' < x < S_t'$, the "ordering up to S_t'" is optimal, therefore, the base-stock policy is not optimal. Now, we can show that the convexity of the optimal value function (structured value function) can ensure that the optimal inventory policy is a base-stock policy.

Proposition 10.2　*If V_{t+1}^* is convex, then*

1. G_t is convex.

    2. A base-stock policy is optimal in period t. Any minimizer of G_t is an optimal base stock level.
    3. V_t^* is convex.

Before we prove this proposition, the following lemma is established.

Lemma 10.1 *Suppose that g is a convex function defined on $\mathbb{R}$. Let $G(y) := E_D[g(y - D)]$, where D is a random variable with a p.d.f. $f_D(\cdot)$. Then G is convex on $\mathbb{R}$.*

Proof. Consider $x, z \in \mathbb{R}$ such that $x < z$. For $0 < t < 1$, we have

$$G(tx + (1-t)z) = E_D[g(tx + (1-t)z - D)] = \int_0^\infty g(tx + (1-t)z - v)f_D(v)dv$$

$$= \int_0^\infty g(t(x-v) + (1-t)(z-v))f_D(v)dv$$

$$\leq \int_0^\infty [tg(x-v) + (1-t)g(z-v)]f_D(v)dv$$

$$= tG(x) + (1-t)G(z),$$

where the inequality follows from the convexity of g. This completes the proof. ∎

    Now we can prove Proposition 10.2.
Proof. (1) G_t can be written as

$$G_t(y) = cy + E[\mathcal{H}(y - D)] + \gamma E[V_{t+1}^*(y - D)]. \qquad (10.11)$$

It follows from $c_b > (1 - \gamma)c$ and $c_h + (1 - \gamma)c$ that $c_b + c_h > 0$ which leads to $\mathcal{H}$ being convex. It follows from the convexity of V_{t+1}^* (assumption), the convexity of $\mathcal{H}$, and Lemma 10.1 that the second and third term on the RHS in (10.11) are convex. Thus, G_t is the sum of three convex functions and hence is convex.

    (2) Denote by S_t the minimizer of $G_t(y)$ over all real y. If $x < S_t$, then the minimizing $y \geq x$ is at $y = S_t$. If $x \geq S_t$, then minimizing y is at $y = x$. That is the base-stock policy with base-stock level S_t is optimal for period t.

    (3) To show the convexity of V_t^*, for an interval $[x, z]$, we consider three cases by using (10.10). Case (i) $[x, z]$ is to the right of S_t as shown in Figure 10.3; Case (ii) $[x, z]$ is to the left of S_t as shown in Figure 10.4; and Case (iii) $[x, z]$ includes S_t as shown in Figures 10.5–10.6. These figures reveals that V_t^* is convex. This completed the proof of the Proposition. ∎

Exercise 10.5 *Prove Proposition 10.2 using a different approach if g is differentiable everywhere.*

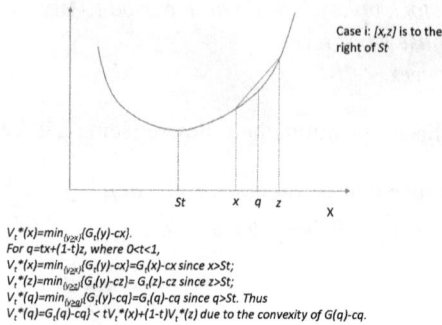

$V_t^*(x)=min_{\{y \geq x\}}\{G_t(y)-cx\}.$
For $q=tx+(1-t)z$, where $0<t<1$,
$V_t^*(x)=min_{\{y \geq x\}}\{G_t(y)-cx\}=G_t(x)-cx$ since $x>St$;
$V_t^*(z)=min_{\{y \geq z\}}\{G_t(y)-cz\}= G_t(z)-cz$ since $z>St$;
$V_t^*(q)=min_{\{y \geq q\}}\{G_t(y)-cq\}=G_t(q)-cq$ since $q>St$. Thus
$V_t^*(q)=G_t(q)-cq) < tV_t^*(x)+(1-t)V_t^*(z)$ due to the convexity of $G(q)-cq$.

Figure 10.3   Case i: $[x, z]$ is to the right of S_t.

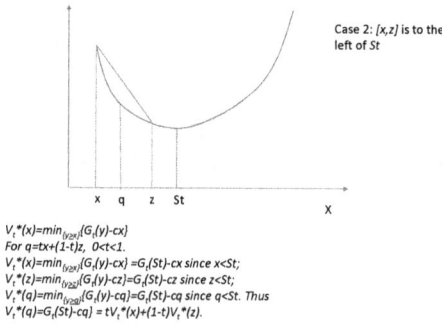

$V_t^*(x)=min_{\{y \geq x\}}\{G_t(y)-cx\}$
For $q=tx+(1-t)z$, $0<t<1$.
$V_t^*(x)=min_{\{y \geq x\}}\{G_t(y)-cx\} =G_t(St)-cx$ since $x<St$;
$V_t^*(z)=min_{\{y \geq z\}}\{G_t(y)-cz\}=G_t(St)-cz$ since $z<St$;
$V_t^*(q)=min_{\{y \geq q\}}\{G_t(y)-cq\}=G_t(St)-cq$ since $q<St$. Thus
$V_t^*(q)=G_t(St)-cq) = tV_t^*(x)+(1-t)V_t^*(z)$.

Figure 10.4   Case ii: $[x, z]$ is to the left of S_t.

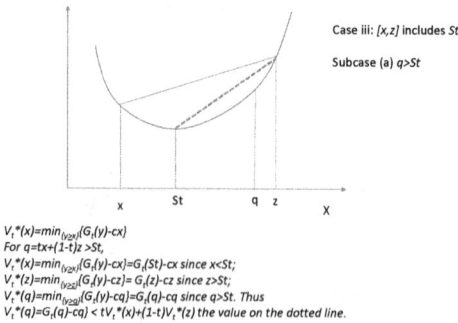

$V_t^*(x)=min_{\{y \geq x\}}\{G_t(y)-cx\}$
For $q=tx+(1-t)z >St$,
$V_t^*(x)=min_{\{y \geq x\}}\{G_t(y)-cx\}=G_t(St)-cx$ since $x<St$;
$V_t^*(z)=min_{\{y \geq z\}}\{G_t(y)-cz\}= G_t(z)-cz$ since $z>St$;
$V_t^*(q)=min_{\{y \geq q\}}\{G_t(y)-cq\}=G_t(q)-cq$ since $q>St$. Thus
$V_t^*(q)=G_t(q)-cq) < tV_t^*(x)+(1-t)V_t^*(z)$ the value on the dotted line.

Figure 10.5   Case iii: q is to the right of S_t.

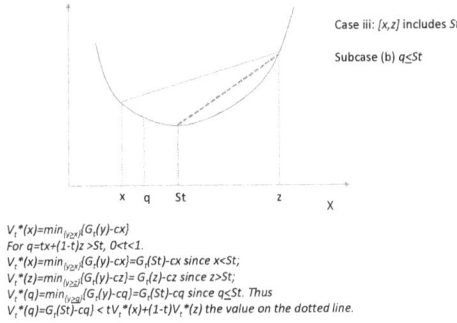

Case iii: $[x,z]$ includes St

Subcase (b) $q \leq St$

$V_t^*(x) = \min_{(y \geq x)} \{G_t(y) - cx\}$
For $q = tx + (1-t)z > St$, $0 < t < 1$.
$V_t^*(x) = \min_{(y \geq x)} \{G_t(y) - cx\} = G_t(St) - cx$ since $x < St$;
$V_t^*(z) = \min_{(y \geq z)} \{G_t(y) - cz\} = G_t(z) - cz$ since $z > St$;
$V_t^*(q) = \min_{(y \geq q)} \{G_t(y) - cq\} = G_t(St) - cq$ since $q \leq St$. Thus
$V_t^*(q) = G_t(St) - cq) < tV_t^*(x) + (1-t)V_t^*(z)$ the value on the dotted line.

Figure 10.6   Case iii: q is to the left of S_t.

With this proposition, the convexity of the optimal value function in the terminal period leads to the convexity of the optimal value functions for all periods backward via the convexity of G_t, recursively. The convexity of G_t then results in the base-stock policy being optimal for each period.

Theorem 10.4 *A base-stock policy is optimal in each period of a finite-horizon inventory control problem.*

Proof. It follows from the assumption that the terminal value function (treated as the optimal value function $V_{N+1}^* = V_T$) is convex. By Proposition 10.2, G_N is convex and a base-stock policy is optimal for period N. By (3) of Proposition 10.2, V_N^* is convex as well. Therefore, this argument iterates backward through the periods in sequence $t = N, N-1, \dots 1$. This completes the proof. ∎
Furthermore, assume that the terminal value function is given by

$$V_T(x) = -cx,$$

which means that we can get the reimbursement c for any leftover unit and must pay c for each unit backlogged (in addition to any unit shortage cost incurred in the previous period). Then, the base-stock level for the last period can be obtained by solving a one-period problem like a newsvendor problem. Suppose that the final period starts with inventory x_0 and the order-up-to y. Then, the expected total cost is given by

$$c(y - x_0) + H(y) - \gamma \int_{v=0}^{\infty} c(y-v)f_D(v)dv = c(1-\gamma)y + H(y) + (\gamma\mu - x_0)c$$
$$= g(y) + (\gamma\mu - x_0)c,$$

where

$$g(y) := c(1-\gamma)y + H(y).$$

Since the base-stock policy is optimal for this one-period problem, the base-stock level S should satisfy $g'(S) = 0$. To solve for the optimal S, we write

$$g(y) = c(1-\gamma)y + \int_0^y c_h(y-v)f_D(v)dv + \int_y^\infty c_b(v-y)f_D(v)dv$$

$$= c(1-\gamma)y + \int_0^y c_h(y-v)f_D(v)dv - \int_0^y c_b(v-y)f_D(v)dv$$

$$+ \int_0^\infty c_b(v-y)f_D(v)dv$$

$$= c(1-\gamma)y + \int_0^y c_h(y-v)f_D(v)dv - \int_0^y c_b(v-y)f_D(v)dv + c_b\mu - c_by.$$

$$\tag{10.12}$$

Taking the first-order derivative of (10.12) and setting it to zero yields

$$\frac{dg(y)}{dy} = c(1-\gamma) + c_h \int_0^y f_D(v)dv + c_b \int_0^y f_D(v)dv - c_b = 0$$

$$\Rightarrow c(1-\gamma) + c_h F_D(y) + c_b F_D(y) - c_b = 0$$

$$\Rightarrow F_D(y) = \frac{c_b - c(1-\gamma)}{c_h + c_b}.$$

Thus, the optimal base-stock level is given by

$$S = F_D^{-1}\left(\frac{c_b - c(1-\gamma)}{c_h + c_b}\right). \tag{10.13}$$

The key to be able to obtain the explicit result for the optimal base-stock level S is due to the fact that the terminal value function has a slope of $-c$. If we can show that all the optimal value functions have a slope of $-c$, then the optimal base-stock level for all periods can be determined by (10.13). The proof of this result is left as an exercise. Note that this result further demonstrates that the more the structural policy is characterized, the more structural properties the optimal value function must possess.

Exercise 10.6 *Consider a finite horizon inventory control system described above. If V_{t+1}^* is convex and $dV_{t+1}^*(x)/dx = -c$ for $x < S$, where S is defined by (10.13), show the following:*

(a) S minimizes $G_t(y)$ over all real y.

(b) The optimal base-stock level in period t is also S.

(c) V_t^* is convex and $dV^*(x)/dx = -c$ for $x < S$.

Remark 10.1 *The base-stock policy is considered to be one of the simplest inventory control policies and the "one-period optimal policy" being used for all periods is called "myopic policy". We have demonstrated that some conditions* *on the optimal value functions are required for myopic policy to be optimal in the base-stock policy example. More general conditions for optimality of my-opic policy are discussed in Chapter 6 of Porteus (2002). Furthermore, if we consider a more complex cost structure that includes the setup cost K of mak-ing an order (a fixed cost), then a particular property called K-convexity is required for the optimal value functions to make (s, S) policy optimal. The de-tails for this development are provided in Chapter 7 of Porteus (2002). Along the same line of characterizing the optimal policy in a stochastic system, we may also be interested in the question of when optimal policies are monotone functions of the state. The answer to this question depends on a class of prop-erties for functions, called submodularity and/or supermodularity, rather than convexity and/or concavity (see Topkis (1998) for a comprehensive presen-tation). For the details about the monotonicity of optimal inventory policies, interested readers are referred to Chapter 8 of Porteus (2002).*

Exercise 10.7 *(Choosing a parking space) Suppose that George is driving down a one-way road to the destination building with angle parking spots along the road. As he drives along, he can observe only one parking space at a time, the one right next to him, which is either available or occupied (with a parked car or reserved). If the one observed is available, he may either (1) stop and park there or (2) drive on to the next space. If it is occupied, then he must drive on to the next space. He cannot turn around to the available spaces passed. Suppose that the available spaces occur independently and with prob-ability p for each space (where $0 < p < 1$). If he has not parked by the time he reaches the building, he must park the car in the underground pay parkade at a cost $c > 1$. If he parks x spaces away from the destination building, then he considers that a parking cost of x has been incurred. The closest space (before reaching the pay parking parkade) is $x = 1$. Since there are possible parking spaces that are better than driving straight to the pay parkade, George wants to determine an optimal parking strategy (when to stop and park) to minimize his total expected parking cost. Find the optimal parking strategy for George.*

For an infinite horizon MDP or an episodic MDP, we utilize the long-run return (expected present value of all returns received over an infinite horizon) as decision criterion and perform the stationary analysis. This approach is sim-ilar to that for DTMCs or DTMPs except now a strategy is implemented that controls the dynamics of the process. In addition, the value function plays the same role as the marginal probability function for a state in DTMCs or DTMPs. Hence, the value function and optimal value function become time-independent when the steady state is reached. Stationarity is guaranteed for a DTMDP with finite state space and action space under a stationary policy (for the infinite horizon MDP, a strategy is often called a policy). The infi-nite horizon problem can be considered as a limiting case of the finite horizon problem as the ending period n goes to infinity. To justify the existence of the stationary value function and policy, we need some basics on metric space. Here, we provide a brief review (for more details see Chapter G).

Let V be the set of value functions of a DTMDP endowed with the so-called L^∞ metric ρ which is defined as

$$\rho(u,v) := \sup_{s \in \mathscr{S}} |u(s) - v(s)| = max(b - a),$$

where $a := \inf_s[u(s) - v(s)]$ and $b := \sup_s[u(s) - v(s)]$. Such a metric represents the maximum absolute distance between the two functions/vectors u and v. Thus, a metric is a relative measure of how far the two vectors are. The pair (ρ, V) is called a metric space. An operator T is an α-contraction if $\alpha > 0$ and for all $u, v \in V$, we have

$$\rho(Tu, Tv) \le \alpha\rho(u,v).$$

For the case of $\alpha \in [0,1)$, T is a contraction. If we apply the contraction operator to two different vectors, the resulting distance is smaller. We can show that the isotone operator $T(\delta)\mathbf{V} := \mathbf{R}(\delta) + \gamma\mathbf{P}(\delta)\mathbf{V}$ (or the optimal operator M) is an α-contraction for each decision rule $\delta \in \Delta$. Consider a and b are the smallest and largest differences (not in absolute value) between two value functions $\mathbf{U}, \mathbf{V} \in V$ in an MDP.

Proposition 10.3 (a) If $a\mathbf{e} \le \mathbf{U} - \mathbf{V} \le b\mathbf{e}$, then for each $\delta \in \Delta$,

$$\gamma a\mathbf{e} \le T(\delta)\mathbf{U} - T(\delta)\mathbf{V} \le \gamma b\mathbf{e}.$$

(b) $T(\delta)$ is a γ-contraction for each $\delta \in \Delta$.
(c) The optimal value operator M is a γ-contraction.

Proof. (a) It follows from $\mathbf{U} - \mathbf{V} \le b\mathbf{e}$ that

$$T(\delta)\mathbf{U} - T(\delta)\mathbf{V} = \gamma\mathbf{P}(\delta)(\mathbf{U} - \mathbf{V})$$
$$\le \gamma\mathbf{P}(\delta)b\mathbf{e} = \gamma b\mathbf{e}.$$

Similarly, it can be shown that $T(\delta)\mathbf{U} - T(\delta)\mathbf{V} \ge \gamma a\mathbf{e}$.

(b) Let $a := \inf_s[\mathbf{U}(s) - \mathbf{V}(s)]$ and $b := \sup_s[\mathbf{U}(s) - \mathbf{V}(s)]$. Then, $a\mathbf{e} \le \mathbf{U} - \mathbf{V} \le b\mathbf{e}$, and $\rho(\mathbf{U}, \mathbf{V}) = max(b, -a)$. By (a), we have

$$\rho(T(\delta)\mathbf{U}, T(\delta)\mathbf{V}) \le max(\gamma b, -\gamma a) = \gamma max(b, -a) = \gamma\rho(\mathbf{U}, \mathbf{V}).$$

(c) Suppose that for each $\mathbf{V} \in V$, there exists $\delta \in \Delta$ such that $T(\delta) = M\mathbf{V}$ (i.e., a decision rule attains the optimization for each $\mathbf{V}$). Then, for a given $\mathbf{U}$, select a decision rule δ such that $M\mathbf{U} = T(\delta)\mathbf{U}$. Hence, we have

$$M\mathbf{U} = T(\delta)\mathbf{U}$$
$$\leq T(\delta)\mathbf{V} + \gamma b\mathbf{e}$$
$$\leq M\mathbf{V} + \gamma b\mathbf{e}.$$

Similarly, it can be shown that $M\mathbf{U} \geq M\mathbf{V} + \gamma a\mathbf{e}$. Thus, $\gamma a\mathbf{e} \leq M\mathbf{U} - M\mathbf{V} \leq \gamma b\mathbf{e}$, which implies M is a γ contraction. ∎

The existence of stationary value function for a DTMDP is based on the Banach fixed-point theorem, also called the contraction mapping theorem. This theorem can be found in most textbooks on functional analysis (e.g., Kolmogorove and Fomin (1957)). The value function space is a complete metric space in which all Cauchy sequences converge (see Chapter G). A Cauchy sequence is a sequence of elements in V such that $\rho(v_n, v_m) \to 0$ as $n, m \to \infty$. Since our value function space for the finite MDP is $\mathbf{R}^{|\mathscr{J}|}$, it is always a complete metric space. For the infinite-dimensional normed vector space situation such as the MDP with infinite state space, the complete metric space, not always guaranteed, is required and called a Banach space. For a given Cauchy sequence of elements in V (complete metric space), there exists a $v \in V$ such that

$$\lim_{n \to \infty} \rho(v_n, v) = 0.$$

Using the contraction property of T, recursion of the value function, triangular inequality for metric, and discount factor $\alpha < 1$, we can prove the following Banach fixed-point Theorem.

Theorem 10.5 *Assume that T is a γ-contraction with $\gamma \in [0,1)$ defined on a complete metric space (ρ, V). For $\mathbf{V} \in V$, define $T^n\mathbf{V} := T(T^{n-1}\mathbf{V})$ for $n \geq 1$ with $T^0\mathbf{V} := \mathbf{V}$. The followings hold:*

1. For each $\mathbf{V} \in V$, there exists a $\mathbf{V}^* \in V$ such that

$$\mathbf{V}^* = \lim_{n \to \infty} T^n\mathbf{V}.$$

2. Each such limit $\mathbf{V}^*$ is a fixed point of T, i.e., $\mathbf{V}^* = T\mathbf{V}^*$.
3. The operator T has a unique fixed point in V. $\mathbf{V}^* = \lim_{n \to \infty} T^n\mathbf{V}$ is independent of $\mathbf{V}$.
4. For the unique fixed point of T, $\mathbf{V}^*$,

$$\rho(\mathbf{V}^*, T^n\mathbf{V}) \leq \gamma^n \rho(\mathbf{V}^*, \mathbf{V}),$$

for every $n = 1, 2, \ldots$

Proof. (1) Select a $\mathbf{V} \in V$ and define a sequence as $\mathbf{V}_0 := \mathbf{V}$ and $\mathbf{V}_n = T\mathbf{V}_{n-1}$ for $n \geq 1$. In this sequence, for $n \leq m$, we have

$$
\begin{aligned}
\rho(\mathbf{V}_n, \mathbf{V}_m) &= \rho(T\mathbf{V}_{n-1}, T\mathbf{V}_{m-1}) \\
&\leq \gamma\rho(\mathbf{V}_{n-1}, \mathbf{V}_{m-1}) \text{ due to } \gamma\text{- contraction of } T \\
&\leq \cdots \leq \gamma^n \rho(\mathbf{V}, \mathbf{V}_{m-n}) \text{ due to recursion} \\
&\leq \gamma^n [\rho(\mathbf{V}, \mathbf{V}_1) + \rho(\mathbf{V}_1, \mathbf{V}_2) + \cdots + \rho(\mathbf{V}_{m-n-1}, \mathbf{V}_{m-n})] \\
&\quad \text{due to triangle inequality} \\
&\leq \gamma^n \frac{\rho(\mathbf{V}, \mathbf{V}_1)}{1 - \gamma} \text{ due to geometric series} \\
&\to 0 \text{ as } m, n \to \infty,
\end{aligned}
$$

Thus, $\{\mathbf{V}_n\}$ is a Cauchy sequence. Since (ρ, V) is complete, there exists a limit $\mathbf{V}^* \in V$ such that $\mathbf{V}^* = \lim_{n \to \infty} T^n \mathbf{V}$.

(2) To show that $\mathbf{V}^*$ is a fixed point of T, given an ε, we select N such that for $n \geq N$ the relation $\rho(\mathbf{V}_{n-1}, \mathbf{V}^*) \leq \varepsilon/2$ holds. Thus, when $n \geq N$, we have

$$
\begin{aligned}
\rho(\mathbf{V}^*, T\mathbf{V}^*) &\leq \rho(\mathbf{V}^*, \mathbf{V}_n) + \rho(\mathbf{V}_n, T\mathbf{V}^*) \\
&\leq \frac{\varepsilon}{2} + \rho(T\mathbf{V}_{n-1}, T\mathbf{V}^*) \\
&\leq \frac{\varepsilon}{2} + \gamma\rho(\mathbf{V}_{n-1}, \mathbf{V}^*) \\
&\leq \frac{\varepsilon}{2} + \gamma\left(\frac{\varepsilon}{2}\right) \leq \varepsilon.
\end{aligned}
$$

Since ε is arbitrary, we must have $\rho(\mathbf{V}^*, T\mathbf{V}^*) = 0$ which implies $\mathbf{V}^* = T\mathbf{V}^*$.

(3) Assume that $\mathbf{V}'$ is another fixed point of T. Then we have

$$
\rho(\mathbf{V}^*, \mathbf{V}') = \rho(T\mathbf{V}^*, T\mathbf{V}') \leq \gamma\rho(\mathbf{V}^*, \mathbf{V}'),
$$

which, since $0 \leq \gamma < 1$, implies that $\rho(\mathbf{V}^*, \mathbf{V}') = 0$, hence, $\mathbf{V}^*$ is the unique fixed point of T.

(4) It follows from $\mathbf{V}^* = T^n \mathbf{V}^*$ and n contraction operations that $\rho(\mathbf{V}^*, T^n \mathbf{V}) \leq \gamma^n \rho(\mathbf{V}^*, \mathbf{V})$. ∎

Every item of this theorem has a direct implication in developing the stationary analysis on a DTMDP with infinite horizon. The first one implies that as the time horizon increases, the value function approaches to the limit which is the stationary value function. The second one indicates that the limiting or stationary value function can be solved by an equation system. The third one shows that the limiting value function is independent of the starting state. Finally, the fourth one characterizes the rate of convergence of the value function from a starting state to the limit. For an infinite horizon DTMDP, we should consider the stationary policy which utilizes the same decision rule in each time period. Using the generic decision rule notation, a stationary policy can be denoted by $\delta^\infty := (\delta, \delta, ...)$, meaning that decision rule δ is applied for each period. It can be proved that the value function for an infinite horizon DTMDP approaches to the stationary value function which describes the dynamic and economic behavior of the process in steady-state. This result is summarized below (see Denardo (1967)). The proofs of the propositions and theorems presented above provide readers with some flavor of the techniques used in developing the dynamic optimization theory. The proofs of the following theorems will be left as exercises for readers.

Proposition 10.4 *For every decision rule $\delta \in \Delta$, there exists a vector $\mathbf{V}_\delta \in V$ that makes the followings hold:*

1. $\mathbf{V}_\delta = \lim_{n \to \infty} T^n(\delta)\mathbf{v}$ for any $\mathbf{v} \in V$.
2. $\mathbf{V}_\delta$ is a unique fixed point of $T(\delta)$.
3. If $\pi = \delta^\infty$, then $\mathbf{V}_t(\pi) = \mathbf{V}_\delta$ for all t.

Exercise 10.8 *Prove Proposition 10.4.*

When the DTMDP reaches a steady state under a stationary policy, the transition probabilities and the one-step returns are all stationary. Thus, we can remove the time index in the recursion for the value functions. The time-independent feature makes the analysis for the infinite horizon DTMDP simpler than that for the finite horizon DTMDP. Under a given stationary policy, the vector of value functions for the infinite horizon DTMDP with finite state and action space reduces to solving a finite linear equations systems. Equations (10.2) can be written as

$$\mathbf{V}^\delta = \mathbf{R}(\delta) + \gamma \mathbf{P}(\delta)\mathbf{V}^\delta, \tag{10.14}$$

which can be solved as

$$\mathbf{V}^\delta = (\mathbf{I} - \gamma \mathbf{P}(\delta))^{-1}\mathbf{R}(\delta),$$

as long as the matrix $\mathbf{I} - \gamma \mathbf{P}(\delta)$ is invertible. This matrix invertibility is ensured if $0 < \gamma < 1$. Similarly, the stationary analysis above also applies to the optimal value functions and optimal policies for the infinite horizon DTMDP. The following two theorems are due to Blackwell (1965).

Theorem 10.6 *For an infinite horizon DTMDP with finite state space and action space, we have*
(a) The optimal value functions are stationary and satisfy the optimality equations as

$$V^*(s) = \max_{a \in A(s)} \left\{ R(s,a) + \gamma \sum_{j \in \mathscr{S}} P(j|s,a)V^*(j) \right\}. \tag{10.15}$$

which, in matrix/vector and operator notation, can be written as

$$\mathbf{V}^* = M\mathbf{V}^*.$$

(b) The optimal value function is the unique fixed point of the optimal value operator M and satisfies

$$\mathbf{V}^* = \lim_{n \to \infty} M^n \mathbf{V}$$

for every $\mathbf{V} \in V$.

Exercise 10.9 *Prove Theorem 10.6.*

Note that the optimal value function is independent of the terminal value function for the corresponding finite horizon problem. The next theorem shows that the optimal stationary policy can be determined by solving the one-period optimization problem with the optimal value function as the arrival value function (i.e., greedy or myopic optimal policy is the optimal stationary policy for the infinite horizon DTMDP).

Theorem 10.7 *For an infinite horizon DTMDP,*
  (a) If a decision rule δ *satisfies*

$$T(\delta)\mathbf{V}^* = M\mathbf{V}^*,$$

then the stationary policy δ^∞ *is optimal.*
  (b) If the state space and action space are finite, a stationary optimal policy always exists.

Exercise 10.10 *Prove Theorem 10.7.*

The optimality of structured policies can be addressed similarly as for the finite horizon case.

Theorem 10.8 *If (i) there is a nonempty set of of structured value functions* $V^* \subset V$ *and* (ρ, V^*) *is complete; (ii) if* $\mathbf{v} \in V^*$, *then there exists a structured decision rule* $\delta^* \in \Delta^*$ *such that* $T(\delta^*)\mathbf{v} = M\mathbf{v}$; *and (iii)* $M : V^* \to V^*$, *then*

  • *The optimal value function* $\mathbf{V}^*$ *is structured, and*
  • *There exists an optimal structured stationary policy.*

Exercise 10.11 *Prove Theorem 10.8.*

10.2   COMPUTATIONAL APPROACH TO DTMDP

Although it is desirable to characterize the structure of the optimal policy in dynamic optimization problems, such a structure may not exist. However, we can always determine the optimal policy numerically via computational algorithms. In this section, we utilize a well-known example, called Grid World (written as Gridworld), to demonstrate the computation for the optimal policy in an infinite (or indefinite) horizon. Assume that an agent (or a robot) lives in a grid world as shown in Figure 10.7, where walls block the agent's path. The agent can go from one cell to another. Every cell has a reward (either positive or negative) and it can be considered as the state if the agent occupies that cell. All cells are transient states except for two cells. These two cells are terminal states with one having a large positive reward, called goal state, and the other having a large negative reward, called bad state. Each transient cell has a "living" reward (either positive or negative). The goal of the agent is to maximize the sum of expected rewards during the process moving from a starting transient state to a terminal state. The agent can decide which direction to move. However, the movement is not deterministic. It is subject to some uncertainty (also called noise). For example, if the agent decides to move north, then there is 80% chance that the movement direction is correct and there is 10% chance that the movement is to the west and another 10% chance to the east. The same deviation applies for other decisions. It is worth noting that we can utilize the stationary analysis for this problem. This is because we can repeat this process over and over again. Another thinking is that we can consider the goal state and the bad state as a non-terminal state, then the process is never ending so that a finite state MDP with infinite time horizon fits the situation. Therefore, we can drop the time index in the recursion and focus on the stationary analysis. Based on the fixed point theorem, we can develop either value-iteration (VI) algorithm or policy-iteration (PI) algorithm to determine the optimal policy in the DTMDP with infinite horizon.

10.2.1   VALUE ITERATION METHOD

This approach is based on the optimality equation (10.15). To illustrate this method by using the Gridworld problem, we assume that the following data are available. The one-step (also called immediate) reward for each state $s = (n, m)$ is given in Figure (10.7). Since the one-step reward function is independent of the action a (i.e., the reward does not depend on the moving direction), the optimality equation (10.15) can be written as

$$V^*(s) = R(s) + \gamma \max_{a \in A(s)} \sum_{j \in \mathscr{S}} P(j|s,a)V^*(j).$$

The VI algorithm proceeds as follows:

$R(s,a)=R((n,m))$ one-step reward function

n \ m	1	2	3	4
3	0	0	0	1
2	0		0	-100
1	0	0	0	0

Figure 10.7 One-step reward function Gridworld problem.

1. Initialize an estimate for the stationary value function, denoted by $\hat{V}_0(s)$, arbitrarily (e.g., zero)

$$\hat{V}_0(s) \leftarrow 0, \forall s \in \mathscr{S},$$

and set $k = 0$.

2. Repeat updating for $k \geq 0$:

$$\hat{V}_{k+1}(s) \leftarrow R(s) + \gamma \max_{a \in A(s)} \sum_{j \in \mathscr{S}} P(j|s,a)\hat{V}_k(j), \forall s \in \mathscr{S} \qquad (10.16)$$

until the maximum difference between the value function at the current step, denoted by $k^* + 1$, and that at the previous step, denoted by k^*, is smaller than some small positive constant.

3. Then we have $V^*(s) = \hat{V}_{k^*+1}, \forall s \in \mathscr{S}$. Stop.

The convergence of VI method is guaranteed by the following theorem.

Theorem 10.9 *Value iteration converges to the optimal value function:* $\hat{V} \rightarrow V^*$.

Proof. It follows from the optimal value operator M that

$$M\hat{V}(s) = R(s) + \gamma \max_{a \in A(s)} \sum_{j \in \mathscr{S}} P(j|s,a)\hat{V}(j)$$

First, we show that M is a contraction. For any value function estimates $\hat{V}_1, \hat{V}_2$, we have

$$|M\hat{V}_1(s) - M\hat{V}_2(s)|$$

$$= \gamma \left| \max_{a \in A(s)} \sum_{j \in \mathscr{S}} P(j|s,a)\hat{V}_1(j) - \max_{a \in A(s)} \sum_{j \in \mathscr{S}} P(j|s,a)\hat{V}_2(j) \right|$$

$$\leq \gamma \max_{a \in A(s)} \left| \sum_{j \in \mathscr{S}} P(j|s,a)\hat{V}_1(j) - \sum_{j \in \mathscr{S}} P(j|s,a)\hat{V}_2(j) \right|$$

$$= \gamma \max_{a \in A(s)} \sum_{j \in \mathscr{S}} P(j|s,a)|\hat{V}_1(j) - \hat{V}_2(j)|$$

$$\leq \gamma \max_{s \in \mathscr{S}} |\hat{V}_1(s) - \hat{V}_2(s)|,$$

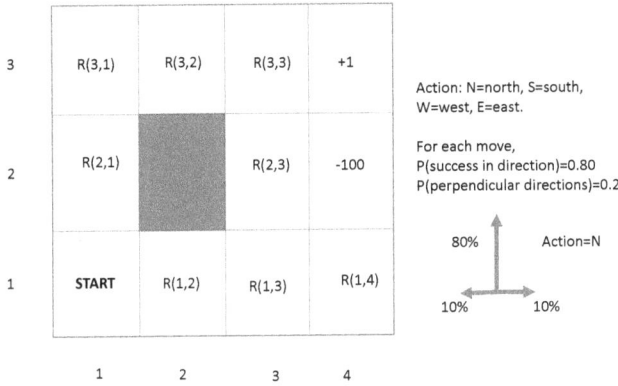

Figure 10.8   Gridworld – An agent tries to move in the maze to a terminal state to achieve the maximum rewards

where the third line follows from the property that

$$\left|\max_s f(x) - \max_s g(x)\right| \leq \max_x |f(x) - g(x)|$$

and the final line follows from the fact that $P(j|s,a)$ are non negative and sum to one. Since $MV^* = V^*$, using the contraction property, we have

$$\max_{s\in\mathscr{S}} |M\hat{V}(s) - V^*(s)| \leq \gamma\max_{s\in\mathscr{S}} |\hat{V}(s) - V^*(s)| \Longrightarrow \hat{V} \to V^*.$$

The theorem is proved. ■

Now we use the Gridworld example as shown in Figure 10.8 to compute the optimal value functions and optimal policy using the VI method. If we assume that the agent (or a robot) can repeat the process of going from the start cell to the terminal cell indefinitely, then we can assume that the state space, action space, transition probability, and value function are all stationary. We use the row and column numbers to indicate the state. That is $s = (i, j)$ and action space is $A(s) = \{N, S, E, W\}$, representing "north", "south", "east", and "west". To implement the VI algorithm, we have two choices for the terminal states. We either treat the value for the terminal cell constant or update the value using the algorithm like a non-terminal cell. Suppose that the immediate reward function $R(s) = 0$ for all cells. We first show how calculations for the VI algorithm are implemented at discount rate $\gamma = 0.9$. First we initialize the value function by assigning zero to all states, $\hat{V}_0(s) \leftarrow 0, \forall s \in S$, which are shown in Figure 10.9. After one iteration, we get the updated value function in Figure 10.10. We show how the value function is updated by taking a few examples. For the

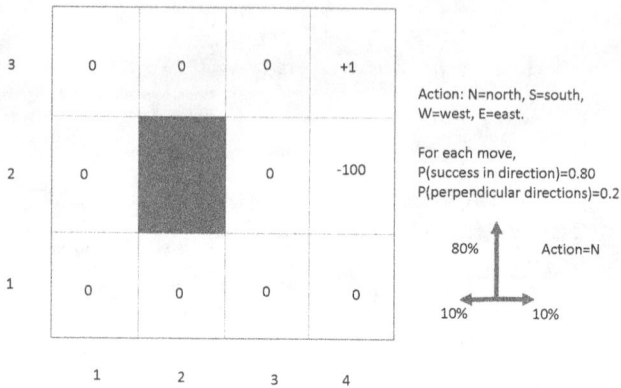

Figure 10.9 Gridworld – Initial Value function – also the immediate reward function

target state $s = (3,4)$, we have

$$\hat{V}_1(3,4) = R(3,4) + \gamma \max_{a \in \{N,S,W,E\}} \sum_{j \in \text{ reachable states of } S} P(j|(3,4),a)\hat{V}_0(j).$$

In this cell, we consider all four actions and compute their value functions as follows:

$$\hat{V}_1^N(3,4) = 1 + 0.9(0.8\hat{V}_0(3,4) + 0.1\hat{V}_0(3,4) + 0.1\hat{V}_0(3,3))$$
$$= 1 + 0.9(0.8 \times 1 + 0.1 \times 1 + 0.1 \times 0) = 1.81,$$
$$\hat{V}_1^S(3,4) = 1 + 0.9(0.8\hat{V}_0(2,4) + 0.1\hat{V}_0(3,4) + 0.1\hat{V}_0(3,3))$$
$$= 1 + 0.9(0.8 \times (-100) + 0.1 \times 1 + 0.1 \times 0) = -70.82,$$
$$\hat{V}_1^W(3,4) = 1 + 0.9(0.8\hat{V}_0(3,3) + 0.1\hat{V}_0(3,4) + 0.1\hat{V}_0(2,4))$$
$$= 1 + 0.9(0.8 \times 0 + 0.1 \times 1 + 0.1 \times (100)) = -7.91,$$
$$\hat{V}_1^E(3,4) = 1 + 0.9(0.8\hat{V}_0(3,4) + 0.1\hat{V}_0(3,4) + 0.1\hat{V}_0(2,3))$$
$$= 1 + 0.9(0.8 \times 1 + 0.1 \times 1 + 0.1 \times (-100)) = -7.19.$$

Thus, the updated value function for this target state is

$$\hat{V}_1(3,4) = 1 + \max(0.81, -8.91, -8.19, -71.82) = 1.81.$$

Another example is for the non-target cell $s = (3,3)$ and the calculations are below:

$$\hat{V}_1^N(3,3) = 0 + 0.9(0.8\hat{V}_0(3,3) + 0.1\hat{V}_0(3,2) + 0.1\hat{V}_0(3,4))$$
$$= 0 + 0.9(0.8 \times 0 + 0.1 \times 0 + 0.1 \times 1) = 0.09,$$
$$\hat{V}_1^S(3,3) = 0 + 0.9(0.8\hat{V}_0(2,3) + 0.1\hat{V}_0(3,4) + 0.1\hat{V}_0(3,2))$$
$$= 0 + 0.9(0.8 \times 0 + 0.1 \times 1 + 0.1 \times 0) = 0.09,$$

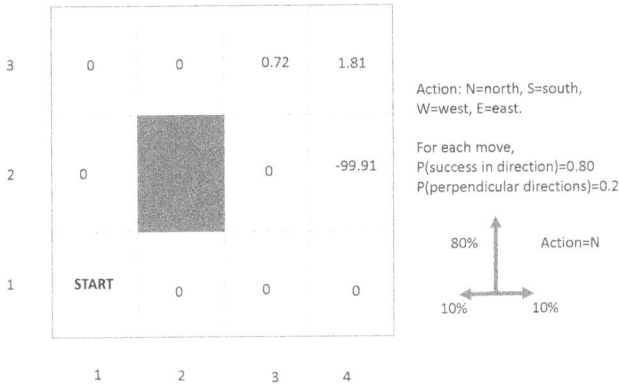

Figure 10.10 Gridworld – Value function after the first itearation

$$\hat{V}_1^W(3,3) = 1 + 0.9(0.8\hat{V}_0(3,2) + 0.1\hat{V}_0(3,3) + 0.1\hat{V}_0(2,3))$$
$$= 0 + 0.9(0.8 \times 0 + 0.1 \times 0 + 0.1 \times 0) = 0.00,$$
$$\hat{V}_1^E(3,3) = 0 + 0.9(0.8\hat{V}_0(3,4) + 0.1\hat{V}_0(3,3) + 0.1\hat{V}_0(2,3))$$
$$= 0 + 0.9(0.8 \times 1 + 0.1 \times 0 + 0.1 \times 0 = 0.72.$$

Thus, the updated value function for this non-target state is

$$\hat{V}_1(3,3) = 0 + \max(0.09, 0.72, 0.09, 0.00) = 0.72.$$

Other states (cells) can be updated in the same way so that updated value function values are shown in Figure 10.10 after the first iteration. The next three Figures 10.11, 10.12, and 10.13 show the results after 5, 10, and 1000 iterations, respectively. The calculations can be done by either Matlab or spreadsheet like Excel. Clearly, the optimal value function converges after 100 iterations as shown in Figure 10.14.

Exercise 10.12 *Verify the value function values in Figures 10.11 to 10.14.*

10.2.2 POLICY ITERATION METHOD

In this section, we discuss another approach to computing the optimal policy and optimal value function. In this approach, the policy is improved after each iteration. When the policy remains the same after an iteration, it becomes the optimal policy. Thus, we call this approach the policy iteration (PI) method. The convergence property of the PI can be proven similarly as in VI by showing that each iteration is also a contraction and policy must be improved each step or be optimal policy (left as an exercise). Below is the PI algorithm:

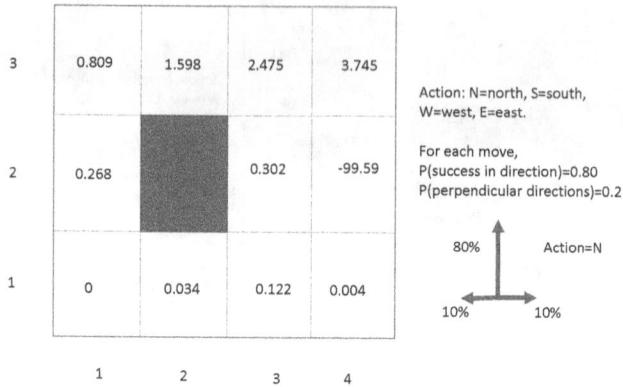

Figure 10.11 Gridworld – Value function after the fifth itearation

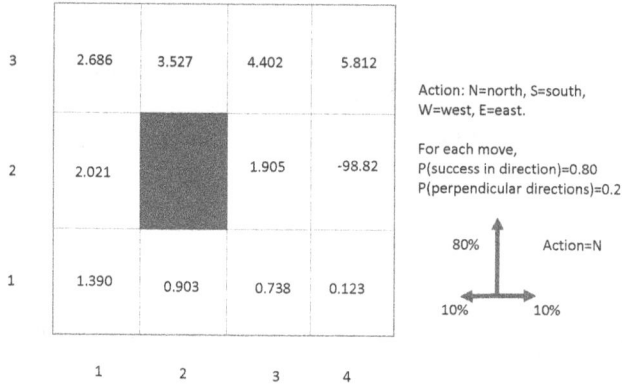

Figure 10.12 Gridworld – Value function after the tenth itearation

1. Initialize policy π (e.g., randomly or arbitrarily chosen).
2. Compute value vector of the policy, $\mathbf{V}^\pi$, via solving a linear equation system (10.14).
3. Update the policy π as a greedy policy with respect to $\mathbf{V}^\pi$

$$\pi(s) \leftarrow \arg\max_a \sum_{s' \in S} P(s'|s,a) V^\pi(s').$$

4. If policy π changes in last iteration, return to step 2. Otherwise, stop.

Using the Gridworld example, we perform the PI by starting with an initial policy which prescribes the same action "N=north" for all states. Figures 10.15, 10.16, and 10.17 show the value function and its associate up-

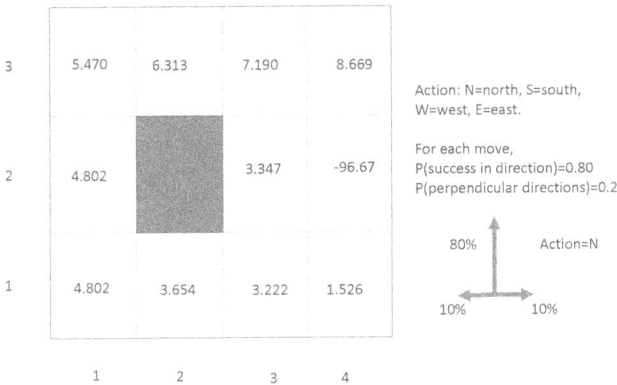

Figure 10.13 Gridworld – Value function after the 1000th itearation

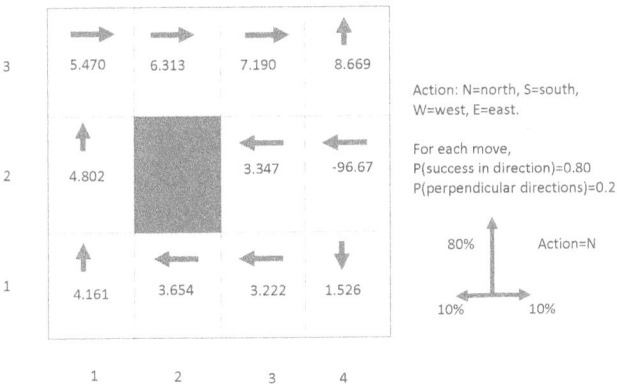

Figure 10.14 Gridworld – Value function and Optimal policy after the 100th itearation

dated policy after each iteration. It only takes three iterations, the policy converges to the optimal one.

After the third iteration, the policy remains the same, indicating that the
optimal policy is obtained. Due to the finite state space, the algorithm can
be implemented by using mathematical software such as Matlab or by using
Excel's solver.

Exercise 10.13 *For the Gridworld example presented above, if the noise level
for making a decision is higher, solve the MDP problem again to see the impact
of the noise level on the optimal policy and the optimal value. Suppose that*

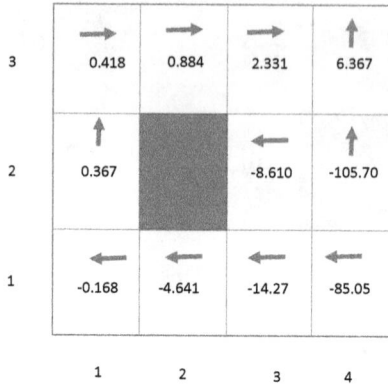

Figure 10.15 Gridworld – Value function values for a given initial policy (Go North for all states) and the updated greedy policy with respect to these values – the first iteration.

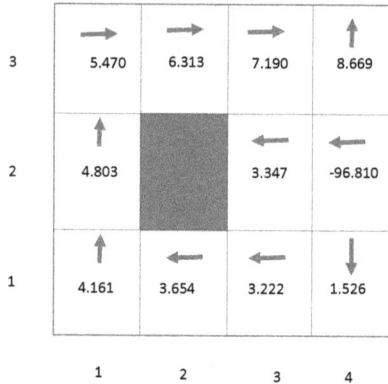

Figure 10.16 Gridworld – Value function values under the policy from the first iteration and the updated greedy policy with respect to these values – the second iteration.

the probability that the movement direction is correct is reduced from 80% to 50% with 25% to each direction perpendicular to the intended direction of movement.

Exercise 10.14 *Prove that the policy iteration algorithm converges to the optimal value function.*

Exercise 10.15 *(Car rental problem). Mr. Smith is a business manager of two locations for a nationwide car rental company. Every day, customers arrive*

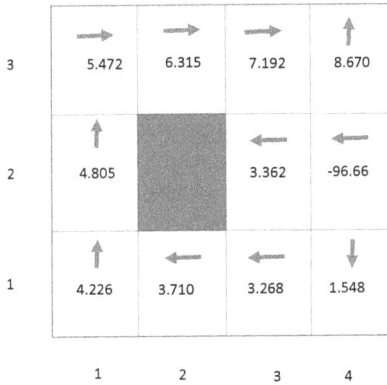

Figure 10.17 Gridworld – Value function values under the policy from the second iteration and the updated greedy policy with respect to these values – the third iteration.

at each location to rent cars. If a car is available at that location, the car is rented out with a credit of £20 by the national company; if no car is available at that location, the business is lost. Cars are available for renting the day after they are returned. To improve the service, Mr. Smith can move cars between two locations overnight, at a cost of £5 per car moved. It is assumed that the number of cars requested and returned at each location follow Poisson distributions with rates λ_i (requests) and μ_i (returns), respectively, with $i = 1, 2$. These parameters are $\lambda_1 = 3, \lambda_2 = 4, \mu_1 = 3$, and $\mu_2 = 2$. Furthermore, it is assumed that there can be no more than 20 cars in each location and a maximum of 5 cars can be moved from one location to the other in one night. Formulate and solve this problem as an MDP with infinite time horizon with $\gamma = 0.9$ to maximize the expected discounted profit. (Hint: Define the state as the number of cars at each location at the end of the day; the time steps are chosen to be days; and the actions are the number of cars moved between the two locations overnight.)

10.2.3 COMPUTATIONAL COMPLEXITY

For the MDPs with finite state space S and finite action sets $A(i)$, $i \in S$, we can compare the VI and PI in terms of computational complexity and the convergence speed. For the VI algorithm, from (10.16) in Step 2, we know that each iteration's requires the number of arithmetic operations that is bounded by $O(|S|^2|A|)$. The VI algorithm may need to take a large number of iterations to converge to the optimal policy. Assume that the maximum reward is R_{max}. Then we have $V^*(s) \leq \sum_{t=1}^{\infty} \gamma^t R_{max} = R_{max}/(1 - \gamma)$. Recall $V_k(s)$ is the value after kth iteration. It follows from the contraction operator and the upper bound of $V^*(s)$ that

$$\max_{s \in S} |V_k(s) - V^*(s)| \le \gamma^k \max_{s \in S} |V_0(s) - V^*(s)| \le \frac{\gamma^k R_{max}}{1 - \gamma}.$$

Hence, we have linear convergence to optimal value function. Clearly, the convergence rate depends on the discount rate γ.

For the PI algorithm, the number of arithmetic operations per iteration is bounded by $O(|S|^3 + |S|^2|A|)$. This is because that the number of arithmetic operations is bounded by $O(|S|^3)$ in Step 2 (solving a linear equation system with $|S|$ equations and S unknowns) and the number of arithmetic operations is bounded by $O(|S|^2|A|)$ in Step 3. The PI algorithm converges faster than the VI algorithm in terms of the number of iterations. It can be shown that the PI has the linear-quadratic convergence.

Exercise 10.16 *Show that the PI algorithm has the linear-quadratic convergence.*

10.2.4   AVERAGE COST MDP WITH INFINITE HORIZON

Another decision criterion is the long-term average cost per unit time for a DT-MDP with infinite horizon. Again, we assume that the state space and action space are finite. Then it can be shown that there exists a stationary optimal policy for the DTMDP if the planning horizon is infinitely long. The rigorous proof of existence of such a policy can be found in Derman (1970) and Puterman (1994). In fact, the issue of the existence of an optimal stationary policy is quite subtle for the MDP with the average cost criterion. This issue becomes more complicated when the state space is not finite but countably infinite. For example, in simple countable-state MDPs, the average cost optimal policies may not exist and, when they do, they may not be stationary. However, these technical difficulties do not occur when the MDP has a finite state space and finite action sets. The interested readers are referred to the book on this topic by Puterman (1994).

For the DTMDP with the average cost criterion, we assume that the system has reached the steady state at time 0. Define the expected n-period cost function $V^\pi(i,n)$ under a stationary policy π and the initial state i by

$$V^\pi(i,n) = \sum_{t=0}^{n-1} \sum_{j \in S} P^{(t)}(j|i,\pi) c_j(\pi(j)), \tag{10.17}$$

where $P^{(t)}(j|i,\pi) = P(X_t = j|X_0 = i, \pi)$ is the t-step transition probability of the DTMDP and $c_j(\pi(j))$ is the one-step cost at state j under the stationary policy π. Then we can define the average cost function $g_i(\pi)$ by

$$g_i(\pi) = \lim_{n \to \infty} \frac{1}{n} V^\pi(i,n), \quad i \in \mathscr{S}. \tag{10.18}$$

This limit exists and represents the long-term average expected cost per time unit when the Markov chain under a stationary policy π is irreducible and positive recurrent. If the initial state i is recurrent under policy π, we can interpret $g_i(\pi)$ as the long-term actual average cost per time unit with probability 1 under policy π. The systems with transient states and/or multiple recurrent subclasses (i.e., disjoint irreducible sets of recurrent states) can be treated by figuring out the average cost rate for each subclass and the probability that the system is absorbed in each recurrent subclass. Here, for simplicity, we focus on the unichain case. In such a case, the Markov chain $\{X_n\}$ under a stationary policy π has a unique stationary distribution $p_j(\pi) = \lim_{m \to \infty}(1/t)\sum_{t=1}^{m} P^{(t)}(j|i, \pi), j \in \mathscr{S}$. Clearly, this stationary distribution can be obtained by solving

$$p_j(\pi) = \sum_{i \in \mathscr{S}} p_i(\pi)P(j|i, \pi(i)), \quad j \in \mathscr{S},$$

$$\sum_{j \in \mathscr{S}} p_j(\pi) = 1.$$

Using the stationary distribution, we have

$$g(\pi) = g_i(\pi) = \sum_{j \in \mathscr{S}} c_j(\pi(j))\pi_j(\pi). \tag{10.19}$$

Now we define the average cost optimal policy.

Definition 10.5 *A stationary policy π^* is said to be average cost optimal if*

$$g_i(\pi^*) \le g_i(\pi)$$

for each stationary policy π, uniformly in initial state i.

It can be shown that such a policy always exists in a DTMDP with a finite state space and finite action sets and furthermore this policy is optimal among the class of all conceivable policies. It follows from the definition of $V^\pi(i,n)$ that

$$V^\pi(i,n) = c_i(\pi(i)) + \sum_{j \in \mathscr{S}} P(j|i, \pi(i))V^\pi(j, n-1), \quad n \ge 1, i \in \mathscr{S} \tag{10.20}$$

with $V^\pi(i,0) = 0$. Due to $\lim_{n \to \infty} V^\pi(i,n)/n = g(\pi)$, for n large, we have

$$V^\pi(i,n) \approx ng(\pi) + v_i(\pi), \tag{10.21}$$

where $v_i(\pi), i \in \mathscr{S}$ is introduced as "bias values". Another name for $v_i(\pi)$ is "relative values". This is because for large n, we have $v_i(\pi) - v_j(\pi) \approx V^\pi(i,n) - V^\pi(j,n)$, which represents the difference in total expected costs for the first n periods when starting in state i rather than in state j, given that policy π is followed. Substituting (10.21) into (10.20) yields

$$g(\pi) + v_i(\pi) \approx c_i(\pi(i)) + \sum_{j \in \mathscr{S}} P(j|i, \pi(i))v_j(\pi), \quad i \in \mathscr{S}. \qquad (10.22)$$

Now, we explain the main idea of the policy improvement which is the key step in the policy iteration algorithm for computing the optimal policy for the DTMDP with the initial state i under the average cost criterion. Define the cost difference ΔC as

$$\Delta C_i(a, \pi) = \lim_{n \to \infty} \left[\left(c_i(a) + \sum_{j \in \mathscr{S}} P(j|i, a)V^{\pi}(j, n-1) \right) \right.$$
$$\left. - \left(c_i(\pi(i)) + \sum_{j \in \mathscr{S}} P(j|i, \pi(i))V^{\pi}(j, n-1) \right) \right],$$

which via (10.21) can be approximated by

$$\Delta C_i(a, \pi) \approx c_i(a) + \sum_{j \in \mathscr{S}} P(j|i, a)v_j(\pi) - c_i(\pi(i)) - \sum_{j \in \mathscr{S}} P(j|i, \pi(i))v_j(\pi)$$
$$= c_i(a) + \sum_{j \in \mathscr{S}} P(j|i, a)v_j(\pi) - g(\pi) - v_i(\pi).$$

$$(10.23)$$

To generate a new improved policy, we need to select an action a to minimize the first three terms of (10.23) for every state i. Thus, we call

$$c_i(a) - g(\pi) + \sum_{j \in \mathscr{S}} P(j|i, a)v_j(\pi)$$

the policy improvement quantity. The main idea is formalized as the following theorem.

Theorem 10.10 *Let g and $v_i, i \in \mathscr{S}$ be a set of fixed numbers. If a stationary policy π' has the property*

$$c_i(\pi'(i)) - g + \sum_{j \in \mathscr{S}} P(j|i, \pi'(i))v_j \leq v_i \text{ for each } i \in \mathscr{S}. \qquad (10.24)$$

Then the long-run average cost of policy π' satisfies

$$g_i(\pi') \leq g, \quad i \in \mathscr{S}. \qquad (10.25)$$

Proof. Let $p_j(\pi')$ be the stationary distribution of the irreducible Markov chain under a stationary policy π'. Multiplying both sides of (10.24) by $p_i(\pi')$ and summing up over i yields

$$\sum_{i \in \mathscr{S}} p_i(\pi')c_i(\pi'(i)) - g + \sum_{i \in \mathscr{S}} p_i(\pi') \sum_{j \in \mathscr{S}} P(j|i, \pi'(i))v_j \leq \sum_{i \in \mathscr{S}} p_i(\pi')v_i.$$

Interchanging the order of summation in the above inequality gives

$$g(\pi') - g + \sum_{j \in \mathscr{S}} p_j(\pi') v_j \le \sum_{i \in \mathscr{S}} p_i(\pi') v_i.$$

Since the two summation terms are the same and cancelled out, the proof is completed. ∎

A formal definition of the relative value for a state i is based on the renewal reward theorem. Consider a recurrent state r in the Markov chain induced by a stationary policy π. Denote by $T_i(\pi)$ the expected time until the first visit to state r and denote by $K_i(\pi)$ the expected costs incurred until the first visit to state r, given that the starting state is i for each $i \in \mathscr{S}$. Clearly, we have

$$g(\pi) = \frac{K_r(\pi)}{T_r(\pi)}.$$

Based on this fact, we define the relative value for state i as

$$w_i(\pi) = K_i(\pi) - g(\pi) T_i(\pi), \quad i \in \mathscr{S}$$

with $w_r(\pi) = 0$. Since $w_i(\pi)$ can be either positive or negative for $i \ne r$, it can be interpreted as the relative value (higher or lower than) to the reference state r. We can show that the average cost $g = g(\pi)$ and the relative values $v_i = w_i(\pi)$ can be determined by a system of linear equations for a given stationary policy π.

Theorem 10.11 *If a stationary policy π induces an irreducible and positive recurrent Markov chain, then we have*

1. The average cost $g(\pi)$ and the relative values $w_i(\pi), i \in \mathscr{S}$, satisfy the following system of linear equations with unknowns g and $v_i, i \in \mathscr{S}$:

$$v_i = c_i(\pi(i)) - g + \sum_{j \in \mathscr{S}} P(j|i, \pi(i)) v_j. \tag{10.26}$$

2. If g and $v_i, i \in \mathscr{S}$ are any solution to (10.26), then $g = g(\pi)$ and $v_i = w_i(\pi) + c$ for some constant c.
3. Let s be any arbitrarily chosen state. Then (10.26) together with $v_s = 0$ have a unique solution.

Proof. (1) By conditioning on the next state after the initial state i, under a stationary policy π, we have

$$T_i(\pi) = 1 + \sum_{j \ne r} P(j|i, \pi(i)) T_j(\pi),$$

$$K_i(\pi) = c_i(\pi(i)) + \sum_{j \ne r} P(j|i, \pi(i)) K_j(\pi)$$

for $i \in \mathcal{S}$. Using these expressions in the definition of relative value yields

$$w_i(\pi) = K_i(\pi) - g(\pi)T_i(\pi) = c_i(\pi(i))$$
$$- g(\pi) + \sum_{j \neq r} P(j|i, \pi(i))[K_j(\pi) - g(\pi)T_j(\pi)]$$
$$= c_i(\pi(i)) - g(\pi) + \sum_{j \in \mathcal{S}} P(j|i, \pi(i))w_j(\pi), \quad i \in \mathcal{S},$$

where the second equality follows from $w_r(\pi) = 0$.

(2) Suppose that $\{g, v_i\}$ is a solution to (10.26). By induction, it can be shown that

$$v_i = \sum_{n=0}^{m-1} \sum_{j \in \mathcal{S}} P^{(n)}(j|i, \pi)c_j(\pi(j)) - mg + \sum_{j \in \mathcal{S}} P^{(m)}(j|i, \pi)v_j$$
$$= V^\pi(i, m) - mg + \sum_{j \in \mathcal{S}} P^{(m)}(j|i, \pi)v_j \quad i \in \mathcal{S}. \tag{10.27}$$

(Proving this expression is left as an exercise.) Note that $\lim_{m \to \infty} V^\pi(i, m)/m = g(\pi)$ for each i. Dividing both sides of (10.27) and letting $m \to \infty$ leads to $g = g(\pi)$. Let $\{g, v_i\}$ and $\{g', v_i'\}$ be any two solutions to (10.26). It follows from $g = g' = g(\pi)$ and (10.27) that

$$v_i - v_i' = \sum_{j \in \mathcal{S}} P^{(m)}(j|i, \pi)(v_j - v_j'), \quad i \in \mathcal{S}, \text{ and } m \geq 1.$$

Summing both sides of the equation above over $m = 1, ..., n$ and then dividing by n yields

$$v_i - v_i' = \frac{1}{n} \sum_{m=1}^{n} \sum_{j \in \mathcal{S}} P^{(m)}(j|i, \pi)(v_j - v_j') = \sum_{j \in \mathcal{S}} \left[\frac{1}{n} \sum_{m=1}^{n} P^{(m)}(j|i, \pi) \right] (v_j - v_j')$$

for $i \in \mathcal{S}$ and $n \geq 1$. By taking the limit of the equation above as $n \to \infty$, we have

$$v_i - v_i' = \sum_{j \in \mathcal{S}} p_j(\pi)(v_j - v_j'), \quad i \in \mathcal{S},$$

where the right-hand side is a constant.

(3) It follows from $\sum_{j \in \mathcal{S}} P(j|i, \pi) = 1$ for each $i \in \mathcal{S}$ that for any constant c, there exist the numbers g and $v_i = w_i(\pi) + c, i \in \mathcal{S}$ that satisfy (10.26). To get a solution with $v_s = 0$ for some s, we need to choose $c = -w_s(\pi)$ which implies that such a solution is unique. ∎

Exercise 10.17 *Prove the expression (10.27).*

Now we can present the policy iteration (PI) algorithm for the average cost DTMDP as follows:

PI Algorithm

1. Initialize policy π.
2. Under the current policy π, solve the following system of linear equations

$$v_i = c_i(\pi(i)) - g + \sum_{j \in \mathscr{S}} P(j|i, \pi(i))v_j,$$

$$v_s = 0$$

   for a set of numbers $\{g(\pi), v_i(\pi)\}, i \in \mathscr{S}$ where s is an arbitrarily chosen state.
3. Update the policy with the numbers determined in the previous step by

$$\pi'(i) \leftarrow \arg \min_{a \in A(i)} \left[c_i(a) - g(\pi) + \sum_{j \in \mathscr{S}} P(j|i, a)v_j(\pi) \right] = \arg \min_{a \in A(i)} T_i(a, \pi)$$

   for $i \in \mathscr{S}$ to generate a new improved policy π'. Here we introduce a notation $T_i(a, \pi)$ for the test quantity for policy improvement under a given policy π.
4. If the policy π' is different from π, return to step 2 with $\pi = \pi'$ (replaced with the new improved policy). Otherwise stop and the optimal policy is obtained as π'.

This algorithm converges after a finite number of iterations. The proof of convergence is discussed briefly later. As mentioned earlier in the discounted cost case, the PI algorithm converges faster than the VI algorithm in terms of the number of iterations although the computational complexity or computational cost can be higher than the VI for a DTMDP with large state space.

The maintenance of a multi-state system is a typical optimization problem in stochastic models. In Chapter 9, we studied the static optimal age replacement policy for a multi-state system that deteriorates with time. Here we present a maintenance problem of a multi-state system in the framework of a DTMDP and solve for the optimal policy under the average cost criterion numerically by using the PI algorithm. The following example is taken from Tijms (2003).

Example 10.1 *Consider a system that is inspected at the beginning of every period. The frequency of inspection determines the length of each period. It can be daily, weekly, or monthly. The system conditions can be classified into N states with 1 representing the best (brand new) and N representing the worst* *(i.e., completely broken and must be replaced with a new one). The states between 1 and N are functional but the efficiency is decreasing in $1 < i < N$. This implies that operating costs in these working states denoted by $C_o(i)$ satisfy $C_o(1) \leq C_o(2) \leq \cdots \leq C_o(N-1)$. For states $2 \leq i \leq N-1$, there are two actions to choose: $a = 0$ do nothing or $a = 1$ do a preventive maintenance. Clearly, at state $i = 1$, the only action is $a = 0$ as there is no need to do preventive maintenance on a brand new system. For state $i = N$, the only action is a replacement, denoted by $a = 2$, which takes two periods of time. We also introduce an auxiliary state to represent that the replacement is in progress already for one day. Thus the state space is*

$$S = \{1, 2, ..., N, N+1\},$$

and action spaces are given by

$$A(1) = \{0\}, \ A(i) = \{0, 1\} \text{ for } 1 < i < N, \ A(N) = A(N+1) = \{2\}.$$

Assume that preventive maintenance and replacement actions will bring the system to state 1. The state transition probabilities under action $a = 0$ (i.e., do nothing) are $P(j|i, a) = P(j|i, 0) = p_{ij}$ for $1 \leq i < N$, which model the stochastic deterioration behavior of the system if $p_{ij} = 0$ for $j < i$. Clearly, with the replacement and preventive maintenance assumption, we have $P(1|i, 1) = 1$ for $1 < i < N$ and $P(j|i, 1) = 0$ for $j \neq 1$; and $P(N+1|N, 2) = P(1|N+1, 2) = 1$. Denote by C_R the replacement cost at state N and $C_P(i)$ the preventive maintenance cost at state $1 < i < N$. The one-step costs $c_i(a)$ are given by

$$c_i(0) = C_o(i), \ c_i(1) = C_P(i), \ c_N(2) = C_R, \ c_{N+1}(2) = 0.$$

Now we have specified all components of a DTMDP. If we assume that this system is in operation indefinitely, the system process can be treated as the DTMDP with infinite horizon under the average cost criterion. We find the optimal stationary policy by using the PI algorithm with the following set of parameters:

$N = 5, C_i(0) = 0$ for $i = 1, 2, 3, 4, C_P(2) = C_P(3) = 7, C_P(4) = 5$ and $C_R = 10$. If $a = 0$ for $i = 1, 2, 3, 4$ and $a = 2$ for $i = 5, 6$ (denoted by a policy π'), the transition probability matrix is given by

$$\mathbf{P}(\pi') = \begin{pmatrix} 0.90 & 0.10 & 0 & 0 & 0 & 0 \\ 0 & 0.80 & 0.10 & 0.05 & 0.05 & 0 \\ 0 & 0 & 0.70 & 0.10 & 0.20 & 0 \\ 0 & 0 & 0 & 0.50 & 0.40 & 0.1 \\ 0 & 0 & 0 & 0 & 0 & 1 \\ 1 & 0 & 0 & 0 & 0 & 0 \end{pmatrix}.$$

Now, we apply the PI algorithm by choosing an initial policy $\pi^{(1)} = (0, 0, 0, 0, 2, 2) = \pi'$. Note that the three boundary states $\pi(1) = 0$ and $\pi(5) = $

$\pi(6) = 2$ are fixed. Thus, we only need to determine the decisions for three states $i = 2,3,4$. If we introduce the vector notations: $\mathbf{v} = (v_1,...v_6)^T$, $\mathbf{c}(\pi) = (c_1(\pi(1)),...,c_6(\pi(6)))^T$ and $\mathbf{g}(\pi) = (g(\pi),...,g(\pi))^T$, then the set of linear equations under policy $\pi^{(1)}$ in Step 2 of PI algorithm can be written as

$$\mathbf{v} = \mathbf{c}(\pi^{(1)}) - \mathbf{g}(\pi^{(1)}) + \mathbf{P}(\pi^{(1)})\mathbf{v} \tag{10.28}$$

with $v_6 = 0$. Now we can start the first iteration:

Iteration 1:

Step 1: Initialize policy with $\pi^{(1)} = (0,0,0,0,2,2)$.

Step 2: Solve the equation system (10.28) under $\pi^{(1)}$ which can be written as

$$v_1 = 0 - g + 0.9v_1 + 0.1v_2$$
$$v_2 = 0 - g + 0.8v_2 + 0.1v_3 + 0.05v_4 + 0.05v_5$$
$$v_3 = 0 - g + 0.7v_3 + 0.1v_4 + 0.2v_5$$
$$v_4 = 0 - g + 0.5v_4 + 0.4v_5 + 0.1v_6$$
$$v_5 = 10 - g + v_6$$
$$v_6 = 0 - g + v_1$$
$$v_6 = 0.$$

The solution is given by

$$g(\pi^{(1)}) = 0.4721, \ v_1(\pi^{(1)}) = 0.4721 \ v_2(\pi^{(1)}) = 5.1931, \ v_3(\pi^{(1)}) = 7.0043,$$
$$v_4(\pi^{(1)}) = 6.6781, \ v_5(\pi^{(1)}) = 9.5279, \ v_6(\pi^{(1)}) = 0.$$

Step 3: Update the policy with the relative values obtained in the previous step. The test quantities for the three states $i = 2,3,4$, in which the decisions must be chosen to improve the policy, are given by

$$T_2(0,\pi^{(1)}) = 5.1931, \ T_2(1,\pi^{(1)}) = 7.0000, \ T_3(0,\pi^{(1)}) = 7.0043,$$
$$T_3(1,\pi^{(1)}) = 7.0000, \ T_4(0,\pi^{(1)}) = 6.6781, \ T_4(1,\pi^{(1)}) = 5.0000.$$

These quantities generate an improved policy $\pi^{(2)} = (0,0,1,1,2,2)$ by choosing the action a that minimizes $T_i(a,\pi^{(1)})$ for each state i with $i = 2,3,4$. Since this updated policy is different from $\pi^{(1)}$, we need to go to the next iteration.

Iteration 2:

Step 2: Value determination under $\pi^{(2)} = (0,0,1,1,2,2)$ by solving linear equation system given by

$$v_1 = 0 - g + 0.9v_1 + 0.1v_2$$
$$v_2 = 0 - g + 0.8v_2 + 0.1v_3 + 0.05v_4 + 0.05v_5$$

$$v_3 = 7 - g + v_1$$
$$v_4 = 5 - g + v_1$$
$$v_5 = 10 - g + v_6$$
$$v_6 = 0 - g + v_1$$
$$v_6 = 0.$$

The solution is given by

$$g(\pi^{(2)}) = 0.4462, \; v_1(\pi^{(2)}) = 0.4462, \; v_2(\pi^{(2)}) = 4.9077, \; v_3(\pi^{(2)}) = 7.0000,$$
$$v_4(\pi^{(2)}) = 5.0000, \; v_5(\pi^{(2)}) = 9.5538, \; v_6(\pi^{(2)}) = 0.$$

Step 3: Update the policy with the relative values obtained in the previous step. The test quantities for the three states $i = 2, 3, 4$, in which the decisions must be chosen to improve the policy, are given by

$$T_2(0, \pi^{(2)}) = 4.9077, \; T_2(1, \pi^{(2)}) = 7.0000, \; T_3(0, \pi^{(2)}) = 6.4092,$$
$$T_3(1, \pi^{(2)}) = 7.0000, \; T_4(0, \pi^{(2)}) = 6.8308, \; T_4(1, \pi^{(2)}) = 5.0000.$$

These quantities generate an improved policy $\pi^{(3)} = (0, 0, 0, 1, 2, 2)$ by choosing the action a that minimizes $T_i(a, \pi^{(2)})$ for each state i with $i = 2, 3, 4$. Since this updated policy is different from $\pi^{(2)}$, we need to go to the next iteration.

Iteration 3:

Step 2: Value determination under $\pi^{(3)} = (0, 0, 0, 1, 2, 2)$ by solving linear equation system given by

$$v_1 = 0 - g + 0.9v_1 + 0.1v_2$$
$$v_2 = 0 - g + 0.8v_2 + 0.1v_3 + 0.05v_4 + 0.05v_5$$
$$v_3 = 0 - g + 0.7v_3 + 0.1v_4 + 0.2v_5$$
$$v_4 = 5 - g + v_1$$
$$v_5 = 10 - g + v_6$$
$$v_6 = 0 - g + v_1$$
$$v_6 = 0.$$

The solution is given by

$$g(\pi^{(3)}) = 0.4338, \; v_1(\pi^{(3)}) = 0.4338, \; v_2(\pi^{(3)}) = 4.7717, \; v_3(\pi^{(3)}) = 6.5982,$$
$$v_4(\pi^{(3)}) = 5.0000, \; v_5(\pi^{(3)}) = 9.5662, \; v_6(\pi^{(3)}) = 0.$$

Step 3: Update the policy with the relative values obtained in the previous step. The test quantities for the three states $i = 2, 3, 4$, in which the decisions *must be chosen to improve the policy, are given by*

$$T_2(0, \pi^{(3)}) = 4.7717, \ T_2(1, \pi^{(3)}) = 7.0000, \ T_3(0, \pi^{(3)}) = 6.5982,$$

$$T_3(1, \pi^{(3)}) = 7.0000, \ T_4(0, \pi^{(3)}) = 5.8927, \ T_4(1, \pi^{(3)}) = 5.0000.$$

These quantities generate an "improved" policy $\pi^{(4)} = (0,0,0,1,2,2)$ by choosing the action a that minimizes $T_i(a, \pi^{(2)})$ for each state i with $i = 2,3,4$. Since this updated policy is the same as $\pi^{(3)}$, the optimal policy is found as $\pi^* = (0,0,0,1,2,2)$.

It is worth noting that although the numerical illustration of using the PI algorithm with a set of simple system parameters, Example 10.1 provides a framework that can be utilized for analyzing and optimizing the maintenance policies for systems that deteriorate with time under more complex cost structures and more sophisticated state spaces.

Exercise 10.18 *For the system considered in Example 10.1, assume that all parameters are the same except for the consequence of the preventive maintenance $a = 1$. Instead of bringing the system to state $i = 1$ (a brand new system), action $a = 1$ for a state i with $i = 2,3,4$ will result in state $i - 1$ (better condition). Such an assumption models the situation where the preventive maintenance is not a replacement with a new system. Determine the optimal maintenance policy with the PI algorithm.*

For the average cost DTMDP, we can also design the VI algorithm based on the recursion (10.20). We write the recursive relation based on the OE as

$$\hat{V}(i,n) = \min_{\pi(i) \in A(i)} \left\{ c_i(\pi(i)) + \sum_{j \in \mathscr{S}} P(j|i, \pi(i)) \hat{V}(j, n-1) \right\}, \quad n \geq 1, i \in \mathscr{S},$$

for $n = 1, 2, \ldots$. Here, $\hat{V}(i,n)$ can be interpreted as the minimum total expected costs with n periods left to the time horizon given that the current state is i and a terminal cost $\hat{V}(i,0)$ is incurred. Starting with an arbitrarily function $\hat{V}(i,0)$, we can recursively compute the optimal value functions as time horizon increases. A big difference between the average cost case and the discounted cost case is the limit that this sequence will converge. For the discounted cost case, we have $\hat{V}(i,n) \to V^*(i)$ as $n \to \infty$. However, for the average cost case, we have $\hat{V}(i,n) - \hat{V}(i,n-1) \to g$ as $n \to \infty$. We also use the optimal value operator M for the average cost case which can be applied to state function $v = (v_i, \in \mathscr{S})$:

$$Mv_i = \min_{a \in A(i)} \left\{ c_i(a) + \sum_{j \in \mathscr{S}} P(j|i,a) v_j \right\}. \tag{10.29}$$

Thus, we have $M\hat{V}(i,n-1) = \hat{V}(i,n)$ and $\hat{V}(i,n) - \hat{V}(i,n-1) = M\hat{V}(i,n-1) - \hat{V}(i,n-1)$. The theorem below provides the upper and lower bounds for the optimal average cost per time unit generated by applying the optimal value operator to value functions.

Theorem 10.12 *Suppose that the Markov chain induced by a given stationary policy in a DTMDP is irreducible and positive recurrent. Let $v = (v_i), i \in \mathcal{S}$ be a vector of given values. Define the v-based stationary policy $\pi(v)$ as a policy that prescribes action $a = \pi(v)$ that minimizes the right-hand side of (10.29). Then we have*

$$\min_{i \in \mathcal{S}}\{Mv_i - v_i\} \le g^* \le g_s(\pi(v)) \le \max_{i \in \mathcal{S}}\{Mv_i - v_i\} \tag{10.30}$$

for any $s \in \mathcal{S}$, where g^* is the minimum long term average cost per time unit and $g_s(\pi(v))$ is the long term average cost per time unit under policy $\pi(v)$ given s as the initial state.

Proof. By the definition of M, we have for any $i \in \mathcal{S}$

$$Mv_i \le c_i(a) + \sum_{j \in \mathcal{S}} P(j|i,a)v_j, \quad a \in A(i), \tag{10.31}$$

which by choosing $a = \pi(i)$ leads to

$$Mv_i \le c_i(\pi(i)) + \sum_{j \in \mathcal{S}} P(j|i,\pi(i))v_j. \tag{10.32}$$

Define the lower bound lb by

$$lb = \min_{i \in \mathcal{S}}\{Mv_i - v_i\}.$$

It follows from $lb \le Mv_i - v_i$ for $i \in \mathcal{S}$ and (10.32) that

$$c_i(\pi(i)) - lb + \sum_{j \in \mathcal{S}} P(j|i,\pi(i))v_j \ge v_i, \quad i \in \mathcal{S}.$$

It follows from Theorem 10.10 and the inequality above that

$$g_i(\pi) \ge lb, \quad i \in \mathcal{S}.$$

This relation holds for every policy π so that $g^* = \min_\pi g_i(\pi) \ge lb$ which is the inequality for the lower bound. Similarly, by defining $ub = \max_{i \in \mathcal{S}}\{Mc_i - v_i\}$, we can prove the upper bound inequality, $g_i(\pi(v)) \le ub$ for all $i \in \mathcal{S}$ (the proof is omitted and left as an exercise). ∎

Exercise 10.19 *Prove the upper bound inequality in Theorem 10.12.*

Now we present the VI algorithm for the average cost DTMDP:
VI Algorithm

1. Initialize the stationary value function $\hat{V}(i,0) = 0$ for all $i \in \mathscr{S}$.
2. Repeat updating the value function for $n \geq 1$ by

$$\hat{V}(i,n) = \min_{a \in A(i)} \left\{ c_i(a) + \sum_{j \in \mathscr{S}} P(j|i,a)\hat{V}(j,n-1) \right\}, \quad i \in \mathscr{S}.$$

Let $\pi(n)$ be any stationary policy at step n such that $a = \pi(i,n)$ minimizes the right-hand side of the above recursion.
3. Compute the bounds

$$lb_n = \min_{i \in \mathscr{S}}\{\hat{V}(i,n) - \hat{V}(i,n-1)\}, \quad ub_n = \max_{i \in \mathscr{S}}\{\hat{V}(i,n) - \hat{V}(i,n-1)\}.$$

4. If $0 \leq ub_n - lb_n \leq \varepsilon lb_n$ with $\varepsilon > 0$ as a pre-specified number holds, then stop with policy $\pi(n)$, otherwise, go to Step 2 with $n := n+1$.

Based on Theorem 10.12, we know that

$$0 \leq \frac{g_i(\pi(n)) - g^*}{g^*} \leq \frac{ub_n - lb_n}{lb_n} \leq \varepsilon, \quad i \in \mathscr{S}$$

when the algorithm is stopped after the nth iteration with policy $\pi(n)$.

Exercise 10.20 *Consider Example 10.1. Use the VI algorithm to compute the optimal policy and its associated minimum average cost rate with the accuracy number $\varepsilon = 0.001$. Compare the VI algorithm with the PI algorithm in terms of number of iterations by varying ε values (e.g., 0.1, 0.01, and 0.0001). (Hint: Use mathematical software package such as Matlab or Mathematica.)*

The main theoretical issue in algorithmic approach to determining the optimal policy for a DTMDP is the convergence of the algorithm developed. Such an issue becomes more complex in the average cost DTMDP than in the discounted cost DTMDP. For example, in the VI algorithm for the average cost DTMDP, the lower bound and upper bound for g^* may not converge to the same limit although the lower bound is non-decreasing and the upper bound is nonincreasing (to be proved later). Consider a special DTMDP with only two states 1 and 2 and a single action a in each state. The one-step costs and the one-step transition probabilities are given by $c_1(a) = 1, c_2(a) = 0, P(2|1,a) = P(1|2,a) = 1$ and $P(1|1,a) = P(2|2,a) = 0$. Under the stationary policy (the only one), the system switches from one state to the other state every period.

Applying the VI algorithm to this trivial DTMDP gives value function sequence of $\hat{V}(1,2k) = \hat{V}(2,2k) = k, \hat{V}(1,2k-1) = k$, and $V(2,2k-1) = k-1$ for all $k \geq 1$. Thus, $lb_n = 0$ and $ub_n = 1$ for all n. This implies that the two bound sequences converge to two different limit. The reason for this oscillating behavior of $\hat{V}(i,n) - \hat{V}(i,n-1)$ is the periodicity of the Markov chain induced by the stationary policy. Such a problem does not occur if the Markov chain is aperiodic. Note that if a recurrent state can be re-visited by a direct transition to itself (i.e., one step transition to itself), then the Markov chain is automatically aperiodic. We can transform the original Markov chain induced by a stationary policy to the one with such a property by a so-called perturbation of the one-step transition probability. This is the same idea as the uniformization technique we presented in Chapter 3 to link between CTMC and DTMC. Suppose that a Markov chain $\{X_n\}$ has the one-step transition probability p_{ij}. These transition probabilities are perturbed as $p'_{ij} = \tau p_{ij}$ for $i \neq j$ and $p'_{ii} = \tau p_{ii} + 1 - \tau$ for some constant $\tau \in (0,1)$. Now the perturbed Markov Chain $\{X'_n\}$ with one-step transition probabilities p'_{ij} is aperiodic. It is easy to check that $\{X'_n\}$ and $\{X_n\}$ have the same stationary distribution. Thus, the periodicity issue can be resolved. We present the following theorem without a proof.

Theorem 10.13 *If the Markov chain induced by the average cost optimal stationary policy in a DTMDP is irreducible, aperiodic, and positive recurrent, then there exist finite parameters $\alpha > 0$ and $0 < \beta < 1$ such that*

$$|ub_n - lb_n| \leq \alpha \beta^n, \quad n \geq 1.$$

Further, we have $\lim_{n \to \infty} ub_n = \lim_{n \to \infty} lb_n = g^*$.

For the proof of this theorem which is quite technical, interested readers are referred to White (1963). However, to give readers some flavor of the proof techniques for convergence properties of computational algorithms, we present a few results with proofs.

Theorem 10.14 *In the VI algorithm for a DTMDP, the lower and upper bounds satisfy*

$$lb_{k+1} \geq lb_k \quad and \quad ub_{k+1} \leq ub_k \text{ for } k \geq 1.$$

Proof. It follows from the definition of $\pi(n)$, the updated stationary policy after the nth iteration, that

$$\hat{V}(i,n) = c_i(\pi(i,n)) + \sum_{j \in \mathscr{S}} P(j|i,\pi(i,n))\hat{V}(j,n-1), \ i \in \mathscr{S}. \tag{10.33}$$

For any stationary policy π, we have

$$\hat{V}(i,n) \leq c_i(\pi(i)) + \sum_{j \in \mathscr{S}} P(j|i,\pi(i))\hat{V}(j,n-1), \ i \in \mathscr{S}. \tag{10.34}$$

Letting $n = k$ in (10.33) and $n = k+1$, $\pi = \pi(k)$ in (10.34) (i.e., in the $(k+1)$st iteration, keeping the kth policy) and subtracting (10.33) from (10.34), we have

$$\hat{V}(i,k+1) - \hat{V}(i,k) \leq \sum_{j \in \mathscr{S}} P(j|i, \pi(i,k))[\hat{V}(j,k) - \hat{V}(j,k-1)], \ i \in \mathscr{S}.$$

(10.35)

Similarly, by taking $n = k+1$ in (10.33) and $n = k$, $\pi = \pi(k+1)$ in (10.34) (i.e., in the kth iteration, using the $(k+1)$st policy) and subtracting (10.34) from (10.33), we get

$$\hat{V}(i,k+1) - \hat{V}(i,k) \geq \sum_{j \in \mathscr{S}} P(j|i, \pi(i,k+1))[\hat{V}(j,k) - \hat{V}(j,k-1)], \ i \in \mathscr{S}.$$

(10.36)

Using the fact that $\hat{V}(j,k) - \hat{V}(j,k-1) \leq ub_k$ for all $j \in \mathscr{S}$ and $\sum_{j \in \mathscr{S}} P(j|i, \pi(i,k)) = 1$, we obtain $\hat{V}(i,k+1) - \hat{V}(i,k) \leq ub_k$ for all $i \in \mathscr{S}$. This gives $ub_{k+1} \leq ub_k$. Similarly, we obtain $lb_{k+1} \geq lb_k$ from (10.36). ∎

Next, we present the convergence property of the PI algorithm for an average cost DTMDP.

Theorem 10.15 *Suppose that the Markov chain induced by the stationary policy is irreducible and positive recurrent. The PI algorithm converges to the optimal policy in finite number of iterations.*

Proof. First, we show that a lexicographical ordering for the average cost and the relative values associated with the policies generated by the PI algorithm can be established. For a stationary policy π, we denote by $R(\pi)$ the class of recurrent states under policy π and number these states as $i = 1, ..., N$ according to their relative values, $w_i(\pi) = K_i(\pi) - g(\pi)T_i(\pi)$, with the largest state being r such that $w_r(\pi) = 0$ (i.e., regeneration state). Let π and π' be immediate successors in the sequence of policies generated by the PI algorithm. We claim that these two policies are different only if

1. $g(\pi') < g(\pi)$, or
2. $g(\pi') = g(\pi)$ and $w_i(\pi') \leq w_i(\pi)$ for all $i \in \mathscr{S}$ with strict inequality for at least one state i.

This implies that a new policy generated either reduces the average cost rate or reduces the relative value of a (transient) state. In other words, the new policy due to (2) is resulted from the different action in at least one transient state (i.e., $i \notin R(\pi)$). Since the number of stationary policies in a DTMDP with a finite state space and finite action sets is finite, this claim indicates that the PI algorithm converges after finite number of iterations. Now we prove this claim. By the construction of policy π', we have

$$c_i(\pi'(i)) - g(\pi) + \sum_{j \in \mathscr{S}} P(j|i, \pi'(i))w_j(\pi) \leq w_i(\pi), \ i \in \mathscr{S} \qquad (10.37)$$

with strict inequality only for those states i with $\pi'(i) \neq \pi(i)$. From Theorem 10.10 and (10.37), we have $g(\pi') \leq g(\pi)$ where the strict inequality holds only if the strict inequality holds in (10.37) for some recurrent state under policy π'. This proves claim (1). Next, consider the case of $g(\pi') = g(\pi)$. In this case, the equality holds in (10.37) for all $i \in R(\pi')$. Then, it follows from the policy improvement step that

$$\pi'(i) = \pi(i), \quad \text{for } i \in R(\pi'), \tag{10.38}$$

which leads to $R(\pi) = R(\pi')$ because the set $R(\pi')$ is closed under policy π'. Since under any stationary policy Q, for the recurrent states $i \in R(Q)$, their value functions do not depend on the actions in transient states $i \notin R(Q)$, this implies that

$$w_j(\pi) = w_j(\pi'), \quad j \in R(\pi'). \tag{10.39}$$

So the only way to make π' to be different from π is to have different action in at least one transient state. To verify this, by iterating (10.37), we have

$$\begin{aligned}
w_i(\pi) &\geq c_i(\pi'(i)) - g(\pi) + \sum_{j \in \mathscr{S}} P(j|i, \pi'(i)) w_j(\pi) \\
&\geq V^{\pi'}(i, m) - m g(\pi) + \sum_{j \in \mathscr{S}} P^{(m)}(j|i, \pi') w_j(\pi),
\end{aligned} \tag{10.40}$$

where the strict inequality holds in the first inequality for each i with $\pi'(i) \neq \pi(i)$. By replacing π with π' and the fact $g(\pi') = g(\pi)$ in (10.40), we have the equality as

$$w_i(\pi') = V^{\pi'}(i, m) - m g(\pi) + \sum_{j \in \mathscr{S}} P^{(m)}(j|i, \pi') w_j(\pi'), \quad i \in \mathscr{S}. \tag{10.41}$$

This equality can be re-written by replacing $w_j(\pi')$ with $w_j(\pi) - [w_j(\pi) - w_j(\pi')]$ as

$$\begin{aligned}
&V^{\pi'}(i, m) - m g(\pi) + \sum_{j \in \mathscr{S}} P^{(m)}(j|i, \pi') w_j(\pi) \\
&= w_i(\pi') + \sum_{j \in \mathscr{S}} P^{(m)}(j|i, \pi') \{w_j(\pi) - w_j(\pi')\}, \quad i \in \mathscr{S} \; m \geq 1.
\end{aligned}$$

Thus, using this equality in (10.40), we obtain

$$\begin{aligned}
w_i(\pi) &\geq c_i(\pi'(i)) - g(\pi) + \sum_{j \in \mathscr{S}} P(j|i, \pi'(i)) w_j(\pi) \\
&\geq w_i(\pi') + \sum_{j \in \mathscr{S}} P^{(m)}(j|i, \pi') \{w_j(\pi) - w_j(\pi')\}, \quad i \in \mathscr{S} \; m \geq 1,
\end{aligned}$$

where the strict inequality holds in the first inequality for each i with $\pi(i) \neq \pi'(i)$. Note that the second term on the left-hand side of the second inequality becomes 0 due to $w_j(\pi) = w_j(\pi')$ if $j \in R(\pi')$ or $P^{(m)}(j|i, \pi'(i)) \to 0$ as $m \to \infty$ if $j \notin R(\pi')$. Thus, we conclude that $w_i(\pi) \geq w_i(\pi')$ for all $i \in \mathscr{S}$. This completes the proof. ∎

As mentioned earlier, the proof of the convergence for the VI algorithm (Theorem 10.13) is quite technical. For the special case under a strong aperiodicity assumption, Tijms (2003) provides a proof of the VI algorithm convergence.

It is worth noting that in addition to the PI and VI algorithms there is the third approach to computing the optimal policy for the average cost DTMDP. That is to solve the optimality equation (10.29) by using linear program. We do not present this approach in this book. Interested readers are referred to Tijms (2003) for details of this approach.

10.3 SEMI-MARKOV DECISION PROCESS

10.3.1 CHARACTERIZING THE STRUCTURE OF OPTIMAL POLICY

Now we consider a continuous-time stochastic system with discrete state space and try to determine the structure of the optimal dynamic policy. We use a simple service system consisting of two separate queues which is modeled as a semi-Markov decision process (SMDP). Consider two single server queues that work in parallel. Both queues have finite waiting rooms of size C_i, and the service times are exponentially distributed with mean $1/\mu_i$ in queue $i, i = 1, 2$. Arrivals occur in this two-queue system according to a Poisson process with rate λ. At every arrival, a scheduler observes the number in the system in each queue and decides to take one of three control actions: reject the arrival, send the arrival to queue 1, or send the arrival to queue 2. Assume that the control actions happen instantaneously and customers cannot jump from one queue to the other or leave the system before their service is completed. The system earns a reward r dollars for every accepted customer and incurs a holding cost h_i dollars per unit time per customer held in queue i (for $i = 1, 2$). Assume that the reward and holding cost values are such that the scheduler rejects an arrival only if both queues are full. We want to determine the structure of the scheduler's optimal policy. Such a system can represent a make-to-order system which offers either tangible products or services. Customer orders arrive at a company's order processing office (OPO). The company has two service facilities in two different locations. These two facilities are heterogeneous in parameters h_i, μ_i, and C_i for $i = 1, 2$. h_i can be used as a proxy for the service quality of the facility; μ_i reflects the service capacity of the facility; and C_i can be set to ensure that the maximum expected waiting time is below a threshold.

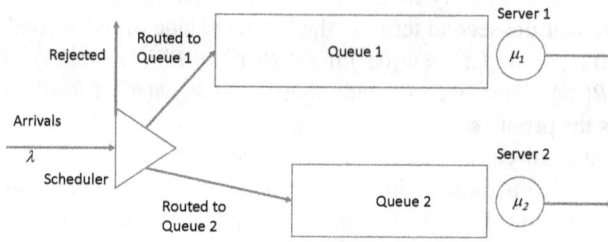

Figure 10.18 Customer allocation in a two queue system.

The optimization problem is to determine the best order allocation policy to maximize the total expected reward of the company. The system is depicted in Figure 10.18.

For $i = 1, 2$, let $X_i(t)$ be the number of customers in the system in queue i at time t (including the customer in service). If an arrival occurs at time t, the OPO looks at $X_1(t)$ and $X_2(t)$ to decide whether the arriving order should be rejected or allocated to facility 1 or facility 2. Note that because of the assumption that the OPO rejects a customer order only if both queues are full, the OPO's action in terms of whether to accept or reject a customer order is already made in such a full-buffer state. Also, if only one of the queues is full, then the assumption requires sending the arrival to the nonfull queue. Therefore, the problem is simplified to the one that decides which queue to send an arrival only when both buffers are not full (we also call such a policy as a routing policy, i.e., decision to send to queue 1 or 2 depending on the number in each queue). Intuitively, the optimal policy seems to be the one that sends an arriving request to queue i if it is "shorter" than queue $3 - i$ for $i = 1, 2$, also called "routing to the shortest queue". If $\mu_1 = \mu_2, C_1 = C_2$, and $h_1 = h_2$, then it can be shown that routing to the shorter queue is optimal. However, in the generic case, we expect a state-dependent threshold policy, or called switching curve, as the optimal policy (with joining the shorter queue being a special case of that).

Before deriving this result, we first illustrate the structure of the optimal policy in Figure 10.19. As shown in the figure, there are three regions. If an arrival occurs at time t when there are $X_1(t)$ in queue 1 and $X_2(t)$ in queue 2, then the optimal action taken depends on the values of $(X_1(t), X_2(t))$, which is the coordinates on the $X_1(t)$–$X_2(t)$ plane. In particular, if $(X_1(t), X_2(t))$ is in region 1, 2, or 3, then the optimal action is to send the arrival at time t to queue 1, queue 2, or reject, respectively. Note that region 3 is the single point (C_1, C_2). Although the threshold policy or switching curve in Figure 10.19 is quite intuitive, showing the optimality of this policy structure is not trivial. Since this is a typical optimization problem in queueing setting, we will present the main

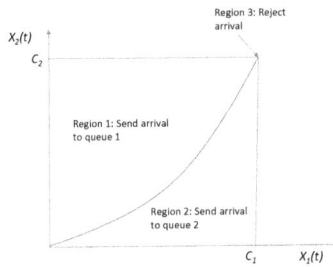

Figure 10.19   Structure of the optimal policy in a two queue system.

idea of the mathematical proof. First, we need some notations. Let $\mathbf{x}$ denote a vector $\mathbf{x} = (x_1, x_2)$ for an actual observed state at an arrival instant where x_1 and x_2 are the number of customers in queue 1 and queue 2, respectively. In essence, suppose $\mathbf{x}$ denotes the state of the system as seen by an arrival. Let e_i denote a unit vector in the ith dimension, for $i = 1, 2$. Obviously, $e_1 = (1, 0)$ and $e_2 = (0, 1)$. If the arriving customer is allocated to queue i, then the new state becomes $\mathbf{x} + e_i$. To show the monotonic switching curve in Figure 10.19 is optimal, all we need to show is the following:

- If the optimal policy in state $\mathbf{x}$ is to allocate to queue 1, then the optimal policy in states $\mathbf{x} + e_2$ and $\mathbf{x} - e_1$ would also be to allocate to queue 1.
- If the optimal policy in state $\mathbf{x}$ is to route to queue 2, then the optimal policy in states $\mathbf{x} + e_1$ and $\mathbf{x} - e_2$ would also be to allocate to queue 2.

These properties simply say that (i) if you route a customer order to queue 1 at state $\mathbf{x}$, then you will also do so for the state with even longer queue 2 or the state with even shorter queue 1; and (ii) if you route a customer order to queue 2 at state $\mathbf{x}$, then you will also do so for the state with even longer queue 1 or the state with even shorter queue 2.

To show these properties, we need to first formulate the problem as a semi-Markov decision process (SMDP) and then investigate the optimal policy in various states. Using the normalization method, we can treat an SMDP similarly as a DTMDP discussed in previous sections. We first define the value function $V(\mathbf{x})$, which is the maximum expected total discounted reward over an infinite horizon, starting from state $\mathbf{x} = (x_1, x_2)$. We also use the term "discounted value" because we consider a discount factor γ and $V(\mathbf{x})$ denotes the expected present value. It is common in the SMDP analysis to select appropriate time units so that $\gamma + \lambda + \mu_1 + \mu_2 = 1$ and therefore in our analysis to compute $V(\mathbf{x})$, we do not have to be concerned about γ (this procedure is also

called normalization and is explained below). To obtain $V(\mathbf{x})$, we first introduce some notations. Let $h(\mathbf{x})$ be the holding cost incurred per unit time in
state $\mathbf{x}$, that is, $h(\mathbf{x}) = h_1 x_1 + h_2 x_2$. Let a^+ denote $\max\{a, 0\}$ if a is a scalar
and $\mathbf{x}^+ = (x_1^+, x_2^+)$. We start with the optimality equation or Bellman equation.
The value function $V(\mathbf{x})$ satisfies the following optimality equation: for $x_1 \in$
$[0, C_1)$ and $x_2 \in [0, C_2)$.

$$
\begin{aligned}
V(\mathbf{x}) = &-h(\mathbf{x}) + \lambda \max\{r + V(\mathbf{x} + e_1), r + V(\mathbf{x} + e_2)\} \\
&+ \mu_1 V((\mathbf{x} - e_1)^+) + \mu_2 V((\mathbf{x} - e_2)^+).
\end{aligned}
\tag{10.42}
$$

In this SMDP, the state-change happens at either a customer arrival instant or
a customer departure instant (service completion instant). These state-change
instants are called embedded points. We are interested in the process on these
embedded points which forms a Markov chain. Since some of these embedded
points are decision epochs, the SMDP is also called a controlled Markov chain.
The optimality equation indicates the dynamic change of the value function
from the current state to a new state. Here we consider an infinite time horizon
and the optimality equation is written in a recursion. Since we focus on the
steady state of the system, we use the same value function at these consecutive
state change instants. Thus, if the left-hand side (LHS) represents the present
value at time instant n, then the right-hand side shows how this value can be
computed by using the value function at time instant $n + 1$. It is worth noting
that the optimality equation can be also written backward in a recursive way.
That is the LHS at time instant n can be written in terms of value function at
time $n - 1$. In (10.42), the first term is the holding cost of $h(x)$ per unit time on
the right-hand side (RHS). The second term represents the value of taking the
best of the two actions at each decision epoch (i.e., an arrival instant), either
routing to queue 1 or queue 2. The last two terms indicate the values of moving
to the next states at service completion instants. Since inter-arrival and service
times are all exponentially distributed, the time interval between two consecutive embedded points is also exponentially distributed. This is because that
the minimum of exponentially distributed random variables is also exponentially distributed. Note that the value of the SMDP is continuously discounted
with rate γ for any time interval. In fact, the RHS of the optimality equation
should be multiplied by a factor of $1/(\gamma + \lambda + \mu_1 + \mu_2)$, which becomes 1 if
$\gamma + \lambda + \mu_1 + \mu_2 = 1$ is set by normalization. This is because the time interval
from the current state x to the next embedded point is the minimum of three
exponential random variables, namely, the inter-arrival time, the service time
in queue 1, and the service time in queue 2. Thus, this time interval, denoted by
T, follows an exponential distribution with rate $\lambda + \mu_1 + \mu_2$. Consequently, the
expected cost or reward of the SMDP over this interval under a continuously
discounted factor can be computed easily if the cost rate or reward rate remains
constant or the cost or reward occurs as a lump sum at a state transition instant.

The following results are needed for our formulation.

Lemma 10.2 *If costs are incurred at rate of h per unit time over the time interval $[0, T]$ and T follows an exponential distribution with rate v, then the expected present value of the cost incurred over time $[0, T]$ at discount rate of γ is given by*

$$E \int_{t=0}^{T} e^{-\gamma t} h\, dt = \int_{t=0}^{\infty} \left(\int_{\tau=0}^{t} e^{-\gamma \tau} h\, d\tau \right) v e^{-vt}\, dt = \frac{h}{\gamma + v}.$$

Lemma 10.3 *Assume that $T_i, i = 1, 2, ..., n$ is a set of n independent and exponentially distributed time intervals, with rates v_i, respectively. Let $T := \min_i T_i$, be the minimum of these time intervals. Then (i) T follows an exponential distribution with rate $v = \sum_i v_i$; (ii) $P(T = T_i) = P(T = T_i | T = t) = v_i / v$ for every $t \geq 0$; (iii) If a lump sum cost C_i is incurred at time T if $T = T_i$, then the expected present value of the costs to be incurred at time T, denoted by C, at discount rate γ is*

$$E[e^{-\gamma T} C] = \frac{\sum_i v_i C_i}{\gamma + \sum_i v_i}.$$

The proofs of these results are left as exercises.

Exercise 10.21 *Prove Lemmas 10.2 and 10.3.*

For the SMDP with a finite state space, we need to take extra care of these boundary states for which the optimality equation may have different forms. In our model, if $x_i = 0$ for $i = 1$ or 2, then the actual equation is

$$V(\mathbf{x}) = -\frac{h(\mathbf{x})}{\gamma + \lambda + \mu_{3-i}} + \frac{\lambda}{\gamma + \lambda + \mu_{3-i}} \max\{r + V(\mathbf{x} + e_1),$$
$$r + V(\mathbf{x} + e_2)\} + \frac{\mu_{3-i}}{\gamma + \lambda + \mu_{3-i}} V((\mathbf{x} - e_{3-i})^+),$$

since server i cannot complete service as there are no customers. Thus, we cannot utilize the normalization condition $\gamma + \lambda + \mu_1 + \mu_2 = 1$ in this equation at the present form. This is mainly because the total transition rate for the boundary state is different from that of non-boundary state. To solve this problem, we utilize the uniformization technique (same idea as the one presented for converting a CTMC into a DTMC in Chapter 3). Assume that there is an "artificial transition" rate μ_i and such an transition is to the present state (i.e., no real transition made). Multiply both sides of the equation above by $(\gamma + \lambda + \mu_{3-i})$, we obtain $(\gamma + \lambda + \mu_{3-i})V(\mathbf{x}) = -h(\mathbf{x}) + \lambda \max\{r + V(\mathbf{x} + e_1), r + V(\mathbf{x} + e_2)\} + \mu_{3-i}V((\mathbf{x} - e_{3-i})^+)$. Now we can add $\mu_i V(\mathbf{x})$ to the LHS and $\mu_i V((\mathbf{x} - e_i)^+)$ to the RHS of this equation because of $\mu_i V(\mathbf{x}) = \mu_i V((\mathbf{x} - e_i)^+)$ for $x_i = 0$. Then we get the same form as (10.42) by applying the normalization condition. We can treat the optimality equations for other boundary states similarly. The general approach (also called Lippman's transformation, see Lippman (1975)) is to make the total transition rate (sum of all transition rates) for every state equal to the maximum of all possible total transition rates for all states by introducing artificial transitions to the current state. If we denote such a uniform total rate by v, then the effective discount factor is equal to $\gamma/(\gamma+v)$. In our example, $v = \lambda + \mu_1 + \mu_2$. To simplify the optimality equation for the structural analysis of the optimal policy, we can normalize $\gamma+v = 1$ by appropriate time scaling. For example, assume that γ_0 is a given discount rate per year (i.e., the time unit is one year). If we use $1/n$ year as the new time unit such that $n = \gamma_0/[1 - (\lambda + \mu_1 + \mu_2)]$, then $\gamma = \gamma_0/n$ will satisfy the normalization condition. Thus, we can focus on (10.42) to characterize the structure of the optimal policy. There are sufficient conditions as follows that $V(\mathbf{x})$ must satisfy for the switching curve policy:

$$V(\mathbf{x}+e_2+e_1) - V(\mathbf{x}+e_2+e_2) \geq V(\mathbf{x}+e_1) - V(\mathbf{x}+e_2), \qquad (10.43)$$
$$V(\mathbf{x}+e_1+e_2) - V(\mathbf{x}+e_1+e_1) \geq V(\mathbf{x}+e_2) - V(\mathbf{x}+e_1), \qquad (10.44)$$
$$V(\mathbf{x}+e_2) - V(\mathbf{x}+e_1+e_2) \geq V(\mathbf{x}) - V(\mathbf{x}+e_1). \qquad (10.45)$$

It is relatively straightforward to check that if these conditions are met, then: (i) if the optimal policy in state $\mathbf{x}$ is to route to queue 1, then the optimal policy in states $\mathbf{x}+e_2$ and $\mathbf{x}-e_1$ would also be to route to queue 1; (ii) if the optimal policy in state $\mathbf{x}$ is to route to queue 2, then the optimal policy in states $\mathbf{x}+e_1$ and $\mathbf{x}-e_2$ would also be to route to queue 2. For example Condition (10.43) ensures that if the optimal policy in state $\mathbf{x}$ is to route to queue 1, then the optimal policy in state $\mathbf{x}+e_2$ would also be to route to queue 1; if the optimal policy in state $\mathbf{x}$ is to route to queue 2, then the optimal policy in state $\mathbf{x}-e_2$ would also be to route to queue 2 (it follows by reversing the inequality of (10.43)). While conditions (10.43) and (10.44) are straightforward, condition (10.45) is not intuitive. The reason that it is needed is because this third condition is used in the proof of the two intuitive conditions. Such a proof is left as an exercise. We can use mathematical induction to prove that the value function satisfies these conditions.

Denote by V^* the set that the optimal value function $V(\mathbf{x})$ belongs to. Conditions (10.43), (10.44), and (10.45) imply that V^* is a set of structured value functions. Hence, Theorem 10.8 applies. The proof below verifies the preservation and attainment conditions in the theorem.

This proof is based on the value iteration presented in the previous section for numerically compute $V(\mathbf{x})$. Let $V_0(\mathbf{x}) = 0$ for all $\mathbf{x}$ as the initial values. The value function at the $(n+1)$th iteration is given by

$$\begin{aligned} V_{n+1}(\mathbf{x}) = {}&-h(\mathbf{x}) + \lambda \max\{r+V_n(\mathbf{x}+e_1), r+V_n(\mathbf{x}+e_2)\} \\ &+ \mu_1 V_n((\mathbf{x}-e_1)^+) + \mu_2 V_n((\mathbf{x}-e_2)^+), \end{aligned} \qquad (10.46)$$

for $n = 0, 1, 2, \ldots.$ Clearly, $V_0(\mathbf{x})$ satisfies conditions (10.43),(10.44), and (10.45). Next, we assume that $V_n(\mathbf{x})$ satisfies all these conditions. Based on this, we need to show that $V_{n+1}(\mathbf{x})$ also satisfies these conditions. Therefore, $V_k(\mathbf{x})$ satisfies these conditions for all k so that $V(\mathbf{x}) = \lim_{k \to \infty} V_k(\mathbf{x})$ will also satisfy these conditions.

Assuming that $V_n(\mathbf{x})$ satisfies conditions (10.43), (10.44), and (10.45), we need to show that all terms on the RHS of (10.46) satisfy these conditions. These terms involve the following functions: (i) the holding cost function $h(x)$; (ii) the max function, denoted by $g(\mathbf{x}) = \max\{r + V_n(\mathbf{x} + e_1), r + V_n(x + e_2)\}$; and (iii) the value function after a service completion, $V_n(\mathbf{x} - e_i)^+)$ for $i = 1, 2$. Note that showing that (iii) is satisfied is relatively easy and left as an exercise for readers (Note that condition (10.45) is needed to prove (iii)). It is also quite easy to show (i) satisfies all conditions. For example, to show $h(\mathbf{x})$ satisfies (10.43), we write

$$h(\mathbf{x} + e_2 + e_1) - h(\mathbf{x} + e_2 + e_2) = h_1(x_1 + 1) + h_2(x_2 + 1) - h_1 x_1 - h_2(x_2 + 2)$$
$$= h_1 - h_2,$$

and

$$h(\mathbf{x} + e_1) - h(\mathbf{x} + e_2) = h_1(x_1 + 1) + h_2 x_2 - h_1 x_1 - h_2(x_2 + 1) = h_1 - h_2.$$

Combining the equations above yields that $h(\mathbf{x})$ satisfies condition (10.43). Other conditions satisfied by $h(\mathbf{x})$ can be shown similarly. Now we only need to prove that (ii) satisfies all conditions and $V_n(\mathbf{x})$ is non-increasing. While the latter is easy to show by mathematical induction (showing $V_n(x)$ is non-increasing for all n is left as an exercise for readers), the former needs more work. To show $g(\mathbf{x})$ satisfies all conditions given $V_n(\mathbf{x})$ satisfies them for all $\mathbf{x}$, we start with condition (10.45) and consider four cases representing all possible actions in states $\mathbf{x}$ and $\mathbf{x} + e_1 + e_2$ (i.e., (x_1, x_2) and $(x_1 + 1, x_2 + 1)$).

Case 1: Action in $\mathbf{x}$ is to allocate to Q1 and in $\mathbf{x} + e_1 + e_2$ is to allocate to Q1. This implies $g(\mathbf{x}) = r + V_n(\mathbf{x} + e_1)$ and $g(\mathbf{x} + e_1 + e_2) = r + V_n(\mathbf{x} + 2e_1 + e_2)$. To prove (10.45), we also need $g(\mathbf{x} + e_1)$ and $g(\mathbf{x} + e_2)$. Based on the definition of $g(\mathbf{x})$, we can obtain two inequalities for these terms. They are $g(\mathbf{x} + e_1) = \max\{r + V_n(\mathbf{x} + 2e_1), r + V_n(\mathbf{x} + e_1 + e_2)\} \geq r + V_n(\mathbf{x} + 2e_1)$ and $g(\mathbf{x} + e_2) = \max\{r + V_n(\mathbf{x} + e_1 + e_2), r + V_n(\mathbf{x} + 2e_2)\} \geq r + V_n(\mathbf{x} + e_1 + e_2)$. Using these relations, we can verify that $g(\mathbf{x})$ satisfies (10.45)) as follows:

$$g(\mathbf{x} + e_2) - g(\mathbf{x} + e_1 + e_2) - g(\mathbf{x}) + g(\mathbf{x} + e_1)$$
$$= g(\mathbf{x} + e_2) - r - V_n(\mathbf{x} + 2e_1 + e_2) - r - V_n(\mathbf{x} + e_1) + g(\mathbf{x} + e_1)$$
$$\geq r + V_n(\mathbf{x} + e_1 + e_2) - r - V_n(\mathbf{x} + 2e_2 + e_2) - r - V_n(\mathbf{x} + e_1) + r$$
$$+ V_n(\mathbf{x} + 2e_1)$$
$$= V_n(\mathbf{x} + e_1 + e_2) - V_n(\mathbf{x} + 2e_1 + e_2) - V_n(\mathbf{x} + e_1) + V_n(\mathbf{x} + 2e_1) \geq 0.$$

The last inequality follows from the fact that $V_n(\mathbf{x}+e_1)$ satisfies condition (10.45). Thus, $g(\mathbf{x})$ satisfies (10.45).

Case 2: Action in $\mathbf{x}$ is to allocate to Q1 and in $\mathbf{x}+e_1+e_2$ is to allocate to Q2. This implies $g(\mathbf{x}) = r + V_n(\mathbf{x}+e_1)$ and $g(\mathbf{x}+e_1+e_2) = r + V_n(\mathbf{x}+e_1+2e_2)$. Again, we have $g(\mathbf{x}+e_1) = \max\{r + V_n(\mathbf{x}+2e_1), r + V_n(\mathbf{x}+e_1+e_2)\} \geq r + V_n(\mathbf{x}+e_1+e_2)$ and $g(\mathbf{x}+e_2) = \max\{r + V_n(\mathbf{x}+e_1+e_2), r + V_n(\mathbf{x}+2e_2)\} \geq r + V_n(\mathbf{x}+e_1+e_2)$. Using these relations, we can verify that $g(\mathbf{x})$ satisfies (10.45)) as follows:

$$
\begin{aligned}
&g(\mathbf{x}+e_2) - g(\mathbf{x}+e_1+e_2) - g(\mathbf{x}) + g(\mathbf{x}+e_1) \\
&= g(\mathbf{x}+e_2) - r - V_n(\mathbf{x}+e_1+2e_2) - r - V_n(\mathbf{x}+e_1) + g(\mathbf{x}+e_1) \\
&\geq r + V_n(\mathbf{x}+e_1+e_2) - r - V_n(\mathbf{x}+e_1+2e_2) - r - V_n(\mathbf{x}+e_1) + r \\
&\quad + V_n(\mathbf{x}+e_1+e_2) \\
&= V_n(\mathbf{x}+e_1+e_2) - V_n(\mathbf{x}+e_1+2e_2) - V_n(\mathbf{x}+e_1) + V_n(\mathbf{x}+e_1+r_2) \geq 0.
\end{aligned}
$$

The last inequality follows by adding (10.43) and (10.45) at state $\mathbf{x}+e_1$. This also shows that $V_n(\mathbf{x}+e_1)$ is cancave in x_2 for a fixed x_1+1. Thus, $g(\mathbf{x})$ satisfies (10.45).

Case 3: Action in $\mathbf{x}$ is to allocate to Q2 and in $\mathbf{x}+e_1+e_2$ is to allocate to Q1. This implies $g(\mathbf{x}) = r + V_n(\mathbf{x}+e_2)$ and $g(\mathbf{x}+e_1+e_2) = r + V_n(\mathbf{x}+2e_1+e_2)$. This situation is a symmetric to Case 2. Thus, we can show that $g(\mathbf{x})$ satisfies (10.45) similarly.

Case 4: Action in $\mathbf{x}$ is to allocate to Q2 and in $\mathbf{x}+e_1+e_2$ is to allocate to Q2. This implies $g(\mathbf{x}) = r + V_n(\mathbf{x}+e_2)$ and $g(\mathbf{x}+e_1+e_2) = r + V_n(\mathbf{x}+e_1+2e_2)$. This situation is a symmetric to Case 1. Thus, we can show that $g(\mathbf{x})$ satisfies (10.45) similarly.

Next, we show that $g(\mathbf{x}) = \max\{r + V_n(\mathbf{x}+e_1), r + V_n(\mathbf{x}+e_2)\}$ satisfies (10.44) if $V_n(\mathbf{x})$ satisfies (10.43), (10.44), and (10.45) for all $\mathbf{x}$. We again consider all possible actions in the following states $\mathbf{x}+e_2 = (x_1, x_2 + 1)$ and $\mathbf{x}+2e_1 = (x_1 + 2, x_2)$.

Case 1: Action in $\mathbf{x}+e_2$ is to allocate to Q1 and in $\mathbf{x}+2e_1$ is to allocated to Q1. This implies $g(\mathbf{x}+e_2) = r + V_n(\mathbf{x}+e_1+e_2)$ and $g(\mathbf{x}+2e_1) = r + V_n(\mathbf{x}+3e_1)$. To prove (10.44), we also need $g(\mathbf{x}+e_1+e_2)$ and $g(\mathbf{x}+e_1)$. Based on the definition of $g(\mathbf{x})$, we can obtain two inequalities for these terms. They are $g(\mathbf{x}+e_1+e_2) = \max\{r + V_n(\mathbf{x}+2e_1+e_2), r + V_n(\mathbf{x}+e_1+2e_2)\} \geq r + V_n(\mathbf{x}+2e_1+e_2)$ and $g(\mathbf{x}+e_1) = \max\{r + V_n(\mathbf{x}+2e_1), r + V_n(\mathbf{x}+e_1+e_2)\} \geq r + V_n(\mathbf{x}+2e_1)$. Using these relations, we can verify that verify that $g(\mathbf{x})$ satisfies

(10.44)) as follows:

$$
\begin{aligned}
& g(\mathbf{x}+e_1+e_2) - g(\mathbf{x}+e_1+e_1) - g(\mathbf{x}+e_2) + g(\mathbf{x}+e_1) \\
&= g(\mathbf{x}+e_1+e_2) - r - V_n(\mathbf{x}+3e_1) - r - V_n(\mathbf{x}+e_1+e_2) + g(\mathbf{x}+e_1) \\
&\geq r + V_n(\mathbf{x}+2e_1+e_2) - r - V_n(\mathbf{x}+3e_1) - r - V_n(\mathbf{x}+e_1+e_2) + r \\
&\quad + V_n(\mathbf{x}+2e_1) \\
&= V_n(\mathbf{x}+2e_1+e_2) - V_n(\mathbf{x}+3e_1) - V_n(\mathbf{x}+e_1+e_2) + V_n(\mathbf{x}+2e_1) \geq 0.
\end{aligned}
$$

The last inequality follows from the fact that $V_n(\mathbf{x}+e_1)$ satisfies condition (10.44). Thus, $g(\mathbf{x})$ satisfies (10.44).

Case 2: Action in $\mathbf{x}+e_2$ is to allocate to Q1 and in $\mathbf{x}+2e_1$ is to allocate to Q2. This implies $g(\mathbf{x}+e_2) = r + V_n(\mathbf{x}+e_1+e_2)$ and $g(\mathbf{x}+2e_1) = r + V_n(\mathbf{x}+2e_1+e_2)$. In addition, we can obtain two inequalities. They are $g(\mathbf{x}+e_1+e_2) = \max\{r + V_n(\mathbf{x}+2e_1+e_2), r + V_n(\mathbf{x}+e_1+2e_2)\} \geq r + V_n(\mathbf{x}+2e_1+e_2)$ and $g(\mathbf{x}+e_1) = \max\{r + V_n(\mathbf{x}+2e_1), r + V_n(\mathbf{x}+e_1+e_2)\} \geq r + V_n(\mathbf{x}+e_1+e_2)$ (choosing these terms with reference of $g(\mathbf{x}+e_2)$, $g(\mathbf{x}+2e_1)$, and (10.44)). Using these these relations, we can verify that $g(\mathbf{x})$ satisfies (10.44)) as follows:

$$
\begin{aligned}
& g(\mathbf{x}+e_1+e_2) - g(\mathbf{x}+e_1+e_1) - g(\mathbf{x}+e_2) + g(\mathbf{x}+e_1) \\
&= g(\mathbf{x}+e_1+e_2) - r - V_n(\mathbf{x}+2e_1+e_2) - r - V_n(\mathbf{x}+e_1+e_2) + g(\mathbf{x}+e_1) \\
&\geq r + V_n(\mathbf{x}+2e_1+e_2) - r - V_n(\mathbf{x}+2e_1+e_2) - r - V_n(\mathbf{x}+e_1+e_2) + r \\
&\quad + V_n(\mathbf{x}+e_1+e_2) = 0.
\end{aligned}
$$

Thus, $g(\mathbf{x})$ satisfies (10.44).

Case 3: Action in $\mathbf{x}+e_2$ is to allocate to Q2 and in $\mathbf{x}+2e_1$ is to allocate to Q1. This implies $g(\mathbf{x}+e_2) = r + V_n(\mathbf{x}+2e_2)$ and $g(\mathbf{x}+2e_1) = r + V_n(\mathbf{x}+3e_1)$. This is symmetric to case 2. Thus, we can show that $g(\mathbf{x})$ satisfies (10.44) similarly.

Case 4: Action in $x+e_2$ is to allocate to Q2 and in $x+2e_1$ is to allocate to Q2. This implies $g(\mathbf{x}+e_2) = r + V_n(\mathbf{x}+2e_2)$ and $g(\mathbf{x}+2e_1) = r + V_n(\mathbf{x}+2e_1+e_1)$. This is symmetric to case 1. Thus, we can show that $g(\mathbf{x})$ satisfies (10.44) similarly.

Thus, $g(\mathbf{x})$ satisfies condition (10.44) if $V_n(\mathbf{x})$ satisfies conditions (10.43), (10.44), and (10.45). Finally, we can use the same approach to show that $g(\mathbf{x})$ satisfies condition (10.43) if $V_n(\mathbf{x})$ satisfies (10.43), (10.44), and (10.45) for all $\mathbf{x}$. This concludes the mathematical induction. Using the limit argument, this result implies that $V(\mathbf{x})$ satisfies all these conditions. Thus, the optimal policy has the monotonic switching curve structure. Although we have confirmed the structure of the optimal policy, we do not know how the exact optimal policy is computed. A benefit of characterizing the structure of the optimal policy is to simplify the computation of the optimal policy.

The analysis presented in this section indicates that to characterize the structure of the optimal policy, the value function must possess certain properties.

Another important issue in stochastic optimization is to investigate the monotonicity of the optimal policies as mentioned earlier in treating inventory models. The conditions required to prove that optimal policies are monotone with system state also depend on properties of value functions. These properties are submodularity and/or supermodularity which have the economic interpretations based on the definition of economic complements and economic substitutes. Combined with the concavity and convexity properties, submodularity and supermodularity are required to analyze the monotonicity of optimal policies in some stochastic inventory systems. Since this is a rich area of research, interested readers are referred to Topkis (1998) for a comprehensive presentation on supermodularity and complementarity and Porteus (2002) for more discussion on monotonicity and structure of optimal policeis in stochastic inventory models.

Exercise 10.22 *For the two queue service system discussed in the section above, show (i) $V_n(x)$ is non-increasing for all n; (ii) if $V_n(x)$ satisfies conditions (10.43), (10.44) and (10.45), then $V_n((x-e_i)^+)$ for $i = 1, 2$ also satisfies these conditions.*

Exercise 10.23 *To determine the switching curve numerically for a two queue system with finite buffers discussed above, one can model the system under a policy with the switching curve structure as a CTMC and evaluate its performance using the stationary distribution. Assume that the system with Poisson arrival process with rate λ and the exponentially distributed service times for the two queues with rates μ_1 and μ_2, respectively has the following parameters: $C_1 = 5, C_2 = 5$ (buffer sizes), $\lambda = 5, \mu_1 = 2, \mu_2 = 3, r = 10, h_1 = 2.0$, and $h_2 = 1.5$. Formulate a CTMC and determine the optimal switching curve numerically. (Hint: An algorithm could be to start with the switching curve being the straight line from (0,0) to (C_1, C_2) and evaluate the expected discounted net benefit (via stationary distribution of the CTMC). Then try all possible neighbors to determine the optimal switching curve that would maximize the expected discounted net benefit.)*

10.3.2   COMPUTATIONAL APPROACH TO SMDP

Similar to the DTMDP, for some SMDPs, the structure of the optimal policy may not be characterized. Then, the optimal policy can be determined numerically by using either the PI or VI algorithm. Here, we present another "customer-routing" problem which is modeled as a SMDP and solved by using the VI algorithm. The problem setting here is different from that of the previous section in several aspects. (i) There are two types of customers; (ii) There are $c > 2$ *identical* servers and each server can serve both types of customers; (iii) There is no waiting space (no buffer) for customers; and (iv) The "routing decision" is not to send an arriving customer to which server but to either accept or reject the arriving customer due to the homogeneous servers.

Suppose that customers of types 1 and 2 arrive at the system with c identical servers without buffer according to two independent Poisson processes with rates λ_1 and λ_2, respectively. The service times for type 1 (type 2) customers are i.i.d. exponential random variables with rate μ_1 (μ_2). If a customer arrives at the system with all c servers being busy, he must be rejected; otherwise, he can be either accepted or rejected. Thus, the problem is to determine the optimal acccept/reject decisions for non-full states (i.e., fewer than c busy server states) to minimize the average rejection rate or, equivalently, to maximize the average throughput rate. This problem can be solved numerically by formulating it as a SMDP. Clearly, we use the customer arrival instants as the only decision instants. Due to the multi-server feature, the one-step transition probability matrix under a stationary policy will have many non-zero elements. To simplify the transition probability matrix, we include the service completion instants as additional decision instants that are fictitious as no real decision is made at these instants. With both customer arrival and service completion instants as decision epochs, the state transition is always to one of at most four neighboring states. Define the state space as

$$ \mathscr{S} = \{(i_1, i_2, k) | i_1, i_2 = 0, 1, ..., c; \ i_1 + i_2 \leq c; \ k = 0, 1, 2\}, $$

where i_1 and i_2 are the number of type 1 and type 2 customers, respectively, currently in service at a decision instant and k represents the type of decision epoch with 1 and 2 for type 1 and type 2 customer arrival instants, respectively, and 0 for a customer departure (or service completion) instant. Denote by $a = 0$ and $a = 1$ the rejecting and accepting decisions, respectively. Note that $a = 0$ decision is the only feasible decision for the states of $i_1 + i_2 = c$ (i.e., all servers are busy) and the states at service completion instants $(i_1, i_2, 0)$. Let $v(i_1, i_2) = \lambda_1 + \lambda_2 + i_1 \mu_1 + i_2 \mu_2$. Since inter-arrival times and service times are all exponentially distributed, we can write the transition probabilities for the states at decision instants (embedded points), denoted by $P(y|x, a)$, and the expected time until next decision instant, denoted by $\tau_x(a)$, as follows:

If action $a = 0$ is taken in state $x = (i_1, i_2, k)$, then

$$ P(y|x, 0) = \begin{cases} \lambda_1/v(i_1, i_2), & y = (i_1, i_2, 1), \\ \lambda_2/v(i_1, i_2), & y = (i_1, i_2, 2), \\ i_1 \mu_1/v(i_1, i_2), & y = (i_1 - 1, i_2, 0), \\ i_2 \mu_2/v(i_1, i_2), & y = (i_1, i_2 - 1, 0), \end{cases} $$

and $\tau_x(0) = 1/v(i_1, i_2)$.

If action $a = 1$ is taken in state $x = (i_1, i_2, 1)$, then

$$P(y|x, 1) = \begin{cases} \lambda_1/v(i_1+1, i_2), & y = (i_1+1, i_2, 1), \\ \lambda_2/v(i_1+1, i_2), & y = (i_1+1, i_2, 2), \\ (i_1+1)\mu_1/v(i_1+1, i_2), & y = (i_1, i_2, 0), \\ i_2\mu_2/v(i_1+1, i_2), & y = (i_1+1, i_2-1, 0), \end{cases}$$

and $\tau_x(1) = 1/v(i_1+1, i_2)$. Similarly, we can write the transition probabilities and the expected time until next decision instant if action $a = 1$ is taken in state $x = (i_1, i_2, 2)$. To minimize the rejection rate, it is assumed that the one-step expected costs $c_x(a)$ are given by

$$c_x(a) = \begin{cases} 1, & x = (i_1, i_2, 1) \text{ and } a = 0, \\ 1, & x = (i_1, i_2, 2) \text{ and } a = 0, \\ 0, & \text{otherwise.} \end{cases}$$

To determine the optimal policy of a SMDP by using the VI algorithm, we can transform the original SMDP to a DTMDP via the uniformization technique. In such a transformation, the one-step expected cost in the DTMDP is given by $\bar{c}_i(a) = c_i(a)/\tau_i(a)$. The VI algorithm can be given as follows:

1. Initialize stationary value function $0 \le \hat{V}(i, 0) \le \min_a\{\bar{c}_i(a)\}$ for all i. Choose a number τ with $0 < \tau \le \min_{i,a} \tau_i(a)$. Let $n := 1$.
2. Compute the function $\hat{V}(i, n), i \in \mathscr{S}$, from

$$\hat{V}(i, n) = \min_{a \in A(i)} \left[\frac{c_i(a)}{\tau_i(a)} + \frac{\tau}{\tau_i(a)} \sum_{j \in \mathscr{S}} P(j|i, a)\hat{V}(j, n-1) \right.$$
$$\left. + \left(1 - \frac{\tau}{\tau_i(a)}\right)\hat{V}(i, n-1) \right]. \tag{10.47}$$

Let $\pi(n)$ be a stationary policy whose actions minimize the right-hand side of (10.47).
3. Compute the bounds

$$lb_n = \min_{j \in \mathscr{S}}\{\hat{V}(j, n) - V(j, n-1)\}, \quad ub_n = \max_{j \in \mathscr{S}}\{\hat{V}(j, n) - \hat{V}(j, n-1)\}.$$

The algorithm is stopped with policy $\pi(n)$ when $0 \le (ub_n - lb_n) \le \varepsilon lb_n$, where ε is a pre-specified accuracy number. Otherwise, let $n := n+1$ and go to Step 1.

To apply this algorithm in the "customer routing" example, we set $\tau = 1/(\lambda_1 + \lambda_2 + c\mu_1 + c\mu_2)$. The recursive relations can be written for two classes of decision instants. For the states at customer departure instants, there is no choice to make and the recursion is simply given by

$$\hat{V}((i_1,i_2,0),n) = \tau\lambda_1\hat{V}((i_1,i_2,1),n-1) + \tau\lambda_2\hat{V}((i_1,i_2,2),n-1)$$
$$+ \tau i_1\mu_1\hat{V}((i_1-1,i_2,0),n-1) + \tau i_2\mu_1\hat{V}((i_1,i_2-1,0),n-1)$$
$$+ [1-\tau\nu(i_1,i_2)]\hat{V}((i_1,i_2,0),n-1),$$

where $\hat{V}((i_1,i_2,1),n-1) = 0$ when $i_1 < 0$ or $i_2 < 0$. For the states at customer arrival instants, a decision of rejecting $a = 0$ or accepting $a = 1$ has to be made. The recursion for a state with a type 1 customer arrival is given by

$$\hat{V}((i_1,i_2,1),n) = \min\big\{v(i_1,i_2) + \tau\lambda_1\hat{V}((i_1,i_2,1),n-1)$$
$$+ \tau\lambda_2\hat{V}((i_1,i_2,2),n-1)$$
$$+ \tau i_1\mu_1\hat{V}((i_1-1,i_2,0),n-1) + \tau i_2\mu_1\hat{V}((i_1,i_2-1,0),n-1)$$
$$+ [1-\tau\nu(i_1,i_2)]\hat{V}((i_1,i_2,0),n-1),$$
$$\tau\lambda_1\hat{V}((i_1+1,i_2,1),n-1) + \tau\lambda_2\hat{V}((i_1+1,i_2,2),n-1)$$
$$+ \tau(i_1+1)\mu_1\hat{V}((i_1,i_2,0),n-1)$$
$$+ \tau i_2\mu_2\hat{V}((i_1+1,i_2-1,0),n-1)$$
$$+ [1-\tau\nu(i_1+1,i_2)]\hat{V}((i_1,i_2,1),n-1)\big\},$$

where the first part inside the curly brackets of the right-hand side is for the value function of taking $a = 0$ and the second part for $a = 1$. To exclude the infeasible decision $a = 1$ for the states $(i_1,i_2,1)$ with $i_1 + i_2 = c$, we set $\hat{V}((i_1,i_2,1),n-1) = \hat{V}((i_1,i_2,2),n-1) = \infty$ when $i_1 + i_2 = c+1$. Similarly, we can write the value function recursion for a state at a type 2 customer arrival instant. With these recursions, we can implement the VI algorithm to compute the optimal policy for this queueing system. Numerical examples reveal that the optimal policy has also a "switching curve" structure. Under an optimal policy, without loss of generality, the customers with higher priority are labeled type 1 and the customers with lower priority are labeled type 2. Type 1 customers are always accepted as long as not all servers are busy. However, a type 2 arriving customer is only accepted when the number of type 2 customers in the system is less than a threshold $i_2^*(i_1)$, which is a function of the number of type 1 customers present at this arrival instant i_1. We present a numerical example (taken from Tijms (2003)) with $c = 10, \lambda_1 = 10, \lambda_2 = 7, \mu_1 = 10$, and $\mu_2 = 1$. Using the VI algorithm, we obtain the optimal policy in terms of the set of thresholds for type 2 customers:

$$i_2^*(0) = i_2^*(1) = 8, \ i_2^*(2) = i_2^*(3) = 7, \ i_2^*(4) = 6,$$
$$i_2^*(5) = 5, \ i_2^*(6) = 4, \ i_2^*(7) = 3, \ i_2^*(8) = 2, \ i_2^*(9) = 1.$$

Graphically, these thresholds exhibit two regions separated by a "switching curve" with one region for "accepting" and the other for "rejecting" of type 2

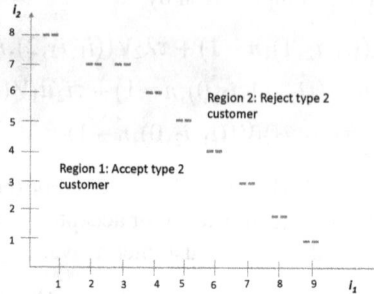

Figure 10.20   Structure of the optimal policy in two queue system.

customers (see Figure 10.20). However, there is no theoretical proof for this optimal policy structure at this time.

Exercise 10.24 *For the multi-server system with two types of customers and no buffer, use a mathematical software package such as Matlab or Mathematica to implement the VI algorithm. (You can use the parameters given in the numerical example illustrated in Figure 10.20).*

Exercise 10.25 *Develop the PI algorithm for a SMDP by using the uniformization technique. (Hint: Argue that the following optimality equation holds:*

$$v_i^* = \min_{a \in A(i)} \left\{ c_i(\pi(i)) - g^* \tau_i(a) + \sum_{j \in \mathscr{S}} P(j|i, \pi(i)) v_j^* \right\}, \quad i \in \mathscr{S}.$$

Then, design the algorithm.

10.4   STOCHASTIC GAMES – AN EXTENSION OF MDP

A stochastic game can be viewed as an extension of an MDP in which there are multiple decision makers with possibly conflicting goals, and their joint actions (decisions) determine state transitions and one-step rewards. In this sense, the traditional MDP discussed in this chapter can be considered as a one decision maker MDP. In this section, we provide a brief introduction to the stochastic game. A stochastic game with N players is defined by:

- A set of states $\mathscr{S}$ with a stage game defined for each state.
- A set of actions $\mathscr{A}_i(s)$ for each player i with $i = 1, 2, ..., N$ at each state $s \in \mathscr{S}$.

- A transition probability matrix **P**, which defines the probability distribution over next states given the current state and actions of the players. That is $P(S_{t+1}|S_t,A_{1t},A_{2t},...,A_{Nt})$ where $S_{t+1} = s' \in \mathscr{S}$, $S_t = s \in \mathscr{S}$ and $A_{it} = a_i \in \mathscr{A}_i(s)$ with $i = 1,2,...N$.
- A one-step reward/cost function $R_i(S_{t+1}|S_t,A_{1t},A_{2t},...,A_{Nt})$. If not dependent on S_{t+1}, it can be written as $R_i(S_t,A_{1t},A_{2t},...,A_{Nt})$.
- Each stage game is played at a set of discrete times t (i.e., discrete-time MDP).

We consider the simplest case by making the following assumptions:

1. Two player game (i.e., $N = 2$).
2. The length of the game is not known (infinite time horizon or episodic).
3. The objective functions for players are the discounted rewards.
4. The transition probabilities and one-step reward functions are independent.

We work on a numerical example of this simple stochastic game as follows. Assume that two players play a game with two states $\mathscr{S} = \{x,y\}$. The action sets for these two players in states x and y are given by

$$\mathscr{A}_1(x) = \{a,b\} \text{ and } \mathscr{A}_2(x) = \{c,d\}$$
$$\mathscr{A}_1(y) = \{e\} \text{ and } \mathscr{A}_2(y) = \{f\}.$$

The transition probabilities from state x are given by

$$[P(S_{t+1} = (x,y)|S_t = x, A_{1t} = \{a,b\}, A_{2t} = \{c,d\})]_{2\times2} = \begin{matrix} & c & d \\ a & (0.5,0.5) & (1,0) \\ b & (1,0) & (1,0) \end{matrix}. \tag{10.48}$$

The transition probabilities from state y are given by

$$[P(S_{t+1} = (x,y)|S_t = y, A_{1t} = \{e\}, A_{2t} = \{f\})]_{1\times1} = \begin{matrix} f \\ e & ((0,1)) \end{matrix}. \tag{10.49}$$

The one-step reward function from state x is given by

$$[R(S_t = x, A_{1t} = \{a,b\}, A_{2t} = \{c,d\}) = (r_1,r_2)]_{2\times2} = \begin{matrix} & c & d \\ a & (8,4) & (5,3) \\ b & (1,5) & (2,6) \end{matrix}. \tag{10.50}$$

The one-step reward function from state y is given by

$$[R(S_t = y, A_{1t} = \{e\}, A_{2t} = \{f\}) = (r_1,r_2)]_{1\times1} = \begin{matrix} f \\ e & ((0,0)) \end{matrix}. \tag{10.51}$$

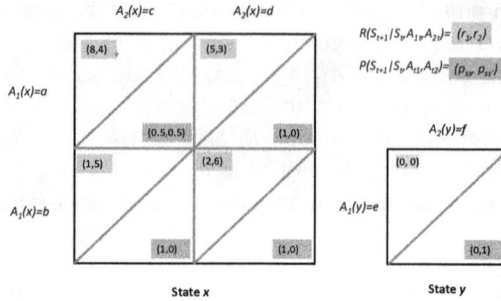

Figure 10.21 Transition probabilities and one-step reward function for the two-player stochastic game.

Figure 10.21 summarizes the information given above. Imagine that the two players may play a series of games which terminate by chance. We can call x "game on" state and y "game off" state. Starting from state x, each player has two actions to choose: "certainty play next game" or "chance play next game". That is for player 1 (2), a (c) is the action of chance play next game and b (d) is the action of certainty play next game. As long as one player chooses the certainty play action, both players will surely play the next game. If both players choose the chance play the next game (i.e., $(A_1(x), A_2(x)) = (a, c)$), then with 50% probability, these two players can play the next game and with 50% probability that the game is terminated (i.e., reaching state y). Based on the one-step rewards, for the one stage game, player 1 prefers the chance play action (i.e., he or she prefers a to b). Although player 2 prefers d (certainty play) to c (chance play) if player 1 chooses b (certainty play), player 2 also prefers c (chance play) to d (certainty play) if player 1 chooses a (chance play). Since player 1 chooses a, then player 2 must choose c. Thus, for the stage game, strategy (a, c) is a Nash equilibrium. However as soon as the players implement strategy pair (a, c) they will potentially go to state y (with 50% probability), the game terminating state at which players gain no further reward. Thus, (a, c) may not be the Nash equilibrium for the stochastic game with multiple stages. This claim is confirmed next. To calculate the total rewards for players in infinite horizon or episodic stochastic games we use a discount factor γ. Denote player 1's policy by π_1 and player 2's policy by π_2. Under a policy pair (π_1, π_2), the value function of player i at state x satisfies

$$V_i^{\pi_1, \pi_2}(x) = R_i(x, \pi_1(x), \pi_2(x)) + \gamma \sum_{x' \in \mathscr{S}} P(x'|x, \pi_1(x), \pi_2(x)) V_i^{\pi_1, \pi_2}(x').$$

(10.52)

Let $V_i^*(x)$ be the expected total value to player i starting from state x if both players are following the optimal policies, also called Nash strategies.

The Nash strategies for the two-player case satisfy

$$V_1^*(x) = \max_{\pi_1(x) \in \mathscr{A}_1(x)} \left(R_1(x, \pi_1(x), \pi_2^*(x)) + \gamma \sum_{x' \in \mathscr{S}} P(x'|x, \pi_1(x), \pi_2^*(x)) V_1^*(x') \right)$$

$$V_2^*(x) = \max_{\pi_2(x) \in \mathscr{A}_2(x)} \left(R_2(x, \pi_1^*(x), \pi_2(x)) + \gamma \sum_{x' \in \mathscr{S}} P(x'|x, \pi_1^*(x), \pi_2(x)) V_2^*(x') \right).$$

Solving these equations is not trivial. Due to the simplicity of the example, we can show how to determine the equilibria in this small stochastic game with $\gamma = 2/3$. Since y is a terminating state without any value to either of the two players, we can focus on state x. Denote the future gains to player i by v_i with $i = 1, 2$. Due to the fact that there are only two possible actions for each player, we can use a 2×2 matrix to represent the expected total values for the two players, respectively, under all policy combinations.

$$[V^{\pi_1, \pi_2}(x) = (V_1, V_2)]_{2 \times 2} = \begin{matrix} & c & d \\ a & \left(\begin{matrix} (8 + \frac{1}{3}v_1, 4 + \frac{1}{3}v_2) & (5 + \frac{2}{3}v_1, 3 + \frac{2}{3}v_2) \\ (1 + \frac{2}{3}v_1, 5 + \frac{2}{3}v_2) & (2 + \frac{2}{3}v_1, 6 + \frac{2}{3}v_2) \end{matrix} \right) \end{matrix}.$$

$$(10.53)$$

For example, for the action pair (a, c) (i.e., both choose the chance play action), the total expected value for player 1 is given by

$$V_1^{(a,c)}(x) = R_1(x, a, c) + \gamma(P(x|x, a, c)V_1^{(a,c)}(x) + P(y|x, a, c) \cdot 0)$$
$$= 8 + \frac{2}{3}(\frac{1}{2}v_1 + \frac{1}{2}0) = 8 + \frac{1}{3}v_1,$$

and the total expected value for player 2 is given by

$$V_2^{(a,c)}(x) = R_2(x, a, c) + \gamma(P(x|x, a, c)V_2^{(a,c)}(x) + P(y|x, a, c) \cdot 0)$$
$$= 4 + \frac{2}{3}(\frac{1}{2}v_2 + \frac{1}{2}0) = 4 + \frac{1}{3}v_2.$$

Similarly, the total expected values for the two players under other three policy combinations in (10.53) can be obtained. Based on (10.53), we can derive the Nash equilibrium (NE) condition for each policy combination as follows:

1. The condition for (a, c) to be NE: $v_1 \leq 21$ and $v_2 \leq 3$.
2. The condition for (a, d) to be NE: $v_2 \geq 3$.
3. The condition for (b, c) to be NE: $v_1 \geq 21$ and $5 \geq 6$.
4. The condition for (b, d) to be NE: $2 \geq 5$.

These conditions can be easily obtained by following the definition of NE. For example, based on (10.53) the NE condition for (a,c) is derived by

$$\text{for player 1}: 8+\frac{1}{3}v_1 \geq 1+\frac{2}{3}v_1 \text{ (i.e., } a \text{ is better than } b) \Rightarrow v_1 \leq 21,$$

$$\text{for player 2}: 4+\frac{1}{3}v_2 \geq 3+\frac{2}{3}v_2 \text{ (i.e., } c \text{ is better than } d) \Rightarrow v_2 \leq 3.$$

Other conditions can be verified similarly. Next, we need to examine each condition to see which one generates the NE strategy for this stochastic game. Clearly, conditions 3 and 4 are not possible so that (b,c) and (b,d) are not NE. For condition 1, when the equilibrium is reached, we have

$$8+\frac{1}{3}v_1 = v_1 \Rightarrow v_1 = 12, \text{ and}$$

$$4+\frac{1}{3}v_2 = v_2 \Rightarrow v_2 = 6,$$

where $v_2 = 6$ contradicts the condition $v_2 \leq 3$. Thus, condition 1 fails in equilibrium and (a,c) is not the NE for the stochastic game. Now the only condition left is for (a,d) (i.e., condition 2). Under this condition, at the equilibrium, we have

$$5+\frac{2}{3}v_1 = v_1 \Rightarrow v_1 = 15, \text{ and}$$

$$3+\frac{2}{3}v_2 = v_2 \Rightarrow v_2 = 9,$$

which is consistent with condition 2: $v_2 \geq 3$ (in fact, the condition for v_1 is not required). So we can conclude that the unique NE for this stochastic game is (a,d) (i.e., player 1 chooses the chance play action and player 2 choose the certainty play action). Note that the NE for the stochastic game is not the NE for the stage game (a,c). As mentioned earlier, this section only provides an easy introduction to stochastic games. For more details about this class of stochastic models, interested readers are referred to Neyman and Sorin (2003).

REFERENCE NOTES

The dynamic optimization problems in stochastic models are well addressed by classical Markov decision process and its extension, Semi-Markov decision process. There are two major issues in this class of problems. The first one is the characterization of the optimal policy structure and the second one is the improvement of computational algorithms or the reduction of computational complexity. The first section addresses the first issue by following the excellent textbook by Porteus [9] and provides readers with an introduction to the general approach. An example of SMDP presented in Section 10.3 is based on [6]. The MDP with average cost criterion and infinite horizon and numerical example are based on [11]. The example of the stochastic game is taken from [4].

REFERENCES

1. D. Blackwell, Discounted dynamic programming. *Annals of Mathematical Statistics*, 36, 226-235, 1965.
2. E.V. Denardo, Contraction mappings in the theory underlying dynamic programming, *SIAM Review*, 9, 165-177, 1967.
3. C. Derman, "Finite State Markovian Decision Processes", Academic Press, New York, 1970.
4. J. Hu and M.P. Wellman, Nash Q-Learning for general-sum stochastic games, *Journal of Machine Learning Research*, 4, 1039-1069, 2003.
5. A. Kolmogorov and S. Fomin, "Elements of Throey of Functions and Functional Analysis", Volumes I and II. Graylock Press, Rochester, N.Y., 1957.
6. N. Gautam, " Analysis of Queues: Methods and Applications", CRC Press, London, 2012.
7. S. Lippman, Applying a new device in the optimization of exponential queueing systems, *Operations Research*, 23, 687-710, 1975.
8. A. Neyman and S. Sorin, "Stochastic Games and Applications", Kluwer Academic Publishers, Boston, 2003.
9. E.L. Porteus, "Fundamentals of Stochastic Inventory Theory", Stanford University Press, Stanford, 2002.
10. M.L. Puterman, "Markov Decision Processes: Discrete Stochastic Dynamic Programming", John Wiley and Sons, Inc., 1994.
11. H. Tijms, "A First Course in Stochastic Models", John Wiley and Sons, 2003.
12. D.M. Topkis, "Supermodularity and Complementarity", Princeton University Press, Princeton, N.J., 1998.
13. D.j. White, Dynamic programming, Markov chains and the method of successive approximations. *Journal of Mathematical Analysis and Applications*, 6, 373–376, 1963.

11 Learning in Stochastic Models

In practice, some assumptions in stochastic models may not be satisfied. For example, the states may not be fully observable or transition models and reward functions are unknown. In other words, information about the future state and/or reward/cost is not available for decision makers. To study such systems, we have to rely on historical data or simulated data for states, actions, and rewards/costs under feasible and reasonable policies. Here, learning processes occur as time goes on. In machine learning theory, this type of learning is called reinforcement learning, where interacting with the environment enables the agent to gain the knowledge about the system dynamics so that better (or optimal) decision can be made. In fact, the optimization in a Markov decision process (either DTMDP or CTMDP) discussed in Chapter 10 belongs to a class of reinforcement learning problems if some parameters of the MDP such as the transition probabilities and/or reward functions are unknown. In such a problem, an agent (decision maker) selects actions and the environment responds to those actions in terms of numerical rewards or costs observed at a sequence of decision epochs. Based on these observed values, the agent estimates these parameters and uses the converged estimates to determine the optimal policy. In this chapter, we present four sections, multi-arm bandits problem, Monte Carlo-based MDP models, hidden Markov models, and partially observable MDPs to illustrate the reinforcement learning in stochastic models.

11.1 MULTI-ARM BANDITS PROBLEM

We consider a situation where a sequence of decisions must be made at some discrete instants over a period of time. At each decision instant, an action from a set of finite actions must be chosen. After each decision we receive a numerical reward which follows a stationary probability distribution that depends on the action selected. The objective is to maximize the expected total reward over this time period, for example, over 1000 time steps. This is the original form of the so-called multi-armed bandit (MAB) problem, so named by analogy to a slot machine, or one-armed bandit, except that it has n levers instead of one. Each action selection is like a play of one of the slot machine's levers, and the rewards are the payoffs for hitting the jackpot. Through repeated action selections we try to maximize the total winnings by concentrating our actions on the best levers. Other sequential decision-making problems exist. Another

DOI: 10.1201/9781003150060-11

example is the problem of a doctor choosing between experimental treatments for a series of seriously ill patients. Each selection is a treatment selection, and each reward is the survival or well-being of the patient. Today the term n-armed bandit problem is sometimes used for a generalization of the problem described above. Here, we utilize this example to demonstrate the main features of a stochastic/sequential decision process with learning. The MAB problem can be considered as a special case of the Markov Decision process (MDP) with a single-state and the learning occurred in such a model belongs to the reinforcement learning as mentioned earlier, which is one of the three categories in machine learning (the other two categories are the unsupervised and supervised learning, respectively).

In an n-armed bandit problem, since payoffs are i.i.d. random variables, we utilize its expected payoff as the value for an action selected. If we knew this true value of each action, then it would be trivial to solve the n-armed bandit problem: we would always select the action with highest value. The main issue is that we do not know the action values with certainty. In fact, we even do not know its distribution! Thus, we have to estimate the mean or the distribution of the payoff. If we maintain estimates of the action values, then at any time step there is at least one action, called a greedy action, whose estimated value is the greatest. Choosing the action with the greatest estimated value is called a greedy policy. Following a greedy policy is said to "*exploiting*" the current knowledge about the values of the actions. If we do not follow the greedy policy and select a non-greedy policy or select a policy randomly, then we say we are "*exploring*". The advantage of choosing non-greedy policy is to improve the estimate of the non-greedy action's value. Exploitation is the right thing to do to maximize the expected reward on the one step (also called myopic), but exploration may produce the greater total reward in the long run at an expense of not achieving the one-step maximum expected reward. Note that the greedy policy is based on the estimated action values with different accuracies. In the MAB problem, the arm being chosen many times has more accurate estimate value than the ones rarely chosen. Thus, exploring those with insufficient number of selections may improve the accuracy of their estimated values. Intuitively, for a finite time horizon, if there are many time steps to the ending of the horizon, exploring these non-greedy actions can be beneficial as the higher value actions may be discovered. At a time instant, we can only select one of exploitation and exploration, these two decisions are mutually exclusive or in conflict. How to make decision at each time step depends in a complex way on multiple factors: the current estimates of action values, the uncertainties, and the number of remaining steps. There are many alternative methods for balancing exploration and exploitation for the MAB problem. However, most of these methods make strong assumptions about parameter stationarity. How to update the information about each action after each exploration determines the policy. In this section, we discuss several methods ranging from very simple sample average approach to Bayesian updating approach (also called Thompson Sampling method).

11.1.1 SAMPLE AVERAGE METHODS

We start with a very simple exploring approach that estimates the value of an action in MAB based on sample average. The only information at the beginning is that the reward (or payoff) for selecting arm a is an i.i.d. random variable, denoted by $X(a)$. Assume that the goal is to maximize the expected total reward over a finite time horizon. Denote the true expected value of selecting arm a by $E(X(a)) = \mu(a)$ and the sample average at time $t > 0$ by $\bar{x}_t(a)$. Then we have

$$\bar{x}_t(a) = \frac{x_1(a) + x_2(a) + \cdots + x_{N_t(a)}(a)}{N_t(a)}, \tag{11.1}$$

where $x_j(a)$ is the actual reward obtained by the jth selecting arm a and $N_t(a)$ is the number of times arm a is selected by time t. If $N_t(a) = 0$, $\bar{x}_t(a)$ is assumed to be some initial value such as 0 or a positive value. As $N_t(a) \to \infty$, by the law of large numbers, $\bar{x}_t(a)$ converges to $\mu(a)$. Under the assumption of stationarity for the mean value of action, this method is considered to be the simplest one. As a result, a simple rule is to select the action at t, denoted by A_t, with the highest estimated mean reward value. Such a policy is the "greedy policy", denoted by A_t^*, where

$$A_t^* = \arg\max_a \bar{x}_t(a). \tag{11.2}$$

In other words, the greedy policy always exploits current information to maximize immediate reward estimated and never tries to improve the mean value estimates of the actions with smaller current sample averages. The disadvantage of the greedy policy is to get stuck in the sub-optimal policy in which the actions with true higher expected values are not selected. To overcome such a weakness, we can mix the exploiting and exploring actions in a simple way. Let ε be a small number less than 1. If with probability of $1 - \varepsilon$ we exploit and with probability of ε we explore, the policy is called ε-greedy policy. Here, for an exploration action, the arm selection can be made either randomly or according a certain rule. We will discuss some methods for exploration later. Now we define the maximum value of action by $\mu^* = \max_{a \in A} \mu(a)$. Then the one-step regret at time instant t is the opportunity loss and is given by

$$l_t = E(\mu^* - \bar{x}_t(a)). \tag{11.3}$$

Further, we define the gap for an action a as $\Delta_a = \mu^* - \mu(a)$, By using the expected count $E(N_t(a))$ (the expected number of times arm a selected by time t), we can define the total regret as

$$
\begin{aligned}
L_t &= E\left[\sum_{\tau=1}^{t}(\mu^* - \bar{x}_\tau(a))\right] \\
&= \sum_{a\in A}E(N_t(a))(\mu^* - \mu(a)) = \sum_{a\in A}E(N_t(a))\Delta_a.
\end{aligned}
\tag{11.4}
$$

Here, we utilize the fact that $E(\bar{x}_\tau(a)) = \mu(a)$. Intuitively, a good policy (or algorithm) ensures small expected counts for large gaps. However, the issue is that these gaps are unknown. As time horizon t increases to infinity, if an algorithm forever explores, it will have a linear and increasing total regret with time. On the other hand, if an algorithm never explores, most likely, it will get stuck in the sub-optimal action and will also have a linear and increasing total regret with time. Our hope is to find a policy or an algorithm that has a decaying total regret with time. We can show that by simulation this decaying total regret behavior happens in an ε-greedy policy.

Exercise 11.1 *Show the details of derivation of total regret formula (11.4).*

To improve the computational efficiency, we develop the recursion for the sample averages as time goes on. That is the sample average for a particular arm based $k+1$ selections is expressed in terms of that based on k selections. That is

$$
\begin{aligned}
\bar{x}_{k+1}(a) &= \frac{1}{k}\sum_{i=1}^{k}x_i(a) \\
&= \frac{1}{k}\left[x_k(a) + \sum_{i=1}^{k-1}x_i(a)\right] \\
&= \frac{1}{k}[x_k(a) + (k-1)\bar{x}_k(a)] \\
&= \frac{1}{k}[x_k(a) + k\bar{x}_k(a) - \bar{x}_k(a)] \\
&= \bar{x}_k(a) + \frac{1}{k}[x_k(a) - \bar{x}_k(a)].
\end{aligned}
\tag{11.5}
$$

Such a recursion holds for arbitrary $x_1(a)$, since $\bar{x}_1(a) = x_1(a)$ and $\bar{x}_2(a) = \bar{x}_1(a) = x_1(a)$. Thus, this recursive calculation can get started with any reasonable guessed value for $x_1(a)$. Note that this recursion applies to the situation where the environment is stationary. For the non-stationary situation, we utilize the constant step-size parameter for the modification term which makes the weights for older values smaller. This is the same idea as the exponential smoothing model in the time-series analysis.

$$
\begin{aligned}
\bar{x}_{k+1}(a) &= \bar{x}_k(a) + \alpha\left(x_k(a) - \bar{x}_k(a)\right) \\
&= \alpha x_k(a) + (1-\alpha)\bar{x}_k(a) \\
&= \alpha x_k(a) + (1-\alpha)[\alpha x_{k-1}(a) + (1-\alpha)\bar{x}_{k-1}(a)] \\
&= \alpha x_k(a) + (1-\alpha)\alpha x_{k-1}(a) + (1-\alpha)^2 \bar{x}_{k-1}(a) \\
&= \alpha x_k(a) + (1-\alpha)\alpha x_{k-1}(a) + (1-\alpha)^2 \alpha x_{k-2}(a) \\
&\quad + \cdots + (1-\alpha)^{k-1}\alpha x_1(a) + (1-\alpha)^k \bar{x}_1 \\
&= (1-\alpha)^k \bar{x}_1(a) + \sum_{i=1}^{k} \alpha(1-\alpha)^{k-i} x_i(a).
\end{aligned}
\tag{11.6}
$$

A more general update rule is to make the step-size depend on the kth selection of action a. Denote such a step-size by $\alpha_k(a)$. Clearly, the sample-average method is the special case of $\alpha_k(a) = 1/k$. Some well-known conditions for convergence with probability one in stochastic approximation theory is as follows:

$$
\sum_{k=1}^{\infty} \alpha_k(a) = \infty \quad \text{and} \quad \sum_{k=1}^{\infty} \alpha_k^2(a) < \infty.
\tag{11.7}
$$

Here, the first condition is required to ensure that the steps are large enough to overcome initial conditions or random fluctuations while the second condition ensures that eventually steps become small enough to achieve convergence. Clearly, the step-size for the sample-average method satisfies both conditions and is appropriate for the stationary environment. On the other hand, the constant step-size for the exponential weight method does not satisfy the second condition. This implies that the estimates never completely converge and continue to vary in response to the most recent rewards. In fact, this non-convergence property is desirable for the non-stationary environment.

11.1.2   EFFECT OF INITIAL VALUES

The recursions developed above depend on the initial action-value estimates which must be set as the best "guessed values" (i.e., $x_1(a)$). In statistical theory, these methods are biased by their initial values. Thus, the initial value effect is called a "bias effect". For the sample-average method, as long as each action is selected at least once, the bias effect disappears. Thus selecting any initial action-value should be fine as long as each action will be explored. For the exponential weight method, the initial bias effect is permanent but decreasing exponentially over time. Therefore, in practice, this initial bias effect is not an issue. In fact, we can balance exploiting and exploring actions by setting the initial action-value estimates. For example, if we set the initial action-value very high (optimistic about each arm), then even we follow a greedy policy, we are more likely to explore every arm in the MAB problem. Thus, the optimistic initial action-values lead to the case where all actions are selected several times before the value estimates converge. By adjusting the initial action-values, we may optimally trade-off between exploitation and exploration so that the average reward over a finite-time horizon is maximized.

11.1.3 UPPER CONFIDENCE BOUNDS

With the exploration on a particular arm a, we can develop some upper bound on the probability that $\bar{x}_t(a)$ is far from its true mean $\mu(a)$. Assume that the payoffs of arm a are i.i.d. random variables with mean $\mu(a)$ and variance $\sigma^2(a)$ and $N_t(a) = n$. For example, using the Chebyshev's inequality, we have

$$P(|\bar{x}_t(a) - \mu(a)| \geq \varepsilon) \leq \frac{Var(\bar{x}_t(a))}{\varepsilon^2} = \frac{\sigma^2(a)}{n\varepsilon^2}.$$

This bound implies that the probability of having a "bad" estimate goes to zero as the number of times of selecting arm a increases. Denote by $S_n(a) = \sum_{i=1}^{n}(x_i(a) - \mu(a)) = n(\bar{x}_t(a) - \mu(a))$. In fact, a better (tighter) upper bound on such a probability can be obtained from the central limit theorem (CLT) which states $S_n(a)/\sqrt{n\sigma^2(a)} \to N(0,1)$, the standard normal distribution as $n \to \infty$. This implies that

$$P(\bar{x}_t(a) - \mu(a) \geq \varepsilon) = P(S_n(a) \geq n\varepsilon) = P\left[\frac{S_n(a)}{\sqrt{n\sigma^2(a)}} \geq \sqrt{\frac{n}{\sigma^2(a)}}\varepsilon\right]$$

$$\approx \int_{\varepsilon\sqrt{\frac{n}{\sigma^2(a)}}}^{\infty} \frac{1}{2\pi} \exp\left(-\frac{x^2}{2}\right) dx,$$

$$(11.8)$$

where the approximation follows from the CTL. The integral in (11.8) cannot be computed with a closed form. However, we could get an upper bound on it. Here we need a simple inequality for a given $b > 0$ as

$$\int_{b}^{\infty} \exp\left(-\frac{x^2}{2}\right) dx = \int_{b}^{\infty} \frac{x}{x} \exp\left(-\frac{x^2}{2}\right) dx \leq \frac{1}{b}\int_{b}^{\infty} x \exp\left(-\frac{x^2}{2}\right) dx$$

$$= \frac{1}{b} \exp\left(-\frac{b^2}{2}\right).$$

Using this inequality in (11.8) yields

$$P(\bar{x}_t(a) - \mu(a) \geq \varepsilon) \lessapprox \sqrt{\frac{\sigma^2(a)}{2\pi\varepsilon^2 n}} \exp\left(-\frac{n\varepsilon^2}{2\sigma^2(a)}\right). \qquad (11.9)$$

To obtain an exact bound with a similar asymptotic rate, we define a class of random variables called subgaussian random variables.

Definition 11.1 *A random variable X is σ-subgaussian if for all $\lambda \in \mathbb{R}$, it holds that $E[\exp(\lambda X)] \leq \exp(\lambda^2 \sigma^2 / 2)$.*

For example, a normal distributed random variable with mean 0 and variance σ^2 is σ-subgaussian. An alternative expression for the condition of σ-subgaussian random variable is to define the cumulant-generating function $\psi(\lambda) = \log M_X(\lambda) \leq (1/2)\lambda^2 \sigma^2$ where $M_X(\lambda) = E[\exp(\lambda X)]$, which is the moment-generating function. A random variable is said to be heavy tailed if the moment generating function is infinite for $\lambda \in \mathbb{R}$ (i.e., $M_X(\lambda) = \infty$).

Exercise 11.2 *Check if Pareto distribution and Weibull distribution are heavy tailed. If yes, under what conditions?*

The reason that we use the term "subgaussian" is because the tail of a σ-subgaussian random variable decay approximately as fast as that of a Normal distribution (also called Gaussian) with zero mean and the same variance. This is formally stated in the following theorem.

Theorem 11.1 *If X is σ-subgaussian, then for any $\varepsilon \geq 0$,*

$$(a) \ P(X \geq \varepsilon) \leq \exp\left(-\frac{\varepsilon^2}{2\sigma^2}\right), \quad (b) \ P(|X| \geq \varepsilon) \leq 2\exp\left(-\frac{\varepsilon^2}{2\sigma^2}\right). \quad (11.10)$$

Proof. (a) For $\lambda > 0$, we have

$$P(X \geq \varepsilon) = P(\exp(\lambda X) \geq \exp(\lambda \varepsilon))$$
$$\leq E[\exp(\lambda X)] \exp(-\lambda \varepsilon)$$
$$\leq \exp\left(\frac{\lambda^2 \sigma^2}{2} - \lambda \varepsilon\right),$$

where the first inequality follows from the Markov's inequality and the second inequality is due to the definition of subgaussianity. Choosing $\lambda = \varepsilon/\sigma^2$ yields the result (a).

(b) Using the same approach, we can obtain

$$P(X \leq -\varepsilon) \leq \exp\left(-\frac{\varepsilon^2}{2\sigma^2}\right).$$

Using the two bounds, we have

$$P(|X| \geq \varepsilon) = P(\{X \geq \varepsilon\} \cup \{X \leq -\varepsilon\})$$
$$\leq P(X \geq \varepsilon) + P(X \leq -\varepsilon) = 2\exp\left(-\frac{\varepsilon^2}{2\sigma^2}\right),$$

which is the result (b). ∎

By letting $\varepsilon = \sqrt{2\sigma^2 \log(1/\delta)}$ in (a) and $\varepsilon = \sqrt{2\sigma^2 \log(2/\delta)}$ in (b) of (11.10), we obtain

$$P(X \geq \sqrt{2\sigma^2 \log(1/\delta)}) \leq \delta, \quad P(|X| \geq \sqrt{2\sigma^2 \log(2/\delta)}) \leq \delta, \quad (11.11)$$

respectively. These bounds are useful. For example, it follows from the second bound in (11.11) that the probability that X falls in the interval

$$\left(-\sqrt{2\sigma^2 \log(2/\delta)}, \ \sqrt{2\sigma^2 \log(2/\delta)} \right)$$

is at least $1 - \delta$. Next, we present more properties of σ-subgaussian random variables.

Lemma 11.1 *Suppose that X is σ-subgaussian and X_1 and X_2 are two independent and σ_1 and σ_2-subgaussian, respectively, then*
(a) $E[X] = 0$ and $Var(X) \leq \sigma^2$.
(b) cX is $|c|\sigma$-subgaussian for all $c \in \mathbb{R}$.
(c) $X_1 + X_2$ is $\sqrt{\sigma_1^2 + \sigma_2^2}$-subgaussian.

Proof. It follows from the definition of σ-subgaussianity that

$$E[\exp(\lambda X)] = \sum_{n \geq 0} \lambda^n \frac{E[X^n]}{n!} \leq \exp\left(\frac{\lambda^2 \sigma^2}{2} \right)$$
$$= \sum_{n \geq 0} \left(\frac{\lambda^2 \sigma^2}{2} \right)^n \frac{1}{n!}.$$

By keeping only the terms up to order 2 of the LHS expansion and collecting all terms with order higher than 2 and expressing it as $g(\lambda)$, we have

$$1 + \lambda E[X] + \frac{\lambda^2}{2} E[X^2] \leq 1 + \frac{\lambda^2 \sigma^2}{2} + g(\lambda), \quad (11.12)$$

where $g(\lambda)/\lambda \to 0$ as $\lambda \to 0$. Now, by dividing both sides of (11.12) by λ and taking the limit when $\lambda \to 0_+$ (from the positive side), we have $E[X] \leq 0$. By taking limit $\lambda \to 0_-$, we show $E[X] \geq 0$. Thus, $E[X] = 0$. By dividing both sides of (11.12) by λ^2 and taking the limit $\lambda \to 0$, we obtain $E[X^2] \leq \sigma^2$.
(b) and (c) are left as exercises. ∎

Exercise 11.3 *Prove (b) and (c) of Lemma 11.1.*

Combining Lemma (11.1) and Theorem 11.1 yields a bound on the tails of $\bar{x}_t(a) - \mu(a)$.

Proposition 11.1 *Suppose that* $x_i(a) - \mu(a)$ *with* $i = 1, ..., n$ *are independent and* σ-*subgaussian random variables. Then for any* $\varepsilon \geq 0$,

$$P(\bar{x}_t(a) - \mu(a) \geq \varepsilon) \leq \exp\left(-\frac{n\varepsilon^2}{2\sigma^2}\right) \text{ and } P(\bar{x}_t(a) - \mu(a) \leq -\varepsilon) \leq \exp\left(-\frac{n\varepsilon^2}{2\sigma^2}\right),$$

where $\bar{x}_t(a) = (1/n)\sum_{i=1}^{n} x_i(a)$.

Proof. It follows from Lemma 11.1 that $\bar{x}_t(a) - \mu(a) = (1/n)\sum_{i=1}^{n}(x_i(a) - \mu(a))$ is $\sigma/\sqrt{n}$-subgaussian. Then applying Theorem 11.1 gives the result. ∎
By choosing $\varepsilon = \sqrt{(2\sigma^2/n)\log(1/\delta)}$, with probability at least $1 - \delta$, we obtain

$$\mu(a) \leq \bar{x}_t(a) + \sqrt{\frac{2\sigma^2\log(1/\delta)}{n}}, \tag{11.13}$$

and symmetrically with probability at least $1 - \delta$, we have

$$\mu(a) \geq \bar{x}_t(a) - \sqrt{\frac{2\sigma^2\log(1/\delta)}{n}}. \tag{11.14}$$

Exercise 11.4 *Derive a two-sided inequality by using a union bound for* $\mu(a)$.

In an MAB problem, the variance of the payoffs for an arm may not be known. Thus, we need to establish a bound on the variance.

Lemma 11.2 *Let* u *be a positive real number. Let* X *be a random variable that has a finite support* $[0, u]$, *i.e.,* $P(0 \leq X \leq u) = 1$. *Then the variance is upper bounded by* $u^2/4$.

Proof. It follows from $0 \leq X \leq u$ that

$$Var(X) = E[X^2] - (E[X])^2 \leq uE(X) - (E[X])^2. \tag{11.15}$$

Let $z = E[X]$. Then the upper bound in (11.15) can be written as

$$-z^2 + uz = -z(z - u),$$

which is a parabola that opens downward and reaches the maximum at $z = u/2$. This implies that

$$Var(X) \leq u\left(\frac{u}{2}\right) - \left(\frac{u}{2}\right)^2 = \frac{u^2}{4}.$$

∎

By choosing $t^{-\alpha} = \delta$ and applying Lemma 11.2 to (11.13) and (11.14), we have for $u = 1$ case

$$\mu(a) \leq UCB_t(a) = \bar{x}_t(a) + \sqrt{\frac{\alpha}{2}} \sqrt{\frac{\log(t)}{n}} = \bar{x}_t(a) + c \sqrt{\frac{\log(t)}{N_t(a)}} \quad (11.16)$$

with $c = \sqrt{\alpha/2}$ called confidence coefficient and UCB representing the upper confidence bound. Note that the relation $t^{-\alpha} = \delta$ reflects the desired property that as time increases the chance that our estimate for $\mu(a)$ is above the upper bound, δ, is decreasing for a given α value (i.e., the confidence that the estimate is below the upper bound $1 - \delta$ increases). In practice, $\alpha > 1$ is often selected.

Auer et al. (2002) proposed an action-selection policy based on such an upper confidence bounds (UCBs) in (11.16) for a MAB problem with any reward distribution over a bounded support $[0, 1]$. This policy, called an UCB policy, prescribes selecting an arm with the maximum upper confidence bound for the mean reward.

Using Hoeffding's Inequality and Chernoff Bounds (see Chapter C), we can easily verify that the confidence level for this UBC is $1 - t^{-\alpha}$ (this is shown in the proof of the following theorem about the regret bound).

Theorem 11.2 *Regret bound for the UCB policy for horizon $t > 1$ in a MAB problem with the reward distribution over $[0, 1]$ is*

$$L_t \leq \sum_{a:\Delta_a > 0} \left(2\alpha \Delta_a^{-1} \log(t) + \frac{\alpha}{\alpha - 1} \Delta_a \right).$$

Proof. Without loss of generality, we assume that arm 1 is optimal. Thus, arm $a \neq 1$ will only be chosen in two cases:

Case 1: Either arms 1 and $a \neq 1$ have been sampled insufficiently to distinguish between their true means; Case 2: The UCB provided by Hoeffding's inequality does not hold for either arm 1, or arm a.

Similar to the UCB in (11.16), let's consider the lower confidence bound (LCB) for the true mean reward of arm a as an event

$$LCB_a = \left\{ \mu(a) \geq \bar{x}_t(a) - \sqrt{\frac{\alpha}{2}} \sqrt{\frac{\log(t)}{N_t(a)}} \right\}$$

$$= \left\{ \bar{x}_t(a) \leq \mu(a) + \sqrt{\frac{\alpha}{2}} \sqrt{\frac{\log(t)}{N_t(a)}} \right\},$$

and the UCB for the true mean of arm 1 as another event

$$UCB_1 = \left\{ \mu(1) \le \bar{x}_t(1) + \sqrt{\frac{\alpha}{2}} \sqrt{\frac{\log(t)}{N_t(1)}} \right\}$$

$$= \left\{ \bar{x}_t(1) \ge \mu(1) - \sqrt{\frac{\alpha}{2}} \sqrt{\frac{\log(t)}{N_t(1)}} \right\}.$$

Clearly, when either the complement of LCB_a or the complement of UCB_1 occurs (i.e., $\bar{x}_t(a)$ is above its UCB or $\bar{x}_t(1)$ is below its LCB), we may select arm a. We use Hoeffding's inequality to bound these probabilities. That is

$$P(LCB_a^c) = P(\bar{x}_t(a) - \mu(a) > \varepsilon) \le exp\left(\frac{-2\varepsilon^2 t^2}{\sum_{i=1}^{t}(b_i - a_i)^2}\right)$$

$$= exp\left(\frac{-2\varepsilon^2 t^2}{\sum_{i=1}^{t}(1-0)^2}\right) = exp(-2\varepsilon^2 t),$$

which can be re-written with $\varepsilon = \sqrt{\frac{\alpha}{2}} \sqrt{\frac{\log(t)}{N_t(a)}}$ as

$$P\left(\bar{x}_t(a) - \mu(a) > \sqrt{\frac{\alpha}{2}} \sqrt{\frac{\log(t)}{N_t(a)}}\right) \le exp\left(\frac{-2t\alpha\log t}{2N_t(a)}\right)$$

$$= exp\left(\frac{-t\alpha\log t}{N_t(a)}\right) \le exp\left(\frac{-t\alpha\log t}{t}\right)$$

$$= e^{-\alpha\log t} = t^{-\alpha}.$$

Similarly, we can obtain the probability bound for the complement of UCB_1. Next we find the upper bound for the $E[N_t(a)]$, where arm $a \ne 1$ is suboptimal. Note that arm a is selected under the UCB policy is when the following condition

$$\bar{x}_t(a) + \sqrt{\frac{\alpha}{2}} \sqrt{\frac{\log(t)}{N_t(a)}} > \bar{x}_t(1) + \sqrt{\frac{\alpha}{2}} \sqrt{\frac{\log(t)}{N_t(1)}} \qquad (11.17)$$

is satisfied and both confidence intervals are valid ones (i.e., LCB_a and UCB_1 estimated at time t are correct). Thus, such a condition holding at time t is due to insufficient sampling (i.e., large variability reflected by the square root term). Event LCB_a implies

$$\mu(a) + 2\sqrt{\frac{\alpha}{2}} \sqrt{\frac{\log(t)}{N_t(a)}} \ge \bar{x}_t(a) + \sqrt{\frac{\alpha}{2}} \sqrt{\frac{\log(t)}{N_t(a)}}. \qquad (11.18)$$

Event UCB_1 implies

$$\bar{x}_t(1) + \sqrt{\frac{\alpha}{2}} \sqrt{\frac{\log(t)}{N_t(1)}} \ge \mu(1). \qquad (11.19)$$

Using (11.17), (11.18), and (11.19), we obtain

$$\mu(a) + 2\sqrt{\frac{\alpha}{2}}\sqrt{\frac{\log(t)}{N_t(a)}} \geq \mu(1),$$

which can be written as

$$\sqrt{\frac{\alpha}{2}}\sqrt{\frac{\log(t)}{N_t(a)}} \geq \frac{\mu(1) - \mu(a)}{2} = \frac{\Delta_a}{2}.$$

Here, the last equality holds due to the fact that arm 1 is optimal and $\Delta_a = \max_j \mu(j) - \mu(a)$. Solving this inequality for $N_t(a)$ yields

$$N_t(a) \leq 2\Delta_a^{-2}\alpha\log(t).$$

Denote by I_i the arm number selected at time i. We have the expected number of times that arm a is selected by time t under a UCB policy as

$$E[N_t(a)] = \sum_{i=1}^{t} E[\mathbf{1}(I_i = a)]$$

$$\leq 2\alpha\Delta_a^{-2}\log(t) + \sum_{i=1}^{t} E[\mathbf{1}(LCB_{ai}^c \cup UCB_{1i}^c)]$$

$$\leq 2\alpha\Delta_a^{-2}\log(t) + \sum_{i=1}^{t} (E[\mathbf{1}(LCB_{ai}^c)] + E[\mathbf{1}(UCB_{1i}^c)])$$

$$\leq 2\alpha\Delta_a^{-2}\log(t) + \sum_{i=1}^{t} (i^{-\alpha} + i^{-\alpha})$$

$$= 2\alpha\Delta_a^{-2}\log(t) + 2\sum_{i=1}^{t} i^{-\alpha}.$$

To bound the second term, we use

$$\sum_{i=1}^{t} i^{-\alpha} \leq 1 + \int_1^\infty x^{-\alpha}dx = 1 + \frac{-1}{1-\alpha} = \frac{-\alpha}{1-\alpha}.$$

Thus, we have

$$E[N_t(a)] \leq 2\alpha\Delta_a^{-2}\log(t) + \frac{2\alpha}{\alpha - 1}.$$

Using this inequality in the definition of the total regret yields

$$L_t = \sum_{a \neq 1} E(N_t(a))\Delta_a$$

$$\leq \sum_{a \neq 1} \left(2\alpha\Delta_a^{-1}\log(t) + \frac{\alpha}{\alpha - 1}\Delta_a\right).$$

This completes the proof. ∎

The action taken by the UCB policy is

$$A_t = \arg\max_a \left[\bar{x}_t(a) + c \sqrt{\frac{\log(t)}{N_t(a)}} \right].$$

It is easy to observe that each time arm a is selected, the uncertainty (i.e., the margin of error) is decreased as $N_t(a)$ is increased. In contrast, each time an action other than a is selected, $\log(t)$ is increased while $N_t(a)$ remains the same. This implies that the uncertainty is increased. Thus, we are motivated to explore arm a.

Exercise 11.5 *Consider a MAB problem with k-arms and 1-subgaussian reward. If $\alpha = 2$, show the regret bound for horizon $t > 1$ is*

$$L_t \leq 3 \sum_{a=1}^{k} \Delta_a + \sum_{a:\Delta_a>0} \frac{16 \log(t)}{\Delta_a}.$$

11.1.4 ACTION PREFERENCE METHOD

An alternative to the action-value estimation method is the action-preference method. In this method, we estimate the action-preference, denoted by $H_t(a)$ by recursions. The larger the preference, the more likely that action is selected. However, the preference has no interpretation in terms of absolute rewards. If each action's reward is increased by a constant, the preference for each action or the likelihood of selecting each action remains the same. Thus, a so-called soft-max distribution (also called Gibbs or Boltzmann distribution) is utilized to quantify the probability that arm a is selected. That is

$$P(A_t = a) = \frac{e^{H_t(a)}}{\sum_{b=1}^{n} e^{H_t(b)}} = \pi_t(a). \tag{11.20}$$

Based on the concept of stochastic gradient ascent, an updating rule exists. Such a rule can be considered as a learning algorithm, also called gradient-bandit algorithm, and implemented as follows. At each step i, after action $A_i = a$ is selected and a reward $x_i(a)$ is received, the preference is updated as

$$\begin{aligned}
H_{i+1}(a) &= H_i(a) + \alpha(x_i(a) - \bar{x}_i(a))(1 - \pi_i(a)), \\
H_{i+1}(b) &= H_i(b) - \alpha(x_i(b) - \bar{x}_i(b))\pi_i(b), \quad \forall b \neq a,
\end{aligned} \tag{11.21}$$

where $\alpha > 0$ is a step-size parameter. Intuitively, we see that if the reward obtained in ith step is higher than the current average, the probability of selecting $A_t = a$ is incremented, and if the reward in ith step is below the current average, then the probability is reduced. The non-selected action's preference moves in the opposite direction. The proof of these recursions is presented below. The algorithm can be considered as a stochastic approximation to gradient ascent. In terms of gradient ascent, the action preference can be expressed as

$$H_{i+1}(a) = H_i(a) + \alpha \frac{\partial E[x_i]}{\partial H_i(a)}, \tag{11.22}$$

where $E[x_i] = \sum_b \pi_i(b)\mu(b)$. Note that

$$\frac{\partial E[x_i]}{\partial H_i(a)} = \frac{\partial}{\partial H_i(a)} \left[\sum_b \pi_i(b)\mu(b) \right]$$

$$= \sum_b \mu(b) \frac{\partial \pi_i(b)}{\partial H_i(a)} = \sum_b (\mu(b) - Y_i) \frac{\partial \pi_i(b)}{\partial H_i(a)}, \tag{11.23}$$

where Y_i is a constant and independent of b. The last equality holds because of $\sum_b \frac{\partial \pi_i(b)}{\partial H_i(a)} = 0$. The performance gradient can be re-written as

$$\frac{\partial E[x_i]}{\partial H_i(a)} = \sum_b \pi_i(b)(\mu(b) - Y_i) \frac{\partial \pi_i(b)}{\partial H_i(a)} / \pi_i(b)$$

$$= E \left[(\mu(A_i) - Y_i) \frac{\partial \pi_i(A_i)}{\partial H_i(a)} / \pi_i(A_i) \right] \tag{11.24}$$

$$= E \left[(x_i - \bar{x}_i) \frac{\partial \pi_i(A_i)}{\partial H_i(a)} / \pi_i(A_i) \right],$$

where the last equality is due to our choosing $Y_i = \bar{x}_i$ and the fact of $E[x_i] = E[\mu(A_i)]$. Next we derive a relation of $\partial \pi_i(b)/\partial H_i(a) = \pi_i(b)(\mathbf{I}_{a=b} - \pi_i(a))$, where $\mathbf{I}_{a=b} = 1$ if $a = b$ and zero otherwise. Using the quotient rule for derivatives, we have

$$\frac{\partial \pi_i(b)}{\partial H_i(a)} = \frac{\partial}{\partial H_i(a)} \left[\frac{e^{H_i(b)}}{\sum_{c=1}^n e^{H_i(c)}} \right]$$

$$= \frac{\frac{\partial e^{H_i(b)}}{\partial H_i(a)} \sum_{c=1}^n e^{H_i(c)} - e^{H_i(b)} \frac{\partial \sum_{c=1}^n e^{H_i(c)}}{\partial H_i(a)}}{\left(\sum_{c=1}^n e^{H_i(c)} \right)^2}$$

$$= \frac{\mathbf{I}_{a=b} e^{H_i(a)} \sum_{c=1}^n e^{H_i(c)} - e^{H_i(b)} e^{H_i(a)}}{\left(\sum_{c=1}^n e^{H_i(c)} \right)^2} \tag{11.25}$$

$$= \frac{\mathbf{I}_{a=b} e^{H_i(b)}}{\sum_{c=1}^n e^{H_i(c)}} - \frac{e^{H_i(b)} e^{H_i(a)}}{\left(\sum_{c=1}^n e^{H_i(c)} \right)^2}$$

$$= \mathbf{I}_{a=b} \pi_i(b) - \pi_i(b)\pi_i(a)$$

$$= \pi_i(b)(\mathbf{I}_{a=b} - \pi_i(a)).$$

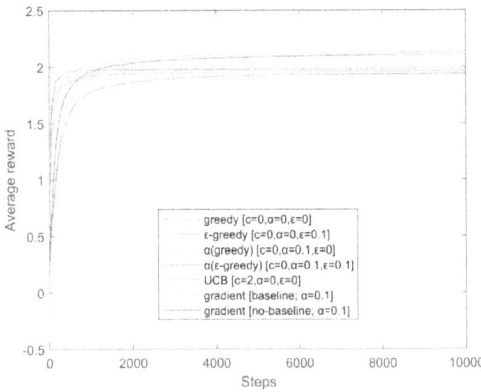

Figure 11.1   10-arm MAB problem simulations under different algorithms.

Using (11.25) in (11.24), we obtain

$$\frac{\partial E[x_i]}{\partial H_i(a)} = E\left[(x_i - \bar{x}_i)\pi_i(A_t)(\mathbf{I}_{a=b} - \pi_i(a))/\pi_i(A_i)\right]$$
$$= E\left[(x_i - \bar{x}_i)(\mathbf{I}_{a=b} - \pi_i(a))\right].$$
(11.26)

Substituting this relation into (11.22) yields

$$H_{i+1}(a) = H_i(a) + \alpha(x_i - \bar{x}_i)(\mathbf{I}_{a=b} - \pi_i(a)), \quad \forall a,$$
(11.27)

which is the algorithm proposed. Note that this algorithm only needs that the reward baseline does not depend on the selected action. The choice of the baseline does not affect the expected update of the algorithm. However, it does affect the convergence rate of the algorithm as it determines the variance of the update. Usually, we choose the baseline as the average of the rewards observed which works reasonably well in practice. As a conclusion of this section, we present in Figure 11.1 the simulation results for different algorithms developed for a 10-armed MAB problem. In this problem, the reward (or payoff) for selecting arm a is an i.i.d. random variable $X(a) \sim N(0,1)$ with an optimistic action-value estimate $\bar{x}_1(a) = 5$. The results are in average rewards that are the averages based on 20 replications with each replication having 10000 steps (other combination of the length of each replication and the number of replications can be chosen, e.g., 200 replications with each replication having 1000 steps).

Exercise 11.6 *Evaluate and compare the ε-greedy, UCB, and gradient bandit algorithm by simulating the 10-armed MAB presented in Figure 11.1.*

By varying the parameter of each algorithm (ε, c, and α) in a reasonable range, show the average reward (performance) curve as a function of the parameter. Create either a figure or a table to compare these algorithms and rank these algorithms for this particular case. (Hint: Use Matlab or Mathematica to simulate the MAB process with 100 replications in 1000 steps.)

Exercise 11.7 *(Thompson Sampling for the Bernoulli Bandit) Consider a K-arm MAB that yields either a success with a fixed probability θ_a or a failure with probability $1 - \theta_a$ when arm a is played ($a \in \{1, 2, ..., K\}$. The success probabilities $(\theta_1, ..., \theta_K)$ are unknown to the player. Assume that the rewards for success and failure are 1 and 0, respectively. In the first period, arm x_1 is played and a reward $r_1 \in \{0, 1\}$ is received with success probability $P(r_1 = 1 | x_1, \theta) = \theta_{x_1}$. After observing r_1, the player applies another action x_2, observe a reward r_2, and this process continues for T periods. The player starts with an independent prior belief over each θ_k. Assume that these priors are beta-distributed with parameters $\alpha = (\alpha_1, ..., \alpha_K)$ and $\beta = (\beta_1, ..., \beta_K)$. That is for action k, the prior probability density function of θ_k is given by*

$$p(\theta_k) = \frac{\Gamma(\alpha_k + \beta_k)}{\Gamma(\alpha_k)\Gamma(\beta_k)} \theta_k^{\alpha_k - 1} (1 - \theta_k)^{\beta_k - 1},$$

where Γ denotes the gamma function. As observations are generated, the distribution is updated according to Bayes' theorem. The reason for assuming Beta distributions is because their conjugacy properties. That is each action's posterior distribution is also Beta with parameters that can be updated according to a simple rule:

$$(\alpha_k, \beta_k) \leftarrow \begin{cases} (\alpha_k, \beta_k) & \text{if } x_t \neq k, \\ (\alpha_k, \beta_k) + (r_t, 1 - r_t) & \text{if } x_t = k. \end{cases}$$

Note that for the special case of $\alpha_k = \beta_k = 1$, the prior $p(\theta_k)$ is uniform over [0,1]. Now consider a 3-armed Beta-Bernoulli bandit problem with mean rewards $\theta_1 = 0.9, \theta_2 = 0.8$, and $\theta_3 = 0.7$. Assume that the prior distribution over each mean reward is uniform. Compare simulated behavior of Thompson sampling with that of greedy algorithm (using Matlab and Mathematica is recommended). NOTE: For a comprehensive discussion on Thompson sampling algorithm, interested readers are referred to Russo et al. (2018).

11.2 MONTE CARLO-BASED MDP MODELS

The idea of "exploring" or "learning", i.e. performing actions in the world to find out and collect samples, in the MAB problem can be applied to the stochastic models discussed in this book when some or all parameters are unknown. In this section, we consider a smaller version of the classical MDP

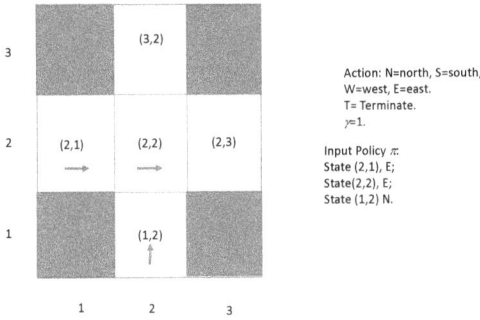

Figure 11.2 Gridworld problem without transition probability and one-step reward information.

model, the Gridworld problem presented in Chapter 10. Assume that the agent tries to move in a maze of 3×3 with four corner cells blocked as shown in Figure 11.2 with same notations as in Chapter 10. The main difference between this Gridworld problem and the one discussed in Chapter 10 is that the transition probabilities and the one-step reward function are unknown. In this section, we discuss several methods to estimate these unknown parameters in stochastic models.

11.2.1 MODEL-BASED LEARNING

For the Gridworld problem with unknown transition probabilities and one-step reward function, we assume that the model structure remains the same as the one considered in Chapter 10. Since there exists terminating states, under a given policy, a path of $n < \infty$ steps (i.e., a sequence of state-action-reward's) is called an episode. Such a path can be denoted by

$$s_0; a_1, r_1, s_1; a_2, r_2, s_2; a_3, r_3, s_3; ...; a_n, r_n, s_n.$$

Recall the recursive relation for the value function introduced in Chapter 10. It can be re-written as

$$V^\pi(s) = \begin{cases} 0 & \text{if } s = s_n \text{ (ending state)}, \\ Q^\pi(s, \pi(s)), & \text{otherwise}, \end{cases}$$

where

$$Q^\pi(s,a) = \sum_j P(j|s,a)[R(j|s,a) + \gamma V^\pi(j)]$$
$$= R(s,a) + \gamma \sum_j P(j|s,a) V^\pi(j).$$

A Small Sample, *n*=4, Example

Observed Episodes (Training Stage)		Learned Model

Episode 1
(2,1), E, -1, (2,2),
(2,2), E, -1, (2,3),
(2,3), +10, T.

Episode 2
(2,1), E, -1, (2,2),
(2,2), E, -1, (2,3),
(2,3), +10, T.

$\hat{P}(j|s,a)$
$\hat{P}\{(2,2)|(2,1), E\}= 1.00$
$\hat{P}\{(2,3)|(2,2), E\}= 0.75$
$\hat{P}\{(3,2)|(2,2), E\}= 0.25$
......

Episode 3
(1,2), N, -1, (2,2),
(2,2), E, -1, (2,3),
(2,3), +10, T.

Episode 4
(1,2), N, -1, (2,2),
(2,2), E, -1, (3,2),
(3,2), -10, T.

$\hat{R}(j|s,a)$
$\hat{R}\{(2,2)|(2,1),E\}= -1$
$\hat{R}\{(2,3)|(2,2),E\}= -1$
$\hat{R}\{T|(2,3), X\}= +10$
$\hat{R}\{T|(3,2), X\}= -10$

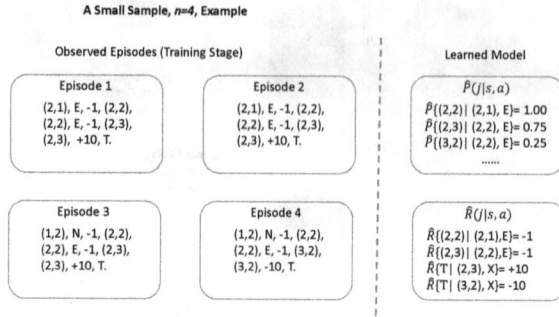

Figure 11.3 Training the MDP model by four episodes (replications) via model-based estimation.

Here, $Q^\pi(s,a)$ is introduced and is called Q-value function which is the expected value if action a is taken in state s and policy π is followed after. Furthermore, we write the one-step reward function in a more general form, which is the expectation on the next (future) state. Now we need to solve the MDP problem in two stages due to the unknown parameters. In the first stage, we estimate the transition probabilities and one-step reward function by using the Monte Carlo approach. This stage is called "learning or training the model". Denote the transition probability estimate by $\hat{P}(j|s,a)$. Such an estimate can be obtained by simulating the system dynamics under a given policy a large number of replications (episodes). Specifically, we count the outcomes j for each s,a pair and normalize it to get $\hat{P}(j|s,a)$. Similarly, we can also estimate the one-step reward function, denoted by $\hat{R}(j|s,a)$. The result of the first stage is a learned MDP model with the estimated Q-value function given by

$$\hat{Q}^\pi(s,a) = \hat{R}(s,a) + \gamma\sum_j \hat{P}(j|s,a)\hat{V}^\pi(j).$$

In the second stage, we can solve the learned MDP via the policy iteration or value iteration approach as we did in Chapter 10. This method is also called "model-based Monte Carlo". We demonstrate the basic idea of "training the model" using the Gridworld problem in Figure 11.3 for a given policy (input policy π) illustrated in Figure 11.2. To make it simple, we only use a very small sample of $n = 4$ episodes. It is easy to check that these two estimates are computed by

$$\hat{P}(j|s,a) = \frac{\text{number of times } (j|s,a) \text{ occurs}}{\text{number of times } (s,a) \text{ occurs}}$$

$$\hat{R}(j|s,a) = r \text{ in } (s,a,r,j),$$

where s is the current state and j is the next state to be visited. In practice, a large number of episodes are required to obtain the good estimates as these unbiased estimators converge to the true limits ($P(j|s,a)$ and $R(j|s,a)$) according to the law of large numbers. A disadvantage of this approach is that it may not scale to large state spaces (i.e., it can be expensive to estimate the parameters for each state-action pair). Since the goal is to estimate the key components of the MDP model, this procedure is called "model-based learning". Here, the data used to estimate the model is the sequence of states, actions, and rewards evolved according to a Markov model in the episode. Note that the samples being averaged within an episode (that happens for long episode cases) are not independent, but they do come from a Markov chain, so it can be shown that these estimates converge to the expectations by the ergodic theorem (a generalization of the law of large numbers for Markov chains). A potential problem exists in the model-based learning approach if we only focus on the deterministic policies, which always map s to the unique action $\pi(s)$, to generate data. That is under a certain policy, there are certain state-action pairs (s,a) that we will never see and thus unable to estimate their Q-value and never know what the effect of those actions are. If these state-action pairs are obviously non-optimal or extremely unlikely in practice, not being able to estimate their Q-values may not be an issue. Otherwise, the need for exploration of all possible state-action pairs motivates us to consider non-deterministic policies which allow us to evaluate each state-action pair infinitely often (possibly over multiple episodes). According to the law of large numbers, these estimates of transition probabilities and one-step reward functions will converge.

Exercise 11.8 *Use the law of large numbers to argue that estimates $\hat{P}(j|s,a)$ and $\hat{R}(j|s,a)$ converge to the limits which are true transition probabilities and one-step reward functions.*

11.2.2 MODEL-FREE LEARNING

An alternative approach, called "model-free learning", is to focus on estimating the optimal Q-value function directly. Thus, this approach does not require to have the transition probabilities and one-step reward function estimated. Based on a sequence of state-action-reward's generated by following a policy π

$$s_0; a_1, r_1, s_1; a_2, r_2, s_2; a_3, r_3, s_3; ...; a_n, r_n, s_n,$$

we can compute the discounted value starting at (s,a), denoted by $q(s,a)$, as

$$q(s,a) = r_0 + \gamma r_1 + \gamma^2 r_2 + \cdots$$

with $s_0 = s$ and $a_1 = a$. Denote by $q_i(s,a)$ the discounted value of the ith episode where $i = 1, ..., n$. Thus, the direct estimate of the Q value for the

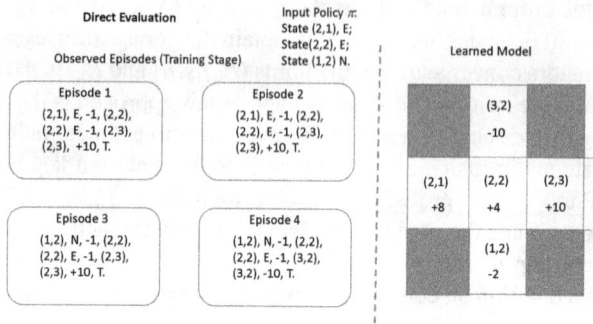

Figure 11.4 Training the MDP model by four episodes (replications) via model-free estimation.

state-action pair (s, a) based on n episodes is given by

$$\hat{Q}^{\pi}(s,a) = \frac{\sum_{i=1}^{n} q_i(s,a)}{n}. \tag{11.28}$$

As an unbiased estimator, we have $E[\hat{Q}^{\pi}(s,a)] = Q^{\pi}(s,a)$ under a given policy π. Now, we show this approach by estimating the Q-value function for the Gridworld problem under the same input policy π with $\gamma = 1$ and $n = 4$ as shown in Figure 11.2. The result is illustrated in Figure 11.4. The state values directly estimated based on the four episodes in Figure 11.4 are easy to compute. Starting from state $(2, 1)$, episodes 1 and 2 both provide the cumulative value of 8 (-1-1+10) which gives the average of 8 for $(2, 1)$. Starting from state $(2, 2)$, all the four episodes provide a total of $(-1 + 10) + (-1 + 10) + (-1 + 10) + (-1 - 10) = 16$. Thus, the average value for $(2, 2)$ is $16/4 = 4$. Starting from state $(1, 2)$, episodes 3 and 4 provide a total of $(-1 - 1 + 10) + (-1 - 1 - 10) = -4$ which leads to the average of $-4/2 = -2$ for $(1, 2)$. Finally, it is clear that the value for the terminal state $(2, 3)$ is $+10$ and the value for the terminal state $(3, 2)$ is -10. Note that in this process we do not need to consider transition probabilities or one-step rewards and the only thing that matters is total rewards resulting from starting with $(s; a)$ pairs. For simplicity, in such a tiny example, we only consider (s, a) which does not show up later in each episode. This is not necessary for the algorithm to work in a long episode for a bigger problem. Since the model-free estimation approach depends strongly on the policy π that is followed, the computational procedure is also called an *on-policy* algorithm. In contrast, the model-based estimation procedure does not depend on the exact policy (as long as it was able to explore all (s, a) pairs), it is called an *off-policy* algorithm.

We can interpret the model-free estimation approach from different perspectives. For a given state-action pair (s, a), let $q(s, a)$ be state value for an episode and k be the number of updates to (s,a). Then, an equivalent expression to (11.28) can be written as a recursive relation

$$\hat{Q}^{\pi(new)}(s,a) = (1-w)\hat{Q}^{\pi(old)}(s,a) + wq(s,a), \qquad (11.29)$$

where $w = 1/(1+k)$ and is called "learning rate". This recursion indicates that averaging as a batch operation is equivalent to interpolating between the old value $\hat{Q}^{\pi}(s,a)$ (current estimate) and the new value $q(s,a)$ (data). Such a weighted average of two values is also called a convex combination of the two values. Note that (11.29) can be re-written as

$$\hat{Q}^{\pi(new)}(s,a) = \hat{Q}^{\pi(old)}(s,a) - w[\underbrace{\hat{Q}^{\pi(old)}(s,a)}_{\text{prediction}} - \underbrace{q(s,a)}_{\text{target}}], \qquad (11.30)$$

which makes the update look like a stochastic gradient update. The recursion (11.30) implies the objective of the least squares regression (i.e., $\min(\hat{Q}^{\pi}(s,a) - q(s,a))^2$). This model-free estimation approach is also called "model-free Monte Carlo" method as the Q-value estimator is updated by the data collected for the entire episode (a replication of a simulation).

Exercise 11.9 *Prove the equivalence of (11.28) and (11.29).*

11.2.3 MODEL-FREE LEARNING WITH BOOTSTRAPPING

Note that the target of model-free Monte Carlo algorithm is $q(s,a)$, the discounted sum of rewards for an episode after taking an action a at state s. However, $q(s,a)$ itself is just an estimate of $Q^{\pi}(s,a)$. If the episode is long, $q(s,a)$ will not be a good estimate. This is because $q(s,a)$ only corresponds to one of many episodes; so as the episode lengthens, it becomes an increasingly less representative sample of what could happen. An alternative to model-free Monte Carlo is to use the previous estimate of the Q-value and sampled immediate reward to develop the new estimate of the Q-value. Since this approach is based on a sequence of State, Action, Reward, State, and Action $(s; a, r, s', a')$, it is called SARSA. In SARSA, the Q-value is updated by the recursive relation with the same structure as in model-free Monte Carlo. The target in SARSA is $r + \gamma \hat{Q}^{\pi}(s', a')$. That is

$$\hat{Q}^{\pi(new)}(s,a) = (1-w)\hat{Q}^{\pi(old)}(s,a) + w[\underbrace{r}_{\text{data}} + \underbrace{\hat{Q}^{\pi(old)}(s',a')}_{\text{estimate}}]. \qquad (11.31)$$

Since the Q-value estimate update is based on the previous estimate, SARSA is said to have the bootstrapping feature. The main advantage that SARSA offers over model-free Monte Carlo is that we don't have to wait until the end of the episode to update the Q-value. If the estimates are already pretty good, then

SARSA will be more reliable since $q(s,a)$ is based on only one path whereas $\hat{Q}^{\pi}(s',a')$ is based on all the ones that the learner has seen before. In terms of the statistical behavior, the estimator $\hat{Q}$ of model-free Monte Carlo, which is based on one entire episode, is unbiased and has large variance. However, the estimator $\hat{Q}$ of SARSA, which starts with the observed value as the initial estimate, is biased and has small variance. Note that such a bias in SARSA converges to zero asymptotically. We can also develop similar algorithms to estimate the value functions $V^{\pi}(s)$ which are presented as the main idea of so-called "temporal-difference learning" in a later subsection. The temporal-difference learning is a combination of the Monte Carlo approach and the MDP approach. The reason that we do not present these value function estimate algorithms here is because that they cannot be used to choose actions in model-free settings.

We can actually interpolate between model-free Monte Carlo (all rewards) and SARSA (one reward). For example, we could update toward $r_t + \gamma r_{t+1} + \gamma^2 \hat{Q}^{\pi}(s_{t+1}, a_{t+2})$ (two rewards). We can even combine all of these updates, which results in an algorithm called SARSA(λ), where λ determines the relative weighting of these targets. See the Sutton and Barto (2020) (Chapter 7) for an introduction.

Exercise 11.10 *Which of the following algorithms can be used to approximately determine the optimal Q-value, denoted by $Q^*(s,a)$:*
   (a) Model-based Monte Carlo.
   (b) Model-free Monte Carlo.
   (c) SARSA

11.2.4 *Q*-LEARNING

To determine the optimal policy under a model-free scenario, we need to estimate the optimal Q-value $Q^*(s,a)$ directly. This can be done by using a so-called Q-learning algorithm. On each (s,a,r,s'), the update on estimating the optimal Q-value is performed by

$$\hat{Q}^*(s,a) \leftarrow (1-w)\hat{Q}^*(s,a) + w(r + \gamma \hat{V}^*(s'))$$
$$= \hat{Q}^*(s,a) + w(r + \gamma \hat{V}^*(s') - \hat{Q}^*(s,a)), \tag{11.32}$$

where $\hat{V}^*(s') = \max_{a' \in A(s')} \hat{Q}^*(s',a')$. The goal of Q-learning is to determine the optimal policy by estimating Q^* in a model-free manner. The Q-learning is an off-policy algorithm. The algorithm of Q-learning can be derived by modifying the value iteration algorithm of DTMDP presented in Chapter 10. Here are a few changes that we can make to convert the VI algorithm of MDP (based on the optimality equation) to the Q-learning algorithm. First, the expectation over the next state s' should be replaced with one sample s'. Second, instead

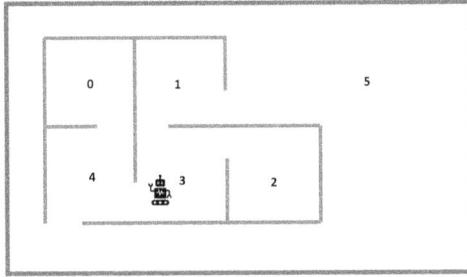

Figure 11.5 A robot trying to exit to the house.

of updating the estimated optimal value function based on the optimality equation, the estimated optimal Q-value is updated by interpolating between old value, called prediction, and the new value, called target. Third, the optimal value function $V^*(s')$ is replaced with $\hat{V}^*(s')$. Importantly, the estimated optimal value $\hat{V}^*(s')$ involves a maximum over actions rather than taking the action of a given policy. This max over a' rather than taking the a' based on the current policy is the principle difference between Q-learning and SARSA.

To demonstrate the calculations, we use a classical example of a robot trying to exit a house with five rooms as shown in Figure 11.5. These five rooms are numbered as 0, 1, 2, 3, and 4. For convenience, we consider "outside the house" as room 5 or "exit" room. A robot needs to learn how to leave the house with the best path possible. Thus, the goal of the robot is to get to Room 5 as fast as it can. To motivate the robot's behavior, the reward value for rooms in the house is 0 and the reward value for the "exit" room is 100. By putting a negative value for infeasible state-action pairs, we can use a reward matrix to represent rewards in this problem as follows:

$$
\mathbf{R} =
\begin{array}{c}
\\
s=0 \\
s=1 \\
s=2 \\
s=3 \\
s=4 \\
s=5
\end{array}
\begin{array}{c}
a=0 \quad a=1 \quad a=2 \quad a=3 \quad a=4 \quad a=5 \\
\left(
\begin{array}{cccccc}
-1 & -1 & -1 & -1 & 0 & -1 \\
-1 & -1 & -1 & 0 & -1 & 100 \\
-1 & -1 & -1 & 0 & -1 & -1 \\
-1 & 0 & 0 & -1 & 0 & -1 \\
0 & -1 & -1 & 0 & -1 & 100 \\
-1 & 0 & -1 & -1 & 0 & 100
\end{array}
\right)
\end{array}
\tag{11.33}
$$

with the entry denoted by $r(s,a)$ (Note that if the one-step reward also depends on the next state s', notation $r(s,a,s')$ can be used. Such a case is presented in the next subsection). From this matrix, we can write the action set for each state as the non-negative value represents a feasible action. For example, $A(4) = \{0,3,5\}$ meaning that in Room 4, the robot can choose to go to Room 0, Room 3, or Room 5 with rewards $r(4,0) = 0, r(4,3) = 0$, and $r(4,5) = 100$, respectively. The main idea of Q-learning is that the robot does not know the reward values in the matrix **R** and needs to learn the state-action values by randomly selecting a feasible action in each state in many episodes. The implementation of the Q-learning algorithm can be illustrated in terms of a matrix $\hat{\mathbf{Q}}^*$. The matrix $\hat{\mathbf{Q}}^*$ is updated by (11.32). To make it simple, we denote the kth updated optimal Q-value by $\hat{Q}^*_k(s,a)$ for state-action pair (s,a), set learning rate $w = 1$ and (11.32) is reduced to

$$\hat{Q}^*_k(s,a) = r(s,a) + \gamma \max_{a' \in A(s')} \hat{Q}^*_{k-1}(s',a') \qquad (11.34)$$

for $k = 1,2,....$ Such a case is an analogy to the "naive" approach in the exponential smoothing forecast model. At the beginning, the robot has no information about the optimal Q-value matrix. Thus, it is initialized with zero as

$$
\hat{\mathbf{Q}}^* =
\begin{array}{c}
\\
s=0 \\
s=1 \\
s=2 \\
s=3 \\
s=4 \\
s=5
\end{array}
\begin{array}{c}
a=0 \quad a=1 \quad a=2 \quad a=3 \quad a=4 \quad a=5 \\
\left(
\begin{array}{cccccc}
0 & 0 & 0 & 0 & 0 & 0 \\
0 & 0 & 0 & 0 & 0 & 0 \\
0 & 0 & 0 & 0 & 0 & 0 \\
0 & 0 & 0 & 0 & 0 & 0 \\
0 & 0 & 0 & 0 & 0 & 0 \\
0 & 0 & 0 & 0 & 0 & 0
\end{array}
\right).
\end{array}
$$

$$(11.35)$$

Then, we can implement the Q-learning algorithm summarized as follows:

1. Start with $\hat{\mathbf{Q}}^*$ in (11.35).
2. Randomly choose an initial state.
3. For each episode (a sequence of state-action-reward's starting with the initial state and ending with the "exit" state), while a state is not the goal state
   (a) Randomly select a feasible action for the current state;
   (b) Consider all feasible actions for the next state;
   (c) Get maximum Q-value for the next state by selecting one action from the feasible action set for the next state.
   (d) Use (11.34) to update $\hat{\mathbf{Q}}^*$.

Similar to the MAB problem, the method of selecting a feasible action in Step 3(a) of the algorithm above can be designed to trade off between exploitation and exploration. Using random selection can ensure that each state-action pair is explored.

While such an algorithm can be easily implemented by some mathematical software package such as Matlab or Mathematica, we show the details of the calculations by doing some specific episodes manually. Assume that discount factor $\gamma = 0.8$. We start with the initial value $\hat{\mathbf{Q}}^*$ in (11.35). First, we randomly select initial state as $s = 1$ (Room 1). At this initial state, from matrix **R**, there are only two feasible actions $a = 3$ and $a = 5$. Assume that $a = 5$ is randomly chosen. In this new (next) state $s = 5$, there are three feasible actions $a = 1, a = 4$, and $a = 5$. Basically, these are the columns in matrix $\mathbf{R}$ with non-negative values in row 5 and we try to select the action with biggest value based on $\hat{\mathbf{Q}}^*$. That is to select the action with $\max\{\hat{Q}^*(5,1),\hat{Q}^*(5,4),\hat{Q}^*(5,5)\}$. However, at this point, all these Q-values are still zero. Now it follows from (11.34) that

$$\hat{Q}^*(1,5) = r(1,5) + 0.8\max\{\hat{Q}^*(5,1),\hat{Q}^*(5,4),\hat{Q}^*(5,5)\}$$
$$= 100 + 0.8(0) = 100.$$

Since the new state is the goal state, this first episode is finished and the $\hat{\mathbf{Q}}^*$ is updated as follows:

$$\hat{Q}^* = \begin{array}{c} \\ s=0 \\ s=1 \\ s=2 \\ s=3 \\ s=4 \\ s=5 \end{array}\begin{array}{cccccc} a=0 & a=1 & a=2 & a=3 & a=4 & a=5 \\ \left(\begin{array}{cccccc} 0 & 0 & 0 & 0 & 0 & 0 \\ 0 & 0 & 0 & 0 & 0 & 100 \\ 0 & 0 & 0 & 0 & 0 & 0 \\ 0 & 0 & 0 & 0 & 0 & 0 \\ 0 & 0 & 0 & 0 & 0 & 0 \\ 0 & 0 & 0 & 0 & 0 & 0 \end{array} \right) \end{array}. \qquad (11.36)$$

Now we consider the second episode. With the current $\hat{\mathbf{Q}}^*$ in (11.36), we randomly select the initial state $s = 3$ (i.e., Room 3). Again, from matrix $\mathbf{R}$, there are three feasible actions $a = 1, a = 2$, and $a = 4$ and $a = 1$ is randomly selected. Next, we will be in Room 1 where there are only two feasible actions $a = 3$ and $a = 5$. Based on the current $\hat{\mathbf{Q}}^*$, we obtain the updated Q-value for the state action pair $(3,1)$ as follows:

$$\hat{Q}^*(3,1) = r(3,1) + 0.8\max\{\hat{Q}^*(1,3),\hat{Q}^*(1,5)\}$$
$$= 0 + 0.8(100) = 80.$$

Since state $s = 1$ is not the goal state, the process continues. Assume that by chance, action $a = 5$ is selected. At the next state $s = 5$, there are three feasible actions $a = 1, a = 4$ and $a = 5$. However, all current $\hat{Q}^*(5,1),\hat{Q}^*(5,4)$, and $\hat{Q}^*(5,5)$ are zeros so that no updates occur for this extra step. Thus, the updated $\hat{\mathbf{Q}}^*$ after the two episodes is given by

$$\hat{Q}^* = \begin{array}{c} \\ s=0 \\ s=1 \\ s=2 \\ s=3 \\ s=4 \\ s=5 \end{array}\begin{array}{cccccc} a=0 & a=1 & a=2 & a=3 & a=4 & a=5 \\ \left(\begin{array}{cccccc} 0 & 0 & 0 & 0 & 0 & 0 \\ 0 & 0 & 0 & 0 & 0 & 100 \\ 0 & 0 & 0 & 0 & 0 & 0 \\ 0 & 80 & 0 & 0 & 0 & 0 \\ 0 & 0 & 0 & 0 & 0 & 0 \\ 0 & 0 & 0 & 0 & 0 & 0 \end{array} \right) \end{array}. \qquad (11.37)$$

After running a large number of episodes (e.g., 100000 episodes implemented by Matlab or Python), we have

$$
\hat{Q}^* =
\begin{array}{c}
\\ s=0 \\ s=1 \\ s=2 \\ s=3 \\ s=4 \\ s=5
\end{array}
\begin{array}{cccccc}
a=0 & a=1 & a=2 & a=3 & a=4 & a=5 \\
\left(\begin{array}{cccccc}
0 & 0 & 0 & 0 & 400 & 0 \\
0 & 0 & 0 & 320 & 0 & 500 \\
0 & 0 & 0 & 320 & 0 & 0 \\
0 & 400 & 256 & 0 & 400 & 0 \\
320 & 0 & 0 & 320 & 0 & 500 \\
0 & 400 & 0 & 0 & 400 & 500
\end{array} \right)
\end{array},
$$

(11.38)

which can be normalized by dividing all positive entries by the maximum entry (i.e., 500) and multiplying them by 100. Thus, the normalized $\hat{Q}^*$ is given by

$$
\hat{Q}^* =
\begin{array}{c}
\\ s=0 \\ s=1 \\ s=2 \\ s=3 \\ s=4 \\ s=5
\end{array}
\begin{array}{cccccc}
a=0 & a=1 & a=2 & a=3 & a=4 & a=5 \\
\left(\begin{array}{cccccc}
0 & 0 & 0 & 0 & 80 & 0 \\
0 & 0 & 0 & 64 & 0 & 100 \\
0 & 0 & 0 & 64 & 0 & 0 \\
0 & 80 & 51 & 0 & 80 & 0 \\
64 & 0 & 0 & 64 & 0 & 100 \\
0 & 80 & 0 & 0 & 80 & 100
\end{array} \right)
\end{array}.
$$

(11.39)

With this optimal Q-value matrix, we can determine the optimal policy from any starting state. Specifically, starting from any non-goal state, we keep choosing the action that gives the maximum value for each state until reaching the goal state. For example, if the starting state is $s = 2$ (i.e., Room 2), we select $a = 3$ which has the maximum value 64 (blue colored). Then we are in state $s = 3$ (i.e., Room 3) where two feasible actions ($a = 1$ and $a = 2$) attain the maximum value of 80. Thus, we arbitrarily select $a = 1$ with blue colored value that brings us to $s = 1$ (i.e., Room 1). Finally, in Room 1, $a = 5$ with the maximum value 100 (blue colored) is selected and the goal state with the maximum value 100 (red colored) is reached. To summarize, if the robot starts from Room 2, its optimal policy is to follow the state-action sequence $(2,3),(3,1),(1,5)$ which reaches the goal state $s = 5$ in three steps with the cumulative reward 0+0+100=100.

Exercise 11.11 *Implement the Q-learning algorithm by using a mathematical package such as Matlab for the robot problem as shown in Figure 11.5.*

11.2.5 TEMPORAL-DIFFERENCE LEARNING

The basic idea of temporal-difference (TD) learning is to learn from every experience gained in each time step. That is the estimated value function $\hat{V}^\pi(S_t)$ is updated each time a sequence of state-action-reward-next state, $(S_t = s, \pi(S_t) = a, r_{t+1}(S_t, \pi(S_t), S_{t+1}) = r, S_{t+1} = s')$, is observed. Similar to estimating Q-value function, we can estimate the state value function based on the updating algorithm for a given policy π as follows:

$$
\begin{aligned}
\hat{V}^\pi(S_t) &\leftarrow (1-w)\hat{V}^\pi(S_t) + w(r_{t+1}(S_t, \pi(S_t), S_{t+1}) + \gamma\hat{V}^\pi(S_{t+1})) \\
&= \hat{V}^\pi(S_t) + w([r_{t+1}(S_t, \pi(S_t), S_{t+1}) + \gamma\hat{V}^\pi(S_{t+1})] - \hat{V}^\pi(S_t))
\end{aligned} \tag{11.40}
$$

with $0 < w \leq 1$ as the learning rate. Here, we use the time index t for the state variable to represent the non-stationary environment. By considering $r_{t+1}(s, \pi(s), s') + \gamma\hat{V}^\pi_{t+1}(s')$ as a sample of value function obtained after time t, we can write (11.40) as

$$
\hat{V}^\pi(S_t) \leftarrow (1-w)\hat{V}^\pi(S_t) + w \cdot sample_t = \hat{V}^\pi(S_t) + w(sample_t - \hat{V}^\pi(S_t)). \tag{11.41}
$$

Estimating the value function under a given policy π is called the "prediction problem". The "sample" in (11.41) can be considered as the target for the estimate so that as more samples are collected the estimated value function is continuously adjusted (improved) by the "estimation error" term. In fact, (11.41) is a general form of learning algorithm in stochastic models. For the TD case, the sample at time t is based on the observed immediate reward r_{t+1} and the previous estimated state value $\hat{V}^\pi(S_{t+1})$ (bootstrapping). For the model-free Monte Carlo case, we have the sample, denoted by $sample_t = q_t(s, \pi(s)) = q_t^\pi(s)$, as the target in the prediction problem. Here, $q_t^\pi(s)$ is the sample value of an episode starting from state s at time t (i.e. In an every-visit Monte Carlo, it is the total rewards starting from s at time t until the end of the episode). Thus, the value function is estimated by the following updating algorithm

$$
\hat{V}^\pi(S_t) \leftarrow \hat{V}^\pi_t(S_t) + w(q_t^\pi(S_t) - \hat{V}^\pi(S_t)). \tag{11.42}
$$

In a model-free Monte Carlo case, the sample $q_t^\pi(s)$ can be obtained by either "first-time visit" or "every-visit" Monte Carlo method (denoted by FVMC or EVMC). While in FVMC, the sample value for state s is the immediate reward after first visit to s in an episode (i.e., the second or later visit to s in the same episode is not counted), in EVMC, the sample value for state s is the total of immediate reward of every visit to s in an episode. Here as mentioned earlier, we assume that EVMC is utilized. Comparing (11.42) with (11.40), we notice that while each update on the estimated value function can be done only after the completion of each episode in the model-free Monte Carlo, the update can be done at every state transition (after every time step) during the progress of the episode in the TD method as shown in Figure 11.6. We can define the TD error e_t^{TD} as

$$
e_t^{TD} = [r_{t+1} + \gamma\hat{V}^\pi(S_{t+1})] - \hat{V}^\pi(S_t),
$$

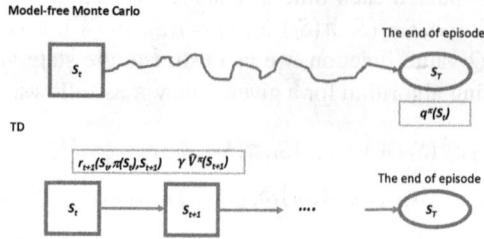

Figure 11.6 Model-free Monte Carlo versus TD learning.

where the arguments of r_{t+1} function are omitted for the simplicity of notation. Similarly, the error for the model-free Monte Carlo approach, denoted by e_t^{MC} can be defined as $e_t^{MC} = q_t^\pi(S_t) - \hat{V}^\pi(S_t)$. If the estimated value function $\hat{V}^\pi$ (an array function) does not change during the episode, then a relation between e_t^{MC} and e_t^{TD} can be established as follows:

$$
\begin{aligned}
e_t^{MC} &= q_t^\pi(S_t) - \hat{V}^\pi(S_t) \\
&= r_{t+1} + \gamma q_{t+1}^\pi - \hat{V}^\pi(S_t) + \gamma \hat{V}^\pi(S_{t+1}) - \gamma \hat{V}^\pi(S_{t+1}) \\
&= [r_{t+1} + \gamma \hat{V}^\pi(S_{t+1}) - \hat{V}^\pi(S_t)] + \gamma(q_{t+1}^\pi - \hat{V}^\pi(S_{t+1})) \\
&= e_t^{TD} + \gamma e_{t+1}^{MC} \\
&= e_t^{TD} + \gamma(e_{t+1}^{TD} + \gamma e_{t+2}^{MC}) \\
&= e_t^{TD} + \gamma e_{t+1}^{TD} + \gamma^2 e_{t+2}^{TD} + \cdots + \gamma^{T-t-1} e_{T-1}^{TD} + \gamma^{T-t}(q_T^{pi} - \hat{V}^\pi(S_T)) \\
&= e_t^{TD} + \gamma e_{t+1}^{TD} + \gamma^2 e_{t+2}^{TD} + \cdots + \gamma^{T-t-1} e_{T-1}^{TD} + \gamma^{T-t}(0 - 0) \\
&= \sum_{j=t}^{T-1} \gamma^{j-t} e_j^{TD}.
\end{aligned}
$$

Note that this relation does not hold exactly if $\hat{V}^\pi$ is updated during the episode. Although the TD learning algorithm (11.40) does update $\hat{V}^\pi$ during the episode, this relation can approximately hold if the step size or learning rate w is small.

It is worth noting that (11.40) can be considered as the simplest TD updating algorithm as the updating can be done immediately upon transition to S_{t+1} and receiving r_{t+1}. It is called *one-step* TD and can be generalized to a multi-step or so-called *n*-step TD in which the update is done after $n \geq 2$ state transitions occurred. We demonstrate the details of calculations of TD by considering the Gridworld example presented in Chapter 10 with different immediate reward values for non-terminal and terminal states. Figure 11.7 shows the progress of an episode on which the TD learning is based. Denote each state by $s = (j, i)$

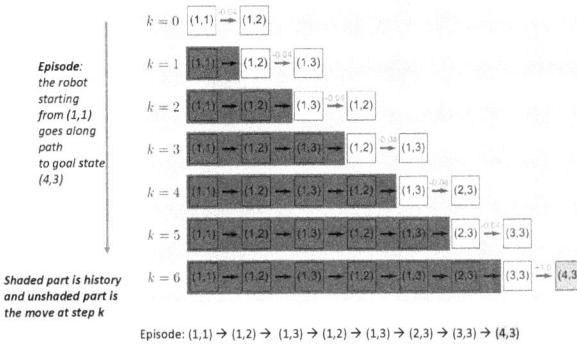

Figure 11.7 is illustrated with the following text elements:

Episode: the robot starting from (1,1) goes along path to goal state (4,3)

Shaded part is history and unshaded part is the move at step k

Episode: (1,1) → (1,2) → (1,3) → (1,2) → (1,3) → (2,3) → (3,3) → (4,3)

Figure 11.7 The progress of an episode in Gridworld problem.

where j is the column number and i is the row number. In this episode, the robot starts from $(1,1)$ and reaches the goal state $(4,3)$ after 7 steps as shown in Figure 11.8. We assume that the goal state $(4,3)$ (colored in green) has a value of $+1$ and the trapping state $(4,2)$ (colored in red) has a value of -1. Each non-terminal state has a value of -0.04. Let $\gamma = 0.9$ and $w = 0.1$. The state values are initialised with zeros. At $k = 1$ the state $(1,1)$ is updated since the robot is in the state $(1,2)$ and the first reward (-0.04) is available. The calculation for updating the state value at $(1,1)$ is given by

$$\hat{V}^{\pi}(S_t) + w([r_{t+1}(S_t, \pi(S_t), S_{t+1}) + \gamma \hat{V}^{\pi}(S_{t+1})] - \hat{V}^{\pi}(S_t))$$
$$= 0.0 + 0.1(-0.04 + 0.9(0.0) - 0.0) = -0.004.$$

Similarly to $(1,1)$ the algorithm updates the state at $(1,2)$. At $k = 3$ the robot goes back and the calculation takes the form: $0.0 + 0.1\ (-0.04 + 0.9\ (-0.004) - 0.0) = -0.00436$. At $k = 4$ the robot changes again its direction. In this case the algorithm update for the second time the state $(1,2)$ as follow: $-0.004 + 0.1\ (-0.04 + 0.9\ (-0.00436) + 0.004) = -0.0079924$. The same process is applied until the end of the episode. To estimate the value function by TD for a given policy, a large number of episodes (iterations) should be generated until these state values converge. We can evaluate the effectiveness of TD learning by using the optimal policy of this problem. As in Chapter 10, we assume that the transition probability model is known with $\gamma = 0.999$. Then the optimal value function and policy can be obtained by the MDP approach (either VI algorithm or PI algorithm) as shown in Figure 11.9. Now we focus on this optimal policy and apply the TD algorithm to estimate the value function. Using Matlab or Python, a program can be developed to implement this algorithm. After 300000 iterations, we obtain the estimated value function as shown in Figure 11.10, which is not too far from the one determined by the MDP with

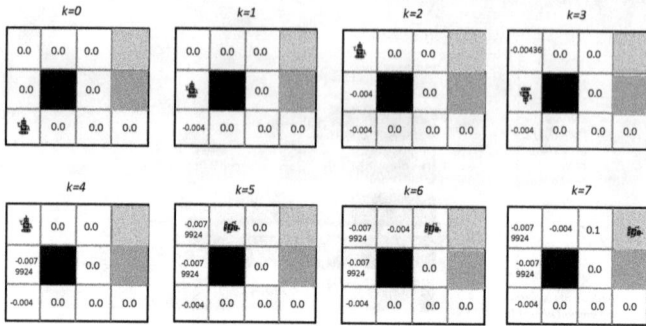

Figure 11.8 TD learning based on an episode in Gridworld problem.

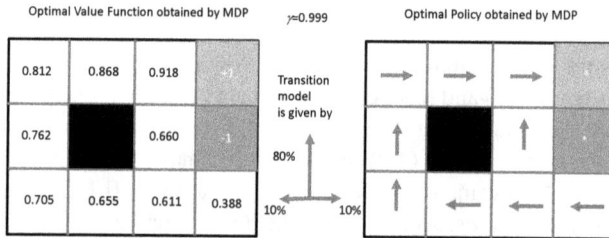

Figure 11.9 Optimal value function and policy determined by MDP for Gridworld problem.

the given transition probability model. Note that the values for the two terminal states cannot be estimated by the TD learning.

Exercise 11.12 *Use Matlab or Python to implement the TD learning algorithm for the Gridworld problem shown in Figure 11.8.*

Whereas the goal of the TD prediction problem is to estimate the state value function under a given fixed policy, the TD control problem is to find either an appropriate policy that balances between exploration and exploitation or the optimal policy that maximizes the total reward. In fact these TD control problems can be solved by the SARSA and Q-learning algorithms presented in the previous subsections. The key difference between the "prediction" problem and "control" problem in learning is that there is a policy updating step in each iteration of the TD control algorithm. Thus, it is obvious that Q-learning algorithm presented in Subsection 11.2.4 is to solve a control problem as the improved policy is generated in each iteration. For SARSA presented in Subsection 11.2.3, it addresses the prediction problem as the policy remains fixed.

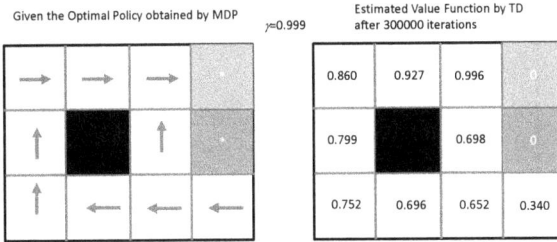

Figure 11.10 The value function estimated for the optimal policy by TD for Gridworld problem.

However, if the policy updating feature is introduced, SARSA can solve the control problem. In fact, the SARAR algorithm with the ε-greedy updating rule, called on-policy TD control, can determine an appropriate policy that balances between exploration and exploitation and is called ε-greedy policy. Meanwhile, the Q-learning algorithm, called off-policy TD control, can determine the optimal policy. The reason for SARSA being called on-policy learning is because a new action a' is chosen using the same ε-greedy policy that generated s'. On the other hand, Q-learning is called off-policy learning because the new action $a_g = \arg\max_{a \in A(s)} Q(s,a)$ is taken as greedy, not using the current policy. Note that under the ε-greedy rule, the probability of exploitation (choosing the action that maximizes the current updated Q-value) is $1 - \varepsilon$ and the probability of exploration (choosing any other action) is ε. Therefore, if we set $\varepsilon = 0$ in the SARSA algorithm, it becomes the Q-learning algorithm. We have presented a detailed numerical example for Q-learning in Subsection 11.2.4. Similarly, the ε-greedy policy for this example can be determined similarly by using the SARSA algorithm (left as an exercise). We use another Gridworld example to show the main difference between the Q-learning and SARSA methods. This example is taken from Sutton and Barto (2020). Consider a Gridworld with three rows and 12 columns. Denote by $s = (i, j)$ the state where i is row number and j is column number. Assume 10 out of 12 states in row 1 represent the cliff with reward -100 and other states have reward -1 as shown in Figure 11.11. By taking $\varepsilon = 0.1$, we compare the ε-greedy optimal policy derived by SARSA with the optimal policy obtained by Q-learning. SARSA generates a ε-greedy optimal policy, and this policy is represented by the green (safer) path in the figure. On the other hand, Q-learning gives the optimal policy represented by the blue path, which is the row next to the cliff. We can explain why the robot prefers to go further away from the cliff in the ε-greedy policy of SARSA? Suppose that the robot follows the ε-greedy policy and is on the row just above the cliff (the same path as the Q-learning's optimal policy), then **LEFT** and **RIGHT** action will make it still stay on this row and

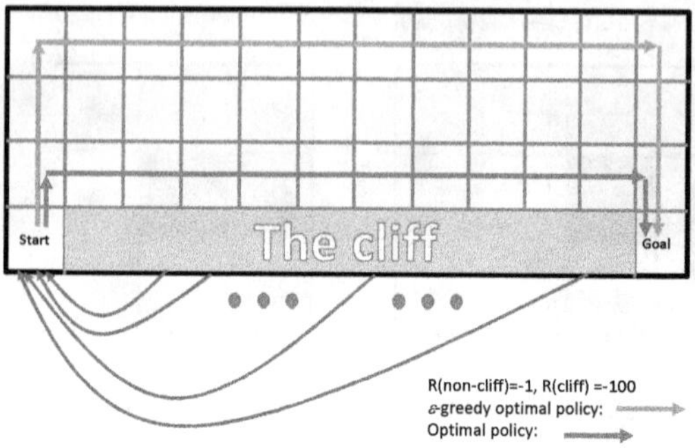

Figure 11.11 The Gridworld with cliff problem.

there is $\varepsilon/4 = 10\%/4 = 2.5\%$ chance that it will go into the Cliff (i.e., going DOWN action), which is a disaster immediately. Remember that in SARSA the robot chooses the next action based on the same ε-greedy policy (so-called on-policy). This implies that for an ε-greedy policy with $\varepsilon > 0$ not so small, the states just above the cliff (blue path) are dangerous and should have small value function and thus the robot prefers to choose the path farther away from the cliff (green path).

On the other hand, Q-learning is an off-policy algorithm. It estimates the reward for state-action pairs based on the optimal (greedy) policy, independent of the agent's actions. An off-policy algorithm approximates the optimal action-value function, independent of the policy. In this case, the Q-learning algorithm can explore and benefit from actions that did not happen during the learning phase. As a result, Q-learning is a simple and effective learning algorithm. However, due to the greedy action selection, the algorithm (usually) selects the next action with the best reward, making it a short-sighted learning algorithm. In the Q-learning process, since the robot always explores and likes to stay closer to the goal state, the blue path (the greedy direction), it has more chances of going into the cliff. Thus, its online performance is worse than that of SARSA. However, Q-learning still learns the optimal value function and optimal policy.

Exercise 11.13 *Based on the updating equations of SARSA and Q-learning, develop the pseudo codes (procedural forms) for the on-policy TD control and the off-policy TD control respectively, and point the major differences between them.*

Exercise 11.14 *Argue that as $\varepsilon \to 0$, the ε-greedy policy generated by SARSA will converge to the overall optimal policy.*

Exercise 11.15 *Instead of taking the maximum over the next state-action pairs in Q-learning, one can use the expected value, taking into account how likely each action is under the current policy. That is called Expected SARSA. Develop the updating equation for this learning algorithm. What is the main advantage and disadvantage of Expected SARSA?*

11.2.6 CONVERGENCE OF LEARNING ALGORITHMS

We have studies several learning algorithms and have assumes that they converge to the limits (e.g., the true state values or Q-values) as the number of iterations increases. In fact, the rigorous proofs of these algorithms' convergence are not trivial and can be quite technical. Here, we present the proof of a very simple case to provide readers some basic idea of proving the convergence for a learning algorithm. For more general and complex cases, interested readers are referred to some studies in this area (see Sutton and Barto (2020)).

Consider a deterministic MDP model, a special case of general MDP model, in which the immediate reward function and the transition function are given by

$$r_t = r(s_t, a_t),$$
$$s_{t+1} = f(s_t, a_t).$$

Accordingly, given initial state $s_0 = s$, we have the value function under a policy π for an infinite horizon problem can be written as

$$V^\pi(s) = \sum_{t=0}^\infty \gamma^t r(s_t, a_t) = \sum_{t=0}^\infty \gamma^t r(s_t, \pi(s_t))$$

with the optimal value function $V^*(s) = \max_\pi V^\pi(s)$. Due to the deterministic nature, we set $w = 1$ and the Q-value function is given by

$$\hat{Q}^*(s,a) = r(s,a) + \gamma \max_{a'} \hat{Q}^*(f(s,a), a'). \tag{11.43}$$

Denote by $\hat{Q}_n^*$ the estimated Q-value function updated on the nth iteration. Then the learning algorithm can be designed as

- Initialize: Set $n = 0, \hat{Q}_n^*(s,a) = q_0(s,a)$ for all state-action pairs (s,a).
- Iteration:
  - For each (s,a), observe $s,a,r,s' = f(s,a)$.
  - Update $\hat{Q}_{n+1}^*(s,a)$ by $\hat{Q}_{n+1}^*(s,a) := r_n + \gamma \max_{a'} \hat{Q}_n^*(s',a')$.
  - If $|\hat{Q}_{n+1}^*(s,a) - \hat{Q}_n^*(s,a)| < \varepsilon$ for every state-action pair (s,a), STOP. Otherwise, set $n = n+1$ and start the next iteration.

The convergence of such an algorithm can be established.

Theorem 11.3 *In a deterministic MDP model, if each state-action pair is visited infinitely often, then the estimated Q-value function converges. That is*

$$\lim_{n \to \infty} \hat{Q}_n^*(s,a) = Q^*(s,a)$$

for all (s,a).

Proof. Define

$$\Delta_n := \|\hat{Q}_n^* - Q^*\| = \max_{s,a} |\hat{Q}_n^*(s,a) - Q^*(s,a)|.$$

At every stage n, for $\hat{Q}_n^*(s,a)$ updated on the $(n+1)$st iteration, the deviation from the true optimal Q-value for this state-action pair is given by

$$|\hat{Q}_{n+1}^*(s,a) - Q^*(s,a)| = |r_n + \gamma\max_{a'} \hat{Q}_n^*(s',a') - (r_n + \gamma\max_{a''} Q^*(s',a''))|$$

$$= \gamma|\max_{a'} \hat{Q}_n^*(s',a') - \max_{a''} Q^*(s',a'')|$$

$$\leq \gamma\max_{a'} |\hat{Q}_n^*(s',a') - Q^*(s',a')| \leq \gamma\Delta_n,$$

where the first inequality follows from the fact that $|\max_a f_1(a) - \max_a f_2(a)| \leq \max_a |f_1(a) - f_2(a)|$ and the second inequality is due to the definition of Δ_n (i.e., the maximum is taken over both s and a). Consider now some interval $[n_1, n_2]$ over which all state-action pairs (s,a) appear at least once. Such an interval always exists due to the condition that each state-action pair is visited infinitely often. Using the above relation and simple induction, it follows that $\Delta_{n_2} \leq \gamma\Delta_{n_1}$. Since $\gamma < 1$ and there is an infinite number of such intervals by assumption, it follows that $\gamma\Delta_n \to 0$ as $n \to \infty$. ∎

It is clear that for the stochastic MDP, the state value function and Q-value functions are defined in terms of the expected values. The convergence of the learning algorithm for the non-deterministic case can also be proved. However, the proof is more technical and involved. Interested readers are referred to Watkins and Dayan (1992). Here we only present the main results. The convergence can be proved under the following two assumptions:

Assumption 11.1 *Every pair of state $s \in \mathcal{S}$ and action $a \in \mathcal{A}(s)$ is visited infinitely often.*

Assumption 11.2 *Let $w_t(s,a)$ denote the learning rate at time t for every state-action pair (s,a). Then the non-negative learning rates satisfy*

$$\sum_{t=0}^{\infty} w_t(s,a) = \infty, \quad \sum_{t=0}^{\infty} w_t(s,a)^2 < \infty \qquad (11.44)$$

with probability 1 and are decreasing monotonically with t.

Under these assumptions, the following theorem can be established.

Theorem 11.4 *Suppose that Assumptions 11.1 and 11.2 hold, then we have*

$$\lim_{k \to \infty} \hat{Q}_k^*(s,a) = Q^*(s,a)$$

with probability 1 for every pair of (s,a).

This theorem is proved by constructing a notional Markov decision process called Action-Replay Process (ARP), which is similar to the real process. Then it is shown that the Q-Values produced by the one step Q-learning process after "n" training examples, are the exact optimal action values for the start of the action-replay process for "n" training examples. The details of the proof can be found in Watkins and Dayan (1992). Note that these assumptions are sufficient conditions for the convergence of Q-learning. For example, for the learning algorithms of episodic processes, $\sum_{t=0}^{\infty} w_t(s,a)^2 < \infty$ is automatically satisfied.

11.2.7 LEARNING IN STOCHASTIC GAMES – AN EXTENSION OF Q-LEARNING

As shown in Chapter 10, the MDP model can be extended to a stochastic game with multi-agents. Accordingly, the Q-learning can be extended to multi-agent environments. We utilize the framework of general-sum stochastic games. General-sum games allow the agents' rewards to be arbitrarily related. As special cases, "zero-sum games" are instances where agents' rewards are always negatively related, and in "coordination games" rewards are always positively related. In this framework, we define optimal Q-values as Q-values received in a Nash equilibrium, and call them "Nash Q-values". The goal of learning is to find Nash Q-values through repeated plays. Based on learned Q-values, our agent can then derive the Nash equilibrium and choose its actions accordingly. Again, we adopt the Gridworld games to demonstrate the basic ideas of learning in stochastic games. This example is taken from Hu and Wellman (2003).

We consider two Gridworld games as shown in Figure 11.12. Game 1 has the deterministic moves and Game 2 has probabilistic transitions. To simplify the notation for state, we number the cells from 0 to 8. In both games, the two players start from their respective starting point (cell 0 and cell 2, respectively). Each player can move only one cell a time with four possible directions: L (left), R (right), U (up), and D (down). If two players attempt to move into the same cell (excluding a goal cell), they are bounced back to their previous cells. They choose their actions simultaneously. The game is over as soon as one player reaches its goal cell and becomes the winner with a positive reward. If both players reach their goal cells at the same time, both are winners (a tie). The goal of each player is to reach its goal cell with a minimum number of steps

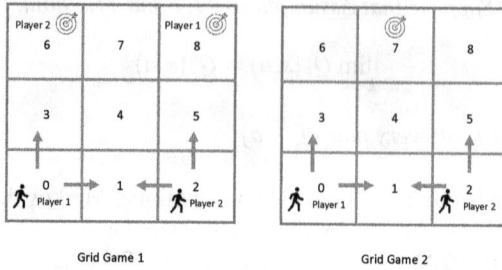

Figure 11.12   Two Gridworld games.

(i.e., the fastest way). It is assumed that the players do not know the locations of their goal cells and their own and other player's reward functions at the beginning of the learning period. However, they can observe the previous actions of both players, the current state (positions of both players), and the immediate rewards after each move (actions of both players). Let's first describe this gridworld problem as a stochastic game. Denote a state by $s = (l^1, l^2)$ where l^i is the position (cell number) of player i with $i = 1, 2$. Then the state space for this problem is $\mathscr{S} = \{(0,1),(0,2),...,(8,7)\}$. Since two players cannot stay in the same cell and the cases with at least one player reaching his goal cell should be excluded, the number of possible states (joint positions) is 56 for these grid games. The payoff structure for these two games are as follows: (1) When a player reaches the goal cell, he received a reward of 100; (2) If a player moves into another non-goal cell without colliding with the other player, he receives a reward of 0; and (3) If a player collides with the other player, he receives a reward of -1 and both players are bounced back to their previous positions. Denote by $L(l_t^i, a_t^i)$ the potential new location for player i if he chooses action a_t^i in location l_t^i at time t and by r_t^i the immediate reward for player i at time t. Then the reward function is given by

$$
r_t^i = \begin{cases} 100 & \text{if } L(l_t^i, a_t^i) = Goal_i, \\ -1 & \text{if } L(l_t^1, a_t^1) = L(l_t^2, a_t^2) \text{ and } L(l_t^2, a_t^2) \neq Goal_j, \ j = 1,2, \\ 0 & \text{otherwise.} \end{cases}
$$

All state transitions are deterministic in Grid Game 1. In Grid Game 2, only moving up from state 0 or 2 is not deterministic (all other moves remain deterministic). That is in state 0 or 2 the player moves up with probability 0.5 and remains in the previous position with probability 0.5. Thus, if both players decide to move up from state (0,2), then the next state is equally likely to be in one of the four states (0,2), (3,2), (0,5), and (3,5) (i.e., with 0.25 probability in each state). Similarly, from state (0,2), the transition probabilities under other

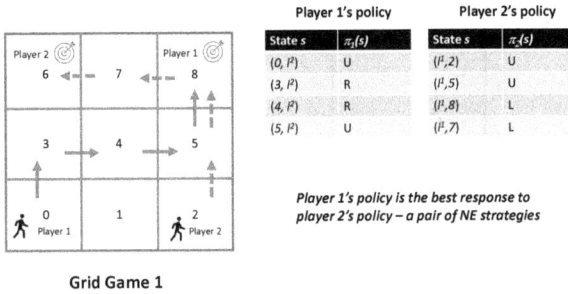

Figure 11.13 A pair of NE strategies for two players.

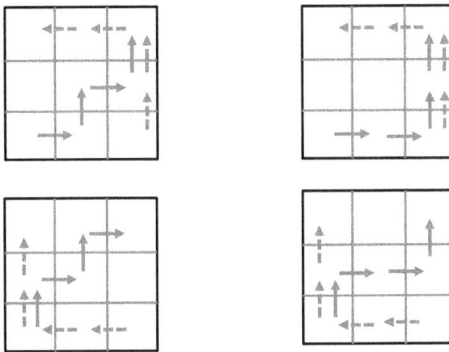

Figure 11.14 Other possible NE policy pairs for two players in Grid Game 1.

action combinations, $P(s'|s, A_1, A_2)$ are given by

$$P((0,1)|(0,2), U, L) = 0.5, \quad P((3,1)|(0,2), U, L) = 0.5,$$
$$P((1,2)|(0,2), R, U) = 0.5, \quad P((1,5)|(0,2), R, U) = 0.5.$$

Now we can discuss the Nash Q-values for these games. A Nash equilibrium (NE) in a grid game consists of a pair of strategies (π_1^*, π_2^*) in which each strategy is a best response to the other. We limit our study to stationary policeis (strategies). Clearly, two shortest paths that do not collide with each other generate an NE equilibrium since each policy is the best response to the other. An example of such an NE equilibrium is shown in Figure 11.13. Other possible NE policy pairs are shown in Figure 11.14. Note that the symmetric policy pairs (i.e., two players interchange their policies) are also NE equilibria. Similarly, the NE equilibria for Grid Game 2 are shown in Figure 11.15. The

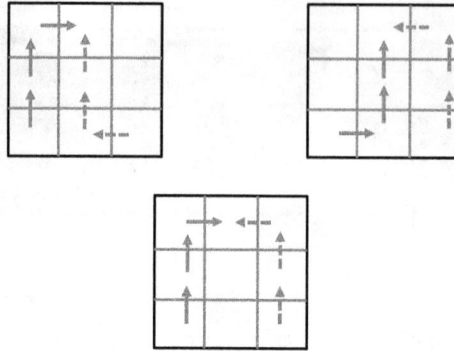

Figure 11.15 Possible NE policy pairs for two players in Grid Game 2.

optimal value function for player 1 is given by

$$V_1^*(s_0) = \sum_t \gamma^t E(R_{1t}(s_0, \pi_1^*, \pi_2^*)),$$

which can be computed easily due to the simplicity of the problem. In Grid Game 1, with $s_0 = (0,2)$ and $\gamma = 0.99$, we have

$$V_1^*(s_0) = 0 + 0.99 \cdot 0 + 0.99^2 \cdot 0 + 0.99^3 \cdot 100 = 97.0.$$

Based on the optimal value function, we can compute the Nash Q-values for player 1 in state s_0 as

$$Q_1^*(s_0, \pi_1(s_0), \pi_2(s_0)) = R_1(s_0, \pi_1(s_0), \pi_2(s_0))$$
$$+ \gamma \sum_{s'} P(s'|s_0, \pi_1(s_0), \pi_2(s_0))V_1^*(s').$$

For example, for $s_0 = (0,2)$ we can compute the Q-values of player 1 for action pairs $(A_1(s_0) = R, A_2(s_0) = L)$ and $(A_1(s_0) = U, A_2(s_0) = U)$ as follows:

$$Q_1^*(s_0, R, L) = -1 + 0.99V_1^*((0,2)) = 95.1,$$
$$Q_1^*(s_0, U, U) = 0 + 0.99V_1^*((3,5)) = 97.0.$$

The Q-values for other action pairs at s_0 for both players can be computed similarly and are listed in Table 11.1. It is clear that there are three NEs for this stage game $(Q_1(s_0), Q_2(s_0))$ and each is a global optimal action pair with value (97.0, 97.0).

For Grid Game 2, we can also work out the Q-values for both players similarly. Again, we start with computing some optimal state values for player 1,

TABLE 11.1

Nash Q-values in state (0,2) of Grid Game 1.

		$A_2(s_0)$	
		L	U
$A_1(s_0)$	R	95.1, 95.1	97.0, 97.0
	U	97.0, 97.0	97.0, 97.0

which will be used for determining the Q-values. These are

$$V_1^*((0,1)) = 0 + 0.99 \cdot 0 + 0.99^2 \cdot 0 = 0,$$
$$V_1^*((0,l^2)) = 0, \text{ for } l^2 = 3,...,8,$$
$$V_1^*((1,2)) = 0 + 0.99 \cdot 100 = 99,$$
$$V_1^*((1,3)) = 0 + 0.99 \cdot 100 = 99 = V_1^*((1,5)),$$
$$V_1^*((1,l^2)) = 0, \text{ for } l^2 = 4,6,8.$$

Due to the probabilistic transitions, the optimal value for the starting state $s_0 = (0,2)$ can be obtained by analyzing the stage game $(Q_1^*(s_0), Q_2^*(s_0))$. We only demonstrate the calculations for player 1 and leave the calculations for player 2 as an exercise for readers.

$$Q_1^*(s_0, R, L) = -1 + 0.99V_1^*(s_0),$$
$$Q_1^*(s_0, R, U) = 0 + 0.99(0.5V_1^*((1,2)) + 0.5V_1^*((1,3))) = 98,$$
$$Q_1^*(s_0, U, L) = 0 + 0.99(0.5V_1^*((0,1)) + 0.5V_1^*((3,1))) = 0.99(0 + 0.5 \cdot 99) = 49,$$
$$Q_1^*(s_0, U, U) = 0 + 0.99(0.25V_1^*((0,2)) + 0.25V_1^*((0,5))$$
$$+ 0.25V_1^*((3,2)) + 0.25V_1^*((3,5)))$$
$$= 0.99(0.25V_1^*((0,2)) + 0 + 0.25 \cdot 99 + 0.25 \cdot 99)$$
$$= 0.99 \cdot 0.25V_1^*(s_0) + 49.$$

$$(11.45)$$

Similar analysis can be done for the Q-values of player 2. These Q-values for both players are summarized in Table 11.2. Next, we consider two potential pure strategy Nash equilibria: (U,L) and (R,U) and a mixed strategy Nash equilibrium for the stage game, which is $(\pi_1(s_0), \pi_2(s_0)) = (\{P(R) = 0.97, P(U) = 0.03\}, \{P(L) = 0.97, P(U) = 0.03\})$. The Nash Q-values for these NEs are summarized as Tables in Figure 11.16. We show how to compute the Q-values for the first two NEs with pure strategies. NE1 is the case that $(\pi_1(s_0), \pi_2(s_0)) = (U,L)$ is the first move of the policy pair. Based on the third equation of (11.45), we have $V_1^*(s_0) = Q_1^*(s_0, U, L) = 49$ for player 1.

TABLE 11.2

Nash Q-values in state (0,2) for Grid Game 2.

Player 2
Player 1

	L	U
R	$-1+0.99R_1$, $-1+0.99R_2$	98, 49
U	49, 98	$49+0.99\cdot0.25R_1$, $49+0.99\cdot0.25R_2$

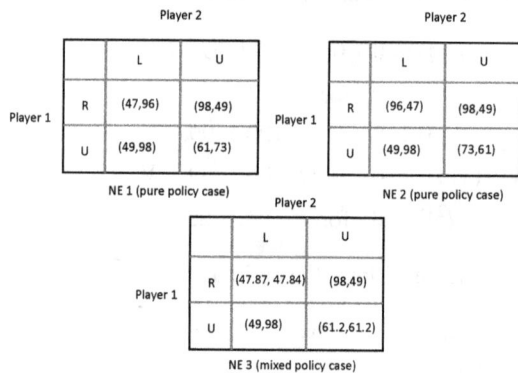

Player 2

	L	U
R	(47,96)	(98,49)
U	(49,98)	(61,73)

Player 1

NE 1 (pure policy case)

Player 2

	L	U
R	(96,47)	(98,49)
U	(49,98)	(73,61)

Player 1

NE 2 (pure policy case)

Player 2

	L	U
R	(47.87, 47.84)	(98,49)
U	(49,98)	(61.2,61.2)

Player 1

NE 3 (mixed policy case)

Figure 11.16 Q-values for three NEs in Grid Game 2.

With this value for $V_1^*(s_0)$, we can compute

$$Q_1^*(s_0,R,L) = -1+0.99\cdot49 = 47.51,$$
$$Q_1^*(s_0,U,U) = 0.99\cdot0.25\cdot49+49 = 61.1275,$$

which along with $Q_1^*(s_0,R,U) = 98$ yields the Q-values for player 1 in Table 1 of Figure 11.16. Similarly, the Q-values for player 2 can be computed for NE1. For NE2, the first move of the policy pair is $(\pi_1(s_0), \pi_2(s_0)) = (R,U)$. Based on the second equation of (11.45), we have $V_1^*(s_0) = Q_1^*(s_0,R,U) = 98$ for player 1. From this $V_1^*(s_0)$ value, we can compute other Q-values for player 1 as

$$Q_1^*(s_0,R,L) = -1+0.99\cdot98 = 96.02,$$
$$Q_1^*(s_0,U,U) = 0.99\cdot0.25\cdot98+49 = 73.255,$$

which along with $Q_1^*(s_0, U, L) = 49$ yields the Q-values for player 1 in Table 2 of Figure 11.16. Again, we can also figure out the Q-values for player 2 in this NE. The computation of Q-values for NE3 is left as an exercise. In contrast to Grid Game 1 where the Q-values for multiple NEs are the same, the three sets of Q-values for the three different NEs in Grid Game 2 are different. This feature revealed in Grid Game 2 makes the convergence of learning process problematic for the multi-NEs situations. For example, none of the NEs of the stage game of Grid Game 2 is a global optimal or saddle point, whereas in Grid Game 1 the NE is a global optimum.

Exercise 11.16 *Compute the Q-values for the NEs in Grid Game 1 discussed for player 2 to verify these numbers given in Table 11.1.*

Exercise 11.17 *Compute all the Q-values for three NEs in Grid Game 2 discussed for both players to verify these numbers given in Figure 11.16.*

Now we consider the learning process in these Grid games. Consider player 1 who initializes the estimated Nash Q-values as zeros. That is $\hat{Q}_{10}^{*1}(s, a_1, a_2) = 0$ and $\hat{Q}_{20}^{1*}(s, a_1, a_2) = 0$ for all s, a_1, a_2 (state-action tuples). Note that the learning process of player 2 can be conducted similarly. Since this example is taken from Hu and Wellman (2003), the following discussion on learning process is based on Hu and Wellman (2003). A game starts from the initial state $(0, 2)$. From the current state, the two players choose their actions simultaneously. Then they observe the new state, rewards, and actions of both players. The learning player, player i, updates its estimated Q-value functions for player j, with $i = 1, 2$ and $j = 1, 2$, based on

$$\hat{Q}_{jt}^{*i}(s, a_1, a_2) = (1 - w_{t-1})\hat{Q}_{jt-1}^{*i}(s, a_1, a_2) + w_{t-1}[r_{jt-1} + \gamma\hat{Q}_{jt-1}^{*i}(s')]. \quad (11.46)$$

In the new state, the two players repeat the process above. When at least one player reaches its goal cell, the current game terminates and a new game restarts with each player randomly assigned to a non-goal cell (i.e., a new episode starts). The learning player keeps the estimated Q-values learned from previous episodes. In Hu and Wellman (2003), a total of 5000 episodes are utilized in the learning process (also called "training stage"). Since each episode takes about eight steps on average, one experiment run will have about 40,000 steps. The total number of state-action tuples in Grid Game 1 is 421 (left as an exercise). Thus, each tuple is visited $n_t(s, a_1, a_2) = 95$ times on average (taking the integer). The learning rate can be defined as $w_t = 1/n_t(s, a_1, a_2)$. If $w_t = 1/95 = 0.01$ which satisfies the assumptions for the convergence of the learning algorithm (see Assumptions 11.1 and 11.2). The stochastic game version of Assumption 11.2 can be stated as

Assumption 11.3 *The learning rate w_t satisfies the following conditions for every (s, a_1, a_2) at t.*

1. $0 \leq w_t(s, a_1, a_2) < 1$, $\sum_{t=0}^{\infty} w_t(s, a_1, a_2) = \infty$, $\sum_{t=0}^{\infty} w_t(s, a_1, a_2)^2 < \infty$ and the latter two hold uniformly with probability 1.

2. $w_t(s, a_1, a_2) = 0$ if $(s, a_1, a_2) \neq (s_t, a_{1t}, a_{2t})$.

Note that the second condition in Assumption 11.3 implies that the player updates only the Q-function element corresponding to current state s_t and action a_{1t}, a_{2t}. Under these assumptions, the convergence of Q-learning algorithms for stochastic games can be proved. There are extensive studies in this area and interested readers are referred to Hu and Wellman (2003) for more discussion on this topic and references therein. In fact, the convergence of Q-learning algorithms in stochastic games is more complex and the proof is more challenging. When multiple Nash equilibria exist, the convergence proof requires that the stage games encountered during learning have global optima, or alternatively, that they all have saddle points. Moreover, it mandates that the learner consistently chooses either global optima or saddle points in updating its Q-values. Thus, in addition to the previous assumptions, the third assumption is made as follows.

Assumption 11.4 *One of the following conditions holds during learning:*

- *Condition A. Every stage game $(Q_{1t}(s), Q_{2t}(s))$, for all t and s, has a global optimal point and player's rewards in this equilibrium are used to update their Q-functions.*
- *Condition B. Every stage game $(Q_{1t}(s), Q_{2t}(s))$, for all t and s, has a saddle point and player's rewards in this equilibrium are used to update their Q-functions.*

During the learning process, if Assumption 11.4 is not satisfied, the player has to choose one of many NEs to update the Q-values. In our Grid Game examples, multiple NEs exist. When updating the Q-values during the learning process, the player has to choose among the multiple NEs from stage games. Here an approach called Lemke–Howson method can be used to generate equilibria in a fixed order (Cottle et al., 1992). There are several options for each player. A *first-Nash learning player* is defined as the one that updates its Q-values using the first NE generated. Similarly, a so-called *second Nash learning player* uses the second NE generated to update its Q-values. A *best-expected-Nash player* selects the NE that generates the highest payoff to itself. By implementing the Q-learning process, we notice that both Grid Games 1 and 2 violate Assumption 11.4. The table on the left in Figure 11.17 shows the results for using the first-Nash learning from state $s_0 = (0, 2)$ in Grid Game 1 after a certain number of episodes. This only NE in this stage game, (R, L), is neither a global optimum nor a saddle point. Similarly, in Grid Game 2, Assumption 11.4 is violated and the unique NE, (U, U) in the stage game shown in the table on the right in Figure 11.17, is neither a global optimum nor a saddle point.

	L	U
R	(-1,-1)	(49,0)
U	(0,0)	(0,97)

Player 1

	L	U
R	(31,31)	(0,65)
U	(0,0)	(49,49)

Player 1

Grid Game 1: Q-values in state (0 2) after 20 episodes if always choosing the first Nash

Grid Game 2: Q-values in state (0 2) after 61 episodes if always choosing the first Nash

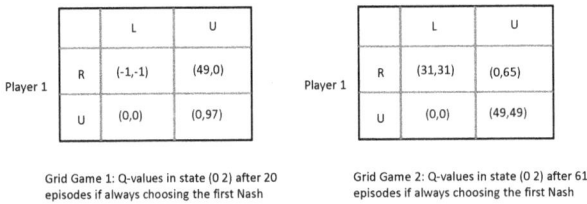

Figure 11.17 *Q*-values updating during the learning process for Grid Games 1 and 2.

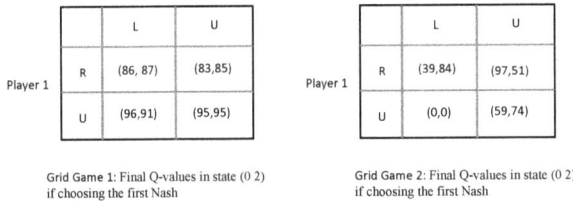

	L	U
R	(86, 87)	(83,85)
U	(96,91)	(95,95)

Player 1

	L	U
R	(39,84)	(97,51)
U	(0,0)	(59,74)

Player 1

Grid Game 1: Final Q-values in state (0 2) if choosing the first Nash

Grid Game 2: Final Q-values in state (0 2) if choosing the first Nash

Figure 11.18 *Q*-values updating during the learning process for Grid Games 1 and 2.

After 5000 episodes of learning, the players' Q-values are stabilized at certain values. Some results are presented in Figure 11.18 (taken from Hu and Wellman 2003). Note that a learning player always use the same NE value to update his own and other player's Q-functions. In Figure 11.18, we observe that the results in the table on the left are close to Table 11.1 and the results in the table on the right are close to the first table (NE 1) in Figure 11.16. Although it is not possible to validate theoretical asymptotic convergence with a finite number of episodes, these examples confirm that the learned Q-functions possess the same equilibrium strategies as Q^*. For more discussion on the offline and online performances in these two Grid Games, we refer readers to Hu and Wellman (2003).

In this section, we have presented a brief introduction to the learning algorithms in MDP models. This is the topic of reinforcement learning, a rich and active research area. The aim of this introduction is to provide readers with an easy and intuitive description of most popular learning models in stochastic processes. Our hope is to stimulate readers' interest in this area. We offer some references in the bibliographical notes at the end of the chapter for interested readers to explore.

11.3 HIDDEN MARKOV MODELS

As another model-based learning model, we introduce the hidden Markov model by using a simple example. Suppose we want to determine the average annual rainfall at a particular location on earth over a series of years. To make it non-trivial and interesting, we assume that this series of years lie in the distant past, before rainfalls were recorded (i.e., before human civilization). Since we cannot go back in time, we instead look for indirect evidence of the rainfall. To simplify the situation, we only consider the two categories of "abundant rainfall" and "drought" for the average annual rainfall measure. Suppose that scientific study indicates that the probability of an abundant rainfall year followed by another abundant rainfall year is 0.8 and the probability that a drought year is followed by another drought year is 0.7. We'll assume that these probabilities also held in the distant past. This information can be summarized as

$$
\begin{array}{cc} & R \quad D \end{array} \\
A = \begin{array}{c} R \\ D \end{array} \begin{pmatrix} 0.8 & 0.2 \\ 0.3 & 0.7 \end{pmatrix}, \tag{11.47}
$$

where R represents "abundant rainfall" and D represents "drought". It is known that there is a correlation between the size of tree growth rings and the rainfall amount per year. For most locations, tree rings will be wider during years of R and narrower during years of D. For simplicity, we only consider three different tree ring sizes, small, medium, and large, denoted S, M, and L, respectively. Furthermore, suppose that based on currently available evidence, the probabilistic relationship between annual rainfall amount and tree ring sizes is given by

$$
\begin{array}{ccc} & S \quad M \quad L \end{array} \\
B = \begin{array}{c} R \\ D \end{array} \begin{pmatrix} 0.2 & 0.3 & 0.5 \\ 0.7 & 0.2 & 0.1 \end{pmatrix}. \tag{11.48}
$$

For this system, the state is the average annual rainfall amount, either R or D. The transition from one state to the next is a Markov chain based on the transition matrix. However, these actual states are "hidden" since we cannot directly observe the rainfall amount in the past. Although we cannot observe the state (rainfall amount) in the past, we can observe the size of tree rings. From (11.48), tree rings provide us with probabilistic information regarding the rainfall amount. Since the underlying states are hidden, this type of system is known as a hidden Markov model (HMM). We denote the HMM's state space by $\Omega = \{R, D\}$ and the observation set by $V = \{S, M, L\} = \{0, 1, 2\}$ where numerical codes for these observations are used. Our goal is to utilize the observable information appropriately to gain insight into various aspects of the Markov process. In addition to the two probability matrices of A and B, which are row stochastic (i.e., stochastic matrices), we assume that the initial state

A Hidden Markov process with T periods

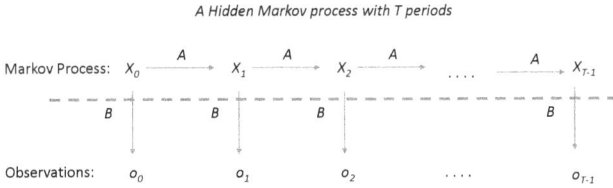

Figure 11.19 A hidden Markov process with T periods.

distribution, denoted by π, over the system states is available for this location as

$$\pi = (0.7, 0.3), \tag{11.49}$$

which means that the chance for state R is 0.7 and for state D is 0.3 (it is a rainforest region on earth). If we denote by N the number of states of the MC and by M the number of observations, an HMM model can be denoted by a triple tuple $\lambda = (\pi, A, B)$ of size $N \times M$ and is illustrated in Figure 11.19.

Usually, a sequence of finite number, T, of observations, denoted by O, is available as a piece of available information for the HMM. In contrast, the corresponding state sequence is denoted by S, which is unobservable.

For example, we may have four year of tree ring sizes available for a four-year period in distant past for this location. Say $O = (S, M, S, L) = (0, 1, 0, 2)$.

There are three typical problems for an HMM that we most likely are interested in. The first one is called "likelihood" or "evaluating" problem, which is to determine the probability $P(O|\lambda)$, given the model $\lambda = (\pi, A, B)$, that the observation sequence O occurs. Such a probability is needed for solving the next problem. The second problem, called "decoding" problem, is to discover the most likely state sequence S given an observation sequence O and an HMM $\lambda = (\pi, A, B)$. The third problem, called "learning" problem, is to update the parameters of the HMM based on an observation sequence O. Now we use the rainfall amount example to demonstrate the basic idea of analyzing an HMM by answering these three questions. This is an HMM with $T = 4, N = 2, M = 3, \Omega = \{R, D\}$ and $V = \{0, 1, 2\}$. For the $N \times N$ matrix $A = \{a_{ij}\}$, the element $a_{ij} = P(s_{t+1} = q_j | s_t = q_i)$ where $q_i, q_j \in S = \{R, D\}$. We assume that the transition matrix is stationary (i.e., independent of time t). For the $N \times M$ matrix B, the element $b_{ij} = b_i(o_j) = P(o_j$ at time $t | s_t = q_i)$ which is the probability of state i generating observation j at time t. Again it is assumed to be independent of time t and the notation $b_i(o_j)$ is a conventional symbol used in HMM's. Suppose that an observation sequence of length four is available and denoted by

$$O = (o_0, o_1, o_2, o_3),$$

which was generated by a sequence of (hidden) states

$$S = (s_0, s_1, s_2, s_3).$$

In addition, the initial distribution over the hidden states is π_{s_0}. Thus, the joint distribution of these two sequences can be given by

$$P(S, O) = \pi_{s_0} b_{s_0}(o_0) a_{s_0, s_1} b_{s_1}(o_1) a_{s_1, s_2} b_{s_2}(o_2) a_{s_2, s_3} b_{s_3}(o_3). \qquad (11.50)$$

Note that here s_i represents the indices in the A and B matrices. In our example, since the observation sequence is $O = (S, M.S, L) = (0, 1, 0, 2)$ is given, we can compute all possible joint sequence probabilities. For example, we can compute the joint probability for a possible state sequence (R, R, D, D) as

$$
\begin{aligned}
P(RRDD, O) &= \pi_R b_R(S) a_{R,R} b_R(M) a_{R,D} b_D(S) a_{D,D} b_D(L) \\
&= 0.7(0.2)(0.8)(0.3)(0.2)(0.7)(0.7)(0.1) = 0.000329.
\end{aligned}
$$

Similarly, we can compute all possible joint probabilities $P(S, O)$ for a given observation sequence O. Table 11.3 lists all these probabilities with the last column containing the normalized probabilities. From Table 11.3, we can sum the probabilities to obtain the probabilities that R and D are in certain positions. These probabilities are presented in Table 11.4.

With all these probabilities, we can obtain some information about this HMM. For example, we can easily find the hidden state sequence with the highest probability. While the computation of this approach is straightforward, the computational complexity is high. In fact, this enumeration approach requires $2TN^T$ multiplications as there are N^T possible sequences and each sequence requires multiplication of $2T$ probabilities. Therefore, we have to utilize the dynamic programming (DP) approach to answer the three questions about an HMM.

Exercise 11.18 *Explain that the computational complexity of computing all possible joint probabilities of the state and observation sequences for T periods in an HMM with N states is $O(TN^T)$.*

Let's first answer the likelihood question. That is to find $P(O|\lambda)$. Consider a state sequence $S = (s_0, s_1, ..., s_{T-1})$. Conditioning on this sequence and the HMM model, we have

$$P(O|S, \lambda) = b_{s_0}(o_0) b_{s_1}(o_1) \cdots b_{s_{T-1}}(o_{T-1}).$$

It follows from the initial distribution and transition matrix A of the HMM that

$$P(S|\lambda) = \pi_{s_0} a_{s_0, s_1} a_{s_1, s_2} \cdots a_{s_{T-2}, s_{T-1}}.$$

TABLE 11.3
State sequence probabilities

state	probability	normalized probability
RRRR	0.002150	0.219429
RRRD	0.000108	0.010971
RRDR	0.000706	0.072000
RRDD	0.000329	0.033600
RDRR	0.000134	0.013714
RDRD	0.000007	0.000686
RDDR	0.000412	0.042000
RDDD	0.000192	0.019600
DRRR	0.001210	0.123429
DRRD	0.000060	0.006171
DRDR	0.000397	0.040500
DRDD	0.000185	0.018900
DDRR	0.000706	0.072000
DDRD	0.000035	0.003600
DDDR	0.002161	0.220500
DDDD	0.001008	0.102900

TABLE 11.4
HMM probabilities

	State Position			
	0	1	2	3
$P(R)$	0.4120	0.5250	0.4500	0.8036
$P(D)$	0.5880	0.4750	0.5500	0.1964

Using the total probability and multiplication formulas, we obtain

$$P(O|\lambda) = \sum_S P(O,S|\lambda) = \sum_S P(O|S,\lambda)P(S|\lambda)$$

$$= \sum_S \pi_{s_0} b_{s_0}(o_0) a_{s_0,s_1} b_{s_1}(o_1) \cdots a_{s_{T-2},s_{T-1}} b_{s_{T-1}}(o_{T-1}).$$

As indicated earlier, the computational complexity is too high (i.e., $O(TN^T)$). To overcome this issue, we develop the so-called forward algorithm (also called α-pass). Define

$$\alpha_t(i) = P(o_0, o_1, ..., o_t, s_t = q_i|\lambda), \tag{11.51}$$

which is the probability of the partial observation sequence up to time t, where the underlying Markov process is in state q_i at time t. Such a probability can be computed recursively as follows:

$$\alpha_0(i) = P(o_0, s_0 = q_i|\lambda) = \pi_i b_i(o_0), \text{ for } i = 0, 1, ..., N-1,$$

$$\alpha_1(i) = P(o_0, o_1, s_1 = q_i|\lambda) = \sum_{j=0}^{N-1} P(o_0, s_0 = q_j, o_1, s_1 = q_i|\lambda)$$

$$= \left[\sum_{j=0}^{N-1} \alpha_0(j) a_{ji} \right] b_i(o_1),$$

$$\vdots$$

$$\alpha_t(i) = P(o_0, o_1, ..., o_t, s_t = q_i|\lambda) = \sum_{j=0}^{N-1} P(o_0, o_1, ...s_{t-1} = q_j, o_t, s_t = q_i|\lambda)$$

$$= \left[\sum_{j=0}^{N-1} \alpha_{t-1}(j) a_{ji} \right] b_i(o_t).$$

It is easy to see that the computational complexity is much less than the enumeration approach. Since for each t, there are about N^2 multiplications, the forward algorithm only requires about $N^2 T$ multiplications which is less than $2TN^T$ multiplications in the enumeration approach. Using the recursion, we can compute

$$P(O|\lambda) = \sum_{i=0}^{N-1} \alpha_{T-1}(i).$$

Exercise 11.19 *In the "rainfall and drought" problem studied, using the forward algorithm based on α-pass, compute the probability $P(O|\lambda) = P(S,M,S,L|\lambda)$.*

Next, we solve the "decoding" problem. That is to find the most likely state sequence for a given observation sequence O. The "most likely" state sequence can be defined in different ways. The most common definitions are the followings: (i) Maximum Individual State Probability (MIP): the state sequence that maximizes the expected number of correct states (i.e., the sequence of the states with maximum individual probabilities for a given observation sequence). (ii) Maximum Sequence Probability (MSP): the state sequence with the highest path probability. Note that these two different definitions may lead to different solutions. Again, we utilize the recursive relation (DP approach) which is based on the backward algorithm, also called β-pass. For $t = 0, 1, ..., T-1$ and $i = 0, 1, ..., N-1$, define

$$\beta_t(i) = P(o_{t+1}, o_{t+1}, ..., o_{t+1} | s_t = q_i, \lambda).$$

The β-pass can be computed recursively as follows:

$$\beta_{T-1}(i) = P(o_T | s_{T-1} = q_i) = 1, \text{ for } i = 0, 1, ..., N-1,$$

$$\beta_{T-2}(i) = P(o_{T-1}, | s_{T-2} = q_i, \lambda) = \sum_{j=0}^{N-1} P(o_{T-1}, s_{T-1} = q_j | s_{T-2} = q_i, \lambda)$$

$$= \sum_{j=0}^{N-1} P(o_{T-1} | s_{T-1} = q_j) P(s_{T-1} = q_j | s_{T-2} = q_i, \lambda)$$

$$= \sum_{j=0}^{N-1} a_{ij} b_j(o_{T-1}),$$

$$\beta_{T-3}(i) = P(o_{T-2}, o_{T-1}, | s_{T-3} = q_i, \lambda)$$

$$= \sum_{j=0}^{N-1} P(o_{T-2}, o_{T-1}, s_{T-2} = q_j | s_{T-3} = q_i, \lambda)$$

$$= \sum_{j=0}^{N-1} P(o_{T-2}, o_{T-1} | s_{T-2} = q_j, \lambda) P(s_{T-2} = q_j | s_{T-3} = q_i, \lambda)$$

$$= \sum_{j=0}^{N-1} P(o_{T-2} | s_{T-2} = q_j, \lambda) P(o_{T-1} | s_{T-2} = q_j, \lambda)$$

$$\times P(s_{T-2} = q_j | s_{T-3} = q_i, \lambda)$$

$$= \sum_{j=0}^{N-1} a_{ij} b_j(o_{T-2}) \beta_{T-2}(j),$$

$$\vdots$$

$$\beta_t(i) = P(o_{t+1}, o_{t+2}, ..., o_{T-1} | s_t = q_i, \lambda) = \sum_{j=0}^{N-1} a_{ij} b_j(o_{t+1}) \beta_{t+1}(j).$$

Now we can obtain the conditional probability that the Markov chain is at a particular state at time t, denoted by $\delta_t(i) = P(s_t = q_i|O,\lambda)$, given a observation sequence as follows.

$$\delta_t(i) = P(s_t = q_i|O,\lambda) = \frac{P((O \cap (s_t = q_i)|\lambda))}{P(O|\lambda)} = \frac{\alpha_t(i)\beta_t(i)}{P(O|\lambda)}. \qquad (11.52)$$

The numerator follows from the Markov property of the process and the fact that $\alpha_t(i)$ is the probability for the observation sequence up to time t and $\beta_t(i)$ is the relevant probability after time t given that $s_t = q_i$. The denominator $P(O|\lambda)$ is obtained from the solution to the first problem (the likelihood problem). Under the MIP definition, the most likely state sequence will consist of the states with maximum $\delta_t(i)$ which can be obtained by taking the maximum over the state index i. For the MSP definition, the most likely state sequence can be determined by using the DP approach. This DP approach is based on the forward algorithm (α-pass) where the summation in the recursion is replaced with the maximum operator. For a given HMM $\lambda = (\pi,A,B)$, the DP algorithm is as follows:

1. Let $\gamma_0(i) = \pi_i b_i(o_0)$, for $i = 0,1,...,N-1$.
2. For $t = 1,2,...,T-1$ and $i = 0,1,...,N-1$, compute

$$\gamma_t(i) = \max_{j \in \{0,1,...,N-1\}} [\gamma_{t-1}(j)a_{ij}b_i(o_t)].$$

3. The probability of the optimal overall path is

$$\max_{j \in \{0,1,...,N-1\}} [\gamma_{T-1}(j)].$$

Based on this maximum probability, the optimal overall path can be determined by tracking back from the final state.

Exercise 11.20 *In the "rainfall and drought" problem studied, using the backward algorithm based on β-pass, determine the "most likely" state sequence probability given the observation $O = \{S,M,S,L\}$ by using either MIP or MSP standard.*

Finally, we solve the third problem (i.e., the learning problem) which is to update the parameters of the HMM based on an observation sequence. We start with an HMM with initial parameters of A,B, and π and a given observation sequence O. These initial parameters can be a set of best guessed values. If no reasonable guess is available, we may choose random values such as $\pi_i \approx 1/N$, $a_{ij} \approx 1/N$, and $b_j(k) \approx 1/M$. These parameters should be randomized as using the exact uniform values may result in a local maximum from which the model cannot climb. This learning or training process for the HMM is based on the following conditional probability for a two consecutive state sequence:

$$\delta_t(i,j) = P(s_t = q_i, s_{t+1} = q_j | O, \lambda) = \frac{\alpha_t(i)a_{ij}b_j(o_{t+1})\beta_{t+1}(j)}{P(O|\lambda)}.$$

Clearly, for $t = 0, 1, ..., T - 2$, we have

$$\delta_t(i) = \sum_{j=0}^{N-1} \delta_t(i,j).$$

With the formulas for $\delta_t(i,j)$ and $\delta_t(i)$, we can design the following procedures for re-estimating the parameters A and B based on the observation sequence with the maximum likelihood property.

1. For $i = 0, 1, ..., N - 1$, set

$$\pi_i = \delta_0(i),$$

   and compute $\delta_t(i,j)$ and $\delta_t(i)$ for $t = 0, 1, ..., T - 2$ with the initial A and B based on the given observation sequence O.
2. For $i = 0, 1, ..., N - 1$ and $j = 0, 1, ..., N - 1$, compute the new estimates

$$\hat{a}_{ij} = \frac{\sum_{t=0}^{T-2} \delta_t(i,j)}{\sum_{t=0}^{T-2} \delta_t(i)}.$$

3. For $j = 0, 1, ..., N - 1$ and $k = 0, 1, ..., M - 1$, compute the new estimates

$$\hat{b}_j(k) = \frac{\sum_{t \in \{0,1,...,T-1\} o_t = k} \delta_t(j)}{\sum_{t=0}^{T-1} \delta_t(j)}.$$

The numerator for $\hat{a}_{ij}$ can be considered as the expected number of transitions from state q_i to state q_j and the denominator is the total expected number of transitions from state q_i to any state for the given observation sequence. Thus, this ratio is the new estimated transition probability from state q_i to q_j. Similarly, the numerator of $\hat{b}_j(k)$ is the expected number of times the model is in state q_j with observation k, while the denominator is the expected number of times the model is in state q_j for the given observation sequence. The learning process is an interactive process and continues until the probability $P(O|\lambda)$ stops increasing. This can be summarized as follows:

1. Initialize the HMM $\lambda = (\pi, A, B)$.
2. Compute $\alpha_t(i), \beta_t(i), \delta_t(i,j)$ and $\delta_t(i)$.
3. Re-estimate the model parameters A and B using the procedure above.
4. If $P(O|\lambda)$ increases, go to Step 2. Otherwise, stop.

To make the computational time of the algorithm reasonable, we may set a minimum threshold for the probability increase per iteration. If the increase is less than the threshold, the algorithm stops. Another way is to set a maximum number of iterations.

Exercise 11.21 *In the "rainfall and drought" problem studied, Based on the observation $O = (S,M,S,L)$, given initial distribution π, and matrices A and B, update the parameters of the HMM.*

11.4 PARTIALLY OBSERVABLE MARKOV DECISION PROCESSES

A partially observable Markov decision process, denoted by POMDP, is a controlled HMM with the following components: (i) a state space $\mathscr{S}$, (ii) an action space A, (iii) transition probability matrices that depend on the action taken for each state, (iv) an immediate reward/cost function, (v) a discount factor between 0 and 1, (vi) an observation set O, (vii) observation probability matrices, and (viii) an initial belief, the probability distribution of starting state. The first five components are the same as those in the fully observable MDP. Denote the time horizon by N. If a policy is fixed, the system becomes an HMM. We utilize a simple running example to demonstrate a finite horizon POMDP formulation and solution via value iteration.

Consider a maintenance problem for a highly complex system (e.g., a machine or a biological system). To make it simple, we consider the maintenance process with a finite time horizon and two system states called "good condition", denoted by G, and "working condition" (not good condition but still operating), denoted by W. Due to the system complexity, these conditions (states) are not observable. However, the quality of the product produced by the system in terms of the proportion of defective items, is observable (for a biological system, some health indicator may be observable). Assume that there are three quality levels: low (L), average (A), and high (H). To keep the system running well, for each period, one of two maintenance actions applies based on the observations. One action is corrective maintenance, denoted by C, and the other is preventive maintenance, denoted by S. Here, corrective maintenance has some negative effect on the product quality in good condition. From the perspective of the product quality, we should perform C on a W state and S on a G state.

To further simplify our analysis, we consider the case with a horizon of length $N = 3$. Thus the state space is $\mathscr{S} = (G,W)$, the action space is $A = (C,S)$, and the observation set is $O = (L,A,H)$. The state dynamics forms a DTMC. Assume that the two transition probability matrices corresponding the two actions are as follows:

$$\mathbf{P}(a=C) = \begin{matrix} & G & W \\ G \\ W \end{matrix}\begin{pmatrix} 0.3 & 0.7 \\ 0.6 & 0.4 \end{pmatrix} \quad \mathbf{P}(a=S) = \begin{matrix} & G & W \\ G \\ W \end{matrix}\begin{pmatrix} 0.2 & 0.8 \\ 0.9 & 0.1 \end{pmatrix} \tag{11.53}$$

and the two observation probability matrices are

$$\mathbf{B}(a=C) = \begin{matrix} & L & A & H \\ G \\ W \end{matrix}\begin{pmatrix} 0.7 & 0.2 & 0.1 \\ 0.1 & 0.4 & 0.5 \end{pmatrix} \quad \mathbf{B}(a=S) = \begin{matrix} & L & A & H \\ G \\ W \end{matrix}\begin{pmatrix} 0.1 & 0.4 & 0.5 \\ 0.6 & 0.3 & 0.1 \end{pmatrix}.$$
$$\tag{11.54}$$

In addition, suppose that the initial distribution for the state (belief) is $\pi = (0.75, 0.25)$. A value structure, called a reward matrix, is also imposed as follows:

$$\mathbf{R} = \begin{matrix} & C & S \\ G \\ W \end{matrix}\begin{pmatrix} 0 & 1.5 \\ 1 & 0 \end{pmatrix}. \tag{11.55}$$

Here we assume that the immediate reward only depends on the current state and action taken. If the immediate reward also depends on the next state and observation, then the expected immediate reward can be computed as shown below. To summarize this case, we have a POMDP with a horizon length of 3, two states, two actions, and three observations. It can be shown that a POMDP is equivalent to a classical observable MDP with a continuous state space. We first define the "belief state" as the distribution over the states of the POMDP. It is a vector with dimension of $|\mathscr{S}|$ for each period t. Each element is for each state and is defined as

$$b_t(s_t) = P(s_t|o_t, a_{t-1}, o_{t-1}, a_{t-2}, ..., o_1, a_0) = P(s_t|h_t), \tag{11.56}$$

where $h_t = (o_t, a_{t-1}, o_{t-1}, a_{t-2}, ..., o_1, a_0)$ is the complete history of observations up to and including t as well as history of actions up to and including time $t-1$. Note that if we focus on the actual state s_t to study the system dynamics, it is not only unobservable but also depends on the entire history of actions and observations. Hence, it is not Markovian. For POMDPs, the belief state $\mathbf{b}_t$ (vector) plays the same role as the state in fully observable MDP and is a "sufficient statistic" to keep the Markovian property and define the optimal policy. The key to maintain the Markovian property is to perform Bayesian updating. We show how the Bayesian updating works below. First we write

$$b_t(s_t) = P(s_t|o_t, a_{t-1}, o_{t-1}, a_{t-2}, ..., o_1, a_0)$$
$$= \frac{P(s_t, o_t, a_{t-1}|o_{t-1}, a_{t-2}, ..., o_1, a_0)}{P(o_t, a_{t-1}|o_{t-1}, a_{t-2}, ..., o_1, a_0)}. \tag{11.57}$$

Letting $h_{t-1} = (o_{t-1}, a_{t-2}, ..., o_1, a_0)$, the numerator can be written as

$$P(s_t, o_t, a_{t-1} | h_{t-1}) = \sum_{s_{t-1} \in \mathscr{S}} P(s_t, o_t, a_{t-1}, s_{t-1} | h_{t-1})$$

$$= \sum_{s_{t-1} \in \mathscr{S}} P(o_t | s_t, a_{t-1}, s_{t-1}, h_{t-1}) P(s_t | a_{t-1}, s_{t-1}, h_{t-1})$$

$$\times P(a_{t-1} | s_{t-1}, h_{t-1}) P(s_{t-1} | h_{t-1}) \qquad (11.58)$$

$$= P(a_{t-1} | h_{t-1}) P(o_t | s_t, a_{t-1}) \sum_{s_{t-1} \in \mathscr{S}} P(s_t | a_{t-1}, s_{t-1}) b_{t-1}(s_{t-1})$$

and the denominator can be written as

$$P(o_t, a_{t-1} | h_{t-1}) = \sum_{s_t' \in \mathscr{S}} \sum_{s_{t-1} \in \mathscr{S}} P(s_t', o_t, a_{t-1}, s_{t-1} | h_{t-1})$$

$$= \sum_{s_t' \in \mathscr{S}} \sum_{s_{t-1} \in \mathscr{S}} P(o_t | s_t', a_{t-1}, s_{t-1}, h_{t-1}) P(s_t' | a_{t-1}, s_{t-1}, h_{t-1})$$

$$\times P(a_{t-1} | s_{t-1}, h_{t-1}) P(s_{t-1} | h_{t-1}) \qquad (11.59)$$

$$= P(a_{t-1} | h_{t-1}) \sum_{s_t' \in \mathscr{S}} P(o_t | s_t', a_{t-1}) \sum_{s_{t-1} \in \mathscr{S}} P(s_t' | a_{t-1}, s_{t-1}) b_{t-1}(s_{t-1}).$$

Now everything is in terms of transition probabilities (i.e., $P(a)$) and observation probabilities (i.e., $B(a)$) given by the POMDP, and the prior belief state vector. Thus, we have

$$b_t(s_t) = \frac{P(o_t | s_t, a_{t-1}) \sum_{s_{t-1} \in \mathscr{S}} P(s_t | a_{t-1}, s_{t-1}) b_{t-1}(s_{t-1})}{\sum_{s_t' \in \mathscr{S}} P(o_t | s_t', a_{t-1}) \sum_{s_{t-1} \in \mathscr{S}} P(s_t' | a_{t-1}, s_{t-1}) b_{t-1}(s_{t-1})}. \qquad (11.60)$$

This equation in fact states the relation $P(s_t | o_t) = P(o_t \cap s_t)/P(o_t)$. Now if we treat $\mathbf{b}_t$ as a $|\mathscr{S}|$ dimensional state of the system which will only depend on $\mathbf{b}_{t-1}$, then the system is an MDP with a continuous state space. Since b_t is a complete probability distribution, we can use $|\mathscr{S}| - 1$ dimensional space to show the belief state space graphically if $|\mathscr{S}| \leq 4$. Note that each element of this belief state vector is a probability (i.e., $0 \leq b_t(s) \leq 1$). The initial distribution over the state space $\mathscr{S}$ is treated as $\mathbf{b}_0$. It is worthy to note that $b_t(s_t)$ is a function of o_t, a_{t-1} and $b_{t-1}(s_{t-1})$ and can be written as $b_t(s_t) = f(o_t, a_{t-1}, b_{t-1}(s_{t-1}))$. This means for any belief state $b_{t-1}(s_{t-1})$, if an action a_{t-1} is taken and an observation o_t is made, then a unique belief state is, $b_t(s_t)$, reached.

Now we introduce the matrix notations and express the state transitions of the MDP with continuous belief state space in matrix form. Define

$$\mathbf{B}_{o_t}(a_{t-1}) = diag(B_{1,o_t}(a_{t-1}), \cdots, B_{|S|,o_t}(a_{t-1})),$$

$$\mathbf{P}(a_{t-1}) = \begin{pmatrix} p_{11}^{a_{t-1}} & p_{12}^{a_{t-1}} & \cdots & p_{1|S|}^{a_{t-1}} \\ p_{21}^{a_{t-1}} & p_{22}^{a_{t-1}} & \cdots & p_{2|S|}^{a_{t-1}} \\ \vdots & \vdots & \cdots & \vdots \\ p_{|S|1}^{a_{t-1}} & p_{|S|2}^{a_{t-1}} & \cdots & p_{|S||S|}^{a_{t-1}}, \end{pmatrix}, \qquad (11.61)$$

where $B_{i,o_t}(a_{t-1}) = P(o_t|s_t = i, a_{t-1})$ and $p_{ij}^{a_{t-1}} = P(s_t = j|a_{t-1}, s_{t-1} = i)$ with $i, j \in \mathscr{S}$. Denoting the belief state vector by $\mathbf{b}_t = (b_t(1), b_t(2), \cdots, b_t(|\mathscr{S}|))^T$, we have

$$\mathbf{b}_t = \mathbf{f}(o_t, a_{t-1}, \mathbf{b}_{t-1}) = \frac{\mathbf{B}_{o_t}(a_{t-1})\mathbf{P}'(a_{t-1})\mathbf{b}_{t-1}}{\sigma(o_t, a_{t-1}, \mathbf{b}_{t-1})}, \qquad (11.62)$$

where $\sigma(o_t, a_{t-1}, \mathbf{b}_{t-1}) = \mathbf{1}'\mathbf{B}_{o_t}(a_{t-1})\mathbf{P}'(a_{t-1})\mathbf{b}_{t-1} = P(o_t, a_{t-1}, \mathbf{b}_{t-1})$. Since sequence of action a_{t-1} and observation o_t is fixed, we can also omit the time index by letting $u = a_{t-1}$ and $y = o_t$.

Since an action will generate multiple possible observations, the current belief state may reach different future belief states and the probability of reaching a particular belief state is determined by $P(o_t|s_t, a_{t-1}) = B_{s_t, o_t}(a_{t-1})$. Note that here we keep a_{t-1} in the condition to clearly indicate that the observation probability depends on the action made in the previous period (i.e., the $\mathbf{B}(a)$ matrix) and it may be omitted and written as $P(o_t|s_t)$ if no confusion is caused. We summarize this dynamics and develop the optimality equations for the POMDP. We need to introduce the following notations. The immediate reword row vector, denoted by $\mathbf{r}_t(a_t) = (r_t^{a_t}(1), ..., r_t^{a_t}(|\mathscr{S}|))^T$, is the expected immediate rewards under transitions and observations with each element given by

$$r_t^{a_t}(s_t) = \sum_{o_{t+1} \in O} \sum_{s_{t+1} \in \mathscr{S}} r(s_t, a_t, s_{t+1}, o_{t+1})P(s_{t+1}|s_t, a_t)P(o_{t+1}|s_{t+1}), \quad (11.63)$$

where $r(s_t, a_t, s_{t+1}, o_{t+1})$ is the actual immediate reward that depends on not only the current period state and action but also the next period observation and state. Note that such a computation may not be needed if the immediate reward does not depend on next period such as in our running example where the immediate rewards are given in the reward matrix $\mathbf{R} = (\mathbf{r}(a_1), ..., \mathbf{r}(a_A))$. Denote the optimal value function vector for period t with a belief vector $\mathbf{b}_t$ by $\mathbf{v}_t(\mathbf{b}_t)$, which is defined on the belief state space in POMDPs, where $t =$

$0, 1, \cdots, N$, for $N+1$ period problem. We can write the OE as follows:

$$\mathbf{v}_t^*(\mathbf{b}_t) = \max_{a_t \in A} \left\{ \mathbf{b}_t^T \cdot \mathbf{r}_t(a_t) + \gamma \sum_{o_{t+1} \in O} P(o_{t+1}, a_t, \mathbf{b}_t) \mathbf{v}_{t+1}^*(\mathbf{f}(o_{t+1}, a_t, \mathbf{b}_t)) \right\},$$
(11.64)

with the terminal state condition of $\mathbf{v}_{N+1}^*(\mathbf{b}_{N+1}) = \mathbf{b}_{N+1}^T \cdot \mathbf{r}_{N+1}$. Note that $P(o_{t+1}, a_t, \mathbf{b}_t) = P(o_{t+1} | \mathbf{b}_t, a_t)$ (i.e., two different notations for the same probability).

Now using the running example of maintenance problem, we discuss how to solve this POMDP by using value iteration approach. We will utilize the OE (11.64) to demonstrate the VI approach (i.e., backward approach). Since $|\mathscr{S}| = 2$, the two-dimensional belief vector is $\mathbf{b}_t = (P(s_t^1) 1 - P(s_t^1))^T$. Here, $P(s_t^1)$ can be shown graphically by using a bar of length 1 in Figure 11.20, where $s_t^1 = G$ and $s_t^2 = W$. In the following discussion, we reserve the subscript for the time index and use the superscript as the state, action, or observation number if time index is shown. For example, $(s_0^1 = G, a_0^1 = C, o_1^2 = A)$ represents that we take action 1 (corrective maintenance) in state G (good condition) at time $t = 0$ and observe outcome A (average quality product made) at time $t = 1$. When the time index is not needed, we do not use superscript for state, action, or observation number. For example, $a1, o2, s1$ represent action 1, observation 2, and state 1, respectively, if the time index is not necessary.

Exercise 11.22 *For a POMDP with a three-dimensional state space (i.e., $|\mathscr{S}| = 3$), show the state space of the belief state vector graphically. How about a 4-dimensional state space case?*

Note that in our maintenance problem $s_t^1 = G$ and $s_t^2 = W$. We start with a $N = 1$ (i.e., the first horizon) case. Let's specify more symbols for this case for developing the formulas. Since the immediate reward matrix $\mathbf{R}$ does not depend on time, we can denote the four elements of the matrix as follows: $r(s_t^1, a_t^1) = r_{11}, r(s_t^1, a_t^2) = r_{12}, r(s_t^2, a_t^1) = r_{21}, r(s_t^2, a_t^2) = r_{22}$ where $a_t^1 = C$ and $a_t^2 = S$. Thus, for the horizon 1 problem, we have the optimal value function as

$$\mathbf{v}_0^*(\mathbf{b}_0) = \max_{a1, a2} \{ \mathbf{b}_0^T \cdot \mathbf{r}(a_0^1), \mathbf{b}_0^T \cdot \mathbf{r}(a_0^2) \}$$

$$= \max \{ P(s_0^1) r_{11} + (1 - P(s_0^1)) r_{21}, P(s_0^1) r_{12} + (1 - P(s_0^1)) r_{22} \} \quad (11.65)$$

$$= \max \{ p_0 r_{11} + (1 - p_0) r_{21}, p_0 r_{12} + (1 - p_0) r_{22} \},$$

where $P(s_0^1) = p_0, \mathbf{r}(a_0^1) = (r_{11}, r_{21})'$ and $\mathbf{r}(a_0^2) = (r_{12}, r_{22})'$ are column vectors of $\mathbf{R}$ matrix. It is easy to obtain the critical number, $x^* = (r_{22} - r_{21})/(r_{11} - r_{21} - r_{12} + r_{22})$, such that $\mathbf{v}_0^*(\mathbf{b}_0)$ is the first expression in the maximum operator if $p_0 \geq x^*$, and the second expression, otherwise.

Now look at our running example for numerical and graphical illustration. Suppose that the initial belief state is $\mathbf{b}_0 = (p_0, 1 - p_0)^T = (0.75, 0.25)^T$ (can

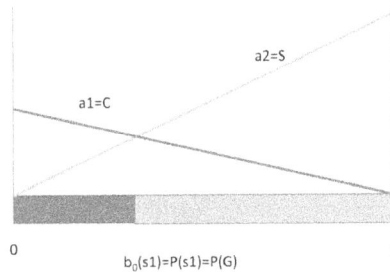

Figure 11.20   Horizon 1 value function.

be changed). Then, in Figure 11.20, the value of taking action $a_0^1 = C$ in this belief state is 0.75 x 0 + 0.25 x 1 = 0.25 (a value determined by the blue line, the first expression of (11.65)). Similarly, taking $a_0^2 = S$ gives 0.75 x 1.5 + 0.25 x 0 = 1.125 (a value determined by the green line, the second expression of (11.65)). We can display these values over belief space $(P(s_0^1))$ as line functions in Figure 11.20 (the graphical representation of OE (11.65)). In fact, the horizon 1 value function is just the immediate reward as there is only one period. Note that this value function is piece-wise linear and convex (PWLC). Such a property holds for all periods when the horizon is more than one period. Since we want to maximize the value for each belief state, the maximum of these two linear functions becomes the optimal value function. The blue region is all the belief states where action a_0^1 is the best strategy to use, and the green region is the belief states where action a_0^2 is the best strategy.

Exercise 11.23 *(The tiger problem) A tiger is put with equal probability behind one of two doors, while treasure is put behind the other one. You are standing in front of the two closed doors and need to decide which one to open. If you open the door with the tiger, you will get hurt (negative reward), say -100. But if you open the door with treasure, you receive a positive reward, say +10. Instead of opening a door right away, you also have the option to wait and listen for tiger noises. But listening is neither free nor entirely accurate. You might hear the tiger behind the left door while it is actually behind the right door and vice versa. The tiger is randomly shuffled between doors after each door is opening. However, the tiger does not move if you decide to listen. This decision problem can be modeled as a POMDP with two states and two observations. The state space is $\mathscr{S} = (TL, TR)$ where $TL = $ tiger behind left door and $TR = $ tiger behind right door. There are three actions $A = (H, L, R)$, with H representing "hold and listen", L representing "open left door", and R representing "open right door". When you choose H (Listen), the door that appears to hide the tiger is denoted by either OTL (hear noise behind left door) or OTR (hear noise behind right door). The transition probability matrices are* *given by*

$$
P(a = H) = \begin{matrix} & TL \quad TR \\ TL \\ TR \end{matrix} \begin{pmatrix} 1.0 & 0.0 \\ 0.0 & 1.0 \end{pmatrix}, \quad P(a = L) = \begin{matrix} & TL \quad TR \\ TL \\ TR \end{matrix} \begin{pmatrix} 0.5 & 0.5 \\ 0.5 & 0.5 \end{pmatrix},
$$

$$
P(a = R) = \begin{matrix} & TL \quad TR \\ TL \\ TR \end{matrix} \begin{pmatrix} 0.5 & 0.5 \\ 0.5 & 0.5 \end{pmatrix},
$$

and the observation probability matrices are given by

$$
B(a = H) = \begin{matrix} & OTL \quad OTR \\ TL \\ TR \end{matrix} \begin{pmatrix} 0.85 & 0.15 \\ 0.15 & 0.85 \end{pmatrix}, \quad B(a = L) = \begin{matrix} & OTL \quad OTR \\ TL \\ TR \end{matrix} \begin{pmatrix} 0.5 & 0.5 \\ 0.5 & 0.5 \end{pmatrix},
$$

$$
B(a = R) = \begin{matrix} & OTL \quad OTR \\ TL \\ TR \end{matrix} \begin{pmatrix} 0.5 & 0.5 \\ 0.5 & 0.5 \end{pmatrix}.
$$

The immediate cost of listening is -1. Solve the $N = 1$ (one time period) problem.

Next we consider the problem with the horizon length of 2. Like observable MDP, we develop the optimal value function by choosing the best action (one out of two options for every belief state in our running example). We start with finding the value of a particular belief state if we take action $a_0^1 = a1$ and receive an observation $o1$ for a horizon 2 problem. This value should be the immediate value of the action plus the value of the next stage (which has been solved as a horizon 1 problem above). Of course, eventually we should also find this function value for taking action $a_0^2 = a2$ and use the maximum.

Figure 11.21 show how the first stage progress if we start with initial belief state $b_0(G)$, denoted by a dot, and take action $a_0^1 = a1$. Then, an immediate reward occurs and the new belief state $b_1(G)$ is reached by the transformation (i.e., Bayesian updating). This happens with probability of $P(o1|a_0^1, b_0) = P(o1|a1, b_0)$ (thanks for the lower dimension of the state space, that is, we use the first row of $\mathbf{B}(a)$ matrix with 0.75 chance and use the second row of $\mathbf{B}(a)$ matrix) with 0.25 chance. As soon as the new belief state $b_1(G)$ is reached, the problem becomes a horizon 1 problem and since it is in the green region, the optimal action is $a_0^2 = a2$. Thus, if we are forced to take action $a1$ in stage 1 $(t = 0)$, and we observed $o1$, then the optimal decision for stage 2 is to take $a2$ $(t = 1)$.

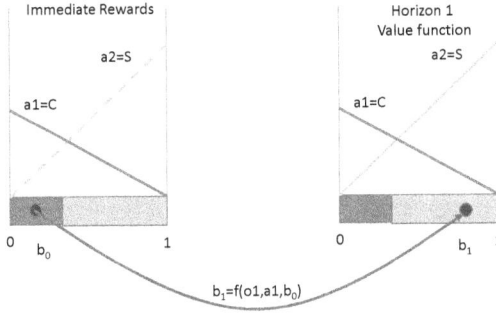

Figure 11.21 Horizon 1 value function.

We next show the detailed calculations for the horizon 2 problem and leave the calculations for the horizon 3 problem as an exercise. Applying (11.61) and (11.62) to our maintenance problem example with two periods ($t = 0, 1$), $a_0^1 = a1 = C$ ($a1$ means action 1) and getting observation $o_1^1 = o1 = L$, we have

$$\mathbf{B}_L(C) = diag(B_{G,L}(C), B_{W,L}(C))$$

$$\mathbf{P}(C) = \begin{pmatrix} p_{11}^C & p_{12}^C \\ p_{21}^C & p_{22}^C \end{pmatrix}$$

$$\mathbf{b}_1 = \mathbf{f}(L, C, \mathbf{b}_0) = \frac{\mathbf{B}_{o_t}(a_{t-1})\mathbf{P}'(a_{t-1})\mathbf{b}_{t-1}}{\sigma(o_t, a_{t-1}, \mathbf{b}_{t-1})} = \frac{\mathbf{B}_L(C)\mathbf{P}'(C)\mathbf{b}_0}{\sigma(L, C, \mathbf{b}_0)}$$

$$= \frac{1}{B_{G,L}(C)(p_{11}^C b_0(G) + p_{21}^C b_0(W)) + B_{W,L}(C)(p_{21}^C b_0(G) + p_{22}^C b_0(W))}$$
$$\times \begin{pmatrix} B_{G,L}(C)(p_{11}^C b_0(G) + p_{21}^C b_0(W)) \\ B_{W,L}(C)(p_{21}^C b_0(G) + p_{22}^C b_0(W)) \end{pmatrix}.$$

Next, we examine the value function of the joint events $a_0^1 = a1 = C$ and $o_1^1 = o1 = L$, denoted by $V(o1 = L, a1 = C)$. It follows from the above equation that

$$V(o1 = L, a1 = C) = \sigma(L, C, \mathbf{b}_0)\mathbf{b}_1^T \cdot \mathbf{r}(a1 = C) = (\mathbf{B}_L(C)\mathbf{P}'(C)\mathbf{b}_0)^T \cdot \mathbf{r}(a1 = C)$$
$$= (B_{G,L}(C)(p_{11}^C b_0(G) + p_{21}^C b_0(W)), B_{W,L}(C)(p_{12}^C b_0(G) + p_{22}^C b_0(W)))$$
$$\times \begin{pmatrix} r_G^C \\ r_W^C \end{pmatrix}$$
$$= B_{G,L}(C)(p_{11}^C b_0(G) + p_{21}^C b_0(W))r_G^C + B_{W,L}(C)(p_{12}^C b_0(G) + p_{22}^C b_0(W))r_W^C$$
$$= 0.1(0.7b_0(G) + 0.4(1 - b_0(G)) \cdot 1 = 0.03p_0 + 0.04.$$

$$(11.66)$$

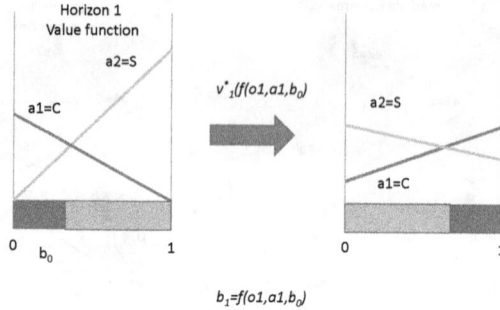

Figure 11.22   Transformed value function for a given $a1$ and $o1$.

Since we can use the same procedure for the value function for taking the action $a_0^2 = a2 = S$ and receiving $o_1^1 = o1 = L$, we will only present result.

$$V(o1 = L, a2 = S) = 0.1(0.2b_0(G) + 0.9(1 - b_0(G)) \cdot 1.5 = 0.135 - 0.105p_0.$$
$$(11.67)$$

The two value functions of (11.66) and (11.67) are illustrated in Figure 11.22 by the blue ($a1$) and green ($a2$) lines, respectively, on the right panel.

We now present more formulas. Given the first stage action is $a1$ ($t = 0$), an observation $o1$ is made ($t = 1$), and the optimal value is utilized for the second stage ($t = 1$), we have the value function as

$$v_0^{(a1,o1)}(b_0(G)) = \mathbf{b}_0 \cdot \mathbf{r}(a1) + \gamma v_1^*(\mathbf{f}(o1, a1, \mathbf{b}_0)), \qquad (11.68)$$

where

$$v_1^*(\mathbf{f}(o1, a1, \mathbf{b}_0)) = \max_{a \in \{a1, a2\}} (V(o1 = L, a1 = C), V(o1 = L, a2 = S))$$

$$= \max_{a \in \{C, S\}} (0.03p_0 + 0.04, 0.135 - 0.105p_0),$$

which can be considered as a transformed value function for this two-period problem with the optimal value for the second period. It is easy to show that $v_0^{(a1,o1)}(b_0(G))$ is again a PWLC function as it is a linear function ($\mathbf{b}_0 \cdot \mathbf{r}(a1)$) plus a PWLC ($\gamma v_1^*(f(o1, a1, b_0))$) as shown in in Figure 11.22). Thus, for a two period problem, the value of a belief state for the fixed action $a1$ and observation $o1$ is simply the immediate reward plus the value from the transformed function $\gamma v_1^*(\mathbf{f}(o1, a1, \mathbf{b}_0))$. The intersection of the two lines in Figure 11.22) can be determined easily by solving

$$0.03p_0 + 0.04 = 0.135 - 0.105p_0.$$

That is $p_0 = 0.7037$. Assume that the first stage ($t = 0$) action taken is $a1 = C$ and observation $o1 = L$ is observed. If the starting belief state is below 0.7037,

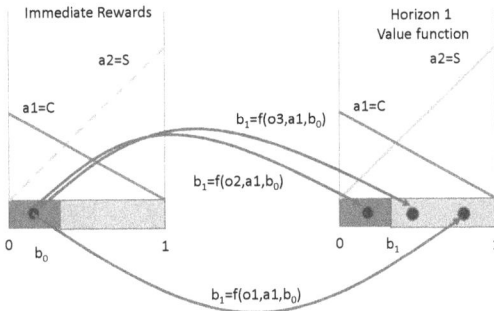

Figure 11.23 Transformed belief states under $a1$.

then the $a2 = S$ is the optimal action for the second stage; otherwise (the start-ing state is above 0.7037), $a1 = C$ is the optimal action for the second stage. In other words, this is the optimal decision rule for the second stage condition-ing on the initial starting state, the first stage action, and the observation made after the first stage action. If we consider all possible observations and their probabilities, we can get the value function of the belief state given only the action $a_0^1 = a1$ in our running example as follows:

$$v_0^{a1}(\mathbf{b}_0) = \mathbf{b}_0 \cdot \mathbf{r}(a1) + \gamma(P(o1, a1, \mathbf{b}_0)v_1^*(f(o1, a1, \mathbf{b}_0))$$
$$+ P(o2, a1, \mathbf{b}_0)v_1^*(f(o2, a1, \mathbf{b}_0)) + P(o3, a1, \mathbf{b}_0)v_1^*(f(o3, a1, \mathbf{b}_0)))$$
$$(11.69)$$

as there are three possible observations which lead to three different trans-formed belief states as shown in Figure 11.23. Each observation, occurring with a certain probability, will lead to a specific transformed value function (as shown above for $o1$). These three transformed value functions are shown in Figure 11.24.

It is now clear that the value of a belief state (starting state), given a partic-ular action $a1$, depends on not only $a1$ but also what action we take next (the second stage where the horizon length will be 1). However, what action we take next will depend on what observation we get. The transformed value function shown in Figure 11.24 generates the best strategy for this remaining period (Horizon 1 problem). This strategy prescribes the best action for each possible observation generated by the starting belief state and the action taken in stage 1, which is shown in Figure 11.25 below. Here, the three stacked bars repre-sent four different second stage strategies (a_2) with green color representing $a2 = S$ and blue color representing $a1 = C$. The color in the top bar represents the maximum value function line color and the corresponding best strategy for the specific value range of the starting belief state. For example, the first range

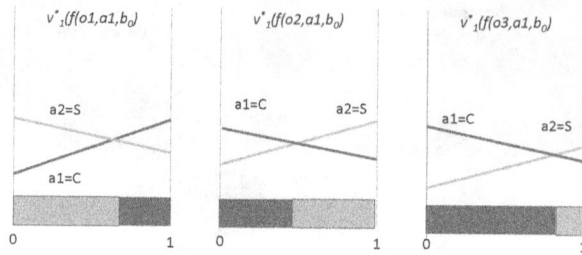

Figure 11.24 Transformed value functions for all observations under $a1$.

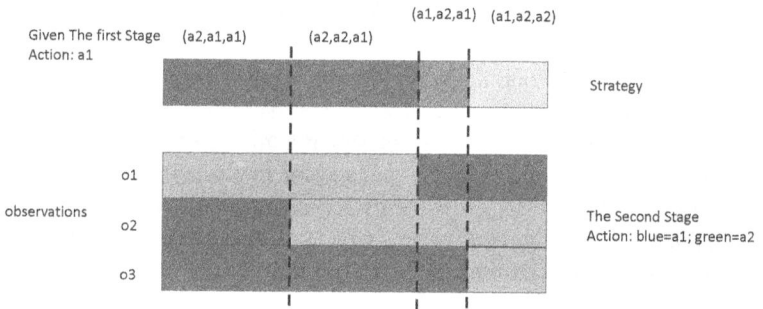

Figure 11.25 Optimal Strategy at the second stage under $a1$ for the first stage.

is blue color in the top bar. This means that if the first stage action is fixed at $a_0^1 = a1 = C$, the maximum value function at stage 2 is the blue line segment and the corresponding strategy for stage 2 is $(o1, o2, o3) \rightarrow (a2, a1, a1)$, meaning that if $o1$ is observed, take action $a_1^2 = a2$; if $o2$ or $o3$ is observed, take action $a_1^1 = a1$. Similarly, we can interpret other color sections in the top bar. Clearly, there are four strategies for stage 2 here. The corresponding value function for stage 2 with the best action taken, given that the first stage action is $a_0^1 = a1$ can be shown in Figure 11.26.

So far, all illustrations are for case where our first stage action is $a_0^1 = a1 = C$ $(t = 0)$. Now we can perform the same analysis for the other case where the first stage action is $a_0^2 = a2 = S$. The results are summarized in Figure 11.27. The value function consists of only two line segments which correspond to the two best strategies (three action bars for the three observations are omitted). Next, we put both value functions for $a1$ and $a2$, respectively, as the first stage action a_0 over the same belief state range $b_0(G)$ as shown in Figure 11.28. Then, we can determine the optimal value function and the optimal action for the first stage according the the value function line segment (represented by color) by taking the maximum value.

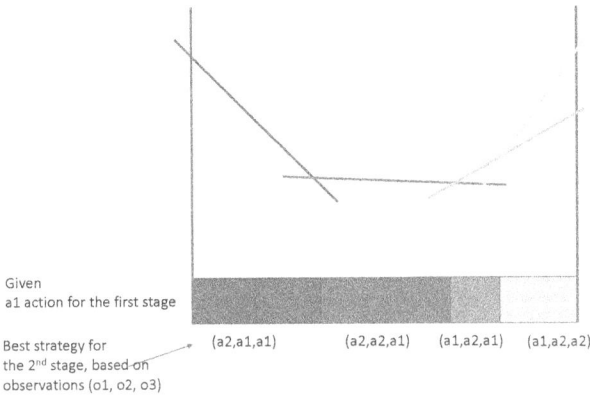

Figure 11.26 Value function for stage 2 under $a_0^1 = a1$.

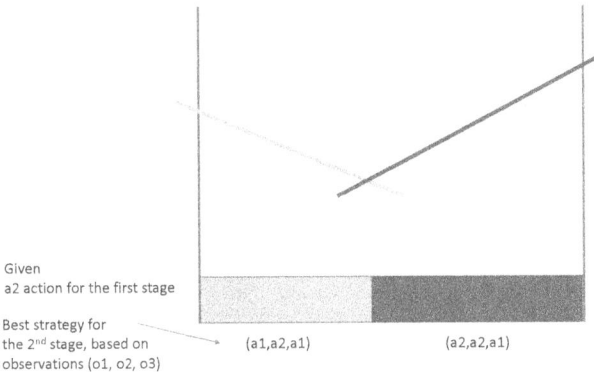

Figure 11.27 Value function for stage 2 under $a_0^2 = a2$.

A more clear value function for horizon 2 problem is illustrated in Figure 11.29.

We can summarize the optimal policy for the two period problem as follows: In the first stage ($t = 0$), based on Figure 11.29, if the initial belief state $b_0(G)$ is in the blue or red range, take action $a_0^1 = a1 = C$; if the initial belief state $b_0(G)$ is in the brown range, take action $a_0^2 = a2 = S$; otherwise (i.e., $b_0(G)$ in yellow range), take action $a_0^1 = a1 = C$. In the second stage, the optimal strategy is determined by the colors in Figures 11.26 (blue, red, yellow) and 11.27 (brown). For example, if the initial belief is in blue range, take action 1 ($a1 = C$) in the first stage ($t = 0$), then take action 2 ($a2 = S$) if observing $o1$;

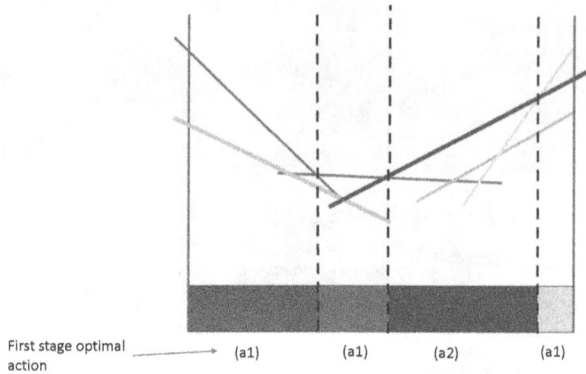

First stage optimal ─────────→   (a1)   (a1)   (a2)   (a1)
action

Figure 11.28   Value functions for both $a1$ and $a2$ as the first stage action, respectively.

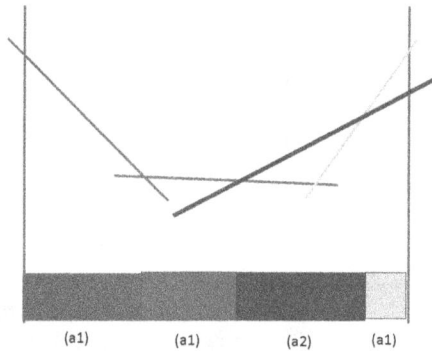

            (a1)   (a1)   (a2)   (a1)

Figure 11.29   Value function for horizon 2 problem.

and take action 1 ($a1 = C$) if observing $o2$ or $o3$ in the second stage ($t = 1$).

Similarly, the optimal value function along with the associated optimal actions/strategies (represented colors) for this horizon 2 problem will be utilized for determining the optimal strategy for the horizon 3 problem.

Now the horizon 2 value function in Figure 11.29 is utilized to transform the value of the state-belief in a three time period (horizon 3) problem. In the same way, for a given action $a_0^1 = a1$ at the beginning, updating the belief state from $\mathbf{b}_0$ to $\mathbf{b}_1$, observing the outcomes, and utilizing the horizon 2 value function to transform the next period value for the possible outcomes shown in Figure 11.30.

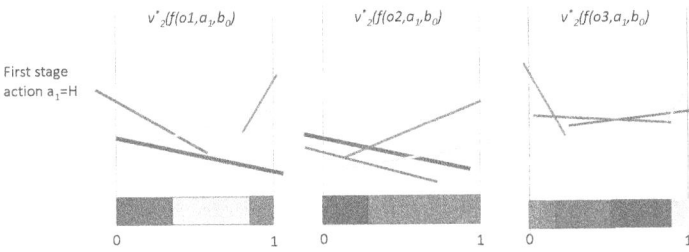

Figure 11.30 Transformed Value functions for horizon 2 problem given $a1$ at the beginning and different outcomes.

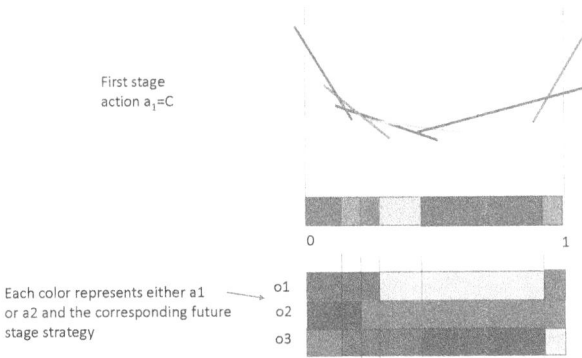

Figure 11.31 Value function for horizon 2 problem given $a1$ at the beginning.

Putting these three cases together generates best strategies for the next period and on. Figure 11.31 shows six strategies (in 6 colors) on the horizontal axis and the corresponding value function (line segments with corresponding colors). Below the value function chart shows the details of these strategies in terms of the best strategy colors for the horizon 2 problem (four colors).

Similarly, we can obtain the results for the case where we start with action $a_0^2 = a2$ which is shown in Figure 11.32.

Finally, we put both the value functions under starting actions of $a1$ and $a2$, respectively, in Figure 11.33. Then we obtain the optimal starting action for the horizon 3 problem.

Figure 11.34 shows the value function for the horizon 3 problem which can be utilized for solving the horizon 4 problem in the same way. In theory, we have solved the finite horizon POMDP problem by value iteration (or backward approach).

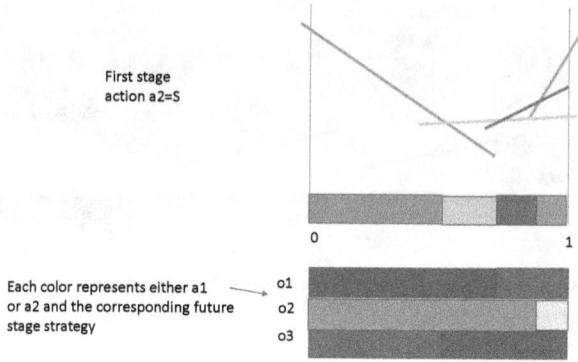

First stage
action a2=S

Each color represents either a1
or a2 and the corresponding future
stage strategy

o1
o2
o3

0   1

Figure 11.32   Value function for horizon 2 problem given $a2$ at the beginning.

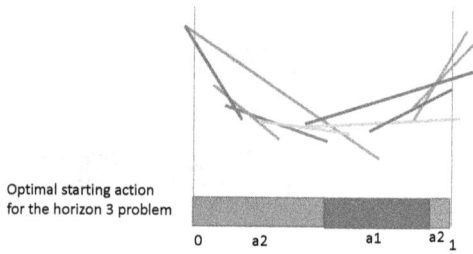

Optimal starting action
for the horizon 3 problem

0   a2   a1   a2 1

Figure 11.33   Optimal starting action for a horizon 3 problem.

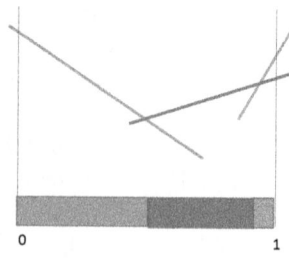

0   1

Figure 11.34   Value function for a horizon 3 problem.

However, solving a POMDP is computationally expensive. In fact, the computational time increases exponentially with the number of actions, the number of observations for each action, and the number of decision epochs (number of periods in the time horizon). Thus, to solve real-world problems with POMDP, we have to reply on approximation algorithms. In fact, this is a very rich research area and there are many excellent books and research papers on this topic. We list some references on the POMDP and the approximation methods at the end of this chapter. Interested readers can further explore this active research area.

Exercise 11.24 *Summarize the optimal strategies for the three period problem of the running example of machine maintenance problem based on the colors in the value function figures.*

Exercise 11.25 *(The tiger problem continues). For the tiger problem, solve the $N = 2$ problem.*

Exercise 11.26 *For a finite horizon maximization POMDP with finite action space and finite observation space, prove that the optimal value function in each stage is piecewise linear and convex (or concave for minimization POMDP) with respect to the initial belief state vector.*

REFERENCE NOTES

To introduce the learning process in a stochastic model, we present the multi-arm bandits problem which is adopted from [11]. The Thompson sampling for the Bernoulli bandit described in Exercise 11.7 is a typical example that contains the main features in machine learning problems. For more discussion on this example, readers are referred to [7]. Monte Carlo-based MDP models can be classified into model-based learning, model-free learning, and model-free learning with bootstrapping types. To determine the optimal policy in a model-free learning situation, Q-learning is introduced. The example of a robot trying to exit a house that demonstrates the Q-learning is taken from [8]. The grid-world example showing the temporal-difference learning is taken from [6]. The comparison between SARSA and Q-learning is demonstrated by a gridworld example with cliff which is based on [5]. The learning in stochastic games is demonstrated by the example from [4]. The hidden Markov model is based on the example from [10]. For more discussion on POMDP with finite horozon, see Smallwood and Sondik [9]. Finally, the graphical illustrations of the value functions for finite-horizon POMDP is based on [3].

REFERENCES

1. P. Auer, N. Cesa-Bianchi, and P. Fischer, Finite-time analysis of the multiarmed bandit problem. Machine learning, 47(2-3), 235–256, 2002.
2. R. W. Cottle, J.-S. Pang, and R. E. Stone, "The Linear Complementarity Problem". Academic Press, New York, 1992.
3. G. Hollinger, Partially Observable Markov Decision Processes (POMDPs), lecture notes, Spring 2011. https://robotics.usc.edu/geoff/cs599/POMDP.pdf
4. J. Hu and M.P. Wellman, Nash Q-Learning for general-sum stochastic games. *Journal of Machine Learning Research*, 4, 1039-1069, 2003.
5. T.C. Nguyen, "Notes on SARSA v.s. Q-Learning", Online notes, 2017. `https://tcnguyen.github.io/reinforcement_learning/sarsa_vs_q_learning.html`
6. M. Patacchiola, "Dissecting Reinforcement Learning-Part.3", online post, 2017. `https://mpatacchiola.github.io/blog/2017/01/29/dissecting-reinforcement-learning-3.html`
7. D.J. Russo, B. Van Roy, A. Kazerouni, I. Osband, and Z. Wen. A Tutorial on Thompson Sampling. *Foundations and Trends in Machine Learning*, 11(1), 1–96, 2018.
8. L.A.D. Santos, "Artificial intelligence", Interactive Book on Artificial Intelligence - [Gitbook], `https://leonardoaraujosantos.gitbook.io/artificial-inteligence/`
9. R. D. Smallwood and E. J. Sondik, The optimal control of partially observable Markov processes over a finite horizon. *Operations Research*, 21(5), 1071–1088, 1973.
10. M. Stamp, "Introduction to Machine Learning with Applications in Information Security", Chapman and Hall/CRC, 2017.
11. R.S. Sutton and A.G. Barto, "Reinforcement Learning: An Introduction",The MIT Press, Cambridge, Massachusetts, London, Englan 2020.
12. C.J.C.H. Watkins and P. Dayan, Q-learning. *Machine Learning*, 8, 279-292, 1992.

Part II

Appendices: Elements of Probability and Stochastics

Elements of Probability and Stochastics

Part II of this book contains a set of appendices. The foundation of stochastic modeling is the probability and stochastic process theory. Some basic concepts and theories are reviewed here. The topics selected here are relevant to the materials presented in this book. These reviews are by no means comprehensive. Thus, these appendices cannot be used as a replacement of the textbooks on these topics. Instead, they serve as a quick reference to and refresher of some key and fundamental results in probability and stochastics as we assume that readers have the knowledge of calculus and elementary probability theory. For the details of each topic, we provide some references (books and research papers) for interested readers. A common feature of these reviews is that a simplified case or scenario is utilized whenever possible to develop the key results which also hold in a more general setting. An advantage of this feature is that the presentation of the focused topic can be made more intuitive and accessible. This approach is in sharp contrast to those of the traditional textbooks which usually develop the theoretical results with rigor for more general settings. Since this book is focused on applied stochastic modeling, we do not emphasize the mathematical rigor which requires more advanced mathematical treatments. Hence, we refer interested readers to those traditional textbooks for more complete and rigorous discussions on these topics. The appendices in this part are organized into 10 Chapters A to J. Each Chapter focuses on a specific topic.

A Basics of Probability Theory

A.1 PROBABILITY SPACE

Probability theory is about studying uncertainties. We start with a random experiment. A random experiment is an experiment whose outcomes cannot be determined in advance. Flipping a coin once is a simple random experiment as we do not know for sure if we will get tails or heads.

Definition A.1 *Sample Space: is the set of all possible outcomes when a random experiment is performed.*

Sample space, denoted by Ω, can be either discrete or continuous. For example, rolling a die yields a sample space with six discrete outcomes $\Omega = \{1, 2, 3, 4, 5, 6\}$ and choosing randomly a real number between 0 and 1 generates a sample space which is continuous and has an infinite number of outcomes (i.e., Ω is the entire segment [0, 1]). All subsets of the sample space form a set denoted by 2^Ω. The reason for this notation is that the elements of sample space can be put into a bijective correspondence with the set of binary functions $f : \Omega \to \{0, 1\}$. That is each element of Ω can either occur or not occur. The number of elements of this set is $2^{|\Omega|}$, where $|\Omega|$ denotes the cardinal of Ω. If the set is finite, i.e., $|\Omega| = n$, then $2^{|\Omega|}$ has 2^n elements. If Ω is infinitely countable (i.e., can be put into a bijective correspondence with the set of natural numbers), then 2^Ω is infinite and its cardinal is the same as that of the real number set $\mathbb{R}$. For example, if Ω represents all possible states of the financial world, then 2^Ω describes all possible events, which might happen in the market; this should be a fully description of the total information of the financial world. The following two examples provide instances of sets 2^Ω in the finite and infinite cases.

Example A.1 *Flip a coin once. The sample space contains two outcomes. i.e., $\Omega = \{H, T\}$. For any element of the sample space, assign 0 if it does not occur and a 1 otherwise. Then we have the following four possible assignments:*

$$\{H \to 0, T \to 0\}, \quad \{H \to 0, T \to 1\}, \quad \{H \to 1, T \to 0\}, \quad \{H \to 1, T \to 1\}.$$

Thus, the set of all subsets of H, T can be represented as 4 sequences of length 2 formed with 0 and 1: $\{0, 0\}, \{0, 1\}, \{1, 0\}, \{1, 1\}$. These correspond to sets $\emptyset, \{T\}, \{H\}, \{H, T\}$, which is the set $2^{\{H, T\}}$.

DOI: 10.1201/9781003150060-A

The next example is a case with infinite sample space.

Example A.2 *Pick a natural number at random. The sample space is natural number set, i.e., $\Omega = \mathbb{N}$. Any subset of the sample space can be associated to a sequence formed with 0 and 1. For instance, the subset $\{1,3,4,7\}$ corresponds to the sequence 101100100000 . . . having 1 on the 1st, 3rd, 4th and 7th places and 0 in rest. It is known that the number of these sequences is infinite and can be put into a bijective correspondence with the real number set $\mathbb{R}$. This can be also written as $|2^{\mathbb{N}}| = |\mathbb{R}|$, and stated by saying that the set of all subsets of natural numbers $\mathbb{N}$ has the same cardinal as the real numbers set $\mathbb{R}$.*

Since 2^{Ω} contains all subsets of the sample space, it is called the power set of the sample space. Note that the power set satisfies the following properties:

1. It contains the empty set $\emptyset$;
2. If it contains a set A, then it also contains its complement $A^c = \Omega \setminus A$;
3. It is closed with regard to unions, i.e., if $A_1, A_2, ...$ is a sequence of sets, then their union $A_1 \cup A_2 \cup \cdots$ also belongs to the power set.

Definition A.2 *σ-algebra and Events: Any subset $\mathscr{F}$ of 2^{Ω} that satisfies the previous three properties is called a σ-algebra (or called σ-field). The sets belonging to $\mathscr{F}$ are called events.*

We say that an event occurs if the outcome of the experiment is an element of that event.

Definition A.3 *Probability Space: The likelihood of occurrence of an event is measured by a probability function $P : \mathscr{F} \to [0,1]$ which satisfies the following two properties:*

    1. $P(\Omega) = 1$,
    2. For any mutually exclusive events $A_1, A_2, \cdots \in \mathscr{F}$, then $P(\cup_{i=1}^{\infty} A_i) = \sum_{i=1}^{\infty} P(A_i)$.

The triplet $(\Omega, \mathscr{F}, P)$ is called a probability space.

We provide two simple examples of probability space.

Example A.3 *For a random experiment of flipping a coin once, the probability space has the following elements: $\Omega = \{H, T\}$, $\mathscr{F} = \{\emptyset, \{H\}, \{T\}, \{H,T\}\}$, and P defined by $P(\emptyset) = 0, P(\{H\}) = 1/2, P(\{T\}) = 1/2, P(\{H,T\}) = 1$.*

Example A.4 *For a random experiment with a finite sample space $\Omega = \{s_1, \cdots, s_n\}$, with the σ-algebra $\mathscr{F} = 2^{\Omega}$, and probability is defined by $P(A) = |A|/n$, $\forall A \in \mathscr{F}$. Then $(\Omega, 2^{\Omega}, P)$ is called the classical probability space.*

A.2 BASIC PROBABILITY RULES

Suppose that two events A and B from the σ-algebra of a probability space with $P(A) > 0$. Denote by $P(A \cap B)$ the probability that both A and B occur, also called joint or intersection probability, and $P(A|B)$ the conditional probability that A occurs given that B has occurred. Then, this conditional probability can be defined as

$$P(A|B) = \frac{P(A \cap B)}{P(B)}. \tag{A.1}$$

Re-writing this formula as

$$P(A \cap B) = P(A|B)P(B) \tag{A.2}$$

provides the meaningful multiplication of two probabilities. That is multiplying a conditional probability by an unconditional probability yields a joint probability. In addition, it naturally generates the definition of independence and its associate product rule. If $P(A|B) = P(A)$, the two events A and B are independent and $P(A \cap B) = P(A)P(B)$. Similarly, we can develop a union formula that can generate the addition rule for probabilities. Denote by $P(A \cup B)$ the probability that at least one of the two events occurs, also called union probability. Then,

$$P(A \cup B) = P(A) + P(B) - P(A \cap B),$$

which is equivalent to

$$P(A) + P(B) = P(A \cup B) + P(A \cap B).$$

Thus, when the two events are mutually exclusive (i.e., disjoint), the sum of the two probabilities is equal to the union probability of the two events, that is $P(A) + P(B) = P(A \cup B)$.

Next, we show that the probability of an event can be expressed as sum of a finite or countably infinite number of probabilities of sub-events. Such a formula is called the law total probability or the probability decomposition formula. This formula is based on a partition of the sample space. A partition of sample space Ω is defined as a set of pairwise disjoint events, denoted by $\{B_n, n = 1, 2...M.\}$, whose union is the entire sample space. That is $\bigcup_{n=1}^{M} B_n = \Omega$ and $B_i \cap B_j = \emptyset$ for $i \neq j$. Note that the partition can be either $M < \infty$ or $M = \infty$ (countably infinite).

$$P(A) = \sum_{n=1}^{M} P(A \cap B_n) = \sum_{n=1}^{M} P(A|B_n)P(B_n). \tag{A.3}$$

For the finite M case, this formula can be justified by using a Venn Diagram. Equations (A.1) and (A.3) are the two fundamental relations in the probability theory. They are used to develop important probability laws and theorems.

For example, combining these two equations yields the well-known Bayes' theorem as follows:

$$P(B_i|A) = \frac{P(B_i \cap A)}{P(A)} = \frac{P(A|B_i)P(B_i)}{\sum_{n=1}^{M} P(A \cap B_n)} = \frac{P(A|B_i)P(B_i)}{\sum_{n=1}^{M} P(A|B_n)P(B_n)} \tag{A.4}$$

for $i = 1, ..., n$.

A.2.0.1 Bayesian Belief Networks

It is worth noting that Bayes' theorem plays a central role in machine learning theory. As an application in this area, we briefly present the Bayesian Belief Network which is an important approach in machine learning. A Bayesian Belief Network (BBN) is a network of connected random events (or random variables) that generate predictions about these random events based on the inter-dependence of these events. The random events are represented by a set of nodes and the relations are represented by directed arcs. As the name indicates, the basis for BBN is the Bayes' theorem. We utilize a simple example to demonstrate the belief propagation and its implication in machine learning.

Consider two commuters, Mr. A and Mr. B, travel from home to work in the event of transit worker strike. There are three events of our interest: (i) Transit worker strike, denoted by S; (ii) Mr. A is late for work, denoted by A; and (iii) Mr. B is late for work, denoted by B. The relationships among these events are represented by a network as shown in Figure A.1. Such a network with associated probability tables is and example of Bayesian Belief Network. An important use of BBNs is in revising the probabilities for some events of interest (also called "beliefs"), which are given in the tables in Figure A.1, based on actual observations. This process, based on Bayes' theorem, is called the belief propagation in a BBN. We first compute the unconditional probability for the "child node" event based on the probabilities given. That is $P(A)$ or $P(B)$ (i.e., the probability that Mr. A (or Mr. B) is late).

$$P(A) = P(A \cap S) + P(A \cap S^c) = P(A|S)P(S) + P(A|S^c)P(S^c)$$
$$= 0.7 \cdot 0.1 + 0.6 \cdot 0.9 = 0.61,$$
$$P(B) = P(B \cap S) + P(B \cap S^c) = P(B|S)P(S) + P(B|S^c)P(S^c)$$
$$= 0.6 \cdot 0.1 + 0.1 \cdot 0.9 = 0.15.$$

Now suppose that we have some observation (additional information). For example, we observe that Mr. A is late for work. However, we do not know if there is a strike. How about the probability that the transit work strike occurs? This is apparently computable with the Bayes' theorem. That is

$$P(S|A) = \frac{P(A|S)P(S)}{P(A)} = \frac{0.7 \cdot 0.1}{0.61} = 0.12,$$

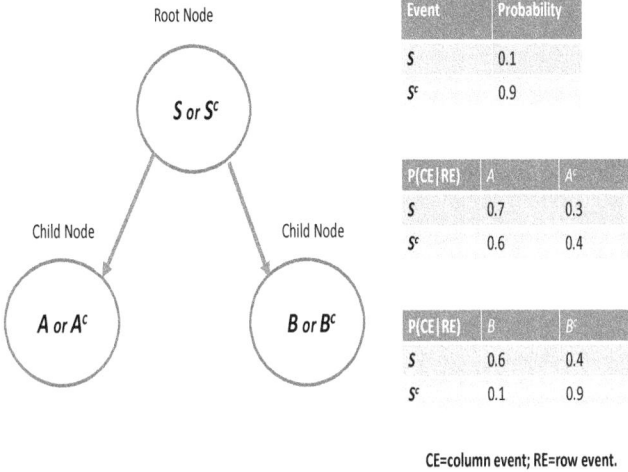

Figure A.1 Bayes' belief network and probability information.

which is a little higher than $P(S) = 0.1$ (called prior probability). Such a revision in probability will also affect the probability that Mr. B is late. Thus, we revise $P(B)$ into $P(B|A)$ as follows:

$$P(B|A) = P((B|A) \cap S) + P((B|A) \cap S^c) = P(B|S,A)P(S|A) + P(B|S^c,A)P(S^c|A)$$
$$= P(B|S)P(S|A) + P(B|S^c)P(S^c|A) = 0.6 \cdot 0.12 + 0.1 \cdot 0.88 = 0.16,$$

where the third equality on the second line follows from the conditional independence (to be defined below). Compared with the probability before knowing that Mr. A is late $P(B)$, the probability that Mr. B is late now $P(B|A)$ is increased from 0.15 to 0.16. So, the evidence that Mr. A is late propagates through the network and changes the belief of the strike and of Mr. B being late.

With this BBN, we can conveniently define the "conditional dependence" and "conditional independence" among random events (or random variables). For example, we can state

- A is conditionally dependent on S with conditional probability $P(A|S)$.
- B is conditionally dependent on S with conditional probability $P(B|S)$.

Since for a given S (or S^c), events A and B have no effect on each other, we can state the conditional independencies as follows:

- A is conditionally independent from B given S. That is $P(A|S,B) = P(A|S)$.
- B is conditionally independent from A given S. That is $P(B|S,A) = P(B|S)$.

It is easy to verify $P(A \cap B|S) = P(A|S)P(B|S)$ which is another way of stating conditional independence. Also we have

$$P(A \cap S \cap B) = P(A|S,B)P(S \cap B) = P(A|S)P(B|S)P(S).$$

The advantage of using BBN is that the joint and marginal probabilities can be computed easily based on the structure of the BBN when the number of events is large and their relations are complex. The results above can be generalized to the followings: Suppose that a BBN has n nodes, denoted by $X_1, X_2, ..., X_n$, representing n multi-outcome random events. Then the joint probability for the n events is given by

$$P(X_1, X_2, ..., X_n) = \prod_{i=1}^{n} P(X_i|Parents(X_i)).$$

Note that for a node without a parent node, its conditional probability on its parent node is the unconditional probability. Using this joint probability, we can compute the marginal (unconditional) probability as

$$P(X_i = x_i)$$
$$= \sum_{x_1 \in S_1} \cdots \sum_{x_{i-1} \in S_{i-1}} \sum_{x_{i+1} \in S_{i+1}} \cdots \sum_{x_n \in S_n} P(X_1 = x_1, ... X_{i-1} = x_{i-1},$$
$$X_i = x_i, X_{i+1} = x_{i+1} ... X_n = x_n),$$

where S_i is the set of values for X_i (the set of outcomes of the random event for node i) with $i = 1, ..., n$. These formulas are basis for developing the learning algorithms in BBNs. There are two types of learning in BBNs: structure learning and parameter learning. The former is for the case we are not sure about the structure of the BBN and have to create a structure based on domain knowledge and expertise or some machine learning algorithms. Then, two approaches, score-based and constraint-based methods, can be applied to learn the structure of the BBN. The parameter learning is for the case that the structure of the BBN is known but the conditional probabilities (given in the tables associated with a BBN) are unknown or need to be updated. BBN learning is an important research area in machine learning and artificial intelligence. Interested readers should be referred to some good textbooks or research papers in this area.

A.3   RANDOM VARIABLES

Since the σ-algebra $\mathscr{F}$ provides the knowledge about which events are possible on a probability space of interest, then $\mathscr{F}$ can be considered as the information component of the probability space $(\Omega, \mathscr{F}, P)$. A random variable X is a function that assigns a numerical value to each element of the sample space (i.e., state of the world), $X : \Omega \to \mathbb{R}$, such that the values taken by X are known to someone who has access to the information $\mathscr{F}$. In other words, the person with access to $\mathscr{F}$ can determine how a particular value is taken by X. More precisely, given any two numbers $a,b \in \mathbb{R}$, then all the states of the world for which X takes values between a and b form a set that is an event (an element of $\mathscr{F}$), i.e.,

$$\{\omega \in \Omega : a < X(\omega) < b\} \in \mathscr{F}.$$

X can also be said to be an $\mathscr{F}$-measurable function. It is worth noting that in the case of the power set as the σ algebra (i.e., $\mathscr{F} = 2^{\Omega}$), the knowledge is maximal, and hence the measurability of random variables is automatically satisfied. In the following discussion, instead of using the word *measurable*, we will also use the more suggestive word *predictable*. This will make more sense in the Chapter B on conditional expectations.

Example A.5 *Consider an experiment of flipping a coin two times and record the sequence of outcomes. Due to the small sample space, it is not too time-consuming to write the three components of the probability space: sample space, power set σ-algebra, probability measure function. Furthermore, we can define a random variable on the sample space and demonstrate that this random variable satisfies the measurable property. The sample space is simply $\Omega = \{HH,HT,TH,TT\}$. Thus, there are four elements in this sample space, i.e., $|\Omega| = 4$. Thus, there are $2^{|\Omega|} = 2^4 = 16$ subsets of Ω in the power set, the σ-algebra. We can list all of them by using the corresponding $|\Omega| = 4$-dimensional vector with binary elements, which can also be written as $\{0,1\}^{|\Omega|}$. All subsets of Ω (i.e., all subsets in $\mathscr{F} = 2^{\Omega}$) can be listed as follows:*

$$\{0,0,0,0\} \Leftrightarrow \emptyset$$
$$\{1,0,0,0\} \Leftrightarrow \{HH\}$$
$$\{0,1,0,0\} \Leftrightarrow \{HT\}$$
$$\{0,0,1,0\} \Leftrightarrow \{TH\}$$
$$\{0,0,0,1\} \Leftrightarrow \{TT\}$$
$$\{1,1,0,0\} \Leftrightarrow \{HH,HT\}$$
$$\{1,0,1,0\} \Leftrightarrow \{HH,TH\}$$
$$\{1,0,0,1\} \Leftrightarrow \{HH,TT\}$$
$$\{0,1,1,0\} \Leftrightarrow \{HT,TH\}$$
$$\{0,1,0,1\} \Leftrightarrow \{HT,TT\}$$
$$\{0,0,1,1\} \Leftrightarrow \{TH,TT\}$$
$$\{1,1,1,0\} \Leftrightarrow \{HH,HT,TH\}$$
$$\{0,1,1,1\} \Leftrightarrow \{HT,TH,TT\}$$
$$\{1,0,1,1\} \Leftrightarrow \{HH,TH,TT\}$$
$$\{1,1,0,1\} \Leftrightarrow \{HH,HT,TT\}$$
$$\{1,1,1,1\} \Leftrightarrow \{HH,HT,TH,TT\} = \Omega$$

The probability function can be defined as $P = |A|/|\Omega|$, where A is any subset of the $\mathscr{F}$ listed above. Thus, we obtain the probability space $(\Omega, \mathscr{F}, P)$ explicitly. Next, we can define a few random variables, denoted by X.

The first one is the number of heads obtained for this experiment. In this case, $X(TT) = 0$, or $X(TH) = 1$. Obviously, the sets

$$\{\omega : X(\omega) = 0\} = \{TT\}, \quad \{\omega : X(\omega) = 1\} = \{HT, TH\}, \quad \{\omega : X(\omega) = 2\} = \{HH\},$$

belong to 2^{Ω}, and hence X is a random variable.

The second one is a binary variable X that is equal to 1 if we get the same outcomes and 0 otherwise. In this case, we have the following sets

$$\{\omega : X(\omega) = 0\} = \{HT, TH\}, \quad \{\omega : X(\omega) = 1\} = \{HH, TT\},$$

that belong to 2^{Ω}, and hence X is a random variable.

The third one is also a binary variable X that is equal to 1 if we get at least one heads and 0 otherwise. In this case, we have the following sets

$$\{\omega : X(\omega) = 0\} = \{TT\}, \quad \{\omega : X(\omega) = 1\} = \{HH, HT, TH\},$$

that belong to 2^{Ω}, and hence X is a random variable.

Obviously, a random variable is a set function from Ω to $\mathbb{R}$ and a non-zero measure function defined on feasible sets is the probability measure. Note that sum of all non-zero probabilities is equal to 1 and all other sets in $\mathscr{F}$ with probability measure of zero are called infeasible sets.

It is worth noting that a random variable is defined as a deterministic function from sample space Ω (domain) to a set in real line (range). However, the measurability of the random variable is defined as the mapping between sets in the σ-algebra of $\mathbb{R}$ (also called Borel σ-algebra) and sets in $\mathscr{F}$, the σ-algebra for the probability space. Saying a random variable is measurable implies that we can assign a probability measure to a set of real values (open or closed interval in $\mathbb{R}$) that the random variable can take. When a random variable is measurable, we can also call that this variable is predictable. Since the key for a variable to be measurable is to know $\mathscr{F}$, a random variable is also called $\mathscr{F}$-measurable as shown in Figure A.2.

A.4 PROBABILITY DISTRIBUTION FUNCTION

Assume that X is a real-valued random variable on the probability space $(\Omega, \mathscr{F}, P)$. A convenient way to characterizing the probability measure of a random variable is to define its cumulative distribution function (cdf). The cdf of X is the function $F_X : \mathbb{R} \to [0, 1]$ defined by $F_X(x) = P(\omega : X(\omega) \leq x)$. It is worth observing that since X is a random variable, then the set $\{\omega : X(\omega) \leq x\}$ belongs to the information set $\mathscr{F}$. This definition is illustrated in Figure A.3.

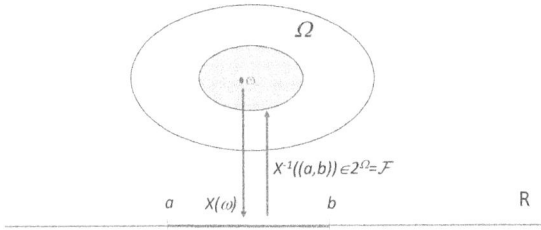

Figure A.2   Measurability of a random variable.

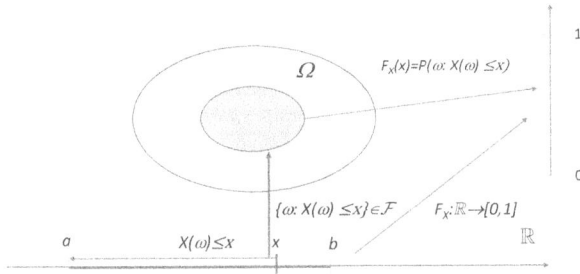

Figure A.3   Distribution function of a random variable.

The distribution function is non-decreasing and satisfies the limits

$$\lim_{x \to -\infty} F_X(x) = 0, \quad \lim_{x \to +\infty} F_X(x) = 1.$$

If we have $dF_X(x)/dx = f(x)$, then $f(x)$ is called the probability density function of X (pdf). A useful property that follows from the fundamental theorem of calculus is

$$P(a < X < b) = P(\omega : a < X(\omega)) < b) = \int_a^b f(x)dx.$$

In the case of discrete random variables the aforementioned integral is replaced by the following sum

$$P(a < X < b) = \sum_{a < x < b} P(X = x) = \sum_{a < x < b} p(x),$$

where $p(x)$ is called the probability mass function (pmf). With pdf or pmf, we can compute the rth moment of a random variable X as

$$E[X^r] = \int_{-\infty}^{\infty} x^r f(x)dx, \ \text{ or } \ E[X^r] = \sum_{x \in S} x^r p(x).$$

The first moment is the well-known mean value $E[X]$ and the variance is computed as $Var[X] = E[X^2] - (E[X])^2$. For more details about properties of random variables, the reader is referred to a traditional probability book, such as Feller (1971).

A.4.1 MULTIVARIATE DISTRIBUTION AND COPULAS

The cdf can be extended to multivariate cases. That is $F_X : \mathbb{R}^d \to [0,1]$ defined by $F_X(x) = P(\omega : X(\omega) \le x)$ where X is the d-dimensional vector. Multivariate cases are more complex and occur in practice when a relationship for several outcomes is required. To study such a relationship, "copula" (a Latin word meaning " a link") is introduced, which describes the relationship via linking the marginal distributions of individual random variables. Copulas have been widely used in mathematical finance and actuarial science. As a special case of $d = 2$, i.e. two continuous random variable vector $X = (X_1, X_2)$, their joint cdf is given by

$$F_X(x_1, x_2) = \int_{-\infty}^{x_1} \int_{-\infty}^{x_2} f(y_1, y_2) dy_2 dy_1, \quad -\infty \le x_1 \le \infty, -\infty \le x_2 \le \infty,$$

where $f(x_1, x_2) \ge 0$ is the joint pdf with $\int_{-\infty}^{\infty} \int_{-\infty}^{\infty} f(y_1, y_2) dy_1 dy_2 = 1$. With these definitions, we can define the marginal distribution of each variable and conditional distribution of a variable given the value of the other variable. The copula can be defined by using the marginal distributions of the data. Intuitively, the marginal distribution describes one random variable without considering the value of the other random variable (or ignoring the relationship or dependence between these variables), the copula is the device that link these marginal distributions or characterizing the relationship or inter-dependence among these variables. Therefore, the joint cdf of multiple random variables can be written in terms of the marginal distributions of the variables and the copula itself. A copula is a multivariate distribution whose marginals are all uniform over $[0,1]$. Any continuous random variable can be transformed, by its probability integral transform (via cdf, see the next subsection), to be uniform over $[0,1]$. Thus, copulas can be used to give inter-dependence information separate from that of the marginal distributions (see Yan (2007)).

Consider a random vector $X = (X_1, ..., X_d)$ with marginal cdfs $F_{X_i}(x) = P(X_i \le x)$ where $i = 1, ..., d$. Transforming these marginal cdfs into uniform cdfs, we obtain vector $U = (U_1, ..., U_d) = (F_{X_1}(x_1), ..., F_{X_d}(x_d))$, which has uniform marginal distributions. This can be easily verified by

$$P(U_i \le y_i) = P(F_{X_i}(x_i) \le y_i) = P(x_i \le F_{X_i}^{-1}(y_i)) = F_{X_i}(F_{X_i}^{-1}(y_i)) = y_i, \quad \text{for } i = 1, ..., d.$$

The copula of X is defined as the joint cdf of U, denoted by C. That is

$$C(u_1, ..., u_d) = P(U_1 \le u_1, ..., U_d \le u_d).$$

This is equivalent to saying that a copula is any function $C : [0, 1]^d \rightarrow [0, 1]$ with the following properties:

- $C(x_1, ..., x_d)$ is increasing in each x_i.
- $C(1, ..., 1, x_i, ..., 1) = x_i$ for $i = 1, ..., d$, and $0 \leq x_i \leq 1$.
- For $a_i \leq b_i$, the nonnegative joint probability $P(a_1 \leq U_1 \leq b_1, ..., a_d \leq U_d \leq b_d)$ results in the rectangular inequality

$$\sum_{i_1=1}^{2} \cdots \sum_{i_d=1}^{2} (-1)^{i_1+\cdots+i_d} C(u_{1i_1}, ..., u_{di_d}) \geq 0.$$

Copulas are useful in describing the joint multivariate distribution of multiple random variables (or data) as it can be decomposed into the individual marginal distributions of these variables and the copula linking them together. Such a property is summarized as the Sklar's Theorem which states that every multivariate cdf of a random vector $\mathbf{X} = (X_1, ..., X_d)$, denoted by $H(x_1, ..., x_d) = P(X_1 \leq x_1, ..., X_d \leq x_d)$, can be expressed in terms of its marginal distributions $F_{X_i}(x_i) = P(X_i \leq x_i)$ and a copula C. That is

$$H(x_1, ..., x_d) = C(F_{X_1}(x_1), ..., F_{X_d}(x_d)). \tag{A.5}$$

If the multivariate pdf exists and is denoted by h, we have

$$h(x_1, ..., x_d) = c(F_{X_1}(x_1), ..., F_{X_d}(x_d)) f_{X_1}(x_1) \cdots f_{X_d}(x_d),$$

where c is the density of the copula and $f_{X_i}(x_i)$ is the marginal pdf of X_i. We can verify (A.5) by writing the RHS as

$$C(F_{X_1}(x_1), ..., F_{X_d}(x_d)) = P(U_1 \leq F_{X_1}(x_1), ..., U_d \leq F_{X_d}(x_d))$$
$$= P(F_{X_1}^{-1}(U_1) \leq x_1, ..., F_{X_d}^{-1}(U_d) \leq x_d)$$
$$= P(X_1 \leq x_1, ..., X_d \leq x_d),$$

which is the LHS of (A.5). If the marginal distributions are continuous, then the copula is unique. In general, copulas can be classified into two categories: Elliptical copulas and Archimediean copulas. Elliptical copulas are based on elliptical distributions which include multivariate Gaussian distributions, multivariate student-t distributions, symmetric multivariate Laplace distributions, symmetric multivariate stable distributions, and so on.

Definition A.4 *The random vector* $\mathbf{X} = (X_1, ..., X_d)$ *is said to have an elliptical distribution if the pdf has the form*

$$f(\mathbf{x}) = k g((\mathbf{x} - \mu)^T \Sigma^{-1} (\mathbf{x} - \mu)),$$

where k is the normalizing constant, $\mathbf{x}$ is a d-dimensional random vector with median or mean vector μ, Σ is a positive definite matrix which is proportional to the covariance matrix, and g is the generator function.

Clearly, multivariate Gaussian distribution is a special case of the elliptical distribution. Next, we present the definitions for the two types of copulas.

Definition A.5 *(Elliptical copula) A family of copula is said to be elliptical if it is written in the form*

$$C(u_1, u_2, ..., u_d; \Sigma) = \Psi_\Sigma^d(\Psi^{-1}(u_1), \Psi^{-1}(u_2), ..., \Psi^{-1}(u_d)).$$

where Ψ_Σ^d is a d-dimensional multivariate ξ distribution with Σ as the correlation matrix; and Ψ^{-1} is an inverse of the univariate ξ distribution. The distribution ξ corresponds to either standard normal distribution resulting in a Gaussian copula or the Student-t distribution $(t_{d.f.})$ resulting in a student-t copula.

The elliptical copulas are also known as inversion-method copulas. The bivariate Gaussian copula is given by

$$C_\rho^\Sigma(u_1, u_2) = \frac{1}{2\pi \sqrt{|\Sigma|}} \int_{-\infty}^{\Phi^{-1}(u_1)} \int_{-\infty}^{\Phi^{-1}(u_2)} \left(-\frac{x_1^2 - 2\rho x_1 x_2 + x_2^2}{2|\Sigma|} \right) dx_1 dx_2,$$

where $u = (u_1, u_2), x = (x_1, x_2)$, and $\rho \in (-1, 1)$ is the correlation parameter in the 2×2 matrix Σ, and Φ^{-1} is the inverse of a univariate standard normal distribution. It is easy to check all properties are satisfied by this Gaussian copula. For example, the second property can be verified as follows:

$$C_\rho^\Sigma(1, u_2) = \Phi_\Sigma^2(\Phi^{-1}(1), \Phi^{-1}(u_2)) = \Phi(\Phi^{-1}(u_2)) = u_2,$$

where Φ_Σ^2 is a bivariate normal distribution and Φ is a univariate standard normal distribution. This copula is for modeling the situation that the tail dependence between the two variables is zero. The bivariate student-t copula can be given similarly and is more appropriate for modeling the situation that the upper and lower tail dependencies are symmetric.

Definition A.6 *(Archimedean copula) A copula family is said to be Archimedean if it is written in the form*

$$C(u_1, u_2, ..., u_d) = \psi^{-1}(\psi(u_1) + \psi(u_2) + \cdots + \psi(u_d)),$$

with the generator of the copula $\psi(x)$, satisfying the following conditions:

- *$\psi(0) = \infty$ and $\psi(1) = 0$.*
- *The inverse function $\psi^{-1}(x)$ corresponds to a probability, such that $\psi^{-1} : [0, \infty] \to [0, 1]$.*
- *The function $\psi(x)$ is convex and decreasing, such that $x \in [0, \infty)$.*

Three typical copulas in this family include Clayton, Gumbel, and Frank copulas. Here, we present the bivariate copulas of these three types. For a bivariate Clayton copula, the generator function is $\psi(k) = \alpha^{-1}(k^{-\alpha} - 1)$, where $\alpha \in [-1, \infty) \setminus \{0\}$. Then the bivariate Clayton copula is defined as

$$C_\alpha(u_1, u_2) = \max\{[u_1^{-\alpha} + u_2^{-\alpha} - 1]^{-1/\alpha}, 0\}. \tag{A.6}$$

It is easy to check that all conditions in the definition of Archimedean copula have been satisfied. For the copula form, we first find the inverse of ψ by solving $\psi(k) = \alpha^{-1}(k^{-\alpha} - 1) = y$ for k. That is $k = \psi^{-1}(y) = (1 + \alpha y)^{-\frac{1}{\alpha}}$. Substituting $y = \psi(u_1) + \psi(u_2) = \alpha^{-1}(u_1^{-\alpha} - 1) + \alpha^{-1}(u_2^{-\alpha} - 1)$ into $\psi^{-1}(y)$ yields

$$\begin{aligned}
C_\alpha(u_1, u_2) &= \psi^{-1}(\psi(u_1) + \psi(u_2)) \\
&= (1 + \alpha[\alpha^{-1}(u_1^{-\alpha} - 1) + \alpha^{-1}(u_2^{-\alpha} - 1)])^{-\frac{1}{\alpha}} \\
&= (u_1^{-\alpha} + u_2^{-\alpha} - 1)^{-\frac{1}{\alpha}}.
\end{aligned}$$

Since the copula cannot be negative, we obtain (A.6). The Clayton copula is appropriate for modeling data points, which are strongly correlated at the low values and weakly correlated at the upper values. Similarly, we can define the bivariate Gumbel copula with the generator function of $\psi(k) = (-\log k)^\alpha$ with $\alpha \geq 1$. The corresponding copula is given by

$$C_\alpha(u_1, u_2) = \exp\{-[(-\log u_1)^\alpha + (-\log u_2)^\alpha]^{\frac{1}{\alpha}}\}, \quad \text{for } \alpha > 1.$$

This copula is more appropriate for modeling data points, which are strongly correlated at the upper values and weakly correlated at the lower values. The bivariate Frank copula has the generator function $\psi(k) = -\log[(e^{-\alpha k} - 1)(e^{-\alpha} - 1)^{-1}]$ with $\alpha \in \mathbb{R} \setminus \{0\}$. Then the copula is given by

$$C_\alpha(u_1, u_2) = -\frac{1}{\alpha}\left[1 + \frac{(e^{-\alpha u_1} - 1)(e^{-\alpha u_2} - 1)}{(e^{-\alpha} - 1)}\right].$$

This Achimedean copula is appropriate for modeling data points, in which the upper and lower tail dependencies are zero.

An application of using Gaussian copula is to simulate data from a bivariate distribution that has different marginal distributions which are correlated. Specifically, we consider to generate random values from a bivariate distribution with the following properties:

- The rank correlation between the variables is ρ.
- The marginal distribution of the first variable, X, is $F_X(x)$.
- The marginal distribution of the second variable, Y, is $F_Y(y)$.

It can be shown that any distribution can be transformed to uniform distribution by using the cdf and the uniform distribution can be transformed to any distribution by using the inverse cdf (see the next subsection). Thus, to simulate the bivariate distributed values via Gaussian copula, we follow the following three steps:

1. Simulate correlated bivariate normal data from a correlation matrix. The marginals are all standard normal.
2. Use the standard normal cdf to transform the normal marginals to uniform distribution.
3. Use the inverse cdfs to transform the unform marginals to whatever distributions desired.

To implement this procedure, the simulated data in Step 1 are put in two columns. Then the transformations in Step 2 and Step 3 are performed on the individual columns of the simulated data. Due to the monotonic property of these transformations, the rank correlation between columns is preserved. Note that the simulation is Step 1 is relatively easy to implement by using R or even Excel and Step 2 is based on the Gaussian copula. Step 1 can be done based on the conditional distribution of the bivariate normal distribution. Recall that the following properties hold for the bivariate normal distribution, denoted by $(X,Y)^T \sim \Phi^2(\mu, \Sigma)$.

1. The marginal distributions are normal. That is

$$X \sim N(\mu_x, \sigma_x^2), \quad Y \sim N(\mu_y, \sigma_y^2).$$

2. The conditional distributions are also normal with

$$X|Y = y \sim N(\mu_{x|y}, \sigma_{x|y}) = N(\mu_x + \rho \frac{\sigma_x}{\sigma_y}(y - \mu_y), \sigma_x^2(1 - \rho^2)),$$

$$Y|X = x \sim N(\mu_{y|x}, \sigma_{y|x}) = N(\mu_y + \rho \frac{\sigma_y}{\sigma_x}(x - \mu_x), \sigma_y^2(1 - \rho^2)).$$

3. If $\rho = 0$, X and Y are independent.
4. Linear combinations of X and Y are also normally distributed. That is

$$aX + bY \sim N(a\mu_x + b\mu_y, a^2\sigma_x^2 + b^2\sigma_y^2 + 2ab\rho\sigma_x\sigma_y),$$

where a, b are constants.

For example, to simulate the data from a bivariate normal distribution via Excel, we can repeat the following:

$$X = NORMINV(RAND(), \mu_x, \sigma_x),$$
$$Y = NORMINV(RAND(), \mu_{y|x}, \sigma_{y|x}).$$

Such a procedure can be extended to higher dimension cases with the pdf of the multivariate normal in matrix form as

$$f_{\mathbf{X}}(x, 1, ..., x_d) = \frac{1}{\sqrt{(2\pi)^d |\Sigma|}} e^{\left(-\frac{1}{2}(\mathbf{x}-\mu)^T \Sigma^{-1}(\mathbf{x}-\mu)\right)},$$

where $\mathbf{x}$ is a real d-dimensional column vector with $d > 2$ and $|\Sigma| \equiv \det\Sigma$ is the determinant of Σ.

A.4.2 TRANSFORMING DISTRIBUTION FUNCTIONS

We can transform a probability density function by transforming a random variable. Such a transformation is rooted from the integration by substitution of a new variable. Consider a continuous random variable X with the pdf of $f(x)$ (the cdf of $F(x)$). Then, for an interval $[a, b]$ in the domain of X, we have $P(a \leq X \leq b) = \int_a^b f(x)dx$. Select a transformation function $y(x)$ to transform X to a new random variable Y and write its inverse as $x(y)$. Thus, we obtain

$$P(a \leq X \leq b) = \int_a^b f(x)dx = \int_{y(a)}^{y(b)} f(x(y))\frac{dx(y)}{dy}dy = P(y(a) \leq Y \leq y(b)),$$

which yields the new pdf for Y as

$$g(y) = f(x(y))\frac{dx(y)}{dy}.$$

Since the pdf must be non-negative, we take the absolute value of the derivative of the inverse function and write

$$g(y) = f(x(y))\left|\frac{dx(y)}{dy}\right|. \tag{A.7}$$

The simplest continuous distribution is the uniform distribution. We utilize the uniform distribution over $[0, 1]$, with the pdf denoted by $f_U(u) = 1$ and the random variable denoted by U, as the benchmark and show the transformation from an arbitrary distribution F to F_U and the reverse. To transform a pdf $f(x)$ to $f_U(u) = 1$, we set $g(y) = f(x(y))\frac{dx(y)}{dy} = 1$, which leads to $dy = f(x(y))dx$. Integrating both sides of this equation yields

$$y = \int_{-\infty}^x f(x)dx = F(x),$$

which implies that if defining a new random variable $Y = F(X)$, the pdf of Y is the uniform distribution over $[0, 1]$ or $U = Y = F(X)$. Obviously, this transformation can be called "probability integral transform" due to using the cdf $F(x)$. Consequently, we can transform the uniform distribution F_U of U to any distribution F of X by the transformation $X = F^{-1}(U)$. Such a transformation is the foundation to generate simulated values of a desirable distribution based on the uniformly distributed numbers between 0 and 1. Another application of using these transforms is to transform a skewed empirical distribution (based on a real data set) to a normal distribution in two steps. The first step is to compute the empirical cdf $\tilde{F}(x)$ by using fractional ranking the data and defining the uniform distributed random variable U by setting $U = \tilde{F}(X)$. The second step is to transform U to a normally distributed new random variable Y via $Y = F_N^{-1}(U)$ where F_N is the normal cdf.

A.5   INDEPENDENT RANDOM VARIABLES

For two random variables X and Y, if for any sets $A, B \in \mathbb{R}$, the events

$$E_A = \{\omega : X(\omega) \in A\}, \quad E_B = \{\omega : Y(\omega) \in B\}$$

are independent (i.e., $P(E_A \cap E_B) = P(E_A)P(E_B)$), then X and Y are called independent random variables.

Proposition A.1 *If X and Y are independent random variables with probability density functions $f_X(x)$ and $f_Y(y)$, then the joint probability density function of (X, Y) is given by $f_{(X,Y)} = f_X(x)f_Y(y)$.*

The proof of this proposition is a good exercise.

A.6   TRANSFORMS FOR RANDOM VARIABLES

Another way of describing a random variable's behavior is to use the transforms of its distribution function. These transforms contain the equivalent information as what the distribution function contains and are defined as the expected value of the function of the random variable. Some of these transforms require fewer conditions for their existence than others. The most general transform, which always exists, is the characteristic function. For a real-valued random variable X with the cdf $F_X(x) = E[\mathbf{1}_{\{X \leq x\}}]$ and pdf $f_X(x)$. The characteristic function, denoted by $\psi_X(t)$, is defined as

$$\psi_X(t) = E[e^{itX}] = \int_{x \in S} e^{itx} dF(x) = \int_{x \in S} e^{itx} f(x)dx. \qquad (A.8)$$

The second transform for a random variable is the Laplace-Stieltjes transform (LST) that is defined as

$$L_X(s) = E[e^{-sX}] = \int_{x \in S} e^{-sx} dF_X(x) = \int_{x \in S} e^{-sx} f_X(x)dx. \qquad (A.9)$$

Note that (1) these transforms are defined for any general functions and here the functions considered are the cdf's (or pdf's) of random variables; and (2) these transforms can be extended to multi-dimensional cases (i.e., random vectors).

The third transform is the moment generating function (mgf) and is defined by

$$M_X(t) = E[e^{tX}] = \int_{x \in S} e^{tx} dF_X(x) = \int_{x \in S} e^{tx} f_X(x) d(x), \qquad (A.10)$$

provided this expectation exists for t in some neighborhood of 0. For the discrete random variables, the integral is replaced with the summation in the definition. There are four important properties for these transforms of random variables as follows:

(1) $\psi_X(0) = L_X(0) = M(0) = 1$ due to the probability distribution function property.

(2) These transforms follow some computational rules for the linear function of a random variable and the linear combination of independent random variables. We use the mgf as an example for this and remaining properties.

- Let a and b be two constant numbers and X is a random variable with a mgf $M_X(t)$. Then, for $Y = a + bX$, we have

$$M_Y(t) = E[e^{(a+bX)t}] = e^{at} E[e^{bXt}] = e^{at} M_X(bt).$$

- Let a_i with $i = 1, ..., n$ be n constant numbers and X_i with $i = 1, ..., n$ be n independent random variables. Then for $S_n = \sum_{i=1}^{n} a_i X_i$, we have

$$M_{S_n}(t) = M_{X_1}(a_1 t) M_{X_2}(a_2 t) \cdots M_{X_n}(a_n t).$$

(3) The mgf, if existed, uniquely determines the distribution and the nth moment of the random variable can be obtained by taking the nth derivative as follows:

$$E[X^n] = M_X^{(n)}(0) = \frac{d^n M_X}{dt^n}\Big|_{t=0}, \quad n = 1, 2, ...$$

(4) The pgf is positive and log-convex and has a lower bound established by Jensen's inequality as $M_X(t) \geq e^{\mu t}$.

In addition to these properties, the mgf can also be used to bound the upper tail probability and moments of a random variable. For example, using the Markov inequality, we have

$$P(X \geq a) = P(e^{tX} \geq e^{ta}) \leq e^{-at} E[e^{tX}] = e^{-at} M_X(t).$$

Similar properties hold for the characteristic functions and LSTs. The relations among these transforms are straightforward from their definitions and are given by

$$M_X(-t) = L_X(t), \quad \text{and} \quad M_X(t) = \psi_X(-it).$$

TABLE A.1
Major Discrete Distributions

distribution	pmf	mean	variance	mgf
Bernoulli	$px + (1-p)(1-x), x = 0,1$	p	$p(1-p)$	$1-p+pe^t$
Binomial	$\binom{n}{x}p^x(1-p)^{n-x}, x = 0,1,...,n.$	np	$np(1-p)$	$(1-p+pe^t)^n$
Poisson	$\frac{e^{-\lambda}\lambda^x}{x!}, x = 0,1,....$	λ	λ	$e^{\lambda(e^t-1)}$
Geometric	$p(1-p)^{x-1}, x = 1,2,....$	$\frac{1}{p}$	$\frac{1-p}{p^2}$	$\frac{pe^t}{1-(1-p)e^t}$
Negative Binomial	$\binom{x-1}{k-1}p^k(1-p)^{x-k}, x = k,k+1,....$	$\frac{k}{p}$	$\frac{k(1-p)}{p^2}$	$\left[\frac{pe^t}{1-(1-p)e^t}\right]^k$
Hypergeometric	$\frac{\binom{k}{x}\binom{N-k}{n-x}}{\binom{N}{n}}, x = 0,1,...,n.$	$\frac{nK}{N}$	$n\frac{k}{N}\left(1-\frac{k}{N}\right)\frac{N-n}{N-1}$	

Another transform, called the probability generating function (pgf), is defined by

$$G_X(z) = E[z^X],$$

which is often used for discrete random variables. Clearly, $G_X(e^t) = E[e^{tX}] = M_X(t)$ holds.

A.7 POPULAR DISTRIBUTIONS IN STOCHASTIC MODELS

We present the popular discrete and continuous distributions used in stochastic models in Tables A.1 and A.2. We list the pmf, mean, variance, and mgf (if exists) for each distribution.

Bernoulli and geometric distributions are considered as two most basic distributions of random variables as other distributions can be developed from them. For example, the sum of n (finite) Bernoulli random variables with parameter p has the Binomial distribution with parameters n and p. By taking appropriate limit, a Binomial distribution can approach to a Poisson distribution or a normal distribution. Verifying these relations can be good exercises. As an example, we show the relation between the geometric distribution and the exponential distribution. Assume that a random variable $X = 1,2,...$ follows a geometric distribution with parameter p. Consider a regime when the probability of success p is very small such that $np = \lambda$ (a constant). Letting

TABLE A.2
Major Continuous Distributions

distribution	pdf	mean	variance	mgf
Uniform	$\begin{cases} \frac{1}{b-a} & \text{if } a \leq x \leq b, \\ 0 & \text{otherwise} \end{cases}$	$\frac{a+b}{2}$	$\frac{(b-a)^2}{12}$	$\frac{e^{tb}-e^{ta}}{t(b-a)}$
Normal	$\frac{1}{\sqrt{2\pi}\sigma}e^{-\frac{1}{2}\left(\frac{x-\mu}{\sigma}\right)^2}, x \in (-\infty, \infty).$	μ	σ^2	$e^{\mu t + \sigma^2 t^2/2}$
Exponential	$\lambda e^{-\lambda x}, x > 0.$	$\frac{1}{\lambda}$	$\frac{1}{\lambda^2}$	$\frac{\lambda}{\lambda-t}$
Gamma	$\frac{1}{\beta^\alpha \Gamma(\alpha)}x^{\alpha-1}e^{-x/\beta}, x > 0.$	$\alpha\beta$	$\alpha\beta^2$	$\left(\frac{1}{1-\beta t}\right)^\alpha$
Weibull	$\binom{x-1}{k-1}p^k(1-p)^{x-k}, x = k, k+1, \dots$	$\frac{k}{p}$	$\frac{k(1-p)}{p^2}$	$\left[\frac{pe^t}{1-(1-p)e^t}\right]^k$
Beta	$\frac{\Gamma(\alpha+\beta)}{\Gamma(\alpha)\Gamma(\beta)}x^{\alpha-1}(1-x)^{\beta-1}, x > 0.$	$\frac{\alpha}{\alpha+\beta}$	$\frac{\alpha\beta}{(\alpha+\beta)^2(\alpha+\beta+1)}$	

$x = k/n$ and taking the limit of $n \to \infty$ in the following identity, we have

$$1 = \sum_{k=1}^{\infty} P(X = k) = \sum_{k=1}^{\infty} p(1-p)^{k-1} = \sum_{k=1}^{\infty} \frac{\lambda}{n}\left(1-\frac{\lambda}{n}\right)^{nx-1}$$

$$= \sum_{k=1}^{\infty} \lambda \frac{\left(\left(1-\frac{\lambda}{n}\right)^n\right)^x}{1-\frac{\lambda}{n}} \frac{1}{n} \xrightarrow{n\to\infty} \int_0^\infty \lambda e^{-\lambda x} dx.$$

We can also use the pmf approach. With the same parameters, taking the limit on the mgf, we have

$$\lim_{n\to\infty} M_X(t) = \lim_{n\to\infty} \frac{\lambda/n}{1 - e^{t/n}(1-\lambda/n)} = \lim_{n\to\infty} \frac{\lambda}{n - e^{t/n}(n-\lambda)}$$

$$= \lim_{n\to\infty} \frac{\lambda}{n(1-e^{t/n}) + \lambda e^{t/n}} = \frac{\lambda}{\lambda - t},$$

which leads to the same result. For a comprehensive review on the relationships among more univariate distributions, interested readers are referred to Song (2005).

The distributions in Table A.3 are derived from the normal distribution. They are chi-square, F, and t distributions, which are most commonly used in statistics. We briefly show the derivation of these useful distributions.

TABLE A.3

Distributions Derived from the Normal Distributions

distribution	pdf	mean	variance	mgf
$\chi^2_{(k)}(x)$ with $x > 0$	$\dfrac{1}{2^{k/2}\Gamma(k/2)}x^{(k-2)/2}e^{-x/2}$	k	$2k$	$\left(\dfrac{1}{1-2t}\right)^{k/2}$
$F_{m,n}(x)$ with $x > 0$	$\dfrac{\Gamma\left(\frac{m+n}{2}\right)m^{m/2}n^{n/2}x^{(m/2)-1}}{\Gamma\left(\frac{m}{2}\right)\Gamma\left(\frac{n}{2}\right)(n+mx)^{(m+n)/2}}$	$\dfrac{n}{n-2}$ for $n > 2$	$\dfrac{2n^2(m+n-2)}{m(n-2)^2(n-4)}$ for $n > 4$	
t distribution	$\dfrac{1}{\sqrt{k}B(1/2,k/2)}$ $\left(1+\frac{t^2}{k}\right)^{-(k+1)/2}$	0	$\dfrac{k}{k-2}$ for $k > 2$	

A.7.1 CHI-SQUARE DISTRIBUTION

To show the chi-square distribution, we consider a Binomial distribution with parameters N (the total number of trials) and p (the probability of a success for each trial). Let the number of successes be k. Then, as $N \to \infty$ with $Np \to \mu$, a constant, we have

$$\chi = \frac{k - Np}{\sqrt{Npq}} \to Z,$$

where $q = 1 - p$ and Z is the standard normal random variable. Squaring both sides of the equation above yields $\chi^2 = (k - Np)^2/(Npq)$, which can be re-written as

$$\chi^2 = \frac{(k - Np)^2}{Np} + \frac{(N - k - Nq)^2}{Nq}.$$

The RHS of the equation above asymptotically approaches to the sum of two independent squared standard normal random variables. This is because the denominator for each term of the RHS approaches to the variance as $N \to \infty$. For example, Np (the denominator of the first term of RHS) approaches to Npq (the variance of k) as $N \to \infty$ since the limit is taken in a way such that $Np = $ constant and hence $p \to 0$ and $q \to 1$ as $N \to \infty$ (i.e., $Npq \to Np$ as $N \to \infty$). Similarly, we also have the denominator of the second term of RHS approaches to the variance of $N - k$. That is $Nqp \to Nq$ as $N \to \infty$ by using the same limiting argument. Note that the results obtained above from the Binomial distribution case (two categories) can be extended to the multinomial distribution case (multiple categories $k > 2$). The RHS of the equation above is consistent with

$$\chi^2 = \sum_{i=1}^{k} \frac{(O_i - E_i)^2}{E_i},$$

where χ^2 is called Pearson's cumulative test statistic, which asymptotically approaches to the sum of independent squared standard normal random variables; O_i is the number of observations of type i outcome, $E_i = Np_i$ is the expected number of type i outcome, asserted by the null hypothesis that the proportion of type i in population is p_i, and the denominator (also E_i) approaches to the variance of O_i. Thus, if $Z_1, Z_2, ..., Z_k$ are i.i.d. standard normal random variables, then the sum of their squares $\sum_{i=1}^{k} Z_i^2$ follows the chi-square distribution with k degrees of freedom, denoted by χ_k^2. Further, when the total number of observations is fixed or the average of k standard normal random variables is used as the center, the degree of freedoms for the χ^2 becomes $k-1$. For example, $\sum_{i=1}^{k}(Z_i - \bar{Z})^2 \sim \chi_{k-1}^2$ where $\bar{Z} = (1/k)\sum_{i=1}^{k} Z_i$. Next, we introduce the F distribution which is the ratio of two chi-squares, each divided by its degrees of freedom. Note that a chi-square random variable divided by its degrees of freedom is a variance estimate.

A.7.2 F DISTRIBUTION

First, we prove a theorem

Theorem A.1 *Let* $\mathbf{X} = \{X_1, X_2, ..., X_n\}$ *be a simple random sample of size n from a population with the mean* μ *and variance* σ^2. *Denote by* $\bar{X}$ *and* S^2 *the sample mean and sample variance, respectively. Then the distribution of* $(n-1)S^2/\sigma^2$ *is the chi-square distribution with* $(n-1)$ *degrees of freedom.*

Proof. First, we note

$$\frac{1}{\sigma^2}\sum_{i=1}^{n}(X_i - \mu)^2 = \sum_{i=1}^{n}\left(\frac{X_i - \mu}{\sigma}\right)^2 = \sum_{i=1}^{n} Z_i^2 \sim \chi_{(n)}^2. \qquad (A.11)$$

Then we write an expression as

$$\frac{1}{\sigma^2}\sum_{i=1}^{n}(X_i - \mu)^2 = \frac{1}{\sigma^2}\sum_{i=1}^{n}(X_i - \bar{X} + \bar{X} - \mu)^2$$

$$= \frac{1}{\sigma^2}\sum_{i=1}^{n}(X_i - \bar{X})^2 + \frac{n}{\sigma^2}(\bar{X} - \mu)^2. \qquad (A.12)$$

$$= \frac{(n-1)S^2}{\sigma^2} + \frac{n}{\sigma^2}(\bar{X} - \mu)^2.$$

where the last equality follows from the definition $S^2 = \sum_{i=1}^{n}(X_i - \bar{X})^2/(n-1)$. From (A.11) and (A.12) and by independence, we have

$$\text{mgf of } \chi_n^2 = \text{mgf of [distribution of } (n-1)S^2/\sigma^2] \times \text{mgf of } \chi_1^2,$$

which is

$$(1-2t)^{-n/2} = M_{(n-1)S^2/\sigma^2}(t) \times (1-2t)^{-1/2}.$$

From this relation, we obtain

$$M_{(n-1)S^2/\sigma^2}(t) = (1-2t)^{-(n-1)/2} \Rightarrow (n-1)S^2/\sigma^2 \sim \chi_{(n-1)}^2.$$

This completes the proof. ∎

Since S^2 is an unbiased estimator for σ^2, $\chi_{(n-1)}^2$ is the distribution for testing hypothesis about or computing confidence interval for population variance. In hypothesis test about two population variances, we form a statistics as the ratio of two variance estimates, which is called F-test statistics and is given by

$$F = \frac{S_1^2}{S_2^2} = \frac{\frac{\sigma_1^2 \chi_{(n_1-1)}^2}{n_1-1}}{\frac{\sigma_2^2 \chi_{(n_2-1)}^2}{n_2-1}} = \frac{\chi_{(n_1-1)}^2/(n_1-1)}{\chi_{(n_2-1)}^2/(n_2-1)}, \tag{A.13}$$

where the null hypothesis of $\sigma_1^2 = \sigma_2^2$ is utilized. The distribution for this statistics is called F distribution that depends on two parameters, the degrees of freedom for the numerator $df_n = n_1 - 1$ and for the denominator $df_d = n_2 - 1$. In general, an F distributed random variable X with $df_n = m$ and $df_d = n$ is obtained by dividing U/m by V/n, where U and V are two independent χ^2 distributed random variables with d.f. m and n, respectively. That is $X = (U/m)(V/n)^{-1}$.

A.7.3 T DISTRIBUTION

It is well known that in testing hypothesis about the population mean μ, the test statistics for the case with the variance of the population σ^2 known is

$$Z = \frac{\bar{X} - \mu}{\sigma/\sqrt{n}},$$

where $\bar{X}$ is the sample mean with sample size n. Clearly, Z is a standard normal random variable. However, in many practical situations, σ^2 is unknown and has to be estimated by sample variance S^2. Since $\bar{X}$ remains to be an unbiased estimator, we can define a new test statistics, called T as

$$T = \frac{\bar{X} - \mu}{S/\sqrt{n}}.$$

It follows from the expressions of Z and T that

$$\frac{Z}{T} = \frac{S}{\sigma} = \sqrt{\frac{\chi^2_{(n-1)}}{n-1}} \Rightarrow T = \frac{Z}{\sqrt{\frac{\chi^2_{(n-1)}}{n-1}}}. \tag{A.14}$$

Here, Theorem A.1 is used. The distribution of statistics T is called t distribution.

A.7.4 DERIVATION OF PROBABILITY DENSITY FUNCTION

Since χ^2 distribution is the basis for both F and t distributions, we show how to derive the pdf for χ^2 distribution. There are multiple ways of developing this pdf expression. For example, it can be obtained by directly integrating the density function of the convolution of i.i.d. random variables or using the moment generating function. The direct integration approach is more involved and needs using the change of variables. Here we present a simpler way that combines both approaches. We start with the case of χ^2_1. Denote by Y the random variable that has $\chi^2_{(1)}$ distribution (d.f.=1) and $Y = Z^2$ where Z is the standard normal random variable. Let $F_Y(y)$ and $f_Y(y)$ be the cdf and pdf of Y, respectively. Then, for $y \geq 0$, we have

$$F_Y(y) = P(Y \leq y) = P(Z^2 \leq y) = P(|Z| \leq \sqrt{y}) = P(-\sqrt{y} \leq Z \leq \sqrt{y})$$
$$= F_Z(\sqrt{y}) - F_Z(-\sqrt{y}) = F_Z(\sqrt{y}) - [1 - F_Z(-\sqrt{y})] = 2F_Z(\sqrt{y}) - 1.$$

Taking the first-order derivative of the above with respect to y yields

$$f_Y(y) = \frac{dF_Y(y)}{dy} = 2\frac{d}{dy}F_Z(\sqrt{y}) = 2\frac{d}{dy}\left(\int_{-\infty}^{\sqrt{y}} \frac{1}{\sqrt{2\pi}} e^{-\frac{t^2}{2}} dt\right)$$
$$= 2\frac{1}{\sqrt{2\pi}} e^{-\frac{y}{2}} \frac{d}{dy}(\sqrt{y}) = 2\frac{1}{\sqrt{2\pi}} e^{-\frac{y}{2}}\left(\frac{1}{2}y^{-\frac{1}{2}}\right) = \frac{1}{2^{\frac{1}{2}}\Gamma\left(\frac{1}{2}\right)} y^{-\frac{1}{2}} e^{-\frac{y}{2}},$$

which is actually a Gamma distribution with $\alpha = 1/2$ and $\beta = 2$ (see Table A.2). Here $\Gamma(p) = \int_0^\infty x^{p-1} e^{-x} dx$ for $p > 0$ is called a Gamma function. It can be verified that Gamma function satisfies a recursion via integration by parts as follows:

$$\Gamma(p) = -e^{-x} x^{p-1}\big|_0^\infty - \int_0^\infty -e^{-x}(p-1)x^{p-2} dx$$
$$= 0 + (p-1) \int_0^\infty e^{-x} x^{p-2} dx = (p-1)\Gamma(p-1).$$

Note that $\Gamma(1/2) = \int_0^\infty x^{-1/2} e^{-x} dx = \sqrt{\pi}$ and $\Gamma(1) = \int_0^\infty e^{-x} dx = 1$. Thus, if p is a positive integer, we have $\Gamma(p) = (p-1)!$. Clearly, the mgf of Y is given by $M_Y(t) = (1 - 2t)^{-1/2}$. Now we consider the case of $\chi^2_{(k)}$ with $k \geq 2$. That is

$$Y = Z_1^2 + Z_2^2 + \cdots + Z_k^2,$$

where $Z_1, ..., Z_n$ are i.i.d. standard normal random variables. Then, in terms of mgf's we have

$$M_Y(t) = E[e^{tY}] = E[e^{t(Z_1^2 + Z_2^2 + \cdots + Z_k^2)}] = E[e^{tZ_1^2}]E[e^{tZ_2^2}] \cdots E[e^{tZ_k^2}]$$

$$= \underbrace{\left(\frac{1}{1-2t}\right)^{\frac{1}{2}} \left(\frac{1}{1-2t}\right)^{\frac{1}{2}} \cdots \left(\frac{1}{1-2t}\right)^{\frac{1}{2}}}_{k \text{ terms}} = \left(\frac{1}{1-2t}\right)^{\frac{k}{2}}.$$

This mgf indicates that Y follows a Gamma distribution with $\alpha = k/2$ and $\beta = 2$. Thus, based on Table A.2, we obtain the pdf of Y as

$$f_Y(y) = \frac{1}{2^{k/2}\Gamma(k/2)} y^{(k-2)/2} e^{-y/2}, \quad \text{for } y > 0. \tag{A.15}$$

Note that we can also use the notation $\chi^2_{(k)} = Y$.

Next, we show how to derive the pdf for the t distribution. Note that from (A.14) $T = Z/\sqrt{Y/k}$ where Z is the standard normal random variable and Y follows the $\chi^2_{(k)}$ distribution with $k = n - 1$.

Theorem A.2 *The pdf of T (t distributed random variable) is given by*

$$f_k(t) = \frac{1}{\sqrt{k}B(1/2, k/2)} \left(1 + \frac{t^2}{k}\right)^{-(k+1)/2},$$

where k is the degrees of freedom and B is the beta function given by

$$B(1/2, k/2) = \frac{\Gamma(1/2)\Gamma(k/2)}{\Gamma((k+1)/2)} = \frac{\sqrt{\pi}\Gamma(k/2)}{\Gamma((k+1)/2)}.$$

Proof. Denote by $f(z, y)$ the joint pdf of Z and Y. Then

$$f(z, y) = f_Z(z)f_Y(y) = \left(\frac{1}{\sqrt{2\pi}} e^{-\frac{1}{2}z^2}\right) \left(\frac{1}{2^{k/2}\Gamma(k/2)} e^{-\frac{1}{2}y} y^{(k-2)/2}\right)$$

$$= \frac{1}{\sqrt{\pi}2^{(k+1)/2}\Gamma(k/2)} e^{-\frac{1}{2}(z^2 + y)} y^{(k-2)/2}. \tag{A.16}$$

Let $t = z/\sqrt{y/k}$ and $u = y$ (change of variables). Then $z = t\sqrt{u/k}$ and $y = u$ and the Jacobian of this transformation J is given by

$$|J| = \left|\frac{\partial(z, y)}{\partial(t, u)}\right| = \begin{vmatrix} \frac{\partial z}{\partial t} & \frac{\partial z}{\partial u} \\ \frac{\partial y}{\partial t} & \frac{\partial y}{\partial u} \end{vmatrix}$$

$$= \begin{vmatrix} \sqrt{\frac{u}{k}} & \frac{1}{2}\left(\frac{u}{k}\right)^{-\frac{1}{2}}\left(\frac{1}{k}\right) \\ 0 & 1 \end{vmatrix} = \sqrt{\frac{u}{k}}.$$

The joint pdf in terms of t and u is given by

$$g(t,u) = f(t,u)|J| = \frac{1}{\sqrt{\pi}2^{(k+1)/2}\Gamma(k/2)} e^{-\frac{1}{2}(t^2\frac{u}{k}+u)} u^{(k-2)/2} \sqrt{\frac{u}{k}}$$

$$= \frac{1}{\sqrt{\pi}2^{\frac{k+1}{2}}\Gamma(\frac{k}{2})\sqrt{k}} e^{-\frac{1}{2}(1+t^2/k)u} u^{(k-1)/2}.$$

Now integrating u over the range $[0,\infty)$, we have

$$f_k(t) = \int_0^\infty g(t,u)du = \frac{1}{\sqrt{\pi}2^{\frac{k+1}{2}}\Gamma(\frac{k}{2})\sqrt{k}} \int_0^\infty e^{-\frac{1}{2}(1+t^2/k)u} u^{(k-1)/2} du. \quad (A.17)$$

Let $T = (1/2)(1+t^2/k)u$. Then we get

$$u = \frac{2T}{(1+t^2/k)}, \quad \text{and} \quad du = \frac{2dT}{(1+t^2/k)}. \quad (A.18)$$

Substituting (A.18) into (A.17) and noting that if $u \to 0$, then $T \to 0$ and $u \to \infty$, then $T \to \infty$, we have

$$f_k(t) = \frac{1}{\sqrt{\pi}2^{\frac{k+1}{2}}\Gamma(\frac{k}{2})\sqrt{k}} \int_0^\infty e^{-T} \left(\frac{2T}{1+t^2/k}\right)^{(k-1)/2} \left(\frac{2}{1+t^2/k}\right) dT$$

$$= \frac{1}{\sqrt{\pi}\Gamma(\frac{k}{2})\sqrt{k}(1+t^2/k)^{\frac{k+1}{2}}} \int_0^\infty e^{-T}(T)^{\frac{n+1}{2}-1} dT$$

$$= \frac{1}{\sqrt{\pi}\Gamma(\frac{k}{2})\sqrt{k}(1+t^2/k)^{\frac{k+1}{2}}} \Gamma\left(\frac{n+1}{2}\right)$$

$$= \frac{1}{\sqrt{k}\frac{\Gamma(\frac{1}{2})\Gamma(\frac{k}{2})}{\Gamma(\frac{n+1}{2})}(1+t^2/k)^{\frac{k+1}{2}}} = \frac{1}{\sqrt{k}B(1/2.k/2)(1+t^2/k)^{\frac{k+1}{2}}}. $$

$$(A.19)$$

This completes the proof. ∎

Finally, we derive the pdf of F distribution which is stated as a theorem.

Theorem A.3 *If X follows an F distribution with the numerator $df_n = m$ and the denominator $df_d = n$, its pdf is given by*

$$f_X(x) = \frac{\Gamma\left(\frac{m+n}{2}\right) m^{m/2} n^{n/2} x^{(m/2)-1}}{\Gamma\left(\frac{m}{2}\right)\Gamma\left(\frac{n}{2}\right)(n+mx)^{(m+n)/2}}, \quad x > 0. \quad (A.20)$$

Proof. Recall that $X = (U/m)/(V/n)$ where U and V are two independent χ^2 distributed random variables with U having the degrees of freedom m and V

having the degrees of freedom n, respectively. For notational simplicity, let $p = m/2$ and $q = n/2$. To compute the pdf of X, we first get the joint pdf of U and V based on (A.15) as

$$f_{U,V}(u,v) = \left(\frac{1}{\Gamma(p)2^p} u^{p-1} e^{-u/2} \right) \left(\frac{1}{\Gamma(q)2^q} v^{q-1} e^{-v/2} \right)$$

$$= \frac{1}{\Gamma(p)\Gamma(q)2^{p+q}} u^{p-1} v^{q-1} e^{-u/2} e^{-v/2}, \quad u,v > 0.$$

(A.21)

Next, we find the pdf of the random variable U/V by starting from the cdf as follows:

$$F_{U/V}(y) = P(U/V \le y) = P(U \le yV).$$

On a $U - V$ plane, the region for this inequality can be shown in a Figure (you can make it). Integrating first with respect to u and then with respect to v, we have

$$P(U \le yV) = \frac{1}{\Gamma(p)\Gamma(q)2^{p+q}} \int_0^\infty \left(\int_0^{yv} u^{p-1} e^{-u/2} du \right) v^{q-1} e^{-v/2} dv. \quad \text{(A.22)}$$

Differentiating the integral in (A.22) with respect to y, we have

$$f_{U/V}(y) = F'_{U/V}(y) = \frac{1}{\Gamma(p)\Gamma(q)2^{p+q}} \int_0^\infty [(vy)^{p-1} e^{-vy/2} v] v^{q-1} e^{-v/2} dv$$

$$= \frac{y^{p-1}}{\Gamma(p)\Gamma(q)2^{p+q}} \int_0^\infty v^{p+q-1} e^{-[(y+1)/2]v} dv.$$

(A.23)

To evaluate the integral above, we notice that the integrand is the main components in the pdf of a Gamma random variable with parameters $\alpha = p+q$ and $\beta = ((y+1)/2)^{-1}$, denoted by $f(v)$. Then, it follows from

$$\int_0^\infty f(v)dv = \int_0^\infty \frac{\left(\frac{y+1}{2} \right)^{p+q}}{\Gamma(p+q)} v^{p+q-1} e^{-[(y+1)/2]v} dv = 1,$$

that

$$\int_0^\infty v^{p+q-1} e^{-[(y+1)/2]v} dv = \frac{\Gamma(p+q)}{\left(\frac{y+1}{2} \right)^{p+q}}.$$

Substituting this result into (A.23) yields

$$f_{U/V}(y) = \frac{y^{p-1}}{\Gamma(p)\Gamma(q)2^{p+q}} \cdot \frac{\Gamma(p+q)}{\left(\frac{y+1}{2} \right)^{p+q}}$$

$$= \frac{\Gamma(p+q)}{\Gamma(p)\Gamma(q)} \cdot \frac{y^{p-1}}{(y+1)^{p+q}}.$$

(A.24)

Now consider $X = (U/m)/(V/n) = (n/m)(U/V) = (n/m)Y$. Substituting $y = (m/n)x$ into (A.24) we can get the pdf for X as follows: $f_X(x) = Cf_{U/V}\left(\frac{m}{n}x\right)$ where C is a constant that can be determined by $\int_0^\infty f_X(x) = 1$. That is

$$\int_0^\infty f_X(x)dx = \int_0^\infty Cf_{U/V}\left(\frac{m}{n}x\right)dx = C\left(\frac{n}{m}\right)\int_0^\infty f_{U/V}\left(\frac{m}{n}x\right)d\left(\frac{m}{n}x\right)$$

$$= C\left(\frac{n}{m}\right) = 1,$$

which yields $C = m/n$. Thus, we obtain

$$f_X(x) = \frac{m}{n}f_{U/V}\left(\frac{m}{n}x\right) = \frac{m}{n}\cdot\frac{\Gamma\left(\frac{m+n}{2}\right)}{\Gamma\left(\frac{m}{2}\right)\Gamma\left(\frac{n}{2}\right)}\cdot\frac{\left(\frac{m}{n}x\right)^{(m/2)-1}}{\left(\frac{m}{n}x+1\right)^{(m+n)/2}}$$

$$= \frac{\Gamma\left(\frac{m+n}{2}\right)m^{m/2}n^{n/2}x^{(m/2)-2}}{\Gamma\left(\frac{m}{2}\right)\Gamma\left(\frac{n}{2}\right)(n+mx)^{(m+n)/2}}.$$

This completes the proof. ∎

Clearly, χ^2, F, and t distributions are developed based on the CLT and random behaviors of unbiased estimators $\bar{X}$ and S^2. In particular, an unbiased estimator is a sample statistics (a random variable) that has the mean value equal to the population mean to be estimated. For example, S^2 has the property that $E[S^2] = \sigma^2$. We show this property as follows:

$$S^2 = \frac{1}{n-1}\left[\sum_{i=1}^n X_i^2 - n\bar{X}^2\right] = \frac{1}{n-1}\left[\sum_{i=1}^n X_i^2 - n\left(\frac{\sum_{i=1}^n X_i}{n}\right)^2\right]$$

$$= \frac{1}{n-1}\sum_{i=1}^n X_i^2 - \frac{1}{n(n-1)}\left(\sum_{i=1}^n X_i\right)^2.$$

Taking the expectation of the equation above, we obtain

$$E[S^2] = E\left(\frac{1}{n-1}\sum_{i=1}^n X_i^2 - \frac{1}{n(n-1)}\left(\sum_{i=1}^n X_i\right)^2\right)$$

$$= \frac{1}{n-1}\sum_{i=1}^n E[X_i^2] - \frac{1}{n(n-1)}E[(\sum_{i=1}^n X_i)^2]$$

$$= \frac{1}{n-1}\sum_{i=1}^n (Var(X_i) + (E[X_i])^2) - \frac{1}{n(n-1)}(Var(\sum_{i=1}^n X_i) + (E[\sum_{i=1}^n X_i])^2)$$

$$= \frac{1}{n-1}\sum_{i=1}^n [\sigma^2 + \mu^2] - \frac{1}{n(n-1)}(Var(\sum_{i=1}^n X_i) + (\sum_{i=1}^n E[X_i])^2)$$

$$= \frac{n}{n-1}(\sigma^2 + \mu^2) - \frac{1}{n(n-1)}[\sum_{i=1}^n Var(X_i) + (n\mu)^2]$$

$$= \frac{n}{n-1}(\sigma^2 + \mu^2) - \frac{1}{n(n-1)}[n\sigma^2 + (n\mu)^2]$$

$$= \frac{n}{n-1}(\sigma^2 + \mu^2) - \frac{1}{n-1}[\sigma^2 + n\mu^2]$$

$$= \frac{n\sigma^2 + n\mu^2 - \sigma^2 - n\mu^2}{n-1} = \sigma^2.$$

Here the property that X_i's are i.i.d. random variables is utilized.

A.7.5   SOME COMMENTS ON DEGREES OF FREEDOM

Another concept that can be explained intuitively is the "degrees of freedom". As the name indicates, it represents the number of random variables (e.g., some of X_i's) that can freely change without affecting some sample statistic values calculated based on the sample (e.g., $\bar{X}$). We use a simple example to explain why the degrees of freedom for computing the sample variance is $n-1$, which also validates the unbiased property S^2 shown above. Consider a small sample with only three values $n = 3$, say $\mathbf{x} = (x_1, x_2, x_3) = (2, 5, 8)$ (e.g., the number of cars sold per day of a car dealer for a period of 3 days). With this sample, we can compute the sample average $\bar{x} = (2 + 5 + 8)/3 = 5$. The concept of degrees of freedom comes in if we want to compute the sample variance s^2 based on this sample that measures the variability or uncertainty of the random variable X. Note that the sample variance can be computed only after sample mean $\bar{x}$ (another sample statistic) is computed. This is because s^2 is based on the squared difference between each observed value x_i and $\bar{x}$. Specifically, $\bar{x}$ is the base for computing the deviation from the center for each observed value and summing these squared deviations up yields a quantity that measures the variability of the sample (an estimate of the variability for the random variable or the population). Now we should ask how many sources of uncertainty contribute to this type of variability (mean-based measure). It turns out to be 2. To answer this question, let's ask another relevant question: "how many sampled values can be changed freely to get the same $\bar{x} = 5$?" The answer should be 2. This is because $x_1 + x_2 + x_3 = 3 \times \bar{x} = 3 \times 5 = 15$ must be satisfied to get $\bar{x} = 5$ so that only two of x_1, x_2 and x_3 can be changed freely and third one must be a determined value to ensure that the sum is 15. For example, if x_1 is changed from 2 to 1 and x_2 is changed from 5 to 10 (indicating a sample with more variability), then x_3 must be fixed at 15–2–10 = 3. The implication of this observation is that in computing the cumulative squared difference $\sum_{i=1}^{3}(x_i - \bar{x})^2$, only the changes in two of x_1, x_2, and x_3 make contribution to the uncertainty measure. This is because if x_1 and x_2 are varied freely and x_3 must be determined by the variations of the x_1 and x_2 for a fixed center $\bar{x}$. In this example, the three squared differences are $(1-5)^2 = 16, (10-5)^2 = 25$, and $(3-4)^2 = 1$. Note that 16 and 25 are the variations made by x_1 and x_2 changes, respectively.

But $(x_3 - \bar{x})^2 = 1$ is the variation caused by the joint changes of x_1 and x_2 (not x_3 itself). This means that the cumulative squared deviation $16 + 25 + 1 = 42$ is the total variability measure due to two sources of uncertainty. Thus to average the squared differences, the sum should be divided by $n - 1 = 3 - 1 = 2$. This explains the degrees of freedom intuitively. In general, if we have computed k sample statistics based on a sample of size n, then the degrees of freedom for computing the $(k+1)$st sample statistics, which are based on the k computed sample statistics, should be $n - k$. Such a rule applies to major statistic inference techniques including hypothesis tests, ANOVA, and regression models. For example, in the ANOVA for multiple regression models, the mean square due to regression (MSR) and mean square due to error (MSE) are two estimators for the variance based on the sum of squares due to regression (SSR) and the sum of squares due to error (SSE). Here SSR+SSE=SST where SST is the sum of squares total. Note that $\text{SST}=\sum_{i=1}^{n}(y_i - \bar{y})^2$, $\text{SSR}=\sum_{i=1}^{n}(\hat{y}_i - \bar{y})^2$, and $\text{SSE}=\sum_{i=1}^{n}(y_i - \hat{y}_i)^2$. Clearly, the degrees of freedom for SST is $n - 1$ since only one parameter $\bar{y}$ is computed and the degrees of freedom for computing MSE is $n - (p+1)$, where p is the number of independent variables in the regression model. Note that $p + 1$ is the number of parameters (coefficients of the regression models) computed from the sample. i.e. $b_1, b_2, ..., b_p$ for p predicting variables and b_0 for intercept. Since the total degrees of freedom is $n - 1$, then the degrees of freedom for MSR is surely p.

A.8 LIMITS OF SETS

The random variable convergence relies on the concept of sequence of events, which requires the definition of limits of sets. We first define the infimum and supremum of a sequence of events. Let $A_n \subset \Omega$. Then,

$$\inf_{k \geq n} A_k := \bigcap_{k=n}^{\infty} A_k, \quad \sup_{k \geq n} A_k := \bigcup_{k=n}^{\infty} A_k.$$

Then we can define the lim inf and lim sup as

$$\liminf_{n \to \infty} A_n = \liminf_{k \geq n} A_n := \bigcup_{n=1}^{\infty}\bigcap_{k=n}^{\infty} A_k, \quad \limsup_{n \to \infty} A_n = \limsup_{k \geq n} A_n := \bigcap_{n=1}^{\infty}\bigcup_{k=n}^{\infty} A_k.$$

$\liminf_{n \to \infty} A_n$ can be considered as the largest infimum of the sequence of the sets and $\limsup_{n \to \infty} A_n$ can be considered as the smallest of the supremum of the sequence of the sets. The limit of a sequence of sets is defined as follows. If for some sequence of subsets $\{A_n\}$

$$\limsup_{n \to \infty} A_n = \liminf_{n \to \infty} A_n = A,$$

then A is called the limit of A_n, i.e., $\lim_{n\to\infty} A_n = A$. It can be shown that

$$\liminf_{n\to\infty} A_n = \lim_{n\to\infty}\left(\inf_{k\geq n} A_k\right), \quad \limsup_{n\to\infty} A_n = \lim_{n\to\infty}\left(\sup_{k\geq n} A_k\right).$$

To get some intuitive feelings about the limit of a sequence of sets, we first explain it using the definition of union and intersection of sets and then show it by using a Bernoulli sequence example. A member of $\cup_{n=1}^{\infty} \cap_{k\geq n}^{\infty} A_k$ is a member of at least one of the sets $D_n = \inf_{k\geq n} A_k = \cap_{k=n}^{\infty} A_k$ for $n \geq 1$, meaning it is member of either $D_1 = A_1 \cap A_2 \cap A_3 \cap \cdots$, or $D_2 = A_2 \cap A_3 \cap A_4 \cap \cdots$ or $D_3 = A_3 \cap A_4 \cap A_5 \cap \cdots$, ..., etc. This implies that it is a member of all except finitely many of the A. On the other hand, a member of $\cap_{n=1}^{\infty} \cup_{k\geq n}^{\infty} A_k$ is a member of all of the sets $E_n = \sup_{k\geq n} A_k = \cup_{k=n}^{\infty} A_k$, meaning it is a member of $E_1 = A_1 \cup A_2 \cup A_3 \cup \cdots$ and of $E_2 = A_2 \cup A_3 \cup A_4 \cup \cdots$ and of $E_3 = A_3 \cup A_4 \cup A_5 \cup \cdots$ and of ... etc. This implies that no matter how far down the sequence you go, it is a member of at least one of the sets that come later. Equivalently, it means that it is a member of infinitely many of them, but there might also be infinitely many that it does not belong to. Clearly, D_n is an increasing sequence of sets as $D_1 \subseteq D_2 \subseteq D_3 \subseteq D_3 \subseteq \cdots$ which exhaust D. Here, each D_i is a subset of D and $D = \cup_{n=1}^{\infty} D_n$. Similarly, E_n is a decreasing sequence of sets as $E_1 \supseteq E_2 \supseteq E_3 \supseteq \cdots$ and $E = \cap_{n=1}^{\infty} E_n$. Now for an arbitrary sequence of sets $\{A_1, A_2, ...\}$, we can squeeze it between an increasing sequence $\{D_n\}$ and a decreasing sequence $\{E_n\}$. This is

$$
\begin{array}{ccccccc}
D_1 = & A_1 \cap A_2 \cap A_3 \cap \cdots & \subseteq & A_1 & \subseteq & E_1 = & A_1 \cup A_2 \cup A_3 \cup \cdots \\
D_2 = & A_2 \cap A_3 \cap \cdots & \subseteq & A_2 & \subseteq & E_2 = & A_2 \cup A_3 \cup \cdots \\
D_3 = & A_3 \cap \cdots & \subseteq & A_3 & \subseteq & E_3 = & A_3 \cup \cdots \\
& & & \vdots & & &
\end{array}
$$

Note that $\{D_n\}$ is the largest increasing sequence that $D_n \subseteq A_n$ for all n and $\{E_n\}$ is the smallest decreasing sequence such that $A_n \subseteq E_n$ for all n. Thus, we can define

$$\liminf_{n\to\infty} A_n = \lim_{n\to\infty} D_n, \quad \limsup_{n\to\infty} A_n = \lim_{n\to\infty} E_n.$$

If these lower and upper limits are the same, $\lim_{n\to\infty} A_n$ exists and equals this common value.

Suppose George tosses two dice and win \$100 if getting a double six. He keeps playing the game without ending. Define a double six as a "success", denoted by 1, and a not double six as a "failure", denoted by 0. Then, the sample space of each play is $S = \{0, 1\}$ and the sample space of playing this game indefinitely is

$$\Omega = S^{\infty} = \{(s_1, s_2, ...); s_i \in S := \{0, 1\}\}.$$

Define $A_n = \{\text{George wins the game on the } n\text{th play}\}$. That is

$$A_n = S \times S \times \cdots \times S \times \{1\} \times S \times \cdots$$

where $\{1\}$ occurs in the nth factor. Now we can use the set operations to represent some events of interest as follows:

(1) Given $n \in \mathbb{N}$, what is the event that George wins at least once in the first n plays? This event is

$$\bigcup_{k=1}^{n} A_k.$$

(2) Given $n \in \mathbb{N}$, what is the event that George does not win any of the first n plays? This event is

$$\left(\bigcup_{k=1}^{n} A_k \right)^c = \{0\} \times \{0\} \times \cdots \times \{0\} \times S \times S \times \cdots,$$

where $\{0\}$'s occur in the first n factors.

(3) What is the event that George eventually wins a play? This event is

$$\bigcup_{k=1}^{\infty} A_k.$$

(4) What is the event that George wins the play infinitely many times? This event is

$$\{A_n ; \text{i.o.}\} = \bigcap_{k=1}^{\infty} \bigcup_{n=k}^{\infty} A_n = \limsup_{n \to \infty} A_n.$$

(5) What is the event that after a finite number of losses, George wins the game all the times? This event is

$$\{A_n ; \text{a.a.}\} = \bigcup_{k=1}^{\infty} \bigcap_{n=k}^{\infty} A_n = \liminf_{n \to \infty} A_n.$$

where "a.a." means almost always or almost surely ("a.s."). We can define the limit for a monotonic sequence of sets:

Definition A.7 *Let $\{A_i\}_{i=1}^{\infty}$ be a collection of subsets in a set $\mathbb{X}$. If $A_1 \subset A_2 \subset A_3 \ldots$ and $\bigcup_{i=1}^{\infty} A_i = A$, then $\{A_i\}_{i=1}^{\infty}$ is a monotonically increasing sequence of sets and A_i converges to A, denoted by $A_i \nearrow A$.*

If $A_1 \supset A_2 \supset A_3 \ldots$ and $\bigcap_{i=1}^{\infty} A_i = A$, then $\{A_i\}_{i=1}^{\infty}$ is a monotonically decreasing sequence of sets and A_i converges to A, denoted by $A_i \searrow A$.

A.9 BOREL-CANTELLI LEMMAS

We present an important tool frequently used in connection with questions concerning almost surely convergence in the stochastic process limits.

Lemma A.1 *(The first Borel–Cantelli lemma) Let $\{A_n, n \geq 1\}$ be a sequence of arbitrary events, then*

$$\sum_{n=1}^{\infty} P(A_n) < \infty \Rightarrow P(\{A_n \ i.o.\}) = 0,$$

where i.o. represents "infinitely-often".

Proof. The claim follows by noting

$$P(\{A_n \ i.o.\}) = P(\limsup_{n \to \infty} A_n) = P\left(\bigcap_{n=1}^{\infty} \bigcup_{m=n}^{\infty} A_m \right)$$

$$\leq P\left(\bigcup_{m=n}^{\infty} A_m \right) \leq \sum_{m=n}^{\infty} P(A_m) \to 0 \text{ as } n \to \infty.$$

The last limit term follows from writing $\sum_{n=1}^{\infty} P(A_n) = \sum_{n=1}^{m-1} P(A_n) + \sum_{n=m}^{\infty} P(A_n)$, taking the limit of $m \to \infty$ on the RHS, and using the condition $\sum_{n=1}^{\infty} P(A_n) < \infty$. ∎

The converse does not hold in general unless an additional assumption of independence is made.

Lemma A.2 *(The second Borel–Cantelli lemma) Let $\{A_n, n \geq 1\}$ be a sequence of independent events, then*

$$\sum_{n=1}^{\infty} P(A_n) = \infty \Rightarrow P(\{A_n \ i.o.\}) = 1.$$

Proof. It follows from the given condition expressed as $\sum_{n=1}^{\infty} P(A_n) = \sum_{n=1}^{k-1} P(A_n) + \sum_{n=k}^{\infty} P(A_n) = \infty$ that for any $k \geq 1$

$$\sum_{n=k}^{\infty} P(A_n) = \infty, \qquad (A.25)$$

which will be used next. Denote by A_n^c the complement of the set A_n. Then,

$$P(\{A_n \ i.o.\}) = P\left(\bigcap_{n=1}^{\infty} \bigcup_{m=n}^{\infty} A_m \right) = 1 - P\left(\bigcup_{n=1}^{\infty} \bigcap_{m=n}^{\infty} A_m^c \right)$$

$$= 1 - \lim_{n \to \infty} P\left(\bigcap_{m=n}^{\infty} A_m^c \right) = 1 - \lim_{n \to \infty} \prod_{m=n}^{\infty} P(A_m^c).$$

To complete the proof, we need to show $P\left(\bigcap_{m=n}^{\infty} A_m^c\right) = 0$. Using the independence and the inequality $1 - x \leq e^x$ with $x \geq 0$, we have

$$P\left(\bigcap_{m=n}^{\infty} A_m^c\right) = \prod_{m=n}^{\infty} P(A_m^c) = \prod_{m=n}^{\infty} (1 - P(A_m))$$

$$\leq \prod_{m=n}^{\infty} e^{-P(A_m)}$$

$$= e^{-\sum_{m=n}^{\infty} P(A_m)} = e^{-\infty} = 0,$$

where the last equality follows from (A.25). This completes the proof. $\blacksquare$

By combining the two lemmas, we note, in particular, that if the events $\{A_n, n \geq 1\}$ are independent, then $P(\{A_n \ i.o.\})$ can only assume the values of 0 or 1, depending on the convergence or divergence of $\sum_{n=1}^{\infty} P(A_n)$. This is summarized as a zero-one law.

Theorem A.4 *(A zero-one law) If the events $\{A_n, n \geq 1\}$ are independent, then*

$$P(\{A_n \ i.o.\}) = \begin{cases} 0, & when \ \sum_{n=1}^{\infty} P(A_n) < \infty, \\ 1, & when \ \sum_{n=1}^{\infty} P(A_n) = \infty. \end{cases}$$

The implication of this theorem is that it suffices to prove that $P(\{A_n \ i.o.\}) > 0$ in order to conclude that the probability equals 1 (and that $P\{A_n \ i.o.\} < 1$ in order to conclude that it equals 0).

A.10 A FUNDAMENTAL PROBABILITY MODEL

Consider an experiment of tossing a coin infinite times, called a Bernoulli experiment. Then an outcome of this experiment will be an infinite sequence of outcomes of Bernoulli trials (H for heads or T for tails) which is called a Bernoulli sequence. We can show that the probability assigned to a set of these outcomes is the Lebesure measure from a unit interval. This example demonstrates the entire process of developing a probability model. First, we collect all possible outcomes of this experiment to form a sample space.

Definition A.8 *The sample space of Bernoulli experiment is the collection of all possible Bernoulli sequences, i.e., $\mathbb{B} = \{all \ Bernoulli \ sequences : a_1 a_2 ... a_k ...\}$ where $a_i = H$ or T.*

For a fair coin case, the chances for H and T are equal. We first show that $\mathbb{B}$ can almost surely be represented by real numbers in $\Omega = (0, 1]$ which implies that $\mathbb{B}$ is uncountable. Ω is called the sample space of the probability model.

Theorem A.5 *Except for a countable subset of* $\mathbb{B}$*, the elements in* $\mathbb{B}$ *can be indexed by the real numbers in* $\Omega = (0, 1]$.

Proof. First, we show that a mapping from Ω to $\mathbb{B}$ is not an onto for a countable subset. Any point $\omega \in \Omega$ can be expressed as a fractional binary expansion,

$$\omega = \sum_{i=1}^{\infty} \frac{a_i}{2^i}, \quad a_i = 0, 1.$$

Each such expansion corresponds to a Bernoulli sequence. To see this, define the ith term of the Bernoulli sequence to be H if $a_i = 1$ and T otherwise. Since some numbers in Ω do not have a unique binary expansion, we cannot define a function from Ω to $\mathbb{B}$. For example, a single point $1/2 = 0.1000\cdots =$ $0.0111\cdots$ in Ω corresponds to two different outcomes $HTTT\cdots \neq THHH\cdots$ in $\mathbb{B}$. To overcome this problem, we adopt the convention that if the real number ω has terminating and non-terminating binary expansions, we utilize the non-terminating one. With this convention, we define a one-one map from Ω to $\mathbb{B}$ that is not onto as it does not produce Bernoulli sequences ending in all T's. Thus, we need to ignore a countable set defined as follows: Let $\mathbb{B}_k^T$ be the finite set of Bernoulli sequences that have only T's after the kth term. Then, we have a countable set as

$$\mathbb{B}^T = \bigcup_{k=1}^{\infty} \mathbb{B}_k^T.$$

Thus, there is a one-one and onto correspondence between Ω and $\mathbb{B} \setminus \mathbb{B}^T$. To use Ω as a model for Bernoulli sequences, we have to ignore $\mathbb{B}^T$. ∎

Now we assign the probability to an event in $\mathbb{B}$ by using the measure of the corresponding subset $I_A \subset \Omega$. Let E_H (E_T) be the event in $\mathbb{B}$ consisting of sequences starting with H (T). The corresponding set in Ω is

$$I_{E_H} = \{\omega \in \Omega; \omega = 0.1a_1a_2... : a_i = 0 \text{ or } 1\} = (0.5, 1].$$

Note that 0.5 is not included as it is a terminating expansion. Similarly, we have $I_{E_T} = (0, 0.5]$. Using the Lebesgue measure for these subsets in Ω, we obtain $P(E_H) = 0.5$ and $P(E_T) = 0.5$ if the "measure" of subset in Ω corresponds to the probability of $E \subset \mathbb{B}$. To make this approach work, we need to assign 0 probability to the countable set $\mathbb{B}_T$. In fact, it is required that any finite or countable subset of Ω needs to be assigned a measure of 0. A measure should have the following properties.

Definition A.9 *A measure* μ *is a real-valued function defined on a collection of subsets of a space* $\mathbb{X}$ *called the measurable sets. If A is a measurable set,* $\mu(A)$ *is the measure of A.* μ *must have the following properties:*

- μ *is non-negative.*

- If $\{A_i\}_{i=1}^m$ is a finite collection of disjoint measurable sets, then $\bigcup_{i=1}^m A_i$ is measurable.
- If $\{A_i\}_{i=1}^m$ is a collection of disjoint measurable sets, then,

$$\mu\left(\bigcup_{i=1}^m A_i\right) = \sum_{i=1}^m \mu(A_i).$$

Lebesgue measure is an example. Suppose that $\mathbb{X}$ is a real number interval and the measurable sets include intervals for which

$$\mu((a,b)) = \mu([a,b]) = \mu((a,b]) = \mu([a,b)) = b - a, \ a,b \in \mathbb{X},$$

we call μ the Lebesgue measure on $\mathbb{X}$ and write $\mu = \mu_{\mathscr{L}}$.

We can assign probabilities to more events in the Bernoulli sequence example. For example, define the events E_{HH}, E_{HT}, E_{TH}, and E_{TT} in $\mathbb{B}$ as the sequences starting with the first two outcomes as HH, HT, TH, and TT, respectively. The corresponding intervals in Ω are

$$I_{E_{HH}} = (0.0.25], I_{E_{TH}} = (0.25, 0.5], I_{E_{HT}} = (0.5, 0.75], I_{E_{TT}} = (0.75, 0.25].$$

Thus, we assign the probability of 0.25 to each of these events. We can continue this argument by considering the events starting with specific the first three outcomes, then the first four outcomes, and so on. Based on the sequences starting with the first m outcomes specified, we obtain 2^m disjoint intervals of equal length in Ω which correspond to the events in $\mathbb{B}$ starting with the first m outcomes specified. Then, we can assign equal probability 2^{-m} to each interval and thus each event. In this way, we can have a sequence of binary partitions, denoted by $\mathscr{T}_m$, for Ω which divide $(0, 1]$ into 2^m disjoint subintervals, denoted by $I_{m,j}$ of equal length. That is

$$\Omega = \bigcup_{j=1}^{2^m} I_{m,j}, \ I_{m,j} = ((j-1)\frac{1}{2^m}, j\frac{1}{2^m}], \ j = 1, \cdots, 2^m.$$

Using the properties of a measure function, we can compute the probabilities of some interesting events.

Example A.6 *Consider the following two events: (i) A is the collection of Bernoulli sequences in which the mth outcome is H; and (ii) B is the collection of Bernoulli sequences in which there are exactly i H's in the first mth outcomes ($i \le m$).*

For event A, we have

$$I_A = \{\omega \in \Omega; \omega = 0.a_1a_2...a_{m-1}1a_{m+1}... : a_i = 0 \ or \ 1 \ for \ i \ne m\}.$$

Let $s = 0.a_1 a_2 ... a_{m-1} 1$, I_A *contains* $(s, s + 2^{-m}]$. *We can choose* $a_1, a_2, ..., a_{m-1}$ *in* 2^{m-1} *different ways and the resulting intervals are disjoint from the others, thus, using the finite additivity, we have*

$$P(A) = \mu_{\mathscr{L}}(I_A) = 2^{m-1} \cdot \frac{1}{2^m} = \frac{1}{2}.$$

For event B, we have

$$I_B = \{\omega \in \Omega; \omega = 0.a_1 a_2 ... a_m a_{m+1} ...$$

$$: \text{exactly } i \text{ of the first } m \text{ digits are } 1 \text{ and remaining are } 0 \text{ or } 1\}.$$

Choose $a_1, a_2, ..., a_m$ *so that exactly* i *digits are* 1 *and set* $s = 0.a_1 a_2 ... a_m$. *Then,* I_B *contains* $(s, s + 2^{-m}]$. *The intervals corresponding to different combinations of* $a_1, ..., a_m$ *are disjoint. Hence, we have*

$$P(B) = \mu_{\mathscr{L}}(I_B) = \binom{m}{i} \cdot \frac{1}{2^m}.$$

We can use the Bernoulli sequence to prove the Law of Large Numbers. This proof can demonstrate several fundamental concepts in developing probability theory. We now show the Weak Law of Large Numbers (WLLN). Consider the Bernoulli sequences by tossing a fair coin and let S_m be the number of H's that occur in the first m tosses. Then, for $\omega \in \Omega$, we define the random variable

$$S_m(\omega) = a_1 + \cdots + a_m, \quad \omega = 0.a_1 a_2 \cdots a_m \cdots,$$

which gives the number of heads in the first m outcomes of the Bernoulli sequence. Given $\delta > 0$, we define an event

$$I_{\delta,m} = \left\{ \omega \in \Omega : \left| \frac{S_m(\omega)}{m} - \frac{1}{2} \right| > \delta \right\}, \tag{A.26}$$

which consists of outcomes for which there are not approximately the same number of H and T after m trials. Now we can establish

Theorem A.6 *(WLLN) For a given* $\delta > 0$,

$$\mu_{\mathscr{L}}(I_{\delta,m}) \to 0, \quad as \ m \to \infty. \tag{A.27}$$

To make the Bernoulli sequence more interesting, we introduce the Rademacher function $R_i(\omega) = 2a_i - 1$, where $\omega = 0.a_1 a_2 \cdots$. Suppose we bet on a sequence of coin tosses such that at each toss, we win \$1 if it is heads and lose \$1 if it is tails. Then $R_i(\omega)$ is the amount won or lost at the ith toss in a sequence of tosses represented by ω. The total amount won or lost after the mth toss in the betting game is given by

$$W_m(\omega) = \sum_{i=1}^{m} R_i(\omega) = 2S_m(\omega) - m, \quad \omega = .a_1 a_2 a_3 \cdots.$$

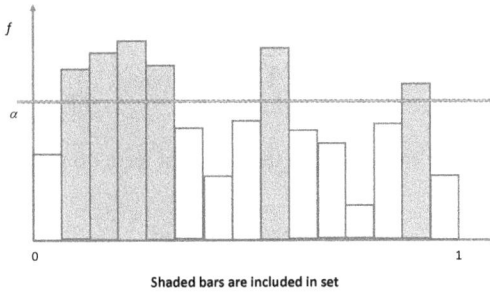

Shaded bars are included in set

Figure A.4 A typical set in Chebyshev's inequality.

We have the equivalent events as follows:

$$\left|\frac{S_m(\omega)}{m} - \frac{1}{2}\right| > \delta \Leftrightarrow |2S_m(\omega) - m| > 2m\delta \Leftrightarrow |W_m(\omega)| > 2\delta m.$$

Since δ is arbitrary, the factor 2 is immaterial, we can define an event

$$D_m = \{\omega \in \Omega : |W_m(\omega)| > m\delta\}.$$

Now we show the WLLN by showing that $\mu_{\mathscr{L}}(D_m) \to 0$ as $m \to \infty$. This can be done by using the special version of Chebyshev's Inequality. Let f be a non-negative, piecewise constant function on Ω and $\alpha > 0$ be a real number. Then

$$\mu_{\mathscr{L}}(\{\omega \in \Omega : f(\omega) > \alpha\}) < \frac{1}{\alpha} \int_0^1 f(\omega)d\omega.$$

This result can be shown in Figure A.4. Since f is piece constant, there is a mesh $0 = \omega_1 < \omega_2 < \cdots < \omega_m = 1$ such that $f(\omega) = c_i$ for $\omega_i < \omega < \omega_{i+1}$ and $1 \leq i \leq m-1$. Then,

$$\int_0^1 f(\omega)d\omega = \sum_{i=1}^m c_i(\omega_{i+1} - \omega_i) \geq \sum_{i=1, c_i > \alpha}^m c_i(\omega_{i+1} - \omega_i)$$

$$> \alpha \sum_{i=1, c_i > \alpha}^m (\omega_{i+1} - \omega_i) = \alpha\mu_{\mathscr{L}}(\{\omega \in I : f(\omega) > \alpha\}).$$

We can re-write the event D_m as

$$D_m = \{\omega \in I : W_m^2(\omega) > m^2\delta^2\},$$

where $W_m^2(\omega)$ is non-negative and piecewise constant. Using the inequality above, we have

$$\mu_{\mathscr{L}}(D_m) < \frac{1}{m^2\delta^2} \int_0^1 W_m^2(\omega)d\omega.$$

We need to compute the RHS of the inequality above as follows:

$$\int_0^1 W_m^2(\omega)d\omega = \int_0^1 \left(\sum_{i=1}^m R_i(\omega)\right)^2 d\omega = \sum_{i=1}^m \int_0^1 R_i^2(\omega)d\omega + \sum_{i,j=1,i\neq j}^m \int_0^1 R_i(\omega)R_j(\omega)d\omega.$$

It follows from $R_i^2(\omega) = 1$ for all ω that the first term on the right is

$$\sum_{i=1}^m \int_0^1 R_i^2(\omega)d\omega = m.$$

To compute the second term on the right, we evaluate $\int_0^1 R_i(\omega)R_j(\omega)d\omega$ when $i \neq j$. Without loss of generality, we assume $i < j$. Set J to be the interval (for a mesh to compute the integral),

$$J = \left(\frac{l}{2^i}, \frac{l+1}{2^i}\right], \quad 0 \leq l \leq 2^i.$$

Due to $j > i$, we notice that R_i is constant on J while R_j oscillates between values -1 and 1 over $2^{(j-i)}$ sub-intervals of interval J. Since this is an even number of sub-intervals with half $+1$ and half -1, cancellation implies

$$\int_J R_i(\omega)R_j(\omega)d\omega = R_i(\omega)\int_J R_j(\omega)d\omega = 0.$$

Thus, $\int_0^1 R_i(\omega)R_j(\omega)d\omega = 0$, which leads to the second term on the right to be 0. We have $\int_0^1 W_m^2(\omega)d\omega = m$ and

$$\mu_{\mathscr{L}}(I_{\delta,m}) < \frac{1}{m^2\delta^2}m = \frac{1}{m\delta^2} \Rightarrow \mu_{\mathscr{L}}(I_{\delta,m}) \to 0 \text{ as } m \to \infty,$$

which proves the WLLN, a fundamental law in probability theory.

A.11   SETS OF MEASURE ZERO

To characterize sets with Lebesgue measure zero, we need some fundamental concepts for metric spaces.

Definition A.10   *Given a subset $A \subset \mathbb{R}^n$, a countable cover of A is a countable collection of sets $\{A_i\}_{i=1}^\infty$ in $\mathbb{R}^n$ such that $A \subset \bigcup_{i=1}^\infty A_i$. If the sets in a countable cover are open, it is called an open cover.*

A set with Lebesgue measure zero can be defined in terms of the property of its countable cover. The following definition is for the metric space $\mathbb{R}$.

Definition A.11 *A set $A \subset \mathbb{R}$ has **Lebesgue measure zero** if for every $\varepsilon > 0$, there is a countable cover $\{A_i\}_{i=1}^{\infty}$ of A, where A_i consists of a finite union of open intervals such that*

$$\sum_{i=1}^{\infty} \mu_{\mathscr{L}}(A_i) < \varepsilon.$$

Then, A has measure zero.

An example for a set with measure zero is for $\mathbb{N}$. Given $\varepsilon > 0$, the set $\mathbb{N}$ has an open cover:

$$\mathbb{N} \subset \bigcup_{i=0}^{\infty} \left(i - \frac{\varepsilon}{2^{i+1}}, i + \frac{\varepsilon}{2^{i+1}} \right).$$

We compute

$$\sum_{i=0}^{\infty} \mu_{\mathscr{L}} \left(\left(i - \frac{\varepsilon}{2^{i+1}}, i + \frac{\varepsilon}{2^{i+1}} \right) \right) = \sum_{i=0}^{\infty} \frac{\varepsilon}{2^i} = \frac{1}{2} \varepsilon < \varepsilon.$$

To show a typical measure theory argument, we prove the following theorem,

Theorem A.7 *The following statements hold.*

> *1. A measurable subset of a set of measure zero has measure zero.*
> *2. If $\{A_i\}_{i=1}^{\infty}$ is a countable collection of sets of measure zero, then $\bigcup_{i=1}^{\infty} A_i$ has measure zero.*
> *3. Any finite or countable set of numbers has measure zero.*

Proof. (1) It follows from the fact that any countable cover of larger set is also a cover of the smaller set. (2) For a given ε, since A_i has measure zero, there exists an open cover $\{B_{i,1}, B_{i,2}, ...\}$ for A_i with

$$\sum_{j=1}^{\infty} \mu_{\mathscr{L}}(B_{i,j}) \leq \frac{\varepsilon}{2^i}.$$

(2) Since the countable union of countable sets is countable, the collection $\{B_{i,j}\}_{i,j=1}^{\infty}$ is countable and covers $\bigcup_{i=1}^{\infty} A_i$. Furthermore,

$$\sum_{i,j=1}^{\infty} \mu_{\mathscr{L}}(B_{i,j}) = \sum_{i=1}^{\infty} \left(\sum_{j=1}^{\infty} \mu_{\mathscr{L}}(B_{i,j}) \right) \leq \sum_{i=1}^{\infty} \frac{\varepsilon}{2^i} = \varepsilon.$$

(3) It follows from (2) and the fact that a point has measure zero. ∎

There are uncountable sets which have measure zero. A famous example is Cantor set.

A.12 CANTOR SET

The Cantor set is constructed by removing the open middle third of each subinterval of the unit interval repeatedly. Such a set possesses several important properties that are relevant in probability theory. Let $C_0 = [0, 1]$. The set C_{k+1} is obtained from C_k by removing the open middle third of each interval in C_k where $k = 0, 1, \ldots$. The Cantor set, denoted by C, is defined as $C = \lim_{k \to \infty} C_k$. The first two steps result in the following sets:

$$C_1 = \left[0, \frac{1}{3}\right] \cup \left[\frac{2}{3}, 1\right], \text{ by removing } \left(\frac{1}{3}, \frac{2}{3}\right),$$

$$C_2 = \left[0, \frac{1}{9}\right] \cup \left[\frac{2}{9}, \frac{3}{9}\right] \cup \left[\frac{6}{9}, \frac{7}{9}\right] \cup \left[\frac{8}{9}, 1\right],$$

by removing the open middle third in each subinterval of C_1.

This process continues indefinitely and can be represented by the following recursive relations:

$$C_k = C_{k-1} \setminus \bigcup_{j=0}^{2^j-1} \left(\frac{1+3j}{3^k}, 2 + 3j3^k\right),$$

or

$$C_k = \frac{C_{k-1}}{3} \cup \left(\frac{2}{3} + \frac{C_{k-1}}{3}\right).$$

The Cantor set has some very interesting properties summarized in the following theorem. Note that the definition of compact or dense set and related concepts can be found in Chapter G.

Theorem A.8 *The Cantor set, C, has the following properties:*

1. *C is closed and compact.*
2. *Every point in C is a limit point in C.*
3. *C has measure zero.*
4. *C is uncountable.*
5. *C is nowhere dense.*
6. *C is a fractal with $\dim(C) = \ln 2 / \ln 3$.*

Proof. (1) The claim is true if we can show the complement of C, denoted by C^c, is open. Note that

$$C^c = (1/3, 2/3) \cup (1/3^2, 2/3^2) \cup (7/3^2, 8/3^2) \cup (1/3^3, 2/3^3)$$
$$= \cup (7/3^3, 8/3^3) \cup (19/3^3, 20/3^3) \cup (25/3^3, 26/3^3) \cup \cdots$$

is open because it is a union of open intervals. Thus C is closed. Also since $C \subseteq [0, 1]$, C is bounded. Then by Heine–Borel theorem, C is compact.

(2) Note that the endpoints of all closed subintervals in $C_k, k \geq 0$, will be contained in the Cantor set. Take any $x \in C = \cap_{k=0}^{\infty} C_k$. Then x is C_k for all k. If x is in C_k, then x must be contained in one of the 2^k intervals that make up the set C_k (as an endpoint of that interval). Let $\delta = 3^{-k}$. Denote by y the same as x except for the $(k+1)^{th}$ digit of the ternary expansion (2 or 0). Then we have $y \in C$ and

$$|y - x| = \frac{2}{3^{k+1}} = \frac{2}{3} \cdot \frac{1}{3^k} = \frac{2}{3}\delta < \delta.$$

Thus, $y \in (x - \delta, x + \delta)$. This completes the proof of this claim. Furthermore, since C is closed, it is also said to be perfect.

(3) Since C_k is a union of disjoint intervals whose lengths sum to $(2/3)^k$, we have

$$\mu(C) = \lim_{k \to \infty} \mu(C_k) = \lim_{k \to \infty} \left(\frac{2}{3}\right)^k = 0.$$

(4) First we show that every point $x \in C$ can be represented by the ternary expansion (i.e., expansion in base 3). Note that an arbitrary number in $[0, 1]$ can be written as

$$x = \sum_{k=1}^{\infty} \frac{a_k}{3^k} = \frac{a_1}{3} + \frac{a_2}{3^2} + \cdots,$$

where a_i takes the values $0, 1$, and 2. For the points in C, we can write it as

$$0.a_1a_2a_3a_4a_5a_6a_7a_8 \cdots .$$

Numbers in $[0, 1]$ with a finite ternary expansion (i.e., such that $a_i = 0$ for $i \geq n$) have two different ternary expansions, a terminating one and a non-terminating one. We adopt the non-terminating one to make the representation unique. Since the points in C are ending points of the closed subintervals, they have a ternary expansion devoid of the number 1:

$$C = \{0.a_1a_2a_3a_4a_5a_6a_7a_8 \cdots \in [0, 1] : a_i = 0 \text{ or } 2\}.$$

Now we prove that C is uncountable. Suppose that C is countable, and we list its elements as $C = \{x_1, x_2, x_3, ...\}$. In terms of the ternary expansions, we can write

$$x_1 = 0.a_{11}a_{12}a_{13}...$$
$$x_2 = 0.a_{21}a_{22}a_{23}...$$
$$\vdots$$
$$x_k = 0.a_{k1}a_{k2}a_{k3}...$$
$$\vdots$$

where $a_{ij} = 0$ or 2 for all i, j. Let $y = 0.b_1b_2b_3...$, where

$$b_i = \begin{cases} 0, & \text{if } a_{ii} = 2 \\ 2, & \text{if } a_{ii} = 0 \end{cases} \tag{A.28}$$

Then $y \neq x_1$ since $b_1 \neq a_{11}$, $y \neq x_2$ since $b_2 \neq a_{22}$, and so on (this is the same diagonalizing proof used for showing $\mathbb{R}$ is uncountable). This implies that $y \notin C$, but this is a contradiction since $b_i \in 0, 2$ for each i, which means $y \in C$. Thus, C is uncountable.

(5) Let $(a, b) \subset [0, 1]$ be an arbitrary interval. This interval contains a number of C^c, say x. Since C^c is open, there is a $\delta > 0$ such that $(x - \delta, x + \delta) \subset C^c$, thus, $(x - \delta, x + \delta) \cap C = \emptyset$ which implies that Cantor set is nowhere dense.

(6) Suppose that we reduce the scale of an object by a factor of k (i.e., we shrink it to $1/k^{th}$ of its original size). Let N be the copies of the reduced object to cover the original one. Then we define a kind of dimension d such that the relation $N = k^d$ holds. When d is non-integer, the object is called a fractal. In the Cantor set case, $N = 2$ and $k = 3$, therefore $\dim(C) = \ln(N)/\ln(k) = \ln 2/\ln 3$. ∎

Knowing the definition of sets with measure zero, we can define the term "almost everywhere" or "almost surely".

Definition A.12 *A property of sets that holds except on a set of measure zero is said to hold "almost everywhere" (a.e.) or "almost surely" (a.s.).*

We present the Strong Law of Large Numbers (SLLN) for general Bernoulli sequences which shows the use of set with Lebesgue measure zero. First, we define the set of normal numbers in I as

$$\mathfrak{N} = \left\{ \omega \in \Omega : \frac{S_m(\omega)}{m} \to p \text{ as } m \to \infty \right\}.$$

Theorem A.9 *(SLLN)* $\mathfrak{N}^c$ *is an uncountable set with Lebesgue measure zero.*

Proof. To simplify the notations, we suppress ω and let $S_m(\omega) = k$. Based on the Borel–Cantelli lemma, we need to show

$$\sum_{m=1}^{\infty} P\left(\left| \frac{k}{m} - p \right| \geq \varepsilon_m \right) < \infty, \tag{A.29}$$

where $\{\varepsilon_m\}$ is a sequence of positive numbers converging to zero. If (A.29) is satisfied, the events $A_m := \left\{ \left| \frac{k}{m} - p \right| \geq \varepsilon_m \right\}$ can occur only for a finite number of indices m in an infinite sequence, or equivalently the events $A_m^c :=$

$\{|\frac{k}{m} - p|\} < \varepsilon$ occur infinitely often. That is the event k/m converges to p almost surely. Now, we prove (A.29). Since

$$\left|\frac{k}{m} - p\right| \geq \varepsilon \Rightarrow |k - mp|^4 \geq \varepsilon^4 m^4.$$

Denote the Binomial distribution by $P_m(k) = \frac{m}{k} p^k q^{m-k}$ where $q = 1 - p$. Then we have

$$\sum_{k=0}^{n} (k - mp)^4 P_m(k) \geq \varepsilon^4 m^4 \left[P\left(\left|\frac{k}{m} - p\right| \geq \varepsilon\right) + P\left(\left|\frac{k}{m} - p\right| < \varepsilon\right) \right],$$

which yields

$$P\left(\left|\frac{k}{m} - p\right| \geq \varepsilon\right) \leq \frac{\sum_{k=0}^{n} (k - mp)^4 P_m(k)}{\varepsilon^4 m^4}. \tag{A.30}$$

Note that $k = S_m = \sum_{i=1}^{m} a_i$ and denote by $Y_i = a_i - p$ (this function is similar to the Rademacher function for $p \neq 1/2$ case). We compute the numerator of (A.30) as

$$\sum_{k=0}^{n} (k - mp)^4 P_m(k) = E\left[\left(\sum_{i=1}^{m} a_i - mp\right)^4\right] = E\left[\left(\sum_{i=1}^{m} (a_i - p)\right)^4\right]$$

$$= E\left[\left(\sum_{i=1}^{m} Y_i\right)^4\right] = \sum_{i=1}^{m}\sum_{k=1}^{m}\sum_{j=1}^{m}\sum_{l=1}^{m} E[Y_i Y_k Y_j Y_l]$$

$$= \sum_{i=1}^{m} E[Y_i^4] + 4m(m-1)\sum_{i=1}^{m}\sum_{j=1}^{m} mE[Y_i^3][Y_j]$$

$$+ 3m(m-1)\sum_{i=1}^{m}\sum_{j=1}^{m} E[Y_i^2]E[Y_j^2]$$

$$= m(p^3 + q^3)pq + 3m(m-1)(pq)^2 \leq [m + 3m(m-1)]pq$$

$$= 3m^2 pq,$$

due to $p^3 + q^3 = (p+q)^3 - 3p^2q - 3pq^2 < 1$ and $pq \leq 1/2 < 1$. Substituting the inequality above into (A.30), we have

$$P\left(\left|\frac{k}{m} - p\right| \geq \varepsilon\right) \leq \frac{3pq}{\varepsilon^4 m^2}.$$

Let $\varepsilon_m = 1/m^{1/8}$ so that

$$\sum_{m=1}^{\infty} P\left(\left|\frac{k}{m} - p\right| \geq \frac{1}{m^{1/8}}\right) \leq 3pq \sum_{m=1}^{\infty} \frac{1}{m^{3/2}} \leq 3pq\left(1 + \int_{1}^{\infty} x^{-3/2} dx\right)$$

$$= 3pq(1+2) = 9pq < \infty. \tag{A.31}$$

Thus, (A.29) has been proved. This implied that the set $\mathfrak{N}^c$ has measure of zero. ∎

A.13 INTEGRATION IN PROBABILITY MEASURE

The notion of expectation is based on integration on measure spaces. In this section we recall briefly the definition of an integral with respect to the probability measure P. We start with introducing the simple function defined on a sample space. Consider a probability space $(\Omega, \mathscr{F}, P)$. A partition of sample space Ω, denoted by $(\Omega_i)_{1 \leq i \leq n}$, is a family of subsets of $\Omega_i \subset \Omega$ satisfying

1. $\Omega_i \cap \Omega_j = \emptyset$, for $i \neq j$;
2. $\bigcup_i^n \Omega_i = \Omega$.

Each Ω_i is an event with probability $P(\Omega_i)$. A simple function is of characteristic functions $f = \sum_i^n c_i \chi_{\Omega_i}$. This means $f(\omega) = c_k$ for $\omega \in \Omega_k$. The integral of the simple function f over the sample space is defined as

$$\int_\Omega f dP = \sum_{i=1}^n c_i P(\Omega_i).$$

If $X : \Omega \to \mathbb{R}$ is a random variable such that there is a sequence of simple functions $(f_n)_{n \geq 1}$ satisfying

1. f_n is Cauchy in probability: $\lim_{n,m \to \infty} P(\omega; |f_n(\omega) - f_m(\omega)| \geq \varepsilon) \to 0, \forall \varepsilon$;
2. f_n converges to X in probability: $\lim_{n \to \infty} P(\omega; |f_n(\omega) - X(\omega)| \geq \varepsilon) \to 0, \forall \varepsilon$,

then the integral of X is defined as the following limit of integrals

$$\int_\Omega X dP = \lim_{n \to \infty} \int_\Omega f_n dP = \int_\Omega X(\omega) dP(\omega).$$

Such an integral with respect to probability measure is defined as a Lebesgue integral. To extend the definition of the Lebesgue integral to arbitrary functions, we approximate them by simple functions. We first show that any function can be expressed as a limit of simple functions. Note that a random variable is a function or mapping from the sample space to a real line. We start with a positive function.

Theorem A.10 *Given a non-negative function f, there exists a sequence of simple functions f_n such that $f_n \to f$ pointwise. On sets of the form $\{x : f(x) \leq M\}$ the convergence is uniform.*

Proof. For each n, define

$$f_n(x) = \begin{cases} \frac{k-1}{2^n}, & \text{if } f(x) \in \left[\frac{k-1}{2^n}, \frac{k}{2^n}\right), \ k = 1, \ldots, n2^n, \\ n, & \text{if } f(x) \geq n. \end{cases} \tag{A.32}$$

Then it is clear that f_n is simple and increasing in n and $0 \leq f_n(x) \leq f(x)$ for all x. Given any x in $A_M = \{x : f(x) \leq M\}$, once $n > M$ (i.e., there is no way that $f(x) \geq n$ holds since n is greater than $f(x)$'s upper bound), we must have $f(x) \in [(k-1)/2^n, k/2^n)$ for some k, so $f(x) - f_n(x) \leq 1/2^n$. Since this bound is independent of x we have $f_n \to f$ uniformly on A_M. Since every x lies in A_M for some M, this proves pointwise convergence everywhere. ∎

A random variable $X : \Omega \to \mathbb{R}$ is said to be integrable if

$$\int_\Omega |X(\omega)| dP(\omega) = \int_\mathbb{R} |x| f(x) dx < \infty,$$

where $f(x)$ is the probability density function of X. Note that this identity follows from changing domain of integration from Ω to $\mathbb{R}$. For an integrable random variable X, we can define the expectation as

$$E[X] = \int_\Omega X(\omega) dP(\omega) = \int_\mathbb{R} x f(x) dx.$$

Further, we can find the expectation for any continuous and measurable function of the random variable X as

$$E[h(X)] = \int_\Omega h(X(\omega)) dP(\omega) = \int_\mathbb{R} h(x) f(x) dx.$$

It is easy to verify that the expectation is a linear operator for the integrable random variables due to the linearity of the integral. That is, for constant numbers a and b and two integrable random variables X and Y, we have

$$E[aX + bY] = aE[X] + bE[X].$$

For the two independent random variables, we have $E[XY] = E[X]E[Y]$. Next, we present the Radon-Nikodym's Theorem which is useful in defining the conditional expectations.

A.13.1 RADON-NIKODYM THEOREM

The Radon–Nikodym theorem (RNT) is one of the three most important theorems in the integration theory. The other two are the Lebesgue's dominated theorem and the Fubini theorem on the product space. Since RNT is the foundation of defining the conditional expectation of random variable, we present this theorem and provide an intuitive proof (also called a heuristic proof). A rigorous proof given by von Neumann can be found in Rudin (1987) and is based on the representation theorem of a linear functional on Hilbert space. There are also other ways of proving this theorem and can be found in some standard textbooks on real analysis or measure theory. Interested readers are referred to these textbooks such as Folland (1999), Halmos (1950), and Bartle (1995). Here, our proof is more intuitive and less technical and based on Cho (2007). Suppose that $(X, \mathscr{F}, \mu)$ is a measurable space. First, we present some preliminary results:

Lemma A.3 *Let λ be a signed measure on a set X, which is a function from $\mathscr{F}$ into the set of real numbers. If a measurable subset A does not contain any positive subset of positive signed measure, then A is a negative set.*

Proof. Assume that A is not a negative set. Then there is a measurable subset E of A such that $\lambda(E) > 0$. Since E cannot be a positive set (the condition of the lemma), there is a measurable subset B_1 of E and a smallest natural number n_1 such that $\lambda(B_1) < -1/n_1$. Then $\lambda(E \cap B_1^c) = \lambda(E) - \lambda(B_1) > \lambda(E) + \frac{1}{n_1} > \lambda(E) > 0$. Thus, the set $E \cap B_1^c$ cannot be a positive set (As if it is a positive set, then $E \cap B_1^c$ is a positive set with positive signed measure which contradicts the condition of the lemma). Similarly, there is a measurable subset B_2 of $E \cap B_1^c$ and a smallest natural number n_2 such that $\lambda(B_2) < -1/n_2$. Then, we can conclude that $E \cap B_1^c \cap B_2^c$ cannot be a positive set. Using the mathematical induction, we can find a measurable set B_k of $E \cap B_1^c \cap \cdots \cap B_{k-1}^c$ and a smallest natural number n_k with $\lambda(B_k) < -1/n_k$. Let $B = \cup_{k=1}^{\infty} B_k$, which is a measurable subset of E. Since all sets B_k are disjoint, we have

$$-\infty < \lambda(B) = \sum_{k=1}^{\infty} \lambda(B_k) < \sum_{k=1}^{\infty} -\frac{1}{n_k} < 0,$$

which implies $\lambda(E \cap B^c) = \lambda(E) - \lambda(B) > \lambda(E) > 0$. Since $E \cap B^c$ cannot be a positive set for the same reason above, there is a measurable subset C of $E \cap B^c$ and a smallest natural number n_j with $\lambda(C) < -1/n_j$, which contradicts to the choices of B_k and natural number n_k (which has exhausted the choices of this type). Thus, A must be a negative set. This completes the proof. ∎

Proposition A.2 *(Hahn Decomposition) Let λ be a signed measure on X. Then there are a positive set P and a negative set N such that $P \cap N = \emptyset$ and $X = P \cup N$.*

The proof of this proposition can be done easily by using the properties of the signed measures.

Let μ be measure on X. We can generate another measure v by using a non-negative measurable function f. For every measurable subset E, v is defined by the integral

$$v(E) = \int_E f d\mu.$$

If $\mu(E) = 0$, then $v(E) = 0$ for every measurable subset E, it is said that v is absolutely continuous with respect to μ. Clearly, this condition follows from the integral. Thus, the measure generated by the integral of a non-negative measurable function with respect to the given measure produces a pair of absolutely continuous measures. The RNT is in fact the converse of this property. The RNT holds for two σ-finite measures (i.e., a measure that can be expressed as a countable sum of finite measures). We provide a proof of the RNT for finite measure cases. A useful technique in such a proof is summarized as a lemma.

Lemma A.4 *Let v and μ be measures on X. Let c_1, c_2 $(0 < c_1 < c_2)$ be real numbers with an average $c_3 = (c_1 + c_2)/2$. Suppose that E is the intersection of a positive set with respect to the signed measure $v - c_1\mu$ and a negative set with respect to the signed measure $v - c_2\mu$. Then, there are two measurable sets E_1 and E_2 such that*

1. $E_1 \cap E_2 = \emptyset, E_1 \cup E_2 = E$,
2. For any subset $F \subset E_1$, we have $c_1\mu(F) \leq v(F) \leq c_3\mu(F)$,
3. For any subset $G \subset E_2$, we have $c_3\mu(G) \leq v(G) \leq c_2\mu(G)$.

Proof. Let N be a negative set with respect to $v - c_3\mu$ (N^c will be the positive set with respect to $v - c_3\mu$ then). Let $E_1 = E \cap N$ and $E_2 = E \cap N^c$. Then, these two measurable sets E_1 and E_2 satisfy the three requirements. Clearly, (1) holds. For (2), a subset F from E_1 is a negative set under $v - c_3\mu$ and also possesses the property made in the assumption (i.e., positive with respect to $v - c_1\mu$ and negative with respect to $v - c_2\mu$). Thus, we have $v(F) - c_3\mu(F) \leq 0 \Rightarrow v(F) \leq c_3\mu(F)$. At the same time, this subset if positive with respect to $v - c_1\mu$. That is $v(F) - c_1\mu(F) \geq 0 \Rightarrow v(F) \geq c_1\mu(F)$. For (3), E_2 is a positive set under $v - c_3\mu$ so that for a subset G from E_2, we have $v(G) - c_3\mu(G_3) \geq 0 \Rightarrow V(G) \geq c_3\mu(G)$. At the same time, this subset if negative under $v - c_2\mu$ so that $V(G) - c_2\mu(G) \leq 0 \Rightarrow V(G) \leq c_2\mu(G)$. This completes the proof. ∎
Now we present the RNT.

Theorem A.11 *(The Radon-Nikodym Theorem). Let v and μ be finite measures on X. Assume that v is absolutely continuous with respect to μ. Then there is a non negative measurable function f such that for every measurable subset E*

$$v(E) = \int_E f d\mu.$$

Proof. We will decompose X in stages. First, we decompose X into pair-wise measurable subset with respect to signed measure $v - k\mu$ for $k = 1, 2, \dots$ Let $X_{1,1}$ be a negative set of a Hahn decomposition of X with respect to the signed measure $v - \mu$ (i.e., $X_{1,1}^c$ is the positive set under the same signed measure). Then for any measurable subset $E \subset X_{1,1}$, we have $0 \cdot \mu(E) \leq v(E) \leq 1 \cdot \mu(E)$. Let $N_{1,2}$ be a negative set of a Hahn decomposition with respect to the signed measure $v - 2\mu$. Let $X_{1,2} = N_{1,2} \cap X_{1,1}^c$, the intersection of a positive set with respect t $v - \mu$ and a negative set with respect to $v - 2\mu$. Then for any measurable subset $E \subset X_{1,2}$, we have $\mu(E) \leq v(E) \leq 2\mu(E)$. Using this approach, suppose that we have constructed pair-wise measurable subsets $X_{1,1}, \cdots, X_{1,n}$ with

$$(k-1)\mu(E) \leq v(E) \leq k\mu(E)$$

for any measurable subset $E \subset X_{1,k}$ with $k = 1, 2, \dots n$. Let $N_{1,n+1}$ be a negative set of a Hahn decomposition with respect to the signed measure $v - (n+1)\mu$ and let $X_{1,n+1} = N_{1,n+1} \cap (X_{1,1} \cup X_{1,2} \cup \cdots \cup X_{1,n})^c$. Then for any measurable subset $E \subset X_{1,n+1}$, we have $n\mu(E) \leq v(E) \leq (n+1)\mu(E)$. Therefore, it follows from the mathematical induction that

$$(n-1)\mu(E) \leq v(E) \leq n\mu(E)$$

for any measurable subset $E \subset X_{1,n}$. Let $B = (\cup_{n=1}^{\infty} X_{1,n})^c$. Then we have $n\mu(B) \leq v(B)$ for all natural number n. Since $v(B) < \infty$, it must be $\mu(B) = 0$. Then since v is absolutely continuous with respect to μ, we have $v(B) = 0$.

Now we define a sequence of non-negative measurable functions. Note that $X = B \cup X_{1,1} \cup \cdots \cup X_{1,k} \cup \cdots$. Define the first measurable function f_1 as

$$f_1(x) = \begin{cases} 0, & x \in B \\ k-1, & x \in X_{1,k}. \end{cases}$$

Thus, for $E \subset X_{1,k}$ we can write

$$\int_E f_1(x) d\mu(x) \leq v(E) \leq \int_E f_1(x) d\mu(x) + \int_E d\mu(x).$$

Next, we divide each $X_{1,k}$ into two subsets by using the Lemma A.3. Treating $c_1 = k - 1, c_2 = k$ and $c_3 = (2k-1)/2$ and the signed measure $v - [(2k-1)/2]\mu$, we can decompose $X_{1,k}$ into two disjoint subsets, denoted by $X_{2,2k-1}$ and $X_{2,2k}$. That is $X_{1,k} = X_{2,2k-1} \cup X_{2,2k}$. Clearly, for any $E \subset X_{2,2k-1}$, we have

$$\frac{2k-2}{2}\mu(E) \leq v(E) \leq \frac{2k-1}{2}\mu(E),$$

and for any $E \subset X_{2,2k}$, we have

$$\frac{2k-1}{2}\mu(E) \leq v(E) \leq \frac{2k}{2}\mu(E).$$

This decomposition motivates us to define the second measurable function f_2 as

$$f_2(x) = \begin{cases} 0, & x \in B \\ \frac{k-1}{2}, & x \in X_{2,k}. \end{cases}$$

Thus, for $E \subset X_{2,k}$ we can write

$$\int_E f_2(x)d\mu(x) \le v(E) \le \int_E f_2(x)d\mu(x) + \frac{1}{2}\int_E d\mu(x).$$

Now, applying Lemma A.3 to $X_{j,k}$ and the signed measure $v - \frac{2k-1}{2^{j-1}}\mu$, we can decompose $X_{j,k}$ into two disjoint measurable subsets $X_{j+1,2k-1}$ and $X_{j+1,2k}$. That is $X_{j,k} = X_{j+1,2k-1} \cup X_{j+1,2k}$. Then, for any $E \subset X_{j+1,2k-1}$, we have

$$\frac{2k-2}{2^j}\mu(E) \le v(E) \le \frac{2k-1}{2^j}\mu(E),$$

and for any $E \subset X_{j+1,2k}$, we have

$$\frac{2k-1}{2^j}\mu(E) \le v(E) \le \frac{2k}{2^j}\mu(E).$$

Accordingly, we define the measurable function f_{n+1} as

$$f_{j+1}(x) = \begin{cases} 0, & x \in B, \\ \frac{k-1}{2^j}, & x \in X_{j+1,k}. \end{cases}$$

Then, for $E \subset X_{j+1,k}$ we can write

$$\int_E f_{j+1}(x)d\mu(x) \le v(E) \le \int_E f_{j+1}(x)d\mu(x) + \frac{1}{2^j}\int_E d\mu(x).$$

By the mathematical induction described above, we can construct a sequence of non-negative measurable functions f_n. It follows from this construction that the function f_n is decreasing in n. Since for each fixed n all $X_{n,k}$ are disjoint, we have

$$\int_E f_n(x)d\mu(x) \le v(E) \le \int_E f_n(x)d\mu(x) + \frac{1}{2^{n-1}}\mu(X).$$

Since $\mu(X) < \infty$, by applying the Monotone convergence theorem, we have for any measurable subset E

$$\int_E f d\mu \le v(E) \le \int_E f d\mu,$$

where $f = \lim_{n \to \infty} f_n$. Thus, we have $v(E) = \int_E f d\mu$ for every measurable subset E. This completes the proof. ∎

Note that $f = dv/d\mu$ is called Radon–Nikodym derivative of v with respect to μ.

One of the most important applications of RNT is to define the conditional expectation of a random variable (its existence and uniqueness).

Definition A.13 *Suppose that X is a random variable defined on a probability space* $(\Omega, \mathscr{F}, P)$ *and* $E[X]$ *is well defined (i.e., X is integrable). Then, a random variable Y is the conditional expectation of X given* $\mathscr{G}$, *where* $\mathscr{G} \subseteq \mathscr{F}$ *is a sub-σ-algebra, if the following two conditions hold:*
  1. $Y : \Omega \to \mathbb{R}$ *is* $\mathscr{G}$-measurable;
  2. $\int_E Y dP = \int_E X dP$, *for all* $E \in \mathscr{G}$.

We can denote by $Y = E[X|\mathscr{G}]$ this conditional expectation. The conditional probability is the special case of $X = \mathbf{1}_A$ for $A \in \mathscr{F}$. That is $P(A|\mathscr{G}) = E[X|\mathscr{F}]$.

The verification of (2) requires the RNT. We present this process as a theorem and its proof.

Theorem A.12 *In a probability space* $(\Omega, \mathscr{F}, P)$, *consider a random variable X on it and a sub-σ-algebra* $\mathscr{G} \subseteq \mathscr{F}$. *If* $E[X]$ *is well-defined, then there exists a* $\mathscr{G}$-measurable function $E[X|\mathscr{G}] : \Omega \to \mathbb{R}$, *unique to P-null sets, such that*

$$\int_E E[X|\mathscr{G}]dP = \int_E X dP, \ \forall E \in \mathscr{G}.$$

Proof. We provide a proof for the case where $X \in L^1$. Suppose that $X \in L^1$, then $X^{\pm} \in L^1$. The measures, defined by

$$\nu^{\pm}(E) = \int_E X^{\pm}dP, \ \forall E \in \mathscr{G},$$

are two finite measures absolutely continuous to P. It follows from the RNT that there exists $\mathscr{G}$-measurable functions $h^{\pm} : \Omega \to [0, \infty)$ such that

$$\nu^{\pm}(E) = \int_E h^{\pm}dP, \ \forall E \in \mathscr{G}.$$

Define $E[X|\mathscr{G}] = h^+ - h^-$. Showing the uniqueness is easy and discussed later. ∎

As a special case of conditional expectation of a random variable, we consider an indicator variable of event A $\mathbf{1}_A$. Define a sub-σ-algebra $\mathscr{G} = \{\emptyset, B, B^c, \Omega\} \subseteq \mathscr{F}$. We compute the conditional expectation $Y = E[\mathbf{1}_A|\mathscr{G}] = P(A|\mathscr{G})$. Since Y is $\mathscr{G}$- measurable by definition, we can write $Y = a\mathbf{1}_B + b\mathbf{1}_{B^c}$ for some constant $a, b \in \mathbb{R}$. However, for any $E \in \mathscr{G} = \{\emptyset, B, B^c, \Omega\}$, we have

$$\int_E Y dP = \int_E \mathbf{1}_A dP \Rightarrow aP(E \cap B) + bP(E \cap B^c) = P(A \cap E).$$

Taking $E = B$ and then $E = B^c$, we can determine

$$a = \frac{P(A \cap B)}{P(B)}, \quad b = \frac{P(A \cap B^c)}{P(B^c)}.$$

With these two parameters, it is easy to check that for every subset $E \in \mathcal{G}$,

$$aP(E \cap B) + bP(E \cap B^c) = P(A \cap E)$$

always holds. Thus, it follows that

$$P(A|\mathcal{G})(\omega) = \begin{cases} P(A|B), & \text{if } \omega \in B, \\ P(A|B^c), & \text{if } \omega \in B^c. \end{cases}$$

Using $Y = P(A|\mathcal{G}) = a\mathbf{1}_B + b\mathbf{1}_{B^c}$, we have the conditional probability

$$P(A|B) = a = \frac{P(A \cap B)}{P(B)}, \quad P(A|B^c) = a = \frac{P(A \cap B^c)}{P(B^c)}.$$

Another special case is to find the condition for $X = 0$ almost surely.

Proposition A.3 *In a probability space* $(\Omega, \mathcal{F}, P)$, *let* $\mathcal{G}$ *be a sub-σ-algebra (i.e., $\mathcal{G} \subseteq \mathcal{F}$). If X is a $\mathcal{G}$-measurable random variable such that*

$$\int_E X dP = 0 \quad \forall E \in \mathcal{G},$$

then $X = 0$ almost surely (a.s.).

Proof. To show that $X = 0$ a.s., it is sufficient to show that $P(\omega; X(\omega) = 0) = 1$. First, we show that X takes values as small as possible with probability 1. That is for all $\varepsilon > 0$, we have $P(|X| < \varepsilon) = 1$. Letting $E = \{\omega; X(\omega) \geq \varepsilon\}$, we have

$$0 \leq P(X \geq \varepsilon) = \int_E dP = \frac{1}{\varepsilon} \int_E \varepsilon dP \leq \frac{1}{\varepsilon} \int_E X dP = 0,$$

and hence $P(X \geq \varepsilon) = 0$. Similarly, we can show $P(X \leq -\varepsilon) = 0$. Thus,

$$P(|X| < \varepsilon) = 1 - P(X \geq \varepsilon) - P(X \leq -\varepsilon) = 1 - 0 - 0 = 1.$$

Taking $\varepsilon \to 0$ yields $P(|X| = 0) = 1$. To justify this fact, we let $\varepsilon = 1/n$ and consider $B_n = \{\omega; |X(\omega)| \leq \varepsilon\}$, with $P(B_n) = 1$. Then we have

$$P(X = 0) = P(|X| = 0) = P(\cap_{n=1}^\infty B_n) = \lim_{n \to \infty} P(B_n) = 1.$$

∎

Corollary A.1 *If X and Y are $\mathcal{G}$-measurable random variables such that*

$$\int_E X dP = \int_E Y dP \quad \forall E \in \mathcal{G},$$

then $X = Y$ a.s..

Proof. Since $\int_E (X - Y)dP = 0$ for all $E \in \mathcal{G}$, it follows from Proposition (A.3) that $X - Y = 0$ a.s. ∎

An application of Corollary (A.1) is to show that Y in Theorem A.12 is unique almost surely. To show this uniqueness, we assume that there are two $\mathcal{G}$-measurable random variables Y_1 and Y_2 with the property stated in Theorem A.12. Then

$$\int_E Y_1 dP = \int_E Y_2 dP, \quad \forall E \in \mathcal{G}.$$

Using Corollary (A.1), we have $Y_1 = Y_2$ a.s..

REFERENCES

1. R.G. Bartle, "The elements of Integration and Lebesgue Measure", John Wiley & Sons, Inc. 1995.
2. S.J. Cho, A Heuristic Proof of the Radon-Nikodym Theorem, https://s-space.snu.ac.kr/bitstream/10371/72845/1/03.pdf, 2007.
3. W. Feller, "An Introduction to Probability Theory and Its Applications", Vol. II. John Wiley & Sons Inc., New York, 2nd ed., 1971.
4. G.B. Folland, "Real Analysis", John Wiley & Sons, Inc., 1999.
5. P.R. Halmos, "Measure Theory", D. Van Nostrand Company Inc., Princeton, N.J, 1950.
6. W. Rudin, "Real and Complex Analysis", McGraw-Hill Book Company, NY, 3rd ed., 1987.
7. W.T. Song, Relationships among some univariate distributions. *IIE Transactions*, 37, 651-656, 2005.
8. J. Yan, Enjoy the joys of copulas. *Journal of Statistical Software*, 21 (4), 2007.

B Conditional Expectation and Martingales

B.1 σ-ALGEBRA REPRESENTING AMOUNT OF INFORMATION

In the world of uncertainty, the information can be considered as the knowledge about the state of nature by observing the value of a random variable, which can help us answer the question of whether or not a relevant event occurs (or some relevant events occur). For example, the event of interest for an investor is that "the ABC stock price is in a certain range, denoted by (p_{min}, p_{max}), at a certain point in time (e.g., on a particular day)". In the probability space setting, all questions can be phrased in terms of the elements of sample space Ω (i.e., whether or not an event occurs). In fact, Ω contains all possible evolutions of the world (including some impossible ones) and the knowledge of the exact $\omega \in \Omega$ (the true state of nature) amounts to the knowledge of everything. Unfortunately, we usually cannot answer the question like "what is the true ω?" as we do not know everything (or we do not have complete or perfect information). The collection of all events A that you have the answer to the question "Is the true ω an element of the event A?" is the mathematical description of your current state of information. This collection, denoted by $\mathscr{F}$, is a σ-algebra. This is because of the following facts:

- First, we always know that the true ω is an element of Ω, thus, $\Omega \in \mathscr{F}$.
- If we know how to answer the question "Is the true ω in A (i.e., $A \in \mathscr{F}$)?", we will also know how to answer the question "Is the true ω in A^c (i.e., $A^c \in \mathscr{F}$)?"
- Let $(A_n)_{n \in \mathbb{N}}$ be a sequence of events that we have the answers to the questions "Is the true ω in A_n for $n \in \mathbb{N}$ (i.e., $(A_n) \in \mathscr{F}$ for all $n \in \mathbb{N}$)?". Then we know the answer to the question: "Is the true ω in $\cup_{n \in \mathbb{N}} A_n$?". In fact, the answer is "No" if we answered "No" to the question " Is the true ω in A_n?" for each n, and it is "Yes" if we answered "Yes" to at least one of these questions. Consequently, $\cup_{n \in \mathbb{N}} A_n \in \mathscr{F}$.

Now we demonstrate the information meaning of a σ-algebra by analyzing a simple random process of discrete time and space with a finite horizon. For a sample space with a finite number of elements, it is easy to show that σ-algebras are generated from the partitions of the sample space resulted from

DOI: 10.1201/9781003150060-B

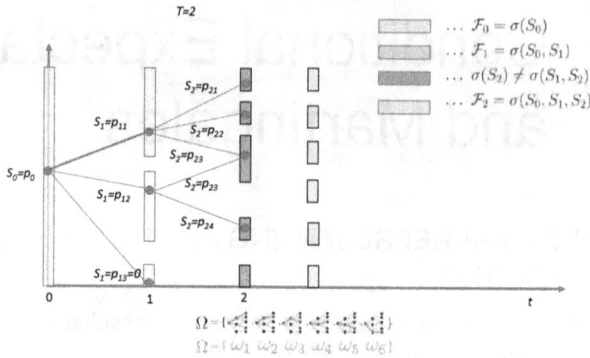

Figure B.1   Two period stock price tree.

observing values of random variables. The finer the partition, the more we know about the true ω or the more information we have. We consider an example about the price evolution of a fictitious stock on a two-day horizon first. As shown in Figure B.1, the stock price on day 0 is $S_0 = p_0$ and moves to one of the three possible values on day 1 and branches further on day 2. Note that the sample space in this two period process has 6 elements. That is $\Omega = \{\omega_1, \omega_2, \omega_3, \omega_4, \omega_5, \omega_6\}$. On day 0, the information available is minimal, the only questions we can answer are the trivial ones: "Is the true ω in Ω?" and "Is the true ω in $\emptyset$?", and this is encoded in the σ-algebra $\{\Omega, \emptyset\}$. Simply, this means we know nothing except whether or not we are in the market (with $\emptyset$ representing "not in the market" and Ω representing "in the market"). On day 1, after having observed the stock price, we know a little bit more. Denote this price by S_1, a random variable. Now we can tell a certain set of elements (events) in Ω occur. In Figure B.1, we know that $\{\omega_1, \omega_2, \omega_3\}$ occur ($S_1 = p_{11}$, thicker arc). But $\{\omega_4, \omega_5\}$ and $\{\omega_6\}$ do not occur. Thus, our information partition is $\{\{\omega_1, \omega_2, \omega_3\}\{\omega_4, \omega_5\}, \{\omega_6\}\}$ and the corresponding σ-algebra $\mathscr{F}_1$, generated by this sample space partition, is a power set of it given by

$$\mathscr{F}_1 = \{\emptyset, \{\omega_1, \omega_2, \omega_3\}, \{\omega_4, \omega_5\}, \{\omega_6\}, \{\omega_1, \omega_2, \omega_3, \omega_4, \omega_5\}, \{\omega_1, \omega_2, \omega_3, \omega_6\},$$
$$\{\omega_4, \omega_5, \omega_6\}, \Omega\}.$$

The σ-algebra $\mathscr{F}_1$ contains some information about predicting the price on day 2. However, the perfect prediction for day 2 is only for the special case when $S_1 = p_{13} = 0$. If this special case occurs (i.e., ω_6), we do not need to wait until day 2 to learn what the stock price at day 2 will be as it is going to remain to 0. On day 2, we know exact what ω occured and the σ-algebra $\mathscr{F}_2$ consists of all subsets of Ω, which is a power set of Ω, consisting of $2^6 = 64$ events including $\emptyset$ and Ω. To demonstrate the concept of "filtration", we examine how we acquire the additional information on each new day. The information is accumulated through the random variables (stock prices) S_0, S_1, and S_2 as their values are gradually revealed. Thus, we can say that the σ-algebra $\mathscr{F}_1$ is generated by random variable S_1 and S_0, which can be denoted by $\mathscr{F}_1 = \sigma(S_0, S_1)$. In other words, $\mathscr{F}_1$ consists of exactly those subsets of Ω, which can be described in terms of the subsets of S_1 values only. That is the collection of events $\{\omega \in \Omega : S_1(\omega) \in A\}$ where A can be $\{p_{11}\}, \{p_{12}\}, \{p_{13}\}, \{p_{11}, p_{12}\}, \{p_{11}, p_{13}\}, \{p_{12}, p_{13}\}, \{p_{11}, p_{12}, p_{13}\}$, and $\emptyset$. Clearly, we have the power set of the sample space partition $\{\{\omega_1, \omega_2, \omega_3\}\{\omega_4, \omega_5\}, \{\omega_6\}\} = \{\{p_{11}\}.\{p_{12}\}, \{p_{13}\}\}$ as follows:

$$\mathscr{F}_1 = \{\emptyset, \{\omega_1, \omega_2, \omega_3\}, \{\omega_4, \omega_5\}, \{\omega_6\}, \{\omega_1, \omega_2, \omega_3, \omega_4, \omega_5\}, \{\omega_1, \omega_2, \omega_3, \omega_6\},$$
$$\{\omega_4, \omega_5, \omega_6\}, \Omega\}$$
$$= \{\emptyset, \{p_{11}\}, \{p_{12}\}, \{p_{13}\}, \{p_{11}, p_{12}\}, \{p_{11}, p_{13}\}, \{p_{12}, p_{13}\}, \{p_{11}, p_{12}, p_{13}\}\}.$$

In the same way, $\mathscr{F}_2$ is a collection of all subsets of Ω which can be described only in terms of values of random variables S_1 and S_2. In our example with 2-day horizon, at day 2, we know exactly what the true state of nature is (all random variable values are revealed and the finest information partition is reached). Hence, $\mathscr{F}_2 = \sigma(S_0, S_1, S_2)$, generated by S_0, S_1, and S_2, contains $2^6 = 64$ subsets of Ω (a power set). A key concept here is that the information is accumulated by the entire history. Suppose a trader who slept through day 1 and woke up on day 2 to observe $S_2 = p_{23}$. If asked about the stock price on day 1, he cannot answer the question as it can be either p_{11} or p_{12}. Thus, the $\sigma(S_0, S_2)$ is strictly smaller (coarser) than $\sigma(S_0, S_1, S_2)$, even though S_2 is observed after S_1. The σ-algebras, $\mathscr{F}_0, \mathscr{F}_1$, and $\mathscr{F}_2$ representing the amounts of information available to the investor is an instance of a so-called "filtration". A filtration is any family of σ-algebras which gets finer and finer (an increasing family of σ-algebras).

Although the two-period example is an over-simplified scenario, the structure can be generalized to a finite horizon setting with $T > 2$ periods. Starting with $S_0 = p_0$, the stock price evolves stochastically from peropd to period. For each period i, it is assumed that the stock price S_i takes a value from a finite set $\{p_{i1}, p_{i2}, ..., p_{in_i}\}$ with $i = 1, 2, ..., T$. The state transitions from period i to period $i + 1$ can be specified by the "tree-diagram" like the one shown in Figure B.1 or a one-step transition probability matrix. For such a discrete-time stochastic process with a finite horizon $S = \{S_i, i = 1, ..., T\}$ on a finite state space, the filtration generated by the process S, denoted by $\mathscr{F}^S$, can be defined as $\mathscr{F}_0^S = \sigma(S_0), \mathscr{F}_1^S = \sigma(S_0, S_1), ..., \mathscr{F}_n^S = \sigma(S_0, S_1, ..., S_n)...$. The filtration $\mathscr{F}^S$ will describe the flow of information of an observer who has access to the value of S_n at time n. In general, we say that a random variable X is measurable with respect to σ-algebra $\mathscr{F}$ if $\sigma(X) \subseteq \mathscr{F}$.

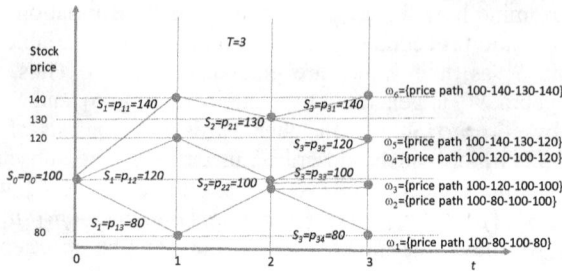

Figure B.2   Three period stock price tree.

B.2   CONDITIONAL EXPECTATION IN DISCRETE TIME

We continue using the stock price model by considering three periods, $T = 3$, as shown in Figure B.2. This example is taken from the lecture notes of G. Zitkovic (2005). Since it is a discrete-time and discrete state model, the sample space for the stock process is also discrete. To make the sample space small, the number of price levels that the process can reach at each period are limited to $n_1 = 3, n_2 = 2$, and $n_3 = 4$. With the restriction on the transitions from period to period, we end up with 6 possible price paths. That is $\Omega = \{\omega_1, \omega_2, \omega_3, \omega_4, \omega_5, \omega_6\}$. For the simplicity, we assume that the branches from the same node have equal probabilities. Stock price at each period, $S_i(\omega)$, forms a discrete-time stochastic process with $i = 0, 1, 2, 3$. The process starts with the initial price $S_0(\omega) = 100$, for all $\omega \in \Omega$. Now we can define the conditional expectation of a random variable, X, at a time point with respect to the σ-algebra $\mathscr{F}$, which represents the information available at that time point, denoted by $E[X|\mathscr{F}]$. We use this simple model to show the concept and important properties of conditional expectation of a random variable.

Now we look at each period. At the beginning (i.e., $i = 0$), since we have no information as $\mathscr{F}_0^S = \{\emptyset, \Omega\}$, the expectation is actually not "conditional". Specifically, we have

$$E[S_1|\mathscr{F}_0^S] = E[S_1] = \frac{1}{3}140 + \frac{1}{3}120 + \frac{1}{3}80 = 113\frac{1}{3},$$

which is a constant number. Similarly, for the next two periods, we can also compute the unconditional expectations

$$E[S_2|\mathscr{F}_0^S] = E[S_2] = \frac{1}{3}130 + \frac{2}{3}100 = 110,$$

$$E[S_3|\mathscr{F}_0^S] = E[S_3] = \frac{1}{6}140 + \frac{2}{6}120 + \frac{2}{6}100 + \frac{1}{6}80 = 110.$$

On day 1, $i = 1$, the stock price $S_1(\omega)$ is known. For the three periods, we have

$E[S_1|\mathcal{F}_1^S] = S_1(\omega),$

$E[S_2|\mathcal{F}_1^S] = \begin{cases} 130, & \text{if } S_1(\omega) = 140, \\ 100, & \text{if } S_1(\omega) = 100 \text{ or } 80 \end{cases} = S_2(\omega)$

$E[S_3|\mathcal{F}_1^S] = \begin{cases} \frac{1}{2}140 + \frac{1}{2}120 = 130, & \text{if } S_1(\omega) = 140 \Rightarrow \omega = \{\omega_5, \omega_6\}, \\ \frac{1}{2}120 + \frac{1}{2}100 = 110, & \text{if } S_1(\omega) = 120 \Rightarrow \omega = \{\omega_3, \omega_4\}, \\ \frac{1}{2}100 + \frac{1}{2}80 = 90, & \text{if } S_1(\omega) = 80 \Rightarrow \omega = \{\omega_1, \omega_2\}. \end{cases}$

On day 2, $i = 2$, the stock price $S_2(\omega)$ is known. For the two periods, we have

$E[S_2|\mathcal{F}_2^S] = S_2(\omega),$

$E[S_3|\mathcal{F}_2^S] = \begin{cases} \frac{1}{2}140 + \frac{1}{2}120 = 130, & \text{if } S_2(\omega) = 130, S_1(\omega) = 140 \\ & \Rightarrow \omega = \{\omega_5, \omega_6\}, \\ \frac{1}{2}120 + \frac{1}{2}100 = 110, & \text{if } S_2(\omega) = 100, S_1(\omega) = 120 \\ & \Rightarrow \omega = \{\omega_3, \omega_4\}, \\ \frac{1}{2}100 + \frac{1}{2}80 = 90, & \text{if } S_2(\omega) = 100, S_1(\omega) = 80 \\ & \Rightarrow \omega = \{\omega_1, \omega_2\}. \end{cases}$

In this example, since the partition of Ω at $t = 2$ is the same as (not finer than) $t = 1$, we have $\mathcal{F}_1^S = \mathcal{F}_2^S$. This is also an expected result due to the deterministic transition from Day 1 to Day 2.

On day 3, $i = 3$, the stock price $S_3(\omega)$ is known. For only period 3, we have

$E[S_3|\mathcal{F}_3^S] = S_3(\omega)$

$= \begin{cases} 140, & \text{if } S_3(\omega) = 140, S_2(\omega) = 130, S_1(\omega) = 140 \\ & \Rightarrow \omega = \{\omega_6\}, \\ 120, & \text{if } S_3(\omega) = 120, S_2(\omega) = 130, S_1(\omega) = 140 \\ & \text{or } S_3(\omega) = 120, S_2(\omega) = 100, S_1(\omega) = 120 \\ & \Rightarrow \omega = \{\omega_5, \omega_4\}, \\ 100 & \text{if } S_3(\omega) = 100, S_2(\omega) = 100, S_1(\omega) = 120 \\ & \text{or } S_3(\omega) = 100, S_2(\omega) = 100, S_1(\omega) = 80 \\ & \Rightarrow \omega = \{\omega_3, \omega_2\}, \\ 80 & \text{if } S_3 = 80, S_2(\omega) = 100, S_1(\omega) = 80 \\ & \Rightarrow \omega = \{\omega_1\}. \end{cases}$

Based on this numerical example, the following properties (labeled as CE for conditional expectation) are intuitive and expected.

- CE1. $E[X|\mathcal{F}] = E[X]$ for all $\omega \in \Omega$ if $\mathcal{F} = \{\emptyset, \Omega\}$. For this smallest (trivial or no information) σ-algebra, the conditional expectation is reduced to the ordinary expectation which is a constant number.

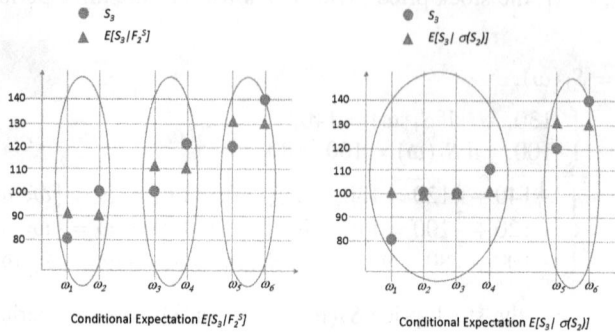

Figure B.3 Atom structure under different σ algebras for the three period stock price tree.

- CE2. $E[X|\mathcal{F}]$ is a random variable which is measurable with respect to $\mathcal{F}$. Here $\mathcal{F}$ is not the largest σ-algebra. For example, $E[S_3|\mathcal{F}_1^S]$ is measurable with respect to $\mathcal{F}_1^S$ since $\{\omega_1,\omega_2\},\{\omega_3,\omega_4\},\{\omega_5,\omega_6\} \in \mathcal{F}_1^S$.
- CE3. $E[X|\mathcal{F}] = X(\omega)$, if X is measurable and $\mathcal{F}$ is the largest σ-algebra. When you know $\mathcal{F}$, you already know the exact value of X. Thus, there is no need for expectation.

Now we introduce the concept of "atom" for a partition of the sample space. An atom is the smallest set of ω's such that $\mathcal{F}$ cannot distinguish between them. Taking an example of $E[S_3|\mathcal{F}_2^S]$, we observe that there are three atoms, $\{\{\omega_5,\omega_6\},\{\omega_3,\omega_4\},\{\omega_1,\omega_2\}\}$, given $\mathcal{F}_2^S$, corresponding to three possible values (130, 110, 90) for $E[S_3|\mathcal{F}_2^S]$ as shown in Figure B.3. These atoms are represented by circles in the figure. We consider a smaller σ-algebra, $\sigma(S_2) \subseteq \mathcal{F}_2^S$ (i.e., we only know the stock price at $i = 2$ but have no idea for $i = 1$). Thus, as shown in Figure B.3, we have

$$E[S_3|\sigma(S_2)] = \begin{cases} 130, & \text{if } S_2(\omega) = 120 \Rightarrow \omega = \{\omega_5,\omega_6\}, \\ \frac{1}{4}80 + \frac{1}{2}100 + \frac{1}{4}120 = 100, & \text{if } S_2(\omega) = 100 \Rightarrow \omega = \{\omega_1,\omega_2,\omega_3,\omega_4\}. \end{cases}$$

Clearly, $Y = E[S_3|\mathcal{F}_2^S]$ is a random variable. We can compute the expected value of this random variable with respect to $\sigma(S_2)$. That is

$$E[Y|\sigma(S_2)] = E[E[S_3|\mathcal{F}_2^S]|\sigma(S_2)]$$
$$= \begin{cases} 130, & \text{if } S_2(\omega) = 120 \Rightarrow \omega = \{\omega_5,\omega_6\}, \\ \frac{1}{2}110 + \frac{1}{2}90 = 100, & \text{if } S_2(\omega) = 100 \Rightarrow \omega = \{\omega_1,\omega_2,\omega_3,\omega_4\}, \end{cases}$$

where the fact that $S_2(\omega) = 100$ is reached from $\{\omega_1,\omega_2\}$ or $\{\omega_3,\omega_4\}$ with equal chance is utilized. This result implies $E[E[S_3|\mathcal{F}_2^S]|\sigma(S_2)] = E[S_3|\sigma(S_2)]$, which is called the tower property of conditional expectation.

- CE4. If X is a random variable and $\mathscr{G}$ and $\mathscr{F}$ are two σ-algebras such that $\mathscr{G} \subseteq \mathscr{F}$, then

$$E[E[X|\mathscr{F}]|\mathscr{G}] = E[X|\mathscr{G}],$$

which is called the tower property of conditional expectation.

Now we generalize the results above to a formula for computing the conditional expectations of a random variable given a σ-algebra in a discrete and finite state space case. Such a case can give you some intuition and the formula structure that holds for a more general case. For a sample space $\Omega = \{\omega_1, \omega_2, ..., \omega_n\}$ with an assigned probability measure $P(\omega) > 0$ for each $\omega \in \Omega$, consider a random variable $X : \Omega \to \mathbb{R}$ and let $\mathscr{F}$ be a σ-algebra on Ω which partitions Ω into m atoms: $A^1 = \{\omega_1^1, \omega_2^1, ..., \omega_{n_1}^1\}, A^2 = \{\omega_1^2, \omega_2^2, ..., \omega_{n_2}^2\},...$ and $A^m = \{\omega_1^m, \omega_2^m, ..., \omega_{n_m}^m\}$. Here, we have $A^1 \cup A^2 \cup \cdots \cup A^m = \Omega, A^i \cap A^j = \emptyset$, for $i \neq j$ and $n_1 + n_2 + ... + n_m = n$. Then we have the conditional expectation of the random variable given a σ algebra as

$$E[X|\mathscr{F}](\omega) = \begin{cases} \frac{\sum_{i=1}^{n_1} X(\omega_i^1)P(\omega_i^1)}{\sum_{i=1}^{n_1} P(\omega_i^1)} = \frac{\sum_{i=1}^{n_1} X(\omega_i^1)P(\omega_i^1)}{P(A^1)} & \text{for } \omega \in A^1, \\ \frac{\sum_{i=1}^{n_2} X(\omega_i^2)P(\omega_i^2)}{\sum_{i=1}^{n_2} P(\omega_i^2)} = \frac{\sum_{i=1}^{n_2} X(\omega_i^2)P(\omega_i^2)}{P(A^2)} & \text{for } \omega \in A^2, \\ \cdots & \cdots \\ \frac{\sum_{i=1}^{n_m} X(\omega_i^m)P(\omega_i^m)}{\sum_{i=1}^{n_m} P(\omega_i^m)} = \frac{\sum_{i=1}^{n_m} X(\omega_i^m)P(\omega_i^m)}{P(A^m)} & \text{for } \omega \in A^m, \end{cases}$$

which indicates that the conditional expectation is constant on the atoms of $\mathscr{F}$ and its value on each atom is computed by taking the weighted average of X using the relative probabilities as weights. Using the indicator function, we can write the formula above as

$$E[X|\mathscr{F}](\omega) = \sum_{i=1}^{m} E[X|A^i]\mathbf{1}_{A^i}(\omega), \tag{B.1}$$

where $E[X|A] = E[X\mathbf{1}_A]/P(A)$. With this formula, we can show all properties of the conditional expectation, those presented before and the followings.

- CE5. For random variables X_1 and X_2 and two real numbers a and b, the following holds:

$$E[aX_1 + bX_2|\mathscr{F}] = aE[X_1|\mathscr{F}] + bE[X_2|\mathscr{F}].$$

Using the fact that the conditional expectation is constant on the atoms of the σ-algebra $\mathscr{F}$, we can establish the following property.

- CE6. Given a σ-algebra $\mathscr{F}$, for two random variables X and Y, If Y is measurable with respect to $\mathscr{F}$, then

$$E[XY|\mathscr{F}](\omega) = Y(\omega)E[X|\mathscr{F}](\omega).$$

This property implies that the random variable measurable with respect to the available information can be treated as a constant in conditional expectations. As an example of using formula B.1 to prove the properties of conditional expectation, we prove CE6 as follows. For $\omega \in A^k$, we have

$$
\begin{aligned}
E[XY|\mathscr{F}](\omega) &= \sum_{i=1}^{m} \frac{1_{A^i}(\omega)}{P(A^i)} \sum_{\omega' \in A^i} X(\omega')Y(\omega')P(\omega') \\
&= \frac{1_{A^k}(\omega)}{P(A^k)} \sum_{\omega' \in A^k} X(\omega')Y(\omega')P(\omega') \\
&= \frac{1}{P(A^k)} \sum_{\omega' \in A^k} X(\omega')y_k P(\omega') = \frac{y_k}{P(A^k)} \sum_{\omega' \in A^k} X(\omega')P(\omega') \\
&= y_k E[X|\mathscr{F}](\omega) = Y(\omega)E[X|\mathscr{F}](\omega),
\end{aligned}
$$

where y_k is the constant for Y on atom A^k. Finally, we point out that conditioning on irrelevant information is the same as conditioning on no information (i.e., unconditional expectation). To formally present this property, we need the following independence definitions.

Definition B.1 *(1) Let $\mathscr{F}$ and $\mathscr{G}$ be two σ-algebras. If for each and every event $A \in \mathscr{F}$ and $B \in \mathscr{G}$, $P(A \cap B) = P(A)P(B)$, then $\mathscr{F}$ and $\mathscr{G}$ are independent.*
(2) If the σ-algebras $\sigma(X)$ and $\mathscr{F}$ are independent , then random variable X and $\mathscr{F}$ are independent.

Now, we can present the following property.

- CE7. If the random variables X and σ-algebra $\mathscr{F}$ are independent, then
$$
E[X|\mathscr{F}](\omega) = E[X], \quad \text{for all } \omega.
$$

B.3   CONDITIONAL EXPECTATION IN CONTINUOUS-TIME

Using the discrete-time stock price model, we have introduced the conditional expectations of random variables with respect to σ-algebra. The concepts and properties presented still hold in the continuous-time model. Although the proofs and derivations are more technical, the basic ideas remain the same. The random variable X_t indexed by a continuous-time t in interval of $[0, \infty)$ or $[0, T]$ for some finite horizon T become a continuous-time stochastic process. The filtration can be defined in the same way as an increasing family of σ-algebras $(\mathscr{F}_t)_{t \in [0, \infty)}$ or $(\mathscr{F}_t)_{t \in [0, T]}$ for continuous-time processes. Of course, a discrete-time filtration $(\mathscr{F}_n)_{n \in \mathcal{N}}$ on infinite Ω's can also be defined similarly. Similar to the discrete-time case, the σ-algebra $\sigma(X)$, generated by X should contain all subsets of Ω of the form $\{\omega \in \Omega : X(\omega) \in (a, b)\}$, for $a < b$ with

$a, b \in \mathbb{R}$. However, such a requirement will not result in a unique σ-algebra. Thus, we define $\sigma(X)$ as the smallest σ-algebra satisfying the requirement of a σ-algebra. This definition can be extended to multiple random variable or infinitely many random variable cases. That is for a continuous-time process $\{X_t\}$ over $[0, T]$ the σ-algebra generated by the process can be defined similarly and denoted by $\sigma((X_s)_{s \in [0,T]})$. It is worth noting that the σ-algebras on infinite Ω's do not necessarily carry the atomic structure as in the discrete-time and finite-state space case. So we cannot describe the σ-algebra by partitions. However, the basic idea of σ-algebra as an information set remains the same. We present the results without proofs and derivations due to the overwhelming technicalities. Interested readers are referred to some books on stochastic analysis such as Harrison (1985). Computing the conditional expectation in continuous time is, in general, more difficult than in discrete-time since there is no formula like (B.1) for the discrete-time case. However, we may work out the conditional expectations by using properties CE1 to CE7. For example, we may compute the conditional expectation of a function of a random variable $g(X)$ given a σ-algebra generated by a random vector $\sigma(X_1, X_2, ..., X_n)$

$$E[g(X)|\mathscr{F}] = E[g(X)|\sigma(X_1, X_2, ..., X_n)]$$

by utilizing the conditional density in two steps:

(i) Compute the conditional density

$$f_{(X|X_1, X_2, ..., X_n)}(x|X_1 = \gamma_1, X_2 = \gamma_2, ..., X_n = \gamma_n)$$

by using the joint density function of the random vector $(X, X_1, X_2, ..., X_n)$. Then use it to obtain the function

$$h(\gamma_1, \gamma_2, ..., \gamma_n) = E[g(X)|X_1 = \gamma_1, X_2 = \gamma_2, ..., X_n = \gamma_n]$$
$$= \int_{-\infty}^{\infty} g(x) f_{(X|X_1, X_2, ..., X_n)}(x|X_1 = \gamma_1, X_2 = \gamma_2, ..., X_n = \gamma_n) dx.$$

(ii) We can get the result by replacing $\gamma_1, \gamma_2, ..., \gamma_n$ with $X_1, X_2, ..., X_n$ in h. That is

$$E[g(X)|\mathscr{F}] = E[g(X)|\sigma(X_1, X_2, ..., X_n)] = h(X_1, X_2, ..., X_n).$$

B.4 MARTINGALES

Based on the concept of conditional expectation, we can define a class of important stochastic processes called *martingales*. We again focus on the discrete-time version martingale that can be used to demonstrate the main concepts and properties which hold in a more general settings. For a discrete-time stochastic process $X = \{X_k\}$, let $X_{m,n}$ denote the portion $X_m, X_{m+1}, ..., X_n$ of the process from time m up to n. We present several definitions for the martingale and start with the simplest one in discrete-time.

Definition B.2 *A process $M_0, M_1, \ldots$ is a martingale if*

$$E[M_{n+1}|M_{0,n}] = M_n \ \text{ for } n \geq 0,$$

and $E[|M_n|] < \infty$ for all $n \geq 0$.

The information set in this definition is based on a part of history of the process itself. Note that the information set can be generalized to a different process as follows:

Definition B.3 *A process $M_0, M_1, \ldots$ is a martingale with respect to process $Y_0, Y_1, \ldots$ if*

$$E[M_{n+1}|Y_{0,n}] = M_n \ \text{ for } n \geq 0,$$

and $E[|M_n|] < \infty$ for all $n \geq 0$.

A martingale can be interpreted as the wealth process of a "fair game". This means that if one plays a fair game, then he expects neither to win nor to lose money on the average. That is, given the history of the player's fortunes up to time n, our expected fortune M_{n+1} at the future time $n+1$ should be the same as the current fortune M_n. If the equality in the condition of the martingale definition is replaced with inequality, then we define a submartingale or a supermartingale.

Definition B.4 *A process $X_0, X_1, \ldots$ is a submartingale (supermartingale) with respect to a process $Y_0, Y_1, \ldots$ if $E[X_{n+1}|Y_{0,n}] \geq (\leq)X_n$ for $n \geq 0$.*

Here we give a classical example of the discrete-time martingale.

Example B.1 *(Polya's Urn). Suppose that we start with one black ball and one white ball in an urn at time 2. Then at each time we draw at random from the urn, and replace it together with a new ball of the same color. Thus, the number of balls at time n is equal to n. Let X_n denote the number of white balls at time $n \geq 2$. Therefore, if $X_n = k$, then with probability k/n a white ball is drawn so that $X_{n+1} = k+1$, and with probability $1 - (k/n)$ a black ball is drawn so that $X_{n+1} = k$. Define $M_n = X_n/n$, the fraction of white balls at time n. Then we can verify that M_n is a martingale as follows:*

$$E[M_{n+1}|X_{2,n}] = E\left(\frac{X_{n+1}}{n+1}\Big|X_n\right) = \frac{1}{n+1}\left[(X_n+1)\frac{X_n}{n} + X_n\left(1 - \frac{X_n}{n}\right)\right]$$

$$= \frac{X_n}{n} = M_n.$$

B.4.1 OPTIONAL SAMPLING

The most important property of martingales is the "conservation of fairness" which is known as optional sampling. For a discrete-time martingale M with respect to Y, this property implies that the expected value of the process at any pre-determined time n (i.e., period n) is the same as that at time 0. Using the condition of a martingale, we have

$$E[M_{n+1}] = E[E[M_{n+1}|Y_{0,n}]] = E[M_n]$$

for all $n \geq 0$. It follows from this relation that $E[M_n] = E[M_0]$. This "fairness" property also holds in other cases. For example, we may inspect the expected value of a martingale at a random time instant T and want to know if this conservation of fairness still holds. That is if $E[M_T] = E[M_0]$ holds. To make this fairness property hold in such a random inspection instant, we need to make two assumptions: (i) T should be upper bounded or $T \leq b$ holds with probability 1 where b is a real number, and (ii) T does not allow the gambler to peek ahead into the future. A random time that satisfies (ii) is called a "stopping time" and its discrete-time version is defined formally as follows:

Definition B.5 *A random variable $T \in \mathbb{N}$ is a stopping time with respect to the process $Y_0, Y_1, \ldots$ if for each integer k, the indicator random variable $I\{T = k\}$ is a function of $Y_{0,k}$.*

The random variable T is usually the time when a random event occurs. Here are two examples.

Example B.2 *(a stopping time case) Suppose that $\{M_n\}$ is a symmetric simple random walk starting at 0. Consider the random variable $T_1 = \min\{n : M_n = 1\}$, the first time the random walk reach 1. Then T_1 is a stopping time. Let $n > 0$. We have $T_1 = n$ if and only if $M_1 < 1, \ldots, M_{n-1} < 1$ and $M_n = 1$. Therefore, $I\{T_1 = n\} = I\{M_1 < 1, \ldots, M_{n-1} < 1, M_n = 1\}$ which is a function of $M_{0,n}$. This means that T_1 is a stopping time.*

Example B.3 *Consider the random variable $T_{max} = \sup\{n \leq 3 : M_n = \max_{0 \leq k \leq 3} M_k\}$. Clearly, T_{max} is not a stopping time. This is because the event $\{T_{max} = 2\}$ depends not only $M_{0,2}$ but also M_3.*

Next, we present the optional sampling theorem.

Theorem B.1 *For $M_0, M_1, \ldots$, a martingale with respect to $W_0, W_1, \ldots$, let T be a bounded stopping time. Then $E[M_T] = E[M_0]$.*

Proof. Assume that T is bounded by n, that is, $T(\omega) \leq n$ holds for all ω. Write M_T as the sum of increments of M plus M_0

$$M_T = M_0 + \sum_{k=1}^{T} (M_k - M_{k-1})$$

$$= M_0 + \sum_{k=1}^{n} (M_k - M_{k-1}) I_{(T \geq k)}.$$

Taking expectations of both sides of the above yields

$$E[M_T] = E[M_0] + \sum_{k=1}^{n} E\left\{ (M_k - M_{k-1}) I_{(T \geq k)} \right\}$$

$$= E[M_0] + \sum_{k=1}^{n} \left\{ E\left(M_k I_{(T \geq k)} \right) - E\left(M_{k-1} I_{(T \geq k)} \right) \right\},$$

where the interchange of sum and expectation is justified by the bounded T. Next, we can show that the sum is zero. Using the fact that $I_{(T \geq k)} = 1 - I_{(T \leq k-1)}$ which is a function of $W_{0,k-1}$, we have

$$E\left(M_k I_{(T \geq k)} \right) = E\left[E\left(M_k I_{(T \geq k)} | W_{0,k-1} \right) \right]$$

$$= E\left[E\left(M_k | W_{0,k-1} \right) I_{(T \geq k)} \right]$$

$$= E\left(M_{k-1} I_{(T \geq k)} \right),$$

which leads to $E\left\{ (M_k - M_{k-1}) I_{(T \geq k)} \right\} = 0$. Thus $E[M_T] = E[M_0]$ is proved. ∎

For the unbounded stopping time cases, this theorem can be applied by using a limiting argument. We demonstrate this approach by considering a simple symmetric random walk, $S_0 = 0, S_1, \dots$. Let a and b be integers with $a < 0 < b$, and define a stopping time $T = \inf\{n : S_n = a \text{ or } S_n = b\}$, the first time that the random walk hits either a or b. In this case, T is not bounded as there is a positive probability that T exceeds any given number. To apply the theorem, we can define a new stopping time

$$T_m = \min\{T, m\} = T \wedge m$$

for each m. Then T_m is a bounded stopping time and the theorem holds so that $E[S_{T_m}] = S_0 = 0$ for all m. Since this random walk is recurrent, $P(T < \infty) = 1$. For each ω such that $T(\omega) < \infty$, we have $T_m(\omega) = T(\omega)$ for sufficiently large m (i.e., $m \geq T(\omega)$). Thus, for each ω satisfying $T(\omega) < \infty$, we have $S_{T_m} = S_T$ for sufficiently large m. This clearly implies that $S_{T_m} \to S_T$ with probability 1. Combining this result with the Bounded Convergence Theorem (see Royden (1988)) yields $E[S_{T_m}] \to E[S_T]$. Therefore, because of $E S_{T_m} = 0$ for all m, we must have $E[S_T] = 0$. With this result, we can obtain the probability that the random walk stops at one of the two boundary values. That is

$$E[S_T] = aP(S_T = a) + bP(S_T = b)$$
$$= a[1 - P(S_T = b)] + bP(S_T = b) = 0,$$

which gives $P(S_T = b) = a/(a - b)$. Now, we give the definition for the continuous-time martingale and the properties presented for the discrete-time martingales also hold for martingales in continuous time.

Definition B.6 *Let $X(t)$ be a continuous-time stochastic process adapted to a right continuous filtration H_t for $0 \le t < \infty$. Then, X is said to be a martingale if $E[|X(t)|] < \infty$ for all t and*

$$E[X(t)|H_s] = X(s)$$

for all $s < t$.

Accordingly, the submartingale and supermartingale can be defined if the equality in the definition above is replaced by $\ge$ or $\le$, respectively.

REFERENCES

1. M. Harrison, "Brownian Motion And Stochastic Flow Systems", John Wiley & Sons, New York, 1985.
2. H.L. Royden, "Real Analysis", Prentice Hall, Englewood Cliffs, NJ, 3rd ed., 1988.
3. G. Zitkovic, Continuous-Time Finance: Lecture Notes, 2005. https://web.ma.utexas.edu/users/gordanz/notes/ctf.pdf

C Some Useful Bounds, Inequalities, and Limit Laws

C.1 MARKOV INEQUALITY

Lemma C.1 *(Markov's Inequality): If Y is a non-negative random variable, then $P(Y > t) \leq \frac{E[Y]}{t}$.*

Proof. We can write the expectation of Y as the integral

$$E[Y] = \int_{x=0}^{\infty} P(Y > x)dx.$$

Note that we can take this integral from 0 since Y is non-negative, and that $P(Y > x)$ is a nonincreasing function. As Figure C.1 illustrates, the nonincreasing nature of $P(Y > x)$ implies the following lower bound: $E[Y] \geq tP(Y > t)$ which leads to the Markov's Inequality. ■

Now we give an application of the Markov's inequality which is to prove the weak law of large numbers (WLLN)

Theorem C.1 *(Weak Law of Large Numbers): For all $\varepsilon > 0$, $\lim_{n \to \infty} P(|\hat{\mu}_n - \mu| > \varepsilon) = 0$.*

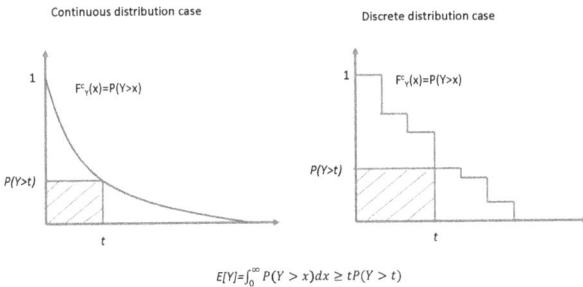

Figure C.1 Markov Inequality.

DOI: 10.1201/9781003150060-C

Proof. Fix $\varepsilon > 0$. Then, $P(|\hat{\mu}_n - \mu| > \varepsilon) = P(|\hat{\mu}_n - \mu|^2 > \varepsilon^2)$. Now, the random variable $|\hat{\mu}_n - \mu|^2$ is non-negative. It follows from the Markov's Inequality that

$$P(|\hat{\mu}_n - \mu| > \varepsilon) = P(|\hat{\mu}_n - \mu|^2 > \varepsilon^2) \leq \frac{E[|\hat{\mu}_n - \mu|^2]}{\varepsilon^2} = \frac{\sigma^2}{n\varepsilon^2},$$

where, in the last equality, we use $E[|\hat{\mu}_n - \mu|^2] = \sigma^2/n$. Taking the limit yields

$$\lim_{n \to \infty} P(|\hat{\mu}_n - \mu| > \varepsilon) \leq \lim_{n \to \infty} \frac{\sigma^2}{n\varepsilon^2} = 0.$$

This completes the proof of the WLLN. ∎

Lemma C.2 *(Chebyshev's Inequality): If X is a random variable, then for t >* 0,

$$P(|X - E[X]| \geq t) \leq \frac{Var(Y)}{t^2}.$$

Proof. This immediately holds by applying Markov's inequality to the non-negative random variable $(X - E[X])^2$. ∎

Next, we present the large deviation estimate (LDE). To demonstrate the relation between WLLN and LDE, we consider the sample proportion estimation problem. Let S_n be the sample count, then the sample proportion is $\hat{p} = S_n/n$, which is an unbiased estimate of p. Note that the variance of S_n is $np(1-p)$. Thus, we have

$$P_n\left(\left|\frac{S_n}{n} - p\right| > \varepsilon\right) = P_n(|S_n - pn| > n\varepsilon)$$

$$\leq \frac{Var(S_n)}{(n\varepsilon)^2} = \frac{p(1-p)}{\varepsilon^2}\frac{1}{n}.$$

This result, which leads to the WLLN, implies that $P_n(|S_n/n - p| > \varepsilon) = O(1/n)$. A more refined estimate of the probability of such a deviation between $\hat{p}$ and p can be obtained and this estimate is the LDE. Given p, we define a function

$$g_+(\varepsilon) = (p + \varepsilon)\ln\left(\frac{p+\varepsilon}{p}\right) + (1 - p - \varepsilon)\ln\left(\frac{1-p-\varepsilon}{1-p}\right).$$

The motivation of defining this function is due to the approximation to Binomial distribution. Let X_n be a Bernoulli random variable with parameter p, then define the Binomial random variable $S_n = X_1 + \cdots + X_n$. Then for $x \in [0, 1]$, we have

$$P_n(S_n = [nx]) = \binom{n}{[nx]} p^{[nx]}(1-p)^{n-[nx]},$$

where $[nx]$ denotes the integer part of nx. We denote by $l = [nx]$ and use Stirling's formula to obtain an approximation to the Binomial probability above

$$P_n(S_n = [nx]) \approx \sqrt{\frac{n}{2\pi l(n-l)}} \exp[n\ln n - l\ln l - (n-l)\ln(n-l) + l\ln p$$

$$+ (n-l)\ln(1-p)]$$

$$= \sqrt{\frac{n}{2\pi l(n-l)}} \exp\left\{n\left[\frac{l}{n}\ln p + \left(1 - \frac{l}{n}\right)\ln(1-p) - \frac{l}{n}\ln\frac{l}{n}\right.\right.$$

$$\left.\left. - \left(1 - \frac{l}{n}\right)\ln\left(1 - \frac{l}{n}\right)\right]\right\}$$

$$= \sqrt{\frac{n}{2\pi l(n-l)}} \exp(-nI(x) + O(\ln n)) = \exp(-nI(x) + O(\ln n)),$$

where

$$I(x) = x\ln\frac{x}{p} + (1-x)\ln\frac{1-x}{1-p}.$$

Now we establish two lemmas by using Taylor expansions of $g_+(\varepsilon)$ and $e^{-ng_+(\varepsilon)}$. These lemmas provide a sense of the asymptotics of these functions.

Lemma C.3

$$g_+(\varepsilon) = \frac{1}{2p(1-p)}\varepsilon^2 + \frac{2p-1}{6p^2(1-p)^2}\varepsilon^3 + \frac{1-3p+3p^2}{12p^3(1-p)^3}\varepsilon^4 + O(\varepsilon^5).$$

Lemma C.4

$$e^{-ng_+(\varepsilon)} = 1 - \frac{n}{2p(1-p)}\varepsilon^2 - \frac{n(2p-1)}{6p^2(1-p)^2}\varepsilon^3$$

$$+ \frac{2n - 6np + 6np^2 - 3n^2p + 3n^2p^2}{24p^3(1-p)^3}\varepsilon^4 + O(\varepsilon^5).$$

Theorem C.2 *(Large Deviations Estimate). For $0 < \varepsilon \leq 1-p$ and $n \geq 1$,*

$$P_n\left(\frac{S_n}{n} \geq p + \varepsilon\right) \leq e^{-ng_+(\varepsilon)}.$$

Proof. For a given t, we have

$$P_n\left(\frac{S_n}{n} \geq p + \varepsilon\right) = P_n\left(e^{t(S_n - np - n\varepsilon)} \geq 1\right).$$

Now apply Markov's Inequality to this positive random variable:

$$P_n\left(e^{t(S_n-np-n\varepsilon)} \geq 1\right) \leq E\left[e^{t(S_n-np-n\varepsilon)}\right]$$

$$= e^{-nt(p+\varepsilon)}E[e^{tS_n}]$$

$$= e^{-nt(p+\varepsilon)}\sum_{k=0}^{n} e^{tk}\binom{n}{k}p^k(1-p)^{n-k}$$

$$= e^{-nt(p+\varepsilon)}(1-p+pe^t)^n$$

$$= e^{-n(t(p+\varepsilon)-\ln(1-p+pe^t))}.$$

To get rid of t in the upper bound, we need to find

$$\sup_{t>0}(t(p+\varepsilon)-\ln(1-p+pe^t)).$$

Let $h(t) = t(p+\varepsilon)-\ln(1-p+pe^t)$ for $0 < p < 1$ and $0 < \varepsilon < 1-p$. Then $h(0) = 0$ and $h'(t) = p+\varepsilon - pe^t/(1-p+pe^t)$. Note that $h'(0) = \varepsilon > 0$ and $\lim_{t\to\infty}h'(t) = p+\varepsilon-1 < 0$. This implies that the supremum of $h(t)$ is attained at some strictly positive value. This value can be determined by the first-order condition $h'(t) = 0$ as follows:

$$t = \ln\left(\frac{-p+p^2-\varepsilon+\varepsilon p}{p(p+\varepsilon-1)}\right) = \ln\left(\frac{(p+\varepsilon)(1-p)}{p(1-p-\varepsilon)}\right).$$

Substituting this critical point into $h(t)$, we have

$$\sup_{t>0} h(t) = (p+\varepsilon)\ln\left(\frac{p+\varepsilon}{p}\right) + (p+\varepsilon)\ln\left(\frac{1-p}{1-p-\varepsilon}\right)$$

$$- \ln\left(1-p+p\frac{(p+\varepsilon)(1-p)}{p(1-p-\varepsilon)}\right)$$

$$= (p+\varepsilon)\ln\left(\frac{p+\varepsilon}{p}\right) + (p+\varepsilon)\ln\left(\frac{1-p}{1-p-\varepsilon}\right)$$

$$- \ln\left((1-p)\left(1+\frac{(p+\varepsilon)}{(1-p-\varepsilon)}\right)\right)$$

$$= (p+\varepsilon)\ln\left(\frac{p+\varepsilon}{p}\right) + (p+\varepsilon)\ln\left(\frac{1-p}{1-p-\varepsilon}\right)$$

$$- \ln\left(\frac{1-p}{1-p-\varepsilon}\right) = g_+(\varepsilon).$$

This completes the proof. ∎

Defining $g_-(\varepsilon) = g_+(-\varepsilon)$ for $0 < \varepsilon < p$, we get

Corollary C.1 *For $0 < \varepsilon < p$ and $n \geq 1$,*

$$P_n\left(\frac{S_n}{n} \leq p - \varepsilon\right) \leq e^{-ng_-(\varepsilon)}.$$

Proof. Interchange the definition of success and failure in the binomial distribution above. Then, "success" occurs with probability $1 - p$ and "failure" occurs with probability p and we have the complementary process $S_n^c = n - S_n$. Thus, we have the equivalent events $\{S_n/n \leq p - \varepsilon\} \Leftrightarrow \{S_n^c/n > 1 - p + \varepsilon\}$. Applying the theorem above, we obtain

$$P_n\left(\frac{S_n^c}{n} > 1 - p + \varepsilon\right) = P_n\left(\frac{S_n}{n} \leq p - \varepsilon\right) \leq e^{-ng_+^c(\varepsilon)},$$

where

$$g_+^c(\varepsilon) = (1 - p - \varepsilon)\ln\left(\frac{1 - p + \varepsilon}{1 - p}\right) + (p - \varepsilon)\ln\left(\frac{p - \varepsilon}{p}\right) = g_+(-\varepsilon) = g_-(\varepsilon).$$

This completes the proof. ∎

Summarizing the above, we have

Corollary C.2 *For $0 < \varepsilon < \min(p, 1 - p)$ and $n \geq 1$,*

$$P_n\left(\left|\frac{S_n}{n} - p\right| \geq \varepsilon\right) \leq e^{-ng_+(\varepsilon)} + e^{-ng_-(\varepsilon)}.$$

Here we also provide a simple proof of the Central Limit Theorem (CLT) using the characteristic function.

Theorem C.3 *(Central Limit Theorem (CLT)):* $\lim_{n \to \infty} \sqrt{n}(\hat{\mu}_n - \mu) \sim \mathcal{N}(0, \sigma^2)$.

Proof. Consider the random variable

$$Z_n = \frac{X_1 + \cdots + X_n - n\mu}{\sqrt{n\sigma^2}}$$

$$= \sum_{j=1}^{n} \frac{X_j - \mu}{\sqrt{n\sigma^2}} = \sum_{j=1}^{n} \frac{1}{\sqrt{n}} Y_j,$$

where $Y_j = (X_j - \mu)/\sigma$.

We will prove the central limit theorem by calculating the characteristic function of this random variable, and showing that, in the limit as $n \to \infty$, it is the same as the characteristic function for $\mathcal{N}(0,1)$. Using the definition of characteristic function, we have

$$\phi_{Z_n}(t) = E[exp(itZ_n)] = E\left[exp\left(it\sum_j \frac{1}{\sqrt{n}}Y_j\right)\right]$$

$$= E\left[\prod_j exp\left(it\frac{1}{\sqrt{n}}Y_j\right)\right] = \prod_j E\left[exp\left(it\frac{1}{\sqrt{n}}Y_j\right)\right]$$

$$= E\left[exp\left(it\frac{1}{\sqrt{n}}Y_j\right)\right]^n = \left(\phi_{Y_1}\left(\frac{t}{\sqrt{n}}\right)\right)^n.$$

Using the Taylor's expansion to approximate the characteristic function, we obtain, for some constant c (maybe complex), as $n \to \infty$ (i.e., $\frac{t}{\sqrt{n}} \to 0$)

$$\phi_{Y_i}\left(\frac{t}{\sqrt{n}}\right) = 1 - \frac{t^2}{2n} + c\frac{t^3}{6n^{\frac{3}{2}}} + o\left(\frac{t^3}{n^{\frac{3}{2}}}\right).$$

Finally, we notice that, as $n \to \infty$, the characteristic function $\phi_{Z_n}(t) \to \left(1 - \frac{t^2}{2n}\right)^n$. Using the identity $e^x = \lim_{n\to\infty}\left(1 + \frac{x}{n}\right)^n$, we have

$$\lim_{n\to\infty}\phi_{Z_n}(t) = e^{-\frac{1}{2}t^2},$$

which is the characteristic function of the standard normal distribution, $\mathcal{N}(0,1)$. ∎

C.2   JENSEN'S INEQUALITY

A classical inequality regarding expectations is given as follows.

Theorem C.4 *(Jensen's inequality) Let $\psi : \mathbb{R} \to \mathbb{R}$ be a convex function and let X be an integrable random variable on the probability space $(\Omega, \mathcal{F}, P)$. If $\psi(X)$ is integrable, then*

$$\psi(E[X]) \leq E[\psi(X)]$$

almost surely.

Proof. Denote the mean of the random variable by $\mu = E[X]$. Since the function ψ is twice differentiable and ψ'' is continuous, expanding ψ in a Taylor series about μ yields

$$\psi(x) = \psi(\mu) + \psi'(\mu)(x - \mu) + \frac{1}{2}\psi''(c)(x - \mu)^2$$

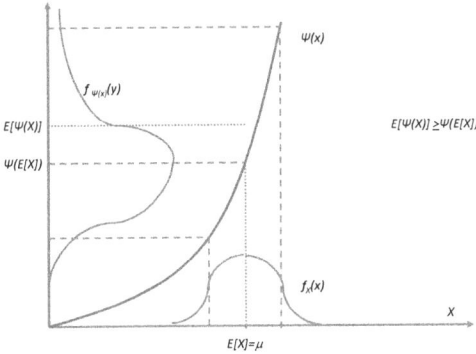

Figure C.2 Jensen's inequality.

with c in between x and μ. Due to the convexity of $\psi'' \geq 0$, we have

$$\psi(x) \geq \psi(\mu) + \psi'(\mu)(x - \mu).$$

Replacing x by the random variable X and taking the expectation in the inequality gives

$$E[\psi(X)] \geq E[\psi(\mu) + \psi'(\mu)(X - \mu)] = \psi(\mu) + \psi'(\mu)(E[X] - \mu)$$
$$= \psi(\mu) = \psi(E[X]),$$

which proves the result. ∎

Figure C.2 shows it intuitively.

C.3 CAUCHY–SCHWARZ INEQUALITY

Another important inequality regarding the expectations of functions of a random variable is the Cauchy–Schwarz inequality stated in the following theorem. Let X be an integrable random variable on the probability space $(\Omega, \mathcal{F}, P)$.

Theorem C.5 *For random variable X and functions and functions $h_1()$ and $h_2()$, it holds that*

$$\{E[h_1(X)h_2(X)]\}^2 \leq E[\{h_1(X)\}^2]E[\{h_2(X)\}^2] \tag{C.1}$$

with equality if and only if either $E[\{h_1(X)\}^2] = 0$ or $E[\{h_2(X)\}^2] = 0$, or $P[h_1(X) = ch_2(X)] = 1$ for some constant $c \neq 0$.

Proof. Let $X_1 = h_1(X)$ and $X - 2 = g_2(X)$ and define

$$Y_1 = aX_1 + bX_2, \quad Y_2 = aX_1 - bX_2.$$

Based on the fact of $E[Y_1^2] \geq 0$ and $E[Y_2^2] \geq 0$, we can write

$$a^2 E[X_1^2] + b^2 E[X_2^2] + 2ab E[X_1 X_2] \geq 0$$
$$a^2 E[X_1^2] + b^2 E[X_2^2] - 2ab E[X_1 X_2] \geq 0.$$

Set $a^2 = E[X_2^2]$ and $b^2 = E[X_1^2]$. Clearly, if either a or b equals 0, the inequality holds. Thus, we only consider $E[X_1^2], E[X_2^2] > 0$ case. The two inequalities can be written as

$$2E[X_1^2]E[X_2^2] + 2\{E[X_1^2]E[X_2^2]\}^{1/2} E[X_1 X_2] \geq 0$$
$$2E[X_1^2]E[X_2^2] - 2\{E[X_1^2]E[X_2^2]\}^{1/2} E[X_1 X_2] \geq 0.$$

Rearranging the terms yields

$$-\{E[X_1^2]E[X_2^2]\}^{1/2} \leq E[X_1 X_2] \leq \{E[X_1^2]E[X_2^2]\}^{1/2},$$

which is equivalent to $\{E[X_1 X_2]\}^2 \leq E[X_1^2]E[X_2^2]$. Thus, (C.1) is proved. The proof of the equality cases is omitted and left as an exercise. ■

C.4 HÖLDER'S INEQUALITY

An inequality regarding the expectations of the product of two random variables is the Hölder's inequality. We start with a lemma

Lemma C.5 *For $a, b > 0$ and $p, q > 1$ satisfying $p^{-1} + q^{-1} = 1$, we have*

$$\frac{a^p}{p} + \frac{b^q}{q} \geq ab$$

with equality if and only if $a^p = b^q$.

Proof. Define a function $h(a,b) = p^{-1}a^p + q^{-1}b^q - ab$. It is easy to show that this function is convex. Thus the first-order condition for the minimum is $a^{p-1} = b$ and the minimum value of $h(a,b)$ is

$$p^{-1}a^p + q-1(a^{p-1})^q - a(a^{p-1}) = p^{-1}a^p + q^{-1}a^p - a^p = 0$$

by the condition $p^{-1} + q^{-1} = 1$. It is concluded that $h(a,b)$ is a nonnegative function with the minimum 0 attained at $a^{p-1} = b$ which is equivalent to $a^p = b^q$. ■

Theorem C.6 *(Hölder's Inequality) Suppose that X and Y are two random variables, and $p, q > 1$ satisfy $p^{-1} + q^{-1} = 1$. Then*

$$|E[XY]| \leq E[|XY|] \leq \{E[|X|^p]\}^{1/p} \{E[|Y|^{1/q}]\}^{1/q}.$$

Proof. We consider absolutely continuous case here (discrete case is similar). Note that

$$E[|XY|] = \int\int |xy| f_{X,Y}(x,y) dx dy \geq \int\int xy f_{X,Y} dx dy = E[XY]$$

and

$$E[XY] = \int\int xy f_{X,Y}(x,y) dx dy \geq \int\int -|xy| f_{X,Y} dx dy = -E[|XY|],$$

which implies $-E[|XY|] \leq E[XY] \leq E[|XY|]$. Thus $|E[XY]| \leq E[|XY|]$. To prove the second inequality, we set

$$a = \frac{|X|}{\{E[|X|^p]\}^{1/p}} \quad b = \frac{|Y|}{\{E[|Y|^q]\}^{1/q}}.$$

Then, it follows from the lemma above that

$$p^{-1} \frac{|X|^p}{E[|X|^p]} + q^{-1} \frac{|Y|^q}{E[|Y|^q]} \geq \frac{|XY|}{\{E[|X|^p]\}^{1/p}\{E[|Y|^q]\}^{1/q}}.$$

Taking expectations of the both sides above, we have the left hand side as

$$p^{-1} \frac{E[|X|^p]}{E[|X|^p]} + q^{-1} \frac{E[|Y|^q]}{E[|Y|^q]} = p^{-1} + q^{-1} = 1,$$

and the right hand side as

$$\frac{E[|XY|]}{\{E[|X|^p]\}^{1/p}\{E[|Y|^q]\}^{1/q}}.$$

Hence, the result follows. ∎

Note that the equality holds if and only if $P(|X|^p = c|Y|^q) = 1$ for some non-zero constant c.

C.5 CHERNOFF BOUNDS AND HOEFFDING'S INEQUALITY

Central limit theorem guarantees are useful for large sample sizes, but if n is small, we would still like to bound the deviation of $\hat{\mu}_n$ from the true mean μ in terms of probability. Chernoff Bounds are for bounding random variables using their moment generating function. We wish to bound the quantity $P(\hat{\mu}_n - \mu > \varepsilon)$. For $\lambda > 0$, we can use the fact that e^x is a monotonically increasing function to obtain a bound as follows:

$$P(\hat{\mu}_n - \mu > \varepsilon) = P(\lambda(\hat{\mu}_n - \mu) > \lambda\varepsilon) = P(e^{\lambda(\hat{\mu}_n - \mu)} > e^{\lambda\varepsilon})$$

$$\leq e^{-\lambda\varepsilon}E\left[e^{\lambda(\hat{\mu}_n - \mu)}\right] = e^{-\lambda\varepsilon}E\left[e^{\lambda(\frac{1}{n}\Sigma_i(X_i - \mu))}\right]$$

$$= e^{-\lambda\varepsilon}E\left[\prod_i e^{\frac{\lambda}{n}(X_i - \mu)}\right] = e^{-\lambda\varepsilon}\prod_i E\left[e^{\frac{\lambda}{n}(X_i - \mu)}\right]$$

$$= e^{-\lambda\varepsilon}E\left[e^{\frac{\lambda}{n}(X_i - \mu)}\right]^n.$$

Next, we introduce a lemma which will be used with the Chernoff bound to get another useful inequality, called Hoeffding's inequality.

Lemma C.6 (*Hoeffding's Lemma*) *Let X be a random variable with a support of* $[a,b]$ *and* $E[X] = 0$. *Then for any real s, the following inequality holds*

$$E[e^{sX}] \leq e^{\frac{s^2(b-a)^2}{8}}.$$

Proof. Since the exponential function e^{sx} is convex in x, we have

$$e^{sX} \leq \frac{b-X}{b-a}e^{sa} + \frac{X-a}{b-a}e^{sb}.$$

Taking the expectation on both sides of the above inequality, we obtain

$$E[e^{sX}] \leq \frac{b-E[X]}{b-a}e^{sa} + \frac{E[X]-a}{b-a}e^{sb} = \frac{b}{b-a}e^{sa} - \frac{a}{b-a}e^{sb}$$

$$= (1-p)e^{sa} + pe^{sb} = \left((1-p) + pe^{s(b-a)}\right)e^{sa}$$

$$= \left(1 - p + pe^{s(b-a)}\right)e^{-sp(b-a)}$$

$$= (1 - p + pe^u)e^{pu},$$

where $p = -a/(b-a)$ and $u = s(b-a)$. Define function ϕ of u as the logarithm of the above expression:

$$\phi(u) = \log((1 - p + pe^u)e^{pu}) = pu + \log(1 - p + pe^u).$$

Thus, we find a bound for $E[e^{sX}] \leq e^{\phi(u)}$. Next, we work out an explicit bound in terms of only parameters a and b. Using Taylor's theorem, we know that there exists some $z \in [0, u]$ so that

$$\phi(u) = \phi(0) + u\phi'(0) + \frac{1}{2}u^2\phi''(z) \leq \phi(0) + u\phi'(0) + \sup_z \frac{1}{2}u^2\phi''(z).$$

We write the first and second derivatives of $\phi(u)$ as

$$\phi'(u) = p + \frac{pe^u}{1 - p + pe^u}, \quad \phi''(u) = \frac{p(1-p)e^u}{(1-p+pe^u)^2}.$$

Using $\phi(0) = \phi'(0) = 0$, we can write the upper bound of $\phi(u)$ as

$$\phi(u) \le \phi(0) + u\phi'(0) + \sup_z \frac{1}{2}u^2\phi''(z) = \sup_z \frac{1}{2}u^2\phi''.$$

Now, we find the maximizer of $\phi''(z)$. Introducing $y = e^u$, we write $\phi''(z)$ as

$$\frac{p(1-p)y}{(1-p+py)^2},$$

which is a linear expression over a quadratic expression of y and thus concave in $y > 0$. Using the first order condition, we can obtain the maximum of $\phi''(u)$. Setting

$$\frac{d}{dy}\frac{p(1-p)y}{(1-p+py)^2} = \frac{p(1-p)(1-p-py)}{(1-p+py)^3} = 0.$$

Solving it gives two critical points: $y = (1-p)/p$ and $y = (p-1)/p$. Note that $E[X] \ge 0$ implies $a \le 0$, which implies $p = \frac{-a}{b-a} \in [0,1]$. Thus, $y = (1-p)/p \ge 0$ and $y = (p-1)/p \le 0$. We therefore select $y = (1-p)/p$. Substituting this critical value in the upper bound expression, we obtain

$$\phi''(u) \le \frac{p(1-p)^{\frac{1-p}{p}}}{(1-p+p^{\frac{1-p}{p}})^2} = \frac{1}{4}.$$

Finally, we have

$$E[e^{sX}] \le e^{\phi(u)} \le e^{\frac{u^2}{8}} = e^{\frac{s^2(b-a)^2}{8}}.$$

This completes the proof. ∎
Now we are ready to present the Hoeffding's Inequality.

Theorem C.7 *(Hoeffding's Inequality) Given independent random variables* $\{X_1, ..., X_m\}$ *where* $a_i \le X_i \le b_i$ *almost surely (with probability 1), we have:*

$$P\left(\frac{1}{m}\sum_{i=1}^{m}X_i - \frac{1}{m}\sum_{i=1}^{m}E[X_i] \ge \varepsilon\right) \le exp\left(\frac{-2\varepsilon^2 m^2}{\sum_{i=1}^{m}(b_i - a_i)^2}\right).$$

Proof. For all $1 \le i \le m$, define a variable $Z_i = X_i - E[X_i]$. This implies that $E[Z_i] = 0$ with support $[a_i - E[X_i], b_i - E[X_i]]$. In particular, we note that the domain must still have length $b_i - a_i$ independent of the expectation of X_i. Let s be some positive value. We have

$$P\left(\sum_{i=1}^{m} Z_i \geq t\right) = P\left(exp\left(s\sum_{i=1}^{m} Z_i\right) \geq e^{st}\right) \overset{\text{Chernoff}}{\leq} \frac{E\left[\prod_{i=1}^{m} e^{sZ_i}\right]}{e^{st}}.$$

Using the independence of Z_i's, we can interchange the expectation and product and obtain

$$\frac{E\left[\prod_{i=1}^{m} e^{sZ_i}\right]}{e^{st}} = \frac{\prod_{i=1}^{m} E\left[e^{sZ_i}\right]}{e^{st}} \overset{\text{hoeffding Lemma}}{\leq} e^{-st} \prod_{i=1}^{m} e^{\frac{s^2(b_i - a_i)^2}{8}}$$

$$= exp\left(-st + \frac{s^2}{8}\sum_{i=1}^{m}(b_i - a_i)^2\right).$$

Selecting a special value of $s = 4t/(\sum_{i=1}^{m}(b_i - a_i)^2)$ and $t = \varepsilon m$, we re-write the expression for the bound as

$$P\left(\sum_{i=1}^{m} Z_i \geq \varepsilon m\right) = P\left(\frac{1}{m}\sum_{i=1}^{m} X_i - \frac{1}{m}\sum_{i=1}^{m} E[X_i] \geq \varepsilon\right)$$

$$\leq exp\left(-\frac{4\varepsilon m}{\sum_{i=1}^{m}(b_i - a_i)^2}\varepsilon m\right.$$

$$+ \frac{1}{8}\left(\frac{4\varepsilon m}{\sum_{i=1}^{m}(b_i - a_i)^2}\right)^2\sum_{i=1}^{m}(b_i - a_i)^2\right)$$

$$= exp\left(\frac{-2\varepsilon^2 m^2}{\sum_{i=1}^{m}(b_i - a_i)^2}\right),$$

which is the result of the theorem. ∎

C.6 THE LAW OF THE ITERATED LOGARITHM

We use a symmetric random walk to demonstrate this law (other versions of Law of Iterated Logarithm (LIL) or Functional LIL (LIL) are similar).

Theorem C.8 *(LIL) For a symmetric random walk, let X_i be the outcome of the ith step, where $X_i = 1$ with probability p and $X_i = -1$ with probability $1 - p$. Let $S_n = \sum_{i=1}^{n} X_i$ denote the value after n steps. Then, almost surely, we have*

$$\limsup_{n\to\infty} \frac{S_n}{\sqrt{2n\log\log n}} = 1$$

and by symmetry,

$$\liminf_{n\to\infty} \frac{S_n}{\sqrt{2n\log\log n}} = -1.$$

The proof of this theorem consists of two stages: preparation and verification. In the verification stage, we accomplish three tasks. First, we prove a lemma about the local CLT and establish a bound on the tail probability of S_n. Based on this lemma, we establish a lower bound on the tail probability of S_n. Then we apply the Borel–Cantelli Lemma to get an upper bound on the probability of a sequence of events regarding the value of S_n. In the verification stage, we apply the results in the first stage to complete the proof.

C.6.1 PREPARATION

Lemma C.7 (a) *Local CLT (LCLT). If* $k = o(n^{3/4})$ *and* $n + k$ *is even, then we have*

$$P(S_n = k) \sim \sqrt{\frac{2}{\pi n}} e^{-\frac{k^2}{2n}}.$$

(b) For all $n, k \geq 0$, *we have* $P(S_n \geq k) \leq e^{-\frac{k^2}{2n}}$.

Proof. (a) Applying Stirling approximation formula to the probability distribution of a symmetric random walk, as $n \to \infty$ we have

$$P(S_n = k) = \binom{n}{\frac{n+k}{2}} \left(\frac{1}{2}\right) = \frac{n! 2^{-n}}{\left(\frac{n+k}{2}\right)! \left(\frac{n-k}{2}\right)!}$$

$$\sim \frac{\sqrt{2\pi n}}{\sqrt{2\pi \frac{n+k}{2}} \sqrt{2\pi \frac{n-k}{2}}} \cdot \frac{\left(\frac{n}{e}\right)^n 2^{-n}}{\left(\frac{n+k}{2e}\right)^{\frac{n+k}{2}} \left(\frac{n-k}{2e}\right)^{\frac{n-k}{2}}} \qquad \text{(C.2)}$$

$$= \sqrt{\frac{2}{\pi n \left(1 - \frac{k^2}{n^2}\right)}} \cdot \frac{n^n}{(n+k)^{\frac{n+k}{2}} (n-k)^{\frac{n-k}{2}}}.$$

Now we can write the second factor on the right-hand side of the equation above as

$$\frac{n^n}{(n+k)^{\frac{n+k}{2}} (n-k)^{\frac{n-k}{2}}} = \exp\left(n\log n - \frac{n+k}{2}\log(n+k) - \frac{n-k}{2}\log(n-k)\right) = \exp(\Delta).$$

Using $\log(1+x) = x - \frac{x^2}{2} + \frac{x^3}{3} - O(x^4)$ as $x \to 0$, we can compute Δ as

$$\Delta = -\frac{n+k}{2}\log\left(1 + \frac{k}{n}\right) - \frac{n-k}{2}\log\left(1 - \frac{k}{n}\right)$$

$$= -\frac{n+k}{2}\left(\frac{k}{n} - \frac{k^2}{2n^2} + \frac{k^3}{3n^3} + O\left(\frac{k^4}{n^4}\right)\right) + \frac{n-k}{2}\left(\frac{k}{n} + \frac{k^2}{2n^2} + \frac{k^3}{3n^3} + O\left(\frac{k^4}{n^4}\right)\right)$$

$$= \frac{k^2}{4n} + \frac{k^2}{4n} - \frac{k^2}{2n} - \frac{k^2}{2n} + O\left(\frac{k^4}{n^3}\right)$$

$$= -\frac{k^2}{2n} + O\left(\frac{k^4}{n^3}\right),$$

where the last equality follows from $n \to \infty$ and $k = o(n^{3/4})$. Substituting this expression into (C.2) gives

$$P(S_n = k) \sim \sqrt{\frac{2}{\pi n \left(1 - \frac{k^2}{n^2}\right)}} e^{-\frac{k^2}{2n} + o\left(\frac{k^4}{n^3}\right)} \sim \sqrt{\frac{2}{\pi n}} e^{-\frac{k^2}{2n}} \quad \left(\begin{matrix} n \to \infty \\ k = o(n^{3/4}) \end{matrix}\right).$$

(b) For any random variable X and a positive θ, it follows from the Markov inequality that

$$P(X \geq t) = P\left(e^{\theta X} \geq e^{\theta t}\right) \leq \frac{E[e^{\theta X}]}{e^{\theta t}}.$$

By taking $X = S_n, t = k$ and comparing Taylor coefficients, we obtain

$$E[e^{\theta X_1}] = \left(\frac{1}{2} e^{\theta} + \frac{1}{2} e^{-\theta}\right) \leq e^{\theta^2/2},$$

$$P(S_n \geq k) \leq \frac{E e^{\theta S_n}}{e^{\theta k}} = \frac{\left(\frac{1}{2} e^{\theta} + \frac{1}{2} e^{-\theta}\right)^n}{e^{\theta k}} \leq \frac{e^{\theta^2 \frac{n}{2}}}{e^{\theta k}} = e^{-\frac{k^2}{2n}},$$

where $\theta = k/n$. ■

Next, we establish a lower bound on the tail probability of S_n.

Lemma C.8 *For* $k \geq \sqrt{n}, k = o(n^{3/4})$, *we have* $P(S_n \geq k) \geq c \frac{\sqrt{n}}{k} e^{-k^2/2n}$ *for some $c > 0$.*

Proof.

$$P(S_n \geq k) \geq P\left(k \leq S_n \leq k + \frac{n}{k}\right) \geq \sqrt{\frac{c}{n}} \sum_{k \leq j \leq \lfloor k + \frac{n}{k} \rfloor} e^{-j^2/2n}.$$

For these j we have

$$e^{-j^2/2n} \geq e^{-\left(k + \frac{n}{k}\right)^2/2n} = e^{-\frac{k^2}{2n} - 1 - \frac{n}{2k^2}} \geq c' e^{-k^2/2n},$$

where the last inequality follows from $k \geq \sqrt{n}$. Hence,

$$P(S_n \geq k) \geq \frac{c''}{\sqrt{n}} e^{-\frac{k^2}{2n}} \cdot \frac{n}{k}.$$

■

Recall from Chapter 6 (Theorem 6.2), we have

$$P(M_n \geq k) = P(S_n = k) + 2P(S_n \geq k + 1), \tag{C.3}$$

where $M_n = \max_{0 \geq j \geq n} S_j$. The Borel–Cantelli Lemma presented in Chapter A is re-stated as follows:

Lemma C.9 *If $\{A_n\}$ is a sequence of events, then (i) If $\sum_n P(A_n) < \infty$, then A_n only finitely many occur almost surely. (ii) If $\sum_n A_n = \infty$ and A_n's are independent, then infinitely many of A_n's occur almost surely.*

Define $A_n = \{S_n \geq (1+\varepsilon)\sqrt{2n \log n}\}$. Using Lemma C.7, we have $P(A_n) \leq \frac{1}{n^{(1+\varepsilon)^2}}$. Since the p-series (or hyperharmonic series), defined as $\sum_{n=1}^{\infty} 1/n^p$ converges with $p > 1$, we have $\sum_n P(A_n) < \infty$. It follows from (i) of Lemma C.9 that A_n only finitely many occur. This implies that $A_n^c = \{S_n \leq (1+\varepsilon)\sqrt{2n \log n}\}$ occur eventually almost surely.

C.6.2 VERIFICATION

For simplicity, we will not be concerned about integrality of indices in the following proof. Actually, we may round the indices appropriately everywhere and argument will still go through. For $\gamma > 0, a > 1, k \in \mathbb{N}$, introducing the notation $u(n) = \sqrt{2n \log \log n}$, and using (C.3), we have

$$P\left(\max_{0 \leq j \leq a^k} S_j \geq (1+\gamma)u\left(a^k\right) \right) \leq P\left(S_{a^k} = (1+\gamma)u\left(a^k\right) \right)$$

$$+ 2P\left(S_{a^k} \geq (1+\gamma)u\left(a^k\right) + 1 \right)$$

$$+ P\left(S_{a^k} = (1+\gamma)u\left(a^k\right) \right)$$

$$\leq 2P\left(S_{a^k} \geq (1+\gamma)u\left(a^k\right) \right) \leq 2e^{-\frac{(1+\gamma)^2 u(a^k)^2}{2a^k}}$$

$$= 2e^{-(1+\gamma)^2 \log\log(a^k)} = \frac{2}{(k \log a)^{(1+\gamma)^2}},$$

where the third inequality follows from (b) of Lemma C.7. Note that these probability estimates are summable in k due to the hyperharmonic series with $p > 1$. It follows from (i) of Lemma C.9 that the event sequence $A_{a^k} = \{\max_{0 \leq j \leq a^k} S_j \geq (1+\gamma)u\left(a^k\right)\}$ only finitely many occur almost surely. This implies

$$P\left(\max_{0 \leq j \leq a^k} S_j \leq (1+\gamma)u\left(a^k\right) \text{ from some } k \text{ on} \right) = 1. \qquad (C.4)$$

Now, for large n, write $a^{k-1} \leq n \leq a^k$. Then, we have

$$\frac{S_n}{u(n)} = \frac{S_n}{u(a^k)} \cdot \frac{u(a^k)}{a^k} \cdot \frac{a^k}{n} \cdot \frac{n}{u(n)} \leq a(1+\gamma)$$

due to (i) $\frac{S_n}{u(a^k)} \leq (1+\gamma)$ which follows from (C.4); (ii) $\frac{a^k}{n} \leq a$ which is from $a^{k-1} \leq n$, and (iii) $\frac{u(a^k)}{a^k} \cdot \frac{n}{u(n)} \leq 1$ because $\frac{u(t)}{t}$ is eventually decreasing. Therefore, we conclude

$$P\left(\limsup_n \frac{S_n}{u(n)} \leq a(1+\gamma)\right) = 1.$$

Since $\gamma > 0$ and $a > 1$ are arbitrary, we have

$$P\left(\limsup_n \frac{S_n}{u(n)} \leq 1\right) = 1 \qquad (C.5)$$

by $\gamma \to 0$ and $a \to 1$. Next, we develop the lower bound by fixing $0 < \gamma < 1, a > 1$. Define $A_k = \{S_{a^k} - S_{a^{k-1}} \geq (1-\gamma)u(a^k - a^{k-1})\}$. We will first show that $\sum_k P(A_k) = \infty$. Denote $n = a^k - a^{k-1}$ and note that for a symmetric random walk we have for $m > n$ $S_m - S_n = S_{m-n}$ in distribution. Thus, for large k, by applying Lemma C.8, we have

$$P(A_k) \geq c\frac{\sqrt{n}}{(1-\gamma)u(n)}e^{-\frac{(1-\gamma)^2 u^2(n)}{2n}} = \frac{c}{1-\gamma}\frac{1}{\sqrt{2\log\log n}}\frac{1}{(\log n)^{(1-\gamma)^2}}.$$

It follows from $\log n \approx k$ that this expression is not summable in k. Thus, from (ii) of Lemma C.9, we get

$$P(\text{ infinitely many } A_k \text{ occur}) = 1.$$

It follows from the upper bound in (C.5) by symmetry that

$$P\left(\liminf_n \frac{S_n}{u(n)} \geq -1\right) = 1.$$

For large k, we have

$$\frac{S_{a^k}}{u(a^k)} \geq (1-\gamma)\frac{u(a^k - a^{k-1})}{u(a^k)} + \frac{S_{a^{k-1}}}{u(a^k)}$$

$$\geq (1-\gamma)\frac{u(a^k - a^{k-1})}{u(a^k)} - \frac{(1+\varepsilon)u(a^{k-1})}{u(a^k)}$$

$$\to (1-\gamma)\sqrt{1+\frac{1}{a}} - \frac{1+\varepsilon}{\sqrt{a}} \quad \text{as } k \to \infty.$$

Thus, we have

$$P\left(\limsup_n \frac{S_n}{u(n)} \geq (1-\gamma)\sqrt{1-\frac{1}{a}} - \frac{1+\varepsilon}{\sqrt{a}}\right) = 1,$$

which leads to the lower bound by taking $\gamma \to 0$ and $a \to \infty$. With these upper and lower bounds, we have

$$\limsup_{n\to\infty} \frac{S_n}{\sqrt{2n\log\log n}} = 1.$$

This completes the proof of the LIL.

For more discussion on the useful bounds, inequalities, and limit laws in probability and statistics, interested readers are referred to some textbooks on probability theory or mathematical statistics such as Taboga (2017).

REFERENCE

1. M. Taboga, "Lectures on Probability Theory and Mathematical Statistics", 3rd ed., CreateSpace Independent Publishing Platform (December 8, 2017).

D Non-Linear Programming in Stochastics

D.1 NON-LINEAR OPTIMIZATION – MULTI-LINEAR REGRESSIONS

Since most optimization issues in stochastic models such as optimizing decision variable(s) or estimating system parameter(s) are involved with the non-linear objective functions, we briefly review some most commonly used conditions for optimality in non-linear programming problems (NLPPs). Note that this review is brief rather than comprehensive. For example, we only present the key and relevant results without proofs and derivations. For more details, we refer to some excellent books on optimization. The standard form of NLPP can be written as

$$\min f(\mathbf{x}) \text{ subject to } g(\mathbf{x}) \leq 0, \ h(\mathbf{x}) = 0,$$

where $\mathbf{x} \in \mathbb{R}^n$. If f, g, and h are continuous functions and one of the two following conditions holds, then this NLPP admits at least one solution.

1. The set of feasible vectors $g(\mathbf{x}) \leq 0, h(\mathbf{x}) = 0$ is a bounded set in $\mathbb{R}^n$;
2. The set of feasible vectors is not bounded, but $\lim_{|\mathbf{x}| \to \infty, g(\mathbf{x}) \leq 0, h(\mathbf{x}) = 0} f(\mathbf{x}) = +\infty$.

We start with the necessary condition for the optimality in the NLPP with only equality constraints.

Theorem D.1 *The necessary conditions of optimality for the NLPP*

$$\min f(\mathbf{x}) \text{ subject to } h(\mathbf{x}) = 0$$

are

$$\nabla f(\mathbf{x}) + \lambda \nabla h(\mathbf{x}) = 0, \ h(\mathbf{x}) = 0,$$

where $\lambda \in \mathbb{R}^m$ is the vector of Lagrange multipliers associated with the m constraints $h(\mathbf{x})$, and ∇ is the gradient of $f(\mathbf{x})$. The definition of the gradient can be found in Chapter J.

Clearly, the optimal solution must satisfy a set of equations. This equation system has $n + m$ equations with the same number of unknowns. The solutions for $\mathbf{x}$ can be extreme points (local or global maximum or minimum) or saddle points. To determine the global optimal solution, the enumeration approach

DOI: 10.1201/9781003150060-D

can be applied over these candidate points. To extend the NLPP to the case with both inequality and equality constraints, we introduce the Karush–Kuhn–Tucker (KKT) optimality conditions.

Theorem D.2 *The optimality conditions for*

$$\min f(\mathbf{x}) \quad subject\ to \quad g(\mathbf{x}),\ h(\mathbf{x}) = 0$$

are

$$\nabla f(\mathbf{x}) + \mu \nabla g(\mathbf{x}) + \lambda \nabla h(\mathbf{x}) = 0,$$
$$\mu g(\mathbf{x}) = 0,$$
$$\mu \geq 0,\ g(\mathbf{x}) \leq 0,\ h(\mathbf{x}) = 0,$$

where $\mu \in \mathbb{R}^m$ is the vector of multipliers associated with the m inequality constraints $g(\mathbf{x})$ and $\lambda \in \mathbb{R}^d$ is the vector of multipliers associated with d equality constraints $h(\mathbf{x})$.

To make these necessary conditions for the optimality sufficient, we need to introduce the concept of convexity.

Definition D.1 *A set S in $\mathbb{R}^n$ is convex if for every pair of vectors $\mathbf{x}, \mathbf{y} \in S$, the segment joining them is also contained in S. That is*

$$t\mathbf{x} + (1-t)\mathbf{y} \in S,\ t \in [0,1].$$

A function $f : S \subset \mathbb{R}^n \to \mathbb{R}$ is said to be convex if S is convex set of vectors and

$$f(t\mathbf{x} + (1-t)\mathbf{y}) \leq tf(\mathbf{x}) + (1-t)f(\mathbf{y}),$$

whenever $\mathbf{x}, \mathbf{y} \in S,\ t \in [0,1]$.

The condition of convexity simplifies the minimization problem as follows.

Theorem D.3 *For a convex function $f : S \subset \mathbb{R}^n \to \mathbb{R}$ where S is a convex set, if $\mathbf{x}^*$ is a local minimum for f in S, then it is also a global minimum for f in S.*

If the function f is differentiable, then we can develop equivalent conditions for verifying the convexity of the continuous function. These conditions are presented as follows.

Proposition D.1 *For a convex function $f : S \subset \mathbb{R}^n \to \mathbb{R}$ where S is a convex and open set,*

    1. If f is differentiable and ∇f is continuous, then f is convex if and only if

$$f(\mathbf{y}) \geq f(\mathbf{x}) + \nabla f(\mathbf{x})(\mathbf{y} - \mathbf{x}),\ \mathbf{x}, \mathbf{y} \in S.$$

2. *If f is twice differentiable and $\nabla^2 f$ is continuous, then f is convex if and only if $\nabla^2 f(\mathbf{x})$, called Hessian matrix, is positive semi-definite for all $\mathbf{x} \in S$.*

The definition of the Hessian matrix can be found in Chapter J. If the $\geq$ is replaced with $>$, the conditions become the one for "strict convexity". For example, f is strictly convex if and only if its Hessian matrix is positive definite at every point in S. To verify the positive definite (or semi-definite) property for the Hessian matrix, we can check if all eigenvalues are strictly positive at every point in S; or even by Sylvester's criterion, if the principal subdeterminants are strictly positive at very point of S. As an example of using NLPP in stochastics, we present the ordinary least squares method and the maximum likelihood method for the multiple linear regression (MLR) models.

D.1.1 MULTIPLE LINEAR REGRESSION

Consider a problem of regression in which the random response variable Y depends on multiple explanatory or independent variables (most likely deterministic) X_i with $i = 1, 2, ..., k$. Such a model is called multiple linear regression model and can be written as

$$Y = \beta_0 + \beta_1 X_1 + \beta_2 X_2 + \cdots + \beta_k X_k + \varepsilon, \tag{D.1}$$

where $\beta_0, \beta_1, ..., \beta_k$ are regression coefficients and ε is the random error term reflecting the randomness of Y. It is assumed that $E[\varepsilon] = 0$. Clearly, $\beta_j = \partial E[Y]/\partial X_j$. Estimating $E[Y]$ by using a sample (data set) turns out to be a non-linear programming problem. The random sample in terms of symbols is given in Table D.1.

D.1.1.1 Least Squares Method

Since the n-tuples of observations are assumed to follow the MLR model of (D.1), we have

$$y_1 = \beta_0 + \beta_1 x_{11} + \beta_2 x_{12} + \cdots + \beta_k x_{1k} + \varepsilon_1$$
$$y_2 = \beta_0 + \beta_1 x_{21} + \beta_2 x_{22} + \cdots + \beta_k x_{2k} + \varepsilon_2$$
$$\vdots$$
$$y_n = \beta_0 + \beta_1 x_{n1} + \beta_2 x_{n2} + \cdots + \beta_k x_{nk} + \varepsilon_n$$

which can be written in matrix form as

$$\mathbf{y} = \mathbf{X}\beta + \varepsilon, \tag{D.2}$$

TABLE D.1
A Sample of Size n for Estimating an MLR Model

Observation number	Response values of Y	Explanatory variables			
		X_1	X_2	$\cdots$	X_k
1	y_1	x_{11}	x_{12}	$\cdots$	x_{1k}
2	y_2	x_{21}	x_{22}	$\cdots$	x_{2k}
$\vdots$	$\vdots$	$\vdots$	$\vdots$	$\ddots$	$\vdots$
n	y_n	x_{n1}	x_{n2}	$\cdots$	x_{nk}

where

$$
\mathbf{y} = \begin{pmatrix} y_1 \\ y_2 \\ y_3 \\ \vdots \\ y_N \end{pmatrix}_{n \times 1}, \quad
\mathbf{X} = \begin{pmatrix} 1 & x_{11} & \cdots & x_{1k} \\ 1 & x_{21} & \cdots & x_{2k} \\ 1 & x_{31} & \cdots & x_{3k} \\ \vdots & \vdots & \cdots & \vdots \\ 1 & x_{n1} & \cdots & x_{nk} \end{pmatrix}_{n \times (k+1)},
$$

$$
\beta = \begin{pmatrix} \beta_0 \\ \beta_1 \\ \vdots \\ \beta_k \end{pmatrix}_{(k+1) \times 1}, \quad
\varepsilon = \begin{pmatrix} \varepsilon_1 \\ \varepsilon_2 \\ \varepsilon_3 \\ \vdots \\ \varepsilon_n \end{pmatrix}_{n \times 1}.
$$

The ordinary least squares is based on some of the following assumptions:

- Linearity. The functional relationship between response variable and explanatory variables is linear in parameters.
- Independence. The observations are independent and identically distributed. That is $\{x_i, y_i\}_{i=1}^n$ are i.i.d.s.
- Exogeneity. (a) $\varepsilon_i | x_i \sim N(0, \sigma_i^2)$; (b) $\varepsilon_i \perp x_i$ (independent); (c) $E[\varepsilon_i | x_i] = 0$ (mean independent); and (d) $Cov(x_i, \varepsilon_i) = 0$ (uncorrelated).
- Error Variance. (a) $Var(\varepsilon_i | x_i) = \sigma^2 < \infty$ (homoscedasticity); (b) $Var(\varepsilon_i | x_i) = \sigma_i^2 < \infty$ (conditional heteroscedastocity).
- Identifiability. The independent variables are not perfectly collinear (i.e., no variable is a linear combination of the others). That is $E[x_i x_i^T] = \mathbf{Q_{XX}}$ is positive definite and finite and $rank(\mathbf{X}) = k + 1 < n$.

The following assumption is needed to study the large sample properties of estimators:

$$\lim_{n \to \infty} \left(\frac{\mathbf{X}^T \mathbf{X}}{n} \right) = \lim_{n \to \infty} \left(\frac{1}{n} \right) \begin{pmatrix} 1 & 1 & 1 & \cdots & 1 \\ x_{11} & x_{21} & x_{31} & \cdots & x_{n1} \\ x_{12} & x_{22} & x_{32} & \cdots & x_{n2} \\ \vdots & \vdots & \vdots & \cdots & \vdots \\ x_{1k} & x_{2k} & x_{3k} & \cdots & x_{nk} \end{pmatrix}_{(k+1) \times n}$$

$$\times \begin{pmatrix} 1 & x_{11} & x_{12} & \cdots & x_{1k} \\ 1 & x_{21} & x_{22} & \cdots & x_{2k} \\ 1 & x_{31} & x_{32} & \cdots & x_{3k} \\ \vdots & \vdots & \vdots & \cdots & \vdots \\ 1 & x_{n1} & x_{n2} & \cdots & x_{nk} \end{pmatrix}_{n \times (k+1)}$$

$$= \lim_{n \to \infty} \left(\frac{1}{n} \right) \begin{pmatrix} n & \sum_{i=1}^{n} x_{i1} & \sum_{i=1}^{n} x_{i2} & \cdots & \sum_{i=1}^{n} x_{ik} \\ \sum_{i=1}^{n} x_{i1} & \sum_{i=1}^{n} x_{i1}^2 & \sum_{i=1}^{n} x_{i1} x_{i2} & \cdots & \sum_{i=1}^{n} x_{i1} x_{ik} \\ \sum_{i=1}^{n} x_{i2} & \sum_{i=1}^{n} x_{i2} x_{i1} & \sum_{i=1}^{n} x_{i2}^2 & \cdots & \sum_{i=1}^{n} x_{i2} x_{ik} \\ \vdots & \vdots & \vdots & \cdots & \vdots \\ \sum_{i=1}^{n} x_{ik} & \sum_{i=1}^{n} x_{ik} x_{i1} & \sum_{i=1}^{n} x_{ik} x_{i2} & \cdots & \sum_{i=1}^{n} x_{ik}^2 \end{pmatrix}$$

$$\text{(D.3)}$$

exists and is a non-stochastic and non-singular matrix (with finite elements). Note that the explanatory variables can be also random. For such a case, the assumption above implies that the first two moments of each explanatory variable and the covariances between these explanatory variables are finite.

Denote by $\hat{Y} = b_0 + b_1 X_1 + b_2 X_2 + \cdots + b_k X_k$ the unbiased estimator of Y. That is $E[\hat{Y}] = E[Y] = \beta_0 + \beta_1 X_1 + \beta_2 X_2 + \cdots + \beta_k X_k$. We use the sample data to compute the estimator for the MLR model. Using the vector notation $\mathbf{b} = (b_0, b_1, ..., b_k)^T$, we can write the squared deviation between the observed value y_i and the estimated value $\hat{y}_i$ as $(y_i - \hat{y}_i)^2 = (y_i - \mathbf{x}_i^T \mathbf{b})^2 = e_i^2$ and $\hat{\mathbf{y}} = \mathbf{Xb}$. The sum of squares, as a measure of the overall deviation, is defined by

$$S(\mathbf{b}) = \sum_{i=1}^{n} (y_i - \mathbf{x}_i^T \mathbf{b})^2 = (\mathbf{y} - \mathbf{Xb})^T (\mathbf{y} - \mathbf{Xb}) = \mathbf{e}^T \mathbf{e}$$

for a given sample of $(\mathbf{y}, \mathbf{X})$. Our problem becomes an NLPP without constraints as $\min_{\mathbf{b}} S(\mathbf{b})$, a special case of NLPP with equality constraints (i.e., $\lambda = 0$). To simplify the notation, we use an alternative symbol $\mathbf{X}' = \mathbf{X}^T$. Furthermore, we utilize the differentiation rule for the product of matrices as follows: If $\mathbf{z}$ is a $m \times 1$ column vector and $\mathbf{A}$ is any $m \times m$ symmetric matrix, then for any matrix function $f(\mathbf{z}) = \mathbf{z}' \mathbf{A} \mathbf{z}$ (a scalar), we have $\partial f(\mathbf{z})/\partial \mathbf{z} = 2\mathbf{A}\mathbf{z}$. Furthermore, if $\mathbf{g}(\mathbf{z}) = \mathbf{z}'\mathbf{X}$ or $\mathbf{h}(\mathbf{z}) = \mathbf{X}\mathbf{z}$, then $\partial \mathbf{g}(\mathbf{z})/\partial \mathbf{z} = \mathbf{X}$ and $\partial \mathbf{h}(\mathbf{z})/\partial \mathbf{z} = \mathbf{X}$.

It is straightforward to verify these derivative rules based on the definitions presented in Section J.5 of this book. Rewrite the sum of squares as

$$S(\mathbf{b}) = \mathbf{y}'\mathbf{y} + \mathbf{b}'\mathbf{X}'\mathbf{X}\mathbf{b} - 2\mathbf{b}'\mathbf{X}'\mathbf{y}. \qquad (D.4)$$

Differentiating $S(\mathbf{b})$ with respect to $\mathbf{b}$ yields

$$\nabla S(\mathbf{b}) = \frac{\partial S(\mathbf{b})}{\partial \mathbf{b}} = 2\mathbf{X}'\mathbf{X}\mathbf{b} - 2\mathbf{X}'\mathbf{y},$$
$$\nabla^2 S(\mathbf{b}) = \frac{\partial^2 S(\mathbf{b})}{\partial \mathbf{b}^2} = 2\mathbf{X}'\mathbf{X}. \qquad (D.5)$$

It is easy to check from the second order derivatives (Hessian matrix) is positive semi-definite. Thus, this objective function is convex and hence the global optimal solution is obtained by solving the first-order condition equation

$$\nabla S(\mathbf{b}) = \frac{\partial S(\mathbf{b})}{\partial \mathbf{b}} = 0 \Rightarrow \mathbf{X}'\mathbf{X}\mathbf{b} = \mathbf{X}'\mathbf{y} \Rightarrow \mathbf{b} = (\mathbf{X}'\mathbf{X})^{-1}\mathbf{X}'\mathbf{y}, \qquad (D.6)$$

which is called ordinary least squares estimator (OLSE) of β. (Note: for the case where $\mathbf{X}$ is not full rank, we need to use the "generalized inverse of $\mathbf{X}'\mathbf{X}$".) Using the OLS $\mathbf{b}$ vector, we can write the unbiased estimator of Y as

$$\hat{\mathbf{y}} = \mathbf{X}\mathbf{b} = \mathbf{X}(\mathbf{X}'\mathbf{X})^{-1}\mathbf{X}'\mathbf{y} = \mathbf{H}\mathbf{y},$$

where $\mathbf{H} = \mathbf{X}(\mathbf{X}'\mathbf{X})^{-1}\mathbf{X}'$ is called "hat matrix". Such a matrix is symmetric, idempotent (i.e., $\mathbf{H}\mathbf{H} = \mathbf{H}$), and with trace of k (i.e., $tr(\mathbf{H}) = k$, for an $n \times n$ matrix A, $tr(\mathbf{A}) = \sum_{i=1}^{n} a_{ii}$). The difference between the observed and estimated values for Y is called residual and it is given by

$$\mathbf{e} = \mathbf{y} - \hat{\mathbf{y}} = \mathbf{y} - \mathbf{X}\mathbf{b} = \mathbf{y} - \mathbf{H}\mathbf{y}$$
$$= (\mathbf{I} - \mathbf{H})\mathbf{y} = \bar{\mathbf{H}}\mathbf{y},$$

where $\bar{\mathbf{H}} = \mathbf{I} - \mathbf{H}$, which is symmetric, idempotent, and with trace of $n - k$. Now we discuss briefly the important properties of the MLR model. First, the estimation error is defined and expressed as follows:

$$\mathbf{b} - \beta = (\mathbf{X}'\mathbf{X})^{-1}\mathbf{X}'\mathbf{y} - \beta = (\mathbf{X}'\mathbf{X})^{-1}\mathbf{X}'(\mathbf{X}\beta + \varepsilon) - \beta$$
$$= (\mathbf{X}'\mathbf{X})^{-1}\mathbf{X}'\varepsilon.$$

With this expression, we can easily verify that $\mathbf{b}$ is the unbiased estimator of β. That is

$$E[\mathbf{b} - \beta] = (\mathbf{X}'\mathbf{X})^{-1}\mathbf{X}'E[\varepsilon] = 0 \Rightarrow E[\mathbf{b}] = \beta.$$

Next, the covariance matrix of $\mathbf{b}$ is given by

$$COV(\mathbf{b},\mathbf{b}) = E[(\mathbf{b}-\beta)(\mathbf{b}-\beta)'] = E[(\mathbf{X}'\mathbf{X})^{-1}\mathbf{X}'\varepsilon\varepsilon'\mathbf{X}(\mathbf{X}'\mathbf{X})^{-1}]$$
$$= \sigma^2(\mathbf{X}'\mathbf{X})^{-1}\mathbf{X}'E[\varepsilon\varepsilon']\mathbf{X}(\mathbf{X}'\mathbf{X})^{-1} = \sigma^2(\mathbf{X}'\mathbf{X})^{-1}\mathbf{X}'\mathbf{I}\mathbf{X}(\mathbf{X}'\mathbf{X})^{-1}$$
$$= \sigma^2(\mathbf{X}'\mathbf{X})^{-1},$$

where $\varepsilon \sim N(0, \sigma^2\mathbf{I}_n)$ and $E[\varepsilon\varepsilon'] = \sigma^2\mathbf{I}_n$. Note that the inverse of the symmetric matrix is also symmetric. The variance of $\mathbf{b}$ can be obtained as

$$Var(\mathbf{b}) = tr(COV(\mathbf{b},\mathbf{b})) = \sum_{i=1}^{k} E[(b_i - \beta_i)^2] = \sum_{i=1}^{k} Var(b_i).$$

The OLS method does not produce the estimate for σ^2 directly. An unbiased estimate can be obtained by performing the residual analysis. This starts with the identity for sum of squares:

$$\underbrace{\sum_{i=1}^{n}(y_i - \bar{y})^2}_{SST} = \underbrace{\sum_{i=1}^{n}(\hat{y}_i - \bar{y})^2}_{SSR} + \underbrace{\sum_{i=1}^{n}(y_i - \hat{y}_i)^2}_{SSE}, \tag{D.7}$$

where SST is the total sum of squares which measures the total variability of $\mathbf{y}$, SSR is the regression sum of squares which measures the variability of $\mathbf{y}$ explained by $\mathbf{x}_1, ..., \mathbf{x}_k$, and SSE is error (residual) sum of squares which measures the variability of $\mathbf{y}$ not explained by $\mathbf{x}$'s. Two properties of residuals are

$$(1)\ \sum_{i=1}^{n} e_i = \sum_{i=1}^{n}(y_i - b_0 - b_1 x_{i1} - \cdots - b_k x_{ik}) = 0,$$

$$(2)\ \sum_{i=1}^{n} x_{ij} e_i = \sum_{i=1}^{n} x_{ij}(y_i - b_0 - b_1 x_{i1} - \cdots - b_k x_{ik}) = 0,\ \ k = 1, 2, ..., k.$$

The first property, called "zero total residual", can be easily verified by using the fact that $(\bar{y}, \bar{x}_1, ..., \bar{x}_k)$ satisfy

$$\hat{Y} = b_0 + b_1 X_1 + b_2 X_2 + \cdots + b_k X_k.$$

That is

$$\sum_{i=1}^{n}(y_i - b_0 - b_1 x_{i1} - \cdots - b_k x_{ik}) = \sum_{i=1}^{n} y_i - \sum_{i=1}^{n} b_0 - b_1 \sum_{i=1}^{n} x_{i1} - \cdots - b_k \sum_{i=1}^{n} x_{ik}$$
$$= n\bar{y} - nb_0 - b_1 n\bar{x}_1 - \cdots - b_k n\bar{x}_k = n(\bar{y} - b_0 - b_1\bar{x}_1 - \cdots - b_k\bar{x}_k) = 0.$$

The second property, called "orthogonality of residuals and explanatory variables", follows from the $\mathbf{b} = (\mathbf{X}'\mathbf{X})^{-1}\mathbf{X}'\mathbf{y}$. That is $\mathbf{X}'(\mathbf{y} - \mathbf{Xb}) = \mathbf{X}'\mathbf{e} = 0$, the matrix form of this property. Based on these two properties, the relation (D.7) can be proved. Note that

$$y_i - \bar{y} = (\hat{y}_i - \bar{y}) + (y_i - \hat{y}_i).$$

Squaring both sides of this equation yields

$$(y_i - \bar{y})^2 = (\hat{y}_i - \bar{y})^2 + (y_i - \hat{y}_i)^2 + 2(\hat{y}_i - \bar{y})(y_i - \hat{y}_i).$$

Summing up over all observations, we have

$$\sum_{i=1}^n (y_i - \bar{y})^2 = \sum_{i=1}^n (\hat{y}_i - \bar{y})^2 + \sum_{i=1}^n (y_i - \hat{y}_i)^2 + 2\sum_{i=1}^n (\hat{y}_i - \bar{y})(y_i - \hat{y}_i),$$

where the last term on the RHS can be shown to be zero. Then we obtain $SST = SSR + SSE$. Based on Section A.7 of Chapter A, we have $SSE/\sigma^2 \sim \chi^2_{n-k-1}$ if the degrees of the freedom for SSE is $n - k - 1$. This is easily verified by noting that the n residuals $e_1, ..., e_n$ cannot all very freely since there are $k+1$ constraints

$$\sum_{i=1}^n e_i = 0 \quad \text{and} \quad \sum_{i=1}^n x_{ji}e_i = 0 \quad \text{for } j = 1, ..., k.$$

Thus, only $n - (k+1)$ of them can be freely changed. Similarly, since the degrees of freedom for SST and SSR are $n - 1$ and k, respectively, we have

$$\frac{SST}{\sigma^2} \sim \chi^2_{n-1}, \quad \frac{SSR}{\sigma^2} \sim \chi^2_k.$$

Note that SSR is independent of SSE and $d.f._{SST} = d.f._{SSR} + d.f._{SSE}$. The mean squares is the sum of squares divided by its degrees of freedom, which is the estimator of variance. Three mean squares are given by

$$MST = \frac{SST}{d.f._{SST}} = \frac{SST}{n-1} = \text{sample variance of } Y,$$

$$MSR = \frac{SSR}{d.f._{SSR}} = \frac{SSR}{k},$$

$$MSE = \frac{SSE}{d.f._{SSE}} = \frac{SSE}{n-k-1} = \hat{\sigma}^2.$$

Clearly, MSE is an unbiased estimator of σ^2 as $E[MSE] = E[\hat{\sigma}^2] = \sigma^2$. To evaluate the goodness of fit of the MLR model, we define the coefficient of determination, called multiple R^2, as

$$R^2 = \frac{SSR}{SST} = 1 - \frac{SSE}{SST}$$

$$= \text{proportion of variability in } Y \text{ explained by } X_1, ..., X_k,$$

which measures the strength of the linear relationship between Y and k independent variables. Note that as more independent variables are added the model, SSE always decreases and hence R^2 increases. However, adding more non-effective independent variables may increase MSE. To evaluate the effectiveness of including an independent variable in MLR model, we introduce the "adjusted R^2 by replacing the sum of squares with mean squares in the definition of R^2 as follows:

$$R^2_{adj} = 1 - \frac{MSE}{MST} = 1 - \frac{SSE/(n-k-1)}{SST/(n-1)}$$

$$= 1 - \frac{n-1}{n-k-1}(1 - R^2).$$

It is easy to show $-k/(n-k-1) \le R^2_{adj} \le R^2 \le 1$ as long as $n > k+1$, a reasonable sample size, and R^2 does not always increases as more independent variables are added to the model. Traditional tests (t and F tests) on the parameters β_i's of MLR and ANOVA can be performed.

D.1.1.2 Maximum Likelihood Method

While the OLS method is to solve a NLPP with a minimization objective function, the maximum likelihood method (MLM) is to solve a NLPP with a maximization objective function. The MLM is widely applied in estimating the parameters of the distribution functions for random variables. For example, in Chapter 4, we utilize the MLM to estimate the PH representation for a PH-distributed random variable. Here we present the application of the MLM in MLR models. In the MLR model $\mathbf{y} = \mathbf{X}\beta + \varepsilon$, the random error terms are assumed to be normal i.i.d. random variables. Thus, the pdf's of these random variables are given by

$$f(\varepsilon_i) = \frac{1}{\sqrt{2\pi}} \exp\left(-\frac{1}{2\sigma^2}\varepsilon_i^2\right), \quad i = 1, 2, ..., n. \tag{D.8}$$

The likelihood function for the MLR model is the joint density of n random error terms $\varepsilon_1, \varepsilon_2, ..., \varepsilon_n$ as follows:

$$L(\beta, \sigma^2) = \prod_{i=1}^{n} f(\varepsilon_i) = \frac{1}{(2\pi\sigma^2)^{n/1}} \exp\left(-\frac{1}{2\sigma^2}\sum_{i=1}^{n}\varepsilon_i^2\right)$$

$$= \frac{1}{(2\pi\sigma^2)^{n/1}} \exp\left(-\frac{1}{2\sigma^2}\varepsilon'\varepsilon\right)$$

$$= \frac{1}{(2\pi\sigma^2)^{n/1}} \exp\left(-\frac{1}{2\sigma^2}(\mathbf{y} - \mathbf{X}\beta)'(\mathbf{y} - \mathbf{X}\beta)\right).$$

Due to the fact that the log transformation is monotonic, we can take the logarithm of $L(\beta, \sigma^2)$. Thus, the MLM problem becomes

$$\max_{\beta, \sigma^2} \ln L(\beta, \sigma^2) = -\frac{n}{2} \ln(2\pi\sigma^2) - \frac{1}{2\sigma^2}(\mathbf{y} - \mathbf{X}\beta)'(\mathbf{y} - \mathbf{X}\beta).$$

The MLM estimators of β and σ^2 can be obtained by solving the first-order derivative condition equation. That is

$$\frac{\partial \ln L(\beta, \sigma^2)}{\partial \beta} = \frac{1}{2\sigma^2} 2\mathbf{X}'(\mathbf{y} - \mathbf{X}\beta) = 0$$

$$\frac{\partial \ln L(\beta, \sigma^2)}{\partial \sigma^2} = -\frac{n}{2\sigma^2} + \frac{1}{2(\sigma^2)^2}(\mathbf{y} - \mathbf{X}\beta)'(\mathbf{y} - \mathbf{X}\beta) = 0.$$

Solving these equations yields the MLM estimators

$$\hat{\beta} = (\mathbf{X}'\mathbf{X})^{-1}\mathbf{X}'\mathbf{y}, \quad \hat{\sigma}^2 = \frac{1}{n}(\mathbf{y} - \mathbf{X}\hat{\beta})'(\mathbf{y} - \mathbf{X}\hat{\beta}). \qquad (\text{D.9})$$

To check the concavity of $L(\beta, \sigma^2)$, we compute the Hessian matrix as follows:

$$\begin{pmatrix} \frac{\partial^2 \ln L(\beta,\sigma^2)}{\partial \beta^2} & \frac{\partial^2 \ln L(\beta,\sigma^2)}{\partial \beta \partial \sigma^2} \\ \frac{\partial^2 \ln L(\beta,\sigma^2)}{\partial \sigma^2 \partial \beta} & \frac{\partial^2 \ln L(\beta,\sigma^2)}{\partial(\sigma^2)^2} \end{pmatrix} = \begin{pmatrix} -\frac{1}{\sigma^2}\mathbf{X}'\mathbf{X} & -\frac{1}{\sigma^4}\mathbf{X}'(\mathbf{y}-\mathbf{X}\beta) \\ -\frac{1}{\sigma^4}(\mathbf{y}-\mathbf{X}\beta)'\mathbf{X} & \frac{n}{2\sigma^4} - \frac{1}{\sigma^6}(\mathbf{y}-\mathbf{X}\beta)'(\mathbf{y}-\mathbf{X}\beta) \end{pmatrix}.$$

This Hessian matrix is negative definite at $\beta = \hat{\beta}$ and $\sigma^2 = \hat{\sigma}^2$. This ensures that the likelihood function is jointly concave with respect to β and σ^2 so that the maximum likelihood is achieved. Note that while the ML estimator $\hat{\beta}$ is unbiased, the ML estimator $\hat{\sigma}^2$ is biased and underestimate the true variance σ^2. In contrast, the OLS estimators $\mathbf{b}$ and $\hat{\sigma}^2$ are all unbiased. While $\hat{\sigma}^2$ is biased, it is consistent and asymptotically efficient.

D.2 ENTROPY AND SUBMODULAR FUNCTIONS OPTIMIZATION

D.2.1 ENTROPY

A measure of uncertainty of a random variable is the entropy which becomes the basis for developing the information theory. Let X be a discrete random variable with a sample space S and the pmf $p(x) = P(X = x), x \in S$. The entropy is defined as follows.

Definition D.2 *The entropy of X, denoted by $H(X)$, is defined by*

$$H(X) = E\left[\log \frac{1}{p(x)}\right] = \sum_{x \in S} p(x) \log \frac{1}{p(x)}$$

$$= -\sum_{x \in S} p(x) \log p(x). \qquad (\text{D.10})$$

For multiple random variable cases, we define the joint entropy and conditional entropy. Let X, Y be a pair of discrete random variables with joint distribution $p(x,y)$ over a state space $(\mathcal{X}, \mathcal{Y})$.

Definition D.3 *The joint entropy $H(X,Y)$ of a pair of random variables (X,Y) is defined as*

$$H(X,Y) = -\sum_{x \in \mathcal{X}} \sum_{x \in \mathcal{Y}} p(x,y) \log p(x,y) = -E[\log p(X,Y)]. \quad (D.11)$$

Similarly, we can define the conditional entropy which is the expectation of the entropies of the conditional distributions.

Definition D.4 *The conditional entropy $H(Y|X)$ is defined as*

$$H(Y|X) = \sum_{x \in \mathcal{X}} p(x) H(Y|X = x) = -\sum_{x \in \mathcal{X}} p(x) \sum_{y \in \mathcal{Y}} p(y|x) \log p(y|x)$$
$$= -\sum_{x \in \mathcal{X}} \sum_{y \in \mathcal{Y}} p(x,y) \log p(y|x) = -E_{p(x,y)}[\log p(Y|X)]. \quad (D.12)$$

Based on the definition, it can be shown that

$$H(X,Y) = H(X) + H(Y|X) = H(Y) + H(X|Y), \quad (D.13)$$

which is called the "chain rule". Now we define the mutual information.

Definition D.5 *For a pair of discrete random variables (X,Y), the mutual information $I(X;Y)$ is defined as a measure of the amount of information that one random variable contains about another random variable. That is*

$$I(X;Y) = H(X) - H(X|Y) = E_{p(x,y)} \log \frac{p(X,Y)}{p(X)p(Y)} \quad (D.14)$$
$$= H(Y) - H(Y|X) = I(Y;X).$$

Clearly, the mutual information is the reduction of uncertainty of one random variable due to the knowledge of the other. We can also define the conditional mutual information as

$$I(X;Y|Z) = H(X|Z) - H(X|Y,Z).$$

It follows from the chain rule that

$$I(X_1,X_2;Y) = H(X_1,X_2) - H(X_1,X_2,|Y)$$
$$= H(X_1) + H(X_2|X_1) - H(X_1|Y) - H(X_2|X_1,Y)$$
$$= I(X_1;Y) + I(X_2;Y|X_1).$$

The mutual information is also called the information gain and denoted by IG. Some well-known properties for the mutual information and entropies include (1) $I(X;Y) \geq 0$; (2) $I(X;Y|Z) \geq 0$; (3) $H(X) \leq \log|\mathscr{X}|$, where $|\mathscr{X}|$ denotes the number of elements in the sample space of the discrete random variable X, with the equality holding if and only if X has a uniform distribution over $\mathscr{X}$; (4) Condition reduces entropy: $H(X|Y) \leq H(X)$ with equality holding if and only if X and Y are independent; (5) $H(X_1, X_2, ..., X_n) \leq \sum_{i=1}^{n} H(X_i)$ with equality holding if and only if the X_i are independent; and (6) $I(X;Y)$ is a concave function of $p(x)$ for fixed $p(y|x)$ and a convext function of $p(y|x)$ for fixed $p(x)$.

D.2.2 MAXIMUM ENTROPY PRINCIPLE IN APPROXIMATING A PROBABILITY DISTRIBUTION

The principle of maximum entropy (PME) is based on the premise that when estimating the probability distribution, the best estimations will keep the largest remaining uncertainty (the maximum entropy) consistent with all the known constraints. This implies that the probability distribution that best represents the current state of knowledge is the one with the largest (information) entropy. If we only know certain statistics about the distribution, such as its mean, then this principle tells us that the best distribution to use is the one with the most surprise (more surprise, means fewer of your assumptions were satisfied). Thus, no more additional assumptions or biases were introduced in the estimation. With PME, we can determine the distribution that makes fewest assumptions about the data. This can be illustrated by a simple example.

Example D.1 *Consider a Bernoulli random variable $X = 0$ or 1 with the pmf $P(X = 0) = 1 - p$ and $P(X = 1) = p$. To estimate p using the maximum entropy principle, we take the first and second order derivative of $H(X) = -p\log p - (1 - p)\log(1 - p)$ with respect to p and obtain*

$$\frac{dH(X)}{dp} = \log\frac{1-p}{p},$$

$$\frac{d^2H(X)}{dp^2} = -\frac{1}{p(1-p)} < 0.$$

The second derivative shows that the entropy function is concave and the global maximum entropy can be determined by solving the first-order condition $dH(x)/dp = 0$ for p. The result is $p = 1/2$. Note that log is the logarithm function with base 2. Thus, at $p = 1/2$, the maximum entropy is 1.

A continuous analogue to discrete entropy is called differential entropy (or continuous entropy) for a continuous random variable X. The differential entropy is given by

$$H(X) = -\int_{-\infty}^{\infty} f(x) \log f(x) dx,$$

where $f(x)$ is the pdf of X and $f(x) \log f(x) = 0$ by definition when $f(x) = 0$. Note that some of the properties of discrete entropy do not apply to differential entropy, for example, differential entropy can be negative. In the next example we derive two continuous distributions with the PME.

Example D.2 (a) *Estimate the continuous distribution with finite support* $[a, b]$ *with* $b > a$ *with PME. This is equivalent to solving the following NLPP:*

$$\max_{f(x)} H(x) = -\int_a^b f(x) \log(f(x)) dx$$

$$\int_a^b f(x) dx = 1.$$

The Lagrangian function is given by

$$L(f(x), \lambda) = -\int_a^b f(x) \log(f(x)) dx + \lambda \left(\int_a^b f(x) dx - 1 \right).$$

Taking the first-order partial derivatives and setting them to zero, we have

$$\frac{\partial L(f(x), \lambda)}{\partial f(x)} = 0 \Rightarrow \log(f(x)) = -1 - \lambda \Rightarrow f(x) = e^{-1-\lambda},$$

$$\frac{\partial L(f(x), \lambda)}{\partial \lambda} = 0 \Rightarrow \int_a^b f(x) dx = 1 \Rightarrow e^{-1-\lambda} \int_a^b dx = 1,$$

which leads to $f(x) = e^{-1-\lambda} = 1/(b-a)$.

(b) *Estimate the continuous distribution with a preassigned standard deviation and mean* μ *with PME. This is equivalent to solving the following NLPP:*

$$\max_{f(x)} H(x) = -\int_{-\infty}^{\infty} f(x) \log(f(x)) dx$$

$$\int_{-\infty}^{\infty} f(x) dx = 1$$

$$\int_{-\infty}^{\infty} (x - \mu)^2 f(x) dx = \sigma^2.$$

The Lagrangian function is given by

$$L(f(x), \lambda_0, \lambda_1) = -\int_{-\infty}^{\infty} f(x) \log(f(x)) dx + \lambda_0 \left(\int_{-\infty}^{\infty} f(x) dx - 1 \right)$$

$$+ \lambda_1 \left(\int_{-\infty}^{\infty} (x - \mu)^2 f(x) dx - \sigma^2 \right).$$

Taking the first-order partial derivative with respect to $f(x)$ and setting it to zero, we have

$$\frac{\partial L(f(x), \lambda_0, \lambda_1)}{\partial f(x)} = 0,$$

$$\Rightarrow -(1 + \log f(x)) + \lambda_0 + \lambda_1 (x - \mu)^2 = 0,$$

$$\Rightarrow f(x) = e^{\lambda_0 + \lambda_1 (x - \mu)^2 - 1}.$$

The other two first-order partial derivative conditions will result in the two constraints. By using the expression of $f(x)$ above, the first constraint is given by

$$\int_{-\infty}^{\infty} e^{\lambda_0 + \lambda_1 (x - \mu)^2 - 1} dx = 1,$$

which after being evaluated by using $\Gamma(1/2) = \sqrt{\pi}$ leads to

$$e^{\lambda_0 - 1} \sqrt{\frac{\pi}{-\lambda_1}} = 1.$$

Similarly, the second constraint

$$\int_{-\infty}^{\infty} (x - \mu)^2 e^{\lambda_0 + \lambda_1 (x - \mu)^2 - 1} dx = \sigma^2,$$

which results in

$$e^{\lambda_0 - 1} = \sqrt{\frac{1}{2\pi} \frac{1}{\sigma}}.$$

Hence, we have $\lambda_1 = -1/(2\sigma^2)$. With the expressions of $e^{\lambda_0 - 1}$ and λ_1, we obtain

$$f(x) = e^{\lambda_0 + \lambda_1 (x - \mu)^2 - 1} = e^{\lambda_0 - 1} e^{\lambda_1 (x - \mu)^2}$$

$$= \sqrt{\frac{1}{2\pi} \frac{1}{\sigma}} e^{-\frac{1}{2\sigma^2}(x - \mu)^2}$$

$$= \frac{1}{\sigma \sqrt{2\pi}} e^{-\frac{1}{2}\left(\frac{x - \mu}{\sigma}\right)^2},$$

which is the normal probability density function.

For more details on information theory, interested readers are referred to some excellent textbooks such as Applebaum (2008).

D.2.3 SUBMODULAR FUNCTION

Submodularity is an important property for a class of set functions. Let f be a set function mapping from the power set of a finite set V to real line.

Definition D.6 $f : 2^V \rightarrow \mathbb{R}$ *is submodular if for any* $A, B \subseteq V$, *we have that*

$$f(A) + f(B) \geq f(A \cup B) + f(A \cap B).$$

An equivalent definition can be established.

Lemma D.1 $f : 2^V \rightarrow \mathbb{R}$ *is submodular if and only if it satisfies the law of diminishing returns. That is*

$$f(A \cup \{e\}) - f(A) \geq f(B \cup \{e\}) - f(B)$$

for all $A \subseteq B \in 2^V$ *and* $e \in V \setminus B$.

If the direction of the inequality changes, we obtain the definition of supermodular function. When the inequality is replaced with equality, then the function is both submodular and supermodular and is said to be modular. It follows from the definition that the following holds.

Lemma D.2 *Let* $f, g : 2^V \rightarrow \mathbb{R}$ *be submodular functions. Then, for* $\lambda, \alpha \geq 0$, $\lambda f + \alpha g$ *is a submodular function.*

Now we give a few examples of submodular functions.

Example D.3 *(Joint entropy) Let* $X_1, ..., X_n$ *with supports* $S_1, ..., S_n$ *(sample spaces) be a finite set of random variables. Their joint entropy is given by*

$$H(X_1, ..., X_n) = - \sum_{x_1 \in S_1} \sum_{x_1 \in S_2} \cdots \sum_{x_1 \in S_n} p(x_1, ..., x_n) \log(p(x_1, ..., x_n)),$$

where $p(x_1, ..., x_n)$ *is the joint pmf and* $H(\emptyset) = 0$. *Note that* $H(\mathcal{X}) \geq 0$ *for all* $\mathcal{X} \subseteq \{X_1, X_2, ..., X_n\}$ *due to the summation of a finite number of non-negative terms. In terms of the conditional entropy* $H(\mathcal{X}|\mathcal{Y}) = H(\mathcal{Y} \cup \mathcal{X}) - H(\mathcal{Y})$ *where* $\mathcal{X}, \mathcal{Y} \subseteq \{X_1, ..., X_n\}$, *we can verify that* $H(\mathcal{X}|\mathcal{Y}) \leq H(\mathcal{X}|\mathcal{Y}')$ *for* $\mathcal{Y}' \subseteq \mathcal{Y}$. *That is the larger the set of variables we condition on, the smaller is the conditional entropy. We can verify then that the joint entropy function is submodular as follows: Consider* $\mathcal{X} \subseteq \mathcal{Y} \subseteq \{X_1, ..., X_n\}$ *and* $X \in \{X_1, ..., X_n\} \setminus \mathcal{Y}$. *Then we have*

$$H(\mathcal{X} \cup X) - H(\mathcal{X}) = H(\{X\}|\mathcal{X}) \geq H(\{X\}|\mathcal{Y}) = H(\mathcal{Y} \cup X) - H(\mathcal{Y}),$$

which shows the submodularity of the joint entropy.

Another example is the famous "sensor placement" problem.

Example D.4 *Consider the following problem: we have to place sensors in a river in such a way to obtain as much information as possible on the status of the river (could be the temperature, the pollution level, etc.). Locations for sensors must chosen from a set V . This problem can be made precise by formulating it using submodular functions. As a first example, assume that, by placing a sensor in location $v \in V$, an area of radius r_v will be covered. Hence define the function $f : 2^V \to \mathbb{R}_+$ as follows:*

$$f(S) = A(\cup_{v \in S} C_v) \ \text{for} \ S \subseteq V,$$

where C_v is the circle of radius r_v centered at v for each $v \in V$, and $A(C)$ is the area of set C. $f(S)$ measures therefore the area covered by sensors placed in locations from S. Using Lemma D.1, one immediately checks that f is submodular.

D.2.4   NAIVE BAYES' MODEL AND FEATURE SECTION PROBLEM

Consider a multiclass classification problem, where the task is to map each input vector **x** to a label y that can take any one of k possible values (classes). As a motivating example of classifying documents into k different types. For example, $y = 1$ might correspond to a sports category, $y = 2$ might correspond to an arts category, $y = 3$ might correspond to a music category, and $y = 4$ might correspond to a science category, and so on. The label $y^{(i)}$ denotes the category of the ith document in the collection. Each component of the vector $x_j^{(i)}$ for $j = 1, ..., d$ might represent the presence or absence of a particular word. For example, $x_1^{(i)}$ can be defined to take $+1$ if the ith document contains the word "Canucks" or -1 otherwise; $x_2^{(i)}$ to take $+1$ if the ith document contains the word "Newton" or -1 otherwise; and so on. Our goal is to model the joint probability

$$P(Y = y, X_1 = x_1, X_2 = x_2, ..., X_d = x_d)$$

for any label y (category) paired with attributes values $x_1, ..., x_d$. The main assumption for developing the Naive Bayes model (NBM) is

$$P(Y = y, X_1 = x_1, X_2 = x_2, ..., X_d = x_d)$$
$$= P(Y = y) \prod_{j=1}^{d} P(X_j = x_j | Y = y), \tag{D.15}$$

where the equality holds due to the independence assumptions. To explain this assumption, we start with the identity

$$P(Y = y, X_1 = x_1, X_2 = x_2, ..., X_d = x_d)$$
$$\doteq P(Y = y) \cdot P(X_1 = x_1, X_2, ..., X_d = x_d | Y = y).$$

Now we focus on the conditional probability

$$P(X_1 = x_1, X_2 = x_2, ..., X_d = x_d | Y = y)$$

$$= \prod_{j=1}^{d} P(X_j = x_j | X_1 = x_1, X_2 = x_2, ..., X_d = x_{j-1}, Y = y) \tag{D.16}$$

$$= \prod_{j=1}^{d} P(X_j = x_j | Y = y).$$

The first equality of (D.16) is based on the multiplication rule as follows

$$P(X_1 = x_1 | Y = y) P(X_2 = x_2 | X_1 = x_1, Y = y)$$
$$\times P(X_3 = x_3 | X_1 = x_1, X_2 = x_2, Y = y)$$
$$\times P(X_4 = x_4 | X_1 = x_1, X_2 = x_2, X - 3 = x_3, Y = y)$$
$$\vdots$$
$$\times P(X_{d-1} = x_{d-1} | X_1 = x_1, ..., X_{d-1} = x_{d-1}, Y = y)$$
$$\times P(X_{d-1} = x_{d-1} | X_1 = x_1, ..., X_{d-1} = x_{d-1}, Y = y),$$

where the product of the first k terms is the joint probability of the k variables X_i with $i = 1, ..., k$ given $Y = y$. The second equality holds by the conditional independence assumption which says that given a value of y, the attribute random variables X_j is mutually independent. This assumption is said to be "Naive" since it is a strong assumption. Based on (D.15), there are two types of parameters for a NBM. We denote them by $q(y) = P(Y = y)$ with $y \in \{1, ..., k\}$ and by $q_j(x|y) = P(X_j = x | Y = y)$ for $j \in \{1, ..., d\}, y \in \{1, ..., k\}$, respectively. Thus, (D.15) can be written as

$$p(y, x_1, ..., x_d) = q(y) \prod_{j=1}^{d} q_j(x_j | y), \tag{D.17}$$

which is defined as an NBM. There are two typical problems: (1) Estimating the parameters $q(y)$ and $q_j(x|y)$ of the NBM; and (2) Selecting most informative features X_j's of the NBM. While the first problem is routinely solved by using the maximum-likelihood method, the second problem can be solved by using the greedy algorithm, a simple heuristic approach. In this section, we describe the basic steps of solving the second problem. For the first problem, we refer readers to some books such as Hastie et al. (2017).

Consider a simple NBM of predicting the probability of "getting sick" based on three indicator variables which are "fever", "age above 60", "travel within 17 days" and "gender". That is a $d = 4$ and $k = 2$ (sick or not) case of the general NBM presented. We want to select the most informative subset $X_A =$

$(X_{i_1},...,X_{i_k})$ to predict $Y = \{0,1\}$. Such a selection is achieved by finding the optimal subset A^* as follows:

$$A^* = \arg\max IG(X_A, Y) \quad \text{s.t.} \quad |A| \leq k,$$

where $IG(X_A, Y) = H(Y) - H(Y|X_A)$, which represents the uncertainty reduction owing to knowing X_A. Let V be a set of d variables X_i with $i = 1,...,d$, called features. Define the utility function $F(A) = IG(X_A, Y)$. The problem is formulated as

$$\text{Find } A^* \subseteq V \text{ such that}$$
$$A^* = \arg\max_{|A| \leq d} F(A),$$

which is NP-hard. A heuristic approach is to apply the following greedy algorithm. It can be shown that the information gain $F(A)$ in NBM is submodular

Greedy Algorithm
Start with $A = \emptyset$
For $i = 1$ to k
$s^* := \arg\max_s F(A \cup \{s\})$
$A := A \cup \{s^*\}$

(see Krause and Guestrin 2009). That is for $A \subseteq B$ and $s \in V \setminus B$, we have $F(A \cup \{s\}) = F(A) \geq F(B \cup \{s\}) - F(B)$. With the submodular property, we can bound the performance of the greedy algorithm.

Theorem D.4 *(Nemhauser et al 1978) Greedy maximization algorithm returns* A_{greedy}:

$$F(A_{greedy}) \geq \left(1 - \frac{1}{e}\right) \max_{|A| \leq k} F(A).$$

This theorem implies that the greedy algorithm can achieve at least about 63% of the performance of the optimal solution.

REFERENCES

1. D. Applebaum, "Probability and Information: An Integrated Approach", Cambridge University Press, 2nd ed., 2008.
2. T. Hastie, R. Tibshirani, and J. Friedman, "The Elements of Statistical Learning: Data Mining, Inference, and Prediction", Springer, 2nd ed., 2017.
3. A. Krause and C. Guerstrin, Optimal value of information in graphical models. *Journal of Artificial Intellegence Research*, (35) 557-591, 2009.
4. G.L. Nemhauser and L.A. Wolsey, Best algorithm for approximating the maximum of a submodular Set Function. *Mathematics of Operations Research*, 3(3) 177-188, 1978.

E Change of Probability Measure for a Normal Random Variable

Consider a random variable X under a standard normal probability measure, $X \sim N(0,1)$. We denote the p.d.f. by $p(x) = \frac{1}{\sqrt{2\pi}} e^{-\frac{1}{2}x^2}$ and $P(x) = P(X \leq x) = \int_{-\infty}^{x} p(s)ds$. Thus, $dP(x) = p(x)dx$. We can write the probability

$$P(\alpha \leq X \leq \beta) = \frac{1}{\sqrt{2\pi}} \int_{\alpha}^{\beta} e^{-\frac{1}{2}x^2} dx.$$

as an example of distribution function. We will show that the change of the distribution on the same random variable is simply the probability measure change. Let's assume that the interval is shifted by a constant μ. That is

$$P(\alpha - \mu \leq X \leq \beta - \mu) = \frac{1}{\sqrt{2\pi}} \int_{\alpha-\mu}^{\beta-\mu} e^{-\frac{1}{2}x^2} dx.$$

Note that we keep the same random variable X. We perform the change of variable to compare these two probabilities to see how much change in the probability of the original interval has been induced by the given shift of the integration interval. Letting $y = x + \mu$ and making the substitutions yields

$$P(\alpha - \mu \leq X \leq \beta - \mu) = \frac{1}{\sqrt{2\pi}} \int_{\alpha-\mu}^{\beta-\mu} e^{-\frac{1}{2}x^2} dx = \frac{1}{\sqrt{2\pi}} \int_{\alpha}^{\beta} e^{-\frac{1}{2}(y-\mu)^2} dy.$$

Now the intervals are the same but the probabilities are shifted. Denote the original interval event by $A = \{\alpha \leq X \leq \beta\}$ and the shifted interval event by $A - \mu$. While event A has the probability of the standard normal distribution in the interval (α, β), event $A - \mu$ has the probability of the normal distribution with mean of μ and standard deviation of 1 in the same interval. We can

DOI: 10.1201/9781003150060-E

re-write the probability

$$P(A - \mu) = P(\alpha - \mu \leq X \leq \beta - \mu) = \frac{1}{\sqrt{2\pi}} \int_\alpha^\beta e^{-\frac{1}{2}(x-\mu)^2} dx$$

$$= \frac{1}{\sqrt{2\pi}} \int_\alpha^\beta e^{-\frac{1}{2}(x^2 - 2x\mu + \mu^2)} dx = \frac{1}{\sqrt{2\pi}} \int_\alpha^\beta e^{-\frac{1}{2}x^2} e^{x\mu - \frac{1}{2}\mu^2} dx$$

$$= \int_\alpha^\beta e^{x\mu - \frac{1}{2}\mu^2} p(x) dx.$$

$$= \int_\alpha^\beta e^{x\mu - \frac{1}{2}\mu^2} dP(x) = \int_A e^{x\mu - \frac{1}{2}\mu^2} dP(x)$$

$$= \int_{-\infty}^\infty e^{x\mu - \frac{1}{2}\mu^2} \mathbf{1}_{[\alpha,\beta]} dP(x) = \int_A e^{x\mu - \frac{1}{2}\mu^2} dP(x)$$

$$= E^P \left[e^{x\mu - \frac{1}{2}\mu^2} \mathbf{1}_A \right] = E^P [Z(x)\mathbf{1}_A].$$

This shift of the interval can be viewed as leading to a change of probability. Now we define a new probability measure. Define $Q(A) = \int_A e^{x\mu - \frac{1}{2}\mu^2} dP(x) = E^P \left[e^{x\mu - \frac{1}{2}\mu^2} \mathbf{1}_A \right] = E^P [z(x)\mathbf{1}_A]$. Consider a left open interval probability

$$Q(X \leq a) = E^P \left[e^{x\mu - \frac{1}{2}\mu^2} \mathbf{1}_{x \leq a} \right] = \frac{1}{\sqrt{2\pi}} \int_{-\infty}^\infty e^{x\mu - \frac{1}{2}\mu^2} \mathbf{1}_{x \leq a} e^{-\frac{1}{2}x^2} dx$$

$$= \frac{1}{\sqrt{2\pi}} \int_{-\infty}^a e^{x\mu - \frac{1}{2}\mu^2} e^{-\frac{1}{2}x^2} dx$$

$$= \frac{1}{\sqrt{2\pi}} \int_{-\infty}^a e^{-\frac{1}{2}(-2x\mu + \mu^2 + x^2)} dx = \frac{1}{\sqrt{2\pi}} \int_{-\infty}^a e^{-\frac{1}{2}(x-\mu)^2} dx,$$

which is the probability of a normal random variable with mean μ and variance of 1 for the interval $(-\infty, a]$ or the distribution function. Thus, Q is a probability measure as it satisfies the requirements of probability measure. We start with a probability measure P and show that shifting the interval induces a new probability measure Q. Let's now look at the relation between these two distributions, i.e., $Q(A) = \int_A e^{x\mu - \frac{1}{2}\mu^2} dP(x)$. We can write

$$Q(\alpha \leq X \leq \beta) = Q(\beta) - Q(\alpha) = \int_\alpha^\beta z(x) dP(x).$$

This is very similar to the fundamental theorem of calculus which establishes a connection between the two central operations, integration and differentiation. Thus the relation between probability measures Q and P through the z function. For such a relation, we do need some conditions regarding the probability measure functions. We need to tighten the continuity definition here. Recall the property spectrum of a function: continuous $->$ uniformly continuous $->$ absolutely continuous $->$ Lipschitz continuous $->$ continuously differentiable. In fact, the relation holds if the probability measure function is absolutely continuous. A measure Q is said to be absolutely continuous with respect to another measure P if whenever P assign 0 probability to an event, then Q also assigns 0 probability to the same event. We write $Q \ll P$ if $P(E) = 0$, then $Q(E) = 0$ or if $P(E) < \delta$, then $Q(E) < \varepsilon$. z is called the derivative of probability measure Q with respect to the probability measure P. This derivative is called Radon-Nikodym derivative. The Rodon-Nikodym theorem (see Chapter A) says: If $Q \ll P$, then there is a z such that:

$$Q(A) = \int_A z(x)dP(x), \quad \text{and} \quad \frac{dQ(x)}{dP(x)} = z(x).$$

Then, we say that these two probability measures are equivalent. We also need to check that z is a probability density function. That is

$$z(x) = e^{x\mu - \frac{1}{2}\mu^2} \geq 0,$$

$$E^P\left[e^{x\mu - \frac{1}{2}\mu^2}\right] = \frac{1}{\sqrt{2\pi}} \int_{-\infty}^{\infty} e^{x\mu - \frac{1}{2}\mu^2} e^{-\frac{1}{2}x^2} dx = \frac{1}{\sqrt{2\pi}} \int_{-\infty}^{\infty} e^{\frac{1}{2}(x-\mu)^2} dx = 1,$$

$$E^P[z(x)] = \int_{-\infty}^{\infty} \frac{dQ(x)}{dP(x)} dP(x) = \int_{-\infty}^{\infty} dQ(x).$$

In general, for a function $h(x)$, we have

$$E^P[z(x)h(x)] = \int_{-\infty}^{\infty} \frac{dQ(x)}{dP(x)} h(x)dP(x) = \int_{-\infty}^{\infty} h(x)dQ(x) = E^Q[h(x)].$$

The probability measure has been shifted to the right. This development can be extended to the Brownian process by taking the limit of the random walk and applying this result to each increment interval. We only provide the parallel results and the derivations can be left as an exercise.

Normal random variable	Change in each step of random walk
$X \sim N(0,1)$ under P	$\Delta X_i \sim (0, \Delta t)$ under P
interval shifted to the left by μ	interval shifted to the left by $\mu_i \Delta t$
$Q(A) = \int_A z(x)dP(x)$	$Q(A) = \int_A z_i(x)dP(x)$
$z(x) = e^{\mu x - \frac{1}{2}\mu^2}$	$z_i(x) = e^{\mu x_i - \frac{1}{2}\mu_i^2 \Delta t}$
X under $Q \sim N(\mu, 1)$	$\Delta X_i \sim N(\mu_i \Delta t, \Delta t)$ under Q
$X - \mu$ under $Q \sim N(0,1)$	$\Delta X_i - \mu_i \Delta \sim N(0, \Delta t)$ under Q

Now consider the summation of the sub-intervals. We have

$\Delta X_i \sim N(0, \Delta t)$ under P	$X_n = \sum_{i=1}^{n} \Delta X_i \sim N(0, t_n)$ under P
$\Delta X_i \sim N(\mu_i \Delta t, \Delta t)$ under Q	$X_n \sim N\left(\sum_{i=1}^{n} \mu_i \Delta t, t_n\right)$ under Q

Now let's look at the exponential term of $z_i(x)$. Taking the limit of $\Delta t \to 0$ or $n \to \infty$, we have

$$\sum_{i=1}^n \mu_i \Delta x_i - \frac{1}{2} \sum_{i=1}^n \mu_i^2 \Delta t \to \int_0^t \mu_i dX_i - \frac{1}{2} \int_0^t \mu_i^2 ds.$$

Consider a SBM B_t under probability measure P. Note that

$$Y(t) = \int_0^t \mu_s ds, \quad Z(t) = \frac{dQ}{dP} = e^{\int_0^t \mu_s dB_s - \frac{1}{2} \int_0^t \mu_s^2 ds}.$$

Then $\tilde{B}_t = B_t - \int_0^t \mu_s ds$ is a BM under Q. If we assume that μ is constant, then, $Y(t) = \mu t, Z(t) = e^{\mu B_t - \frac{1}{2}\mu^2 t}$ and $\tilde{B}_t = B_t - \mu t$ is a BM under Q. Next, we need to show that $Z(t)$ is a valid probability density in continuous time. This can be done by showing that $Z(t)$ is a martingale. That is $E^P[Z(t)|\mathscr{F}_s] = Z(s)$. Since it is bounded, we only need to show it is a local martingale. That is to show it is a stochastic integral with no drift term. To do that we only need to apply Ito's Lemma to the exponential term. Let $X = \mu B_t - (1/2)\mu^2 t$. Recall that the Ito's differential of exponential is $dX = \mu dB_t - (1/2)\mu^2 dt$ due to the constant μ and the quadratic term $dX^2 = \mu^2 dt$ (see Chapter H). Now substituting these terms, we get

$$dZ(t) = e^X \left(dX + \frac{1}{2} dX^2 \right)$$

$$= Z(t) \left(\mu dB_t - \frac{1}{2}\mu^2 dt + \frac{1}{2}\mu^2 dt \right) = \mu Z(t) dB_t.$$

Since B_t is a martingale, it is easy to show that $Z(t)$ is also a martingale. For $0 \le s \le t$, we can write

$$E[Z(t)|\mathscr{F}_s] = E\left[\exp\left(\mu B_t - \frac{1}{2}\mu^2 t \right) | \mathscr{F}_s \right]$$

$$= E[\exp(\mu B_t)|\mathscr{F}_s] \cdot \exp\left(-\frac{1}{2}\mu^2 t \right)$$

$$= E[\exp(\mu(B_t - B_s)|\mathscr{F}_s] \cdot \exp(\mu B_s) \cdot \left(-\frac{1}{2}\mu^2 t \right).$$

Note that

$$E[\exp(\mu(B_t - B_s)|\mathscr{F}_s] = E[\exp(\mu(B_t - B_s))] = \exp\left(-\frac{1}{2}\mu^2(t-s) \right),$$

where the first equality is due to the strong Markov property of a BM and the second equality is due to the moment generating function for the increment of the BM. Combining both above, we have

$$E[Z(t)|\mathscr{F}_s] = \exp\left(-\frac{1}{2}\mu^2(t-s)\right) \cdot \exp(\mu B_s) \cdot \exp\left(-\frac{1}{2}\mu^2 t\right)$$

$$= \exp\left(\mu B_s - \frac{1}{2}\mu^2 s\right) = Z(s).$$

This indicates that $Z(t)$ is a martingale, hence, is a valid density. We can further show that $\tilde{B}_t = B_t - \mu t$ is a BM by checking the conditions of the definition. In classical probability theory, the result, that tells how stochastic processes change under changes in measure, is stated as the Girsanov theorem. This theorem is especially important in the theory of financial mathematics. For more details on this theorem and related applications, interested readers are referred to some good textbooks such as Baxter and Rennie (1996).

REFERENCE

1. M. Baxter and A. Rennie "Financial Calculus - An Introduction to Derivative Pricing", Cambridge University Press, 1996.

F Convergence of Random Variables

Here we provide definitions of different types of convergence and their relationships. Consider a sequence of random variables $\{X_n, n \in \mathbb{N}\}$. This sequence can converge to a random variable X. There are four types of convergence as follows:

1. Convergence in distribution,
2. Convergence in probability,
3. Convergence in mean,
4. Almost sure convergence (also called convergence with probability 1).

F.1 CONVERGENCE IN DISTRIBUTION

Convergence in distribution is in some sense the weakest type of convergence. All it says is that the c.d.f.'s of X_n's, denoted by $F_{X_n}(x)$, converge to the c.d.f. of X, denoted by $F_X(x)$, as n goes to infinity.

Definition F.1 *A sequence of random variables $\{X_n, n \geq 1\}$ converges in distribution to a random variable X, denoted by $X_n \overset{d}{\to} X$, if*

$$\lim_{n \to \infty} F_{X_n}(x) = F_X(x)$$

for all x at which $F_X(x)$ is continuous.

Example F.1 *Suppose that there is a sequence of random variables with distribution functions as*

$$F_{X_n}(x) = \begin{cases} 1 - \left(1 - \frac{1}{n}\right)^{n\lambda x}, & x > 0, \\ 0, & x \leq 0, \end{cases} \quad \text{for } n = 2, 3, \dots.$$

Show that X_n converges in distribution to the exponentially distributed random variable with rate λ. Let $X \sim Exp(\lambda)$. For $x \leq 0$, we have $F_{X_n}(x) = F_X(x) = 0$

for $n = 2, 3, \ldots$. For $x > 0$, we have

$$\lim_{n \to \infty} F_{X_n}(x) = \lim_{n \to \infty} \left(1 - \left(1 - \frac{1}{m} \right)^{n\lambda x} \right)$$

$$= 1 - \lim_{n \to \infty} \left(1 - \frac{1}{n} \right)^{n\lambda x}$$

$$= 1 - e^{-\lambda x} = F_X(x).$$

Thus, we conclude that $X_n \xrightarrow{d} X$.

Another example of convergence in distribution is that a sequence of Binomial random variables, $\{X_n \sim Binom(n, \lambda/n), n \geq 1.\}$ converges to a Poisson distribution with rate λ in distribution. Central Limit Theorem is also an example of the convergence in distribution.

F.2   CONVERGENCE IN PROBABILITY

Convergence in probability is stronger than convergence in distribution.

Definition F.2 *A sequence of random variables $\{X_n, n \geq 1\}$ converges in probability to a random variable X, denoted by $X_n \xrightarrow{P} X$, if*

$$\lim_{n \to \infty} P(|X_n - X| \geq \varepsilon) = 0, \ \ for \ all \ \varepsilon > 0.$$

Example F.2 *Consider a sequence of exponentially distributed random variables $\{X_n, n \geq 1\}$ with rate n. Show that $X_n \xrightarrow{P} 0$. This can be done by checking the definition of this convergence mode as follows:*

$$\lim_{n \to \infty} P(|X_n - 0| \geq \varepsilon) = \lim_{n \to \infty} P(X_n \geq \varepsilon) = \lim_{n \to \infty} e^{-n\varepsilon} = 0$$

for all $\varepsilon > 0$.

The most famous example of convergence in probability is the WLLN.

F.3   CONVERGENCE IN MEAN

We can define the convergence of a sequence of random variables, $\{X_n, n \geq 1\}$ to X in terms of the "distance" between X and X_n. One way to define this distance is by the expected value of the rth power of the absolute distance between X_n and X. That is $E[|X_n - X|^r]$, where $r \geq 1$ is a fixed number. Such a convergence mode requires $E[|X_n^r|] < \infty$. The most common choice is $r = 2$, which is called the mean-square convergence.

Definition F.3 *A sequence of random variables $\{X_n, n \geq 1\}$ converges in the rth mean or in the L^r norm to a random variable X, denoted by $X_n \xrightarrow{L^r} X$, if*

$$\lim_{n\to\infty} E[|X_n - X|^r] = 0.$$

If $r = 2$, it is called the mean-square convergence and denoted by $X_n \xrightarrow{m.s.} X$.

Example F.3 *Consider a sequence of random variables with uniform distributions. That is $X_n \sim Uniform\left(0, \frac{1}{n}\right)$ for $n \geq 1$. Show that $X_n \xrightarrow{L^r} 0$, for any $r \geq 1$. Note that the pdf of X_n is given by*

$$f_{X_n}(x) = \begin{cases} n, & 0 \leq x \leq \frac{1}{n} \\ 0, & otherwise, \end{cases}$$

We have

$$E[|X_n - 0|^r] = \int_0^{\frac{1}{n}} x^r n \, dx = \frac{1}{(r+1)n^r} \to 0, \ as \ n \to \infty$$

for all $r \geq 1$.

We present two theorems regarding this mode of convergence.

Theorem F.1 *For $1 \leq r \leq s$, if $X_n \xrightarrow{L^s} X$, then $X_n \xrightarrow{L^r} X$.*

The proof can be done by using Holder's Inequality.

Theorem F.2 *If $X_n \xrightarrow{L^r} X$, then $X_n \xrightarrow{P} X$.*

Proof. For any $\varepsilon > 0$ and $r \geq 1$, we have

$$P(|X_n - X| \geq \varepsilon) = P(|X_n - X|^r \geq \varepsilon^r)$$
$$\leq \frac{E[|X_n - X|^r]}{\varepsilon^r},$$

where the inequality follows by the Markov's inequality. Since $\lim_{n\to\infty} E[|X_n - X|^r] = 0$, we have $\lim_{n\to\infty} P(|X_n - X| \geq \varepsilon) = 0$, for all $\varepsilon > 0$. ∎

The converse of the theorem is not true in general. Look at the following example.

Example F.4 *Consider a sequence of random variables $\{X_n, n \geq 1\}$ such that*

$$X_n = \begin{cases} n^2 & with \ probability \ \frac{1}{n}, \\ 0 & with \ probability \ 1 - \frac{1}{n}. \end{cases}$$

Show that $X_n \xrightarrow{P} 0$ but X_n does not converge in the rth mean for $r \geq 1$.

We can show $X_n \xrightarrow{P} 0$ by noting that

$$\lim_{n \to \infty} P(|X_n - 0| \geq \varepsilon) = P(X_n = n^2) = \lim_{n \to \infty} \frac{1}{n} = 0.$$

However, for any $r \geq 1$, we have

$$\lim_{n \to \infty} E[|X_n - 0|^r] = \lim_{n \to \infty} \left(n^{2r} \cdot \frac{1}{n} + 0 \cdot \left(1 - \frac{1}{n} \right) \right)$$

$$= \lim_{n \to \infty} n^{2r-1} = \infty.$$

Thus, X_n does not converge in the rth mean.

F.4 ALMOST SURE CONVERGENCE

This is the strongest type of convergence of a sequence of random variables. Consider $\{X_n, n \geq 1\}$ defined on the same sample space Ω. For simplicity, we assume that Ω is a finite set

$$\Omega = \{\omega_1, \omega_2, \cdots, \omega_k\}.$$

Since X_n is a function from Ω to the set of real numbers, we can write

$$X_n(\omega_i) = x_{ni}, \quad \text{for } i = 1, 2, \cdots, k.$$

After this random experiment is performed, one of the ω_i's will be the outcome, and the values of the $\{X_n, n \geq 1\}$ are known (a realization of this random variable sequence is observed). Since this is a sequence of real numbers, we can study its convergence. Does it converge? If yes, what does it converge to? Almost sure convergence is defined based on the convergence of such sequences. If the probability of such a convergence is 100%, it is said to be almost surely convergence or convergence with probability 1. We demonstrate this by using the following example.

Example F.5 *Consider a random experiment of tossing a fair coin once. Then, the sample space is $\Omega = \{H, T\}$. We define a sequence of random variables $X_1, X_2, \cdots$ on this sample space as follows:*

$$X_n(\omega) = \begin{cases} \frac{n}{n+1} & \text{if } \omega = H, \\ (-1)^n & \text{if } \omega = T. \end{cases} \tag{F.1}$$

(i) For each of the possible outcomes (H or T), determine whether the realization of the random variable sequence converges or not.

   If the outcome is H, then the nth value of the sequence is $X_n(H) = \frac{n}{n+1}$. Thus, the sequence is

$$\frac{1}{2}, \frac{2}{3}, \frac{3}{4}, \cdots.$$

This sequence converges to 1 as n goes to infinity. If the outcome is T, then $X_n(T) = (-1)^n$ and the sequence is

$$-1, 1, -1, 1, -1, \cdots .$$

This sequence does not converge as it oscillates between -1 and 1 forever.
    (ii) Find

$$P\left(\left\{\omega_i \in \Omega : \lim_{n\to\infty} X_n(\omega_i) = 1\right\}\right).$$

From part (i), the event $\{\omega_i \in \Omega : \lim_{n\to\infty} X_n(\omega_i) = 1\}$ (convergence of a sequence of real numbers to 1) happens if and only if the outcome is H, thus, we have

$$P\left(\left\{\omega_i \in \Omega : \lim_{n\to\infty} X_n(\omega_i) = 1\right\}\right) = P(H) = \frac{1}{2}.$$

Thus, this is not the almost surely convergence or convergence with probability 1. Here is the definition of this convergence mode.

Definition F.4 *A sequence of random variables* $\{X_n, n \geq 1\}$ *converges almost surely to a random variable* X, *denoted by* $X_n \overset{a.s.}{\to} X$, *if*

$$P\left(\left\{\omega_i \in \Omega : \lim_{n\to\infty} X_n(\omega_i) = X(\omega)\right\}\right) = 1.$$

Example F.6 *Consider the sample space* $S = [0, 1]$ *with a probability measure that is uniform on this space (Lebesgue measure on the unit interval). That is* $P([a, b]) = b - a$, *for all* $0 \leq a \leq b \leq 1$. *Define the sequence of random variables* $\{X_n, n \geq 1\}$ *as follows:*

$$X_n(s) = \begin{cases} 1 & 0 \leq s < \frac{n+1}{2n}, \\ 0 & otherwise. \end{cases}$$

Also define the random variable X *on the same sample space as follows:*

$$X(s) = \begin{cases} 1 & 0 \leq s < \frac{1}{2}, \\ 0 & otherwise. \end{cases}$$

Now we can show that $X_n \overset{a.s.}{\to} X$. *First, we define a set* A *as*

$$A = \left\{s \in S : \lim_{n\to\infty} X_n(s) = X(s)\right\}.$$

We need to show that $P(A) = 1$.

Since $\frac{n+1}{2n} > \frac{1}{2}$, for any $s \in [0, 1/2)$, we have

$$X_n(s) = X(s) = 1.$$

Thus, $[0, 0.5) \subset A$. Now if $s \in (1/2, 0]$, we have $X(s) = 0$. Since $2s - 1 > 0$. Note that $n > \frac{1}{2s-1}$ is equivalent to $s > \frac{n+1}{2n}$ which implies that $X_n(s) = 0$. Therefore, we have

$$\lim_{n \to \infty} X_n(s) = 0 = X(s), \quad \text{for all } s > \frac{1}{2}.$$

We conclude $\left(\frac{1}{2}, 1\right] \subset A$. However, there is a single point in S that is not in A. That is $s = 1/2 \notin A$, since $X_n(1/2) = 1$ for all n but $X(1/2) = 0$. We obtain

$$A = \left[0, \frac{1}{2}\right) \cup \left(\frac{1}{2}, 1\right] = S - \left\{\frac{1}{2}\right\}.$$

Since the point has measure zero, we conclude $P(A) = 1$ and hence $X_n \overset{a.s.}{\to} X$. It is desirable to know some sufficient conditions for almost sure convergence which are easier to verify. Here is a result obtained from the Borel–Cantelli Lemma that is sometimes useful when we would like to prove almost sure convergence.

Theorem F.3 *For a sequence of random variables $\{X_n, n \geq 1\}$, if for all $\varepsilon > 0$,*

$$\sum_{n=1}^{\infty} P(|X_n - X| > \varepsilon) < \infty,$$

then $X_n \overset{a.s.}{\to} X$.

Here is an example of using this theorem.

Example F.7 *Consider a sequence of random variables $\{X_n, n \geq 1\}$ such that*

$$X_n = \begin{cases} -\frac{1}{n} & \text{with probability } \frac{1}{2}, \\ \frac{1}{n} & \text{with probability } \frac{1}{2}. \end{cases} \tag{F.2}$$

Show that $X_n \overset{a.s.}{\to} 0$.

Using the theorem, it suffices to show that $\sum_{n=1}^{\infty} P(|X_n - 0| > \varepsilon) < \infty$. Since $|X_n| = 1/n$, $|X_n| > \varepsilon \Leftrightarrow n < 1/\varepsilon$. Thus, we have

$$\sum_{n=1}^{\infty} P(|X_n| > \varepsilon) \leq \sum_{n=1}^{\lfloor \frac{1}{\varepsilon} \rfloor} P(|X_n| > \varepsilon) = \left\lfloor \frac{1}{\varepsilon} \right\rfloor < \infty.$$

Theorem (F.3) only provides a sufficient condition for almost sure convergence. The following theorem provides a sufficient and necessary condition for almost sure convergence.

Theorem F.4 *For a sequence of random variables $\{X_n, n \geq 1\}$, given any $\varepsilon > 0$, define the set of events*

$$E_m = \{|X_n - X| < \varepsilon, \text{ for all } n \geq m\}.$$

Then $X_n \xrightarrow{a.s.} X$ if and only if $\lim_{m \to \infty} P(E_m) = 1$.

A famous example for almost sure convergence is the SLLN.

We present a version of continuous mapping theorem for a sequence of random variables.

Theorem F.5 *For a sequence of random variables $\{X_n, n \geq 1\}$ and a continuous function $g : \mathbb{R} \mapsto \mathbb{R}$, the following statements hold:*

1. If $X_n \xrightarrow{d} X$, then $g(X_n) \xrightarrow{d} g(X)$.
2. If $X_n \xrightarrow{p} X$, then $g(X_n) \xrightarrow{p} g(X)$.
3. If $X_n \xrightarrow{a.s.} X$, then $g(X_n) \xrightarrow{a.s.} g(X)$.

Proof. (1) Convergence in distribution. The proof is based on the portmanteau theorem which says that the convergence in distribution $X_n \xrightarrow{d} X$ is equivalent to $E[f(X_n)] \to E[f(X)]$ for every bounded continuous functional f.

Thus, it suffices to prove that $E[f(g(X_n))] \to E[f(g(X))]$ for every bounded continuous functional f. Note that $F = f \circ g$ is itself a bounded continuous functional. And so the claim follows from the statement above.

(2) Convergence in probability. We need to show that for any $\varepsilon > 0$,

$$P(|g(X_n) - g(X)| > \varepsilon) \to 0, \text{ as } n \to \infty.$$

For some $\delta > 0$, by the Law of Total Probability, we have

$$P(|g(X_n) - g(X)| > \varepsilon) = P(|g(X_n) - g(X)| > \varepsilon, |X_n - X| < \delta)$$
$$+ P(|g(X_n) - g(X)| > \varepsilon, |X_n - X| \geq \delta).$$

As $\delta \to 0$, the first term on the right hand side converges to 0 by the continuity of g and the second term also converges to 0 by convergence of X_n to X. Thus, $P(|g(X_n) - g(X)| > \varepsilon) \to 0$.

(3) Almost sure convergence. By the definition of the continuity of the function $g()$,

$$\lim_{n \to \infty} X_n(\omega) = X(\omega) \Leftrightarrow \lim_{n \to \infty} g(X_n(\omega)) = g(X(\omega))$$

at each point $X(\omega)$. Thus,

$$P\left(\lim_{n \to \infty} g(X_n) = g(X)\right) = P\left(\lim_{n \to \infty} g(X_n) = g(X), X \notin D_g\right)$$
$$\geq P\left(\lim_{n \to \infty} X_n = X, X \notin D_g\right) = 1,$$

where D_g is the set of discontinuity points with measure 0.

■

Finally, we present the relationships between modes of convergence. Almost sure convergence requires that $\{X_n, n \geq 1\}$ and X be defined on a common probability space. The same is true in general for convergence in probability and in rth mean except when the limit X is a constant. The following relationships are true:

1. $X_n \xrightarrow{a.s.} X$ implies $X_n \xrightarrow{P} X$;
2. $X_n \xrightarrow{P} X$ implies $X_n \xrightarrow{d} X$;
3. $X_n \xrightarrow{L^r} X$ implies $X_n \xrightarrow{P} X$;
4. $X_n \xrightarrow{d} X$ implies $X_n \xrightarrow{P} X$ if X is a constant with probability 1.

Proofs of these relations can be left as exercises for readers or be found in some standard probability textbooks. Here we present a few counter-examples to show that the converse of these relations is not true.

Example F.8 *Convergence in distribution does not imply convergence in probability.*

Let $\Omega = \{\omega_1, \omega_2, \omega_3, \omega_4\}$. Define a sequence of random variables X_n and a random variable X as follows:

$$X_n(\omega_1) = X_n(\omega_2) = 1, X_n(\omega_3) = X_n(\omega_4) = 0, \text{ for all } n$$
$$X(\omega_1) = X(\omega_2) = 0, X(\omega_3) = X(\omega_4) = 1.$$

Moreoever, we assign equal probability to each event (Lebesgue measure in [0,1]). Then we have the c.d.f's as follows:

$$F(x) = \left\{ \begin{array}{ll} 0, & \text{if } x < 0, \\ \frac{1}{2}, & \text{if } 0 \leq x < 1 \\ 1, & \text{if } x \geq 1 \end{array} \right\} \quad F_n(x) = \left\{ \begin{array}{ll} 0, & \text{if } x < 0, \\ \frac{1}{2}, & \text{if } 0 \leq x < 1 \\ 1, & \text{if } x \geq 1 \end{array} \right\}. \quad \text{(F.3)}$$

Since $F_n(x) = F(x)$ for all n, it is trivial that $X_n \xrightarrow{d} X$. However, we also have

$$\lim_{n \to \infty} P\left(\omega : |X_n(\omega) - X(\omega)| \geq \frac{1}{2} \right) = 1,$$

as $|X_n(\omega) - x(\omega)| = 1$ for all n and ω. Thus, $X_n \xrightarrow{P} X$. In fact, it is very easy to find a counter-example for this illustration. For example, let $X_n = U$ for all n and $X = 1 - U$, where U follows a uniform distribution in $[0, 1]$. Then we can show the same conclusion.

Example F.9 *Convergence in probability does not imply almost sure convergence.*

Consider the sequence of random variables $\{X_n, n \geq 1\}$ such that

$$P(X_n = 1) = \frac{1}{n}, \quad P(X_n = 0) = 1 - \frac{1}{n}, \quad n \geq 1.$$

Obviously for any $0 < \varepsilon < 1$ and $X = 0$, we have

$$P(|X_n - X| > \varepsilon) = P(X_n = 1) = \frac{1}{n} \to 0.$$

Thus, $X_n \overset{p}{\to} X$. To show whether or not $X_n \overset{a.s.}{\to} X$, we need to define the following set: For any $\varepsilon > 0$,

$$B_m(\varepsilon) = \bigcup_{n=m}^{\infty} A_n(\varepsilon),$$

where $A_n(\varepsilon) = \{\omega : |X_n(\omega) - X(\omega)| > \varepsilon\}$. Then, we have the following equivalence: $X_n \overset{a.s.}{\to} X \Leftrightarrow P(B_m(\varepsilon)) \to 0$ as $m \to \infty$. An intuitive explanation for $B_m(\varepsilon)$ is that there exist some (or at least one) $A_n(\varepsilon)$ occurring for $n \geq m$ (i.e., the event that X_n is not so close to X occurring). Clearly, $B_m(\varepsilon)$ is a decreasing sequence of events. Thus, $\lim_{m \to \infty} B_m(\varepsilon) = \bigcap_{m=1}^{\infty} \bigcup_{n=m}^{\infty} A_n(\varepsilon) = \limsup_{m \to \infty} A_n(\varepsilon) = A(\varepsilon) = \{\omega : \omega \in A_n(\varepsilon) \ i.o.\}$. Thus, $X_n \overset{a.s.}{\to} X$ or $P(B_m(\varepsilon)) \to 0$ as $m \to \infty$, if and only if $P(A(\varepsilon)) = 0$. Now we show $P(B_m(\varepsilon))$ does not approach to 0 in this example. Note that

$$P(B_m(\varepsilon)) = 1 - \lim_{M \to \infty} P(X_n = 0 \text{ for all } n \text{ such that } m \leq n \leq M)$$

$$= 1 - \lim_{M \to \infty} \left(1 - \frac{1}{m}\right)\left(1 - \frac{1}{m+1}\right) \cdots \left(1 - \frac{1}{M+1}\right)$$

$$= 1 - \lim_{M \to \infty} \frac{m-1}{m} \frac{m}{m+1} \frac{m+1}{m+2} \cdots \frac{M}{M+1}$$

$$= 1 - \lim_{M \to \infty} \frac{m-1}{M+1} = 1,$$

which implies that $P(B_m(\varepsilon)) \to 1$ as $m \to \infty$. Hence, $X_n \overset{a.s.}{\not\to} X$.

Example F.10 *Convergence in probability does not imply convergence in L^r.*

Consider a sequence of random variables $\{X_n, n \geq 1\}$ such that

$$P(X_n = e^n) = \frac{1}{n}, \quad P(X_n = 0) = 1 - \frac{1}{n}.$$

Let $X = 0$. Then, for any $\varepsilon > 0$, we have

$$P(|X_n - X| < \varepsilon) = P(|X_n| < \varepsilon) = 1 - \frac{1}{n} \to 1 \text{ as } n \to \infty.$$

Thus, $X_n \xrightarrow{P} 0$. However, for each $r > 0$, we have

$$E[|X_n - X|^r] = E[X_n^r] = e^{rn}\frac{1}{n} \to \infty \text{ as } n \to \infty.$$

Thus, $X_n \xrightarrow{L^r} 0$.

Example F.11 *Convergence in L^r does not imply almost sure convergence.*

We only need to show a case with $r = 2$ by using the previous example. Consider a sequence of random variables $\{X_n, n \geq 1\}$ such that

$$P(X_n = 1) = \frac{1}{n}, \quad P(X_n = 0) = 1 - \frac{1}{n}, \quad n \geq 1.$$

Then, for $X = 0$,

$$E[|X_n - X|^2] = E[|X_n|^2] = \frac{1}{n} \to 0 \text{ as } n \to \infty,$$

which implies $X_n \xrightarrow{L^2} 0$. As shown in Example F.9, this is the case where $X_n \xrightarrow{a.s.} 0$.

Example F.12 *Almost sure convergence does not imply convergence in L^r.*

We first show the case for $r = 1$. Let the underlying probability space be the unit interval $[0, 1]$ with uniform distribution (which coincides with Lebesgue measure). Consider a sequence of random variables $\{X_n, n \geq 1\}$ such that

$$X_n = \begin{cases} 2^n, & \text{if } \omega \in (a_n, a_n + \frac{1}{2^n}), \\ 0, & \text{otherwise,} \end{cases}$$

where $a_n = 2^{-1} + 2^{-2} + \cdots + 2^{-(n-1)}$ with $a_1 = 0$. Now we check for any $\varepsilon > 0$

$$\sum_{n=1}^{\infty} P(|X_n - 0| > \varepsilon) = \sum_{n=1}^{\infty} P(X_n = 2^n) = \sum_{n=1}^{\infty} \frac{1}{2^n} = 1 < \infty,$$

which implies that $X_n \xrightarrow{a.s.} 0$. However, $E[|X_n - 0|] = E[X_n] = 2^n \cdot \frac{1}{2^n} = 1 \neq 0$ for all n. Thus, we have $X_n \xrightarrow{L^1} 0$. For $r \geq 2$ case, consider a sequence of random variables $\{X_n, n \geq 1\}$ such that

$$P(X_n = n) = \frac{1}{n^r}, \quad P(X_n = 0) = 1 - \frac{1}{n^r}, \quad n \geq 1.$$

We have

$$P(X_n = 0, \text{ for all } n \text{ such that } m \leq n \leq M) = \prod_{n=m}^{M} \left(1 - \frac{1}{n^r}\right).$$

As $M \to \infty$, the infinite product converges to some nonzero quantity, which itself converges to 1 as $m \to \infty$. Hence, $X_n \overset{a.s.}{\to} 0$. However, $E[|X_n - 0|^r] = 1$ and $X_n \overset{L^r}{\not\to} 0$ as $n \to \infty$.

Example F.13 *Convergence in L^r does not imply convergence in quadratic mean.*

Let $\{X_n, n \geq 1\}$ be a sequence of random variables such that

$$X_n = \begin{cases} 0 & \text{with probability } 1 - \frac{1}{n}, \\ \sqrt{n} & \text{with probability } \frac{1}{n}. \end{cases}$$

It is easy to show that $X_n \overset{a.s.}{\to} 0$ and $X_n \overset{L^1}{\to} 0$, however, $X_n \overset{L^2}{\not\to} 0$. For more details on the convergence of random variable sequences, interested readers are referred to many excellent books on probability theory such as Durrett (2010) and Resnick (2005).

REFERENCES

1. R. Durrett, "Probability - Theory and Examples", Cambridge University Press, 4th ed., 2010.
2. S.I. Resnick, "A Probability Path", Springer Science+Business Media, New York, 2005.

G | Major Theorems for Stochastic Process Limits

In this chapter, we present some important theorems for obtaining the limits of the stochastic processes. However, to give some intuitive explanations or justifications, we may only prove the case for the convergence of the sequence of random variables. Thus, the more general sequence of random elements (or random vectors) may replace the sequence of random variables in the statements of theorems. Note that the generalization to a more general setting can be found in some advanced probability and statistics books such as Billingsley (1999) and Chen and Yao (2001).

G.1 SKOROHOD REPRESENTATION THEOREM

Here we present the random variable sequence version of this theorem.

Theorem G.1 *For a sequence of random variables $\{X_n, n \geq 1\}$, if $X_n \xrightarrow{d} X$, then there exist random variables $\{X_n^*, n \geq 1\}$ and X^* defined on a common probability space such that (a) $X_n^* \stackrel{d}{=} X_n$ for all n, and $X^* \stackrel{d}{=} X$ (where $\stackrel{d}{=}$ denotes equality in distribution), and (b) $X_n^* \xrightarrow{a.s.} X^*$.*

This theorem also applies in a more general setting. That is when $\{X_n\}$ and X are random vectors or, in fact, random elements of any separable and complete metric space, the theorem also holds. Here we present a simple and intuitive proof for the random variable case.

Proof. If F_n and F are the distribution functions of X_n and X, we define F_n^{-1} and F^{-1} as their (left-continuous) inverses where, for example, $F^{-1}(t) = \sup\{x : F(x) \geq t\}$ for $0 < t < 1$. Then given a random variable U which has a uniform distribution on $(0, 1)$, we can define $X_n^* = F_n^{-1}(U)$ and $X^* = F^{-1}(U)$. Then, it is easy to verify that $X_n^* \stackrel{d}{=} X_n$, $X^* \stackrel{d}{=} X$ and $X_n^* \xrightarrow{a.s.} X^*$. ∎

Note that this proof cannot be generalized to sequences of random vectors or random elements as it takes advantage of the natural ordering of the real line. For the proof of the more general setting, the readers are referred to Billingsley (1999).

G.2 CONTINUOUS MAPPING THEOREM

The continuous mapping theorem (CMT) states that continuous functions preserve limits even if their arguments are sequences of random variables. A continuous function is such a function that maps convergent sequences into convergent sequences: if $x_n \to x$ then $g(x_n) \to g(x)$. The continuous mapping theorem states that this will also be true if we replace the deterministic sequence $\{x_n\}$ with a sequence of random variables $\{X_n\}$, and replace the standard notion of convergence of real numbers "$\to$" with one of the types of convergence of random variables.

Let $\{X_n\}$, X be random elements defined on a metric space S. Suppose a function $g : S \to S'$ (where S' is another metric space) has the set of discontinuity points D_g such that $P(X \in D_g) = 0$. Then

$$X_n \overset{d}{\to} X \Rightarrow g(X_n) \overset{d}{\to} g(X);$$

$$X_n \overset{p}{\to} X \Rightarrow g(X_n) \overset{p}{\to} g(X);$$

$$X_n \overset{a.s.}{\to} X \Rightarrow g(X_n) \overset{a.s.}{\to} g(X).$$

We present a simple proof of this theorem by starting with the following lemma.

Lemma G.1 *The convergence $X_n \overset{p}{\to} X$ occurs if and only if every sequence of natural numbers $n_1, n_2, \ldots \in \mathbb{N}$ has a subsequence $r_1, r_2, \ldots \in \{n_1, n_2, \ldots\}$ such that $X_{r_k} \overset{a.s.}{\to} X$ as $K \to \infty$.*

A sketch of the proof is as follows:
Proof. We prove the second statement of the CMT by using Lemma G.1. It is sufficient to show that for every sequence $n_1, n_2, \ldots$, we have a subsequence $m_1, m_2, \ldots$ along which $g(X_{m_i}) \overset{a.s.}{\to} g(X)$. A second use of Lemma G.1, shows that we can find a subsequence $m_1, m_2, \ldots$ of $n_1, n_2, \ldots$ along which X_n converges to X with probability 1. Since continuous functions preserve limits this implies that $g(X_n)$ converges to $g(X)$ along that subsequence with probability 1, and the second statement follows. We prove the first statement by using the portmanteau lemma (presented in Section G.0,6 below). It is sufficient to show that for a bounded and continuous function h, we have $E(h(g(X_n))) \to E(h(g(X)))$. Since g, h are continuous with probability 1 and h is bounded, the function $f = h \circ g$ is also continuous with probability 1 and bounded. It follows from the portmanteau lemma that $E(f(X_n)) \to E(f(X))$, proving the first statement.

To prove the third statement, note that we have with probability 1 a continuous function of a convergent sequence. Using the fact that continuous functions preserve limits, we have convergence to the required limit with probability 1. ∎

For more details on the proof of this theorem, interested readers are referred to Billingsley (1999).

G.3 RANDOM TIME-CHANGE THEOREM

Theorem G.2 *Let $\{X_n, n \geq 1\}$ and $\{Y_n, n \geq 1\}$ be two sequences in $\mathscr{D}^J$. Assume that Y_n is non-decreasing with $Y_n(0) = 0$. If as $n \to \infty$, (X_n, Y_n) converges uniformly on compact sets to (X, Y) with X and Y in $\mathscr{C}^J$, then $X_n(Y_n)$ converges uniformly on compact sets to $X(Y)$, where $X_n(Y_n) = X_n \circ Y_n = \{X_n(Y_n(t)), t \geq 0\}$ and $X(Y) = X \circ Y = \{X(Y(t)), t \geq 0\}$.*

Since the proof can be found in many advanced probability books such as Billingsley (1999), we omit the rigorous proof. However, we want to explain why sometimes we need to do random time change for a stochastic process. The basic idea can be demonstrated by the fact that a stochastic process can be transformed to another one by stretching and shrinking the time scale in either deterministic or random manner. The simplest point process is the homogeneous Poisson process with "unit intensity" (i.e., with rate $\lambda = 1$), which can be called a standard Poisson process. In such a process, the inter-arrival time is exponentially distributed with mean of 1. By "time-change technique" we can transform a complicated arrival process into a standard Poisson process. First, we consider a homogeneous Poisson process that does not have unit intensity (i.e., $\lambda \neq 1$). To make a realization of this process look like a realization of the standard Poisson process, we can simply re-scale the time by factor of λ. This is to use the new time variable $\tau(t) = t\lambda$. Thus, the inter-arrival time interval is stretched out when the intensity λ is high or is compressed when λ is low. Thus, $X(\tau)$ looks like a realization of the standard Poisson process with unit intensity. Now consider an non-homogeneous Poisson process with a time varying intensity $\lambda(t)$. The idea of transforming the time variable remains similar. We still want the new time, τ, to run slow and stretch out when the $\lambda(t)$ is high and we want it run fast and compress events when $\lambda(t)$ is low. The time transformation is done by

$$\tau(t) = \int_0^t \lambda(s)ds.$$

If $t_1, t_2, \ldots$ are the arrival instants of events in the original time axis, then the transformed time instants $\tau(t_1), \tau(t_2), \ldots$ come from a standard Poisson process and $\tau(t_{i+1}) - \tau(t_i)$ are i.i.d. and exponentially distributed with mean 1. Next, we consider the case where the intensity function is stochastic and depends on some random factors, including possibly history of the process. in this case, we can write the intensity as $\lambda(t|\mathscr{H}_t)$, where $\mathscr{H}_t$, as a sigma-field, summarizes everything that goes into setting the intensity, and the collection of them over times forms a filtration. The intensity now is a stochastic process adapted to this filtration. If we re-scale time by

$$\tau(t) = \int_0^t \lambda(s|\mathscr{H}_s)ds,$$

then the event arrival process at $\tau(t_1), \tau(t_2), \ldots$ still look like a realization of a standard Poisson process. However, we notice that the way we have re-scaled time is random, and possibly different from one realization to another of the original point process. Hence, we perform a random time change for the stochastic process. In the above examples, let $Y(t) = \tau(t)$, which is non-decreasing random process with $Y(0) = 0$. Consider X as a point process. Then $X(Y) = X \circ Y = \{X(Y(t)), t \geq 0\}$ can be considered as a standard Poisson process in the transformed time Y.

G.4 CONVERGENCE-TOGETHER THEOREM

Theorem G.3 *Let $\{X_n, n \geq 1\}$ and $\{Y_n, n \geq 1\}$ be two sequences in $\mathscr{D}^J$. If as $n \to \infty$, $X_n - Y_n$ converges uniformly on compact sets to zero and X_n converges uniformly on compact sets to X in $\mathscr{C}^J$, then Y_n converges uniformly on compact sets to X.*

Again, we do not provide the rigorous proof of this theorem (see Billingsley (1999) for complete proof). Instead, we present a theorem with the similar flavor for the convergence of sequences of random variables.

Theorem G.4 *For two sequences of random variables $\{X_n, n \geq 1\}, \{Y_n, n \geq 1\}$ and a random variable X, if $|Y_n - X_n| \xrightarrow{P} 0$ and $X_n \xrightarrow{d} X$, as $n \to \infty$, then $Y_n \xrightarrow{d} X$.*

Proof. The proof is based on the Portmanteau Lemma, part B (see Section G.06). Consider any bounded function f which is also Lipschitz. Thus, there exists a $K > 0$ and for all x and y such that $|f(x) - f(y)| \leq K|x - y|$ and $|f(x)| \leq M$. For any $\varepsilon > 0$, we have

$$
\begin{aligned}
|E[f(Y_n)] - E[f(X_n)]| &\leq E[|f(Y_n) - f(X_n)|] \\
&= E[|f(Y_n) - f(X_n)|1_{|Y_n - X_n| < \varepsilon}] \\
&\quad + E[|f(Y_n) - f(X_n)|1_{|Y_n - X_n| \geq \varepsilon}] \\
&\leq E[K|Y_n - X_n|1_{|Y_n - X_n| < \varepsilon}] + E[2M1_{|Y_n - X_n| \geq \varepsilon}] \\
&\leq K\varepsilon P(|Y_n - X_n| < \varepsilon) + 2MP(|Y_n - X_n| \geq \varepsilon) \\
&\leq K\varepsilon + 2MP(|Y_n - X_n| \geq \varepsilon),
\end{aligned}
$$

where the inequality in line three is due to the fact that f is Lipschitz and bounded. Now we have

$$
\begin{aligned}
|E[f(Y_n)] - E[f(X)]| &= |E[f(Y_n)] - E[f(X_n)] + E[f(X_n)] - E[f(X)]| \\
&\leq |E[f(Y_n)] - E[f(X_n)]| + |E[f(X_n)] - E[f(X)]| \\
&\leq K\varepsilon + 2MP(|Y_n - X_n| \geq \varepsilon) + |E[f(X_n)] - E[f(X)]|.
\end{aligned}
$$

Taking the limit of the both sides as $n \to \infty$, the second term on the RHS will go to 0 since $\{Y_n - X_n\}$ converges to 0 in probability; and the third term will also approach to 0 by the Portmanteau Lemma and the fact that X_n converges to X in distribution. Therefore, we have

$$\lim_{n \to \infty} |E[f(Y_n)] - E[f(X)]| \leq K\varepsilon.$$

Since ε is arbitrary, the limit above must be equal to zero. Thus, $E[f(Y_n)] \to E[f(X)]$, which again by the Portmanteau Lemma implies that $\{Y_n\}$ converges to X in distribution. ∎

G.5 DONSKER'S THEOREM – FCLT

Consider an i.i.d. sequence of d-dimensional real-valued random variables, denoted by $\{Z_n; n \geq 1\}$ where $Z_n \in \mathbb{R}^d$. Let $S_n = Z_1 + Z_2 + \cdots + Z_n$ with $S_0 = 0$, which is a random walk. Donsker's theorem describes the limit behavior of this random walk over a long time intervals rather than at some time points in a far future. Note that the limit theorems for the random walk at a certain time point are the classical strong law of large numbers and central limit theorem. If $E\|Z_n\| < \infty$ and $E\|Z_n\|^2 < \infty$, as $n \to \infty$ then

$$\frac{S_n}{n} \to \mu \quad \text{a.s.} \quad \frac{S_n - n\mu}{\sqrt{n}} \to \Sigma^{1/2} N(\mathbf{0}, \mathbf{I}),$$

where $N(\mathbf{0}, \mathbf{I})$ is an $\mathbb{R}^d$-valued multivariate normal random variables with mean vector $\mathbf{0}$ and the identity and covariance matrix, and $\Sigma^{1/2}$ as the square root of the covariance matrix $\Sigma = EZ_n^T Z_n - EZ_n^T \cdot EZ_n$. These limit theorems for the random variable version can be generalized to the limit theorems for the stochastic process version. Naturally, we replace the random variable $n^{-1} S_n$ with the stochastic process $\bar{X}(t) = n^{-1} S_{[nt]}$, where $[x]$ denoted the greatest integer less than or equal to x, and replace the constant μ with the deterministic process $\bar{X}(t) = \mu t$. However, this extension requires more concepts about the convergence of stochastic processes, which are briefly reviewed here. Since the stochastic processes can be considered as random functions of time, we need to focus on a specific function space to deal with the stochastic processes having sample paths that are right continuous with left limits (RCLL). This function space is denoted by $D_E[0, \infty)$.

$$D_E[0, \infty) = \{\omega : [0, \infty) \to E$$
$$\text{such that } \omega(\cdot) \text{ is right continuous for every } t \geq 0$$
$$\text{and has left limits at every } t > 0\}.$$

Insert: Basic Definitions and Concepts Regarding Metric Space.
   A metric space is a set X with metric on it. The metric associates with any pair of elements (points) of X a distance.

Definition G.1 *A metric space is a pair (X,d), where X is a set and d is a metric on X (or distance function on X), that is a function defined as, $d : X \times X \to \mathbb{R}$, such that for all $x, y, z \in X$ the following properties hold:*

> *1. d is real-valued, finite, and non-negative.*
> *2. $d(x,y) = 0$ if and only if $x = y$.*
> *3. $d(x,y) = d(y,x)$ (Symmetry).*
> *4. $d(x,y) \le d(x,z) + d(z,y)$ (Triangle inequality)*

The next definition is about the completeness of a metric space.

Definition G.2 *The space X is said to be **complete** if every Cauchy sequence in X converges to a limit which is an element of X.*

Definition G.3 *A sequence (x_n) in a metric space $X = (X,d)$ is said to be Cauchy (or fundamental) if for every $\varepsilon > 0$ there is an $N = N(\varepsilon)$ such that for every $n, m > N$, we have $d(x_m, x_n) < \varepsilon$.*

Definition G.4 *A sequence (x_n) is said to be **convergent** if there is an element x such that, for every given ε, the following condition holds for all but finitely many n: $d(x_n, x) < \varepsilon$. This x is called the limit of the sequence (x_n).*

Another concept is the separability of a metric space

Definition G.5 *The space X is said to be **separable** if it has a countable subset which is dense in X.*

Definition G.6 *A set A is said to be **countable** if A is finite (has finitely many elements) or if we can associate positive integers with the elements of A so that to each element of A there corresponds a unique positive integer and, conversely, to each positive integer 1,2,3,... there corresponds a unique element of A.*

Definition G.7 *A subset A of a space X is said to be **dense** in X if for every $x \in X$ and for every $r > 0$ we have that $B(x,r) \cap A \ne \emptyset$. That is every open ball in (X,d) contains a point of A.*

Let A be a subset of a metric space X. Then a point x_0 of X (which may or may not be a point of A) is called an **accumulation point** (or **limit point**) of A if every neighborhood of x_0 contains at least one point $y \in A$ distinct from x_0, The set consisting of the points of A and the accumulation points of A is called **the closure** of A and is denoted by $\bar{A}$. A neighborhood of x_0 is defined as any subset of X which contains an ε-neighborhood of x_0, which is also called an open ball $B(x_0, \varepsilon) = \{x \in X; d(x, x_0) < \varepsilon\}$.

Definition G.8 *A subset A of X is **closed** if $A = \bar{A}$; A subset B of X is **bounded** if there exists $r > 0$ such that for all s and t in B, we have $d(s,t) < r$.*

In $\mathbb{R}^n$, the closed and bounded subset is defined as a **compact set**. However, this definition does not hold for more general metric space. Here is the definition.

Definition G.9 *Let (X,d) be a metric space and $A \subset X$. Then A is said to be a **compact** set if and only if for all open cover $\mathcal{O}$ of A, there exist a finite number of sets $\{A_1, A_2, ..., A_k\} \subset \mathcal{O}$ such that*

$$A \subset \bigcup_{i=1}^{k} A_i.$$

That is for any open cover, we only need a finite number of elements of it to cover the whole set.

Definition G.10 *A subset A of X is **relatively compact** if the closure $\bar{A} \subset X$ is a compact subset of X.*

Definition G.11 *A metric space is called **sequentially compact** if every sequence in X has a convergent subsequence.*

Definition G.12 *A metric space is called **totally bounded** if for every $\varepsilon > 0$ there is a finite cover of X consisting of balls of radius ε.*

As an application of using the concept of metric space, we present an important theorem.

Theorem G.5 *Let X be a metric space with metric d. Then the following properties are equivalent:*

  1. X is compact.
  2. Every infinite set has an accumulation point.
  3. X is sequentially compact, i.e., every sequence has a convergent subsequence.
  4. X is totally bounded and complete.

Another important theorem using the metric space concept is the Banach fixed-point theorem that is based on the definition of contraction.

Definition G.13 *Let (X,d) be a metric space. A mapping $T : X \to X$ is called a **contraction** on X if there exists a positive constant $K < 1$ such that*

$$d(T(x), T(y)) \le Kd(x,y) \quad \text{for } x,y \in X.$$

Geometrically, this means that the distance between $T(x)$ and $T(y)$ is smaller than the distance between x and y. It is not hard to prove the following result.

Theorem G.6 *(Banach's Fixed Point Theorem). Let (X,d) be a complete metric space and let $T : X \to X$ be a contraction on X. Then T has a unique fixed point $x \in X$ (such that $T(x) = x$).*

To study the stochastic processes covered in this book, we define the metric on the functional space $D_E[0,1]$. First, we consider a class of strictly increasing and continuous functions $\Lambda : [0,1] \to [0,1]$. If $\lambda \in \Lambda$, then $\lambda(0) = 0$ and $\lambda(1) = 1$. Denote by $\|x\| = \sup_t |x(t)| < \infty$ the uniform norm on $[0,1]$.

Definition G.14 *Let $I \in \Lambda$ be the identity function. Then we define the two uniform norms*

$$\|\lambda - I\| = \sup_t |\lambda(t) - t|,$$

$$\|x - y \circ \lambda\| = \sup_t |x(t) - y(\lambda(t))|.$$

For x and y in $D_E[0,1]$, a metric $d(x,y)$ defined by

$$d(x,y) = \inf_{\lambda \in \Lambda} \{\|\lambda - I\| \vee \|x - y \circ \lambda\|\}$$

is called the Skorokhod metric.

It follows from this definition that the distance between x and y is less than ε if there exists $\lambda \in \Lambda$ such that the supremum of $\sup_t |\lambda - I|$ and $\sup_t |x(t) - y(\lambda(t))| < \varepsilon$. It can be verified that the Skorokhod metric satisfies four properties in Theorem G.6. It is easy to see that the Skorokhod metric is more appropriate for measuring the closeness of the two functions with discontinuous points rather than the uniform metric. Such a case can be the sample path for a stochastic process defined in $D_E[0,\infty)$.

Example G.1 *Consider two jump functions over [0,1]*

$$f(t) = \begin{cases} 0, & \text{if } t \in [0,0.5) \\ 8, & \text{if } t \in [0.6,1] \end{cases} \quad ; \quad g(t) = \begin{cases} 0, & \text{if } t \in [0,0.6) \\ 8, & \text{if } t \in [0.6,1] \end{cases}.$$

If these two functions represent two sample paths of two random arrival processes with a fixed batch of 8 customers over the time interval [0,1] with one arrival for each process. Although these two functions $f(t)$ and $g(t)$ coincide on almost all the points in [0,1] (except for the tiny interval [0.5,0.6)). It is easy to see that these two functions are far apart and the uniform distance between them is $d_U(f,g) = \max_{t \in [0,1]} |f(t) - g(t)| = 8$. However, under the Skorokhod *metric, it can be shown that the distance, denoted by $d_S(f,g)$, is much smaller than 8. In fact, by choosing $\hat{\lambda} \in \Lambda$ as*

$$\hat{\lambda}(t) = \begin{cases} 1.2t & \text{if } t \in [0,0.5), \\ 0.8t + 2 & \text{if } t \in [0.5,1], \end{cases}$$

we have

$$d_S(f,g) = \inf_{\lambda \in \Lambda} \{\|\lambda - I\| \vee \|f - g \circ \lambda\|\} \leq 0.1,$$

which is left as an exercise.

Furthermore, the exact value of $d_S(f,g)$ is also 0.1. This can be verified by showing that for all $\lambda \in \Lambda$ the following holds

$$\|\lambda - I\| \vee \|f - g \circ \lambda\| \geq 0.1.$$

The basic idea of replacing the uniform metric by the Skorokhod metric is to say functions are close if they are uniformly close over [0,1] after allowing small perturbations of time (the function argument). For example, consider a sequence of functions $x_n = (1 + n^{-1})\mathbf{1}_{[2^{-1}+n^{-1},1]}, n \geq 3$ and a single jump function of magnitude 1 at 2^{-1}, $x = \mathbf{1}_{[2^{-1},1]}$. Then it is easy to check that as $n \to \infty$ while $d_U(x_n,x) \geq 1$ for all n, $d_S(x_n,x) \to 0$. In our review on the convergence of stochastic processes, we will utilize the continuous function space $C_E[0,\infty)$ and the Skorokhod space $D_E[0,\infty)$. For more discussions on function spaces, interested readers are referred to Whitt (2002). Note that the strong convergence of stochastic processes can be stated in terms of the Skorokhod metric. For example, the convergence of the Skorokhod representation can be expressed as $d_S(X_n^*,X^*) \to 0$ a.s. Another example of using the Skorokhod metric is to express the continuity of a function $h : D_E[0,\infty) \to D_E[0,\infty)$ in proving the continuous mapping theorem. Since the week convergence is central in the theory of stochastic process limits, we state the following theorem which can be taken as the definition of weak convergence.

Theorem G.7 *Let $X_n, X \in D_E[0,\infty)$ be a sequence of stochastic processes and X a stochastic process limit. Then $X_n \xrightarrow{w} X$ as $n \to \infty$ if and only if $Eh(X_n) \to Eh(X)$ as $n \to \infty$ for all bounded continuous functions $h : D_E[0,\infty) \to \mathbb{R}$. $(d = 1$ case).*

If $X \in C_E[0,\infty)$ and $X_n \xrightarrow{w} X$, then for every collection $t_1,...,t_n$ of time indices, it can be shown that

$$(X_n(t_1),...,X_n(t_m)) \xrightarrow{w} (X(t_1),...,X(t_m)), \quad n \to \infty$$

by applying the continuous mapping theorem. However, it is worth noting that the weakly convergence of the finite-dimensional distribution of X_n is not sufficient to lead to the weak convergence of X_n to X. This can be verified by a simple example:

Example G.2 *Let U be a uniformly distributed random variable over $[0,1]$. Consider a sequence of stochastic processes $X_n(t) = \exp[-n(t-U)^2]$ for $t \geq 0$ and a stochastic process limit $X(t) = 0$ for $t \geq 0$. Consider a set of time instants $0 \leq t_1 < \cdots < t_m$. It is easy to check that as $n \to \infty$*

$$(X_n(t_1), ..., X_n(t_m)) \xrightarrow{w} (0, ..., 0) = (X(t_1), ..., X(t_m)),$$

which implies that the finite-dimensional distrbutions of X_n converge to those of X, where $X_n, X \in C_E[0, \infty)$. To see the reverse, we select a continuous function $h(x) = \max\{|x| : 0 \leq t \leq 1\}$. Since $X \in C_E[0, \infty)$, the continuous mapping theorem would assert that $h(X_n) \xrightarrow{w} h(X)$ if $X_n \xrightarrow{w} X$. It follows from $X_n(t) = \exp(-n(t-U)^2)$ that the maximum of $|X_n(t)|$ is 1 (zero exponent). Thus $h(X_n) = 1$ whereas $h(X) = 0$, which contradicts the weak convergence of X_n to X.

Definition G.15 *Given a family $\{P, P_n : n \geq 0\}$ of probability measures on $\Omega = D_E[0, \infty)$, P_n converges weakly to P, denoted by $P_n \xrightarrow{w} P$, if*

$$\int_\Omega h(\omega) P_n(d\omega) \to \int_\Omega h(\omega) P(d\omega)$$

for all bounded continuous functions $h : \Omega \to \mathbb{R}$.

Similar to the strong convergence of the stochastic processes which can be expressed in terms of the Skorokhod metric, the weak convergence of the probability measures P_n to P (equivalent to the weak convergence of $X_n \xrightarrow{w} X$) can be expressed in terms of another metric. To describe this approach, we need to define a probability measure space $\mathscr{P} = \{Q : Q$ is a probability measure on $D_E[0, \infty)\}$. There exists a metric ρ, called Prohorov metric, on $\mathscr{P}$ such that $P_n \xrightarrow{w} P$ as $n \to \infty$ (see Whitt (2002)).

A standard approach to show that a sequence of elements to converge an element in a metric space consists of two steps:

1. Show that a sequence $\{x_n : n \geq 0\}$ is relatively compact. That is that every subsequence $x_{n'}$ has a further convergent subsequence $x_{n''}$.
2. Show that every convergent subsequence $x_{n'}$ of x_n must converge to x.

Theorem G.8 *Consider $\{P, P_n : n \geq 0\}$ where $P_n(\cdot) = P(X_n \in \cdot), P(\cdot) = P(X \in \cdot)$ and $X_n, X \in D_E[0, \infty)$. If the finite-dimensional distribution of X_n converge to those of X, then the only possible limit point of $\{P_n : n \geq 0\}$ is P.*

This theorem indicates that convergence of the finite-dimensional distribution helps identify the set of possible limit points of P_n. To ensure relative compactness of a sequence $\{P_n : n \geq 0\}$ of probability measures in $\mathscr{P}$, a theorem can be established.

Theorem G.9 *Consider $\{P_n : n \geq 0\}$ where $P_n \in P$. Then, $\{P_n : n \geq 0\}$ is relatively compact in ρ if and only if for every $\varepsilon > 0$, there exists a compact set $K_\varepsilon \subset D_E[0,\infty)$ such that*

$$\inf_{n \geq 0} P_n(K_\varepsilon) \geq 1 - \varepsilon.$$

Definition G.16 *A family $\{P_n \in \mathcal{P} : n \geq 0\}$ of probability measures on $D_E[0,\infty)$ is said to be tight if for every $\varepsilon > 0$, there exists a compact set $K_\varepsilon \subset D_E[0,\infty)$ such that $\inf_{n \geq 0} P_n(K_\varepsilon) \geq 1 - \varepsilon$.*

Now we return to Donsker's theorem. First, the classical SLLN can be extended to the functional strong law of large numbers (FCLT) as follows:

Theorem G.10 *Let $\bar{X}_n(t) = S_{[nt]}/n$ and $\bar{X} = \mu t$. Then, as $n \to \infty$, we have*

$$d_S(\bar{X}_n, \bar{X}) \to 0 \quad a.s..$$

To extend the CLT to the functional central limit theorem (FCLT), we scale $S_{[nt]}$ as follows

$$X_n(t) = n^{1/2}\left(\frac{S_{[nt]}}{n} - \mu t\right) = n^{1/2}(\bar{X}_n(t) - \bar{X}(t))$$
$$= \frac{S_{[nt]} - n\mu t}{\sqrt{n}}.$$

Such a scaled process approaches to a stochastic limit which is diffusion process.

Theorem G.11 *If $E\|Z_n\|^2 < \infty$, then $X_n \overset{w}{\to} \Sigma^{1/2}B$ as $n \to \infty$ in $D_E[0,\infty)$, where $B(\cdot)$ is a standard Brownian motion process on $\mathbb{R}^d$.*

The weak convergence result suggests the approximation in distribution as follows:

$$S_{[nt]} \overset{d}{\approx} \mu nt + n^{1/2}\Sigma^{1/2}B(t).$$

G.6   STRONG APPROXIMATION THEOREM

A more general and powerful result due to Komlos, Major, and Tusnady (1975) is in the form of strong approximation theorem for random walk.

Theorem G.12 *Suppose that $\{Z_n : n \geq 1\}$ is a sequence of i.i.d. real-valued random variables satisfying $E \exp(\alpha Z_n) < \infty$ for α in an open neighborhood of zero. Let $S_0 = 0$ and $S_n = Z_1 + \cdots + Z_n$ for $n \geq 1$. Then there exists a probability space $\{\Omega, \mathscr{F}, P\}$ supporting a standard Brownian motion $\{B(t) : t \geq 0\}$ and a sequence $\{S'_n : n \geq 0\}$ such that*

1. $\{S'_n : n \geq 0\} \overset{d}{=} \{S_n : n \geq 0\}$. That is the sequence $\{S'_n\}$ shares the same distribution as $\{S_n\}$.

2. For every x and n,

$$P\left(\max_{0 \leq k \leq n} |S'_k - k\mu - \sigma B(k)| > C\log n + x\right) < Ke^{-\lambda x},$$

where $\mu = EZ_n, \sigma^2 = VarZ_n$ and K, C and λ are positive constants depending only on the distribution of Z_n.

By using the Borel-Cantelli lemma, from (2) of the theorem above, we can get

$$S'_n = n\mu + \sigma B(n) + O(\log n) \quad \text{a.s.}.$$

This result provides an upper bound for $S'_n - (n\mu + \sigma B(n) + O(\log n))$, which implies the Donsker's theorem. This can be shown as follows. Setting $X'_n(t) = n^{1/2}(n^{-1}S'_{[nt]} - \mu)$, we have

$$X'_n(t) = n^{1/2}(n^{-1}S'_{[nt]} - \mu)$$
$$= n^{1/2}[n^{-1}(n\mu + \sigma B(n) + O(\log n)) - \mu]$$
$$= n^{1/2}[n^{-1}\sigma B(n) + n^{-1}O(\log n)]$$
$$= n^{-1/2}\sigma B(n) + \frac{O(\log n)}{\sqrt{n}},$$

which implies

$$\|X'_n(t) - n^{-1/2}\sigma B(nt)\|_T = O(n^{-1/2}\log n) \quad \text{a.s.}$$

for any $T > 0$. Thus, it follows from (1) of the theorem above and the scaling property of BM (i.e., $n^{-1/2}B(nt) \overset{d}{=} B(t)$), that the Downsker's theorem holds. An important application of the strong approximation theorem is to obtain the bounds on the rates of convergence for various limit theorems related to Donsker's theorem.

The following Lemma plays an important role in establishing the stochastic process limits.

Lemma G.2 (Portmanteau Lemma): A sequence $\{X_n\}$ converges in distribution to X if and only if any of the following conditions are met:

- A. $E[f(X_n)] \to E[f(X)]$ for all bounded, continuous functions f;
- B. $E[f(X_n)] \to E[f(X)]$ for all bounded, Lipschitz functions f;
- C. $\limsup\{P(X_n \in C)\} \leq P(X \in C)$ for all closed sets C;
- D. $P(X_n \in A) \to P(X \in A)$ for any Borel set A such that $P(X \in \partial A) = 0$. Here $\partial A = \bar{A} \setminus int(A)$ is the boundary of A.

REFERENCES

1. P. Billingsley, "Convergence of probability measures", John Wiley & Sons, 1999.
2. H. Chen and D.D. Yao, "Fundamentals of Queueing Networks: Performance, Asymptotics, and Optimization", Springer, New York, 2001.
3. J. Komlos, P. Major, and G. Tusnady, An approximation of partial sums of independent rv's and the sample df. I. *Wahrsch verw Gebiete/Probability Theory and Related Fields,* 32, 111–131, 1975.
4. W. Whitt, "Stochastic-Process Limits An Introduction to Stochastic-Process Limits and Their Application to Queues", Springer, New York, 2002.

H A Brief Review on Stochastic Calculus

Since the BM is the fundamental building block of stochastic calculus, we first show the existence of BM by the construction approach.

H.1 CONSTRUCTION OF BM–EXISTENCE OF BM

To show the existence of BM, we can construct a BM by first defining it on dyadic rational numbers. Consider B_t in $0 \leq t \leq 1$. After obtaining this process in this interval, we can take a countable collection of such processes and connect them appropriately to get the BM on $[0, \infty)$. This section is mainly based on the lecture notes by Chang (2007). Let

$$\mathcal{D}_n = \left\{ \frac{k}{2^n} : k = 0, 1, ..., 2^n \right\},$$

denote the dyadic rationals in $[0, 1]$ with denominator 2^n and let $\mathcal{D} = \cup_n \mathcal{D}_n$. Thus, we can index a set of standard normal random variables $\{Z_q : q \in \mathcal{D}\}$. We first define B_0, B_1, and then $B_{1/2}$ and then $B_{1/4}$ and $B_{3/4}$, and so forth by subdividing intervals. Since we know that $B_1 \sim N(0, 1)$, we can generate (i.e., simulate) a random value, Z_1, then we have $B_1 = Z_1$. We denote $X^{(0)}(t)$ as the initial linear interpolation approximation to B_t. Then $X^{(0)}(t) = Z_1 t$, connecting $(0, 0)$ and $(1, Z_1)$, is the starting crude approximation to the sample path of B_t as shown in Figure H.1. Next, we simulate a value for $B_{1/2}$. Given $B_0 = 0$ and $B_1 = Z_1$, we know that $B_{1/2}$ is normally distributed with mean $Z_1/2$ and variance $(1/2)(1/2) = 1/4$. Since $X^{(0)}(1/2) = Z_1/2$, we only need to add a normal random variable with mean 0 and variance of $1/4$ to $X^{(0)}(1/2)$ to get the right distribution. This can be done by generating another $N(0, 1)$ random variable Z_2 and adding it to $X^{(0)}(1/2)$ to get $B_{1/2} = X^{(0)}(1/2) + (1/2)Z_2$. Now we can get the piecewise linear path, denoted by $X^{(1)}(t)$, connecting these three points $(0, 0), (1/2, B_{1/2}), (1, B_1)$. Next, we simulate $B_{1/4}$ and $B_{3/4}$ based on $X^{(1)}(t)$. Again the mean of B_t at these points are determined by $X^{(1)}(t)$ from the conditional distribution of BM. To have the correct variances at these points, we need to generate two independent $N(0, 1)$ random variable values, denoted

DOI: 10.1201/9781003150060-H

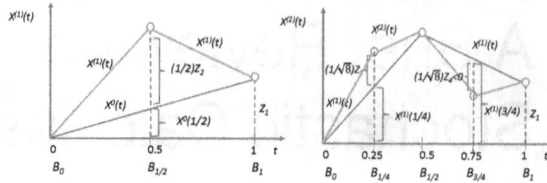

Figure H.1 Construction of a standard Brownian motion.

by Z_3 and Z_4 and then define

$$B_{1/4} = X^{(1)}(1/4) + \frac{1}{\sqrt{8}} Z_3,$$

$$B_{3/4} = X^{(1)}(3/4) + \frac{1}{\sqrt{8}} Z_4.$$

After these points are generated, we can obtain the better approximation $X^{(2)}(t)$ by connecting the five points (see Figure H.1). In the same way, based on $X^{(2)}(t)$, we can simulate B_t for four additional time instants, namely, $B_{1/8}, B_{3/8}, B_{5/8}$, and $B_{7/8}$ by generating four standard normal random variable values multiplied by $\sqrt{1/16} = 1/4$ and adding them to the values $X^{(2)}(1/8), X^{(2)}(3/8), X^{(2)}(5/8)$, and $X^{(2)}(7/8)$. Then we obtain the piecewise linear approximation to B_t by connecting 9 points, which is denoted by $X^{(3)}(t)$. Keep doing this, we can summarize the process as follows: In general, to get from $X^{n)}$ to $X^{(n+1)}$, we need to generate 2^n new standard normal random values, denoted by $Z_{2^n+1}, Z_{2^n+2}, ..., Z_{2^{n+1}}$, multiply them by the appropriate conditional standard deviation $\sqrt{2^{-n-2}} = 2^{-(n/2)-1}$, and add to the values $X^{(n)}(1/2^{n+1}), X^{(n)}(3/2^{n+1}), ..., X^{(n)}(1 - 1/2^{n+1})$ to get the new values $X^{(n+1)}(1/2^{n+1}), X^{(n+1)}(3/2^{n+1}), ..., X^{(n+1)}(1 - 1/2^{n+1})$. It can be shown that with probability 1, the sequence of random functions $X^{(1)}, X^{(2)}, ...$ converges uniformly over the interval $[0, 1]$. The requirement of the uniformity of the convergence stems from the fact that the limit of a uniformly convergent sequence of continuous functions is a continuous function. We first show a property of an i.i.d. sequence of standard normal random variables, $\{Z_n, n \geq 1\}$. Let c be a number greater than 2, then

$$P(|Z_n| \leq \sqrt{c \log n} \text{ for all sufficiently large } n) = 1. \tag{H.1}$$

Recall the tail probability bound for the standard normal random variable

$$P(Z > x) = \frac{1}{\sqrt{2\pi}} \int_x^\infty 1 \cdot e^{-t^2/2} dt \leq \frac{1}{\sqrt{2\pi}} \int_x^\infty \frac{t}{x} e^{-t^2/2} dt = \frac{e^{-x^2/2}}{x\sqrt{2\pi}}.$$

Based on this inequality, we have

$$\sum_{n=1}^{\infty} P(|Z_n| > \sqrt{c\log n}) \le 2\frac{1}{\sqrt{2\pi}}\sum_{n=1}^{\infty}\frac{e^{-(1/2)c\log n}}{\sqrt{c\log n}} = \sqrt{\frac{2}{\pi}}\sum_{n=1}^{\infty}\frac{n^{-(1/2)c}}{\sqrt{c\log n}},$$

which is finite for $c > 2$. Hence, it follows from the Borel–Cantelli lemma that

$$P(|Z_n| > \sqrt{c\log n} \ i.o.;) = 0,$$

which is equivalent to its complement holds for all sufficiently large n.

Define the maximum difference between $X^{(n+1)}$ and $X^{(n)}$, denoted by M_n, as

$$M_n = \max_{t\in[0,1]} |X^{(n+1)}(t) - X^{(n)}(t)|.$$

Note that if $\sum_n M_n < \infty$, then the sequence of functions $X^{(1)}, X^{(2)}, \ldots$ converges uniformly over $[0,1]$ to a continuous function. Thus, we need to show that $P(\sum_n M_n < \infty) = 1$. Since the maximum difference only occurs at the dyadic rationals, we have

$$M_n = 2^{-(n/2)-1} \max\{|Z_{2^n+1}|, |Z_{2^n+2}|, \ldots, |Z_{2^{n+1}}|\}.$$

Taking $c > 2$, it follows from (H.1) that with probability 1

$$M_n \le 2^{-(n/2)-1} \sqrt{c\log(2^{n+1})}$$

holds for all sufficiently large n. That is

$$P(M_n \le 2^{-(n/2)-1} \sqrt{n+1} \sqrt{c\log 2} \text{ eventually}) = 1.$$

Thus, because of $\sum_n 2^{-(n/2)}\sqrt{n+1} < \infty$, we have $\sum_n M_n < \infty$ with probability 1. This implies that the linear interpolation approximations converges to a continuous function over $[0,1]$, denoted by $X = \lim_{n\to\infty} X^{(n)}$. Next we need to check the normality, $X(t) \sim N(0,t)$, and stationary independent increments of the process. Clearly, at any dyadic rational $r = k/2^m$, we have $X(r) \sim N(0,r)$. According to our construction process, for all $n \ge m$, $X^{(n)}(k/2^m) = X^{(m)}(k/2^m)$. Once we assign a value to the process at a dyadic rational time $r = k/2^m$, we will not change it. Thus, $X(r) = X^{(m)}(r)$, Note that $X^{(m)}(k/2^m) \sim N(0, k/2^m)$. Consider $t \in [0,1]$ and a sequence of dyadic rational numbers $\{r_n, n \ge 1\}$ such that $\lim_{n\to\infty} r_n = t$. Then $X(t) = \lim_{n\to\infty} X(r_n)$ with probability 1, by the path continuity of the process X. Since $X(r_n) \sim N(0, r_n)$ and $\lim_{n\to\infty} r_n = t$, we have $X(t) \sim N(0,t)$. This can be justified as follows. We know that $X(r_n) \sim \sqrt{r_n}Z$. Take the limit of both sides as $n \to \infty$. $X(r_n) \to X(t)$ by path continuity, and $\sqrt{r_n}Z \to \sqrt{t}Z \sim N(0,t)$. Similarly, we can verify that X has stationary and independent increments. Consider three time point $s < t < u$. Take three dyadic rational sequences, denoted by

s_n, t_n, and u_n such that $s_n \to s, t_n \to t$, and $u_n \to u$. According to the construction process, we have $X(u_n) - X(t_n) \sim N(0, u_n - t_n)$ which is independent of $X(t_n) - X(s_n) \sim N(0, u_n - t_n)$ for all n. Hence,

$$(X(t_n) - X(s_n), X(u_n) - X(t_n)) \sim (\sqrt{t_n - s_n} Z_1, \sqrt{u_n - t_n} Z_2),$$

where Z_1 and Z_2 are i.i.d. standard normal random variables. Then, taking the limit of both sides and using the path continuity of X, we obtain that $X(u) - X(t) \sim N(0, u - t), X(t) - X(s) \sim N(0, t - s)$, and $X(u) - X(t)$ is independent of $X(t) - X(s)$ for $s < t < u$. Now we have verified that the BM exists by this construction process.

H.2 DIFFUSION PROCESSES AND KOLMOGOROV'S EQUATIONS

A stochastic process that has the strong Markov property and (almost surely) continuous sample paths is called a diffusion process. In this book, it is a CTMP. BM $X(t)$ is an example of diffusion process. Suppose $X(t) = x$ is the state of the process at time t. Then we can define the infinitesimal mean $\mu = \mu(x)$ and infinitesimal variance $\sigma^2 = \sigma^2(x)$ by the following expressions:

$$E[X(t+h) - X(t)|X(t) = x] = \mu(x)h + o(h),$$
$$Var[X(t+h) - X(t)|X(t) = x] = \sigma^2(x)h + o(h),$$

as $h \downarrow 0$. Now we establish the Kolmogorov's backward and forward equations. First we derive the backward equation, which will be used to develop the forward equation. Although the main purpose of these equations is for the transition density, they apply to more general class of functions of the diffusion process X_t. Let $\phi(X_t)$ be a real value function of the diffusion process, which can be interpreted as the "reward" rate earned at state $X(t)$. We are interested in the expected value of this function. Denote $v(x,t) = E_x \phi(X_t)$, the conditional expected reward if we stop the diffusion at time t, given that it started in state x at time 0. To derive the backward equation, we assume that the process starts at state x and consider the expected reward $v(x, t+h)$ at a future time instant $t+h$. Conditioning on the state after a small time increment h from time 0, we have

$$v(x, t+h) = E_x[\phi(X_{t+h})] = E_x[E_x \phi(X_{t+h}|X_h)]] = E_x[v(X_h, t)].$$

Based on this relation, we obtain

$$\frac{\partial}{\partial t} v(x,t) = \lim_{h \to 0} \frac{E_x[v(X_h, t)] - v(x,t)}{h}$$

$$= \lim_{h \to 0} \frac{1}{h} E_x \left[\frac{\partial}{\partial x} v(x,t)(X_h - x) + \frac{1}{2} \frac{\partial^2}{\partial x^2} v(x,t)(X_h - x)^2 + o(h) \right]$$

$$= \mu(x) \frac{\partial}{\partial x} v(x,t) + \frac{1}{2} \sigma^2(x) \frac{\partial^2}{\partial x^2} v(x,t),$$

$$(H.2)$$

which is called the backward equation for v.

If we choose an indicator function $\phi(x) = 1_{x \le y}$ where y is a fixed value. Then

$$v(x,t) = E_x[\phi(X_t)] = P(X_t \le y | X_0 = x) =: F(t,x,y),$$

which is the distribution function of the transition probability.

$$\frac{\partial}{\partial t} F(t,x,y) = \mu(x) \frac{\partial}{\partial x} F(t,x,y) + \frac{1}{2}\sigma^2(x) \frac{\partial^2}{\partial x^2} F(t,x,y). \qquad \text{(H.3)}$$

Next, we derive the forward equation. Denote the probability density of X_t by $f(x,t)$ for all $t \ge 0$. For a homogeneous diffusion X_t, considering a future time s time units from t, we can write

$$v(y,s) = E[\phi(X_{s+t}) | X_t = y].$$

Un-conditioning the state at time t (the initial state relative to time point $t+s$), we have

$$\int v(y,s)f(y,t)dy = \int E[\phi(X_{s+t})|X_t = y]P(X_t \in dy) = E[\phi(X_{s+t})],$$

which is a function of $(s+t)$. Using the chain rule, we obtain

$$\frac{\partial}{\partial s}\int v(y,s)f(y,t)dy = \frac{\partial}{\partial t}\int v(y,s)f(y,t)dy,$$

which is equivalent to

$$\int \left[\frac{\partial}{\partial s}v(y,s)\right]f(y,t)dy = \int v(y,s)\left[\frac{\partial}{\partial t}f(y,t)\right]dy. \qquad \text{(H.4)}$$

Note that the partial derivative term on the left-hand side is given by the backward equation (H.2). Thus the left-hand side of (H.4) is given by

$$\int \mu(y)\left[\frac{\partial}{\partial y}v(y,s)\right]f(y,t)dt + \frac{1}{2}\int \sigma^2(y)\left[\frac{\partial^2}{\partial y^2}v(y,s)\right]f(y,t)dy. \qquad \text{(H.5)}$$

Applying the integration by parts to the first term of (H.5) yields

$$\int \mu(y)\left[\frac{\partial}{\partial y}v(y,s)\right]f(y,t)dt = -\int v(y,s)\left[\frac{\partial}{\partial y}\mu(y)f(y,t)\right]dy. \qquad \text{(H.6)}$$

Here we use the fact $\lim_{y \to \pm\infty} v(y,s)\mu(y)f(v,t) = 0$ under mild assumptions. As $y \to \pm\infty$, $v(y,t)$ will approach zero since it is assumed that ϕ has compact support and $f(y,s)$ also approach to zero as well-behaved probability densities. For the second term of (H.5), we integrate by parts two times and get

$$\frac{1}{2}\int \sigma^2(y)\left[\frac{\partial^2}{\partial y^2}v(y,s)\right]f(y,t)dy = \frac{1}{2}\int v(y,s)\left[\frac{\partial^2}{\partial y^2}\sigma^2(y)f(y,t)\right]dy. \qquad \text{(H.7)}$$

By substituting (H.6) and (H.7) into (H.5), we can write (H.4) as

$$\int v(y,s) \left\{ -\frac{\partial}{\partial y}[\mu(y)f(y,t)]dy + \frac{1}{2}\frac{\partial^2}{\partial y^2}\left[\sigma^2(y)f(y,t)\right] \right\} dy$$

$$= \int v(y,s) \left[\frac{\partial}{\partial t}f(y,t)\right] dy, \tag{H.8}$$

which becomes

$$\int \phi(y) \left\{ -\frac{\partial}{\partial y}[\mu(y)f(y,t)]dy + \frac{1}{2}\frac{\partial^2}{\partial y^2}\left[\sigma^2(y)f(y,t)\right] \right\} dy$$

$$= \int \phi(y) \left[\frac{\partial}{\partial t}f(y,t)\right] dy, \tag{H.9}$$

by letting $s \to 0$. Therefore, we obtain the forward equation for $f(y,t)$ as

$$\frac{\partial}{\partial t}f(y,t) = -\frac{\partial}{\partial y}[\mu(y)f(y,t)]dy + \frac{1}{2}\frac{\partial^2}{\partial y^2}\left[\sigma^2(y)f(y,t)\right]. \tag{H.10}$$

The forward equation gives the more natural description of the evolution of the probability density function for the diffusion process. The probability density is a function of the "forward" variable y, given the initial distribution fixed. In contrast, the backward equation keeps the forward variable y fixed, and describes the density at time t as a function of the "backward" variable x which is the initial state of the process at time 0. An advantage of the backward equation is that it requires weaker assumptions than the forward equation. Such a difference is reflected by the form of the two equations: note that the functions μ and σ appear inside derivatives in the forward equation, while they do not in the backward equation. Thus, one might expect that the forward equation requires more smoothness assumptions on μ and σ. Just like the CTMC, we can use the forward equation to find the stationary distribution of the diffusion process which is a CTMP. Let $\pi(t)$ be the stationary density for the diffusion X_t. That is

$$\lim_{t \to \infty} f(t,y) = \pi(y),$$

which satisfies the forward equation

$$0 = -\frac{d}{dy}[\mu(y)\pi(y)]dy + \frac{1}{2}\frac{d^2}{dy^2}\left[\sigma^2(y)\pi(y)\right]. \tag{H.11}$$

In the next subsection, we present the stochastic differential equations that a diffusion process X_t satisfies. Combined with the stochastic differential equation with the Kolmogorov's equation, we can solve for stationary distribution for a diffusion process if it exists. Here is an example.

Example H.1 *Consider a problem of diffusion in a potential well. Let $g(x)$ be a "potential function" which is differentiable (or at least almost everywhere) and unbounded (i.e., $g : \mathbb{R} \to \mathbb{R}$ and $g(x) \to \infty$ as $|x| \to \infty$. We are interested in a stochastic process X_t that tends to minimize $g(x)$. In particular, we want to determine the stationary distribution of such a process.*

Solution: First, we consider a deterministic model that minimizes $g(x)$ via the gradient decent method. Assume that at time t, the process is at position X_t, then the gradient at this instant is $g'(X_t)$. The process will move a small step in the direction of decreasing g to the position $X_{t+\Delta t} = X_t - g'(X_t)\Delta t$ at time $t + \Delta t$. Taking the limit of $\Delta t \to 0$ gives the differential equation $dX_t = -g'(X_t)dt$ which is the deterministic model for the "potential minimization process". Suppose that the process is subject to random shocks so that some randomness should be added to the small movement. This can be done by modeling X_t as a diffusion with infinitesimal mean $\mu(x) = -g'(X_t)$ and a constant infinitesimal variance $\sigma^2(x) = \sigma^2$. That is

$$dX_t = \mu(X_t)dt + \sigma(X_t)dB_t = -g'(X_t) + \sigma dB_t,$$

which is the standard differential equation that a diffusion process satisfies (see the next section for details). That is (i) X_t is a diffusion process; and (ii) it has the infinitesimal mean and variance given by $-g'(x)$ and σ^2. Applying the stationary distribution equation (H.11) to X_t yields

$$0 = \frac{d}{dx}[g'(x)\pi(x)]dy + \frac{\sigma^2}{2}\frac{d^2}{dx^2}[\pi(x)],$$

which can be written as

$$\pi'(x) + \frac{2}{\sigma^2}g'(x)\pi(x) = B$$

for some constant B. This ordinary differential equation (ODE) can be solved by introducing the integrating factor $\exp[(2/sigma^2)g(x)]$. Thus, the ODE can be rewritten as

$$\left(e^{(2/\sigma^2)g(x)}\pi(x) \right)' = Be^{(2/\sigma^2)g(x)},$$

which gives the solution

$$\pi(x) = e^{-(2/\sigma^2)g(x)}\left[B\int_0^x e^{(2/\sigma^2)g(y)}dy + C \right]. \tag{H.12}$$

Note that the lower limit of the integral can be arbitrary number as it can be compensated by the corresponding change in the constant C. From the assumption that $g(x) \to \infty$ as $|x| \to \infty$ and the fact that $\pi(x)$ is a nonnegative probability measure for all x, we must have $B = 0$. Thus we find

$$\pi(x) = Ce^{-(2/\sigma^2)g(x)},$$

where C is the normalizing constant that makes $\int_{-\infty}^{\infty} \pi(x)dx = 1$. If $g(x) = (1/2)x^2$ and $\sigma^2 = 2$, then $\pi(x) = Ce^{-x^2/2}$ which is the standard Ornstein-Uhlenbeck process (O-U process). Clearly, its stationary distribution is a standard normal distribution.

H.3 STOCHASTIC DIFFERENTIAL EQUATIONS

Now we develop the stochastic differential equation that a stochastic process satisfies. First, consider a deterministic differential equation by modeling the exponential growth of a population. Let $X(t)$ denote the population size at time t. Then a differential equation can be written as $dX(t)/dt = rX(t)$. However, the growth rate may not be completely deterministic due to some random factor (or called "noisy" factor). This random factor can be modeled as a stochastic process $\{R(t), t \geq 0\}$. Then the differential equation can be given by

$$\frac{dX(t)}{dt} = [r + R(t)]X(t) = rX(t) + R(t)X(t),$$

where the right-hand side is the sum of the two functions of the time t and state variable $X(t)$. Compared with the deterministic differential equation $dX(t)/dt = rX(t)$, the second term is introduced by the "noisy" factor $R(t)$. To simplify the notation, we write $X_t = X(t)$ below. Thus, we can write the differential equation in a more general form as

$$\frac{dX_t}{dt} = \mu(X_t, t) + \sigma(X_t, t)R(t), \tag{H.13}$$

where the second term reflects the fact that it is caused by $R(t)$. In many applications, $\{R(t)t \geq 0\}$, called white noise", is assumed to satisfy the followings:

1. $\{R(t), t \geq 0\}$ is a stationary Gaussian process.
2. $E[R(t)] = 0$.
3. $R(s)$ is independent of $N(t)$.

Re-write (H.13) as

$$\begin{aligned} dX_t &= \mu(X_t, t)dt + \sigma(X_t, t)R(t)dt \\ &=: \mu(X_t, t)dt + \sigma(X_t, t)dV(t), \end{aligned} \tag{H.14}$$

where $dV_t = R(t)dt$ indicates that the defined process $\{V_t, t \geq 0\}$ should be a Gaussian process with independent increments. The only process satisfying these requirements should be a standard BM denoted by $B_t = B(t)$. Thus, we can write (H.14) as

$$dX_t = \mu(X_t, t)dt + \sigma(X_t, t)dB_t, \tag{H.15}$$

which is called a stochastic differential equation. Since the paths of $\{B(t), t \geq 0\}$ are not differentiable, a more rigorous way of writing this equation is in the integrated form

$$X_t - X_0 = \int_0^t \mu(X_s, s)ds + \int_0^t \sigma(X_s, s)dB_s. \tag{H.16}$$

The stochastic process that satisfies (H.16) or (H.15) is called a diffusion process. Diffusion process is an important subclass of the class of Itô processes which is introduced by extending the fundamental theorem of calculus (the Newton–Leibnitz formula). Recall this famous theorem in calculus course. To integrate a function $f(x)$ over the interval $[a, b]$, we an find a function F such that $dF(x)/dx = f(x)$ for all x. Then we have

$$F(b) - F(a) = \int_a^b f(x)dx = \int_a^b F'(x)dx. \tag{H.17}$$

Now we consider the integral of $\int_0^T f(B_t)dB_t$ which is resulted from replacing x with the standard BM B_t in the integral in (H.17). It can be shown that an extra term is needed in this new integral (stochastic integration) given by

$$F(B_T) - F(B_0) = \int_0^T F'(B_t)dB_t + \frac{1}{2}\int_0^T F''(B_t)dt, \tag{H.18}$$

provided that the second-order derivative of F is continuous. This formula follows by using the Taylor expansion for function F and the quadratic variation of BM. First, write the approximation based on the Taylor expansion for the function F when $|x - x_0|$ is small as

$$F(x) \approx F(x_0) + F'(x_0)(x - x_0) + \frac{1}{2}F''(x_0)(x - x_0)^2. \tag{H.19}$$

Substituting $x = B_{t_k}$ and $x_0 = B_{t_{k-1}}$ in (H.19) yields

$$F(B_{t_k}) - F(B_{t_{k-1}}) \approx F'(B_{t_{k-1}})(B_{t_k} - B_{t_{k-1}}) + \frac{1}{2}F''(B_{t_{k-1}})(B_{t_k} - B_{t_{k-1}})^2. \tag{H.20}$$

Summing the above expression over $k = 1$ to n, we have

$$F(B_T) - F(B_0) \approx \sum_{k=1}^n F'(B_{t_{k-1}})(B_{t_k} - B_{t_{k-1}}) + \frac{1}{2}\sum_{k=1}^n F''(B_{t_{k-1}})(B_{t_k} - B_{t_{k-1}})^2. \tag{H.21}$$

By taking the limit $n \to \infty$ on both sides of (H.21), we obtain (H.18). This basic Itô formula can be extended to the integrals of the form $\int_0^t f(s, B_s)dB_s$, which is in the non-homogeneous case. Accordingly, $F(\cdot, \cdot)$ is a function of two arguments of t and x. Using the same approximation procedure, we have

$$F(t+\Delta t, B_t + \Delta B_t) - F(t, B_t) \approx \Delta t \frac{\partial}{\partial t} F(t, B_t) + \Delta B_t \frac{\partial}{\partial x} F(t, B_t)$$

$$+ \frac{1}{2} \left((\Delta t)^2 \frac{\partial^2}{\partial t^2} F(t, B_t) + 2\Delta t \Delta B_t \frac{\partial^2}{\partial x \partial t} F(t, B_t) + (\Delta B_t)^2 \frac{\partial^2}{\partial x^2} F(t, B_t) \right),$$

which is based on the multi-dimensional version of Taylor expansion by truncating the third-order or higher terms. Again, by summing and taking the limit $n \to \infty$, we obtain

$$F(t, B_t) = F(0, B_0) + \int_0^t \left(\frac{\partial}{\partial t} F(t, B_t) + \frac{1}{2} \frac{\partial^2}{\partial x^2} F(t, B_t) \right) dt + \int_0^t \frac{\partial}{\partial x} F(t, B_t) dB_t$$

$$\text{(H.22)}$$

for a function F of two arguments t and x, continuously differentiable in the first and the second order. This is called the Itô formula in non-homogeneous form.

The processes X_t which can be written as

$$X_t = X_0 + \int_0^t \mu_s ds + \int_0^t \sigma_s dB_s \tag{H.23}$$

for some adapted processes μ_s and σ_s are called Itô processes. The process μ_s is called the drift process and the process σ_s the volatility process of the Itô process X_t. By comparing (H.22) and (H.23), we observe that process $F(t, B_t)$ is an Itô process with

$$\mu_s = \frac{\partial}{\partial t} F(s, B_s) + \frac{1}{2} \frac{\partial^2}{\partial x^2} F(s, B_s),$$

$$\sigma_s = \frac{\partial}{\partial x} F(s, B_s).$$

The Itô process X_t is a martingale if $\mu_t = 0$ for all t, is a submartingale is $\mu_t \geq 0$ for all t, and is a supermartingale if $\mu_t \leq 0$ for all t. In the case where μ_t changes signs, X_t is neither a supermartingale nor a submartingale. Itô processes and their associated SDEs are fundamental in analyzing diffusion processes. To apply these SDEs, some special rules for integrations and differentiations are needed. We focus on some fundamental rules and discuss them briefly. Then, we discuss the Ito formulas. For more details on SDE's, interested readers are referred to some excellent books on stochastic calculus such as Oksendal (2013).

H.4   SOME STOCHASTIC CALCULUS RULES

H.4.1   VARIATION OF BM

Consider a BM over a finite time period $[0, T]$. With an equidistant partition $0 = t_0 < t_1 < \cdots < t_{n-1} < t_n = T$ with $\Delta_n = T/n$, we define the quadratic variation of the BM as

$$\lim_{n \to \infty} \sum_{k=0}^{n-1} (B_{t_{k+1}} - B_{t_k})^2, \tag{H.24}$$

which can be shown to be equal to T in the mean square sense (see the convergence modes in Chapter F). Denote $X_n = \sum_{k=1}^{n} (B_{t_{k+1}} - B_{t_k})^2$. Compute the mean and variance of X_n as follows:

$$E[X_n] = \sum_{i=0}^{n-1} E[(B_{t_{i+1}} - B_{t_i})^2] = \sum_{i=0}^{n-1} (t_{i+1} - t_i)$$

$$= t_n - t_0 = T,$$

$$Var(X_n) = \sum_{i=0}^{n-1} Var[(B_{t_{i+1}} - B_{t_i})^2] = \sum_{i=0}^{n-1} 2(t_{i+1} - t_i)^2$$

$$= n \cdot 2 \left(\frac{T}{n}\right)^2 = \frac{2T^2}{n}.$$

It follows from $E[X_n] = T$ for all $n \geq 1$ and $Var(X_n) \to 0$ as $n \to \infty$ that

$$\lim_{n \to \infty} \sum_{i=0}^{n-1} X_n = \lim_{n \to \infty} \sum_{i=0}^{n-1} (B_{t_{i+1}} - B_{t_i})^2 = T \quad \text{in } L^2. \tag{H.25}$$

Furthermore, under the condition of $\sum_n ||\Delta_n|| < \infty$, the quadratic variation of BM converges to T almost surely as well. Define $Y_n = X_n - T$. Note that $E[Y_n] = 0$ and $Var(Y_n) = E[Y_n^2] = 2\sum_{i=1}^{n} (t_i - t_{i-1})^2 \leq 2T||\Delta_n||$. Using the Chebychev's inequality, we have

$$P(|Y_n| > \varepsilon) \leq \frac{Var(Y_n)}{\varepsilon^2} = \frac{2\sum_{i=1}^{n}(t_i - t_{i-1})^2}{\varepsilon^2} \leq \frac{2T||\Delta_n||}{\varepsilon^2} < \infty.$$

Summing up the inequality above over n yields

$$\sum_{n=1}^{\infty} P(|Y_n| > \varepsilon) \leq \frac{2T}{\varepsilon^2} \sum_{n=1}^{\infty} ||\Delta_n|| < \infty,$$

if $\sum_{n=1}^{\infty} ||\Delta_n|| < \infty$. By using Borel-Cantelli Lemma, we conclude that $Y_n \to 0$ or $X_n \to T$ almost surely. For example, for a time interval $[0, 1]$, the dyadic rational partition

$$\Delta_n = \left\{ \frac{k}{2^n}, k = 0, ..., 2^n \right\},$$

satisfies

$$\sum_{n=1}^{\infty} ||\Delta_n|| = \sum_{n=1}^{\infty} \frac{1}{2^n} < \infty,$$

so that the quadratic variation converges to $T = 1$ almost surely. We can also show that the total variation of the BM, denoted by TV, is unbounded. That is $TV = \sum_{i=1}^{n} |B_{t_i} - B_{t_{i-1}}| \to \infty$ almost surely as $n \to \infty$ (i.e., $||\Delta_n|| \to 0$). We show this fact by contradiction. Suppose that this total variation is bounded. Note that

$$\sum_{i=1}^{n} |B_{t_i} - B_{t_{i-1}}|^2 \le \max_{1 \le i \le n} |B_{t_i} - B_{t_{i-1}}| \sum_{i=1}^{n} |B_{t_i} - B_{t_{i-1}}| = \max_{1 \le i \le n} |B_{t_i} - B_{t_{i-1}}| TV.$$

Since B_t is continuous almost surely on $[0, T]$, it is necessarily uniformly continuous $[0, T]$. Therefore,

$$\max_{1 \le i \le n} |B_{t_i} - B_{t_{i-1}}| \to 0 \quad \text{as} \quad ||\Delta_n|| \to 0,$$

which implies that

$$\sum_{i=1}^{n} |B_{t_i} - B_{t_{i-1}}|^2 \to 0$$

almost surely. This is a contradiction to the fact that the quadratic variation is non-zero. Thus, we conclude that TV for the BM is infinite. The variation behavior of the BM is due to its non-differentiable property. In calculus, for a differentiable and bounded function over a finite interval, the total variation is bounded and the quadratic variation is zero. Such a difference is the key factor that results in different rules in stochastic calculus from classical calculus. We can re-write (H.25) as

$$\int_{0}^{T} (dB_t)^2 = \lim_{t \to \infty} \sum_{i=0}^{n-1} (B_{t_{i+1}} - B_{t_i})^2 = T = \int_{0}^{T} dt,$$

which results in the fundamental relation

$$(dB_t)^2 = dt. \tag{H.26}$$

H.4.2   STOCHASTIC INTEGRATION

We introduce the Itô integral which is similar to the Riemann integral. The Itô integral is taken with respect to infinitesimal increments of a BM, dB_t, which are random variables. Thus, the Itô integral is a random variable. This is the fundamental difference from the Riemann integral which is a real number. Let $F_t = f(B_t, t)$ be a non-anticipating process which means that F_t is independent of any future increment $B_s - B_t$ for any t and s with $t < s$. For $0 \le a \le b$, we assume

$$E\left[\int_{a}^{b} F_t^2 dt\right] < \infty,$$

which is the condition needed when we discuss the martingale property f the Itô integral. To define the Itô integral, we form the partial sums that are similar to those in the Riemann integral. Divide the interval $[a, b]$ into n subintervals using the partition points

$$a = t_0 < t_1 < \cdots < t_{n-1} < t_n = b$$

and develop the partial sums

$$S_n = \sum_{i=0}^{n-1} F_{t_i}(B_{t_{i+1}} - B_{t_i}).$$

It should be emphasized that the intermediate points are chosen to be the left endpoints of these subintervals. Since the process F_t is nonanticipative, the random variables F_{t_i} and $B_{t_{i+1}} - B_{t_i}$ are independent.

The Itô integral is defined as the limit of the partial sum S_n

$$ms - \lim_{n \to \infty} S_n = \int_a^b F_t dB_t,$$

provided the limit exists with $t_{i+1} - t_i = (b-a)/n$, $i = 0, 1, ..., n-1$. Note the convergence is in the mean square sense (ms). That is

$$\lim_{n \to \infty} E\left[\left(S_n - \int_a^b F_t dB_t\right)^2\right] = 0.$$

The conditions for the existence of the Ito integral can be summarized as follows:

1. The paths $t \to F_t(\omega)$ are continuous on $[a, b]$ for any $\omega \in \Omega$;
2. The process F_t is non-anticipating for $t \in [a, b]$;
3. $E\left[\int_a^b F_t^2 dt\right] < \infty.$

Next we present a few examples of the Ito integral.

Example H.2 *The simplest example is the case of F_t is a constant c.*

In this case, we have

$$S_n = \sum_{i=0}^{n-1} F_{t_i}(B_{t_{i+1}} - B_{t_i}) = \sum_{i=0}^{n-1} c(B_{t_{i+1}} - B_{t_i})$$
$$= c(B_b - B_a),$$

which implies $\int_a^b c dB_t = c(B_b - B_a)$. Setting $c = 1, a = 0$, and $b = T$ and noting that $B_0 = 0$, we get

$$\int_0^T dB_t = B_T.$$

Example H.3 *The next example is the case of $F_t = B_t$. That is to compute* $\int_0^T B_t dB_t$.

With the same equidistant partition of the $[0, T]$, we have

$$\int_0^T B_t dB_t = \lim_{n \to \infty} \sum_{k=1}^n B_{t_{k-1}}(B_{t_k} - B_{t_{k-1}})$$

$$= \frac{1}{2}B_T^2 - \frac{1}{2}\lim_{n \to \infty} \sum_{k=1}^n (B_{t_k} - B_{t_{k-1}})^2$$

$$= \frac{1}{2}B_T^2 - \frac{1}{2}T.$$

The second equality follows from the following identity

$$B_T^2 = B_T^2 - B_0^2 = \sum_{k=1}^n (B_{t_k}^2 - B_{t_{k-1}}^2) = \sum_{k=1}^n (B_{t_k} - B_{t_{k-1}})(B_{t_k} + B_{t_{k-1}})$$

$$= \sum_{k=1}^n (B_{t_k} - B_{t_{k-1}})(B_{t_k} - B_{t_{k-1}} + 2B_{t_{k-1}})$$

$$= \sum_{k=1}^n (B_{t_k} - B_{t_{k-1}})^2 + 2\sum_{k=1}^n B_{t_{k-1}}(B_{t_k} - B_{t_{k-1}}),$$

which leads to an expression for $\sum_{k=1}^n B_{t_{k-1}}(B_{t_k} - B_{t_{k-1}})$ which is S_n. The third equality comes from the quadratic variation result for the BM. Note that an extra term $-(1/2)T$ occurs in this integral compared with the classical integral for a deterministic function. As a random variable, the mean and variance of this Ito integral are $E[\int_0^T B_t dB_t] = 0$ and $Var[\int_0^T B_t dB_t] = T^2/2$, respectively. This result can be re-written as

$$\int_a^b B_t dB_t = \frac{1}{2}(B_b^2 - B_a^2) - \frac{1}{2}(b - a).$$

The properties of the Ito integral can be established for the general non-anticipating process by using the simple process. A process $A(t)$ is a simple process if there exist times

$$0 = t_0 < t_1 < \cdots < t_n < \infty$$

and a random variable $Y_j, j = 0, 1, ..., n$ that are $\mathscr{F}_{t_j}$-measurable such that

$$A(t) = T_j, \quad t_j \leq t < t_j + 1.$$

Then, the stochastic integral regarding the simple process can be obtained as $\int_0^t A(s)dB_s$ and its properties can be easily verified. Then these properties can be extended to the more general non-anticipating process $f(B_t, t)$ case. Some of these properties are similar to those for the Riemann integral and others are random variable related. We present some of these properties without proofs.

1. Linearity: If $a, b \in \mathbb{R}$ are constants, $f(B_t, t), g(B_t, t)$ are non-anticipating processes, then

$$\int_0^T [af(B_t, t) + bg(B_t, t)] dB_t = a \int_0^T f(B_t, t) dB_t + b \int_0^T g(B_t, t) dB_t.$$

2. Partition property:

$$\int_0^T f(B_t, t) dB_t = \int_0^u f(B_t, t) dB_t + \int_u^T f(B_t, t) dB_t, \quad 0 < u < T.$$

3. Zero mean:

$$E\left[\int_a^b f(B_t, t) dB_t \right] = 0.$$

4. Variance:

$$Var\left[\int_a^b f(B_t, t) dB_t \right] = E\left[\left(\int_a^b f(B_t, t) dB_t \right)^2 \right] = E\left[\int_a^b f(B_t, t)^2 dB_t \right].$$

5. Covariance:

$$E\left[\left(\int_a^b f(B_t, t) dB_t \right) \left(\int_a^b g(B_t, t) dB_t \right) \right] = E\left[\int_a^b f(B_t, t) g(B_t, t) dt \right].$$

6. $\mathscr{F}_t$-predictability:

$$\int_0^t f(B_s, s) dB_s$$

is $\mathscr{F}_t$-predictable.
7. Path Continuity: With probability one, the function $t \to X_t = \int_0^t f(B_s, s) dB_s$ is continuous.
8. Martingale property: The process $X_t = \int_0^t f(B_s, s) dB_s$ is a martingale with respect to $\{\mathscr{F}_t\}$.

For the proofs of all these properties, interested readers are referred to some excellent books such as Calin (2022).

As an example, we give the brief proof of the martingale property (8) as follows. To establish such a property for the Ito integral, we need to check the four properties of a martingale.

(i) Integrability: Using the properties of Ito ingerals, we have

$$E[X_t^2] = E\left[\left(\int_0^t f(B_s, s) dB_s \right)^2 \right] = E\left[\int_0^t f^2(B_s, s) ds \right]$$

$$< E\left[\int_0^\infty f^2(B_s, s) ds \right] < \infty,$$

and it follows from the $E[|X_t|]^2 \leq E[X_t^2]$ that $E[|X_t|] < \infty$, for all $t \geq 0$.

(ii) Predictability: X_t is $\mathscr{F}_t$-predictable from Property 6.

(iii) $E[X_t|\mathscr{F}_s] = X_s$. This relation can be verified as follows: Using the partition property, we have

$$
E\left[\int_0^t f(B_u,u)dB_u | \mathscr{F}_s\right]
$$

$$
= E\left[\int_0^s f(B_u,u)dB_u + \int_s^t f(B_u,u)dB_u | \mathscr{F}_s\right] \tag{H.27}
$$

$$
= E\left[\int_0^s f(B_u,u)dB_u | \mathscr{F}_s\right] + E\left[\int_s^t f(B_u,u)dB_u | \mathscr{F}_s\right].
$$

Since $\int_0^s f(B_u,u)dB_u$ is $\mathscr{F}_s$-predictable, we get

$$
E\left[\int_0^s f(B_u,u)dB_u | \mathscr{F}_s\right] = \int_0^s f(B_u,u)dB_u.
$$

Note that $\int_s^t f(B_u,u)dB_u$ contains only information between s and t, it is independent of the information set $\mathscr{F}_s$. Thus, we can drop the condition in the expectation and use the mean property of the Ito integral to get

$$
E\left[\int_s^t f(B_u,u)dB_u | \mathscr{F}_s\right] = E\left[\int_s^t f(B_u,u)dB_u\right] = 0.
$$

Substituting these two expectation terms into (H.27) yields the desired result.

(iv) Continuity: This follows immediately from property (7).

A special case of the Ito's integral is when the non-anticipating stochastic process becomes a deterministic function of only time t, i.e., $f(B_t,t) = f(t)$. Then, $\int_a^b f(t)dB_t$ is called the Wiener integral and is the mean square limit of the partial sums

$$
S_n = \sum_{i=0}^{n-1} f(t_i)(B_{t_{i+1}} - B_{t_i}).
$$

All properties of Ito integral hold for Wiener integrals. Moreover, we can obtain its distribution, which is normally distributed with mean and variance

$$
E\left[\int_a^b f(t)dB_t\right] = 0, \quad E\left[\left(\int_a^b f(t)dB_t\right)^2\right] = \int_a^b f(t)^2 dt,
$$

respectively. The proof is left as an exercise.

There are other types of stochastic integrals. The integration with respect to the compensated Poisson process (a martingale), denoted by $M_t = N_t - \lambda t$, where N_t is a Poisson process with rate λ. Let $F_t = F(M_t, t)$ be a non-anticipating process with $E[\int_a^b F_t^2 dt] < \infty$ for an interval $[a, b]$. The integral of F_{t-} with respect to M_t is defined as the limit of the partial sum

$$
ms \lim_{n \to \infty} S_n = ms \lim_{n \to \infty} \sum_{i=0}^{n-1} F_{t_{i-}} (M_{t_{i+1}} - M_{t_i}) = \int_a^b F_{t-} dM_t,
$$

provided the limit exists. More precisely, this limit means

$$
\lim_{n \to \infty} E\left[\left(S_n - \int_a^b F_{t-} dM_t \right)^2 \right] = 0.
$$

Similar properties to those for the Ito integral can be established for the Poisson integration.

H.4.3 STOCHASTIC DIFFERENTIATION

Since most stochastic processes are not differentiable, we will focus on the infinitesimal changes of the process such as dB_t. For a process X_t, the change in the process between t and $t + \Delta t$ is given by $\Delta X_t = X_{t+\Delta t} - X_t$. As $\Delta t \to 0$, we obtain the infinitesimal change of the process X_t

$$
dX_t = X_{t+dt} - X_t,
$$

which can be also written as $X_{t+dt} = X_t + dX_t$.

Some basic differentiation rules are listed below:

1. The constant multiple rule: If X_t is a stochastic process and c is a constant, then
$$
d(cX_t) = cdX_t.
$$

2. The sum/difference rule: If X_t and Y_t are two stochastic processes, then
$$
d(X_t \pm Y_t) = dX_t \pm dY_t.
$$

3. The product rule: If X_t and Y_t are two stochastic processes, then
$$
d(X_t Y_t) = X_t dY_t + Y_t dX_t + dX_t dY_t.
$$

4. The quotient rule: If X_t and Y_t are two stochastic processes, then
$$
d\left(\frac{X_t}{Y_t}\right) = \frac{Y_t dX_t - X_t dY_t - dX_t dY_t}{Y_t^2} + \frac{X - t}{Y_t^3} (dY_t)^2.
$$

The proofs of these rules are easy and we only provide one example of verifying the product rule. The key is to write the definition expression in terms of the incremental form of the process. For example,

$$
\begin{aligned}
d(X_tY_t) &= X_{t+dt}Y_{t+dt} - X_tY_t \\
&= X_{t+dt}Y_{t+dt} - X_{t+dt}Y_t - X_tY_{t+dt} + X_tY_t + X_{t+dt}Y_t - X_tY_t + X_tY_{t+dt} - X_tY_t \\
&= (X_{t+dt} - X_t)(Y_{t+dt} - Y_t) + Y_t(X_{t+dt} - X_t) + X_t(Y_{t+dt} - Y_t) \\
&= X_t dY_t + Y_t dX_t + dX_t dY_t.
\end{aligned}
$$

If X_t is replaced by the deterministic function $f(t)$, the product rule becomes

$$
d(f(t)Y_t) = f(t)dY_t + Y_t df(t) + df(t)dY_t.
$$

Note that in most cases, Y_t is an Ito diffusion

$$
dY_t = a(t, B_t)dt + b(t, B_t)dB_t.
$$

It follows from $dtdB_t = dt^2 = 0$ that $df(t)dY_t = f'(t)dtdY_t = 0$. Thus, the product rule with a deterministic function reduces to

$$
d(f(t)Y_t) = f(t)dY_t + Y_t df(t),
$$

which is similar to the classical product rule in calculus. Similar result can be obtained from the quotient rule. With the fundamental relation $(dB_t)^2 = dt$, we can find that some extra term may occur in differentiation of stochastic processes compared with the deterministic cases. We give a few examples here.

Example H.4 *Find the differentiations of B_t^2 and B_t^3.*

It follows from the product rule that

$$
\begin{aligned}
d(B_t^2) &= d(B_t \cdot B_t) = B_t dB_t + B_t dB_t + dB_t dB_t \\
&= 2B_t dB_t + dt. \\
d(B_t^3) &= d(B_t \cdot B_t^2) = B_t(dB_t^2) + B_t^2 dB_t + d(B_t^2)dB_t \\
&= B_t(2B_t dB_t + dt) + B_t^2 dB_t + (2B_t dB_t + dt)dB_t \\
&= 2B_t^2 dB_t + B_t dt + B_t^2 dB_t + 2B_t(dB_t)^2 + dt dB_t \\
&= 3B_t^2 dB_t + 3B_t dt,
\end{aligned}
$$

where the last equality follows from the fundamental relations $(dB_t)^2 = dt$ and $dtdB_t = 0$.

Example H.5 *Find the infinitesimal change of the integrated BM and the average of the BM.*

Let $Z_t = \int_0^t B_u du$. The infinitesimal change of Z_t is given by

$$dZ_t = Z_{t+dt} - Z_t = \int_t^{t+dt} B_s ds = B_t dt,$$

where the last equality holds from the continuity of B_s in s. Let $A_t = (1/t)Z_t = (1/2)\int_0^t B_u du$. Then

$$dA_t = d\left(\frac{1}{t}Z_t\right) = d\left(\frac{1}{t}\right)Z_t + \frac{1}{t}dZ_t + d\left(\frac{1}{t}\right)dZ_t$$

$$= -\frac{1}{t^2}Z_t dt + \frac{1}{t}B_t dt - \frac{1}{t^2}B_t(dt)^2$$

$$= \frac{1}{t}\left(B_t - \frac{1}{t}Z_t\right)dt.$$

H.5   ITO'S FORMULA

In calculus, for a real differentiable function $f(x)$, it follows from the Taylor expansion that

$$\Delta f(x) = f'(x)\Delta x + \frac{1}{2}f''(x)(dx)^2 + O(\Delta x)^3,$$

which, as $\Delta x \to 0$, leads to the differential expression

$$df(x) = f'(x)dx.$$

If we consider f as a function of a deterministic process of t, denoted by $x(t)$ and assume that $x(t)$ is a differentiable function of t, then we have the differential form of the well-known chain rule

$$df(x(t)) = f'(x(t))dx(t) = f'(x(t))x'(t)dt.$$

Now what if $x(t)$ is replaced by a stochastic process X_t? As expected, the differential form of $df(X_t)$ will have an extra term of the second order (i.e., the term $(dX_t)^2$ is not negligible as shown later). That is

$$df(X_t) = f'(X_t)dX_t + \frac{1}{2}f''(X_t)(dX_t)^2, \qquad (H.28)$$

which sometimes is called the general Ito's formula.

We consider a case where

$$dX_t = \mu(B_t,t)dt + \sigma(B_t,t)dB_t \qquad (H.29)$$

and X_t in such a case is called an Ito diffusion. Note that this is a special case of (H.15) in which $\mu(X_t,t) = \mu(B_t,t)$ and $\sigma(X_t,t) = \sigma(B_t,t)$.

Theorem H.1 *(Ito's formula for diffusions) If X_t is an Ito diffusion, then*

$$df(X_t) = \left[\mu(B_t,t)f'(X_t) + \frac{\sigma(B_t,t)^2}{2}f''(X_t)\right]dt + \sigma(B_t,t)f'(X_t)dB_t. \quad \text{(H.30)}$$

Proof. The proof is based on using the relations $dB_t^2 = dt$ and $(dt)^2 = dB_t dt = 0$. First note that

$$(dX_t)^2 = (\mu(B_t,t)dt + \sigma(B_t,t)dB_t)^2$$
$$= \mu(B_t,t)^2(dt)^2 + 2\mu(B_t,t)\sigma(B_t,t)dB_t dt + \sigma^2(B_t,t)^2(dB_t)^2$$
$$= \sigma(B_t,t)^2 dt.$$

Substituting this expression and (H.29) into (H.28) yields

$$df(X_t) = f'(X_t)dX_t + \frac{1}{2}f''(X_t)(dX_t)^2$$
$$= f'(X_t)(\mu(B_t,t)dt + \sigma(B_t,t)dB_t) + \frac{1}{2}f''(X_t)\sigma(B_t,t)^2 dt$$
$$= \left[\mu(B_t,t)f'(X_t) + \frac{\sigma(B_t,t)^2}{2}f''(X_t)\right]dt + \sigma(B_t,t)f'(X_t)dB_t.$$

∎

An application of the Ito's formula is to derive a differentiation rule for B_t^n. First, consider the case $X_t = B_t$ or $f(X_t)$ in the formula. That is the case where $\mu(B_t,t) = 0$ and $\sigma(B_t,t) = 1$. Thus, we have

$$df(B_t) = \frac{1}{2}f''(B_t)dt + f'(B_t)dB_t. \quad \text{(H.31)}$$

Now consider $f(x) = x^n$ where n is constant. Then $f'(x) = nx^{n-1}$ and $f''(x) = n(n-1)x^{n-2}$. Thus, replacing B_t by B_t^n in (H.31), we obtain

$$d(B_t^n) = \frac{1}{2}n(n-1)B_t^{n-2}dt + nB_t^{n-1}dB_t.$$

The Ito formula can be extended to the non-homogeneous case. Consider the time-dependent function $f = f(t,x)$, we have

$$df(t,x) = \partial_t f(t,x)dt + \partial_x f(t,x)dx + \frac{1}{2}\partial_{xx}^2 f(t,x)(dx)^2 + O(dx)^3 + O(dt)^2.$$

Substituting $x = X_t$ gives

$$df(t,X_t) = \partial_t f(t,X_t)dt + \partial_x f(t,X_t)dX_t + \frac{1}{2}\partial_{xx}^2 f(t,X_t)(dX_t)^2. \quad \text{(H.32)}$$

If X_t is an Ito diffusion, we have the Ito's formula for the non-homogeneous case

$$df(t,B_t) = \left[\partial_t f(t,X_t) + \mu(B_t,t)\partial_x f(t,X_t) + \frac{\sigma(B_t,t)^2}{2} \partial_{xx}^2 f(t,X_t) \right] dt \quad \text{(H.33)}$$
$$+ \sigma(B_t,t)\partial_x f(t,X_t) dB_t.$$

Another extension of the Ito's formula is to the multidimensional case. If we consider a stochastic process depending on several Ito diffusions. For example, $f(t,X_t,Y_t)$ is a function of two Ito diffusions. Assume that f is a function whose partial derivatives $f_t, f_x, f_y, f_{xx}, f_{xy}$ and f_{yy} are defined and continuous. Using the similar approach, we can obtain

$$df(t,X_t,Y_t) = \frac{\partial}{\partial t} f(t,X_t,Y_t)dt + \frac{\partial}{\partial x} f(t,X_t,Y_t)dX_t + \frac{\partial}{\partial y} f(t,X_t,Y_t)dY_t$$
$$+ \frac{1}{2} \frac{\partial^2}{\partial x^2} f(t,X_t,Y_t)(dX_t)^2 + \frac{1}{2} \frac{\partial^2}{\partial y^2} f(t,X_t,Y_t)(dY_t)^2 \quad \text{(H.34)}$$
$$+ \frac{\partial^2}{\partial x \partial y} f(t,X_t,Y_t)dX_t dY_t.$$

Furthermore, if X_t and Y_t, as the Ito processes, are defined by

$$dX_t = \mu_t^1 dt + \sigma_t^{11} dB_t^1 + \sigma_t^{12} dB_t^2$$
$$dY_t = \mu_t^2 dt + \sigma_t^{21} dB_t^1 + \sigma_t^{22} dB_t^2.$$

To simplify the expressions, we drop the variable names in function f. That is we write $f = f(t,X_t,Y_t)$. Then,

$$df(t,X_t,Y_t) = f_t dt + f_x dX_t + f_y dY_t + \frac{1}{2} f_{xx} d[X,X]_t + f_{xy} d[X,Y]_t + \frac{1}{2} f_{yy} d[Y,Y]_t$$
$$= f_t dt + f_x(\mu_t^1 dt + \sigma_t^{11} dB_t^1 + \sigma_t^{12} dB_t^2) + f_y(\mu_t^2 dt + \sigma_t^{21} dB_t^1 + \sigma_t^{22} dB_t^2)$$
$$+ \frac{1}{2} f_{xx} d[X,X]_t + f_{xy} d[X,Y]_t + \frac{1}{2} f_{yy} d[Y,Y]_t$$
$$= f_t dt + f_x(\mu_t^1 dt) + f_x(\sigma_t^{11}, \sigma_t^{12}) \begin{pmatrix} dB_t^1 \\ dB_t^2 \end{pmatrix}$$
$$+ f_y(\mu_t^2 dt) + f_y(\sigma_t^{21}, \sigma_t^{22}) \begin{pmatrix} dB_t^1 \\ dB_t^2 \end{pmatrix}$$
$$+ \frac{1}{2} f_{xx}[(\sigma_t^{11})^2 (dB_t^1)^2 + (\sigma_t^{12})^2 (dB_t^2)^2] + f_{xy}[\sigma_t^{11}\sigma_t^{21}(dB_t^1)^2 + \sigma_t^{12}\sigma_t^{22}(dB_t^2)^2]$$
$$+ \frac{1}{2} f_{yy}[(\sigma_t^{21})^2 (dB_t^1)^2 + (\sigma_t^{22})^2 (dB_t^2)^2]$$
$$= f_t dt + f_x \left\{ (\mu_t^1 dt) + f_y(\mu_t^2 dt) \right\}$$

$$+\left\{\frac{1}{2}f_{xx}[(\sigma_t^{11})^2(dB_t^1)^2+(\sigma_t^{12})^2(dB_t^2)^2]+f_{xy}[\sigma_t^{11}\sigma_t^{21}(dB_t^1)^2+\sigma_t^{12}\sigma_t^{22}(dB_t^2)^2]\right.$$

$$\left.+\frac{1}{2}f_{yy}[(\sigma_t^{21})^2(dB_t^1)^2+(\sigma_t^{22})^2(dB_t^2)^2]\right\}$$

$$+\left\{f_x(\sigma_t^{11},\sigma_t^{12})\begin{pmatrix}dB_t^1\\dB_t^2\end{pmatrix}+f_y(\sigma_t^{21},\sigma_t^{22})\begin{pmatrix}dB_t^1\\dB_t^2\end{pmatrix}\right\}$$

$$=f_t dt+f_x\left\{(\mu_t^1 dt)+f_y(\mu_t^2 dt)\right\}$$

$$+\left\{\frac{1}{2}f_{xx}[(\sigma_t^{11})^2+(\sigma_t^{12})^2]dt+f_{xy}[\sigma_t^{11}\sigma_t^{21}+\sigma_t^{12}\sigma_t^{22}]dt\right.$$

$$\left.+\frac{1}{2}f_{yy}[(\sigma_t^{21})^2+(\sigma_t^{22})^2]dt\right\}$$

$$+\left\{f_x(\sigma_t^{11},\sigma_t^{12})\begin{pmatrix}dB_t^1\\dB_t^2\end{pmatrix}+f_y(\sigma_t^{21},\sigma_t^{22})\begin{pmatrix}dB_t^1\\dB_t^2\end{pmatrix}\right\}.$$

The Ito's formula can be extended to the n-dimensional case as follows:

$$df(t,X_t^{(1)},...,X_t^{(n)})=\left[\left(\frac{\partial}{\partial t}+\sum_{i=1}^n\mu^{(i)}\frac{\partial}{\partial X_t^{(i)}}+\frac{1}{2}\sum_{i,j=1}^n\sigma^{(i)}\cdot\sigma^{(j)}\frac{\partial^2}{\partial X_t^{(i)}\partial X_t^{(j)}}\right)f\right]dt$$

$$+\sum_{i=1}^n\frac{\partial}{\partial X_t^{(i)}}f\sigma^{(i)}\cdot dB_t$$

$$=\frac{\partial}{\partial t}f dt+\mathscr{L}f dt+\Sigma dB_t\nabla f,$$

(H.35)

where

$$\mathscr{L}=\sum_{i=1}^n\mu^{(i)}\frac{\partial}{\partial X_t^{(i)}}+\frac{1}{2}\sum_{i,j=1}^n\Sigma_{ij}\frac{\partial^2}{\partial X_t^{(i)}\partial X_t^{(j)}}$$

and

$$\nabla f=(\partial f/\partial X_t^{(1)},...,\partial f/\partial X_t^{(n)})^T.$$

H.6   SOME THEOREMS ON STOCHASTIC INTEGRATION

Similar to the elementary calculus, several relations can be useful in computing stochastic integrals.

Theorem H.2 *(The fundamental theorem of stochastic calculus)*

   1. For any $a<t$,

$$d\left(\int_a^t f(s,B_s)dB_s\right)=f(t,B_t)dB_t.$$

2. If Y_t is a stochastic process, such that $Y_t dB_t = dF_t$, then

$$\int_a^b Y_t dB_t = F_b - F_a.$$

The proof of this theorem is straightforward. Next we show the stochastic integration by parts. Consider the process $F_t = f(t)g(B_t)$ with f and g differentiable. It follows from the product rule that

$$dF + t = df(t)g(B_t) + f(t)dg(B_t)$$

$$= f'(t)g(B_t)dt + f(t)(g'(B_t)dB_t + \frac{1}{2}g''(B_t)dt)$$

$$= f'(t)g(B_t)dt + \frac{1}{2}f(t)g''(B_t)dt + f(t)g'(B_t)dB_t.$$

Integrating both sides of this equation and re-arranging the terms yields

$$\int_a^b f(t)g'(B_t)dB_t = f(t)g(B_t)|_a^b - \int_a^b f'(t)g(B_t)dt - \frac{1}{2}\int_a^b f(t)g''(B_t)dt,$$
(H.36)

which is called the formula of stochastic integration by parts. If X_t and Y_t are two Ito diffusions, using the product formula, we have

$$d(X_t Y_t) = X_t dY_t + Y_t dX_t + dX_t dY_t.$$

Integrating both sides of the above equation with limits a and b, using the fundamental theorem, and re-arranging the terms yields

$$\int_a^b X_t dY_t = X_b Y_b - X_a Y_a - \int_a^b Y_t dX_t - \int_a^b dX_t dY_t.$$
(H.37)

To use this formula, the term $dX_t dY_t$ need to be evaluated by using $(dB_t)^2 = dt$ and $dt dB_t = 0$.

H.7 MORE ON STOCHASTIC DIFFERENTIAL EQUATIONS

For the stochastic differential equation introduced earlier, we can write it for a continuous stochastic process X_t as follows:

$$dX_t = \mu(t, B_t, X_t)dt + \sigma(t, B_t, X_t)dB_t$$

or in the integral form

$$X_t = X_0 + \int_0^t \mu(s, B_s, X_s)ds + \int_0^t \sigma(s, B_s, X_s)dB_s.$$

Note that $\mu(t, B_t, X_t)$ and $\sigma(t, B_t, X_t)$, the drift process and volatility process, now are written as functions of t, B_t, and X_t itself.

In many practical situations, we are interested in the mean and variance of X_t. These characteristic processes can be obtained directly from the SDE in some cases without solving it explicitly.

Taking the expectation in the integral form equation above and using the fact that the Ito integral has a zero mean, we have

$$E[X_t] = X_0 + \int_0^t E[\mu(s, B_s, X_s)]ds, \qquad (H.38)$$

which can be written as

$$\frac{d}{dt}E[X_t] = E[\mu(t, B_t, X_t)].$$

Note that the mean process does not depend on the volatility term. Consider a case where $\mu(t, B_t, X_t) = a(t)X_t + b(t)$, where $a(t)$ and $b(t)$ are continuous deterministic functions. Then, we have

$$\frac{d}{dt}E[X_t] = a(t)E[X_t] + b(t),$$

which has the solution

$$E[X_t] = e^{A(t)}\left(X_0 + \int_0^t e^{-A(s)}b(s)ds\right), \qquad (H.39)$$

where $A(t) = \int_0^t a(s)ds$. For example, if $a(t) = 2$ and $b(t) = e^{2t}$, then we can find that $E[X_t] = e^{2t}(X_0 + t)$. Next, we show that both the mean and variance of X_t can be found in a case in which the SDE is given by

$$dX_t = \alpha(t)X_t dt + \beta(t)dB_t.$$

First, by setting $b(t) = 0$ and replacing $a(t)$ by $\alpha(t)$ in (H.39), we obtain

$$E[X_t] = e^{A(t)}X_0,$$

where $A(t) = \int_0^t \alpha(s)ds$. Furthermore, we can obtain the variance of X_t as follows: First, we compute the second moment of X_t. By noting $(dX_t)^2 = \beta^2(t)dt$, we have

$$\begin{aligned} d(X_t)^2 &= 2X_t dX_t + (dX_t)^2 \\ &= 2X_t(\alpha(t)X_t dt + \beta(t)dB_t) + \beta^2(t)dt \\ &= (2\alpha(t)X_t^2 + \beta^2(t))dt + 2\beta(t)X_t dB_t. \end{aligned}$$

Letting $Y_t = X_t^2$, we get

$$dY_t = (2\alpha(t)Y_t + \beta^2(t))dt + 2\beta(t)\sqrt{Y_t}dB_t.$$

Applying (H.39) to Y_t yields

$$E[Y_t] = e^{2A(t)}\left(Y_0 + \int_0^t e^{-2A(s)}\beta^2(s)ds\right)$$

$$= E[X_t^2] = e^{2A(t)}\left(X_0^2 + \int_0^t e^{-2A(s)}\beta^2(s)ds\right).$$

Thus, the variance is given by

$$Var[X_t] = E[X_t^2] - (E[X_t])^2 = e^{2A(t)}\int_0^t e^{-2A(s)}\beta^2(s)ds.$$

The expressions for the mean and variance of some specific stochastic processes can be obtained from the SDEs. We present two cases of the exponential function of the standard BM and the power function of the standard BM as they are very useful in applications.

Example H.6 *Find the mean and variance of (1) e^{kB_t} with k being a constant, and (2) $B_t e^{B_t}$.*

Using the Ito's formula, we have

$$d(e^{kB_t}) = ke^{kB_t}dB_t + \frac{1}{2}k^2 e^{kB_t}dt.$$

Integrating gives

$$e^{kB_t} = 1 + k\int_0^t e^{kB_s}dB_s + \frac{1}{2}k^2\int_0^t e^{kB_s}ds.$$

Taking the expectations, we obtain

$$E[e^{kB_t}] = 1 + + \frac{1}{2}k^2\int_0^t E[e^{kB_s}]ds.$$

Letting $f(t) = E[e^{kB_t}]$ and differentiating the equation above yields

$$f'(t) = \frac{1}{2}k^2 f(t)$$

with $f(0) = E[e^{kB_0}] = 1$. The solution is $f(t) = e^{k^2 t/2}$. Thus, we have

$$E[e^{kB_t}] = e^{k^2 t/2}.$$

The variance is given by

$$Var(e^{kB_t}) = E[e^{2kB_t}] - (e[e^{kB_t}])^2 = e^{4k^2t/2} - e^{k^2t}$$
$$= e^{k^2t}(e^{k^2t} - 1).$$

We can also find $E[B_t e^{B_t}]$ by using the same procedure. First, using the product rule and Ito's formula, we have

$$d(B_t e^{B_t}) = e^{B_t}dB_t + B_t d(e^{B_t}) + dB_t d(e^{B_t})$$
$$= e^{B_t}dB_t + (B_t + dB_t)(e^{B_t}dB_t + \frac{1}{2}e^{B_t}dt)$$
$$= (e^{B_t} + B_t e^{B_t})dB_t + (\frac{1}{2}B_t e^{B_t} + e^{B_t})dt.$$

Integrating and using $B_0 e^{B_0} = 0$ gives

$$B_t e^{B_t} = \int_0^t \left(\frac{1}{2}B_s e^{B_s} + e^{B_s}\right)ds + \int_0^t (e^{B_s} + B_s e^{B_s})dB_s.$$

Taking the expectation, we have

$$E[B_t e^{B_t}] = \int_0^t \left(\frac{1}{2}E[B_s e^{B_s}] + E[e^{B_s}]\right)ds.$$

Letting $f(t) = E[B_t e^{B_t}]$ and using $E[e^{B_s}] = e^{s/2}$, the previous equation can be written as

$$f(t) = \int_0^t \left(\frac{1}{2}f(s) + e^{s/2}\right)ds.$$

Differentiating $f(t)$ yields

$$f'(t) = \frac{1}{2}f(t) + e^{t/2}$$

with the intial condition $f(0) = 0$. This linear differential equation can be re-written as

$$d(e^{-t/2}f(t)) = dt,$$

which leads to the solution $f(t) = te^{t/2}$. Thus, we have

$$E[B_t e^{B_t}] = te^{t/2}.$$

Another useful stochastic process is the power function of B_t. That is $(B_t)^n$ with $n = 1, 2, ...$ We can show the following result. For any non-negative integer $k \geq 0$

$$E[(B_t)^{2k}] = \frac{(2k)!}{2^k k!}t^k, \quad E[(B_t)^{2k+1}] = 0. \tag{H.40}$$

To derive this result, we apply the Ito's formula to B_t^n

$$d(B_t^n) = nB_t^{n-1}dB_t + \frac{n(n-1)}{2}B_t^{n-2}dt.$$

Integrating this SDE yields

$$B_t^n = n\int_0^t B_s^{n-1}dB_s + \frac{n(n-1)}{2}\int_0^t B_s^{n-2}ds.$$

Taking the expectation, we obtain

$$E[B_t^n] = \frac{n(n-1)}{2}\int_0^t E[B_s^{n-2}]ds.$$

Note that $E[B_t] = 0$ and $E[B_t^2] = t$. Using the mathematical induction, we can prove the result.

Another special case is for a process X_t that satisfied the following SDE

$$dX_t = a(t)dt + b(t)dB_t.$$

It can be shown that such a process is a Gaussian distributed with mean $X_0 + \int_0^t a(s)ds$ and variance $\int_0^t b^2(s)ds$.

Now we consider the case in which the drift and volatility depend on t and B_t only. That is

$$dX_t = \mu(t, B_t)dt + \sigma(t, B_t)dB_t, \quad t \geq 0,$$

which defines an Ito diffusion. The integral form is given by

$$X_t = X_0 + \int_0^t \mu(s, B_s)ds + \int_0^t \sigma(s, B_s)dB_s.$$

There are some cases where both integrals can be computed explicitly.
Here a few examples.

Example H.7 *Solve the following SDEs and find the mean and variance of the corresponding processes:*

(1)

$$dX_t = (B_t - c)dt + B_t^2 dB_t, \quad X_0 = 0, \tag{H.41}$$

where c is a constant.

(2)

$$dX_t = e^{-t/2}dt + e^{B_t - t/2}dB_t, \quad X_0 = 0. \tag{H.42}$$

To obtain the solution to the first differential equation, we need the integrated BM process, $Z_t = \int_0^t B_s ds$. Integrating (H.41) from 0 to t yields

$$X_t = \int_0^t dX_s = \int_0^t (B_s - c)ds + \int_0^t B_s^2 dB_s$$

$$= Z_t - ct + \frac{1}{3}B_t^3 - \int_0^t B_s ds \tag{H.43}$$

$$= \frac{1}{3}B_t^3 - ct.$$

Note that $Z_t \sim N(0, t^3/3)$. Taking the expectation and using (H.40), we get

$$E[X_t] = -ct.$$

$$Var(X_t) = E[X_t^2] - (E[X_t])^2$$

$$= E\left[\left(\frac{1}{3}B_t^3 - ct\right)^2\right] - (-ct)^2$$

$$= E\left[\frac{1}{9}B_t^6 - \frac{2}{3}ct + (ct)^2\right] - (ct)^2$$

$$= \frac{1}{9}E[B_t^6] - \frac{2}{3}ct + (ct)^2 - (ct)^2$$

$$= \frac{15}{9}t^3 - \frac{2}{3}ct.$$

For (H.42), integrating from 0 to t

$$X_t = \int_0^t e^{-s/2}ds + \int_0^t e^{-s/2+B_s}dB_s$$

$$= 2(1 - e^{-t/2}) + e^{-t/2}e^{B_t} - 1$$

$$= 1 + e^{-t/2}(e^{B_t} - 2).$$

Note that e^{B_t} is a special geometric BM where its mean and variance are available. Taking the expectation and computing the variance, we obtain

$$E[X_t] = E[1 + e^{-t/2}(e^{B_t} - 2)] = 1 - 2e^{-t/2} + e^{-t/2}E[e^{B_t}]$$

$$= 2 - 2e^{-t/2},$$

$$Var(X_t) = Var[1 + e^{-t/2}(e^{B_t} - 2)] = Var[e^{-t/2}e^{B_t}] = e^{-t}Var(e^{B_t})$$

$$= e^{-t}(e^{2t} - e^t) = e^t - 1.$$

Furthermore, we may also get the distribution of X_t in some cases. For example, in this second example, the distribution of X_t is given by

$$F_t(x) = P(X_t \leq x) = P(1 + e^{-t/2}(e^{B_t} - 2) \leq x)$$

$$= P(B_t \leq \ln(2 + e^{t/2}(x-1))) = P\left(\frac{B_t}{\sqrt{t}} \leq \frac{1}{\sqrt{t}} \ln(2 + e^{t/2}(x-1))\right)$$

$$= \Phi\left(\frac{1}{\sqrt{t}} \ln(2 + e^{t/2}(x-1))\right),$$

where $\Phi(\cdot)$ is the distribution function of the standard normal random variable.

As shown in the introduction of the SDEs, the drift and volatility may be written as follows:

$$\mu(t,x) = \partial_x f(t,x) + \frac{1}{2}\partial_{xx}^2 f(t,x),$$

$$\sigma(t,x) = \partial_x f(t,x).$$

For more details on stochastic calculus, interested readers are referred to some textbooks such as Lawler (2021) and Calin (2022).

REFERENCES

1. J. Chang, "Stochastic Processes", http://www.stat.yale.edu/pollard/Courses/251.spring2013/Handouts/Chang-notes.pdf, 2007.
2. O. Calin, "An Informal Introduction to Stochastic Calculus with Applications", World Scientific, 2nd ed., 2022.
3. G. F. Lawler, "Introduction to Stochastic Calculus with Applications", Chapman and Hall/CRC, 1st ed., 2021.
4. B. Oksendal, "Stochastic Differential Equations - An Introduction with Applications", Springer, 6th ed., 2013.

Comparison of Stochastic Processes – Stochastic Orders

To compare two random variables or two stochastic processes, we need the concept of stochastic orders. In this section, we provide a brief introduction to this important concept. We start with the definition of the stochastically larger relation and then present the related properties. For more details about stochastic orders, interested readers are referred to many good books (i.e., Ross (1996)).

I.1 BASIC STOCHASTIC ORDERING

Definition I.1 *A random variable X is said to be stochastically larger than another random variable Y, denoted by $X \geq_{st} Y$, if*

$$P(X > a) \geq P(Y > a) \quad \text{for all } a.$$

Assume that the distribution functions for X and Y are F and G. Then the stochastically larger relation can be written as

$$F^c(a) \geq G^c(a) \quad \text{for all } a,$$

where F^c (G^c) is the complement of the distribution function F (G). It is a good exercise to verify the following properties about the stochastic larger relation.

If $X \geq_{st} Y$, then

1. $E[X] \geq E[Y]$,
2. Such an ordering is equivalent to

$$E[f(X)] \geq E[f(Y)] \quad \text{for all increasing function } f.$$

We can also define the stochastically increasing (or decreasing) additional life of a non-negative random variable. Such a random variable is usually used for modeling a life time of an operating system (e.g., a machine). Consider a machine that has random life time (from the brand new state to the failure state) X with distribution function F. Assume that this machine has survived

DOI: 10.1201/9781003150060-I

to time t. We let X_t denote the additional life from t to its future failure. Then its complement of distribution function is

$$F_{X_t}^c(a) = P(X_t > a) = P(X - t > a | X > t)$$
$$= \frac{F^c(t+a)}{F^c(t)}, \tag{I.1}$$

which is useful in presenting the property of the additional life of the machine. Note that the distribution function can be written as

$$F_{X_t}(a) = P(X_t \le a) = P(X - t \le a | X > t)$$
$$= \frac{F(t+a)}{F^c(t)}. \tag{I.2}$$

The failure (or hazard) rate function of X can be defined from $F_{X_t}(a)$. Differentiating (I.2) with respect to a yields

$$dF_{X_t}(a) = \frac{d}{da}\left(\frac{F(t+a)}{F^c(t)}\right) da$$
$$= \frac{f(t+a)}{F^c(t)} da,$$

where the term $f(t+a)/F^c(t)$ represents the probability that a t-unit-old machine fails in the interval $(t+a, t+a+d(t+a))$. Letting $a \to 0$, this term leads to the failure rate function $h(t)$

$$h(t) = \frac{f(t)}{F^c(t)},$$

which represents the probability that a t-unit old machine fails in the interval $(t, t+dt)$. There are two types of random variables based on the property of the failure rate function that have wide applications in reliability theory. A random variable X is said to be an increasing (a decreasing) failure rate (IFR) random variable if $h(t)$ is increasing (decreasing) in t. Clearly, an IFR random variable can model the failure process of an aging item which is more likely to fail as it is getting older. Now we can establish the relation between the failure rate function of the additional life time X_t, denoted by $h_t(t)$ and that of X, denoted by $h(t)$, as follows:

$$h_t(a) = \lim_{\delta \to 0} P(a < X_t < a + \delta | X_t \ge a)/\delta$$
$$= \lim_{\delta \to 0} P(a < X - t < a + \delta | X \ge t, X - t \ge a)/\delta$$
$$= \lim_{\delta \to 0} P(t + a < X < t + a + \delta | X \ge t + a)/\delta$$
$$= h(t+a).$$

Next, we express $F_{X_t}^c(s)$ in terms of the failure rate function of X, i.e. $h(t)$. It follows from

$$dF_{X_t}^c(s) = -f_{X_t}(s)ds = -\frac{f_{X_t}(s)}{F_{X_t}^c(s)}F_{X_t}^c(s)ds$$

$$= -h_t(s)F_{X_t}^c(s)ds,$$

that

$$F_{X_t}^c(s) = \exp\left\{-\int_0^s h_t(y)dy\right\}$$

$$= \exp\left\{-\int_t^{t+s} h(z)dz\right\}.$$

(I.3)

Clearly, from (I.3), it follows that if $h(z)$ is increasing (decreasing) in z, then $F_{X_t}^c(s)$ is decreasing (increasing) in t. Conversely, if $F_{X_t}^c(s)$ is decreasing (increasing) in t, then (I.3) implies that $h(z)$ is increasing (decreasing) in z. Therefore, the following property holds.

· X is IFR $\Leftrightarrow X_t$ is stochastically decreasing in t,
· X is DFR $\Leftrightarrow X_t$ is stochastically increasing in t.

A good example is the random variable that follows a Weibull distribution with scale parameter μ and shape parameter k. Its p.d.f. is given by

$$f(x) = \begin{cases} \frac{k}{\mu}\left(\frac{x}{\mu}\right)^{k-1}e^{-(x/\mu)^k} & x \geq 0, \\ 0 & x < 0. \end{cases}$$

For the Weibull distributed random variable, if $k < 1$, it is DFR; if $k > 1$, it is IRF; and if $k = 1$, it has a constant failure rate and becomes an exponentially distributed random variable. The mixtures of exponentially distributed random variables form a class of random variables which are DFR. The distribution function F for a mixture of distributions $F_\alpha, 0 < \alpha < \infty$ can be written as

$$F(x) = \int_0^\infty F_\alpha(x)dH(\alpha),$$

where α is the characterization parameter and its distribution function is $H(\alpha)$. It can be shown that if $F_\alpha(x) = 1 - e^{-\lambda\alpha x}$, $F(x)$ is DFR distribution function. This implies that the mixture of random variables with constant failure rate will be DRF. Furthermore, in general, if F_α is DRF distribution with $0 < \alpha < \infty$ abd $H(\alpha)$ is a distribution on $(0, \infty)$, then $F(t) = \int_0^\infty F_\alpha(t)dH(\alpha)$ is also a DRF distribution. The proof of this property requires the following famous inequality (the Cauchy-Schwarz Inequality):

For any distribution function F and functions $h(t), g(t), t \geq 0$, we have

$$\left(\int h(t)g(t)dF(t) \right)^2 \leq \left(\int h^2(t)dF(t) \right) \left(\int g^2(t)dF(t) \right)$$

provided these integrals exist.

I.1.1 COUPLING

For two random variables X and Y with basic stochastic ordering relation, a property called "coupling" can be established. This property can be stated as follows: If $X \geq_{st} Y$, then there exist random variables X' and Y' which have the same distributions of X and Y such that

$$P(X' \geq Y') = 1.$$

The proof of this coupling property requires the following proposition.

Proposition I.1 *Let F and H be continuous distribution functions. If X has distribution F, then random variable $Y = H^{-1}(F(x))$ has distribution H.*

The proof is relatively easy as follows:

$$\begin{aligned}
P(Y \leq a) = P(H^{-1}(F(X)) \leq a) &= P(F(x) \leq H(a)) \\
&= P(X \leq F^{-1}(H(a))) = F(F^{-1}(H(a))) \\
&= H(a).
\end{aligned}$$

Now we can justify the coupling property. Suppose that X with distribution function F and Y with distribution function H have the basic ordering relation $X \geq_{st} Y$ or $F^c(a) \geq H^c(a)$. We consider another random variable X' following distribution F and introduce an random variable $Y' = H^{-1}(F(X))$. According the proposition above, we know that Y' follows distribution H. Note that $F \leq H$ which implies that $F^{-1} \geq H^{-1}$ (the c.d.f relation vs. the inverse of the c.d.f. relation). Thus, we have

$$Y' = H^{-1}(F(X)) \leq F^{-1}(F(X')) = X',$$

which implies that $P(X' \geq Y') = 1$. We can extend the basic stochastic ordering relation from the random variable to the random vector and stochastic process.

Definition I.2 *(a) A random vector $\mathbf{X} = (X_1, ..., X_n)$ is stochastically larger than a random vector $\mathbf{Y} = (Y_1, ..., Y_n)$, denoted by $\mathbf{X} \geq_{st} \mathbf{Y}$ if for all increasing function f*

$$E[f(\mathbf{X})] \geq E[f(\mathbf{Y})].$$

(b) A stochastic process $\{X(t), t \geq 0\}$ is stochastically larger than another stochastic process $\{Y(t), t \geq 0\}$ if

$$(X(t_1), ..., X(t_n)) \geq_{st} (Y(t_1), ..., Y(t_n))$$

for all $n, t_1, ..., t_n$.

With this definition, we can show

Proposition I.2 *If* **X** *and* **Y** *are vectors of independent components such that $X_i \geq_{st} Y_i$, then* **X** $\geq_{st}$ **Y**.

Proof. First, we show that if $X_i \geq_{st} Y_i$ for $i = 1, ..., n$, then for any increasing function f,

$$f(X_1, ..., X_n) \geq_{st} f(Y_1, ..., Y_n).$$

Suppose that X_i follows distribution F_i and Y_i follows distribution H_i. For each X_i, we can introduce a random variable $Y_i' = H_i^{-1}(F_i(X_i))$. Then Y_i' has the distribution of H_i and $Y_i' \leq X_i$. Thus, we have $f(X_1, ..., X_n) \geq f(Y_1', ..., Y_n')$ due to the increasing function f. This relation implies for any a

$$f(Y_1', ..., Y_n') > a \Rightarrow f(X_1, ..., X_n) > a,$$

which leads to

$$P(f(Y_1', ..., Y_n') > a) \leq P(f(X_1, ..., X_n) > a).$$

Note that the left-hand side of the above equals $P(f(Y_1, ..., Y_n)$ due to the identical distribution for Y_i and Y_i'. That is

$$P(f(X_1, ..., X_n) > a) \geq P(f(Y_1, ..., Y_n) > a).$$

Integrating both sides of the above inequality, we have

$$E[f(X_1, ..., X_n)) \geq E[f(Y_1, ..., Y_n)),$$

which implies the result from the definition of the stochastic ordering for random vectors. ∎

This proof shows the importance of using the coupling property. Using these definitions and concepts, we can compare and order stochastic processes. For example, consider two renewal processes $N_i = \{N_i(t), t \geq 0\}, i = 1, 2$. Suppose that the inter-arrival times for N_1 (N_2) follow distribution F (H). If $F^c \geq H^c$ or the inter-arrival time (i.i.d.) for N_1 is stochastically greater than that for N_2, then we can show that the first renewal process is stochastically greater than the second renewal process. That is

$$\{N_1(t), t \geq 0\} \leq_{st} \{N_2(t), t \geq 0\}.$$

Again, the proof can be done by using the coupling property. Let $X_1, X_2, \ldots$ be a sequence of i.i.d. random variables following distribution F. Then the renewal process generated by X_i, denoted by N_1^*, has the same distribution as N_1. According the coupling property, we can generate a sequence of i.i.d. random variables $Y_1, Y_2, \ldots$ by using $H^{-1}(F(Y_i))$ for $i = 1, 2, \ldots$. This sequence of i.i.d. random variables have the distribution H and the property $Y_i \leq X_i$. Similarly, the renewal process generated by Y_i, denoted by N_2^*, has the same distribution as N_2. However, since $Y_i \leq X_i$ for all i, it follows that

$$N_1^*(t) \leq N_2^*(t) \quad \text{for all } t,$$

which implies

$$P(N_1(t) > m) \leq P(N_2(t) > m)$$

for all m. This implies that $\{N_1(t), t \geq 0\} \leq_{st} \{N_2(t), t \geq 0\}$.

Using the coupling, we can establish the stochastic monotonic properties for some Markov processes such as the Birth-and-Death processes (BD processes). The first monotonic property is that the BD process $\{X(t), t \geq 0\}$ is stochastically increasing in the initial state $X(0) = i$. The basic idea of proving this property is (i) considering two BD processes $\{X_1(t), t \geq 0\}$ with $X_1(0) = i + 1$ and $\{X_2(t), t \geq 0\}$ with $X_2(0) = i$; and (ii) finding a BD process $\{X_3(t), t \geq 0\}$ with $X_3(0) = i + 1$ that is always greater than the BD process $\{X_2(t), t \geq 0\}$ with $X_2(0) = i$ and has the same distribution as that of $\{X_1(t), t \geq 0\}$ with $X_1(0) = i + 1$; and (3) using the coupling to obtain the conclusion. The key is to construct $\{X_3(t), t \geq 0\}$. Since the BD process can only go up or down by 1 at a state change instant and $X_1(t)$ starts with the state with 1 more than that of $X_2(t)$ (i.e., $X_1(0) > X_2(0)$), $X_1(t)$ will be either always greater than $X_2(t)$ or will be equal to $X_2(t)$ at some time. Denote by τ the first time that these two processes become equal. That is

$$\tau = \begin{cases} \infty & \text{if } X_1(t) > X_2(t) \text{ for all } t, \\ \text{1st } t: & X_1(t) = X_2(t) \text{ otherwise.} \end{cases}$$

If $\tau < \infty$, then the two processes are equal at time τ. Thus, before τ, $X_1(t)$ is greater than $X_2(t)$; from the instant τ and onwards, by the Markovian property, $X_1(t)$ and $X_2(t)$ will have the same probabilistic structure (i.e., the two processes have the same distribution on the functional space over $[\tau, \infty)$). Now we can define a third stochastic process $\{X_3(t), t \geq 0\}$ as follows:

$$X_3(t) = \begin{cases} X_1(t) & \text{if } t < \tau, \\ X_2(t) & \text{if } t \geq \tau. \end{cases}$$

Note that $\{X_3(t)\}$ will also be a BD process with the same parameters as the other two processes and $X_3(0) = X_1(0) = i + 1$. It follows from the definition of τ, the fact $X_1(t) > X_2(t)$ for $t < \tau$, and the definition of $X_3(t)$ that

$$X_3(t) \geq X_2(t) \quad \text{for all } t,$$

which implies that $\{X_1(t), t \geq 0\} \geq_{st} \{X_2(t), t \geq 0\}$ by coupling. Therefore, we conclude that the BD process is stochastically increasing in the initial state $X(0) = i$. Furthermore, using this monotonic property, we can prove that the transition probability function for the BD process $P(X(t) \geq j|X(0) = 0)$ is increasing in t for all j (left as an exercise for readers).

I.2   FAILURE RATE ORDERING

For two random variables X and Y, let $h_X(t)$ and $h_Y(t)$ be the hazard (or failure) rate functions of X and Y, respectively. It is said that X is hazard (or failure) rate larger than Y if

$$h_X(t) \geq h_Y(t) \quad \text{for all } t \geq 0. \tag{I.4}$$

It follows from

$$P(X > t + s | X > t) = \exp\left\{ -\int_t^{t+s} h(y)dy \right\}$$

that (I.4) implies

$$P(X > t + s | X > t) \leq P(Y > t + s | Y > t) \Rightarrow X_t \leq_{st} Y_t \quad \text{for all } t \geq 0,$$

where X_t and Y_t represent the remaining lives (also called residual life or additional life) of t-unit-old item having the same distributions of X and Y, respectively. Thus, we conclude that the hazard larger relation implies the stochastically smaller relation for the residual life. Hazard ordering relation is useful in comparing counting processes (see Ross (1996) for examples).

I.3   LIKELIHOOD RATIO ORDERING

Consider two continuous non-negative random variables X and Y with f and g density functions, respectively. It is said that X is likelihood larger than Y, denoted by $X \geq_{LR} Y$, if

$$\frac{f(x)}{g(x)} \leq \frac{f(y)}{g(y)} \quad \text{for all } x \leq y.$$

I.4   VARIABILITY ORDERING

Here, we need the definition of a convex function. A dunction is said to be convex if for all $0 < \alpha < 1, x_1, x_2,$

$$h(\alpha x_1 + (1 - \alpha)x_2) \leq \alpha h(x_1) + (1 - \alpha)h(x_2).$$

It is said that X is more variable than Y, denoted by $X \geq_v Y$, if

$$E[h(X)] \geq E[h(Y)] \quad \text{for all increasing and convex } h.$$

If these random variables have distributions F and H, respectively, we can also write $F \geq_v H$. An equivalent definition of the stochastically more variable relation is stated as the following proposition.

Proposition I.3 *For two non-negative random variables X with distribution F and Y with distribution H, $X \geq_v Y$, if and only if,*

$$\int_a^\infty F^c(x)dx \geq \int_a^\infty H^c(x)dx, \quad for \ a \geq 0.$$

The proof can be done by designing a convex function h_a as follows:

$$h_a(x) = \begin{cases} 0 & x \leq a, \\ x - a & x > a. \end{cases}$$

and the fact that h is an increasing and convex function. (The detailed proof can be found in Ross (1996)). Another property regarding variability ordering is for the two random variables with the same mean. Suppose that X and Y are two non-negative random variables such that $E[X] = E[Y]$. Then, we can show that $X \geq_v Y$, if and only if, $E[h(X)] \geq E[h(Y)]$ for all convex h. This property can intuitively explain the condition $E[h(X)] \geq E[h(Y)]$ for stochastically larger relation $X \geq_v Y$. This is because a convex function assigns more weight to the extreme values and thus $E[h(X)] \geq E[h(Y)]$ indicates that X is more variable than Y. As a special case, for X and Y with $E[X] = E[Y]$, if we choose $h(x) = x^2$, then $E[h(X)] \geq E[h(Y)]$ implies $Var(X) \geq Var(Y)$.

The following proposition is useful in comparison of multi-variate cases.

Proposition I.4 *If $X_1, ..., X_n$ are independent and $Y_1, ..., Y_n$ are independent, and $X_i \geq_v Y_i$ for $i = 1, ..., n$, then*

$$h(X_1, ..., X_n) \geq_v g(Y_1, ..., Y_n)$$

for all increasing and convex function g that are convex in each argument.

The proof of this proposition can be found in Ross (1996).

Finally, an inequality for expectations involved two increasing functions of a random vector can be established for a set of random variables called "associated random variables". We define a set of random variables to be associated if for all increasing functions f and g, the following inequality holds

$$E[f(\mathbf{X})g(\mathbf{X})] \geq E[f(\mathbf{X})]E[g(\mathbf{X})],$$

where $\mathbf{X} = (X_1, ..., X_n)$.

REFERENCE

1. S. Ross, "Stochastic Processes", John Wiley & Sons, New York, 2nd ed., 1996.

J Matrix Algebra and Markov Chains

In this section, we review some concepts and theory in linear algebra (linear system) that are relevant to Markov chains. Most of the proofs are omitted for the conciseness of the review. These proofs are either left as exercises or can be found in some textbooks on linear algebra such as Kani et al. (2021) on which this section is based on.

J.1 POSITIVE AND NON-NEGATIVE VECTORS AND MATRICES

Definition J.1 *A matrix or vector is positive or elementwise positive (non-negative or elementwise non-negative) if all its entries are positive (or non-negative)*

Notation $\mathbf{x} > \mathbf{y}(\mathbf{x} \geq \mathbf{y})$ means $\mathbf{x} - \mathbf{y}$ is elementwise positive (nonnegative). Matrix multiplication preserves nonnegativity if and only if the matrix is nonnegative.

The normalized form of a positive column vector $\mathbf{x}$ is defined as $\mathbf{d} = (1/\mathbf{e}^T\mathbf{x})\mathbf{x}$, which is also called the distribution of $\mathbf{x}$. That is the component i, $d_i = x_i/(\sum_k x_k)$, gives the proportion x_i of the total of $\mathbf{x}$.

Definition J.2 *A nonnegative square matrix $\mathbf{A} \in \mathbf{R}^{n \times n}$ is called regular if for some integer $k \geq 1$, that $\mathbf{A}^k > 0$.*

We can use the graph with connected nodes $1, ..., n$. Consider a matrix $\mathbf{A} = [\mathbf{A}]_{ij} = [A_{ij}]$ representing connections (directed arcs) for a graph with n nodes. We construct the matrix in such a way that $A_{ij} > 0$ whenever there is a directed arc from i to j. Then $[\mathbf{A}^k]_{ij} > 0$ if and only if there is a path of length k from i to j. A matrix $\mathbf{A}$ is regular if for some k there is a path of length k from every node to every other node. Clearly, any positive matrix is regular (i.e., $k = 1$). For example, the following matrices are not regular

$$\begin{pmatrix} 1 & 1 \\ 0 & 1 \end{pmatrix} \quad \text{and} \quad \begin{pmatrix} 0 & 1 \\ 1 & 0 \end{pmatrix}.$$

DOI: 10.1201/9781003150060-J

However, the following matrix is regular

$$\begin{pmatrix} 1 & 1 & 0 \\ 0 & 0 & 1 \\ 1 & 0 & 0 \end{pmatrix}$$

as

$$\begin{pmatrix} 1 & 1 & 0 \\ 0 & 0 & 1 \\ 1 & 0 & 0 \end{pmatrix}^4 = \begin{pmatrix} 3 & 2 & 1 \\ 1 & 1 & 1 \\ 2 & 1 & 1 \end{pmatrix},$$

which implies $k = 4$. Now we can present the Perron-Frobenius Theorems for regular matrices.

Theorem J.1 *If matrix $\mathbf{A} \in \mathbf{R}^{n \times n}$ is nonnegative and regular, that is, $\mathbf{A}^k > 0$ for some $k \geq 1$, then*

- *there is an eigenvalue λ_{pf} of $\mathbf{A}$ that is real and positive, with positive left and right eigenvectors;*
- *for any other eigenvalue λ, we have $|\lambda| < \lambda_{pf}$;*
- *the eigenvalue λ_{pf} is simple, i.e., has multiplicity one, and corresponds to a 1×1 Jordan block.*

The eigenvalue λ_{pf} is called the Perron-Frobenius eigenvalue of $\mathbf{A}$ and the associated positive (left and right) eigenvectors are called the (left and right) PF eigenvectors (and are unique up to positive scaling).

For nonnegative matrices, we have

Theorem J.2 *If matrix $\mathbf{A} \in \mathbf{R}^{n \times n}$ is nonnegative, that is, $\mathbf{A} \geq 0$, then*

- *there is an eigenvalue λ_{pf} of $\mathbf{A}$ that is real and nonnegative, with associated nonnegative left and right eigenvectors;*
- *for any other eigenvalue λ of $\mathbf{A}$, we have $|\lambda| \leq \lambda_{pf}$.*

λ_{pf} *is called the Perron-Frobenius eigenvalue of $\mathbf{A}$ and the associated nonnegative (left and right) eigenvectors are called the (left and right) PF eigenvectors. In this case, they need not be unique, or positive.*

These theorems can be applied to analyze the discrete-time Markov chains. The one-step transition probability matrix is stochastic. That is $\mathbf{P} \geq 0$, $\mathbf{e}^T \mathbf{P} = \mathbf{e}^T$, and $\mathbf{Pe} = \mathbf{e}$. Thus, $\mathbf{P}$ is nonnegative and $\mathbf{e}^T$ ($\mathbf{e}$) is a left (right) eigenvector with eigenvalue 1, which is in fact the PF eigenvalue of $\mathbf{P}$. It is well known that an equilibrium distribution (also called invariant measure), π, of the Makov chain satisfies $\pi \mathbf{P} = \pi$. Thus, π is a right PF eigenvector of $\mathbf{P}$ with eigenvalue 1. Now suppose that $\mathbf{P}$ is regular, which means for some k, we have $\mathbf{P}^k > \mathbf{0}$.

That is there is positive probability of reaching any state from any other state in k steps (i.e., positive recurrent). Under this condition, there is a unique equilibrium distribution $\pi > 0$ and the eigenvalue 1 is simple and dominant. Thus, we hace the limiting probability $\lim_{t \to \infty} [\mathbf{P}]_{ij} = \pi_j$. One can say that the distribution of a regular Markov chain always converges to the unique equilibrium distribution. The rate of convergence to the equilibrium distribution depends on second largest eigenvalue magnitude, i.e., $\mu = \max\{|\lambda_2|, ..., |\lambda_n|\}$, where $\lambda_i, i = 1, 2, ..., n$ are the eigenvalues of $\mathbf{P}$ and $\lambda_1 = \lambda_{pf} = 1$, which is also called the spectral radius of $\mathbf{P}$, denoted by $sp(\mathbf{P})$. The mixing time of the Markov chain is given by

$$T = \frac{1}{\log(1/\mu)},$$

which roughly means the number of steps over which the deviation from equilibrium distribution decreases by factor e.

Suppose $\mathbf{A}$ is non-negative with PF eigenvalue λ_{pf} and $\lambda \in \mathbf{R}$, then $(\lambda \mathbf{I} - \mathbf{A})^{-1}$ exists and is non-negative, if and only if $\lambda > \lambda_{pf}$ for any square matrix $\mathbf{A}$. Thus, for any square matrix $\mathbf{A}$, the power series expansion (to be justified in the next section on the convergence of power matrix series)

$$(\lambda \mathbf{I} - \mathbf{A})^{-1} = \frac{1}{\lambda} \mathbf{I} + \frac{1}{\lambda^2} \mathbf{A} + \frac{1}{\lambda^3} \mathbf{A}^2 + \cdots$$

converges provided $|\lambda|$ is larger than all eigenvalues of $\mathbf{A}$. This result implies that if $\lambda > \lambda_{pf}$, then $(\lambda \mathbf{I} - \mathbf{A})^{-1}$ is non-negative. It is easy to show the converse. Suppose that $(\lambda \mathbf{I} - \mathbf{A})^{-1}$ exists and is non-negative, and let $v \neq 0, v \geq 0$ be a PF eigenvector of $\mathbf{A}$, then we have

$$(\lambda \mathbf{I} - \mathbf{A})^{-1} v = \frac{1}{\lambda - \lambda_{pf}} v \geq 0,$$

which implies that $\lambda > \lambda_{pf}$.

J.2 POWER OF MATRICES

Now we investigate the limiting behavior of the power of a non-negative matrix. That is $\lim_{t \to \infty} \mathbf{A}^n = \mathbf{A}_\infty$. We start with defining the limit for matrices.

Definition J.3 *Let $\mathbf{A}_1, \mathbf{A}_2, ...$ be a sequence of $m \times n$ matrices. Then the limit of this sequence is defined as*

$$\mathbf{A} = \lim_{k \to \infty} \mathbf{A}_k$$

if $\mathbf{A} = [a_{ij}]$ *and* $\lim_{k\to\infty} a_{ij}^{(k)} = a_{ij}$, *for all* i, j *with* $1 \le i \le m$ *and* $1 \le j \le n$, *where*

$$\mathbf{A}_k = \begin{pmatrix} a_{11}^{(k)} & \cdots & a_{1n}^{(k)} \\ \vdots & \cdots & \vdots \\ a_{m1}^{(k)} & \cdots & a_{mn}^{(k)} \end{pmatrix},$$

and $a_{ij}^{(k)}$ *is a function of* k.

For example, if we have two sequences of matrices as follows:

$$\text{(a)} \quad \mathbf{A}_k = \begin{pmatrix} e^{-k} & 2 \\ \frac{1}{k^3} & \frac{k+1}{k} \end{pmatrix}, \quad \text{for } k = 1, 2, \ldots$$

$$\text{(b)} \quad \mathbf{B}_k = \begin{pmatrix} (-1)^k & 0 \\ 0 & 1 \end{pmatrix}, \quad \text{for } k = 1, 2, \ldots,$$

then, we have the limit for (a) given by

$$\mathbf{A}_\infty = \lim_{k\to\infty} \mathbf{A}_k = \begin{pmatrix} 0 & 1 \\ 0 & 1 \end{pmatrix}.$$

However, the limit for (b) does not exist. Some rules for limits of matrices are summarized below.

Theorem J.3 *Assume that* $\lim_{k\to\infty} \mathbf{A}_k = \mathbf{A}$ *exists and let* $\mathbf{B}, \mathbf{C}$ *be matrices such that the product* $\mathbf{CA}_k\mathbf{B}$ *is defined. Then*

1. $\lim_{k\to\infty}(\mathbf{A}_k\mathbf{B}) = \mathbf{AB}$,
2. $\lim_{k\to\infty}(\mathbf{CA}_k) = \mathbf{CA}$,
3. $\lim_{k\to\infty}(\mathbf{CA}_k\mathbf{B}) = \mathbf{CAB}$,

The proof can be easily done by the definition of matrix multiplication and properties of limits.

To study the discrete-time Markov chains, we are interested in the limit of the matrix powers.

Definition J.4 *A square matrix* $\mathbf{A}$ *is called power convergent if* $\lim_{k\to\infty} \mathbf{A}^k$ *exists.*

Example J.1 *The square matrix* $\mathbf{A} = \begin{pmatrix} \frac{1}{2} & 0 & 0 \\ 0 & \frac{2}{3} & 0 \\ 0 & 0 & \frac{3}{4} \end{pmatrix}$ *is power convergent since*

$$\lim_{k\to\infty} \mathbf{A}^k = \lim_{k\to\infty} \begin{pmatrix} \left(\frac{1}{2}\right)^k & 0 & 0 \\ 0 & \left(\frac{2}{3}\right)^k & 0 \\ 0 & 0 & \left(\frac{3}{4}\right)^k \end{pmatrix} = \begin{pmatrix} 0 & 0 & 0 \\ 0 & 0 & 0 \\ 0 & 0 & 0 \end{pmatrix}.$$

However, the matrix $\mathbf{B} = \begin{pmatrix} 1 & 0 \\ 0 & -1 \end{pmatrix}$ *is not power convergent since the* $\mathbf{B}^k = \begin{pmatrix} 1 & 0 \\ 0 & (-1)^k \end{pmatrix}$ *does not converge as* $k \to \infty$.

For some matrices with special structures, the power convergence can be easily determined. Inspired by the previous example, we have

Theorem J.4 *A diagonal matrix* $\mathbf{A} = Diag(\lambda_1, \lambda_2, ..., \lambda_m)$ *is power convergent if and only if for each i with* $1 \le i \le m$, *we have:*

$$\text{either} \quad |\lambda_i| < 1 \quad \text{or} \quad \lambda_i = 1. \tag{J.1}$$

More generally, a diagonal matrix $\mathbf{A}$ *is power convergent if and only if each eigenvalue* λ_i *of* $\mathbf{A}$ *satisfies (J.1).*

To find the conditions for a general matrix to converge, we first consider the Jordan Block matrix $\mathbf{J} = J(\lambda, m)$.

Theorem J.5 *A Jordan block* $\mathbf{J} = J(\lambda, m)$ *is power convergent if and only if either* $|\lambda| < 1$ *or* $\lambda = 1$ *and* $m = 1$, *where* λ *is the entry on the diagonal and* m *is the dimension of the block.*

Recall that in linear algebra, the Jordan canonical form matrix has the following structure

$$\mathbf{J} = Diag(\mathbf{J}_1, \mathbf{J}_2, ..., \mathbf{J}_n) = \begin{pmatrix} \mathbf{J}_1 & 0 & \cdots & 0 \\ 0 & \mathbf{J}_2 & \ddots & \vdots \\ \vdots & \ddots & \ddots & 0 \\ 0 & \cdots & 0 & \mathbf{J}_n \end{pmatrix},$$

where $\mathbf{J}_i = J(\lambda_i, m_i)$ are Jordan blocks with the form

$$\mathbf{J}_i = J(\lambda_i, m_i) = \begin{pmatrix} \lambda_i & 1 & 0 & \cdots & 0 \\ 0 & \lambda_i & 1 & \ddots & 0 \\ \vdots & \cdots & \ddots & \ddots & 0 \\ \vdots & \cdots & & \ddots & 1 \\ 0 & 0 & \cdots & \cdots & \lambda_i \end{pmatrix}.$$

Two simple examples to show this result are as follows:

Example J.2 *For* $\mathbf{J}_1 = J(\frac{1}{2}, 2)) = \begin{pmatrix} \frac{1}{2} & 1 \\ 0 & 1 \end{pmatrix}$, *then*

$$\lim_{k \to \infty} \mathbf{J}_1^k = \lim_{k \to \infty} \begin{pmatrix} \left(\frac{1}{2}\right)^k & k\left(\frac{1}{2}\right)^{k-1} \\ 0 & \left(\frac{1}{2}\right)^k \end{pmatrix} = \begin{pmatrix} 0 & 0 \\ 0 & 0 \end{pmatrix}.$$

Thus, $\mathbf{J}_1$ *is power convergent and converges to* $\mathbf{0}$ *because of* $|\lambda| = 1/2 < 1$. *For* $\mathbf{J}_2 = J(1,2) = \begin{pmatrix} 1 & 1 \\ 0 & 1 \end{pmatrix}$, *then* $\mathbf{J}_2^k = \begin{pmatrix} 1 & k \\ 0 & 1 \end{pmatrix}$, *so the sequence does not converge. That is* $\mathbf{J}_2$ *is not power convergent since* $\lambda = 1$ *and* $m \neq 1$.

Definition J.5 *An eigenvalue* λ_i *of a matrix* $\mathbf{A}$ *is called regular if its geometric multiplicity is the same as its algebraic multiplicity. An eigenvalue* λ_i *of* $\mathbf{A}$ *is called dominant if we have* $|\lambda_j| < |\lambda_i|$, *for every eigenvalue* $\lambda_j \neq \lambda_i$ *of* $\mathbf{A}$.

Here we need to review the definitions of algebraic and geometric multiplicities of eigenvalues.

Definition J.6 *Let* $\mathbf{A}$ *be an* $m \times m$ *matrix and* $\lambda \in \mathbb{C}$ *be its eigenvalue. Then the algebraic multiplicity of* λ *is the multiplicity of* λ *as a root of the characteristic polynomial of* $\mathbf{A}$, *denoted by* $m_{\mathbf{A}}(\lambda)$, *and the geometric multiplicity of* λ *is the dimension of the* λ-*eigenspace of* $\mathbf{A}$, *denoted by* $v_{\mathbf{A}}(\lambda)$.

Some properties regarding $m_{\mathbf{A}}(\lambda)$ and $v_{\mathbf{A}}(\lambda)$ are as follows:

1. λ is an eigenvalue of $\mathbf{A} \Leftrightarrow m_{\mathbf{A}}(\lambda) \geq 1 \Leftrightarrow v_{\mathbf{A}}(\lambda) \geq 1$.
2. $v_{\mathbf{A}}(\lambda) \leq m_{\mathbf{A}}(\lambda)$.
3. $v_{\mathbf{A}}(\lambda) := \dim Nullsp(\mathbf{A} - \lambda \mathbf{I}) = m - \text{rank}(\mathbf{A} - \lambda \mathbf{I})$.

With this definition, we can now characterize power convergent matrices as follows.

Theorem J.6 *A square matrix* $\mathbf{A}$ *is power convergent if and only if either* $|\lambda_i| < 1$ *for all eigenvalues* λ_i *of* $\mathbf{A}$ *or 1 is regular and dominant eigenvalue of* $\mathbf{A}$.

Further, we can find the limit of the power convergent matrix in the following two theorems.

Theorem J.7 *If* $|\lambda_i| < 1$ *for every eigenvalue of* $\mathbf{A}$, *then* $\mathbf{A}$ *is power convergent and* $\lim_{n \to \infty} \mathbf{A}^n = 0$.

Theorem J.8 *If* $\lambda_1 = 1$ *is a regular, dominant eigenvalue of* $\mathbf{A}$, *then* $\mathbf{A}$ *is power convergent and*

$$\lim_{n \to \infty} \mathbf{A}^n = C_{10} \neq \mathbf{0},$$

where C_{10} is the first constituent matrix of $\mathbf{A}$ associated to $\lambda_1 = 1$. Moreover, the other constituent matrices C_{1k} associated to $\lambda_1 = 1$ are equal to zero. That is $C_{11} = \cdots = C_{1,m_1-1} = 0$.

Now we need to review the concepts related to functions of matrices.

J.2.1 FUNCTIONS OF SQUARE MATRICES

The function of a square matrix $\mathbf{A}$ is defined in the Taylor series manner as

$$f(x) = c_0 + c_1 x + c_2 x^2 + c_3 x^3 + \cdots$$

by replacing x with $\mathbf{A}$. To compute a function of $\mathbf{A}$, we need to find the n eigenvalues of $\mathbf{A}$, say $\lambda_1, \lambda_2, ..., \lambda_n$ (some of them may be complex and some may be multiple) and for each of these, compute the corresponding "constituent matrix", denoted by $C_1, C_2, ..., C_n$. Assuming that all eigenvalues are distinct, we can then evaluate any function f of $\mathbf{A}$ by

$$f(\mathbf{A}) = \sum_{i=1}^{n} f(\lambda_i) C_i.$$

The easiest way to find the constituent matrices is to make the above formula true for $f(x) = x^i, i = 0, 1, ..., n-1$ so that a set of equations are established as follows:

$$i = 0 \ \ f(\mathbf{A}) = \sum_{i=1}^{n} f(\lambda_i) C_i = \mathbf{A}^0 \Rightarrow C_1 + C_2 + C_3 + \cdots + C_n = \mathbf{I}$$

$$i = 1 \ \ f(\mathbf{A}) = \sum_{i=1}^{n} f(\lambda_i) C_i = \mathbf{A}^1 \Rightarrow \lambda_1 C_1 + \lambda_2 C_2 + \lambda_3 C_3 + \cdots + \lambda_n C_n = \mathbf{A}$$

$$i = 2 \ \ f(\mathbf{A}) = \sum_{i=1}^{n} f(\lambda_i) C_i = \mathbf{A}^2 \Rightarrow \lambda_1^2 C_1 + \lambda_2^2 C_2 + \lambda_3^2 C_3 + \cdots + \lambda_n^2 C_n = \mathbf{A}^2$$

$$\vdots \quad = \quad \vdots$$

$$i = n-1 \ \ f(\mathbf{A}) = \sum_{i=1}^{n} f(\lambda_i) C_i = \mathbf{A}^{n-1} \Rightarrow \lambda_1^{n-1} C_1 + \lambda_2^{n-1} C_2 + \lambda_3^{n-1} C_3$$

$$+ \cdots + \lambda_n^{n-1} C_n = \mathbf{A}^{n-1}$$

Solving these equations for the constituent matrices by treating the unknowns and the RHS elements as ordinary numbers. Here us an example.

Example J.3 *Consider a matrix* $\mathbf{A} = \begin{pmatrix} 1 & 3 \\ 1 & -1 \end{pmatrix}$. *Find the constituent matrices of this matrix. It follows from the characteristic polynomial that the eigen-*

values are $\lambda_1 = 2$ *and* $\lambda_2 = -2$. *Then we have*

$$C_1 + C_2 = \mathbf{I}$$
$$2C_1 - 2C_2 = \mathbf{A}$$

which has the solution

$$C_1 = \frac{1}{4}\mathbf{A} + \frac{1}{2}\mathbf{I} = \begin{pmatrix} \frac{3}{4} & \frac{3}{4} \\ \frac{1}{4} & \frac{1}{4} \end{pmatrix}$$

$$C_2 = \frac{1}{2}\mathbf{I} - \frac{1}{4}\mathbf{A} = \begin{pmatrix} \frac{1}{4} & -\frac{3}{4} \\ -\frac{1}{4} & \frac{3}{4} \end{pmatrix}.$$

Now we can evaluate any function $\mathbf{A}$. *That is*

$$f(\mathbf{A}) = f(\lambda_1)C_1 + f(\lambda_2)C_2.$$

Suppose that $f(x) = e^x$. *Then*

$$e^{\mathbf{A}} = e^2 C_1 + e^{-2} C_2 = e^2 \begin{pmatrix} \frac{3}{4} & \frac{3}{4} \\ \frac{1}{4} & \frac{1}{4} \end{pmatrix} + e^{-2} \begin{pmatrix} \frac{1}{4} & -\frac{3}{4} \\ -\frac{1}{4} & \frac{3}{4} \end{pmatrix}$$

$$= \begin{pmatrix} 5.5756 & 5.4403 \\ 1.8134 & 1.9488 \end{pmatrix}.$$

Another function is $f(x) = \sin(x)$. *We have*

$$\sin(\mathbf{A}) = \sin(2)C_1 + \sin(-2)C_2$$

$$= \sin(2) \begin{pmatrix} \frac{3}{4} & \frac{3}{4} \\ \frac{1}{4} & \frac{1}{4} \end{pmatrix} + \sin(-2) \begin{pmatrix} \frac{1}{4} & -\frac{3}{4} \\ -\frac{1}{4} & \frac{3}{4} \end{pmatrix}$$

$$= \begin{pmatrix} 0.4546 & 1.3639 \\ 0.4546 & -0.4546 \end{pmatrix}.$$

For case with eigenvalues of multiplicity greater than 1, we need more complicated formula. Suppose that λ_1, λ_2, and λ_3 have the same value. Then instead
of

$$f(\mathbf{A}) = f(\lambda_1)C_1 + f(\lambda_2)C_2 + f(\lambda_3)C_3 + \cdots$$

we have to use

$$f(\mathbf{A}) = f(\lambda_1)C_1 + f'(\lambda_1)D_2 + f''(\lambda_1)E_3 + \cdots.$$

That is the higher-order derivatives will be used for missing terms for C_i,
which are replaced with D_2 and E_3 etc.. That is $f(\lambda_2)C_2 = f'(\lambda_1)D_2$ and
$f(\lambda_3)C_3 = f''(\lambda_1)E_3$. The same rule applies for any algebraic multiplicities
of eigenvalues.

To determine the constituent matrices, we again use $f(x) = x^i, i = 0, 1, 2...$ to have

$$i = 0 \quad f(\mathbf{A}) = \sum_{i=1}^{n} f(\lambda_i)C_i = \mathbf{A}^0 \Rightarrow C_1 + C_4 + \cdots + C_n = \mathbf{I}$$

$$i = 1 \quad f(\mathbf{A}) = \sum_{i=1}^{n} f(\lambda_i)C_i = \mathbf{A}^1 \Rightarrow \lambda_1 C_1 + D_2 + \lambda_4 C_4 + \cdots + \lambda_n C_n = \mathbf{A}$$

$$i = 2 \quad f(\mathbf{A}) = \sum_{i=1}^{n} f(\lambda_i)C_i = \mathbf{A}^2 \Rightarrow \lambda_1^2 C_1 + 2\lambda_2 D_2 + 2E_3 + \lambda_4^2 C_4$$
$$+ \cdots + \lambda_n^2 C_n = \mathbf{A}^2$$

$$\vdots \quad = \quad \vdots$$

$$i = n-1 \quad f(\mathbf{A}) = \sum_{i=1}^{n} f(\lambda_i)C_i = \mathbf{A}^{n-1} \Rightarrow \lambda_1^{n-1} C_1 + (n-1)\lambda_2^{n-2} D_2$$
$$+ (n-1)(n-2)\lambda_3^{n-3} E_3 + \lambda_4^{n-1} C_4 + \cdots + \lambda_n^{n-1} C_n = \mathbf{A}^{n-1}.$$

Example J.4 *Find the constituent matrices for* $\mathbf{A} = \begin{pmatrix} 4 & 7 & 2 \\ -2 & -2 & 0 \\ 1 & 5 & 4 \end{pmatrix}$ *has the eigenvalue 2 with algebraic multiplicity 3. We have*

$$C = \mathbf{I}$$
$$2C + D = \mathbf{A}$$
$$4C + 4D + 2E = \mathbf{A}^2,$$

which has the solution

$$C = \mathbf{I}$$

$$D = \mathbf{A} - 2C = \mathbf{A} - 2\mathbf{I} = \begin{pmatrix} 2 & 7 & 2 \\ -2 & -4 & 0 \\ 1 & 5 & 2 \end{pmatrix}$$

$$E = \frac{1}{2}\mathbf{A}^2 - 2D - 2C = \frac{1}{2}\begin{pmatrix} 4 & 7 & 2 \\ -2 & -2 & 0 \\ 1 & 5 & 4 \end{pmatrix}^2 - 2\begin{pmatrix} 3 & 7 & 2 \\ -2 & -3 & 0 \\ 1 & 5 & 3 \end{pmatrix}$$

$$= \begin{pmatrix} -4 & -2 & 4 \\ 2 & 1 & -2 \\ -3 & -\frac{3}{2} & 3 \end{pmatrix}.$$

Consider a simple function $f(x) = \frac{1}{x}$. *Then* $f'(x) = -\frac{1}{x^2}$ *and* $f''(x) = \frac{2}{x^3}$. *For* $\lambda_1 = 2$, *we have* $f(\lambda_1) = f(2) = 1/2, f'(\lambda_1) = f'(2) = -1/2^2 = -1/4$ *and*

$f''(\lambda_1) = f''(2) = 2/2^3 = 1/4.$ *Thus,*

$$f(\mathbf{A}) = \mathbf{A}^{-1} = f(\lambda_1)C + f'(\lambda_1)D + f''(\lambda_1)E$$

$$= \frac{1}{2}\begin{pmatrix} 1 & 0 & 0 \\ 0 & 1 & 0 \\ 0 & 0 & 1 \end{pmatrix} - \frac{1}{4}\begin{pmatrix} 2 & 7 & 2 \\ -2 & -4 & 0 \\ 1 & 5 & 2 \end{pmatrix} + \frac{1}{4}\begin{pmatrix} -4 & -2 & 4 \\ 2 & 1 & -2 \\ -3 & -\frac{3}{2} & 3 \end{pmatrix}$$

$$= \begin{pmatrix} -1 & -\frac{9}{4} & \frac{1}{2} \\ 1 & \frac{7}{4} & -\frac{1}{2} \\ -1 & -\frac{13}{8} & \frac{3}{4} \end{pmatrix}.$$

Corollary J.1 $\lim_{n\to\infty} \mathbf{A}^n = \mathbf{0} \Leftrightarrow |\lambda_i| < 1,$ *for all eigenvalues* λ_i *of* $\mathbf{A}$.

Now we can demonstrate how to find the limit for a power convergent matrix.

Example J.5 *Consider* $\mathbf{A} = \frac{1}{4}\mathbf{B}$ *and* $\mathbf{B} = \begin{pmatrix} 3 & 0 & 0 & 1 \\ 1 & 3 & -1 & -1 \\ 1 & -1 & 3 & -1 \\ 1 & 0 & 0 & 3 \end{pmatrix}.$ *Find*

$\lim_{n\to\infty}\mathbf{A}^n.$

It follows from the characteristic polynomial $|\mathbf{A} - \lambda\mathbf{I}| = 0$, *which reduces to* $(t-1)^2(t-\frac{1}{2})^2 = 0$, *that the eigenvalues are* $\lambda_1 = 1$ *and* $\lambda_2 = \frac{1}{2}$. *It is easy to check that the rank of* $\mathbf{A} - \lambda_1\mathbf{I}$ *is 2. Thus the geometric multiplicity* $\nu_{\mathbf{A}}(1) = 4 - 2 = 2$ *is equal to the algebraic multiplicity* $m_{\mathbf{A}}(1) = 2$. *Thus, 1 is a regular eigenvalue. Further, 1 is also dominant due to* $|\lambda_1| = 1 > |\lambda_2| = \frac{1}{2}$. *Thus* $\mathbf{A}$ *is power convergent and* $\lim_{n\to\infty}\mathbf{A}^n = C_{10}$ *and* $C_{11} = 0$. *To find* C_{10}, *we apply the spectral decomposition as follows:*

$$f(\mathbf{A}) = f(1)C_{10} + f'(1)C_{11} + f(\frac{1}{2})C_{20} + f'(\frac{1}{2})C_{21}$$

for every $f \in \mathbb{C}[t]$. *Choose* $f(t) = (2t-1)^2$ *to make* $f(\frac{1}{2}) = f'(\frac{1}{2}) = 0$. *Then the equation above can be written as*

$$(2\mathbf{A} - \mathbf{I})^2 = (2\cdot 1 - 1)^2 C_{10} + 2(2\cdot 1 - 1)\cdot 0 + 0C_{20} + 0C_{21}$$

$$\Rightarrow \quad C_{10} = (2\mathbf{A} - \mathbf{I})^2.$$

Hence, we have

$$\lim_{n\to\infty}\mathbf{A}^n = C_{10} = (2\mathbf{A} - \mathbf{I})^2 = \frac{1}{4}(\mathbf{B} - 2\mathbf{I})^2$$

$$= \frac{1}{4}\begin{pmatrix} 1 & 0 & 0 & 1 \\ 1 & 1 & -1 & -1 \\ 1 & -1 & 1 & -1 \\ 1 & 0 & 0 & 1 \end{pmatrix}^2 = \frac{1}{4}\begin{pmatrix} 2 & 0 & 0 & 2 \\ 0 & 2 & -2 & 0 \\ 0 & -2 & 2 & 0 \\ 2 & 0 & 0 & 2 \end{pmatrix}.$$

Definition J.7 *An eigenvalue λ_i of* **A** *is called simple if its algebraic multiplicity is 1.*

Clearly, each simple eigenvalue is regular as the geometric multiplicity is no larger than the algebraic multiplicity.

Theorem J.9 *If* **A** *is power convergent and 1 is a simple eigenvalue of* **A**, *then*

$$\lim_{n\to\infty} \mathbf{A}^n = C_{10} = \frac{1}{\mathbf{u}^T \cdot \mathbf{v}} \mathbf{u} \cdot \mathbf{v}^T,$$

where $\mathbf{u} \in E_{\mathbf{A}}(1)$ *is any non-zero 1-eigenvector of* **A**, *and* $\mathbf{v} \in E_{\mathbf{A}^T}(1)$ *is any non-zero 1-eigenvector* $\mathbf{A}^T$ *with* $E_{\mathbf{A}}(\lambda)$ *as the eigenspace of eigenvalue* λ.

Definition J.8 *For a square matrix* **A**, *its special radius is the real number defined by*

$$\rho(\mathbf{A}) = \max\{|\lambda| : \lambda \in \mathbb{C} \text{ is an eigenvalue of } \mathbf{A}\}.$$

Example J.6 *Find the special radius of* $\mathbf{A} = Diag(1, i, 1+i) = \begin{pmatrix} 1 & 0 & 0 \\ 0 & i & 0 \\ 0 & 0 & 1+i \end{pmatrix}$.

The spectrum of $\mathbf{A} = \{1, i, 1+i\}$. *Thus,*

$$\rho(\mathbf{A}) = \max\{|1|, |i|, |1+i|\} = max(1, 1, \sqrt{2}) = \sqrt{2}.$$

Now we can state some useful theorems by using the spectral radius.

Theorem J.10 **A** *is power convergent if and only if either* $\rho(\mathbf{A}) < 1$ *or* $\rho(\mathbf{A}) = 1$ *and 1 is a dominant, regular eigenvalue of* **A**.

Theorem J.11 $\lim_{n\to\infty} \mathbf{A}^n = 0 \Leftrightarrow \rho(\mathbf{A}) < 1$.

Note that the spectral radius can be estimated without knowing the eigenvalues explicitly. One approach is to relate the spectral radius $\rho(\mathbf{A})$ of **A** to the number

$$\|\mathbf{A}\| = \max_i \left(\sum_{j=1}^{m} |a_{ij}| \right),$$

which is called the *norm* of the $m \times m$ matrix $\mathbf{A} = [a_{ij}]$. The norm $\|\cdot\|$ satisfies the following relations:

1. $\|\mathbf{A}\| \geq 0, \|\mathbf{A}\| = 0 \Leftrightarrow \mathbf{A} = \mathbf{0}$,
2. $\|\mathbf{A} + \mathbf{B}\| \leq \|\mathbf{A}\| + \|\mathbf{B}\|$,
3. $\|c\mathbf{A}\| = |c|\|\mathbf{A}\|$, if $c \in \mathbb{C}$,

4. $\|\mathbf{A} \cdot \mathbf{B}\| \le \|\mathbf{A}\| \cdot \|\mathbf{B}\|$.

Theorem J.12 *If $\mathbf{A}$ is an $m \times m$ matrix, then*

$$\rho(\mathbf{A}) \le \min(\|\mathbf{A}\|, \|\mathbf{A}^T\|).$$

We show how to use this theorem to check if a matrix is power convergent by the following example.

Example J.7 *Estimate the spectral radius of* $\mathbf{A} = \begin{pmatrix} 0 & \frac{1}{3} & \frac{1}{4} \\ \frac{1}{2} & 0 & \frac{1}{4} \\ \frac{1}{2} & \frac{1}{3} & 0 \end{pmatrix}$. *The sums of the absolute values of the rows are* $\frac{7}{12}, \frac{3}{4}, \frac{5}{6}$, *respectively, so* $\|\mathbf{A}\| = \frac{5}{6}$ *and the sums of the absolute values of the columns are* $1, \frac{2}{3}, \frac{1}{2}$, *respectively, so* $\|\mathbf{A}^T\| = 1$. *Thus, we have*

$$\rho(\mathbf{A}) \le \min(\frac{5}{6}, 1) = \frac{5}{6},$$

which implies that $\mathbf{A}$ is power convergent with limit $\mathbf{0}$.

We can check how good this estimate (an upper bound) is for the spectral radius. We can determine the exact spectral radius by solving the characteristic polynomial of $\mathbf{A}$

$$|\mathbf{A} - t\mathbf{I}| = t^3 - \frac{3}{8}t - \frac{1}{12} = 0.$$

The three roots are all real and irrational numbers

$$-0.420505805, -0.282066739, 0.7025747892.$$

Thus, the spectral radius is $\rho(\mathbf{A}) = 0.7025747892$ which is below and not too far away from the bound 5/6=0.83333....

To state the next theorem, called Gershgorin's theorem, we need the definition of *Gershgorin's discs*.

Definition J.9 *Suppose $\mathbf{A} = [a_{ij}] \in M_m(\mathbb{C})$ is an $m \times m$ matrix. Then for each k with $1 \le k \le m$, let*

$$r_k = \sum_{j=1, j\ne k}^{m} |a_{kj}|,$$

$$D_k = D_k(\mathbf{A}) = \{z \in \mathbb{C} : |z - a_{kk}| \le r_k\}.$$

Thus, D_k is the disc of radius r_k centered at the (diagonal element) $a_{kk} \in \mathbb{C}$ and is called Gershgorin disk.

Note that r_k is the sum of the absolute values of the entries of the k-th row excluding the diagonal element.

Theorem J.13 *(Gersgorin, 1931). Each eigenvalue λ of $\mathbf{A}$ lies in at least one Gersgorin disc $D_k(\mathbf{A})$ of $\mathbf{A}$.*

A standard to check whether a matrix is invertible is stated below.

Corollary J.2 *If the diagonal entries of the matrix $\mathbf{A} = [a_{ij}]$ are much larger than their off-diagonal entries in the sense that*

$$|a_{kk}| > \sum_{j=1, j\neq k}^{m} |a_{kj}| \quad for \ k = 1, ..., m,$$

then $\mathbf{A}$ is invertible.

For real matrices, there are other methods of estimating the spectral radius. One of these is based on the trace of the matrix.

Definition J.10 *Let $\mathbf{A} = [a_{ij}]$ be an $m \times m$ matrix, then its trace is defined as the sum of its diagonal elements:*

$$tr(\mathbf{A}) = a_{11} + a_{22} + \cdots + a_{mm}.$$

Some properties of the trace are as follows: Let $\mathbf{A}, \mathbf{B}$, and $\mathbf{P}$ be $m \times m$ matrices and $\mathbf{P}$ is invertible. Then

1. $tr(\mathbf{AB}) = tr(\mathbf{BA})$;
2. $tr(\mathbf{P}^{-1}\mathbf{AP}) = tr(\mathbf{A})$;
3. $tr(A) = m_{\mathbf{A}}(\lambda_1)\lambda_1 + \cdots + m_{\mathbf{A}}(\lambda_s)\lambda_s$.

Now we give the proof of Theorem J.12 which has been applied to analyzing Markov chain with absorbing states and structure Markov chains (Chapter 4 of this book).

Proof. (The proof of Theorem J.12) We first show

$$\rho(\mathbf{A}) \le \|\mathbf{A}\| := \max_{k}(\sum_{j=1}^{m} |a_{kj}|). \tag{J.2}$$

For this, let λ be an eigenvalue of $\mathbf{A}$ of maximal absolute value, i.e., $|\lambda| = \rho(\mathbf{A})$. According to Gersgorin's theorem, $\lambda \in D_k(\mathbf{A})$ for some k. That is

$$|\lambda - a_{kk}| \le \sum_{j=1, j\neq k}^{m} |a_{kj}|, \quad or$$

$$|\lambda - a_{kk}| + |a_{kk}| \le \sum_{j=1}^{m} |a_{kj}| \le \|\mathbf{A}\|.$$

Also note that since $\rho(\mathbf{A}) = |\lambda| = |(\lambda - a_{kk}) + a_{kk}| \le |\lambda - a_{kk}| + |a_{kk}|$ by the triangle inequality, we have

$$\rho(\mathbf{A}) \le |\lambda - a_{kk}| + |a_{kk}| \le \|\mathbf{A}\|,$$

which proves the claim (J.2).

Next, we apply (J.2) to $\mathbf{A}^T$ to get $\rho(\mathbf{A}^T) \leq \|\mathbf{A}\|$. However, since $\mathbf{A}$ and $\mathbf{A}^T$ have the same eigenvalues, $\rho(\mathbf{A}) = \rho(\mathbf{A}^T) \leq \|\mathbf{A}^T\|$. Thus we have $\rho(\mathbf{A}) \leq \|\rho(\mathbf{A}\|$ and $\rho(\mathbf{A}) \leq \|\rho(\mathbf{A})^T\|$. Thus, it follow that $\rho(\mathbf{A}) \leq \|\mathbf{A}\|$ and $\rho(\mathbf{A}) \leq \|\mathbf{A}^T\|$ so that the theorem is proved. ∎

Theorem J.14 *If $\mathbf{A}$ is a real $m \times m$ matrix, then $\rho(\mathbf{A}) \leq \sqrt{\rho(\mathbf{A}^T\mathbf{A})} \leq \sqrt{tr(\mathbf{A}^T\mathbf{A})}$.*

J.2.2 GEOMETRIC SERIES OF MATRICES

Now we are interested in computing the geometric series of a matrix $\mathbf{T}$. That is

$$S_n = \mathbf{I} + \mathbf{T} + \cdots + \mathbf{T}^{n-1}.$$

Recall that the formula for a geometric series in scalar case is given by

$$S_n = 1 + a + a^2 + \cdots + a^{n-1} = \frac{1-a^n}{1-a}$$

and thus, if $|a| < 1$, the infinite series is as follows:

$$S = \sum_{n=0}^{\infty} a^n := \lim_{n \to \infty}(1 + a + \cdots + a^n) = \frac{1}{1-a}.$$

In fact, similar results exist in the matrix case.

Definition J.11 *Suppose $\mathbf{T}$ is a square matrix. Then the sequence $\{S_n, n \geq 0\}$, called the geometric series generated by $\mathbf{T}$ is defined by*

$$S_n = \mathbf{I} + \mathbf{T} + \cdots + \mathbf{T}^{n-1}, \quad S_0 = \mathbf{T}^0 = \mathbf{I}.$$

The series converges if the sequence $\{S_n, n \geq 0\}$ converges with

$$S = \lim_{n \to \infty} S_n = \sum_{n=0}^{\infty} \mathbf{T}^n.$$

Theorem J.15 *The geometric series generated by $\mathbf{T}$ converges if and only if*

$$\rho(\mathbf{T}) < 1; \quad i.e., \; |\lambda_i| < 1, \; \text{for each eigenvalue } \lambda_i \text{ of } \mathbf{T}. \tag{J.3}$$

If this condition holds, then $\mathbf{I} - \mathbf{T}$ is invertible and we have

$$S_n := \sum_{k=0}^{n-1} \mathbf{T}^k = (\mathbf{I} - \mathbf{T})^{-1}(\mathbf{I} - \mathbf{T}^n), \tag{J.4}$$

and hence the series converges to

$$\sum_{k=0}^{\infty} \mathbf{T}^k = (\mathbf{I} - \mathbf{T})^{-1}. \tag{J.5}$$

Proof. ($\Leftarrow$) First, we assume that the sequence $\{S_n, n \geq 0\}$ converges and let $S = \lim_{n \to \infty} S_n$. Note that

$$S_n - S_{n-1} = (T^n + T^{n-1} + \cdots + I) - (T^{n-1} + T^{n-2} + \cdots + I) = T^n.$$

Then, it follows that $\lim_{n \to \infty} T^n = \lim_{n \to \infty}(S_n - S_{n-1}) = \lim_{n \to \infty} S_n - \lim_{n \to \infty} S_n = S - S = 0$. Thus, T is power convergent with limit 0 and this result implies (J.3) holds. ($\Rightarrow$) Now suppose that (J.3) holds. Then the eigenvalues of $I - T$ are $1 - \lambda_i \neq 0$, so $I - T$ is invertible. Compute

$$\begin{aligned} T \cdot S_n &= T(I + T + \cdots + T^{n-1}) \\ &= T + T^2 + \cdots + T^n \\ &= (I + T + T^2 + \cdots + T^{n-1}) + T^n - I \\ &= S_n + T^n - I, \end{aligned}$$

which can be written as

$$T \cdot S_n - S_n = T^n - I \Rightarrow (I - T)S_n = I - T^n.$$

Since $(I - T)$ is invertible, pre-multiplying both sides of the last equation above by $(I - T)^{-1}$ yields (J.4). Finally, it follows from $\lim_{n \to \infty} T^n = 0$ that

$$S = \lim_{n \to \infty} S_n = (I - T)^{-1}(I - \lim_{n \to \infty} T^n) = (I - T)^{-1},$$

which proves (J.5). $\blacksquare$

J.3 STOCHASTIC MATRICES AND MARKOV CHAINS

In this section, we link the linear algebra to the theory of Markov chains by focusing on a DTMC. Since the probability distribution function can be written as a recursive relation $\pi(k) = \pi(k-1)P(k)$, the DTMC can be considered as a discrete linear system. Here the matrix P is a stochastic matrix (i.e., the sum of each rows equals 1). Stochastic matrices have some properties worth noting.

Theorem J.16 *Suppose that A is a stochastic matrix of size $p \times q$ and B is a stochastic matrix of size $q \times r$. Then the matrix AB is a stochastic matrix of size $p \times r$.*

The eigenvalues of a square stochastic matrix play an important role in analyzing DTMCs.

Theorem J.17 *For a square stochastic matrix P of size $m \times m$, we have*

1. 1 is a regular eigenvalue of P.

2. $|\lambda_i| \leq 1$ *for $i = 1, ..., m$. That is $\rho(\mathbf{P}) = 1$.*

Proof. It is easy to show that 1 is an eigenvalue from the property for the stochastic matrix. To prove that 1 is a regular eigenvalue of $\mathbf{P}$, we use the norm of a matrix. It follows again from the stochastic matrix property that

$$\|\mathbf{P}\| = \max(1, 1, ..., 1) = 1.$$

Note that $\|\mathbf{P}^n\| = 1$ for any $n \geq 1$ since $\mathbf{P}^n$ is also stochastic. Thus, it follows from the property of the matrix norm that for any invertible matrix $\mathbf{A}$

$$\|\mathbf{A}^{-1}\mathbf{P}^n\mathbf{A}\| \leq \|\mathbf{A}^{-1}\| \cdot \|\mathbf{A}\| = c_{\mathbf{A}}, \quad \text{for all } n \geq 1. \tag{J.6}$$

We take $\mathbf{A}$ such that $\mathbf{A}^{-1}\mathbf{P}\mathbf{A} = \mathbf{J}$ is a Jordan matrix. It follows from (J.6) that $\|\mathbf{J}^n\| \leq c_{\mathbf{A}}$, for all $n \leq 1$. This implies that $\mathbf{J}$ cannot cotain a Jordan block $J(1, k)$ with $k \geq 2$. Otherwise, we have

$$J(1,k)^n = \begin{pmatrix} 1 & n & * & * & \cdots \\ 0 & 1 & n & * & \cdots \\ \vdots & \vdots & \vdots & \vdots & \cdots \\ 0 & \cdots & \cdots & \cdots & 1 \end{pmatrix},$$

where $*$'s are positive numbers, and hence $\|\mathbf{J}^n\| \geq \|J(1,k)^n\| \geq 1 + n$, which contradicts (J.6). Therefore, 1 is a regular eigenvalue of $\mathbf{P}$. Part (2) follows from the definition of spectral radius and the theorem (J.12). ∎

J.3.1   POSITIVE SEMI-DEFINITE MATRICES

In this section, we review positive semi-definite matrices and their applications to stochastics.

Definition J.12 *Let $\mathbf{A} = [a_{ij}]$ be a real symmetric $n \times n$ matrix. Then $\mathbf{A}$ is said to be positive semi-definite (positive definite) if $\mathbf{x}^T\mathbf{A}\mathbf{x} \geq 0(> 0)$ for all $\mathbf{x} \in \mathbb{R}^n$ (and $\mathbf{x} \neq \mathbf{0}$).*

Note that the function

$$\mathbf{x}^T\mathbf{A}\mathbf{x} = \sum_{i=1}^{n}\sum_{j=1}^{n} a_{ij}x_ix_j = \sum_{i=1}^{n} a_{ii}x_i^2 + 2\sum_{i>j} a_{ij}x_ix_j.$$

is called a quadratic form. If the inequality signs reverse, we can obtain the definition of negative semi-definite (negative definite) matrix. Clearly if a symmetric $\mathbf{A}$ is positive semi-definite, then $-\mathbf{A}$ will be negative semi-definite. If a matrix is neither positive semi-definite nor negative semi-definite, it is called indefinite. Some properties for positive semi-definite matrix are as follows:

(1) For a positive semi-definite matrix $\mathbf{A}$, we have $\mathbf{Ax} = \mathbf{0} \Leftrightarrow \mathbf{x}^T \mathbf{Ax} = 0$.

(2) If a symmetric matrix $\mathbf{A}$ can be expressed in the product form $\mathbf{A} = \mathbf{BB}^T$, then it is positive semi-definite. This is because

$$\mathbf{x}^T \mathbf{Ax} = \mathbf{x}^T \mathbf{BB}^T \mathbf{x} = (\mathbf{B}^T \mathbf{x})^T (\mathbf{B}^T \mathbf{x}) = \|\mathbf{B}^T \mathbf{x}\|^2 \geq 0.$$

Furthermore, it can be shown that $rank(\mathbf{A}) = rank(\mathbf{B})$

An example of positive semi-definite matrices is to express the variance and covariance in multiple random variable cases. Let $\mathbf{X} = (X_1, X_2, ..., X_n)$ be a row vector of n random variables, with

$$\mu_i = E[X_i], \quad \sigma_i = \sqrt{E[(X_i - \mu_i)^2]}, \quad \sigma_{ij} = E[(X_i - \mu_i)(X_j - \mu_j)] \text{ for } i \neq j$$

for $i = 1, 2, ..., n$ as the mean, standard deviation, and covariance, respectively. Note that if $\sigma_{ij} = 0$, then variables X_i and X_j are uncorrelated and if $\sigma_{ij} \neq 0$ the correlation is defined as $\rho_{ij} = \sigma_{ij}/(\sigma_i \sigma_j)$ for $i \neq j$. The second moment matrix is the symmetric $n \times n$ matrix with (i, j) entry $E[X_i X_j]$ is given by

$$\mathbf{S} = \begin{pmatrix} E[X_1^2] & E[X_1 X_2] & \cdots & E[X_1 X_n] \\ E[X_2 X_1] & E[X_2^2] & \cdots & E[X_2 X_n] \\ \vdots & \vdots & \ddots & \vdots \\ E[X_n X_1] & E[X_n X_2] & \cdots & E[X_n^2] \end{pmatrix} = E[\mathbf{XX}^T].$$

Note that this matrix is symmetric and positive semi-definite since for all $\mathbf{a}$ it follows

$$\mathbf{a}^T \mathbf{Sa} = \mathbf{a}^T E[\mathbf{XX}^T]\mathbf{a} = E[\mathbf{a}^T \mathbf{XX}^T \mathbf{a}] = E[(\mathbf{a}^T \mathbf{X})(\mathbf{a}^T \mathbf{X})^T] = E[\mathbf{a}^T \mathbf{X}]^2 \geq 0.$$

The last term of the equation above is the expected value of the square of the scalar random variable $Y = \mathbf{a}^T \mathbf{X}$. The covariance matrix is the symmetric $n \times n$ matrix given by

$$\Sigma = \begin{pmatrix} \sigma_1^2 & \sigma_{12} & \cdots & \sigma_{1n} \\ \sigma_{21} & \sigma_2^2 & \cdots & \sigma_{2n} \\ \vdots & \vdots & \ddots & \vdots \\ \sigma_{n1} & \sigma_{n2} & \cdots & \sigma_n^2 \end{pmatrix} = E\left[\begin{pmatrix} X_1 - \mu_1 \\ X_2 - \mu_2 \\ \vdots \\ X_n - \mu_n \end{pmatrix} \begin{pmatrix} X_1 - \mu_1 \\ X_2 - \mu_2 \\ \vdots \\ X_n - \mu_n \end{pmatrix}^T\right]$$

$$= E[(\mathbf{X} - \mu)(\mathbf{X} - \mu)^T],$$

where μ is the vector of means

$$\mu = (\mu_1, \mu_2, ..., \mu_n) = (E[X_1], E[X_2], ..., E[X_n]).$$

Clearly, the covariance matrix is positive semi-definite as for all $\mathbf{a}$

$$\mathbf{a}^T \Sigma \mathbf{a} = \mathbf{a}^T E[(\mathbf{X} - \mu)(\mathbf{X} - \mu)^T]\mathbf{a} = E[\mathbf{a}^T (\mathbf{X} - \mu)]^2 \geq 0.$$

The correlation matrix has (i, j) entry $\rho_{ij} = \sigma_{ij}/(\sigma_i \sigma_j)$ for $i \neq j$ and 1 for $i = j$. That is

$$\mathbf{C} = \begin{pmatrix} 1 & \rho_{12} & \cdots & \rho_{1n} \\ \rho_{21} & 1 & \cdots & \rho_{2n} \\ \vdots & \vdots & \ddots & \vdots \\ \rho_{n1} & \rho_{n2} & \cdots & 1 \end{pmatrix}.$$

The relation between the covariance matrix and the correlation matrix is $\mathbf{C} = \mathbf{D}\Sigma\mathbf{D}$ where $\mathbf{D} = diag(\sigma_1^{-1}, \sigma_2^{-1}, ..., \sigma_n^{-1})$ is a diagonal matrix. This relation indicates that $\mathbf{C}$ is positive semi-definite. Also $\mathbf{C}$ is the covariance matrix of the standardized random variables $U_i = (X_i - \mu_i)/\sigma_i$.

Next, we discuss the linear combinations of random vectors. Some mean and covariance rules are summarized as follows: Suppose $\mathbf{X}$ is a random vector with mean vector μ and covariance matrix Σ, then

1. Let α be a scalar. Then random vector $\mathbf{Y} = \alpha\mathbf{X}$ has the mean and covariance matrix

$$E[\mathbf{Y}] = \alpha\mu, \quad E[(\mathbf{Y} - E[\mathbf{Y}])(\mathbf{Y} - E[\mathbf{Y}])^T] = \alpha^2\Sigma.$$

2. Let $\mathbf{X}, \mathbf{Y}$ be two random n-vectors with mean $\mu_{\mathbf{X}}, \mu_{\mathbf{Y}}$, and covariance $\Sigma_{\mathbf{X}}, \Sigma_{\mathbf{Y}}$. Then, the random vector $\mathbf{Z} = \mathbf{X} + \mathbf{Y}$ has mean $E[\mathbf{Z}] = \mu_{\mathbf{X}} + \mu_{\mathbf{Y}}$. Further, if $\mathbf{X}$ and $\mathbf{Y}$ aee uncorrelated. That is

$$E[(X_i - \mu_{xi})(Y_j - \mu_{yj})] = 0, \quad i, j = 1, ..., n,$$

then the covariance matrix of $\mathbf{Z}$ is

$$E[(\mathbf{Z} - E[\mathbf{Z}])(\mathbf{Z} - E[\mathbf{Z}])^T] = \Sigma_{\mathbf{X}} + \Sigma_{\mathbf{Y}}.$$

Now we define the **affine transformation**. Suppose that $\mathbf{Y}$ is a random n-vector with mean $\mu_{\mathbf{Y}}$ and covariance matrix $\Sigma_{\mathbf{Z}}$. Then the affine transformation of $\mathbf{Y}$ is given by

$$\mathbf{X} = \mathbf{A}\mathbf{Y} + \mathbf{B},$$

where $\mathbf{A}$ is an $m \times n$ matrix and $\mathbf{b}$ is an m-vector. It is easy to check that the mean vector and covariance matrix of $\mathbf{X}$ are given by

$$E[\mathbf{X}] = E[\mathbf{A}\mathbf{Y} + \mathbf{b}] = \mathbf{A}\mu_{\mathbf{Y}} + \mathbf{b},$$

and

$$\Sigma = E[(\mathbf{X} - E[\mathbf{X}])(\mathbf{X} - E[\mathbf{X}])^T] = E[(\mathbf{A}\mathbf{Y} - \mathbf{A}\mu_{\mathbf{Y}})(\mathbf{A}\mathbf{Y} - \mathbf{A}\mu_{\mathbf{Y}})^T]$$
$$= \mathbf{A}E[(\mathbf{Y} - \mu_{\mathbf{Y}})(\mathbf{Y} - \mu_{\mathbf{Y}})^T]\mathbf{A}^T = \mathbf{A}\Sigma_{\mathbf{Y}}\mathbf{A}^T.$$

Example J.8 *(Factor Model) Suppose a random n-vector* $\mathbf{X}$ *has covariance matrix*

$$\Sigma = \mathbf{AA}^T + \sigma^2 \mathbf{I}$$

$\mathbf{X}$ *can be interpreted as being generated by a model* $\mathbf{X} = \mu_{\mathbf{X}} + \mathbf{AY} + \mathbf{w}$, *where* Y *is a random vector with mean* $\mathbf{0}$ *and covariance matrix* $\mathbf{I}$ *and* $\mathbf{w}$ *is a random error vector, which is uncorrelated with* $\mathbf{Y}$ *and has* $E[\mathbf{w}] = \mathbf{0}$ *and* $E[\mathbf{ww}^T] = \sigma^2 \mathbf{I}$. *In multi-variate statistics, this is called a factor model where the components of* Y *are common factors in* $\mathbf{X}$. *Note that* $\mathbf{X} - \mu_{\mathbf{X}}$ *is a vector* $\mathbf{AY}$ *in a subspace* range$(\mathbf{A})$ *plus error* $\mathbf{w}$.

Here we need the definition of the range of matrix $\mathbf{A}$. Let $\mathbf{A}$ be an $m \times n$ matrix with real entries. Then there are two important subspaces associated to $\mathbf{A}$. One is a subspace $\mathbb{R}^m$, called the range of $\mathbf{A}$, and the other is a subspace of $\mathbb{R}^n$, called the null space of $\mathbf{A}$.

Definition J.13 *The range of* $\mathbf{A}$ *is a subspace of* $\mathbb{R}^m$, *denoted by* $\mathbb{R}(\mathbf{A})$, *defined by*

$$\mathbb{R}(\mathbf{A}) = \{\mathbf{Y}: \text{ there exists at least one } \mathbf{X} \text{ in } \mathbb{R}^n \text{ such that } \mathbf{AX} = \mathbf{Y}\}.$$

Definition J.14 *The null space of* $\mathbf{A}$ *is a subspace of* $\mathbb{R}^n$, *denoted by* $\mathcal{N}(\mathbf{A})$, *defined by*

$$\mathcal{N}(\mathbf{A}) = \{\mathbf{X}: \mathbf{AX} = \mathbf{0}_m\}.$$

We can use the matrix notations to express the estimates of the mean and co-variance of a random vector based on a data set. Consider m observations (sampled values) for each variable of a random n-vector $\mathbf{X} = (\mathbf{X}_1, \mathbf{X}_2, ..., \mathbf{X}_n)$. Then we can use an $m \times n$ data matrix $\mathbf{x}$ as follows:

$$\mathbf{x} = \begin{pmatrix} x_{11} & x_{12} & \cdots & x_{1n} \\ x_{21} & x_{22} & \cdots & x_{2n} \\ \vdots & \vdots & \ddots & \vdots \\ x_{m1} & x_{m2} & \cdots & x_{mn} \end{pmatrix} = \begin{pmatrix} \mathbf{x_1}^T \\ \mathbf{x_2}^T \\ \vdots \\ \mathbf{x_m}^T \end{pmatrix},$$

where $\mathbf{x}_i^T$ is the row vector (i.e., $\mathbf{x}_i$ is a column vector) for the ith observation (n values for n variables). The sample estimate for the mean $E[\mathbf{X}]$ is denoted by $\bar{\mathbf{x}}$. Based on the data set, we have

$$\bar{\mathbf{x}} = \frac{1}{m} \sum_{i=1}^{m} \mathbf{x}_i = \frac{1}{m} \mathbf{x}^T \mathbf{e}.$$

Subtracting $\bar{\mathbf{x}}^T$ from each row gives the centered data matrix

$$\mathbf{x}_c = \mathbf{x} - \mathbf{e}\bar{\mathbf{x}}^T = \left(\mathbf{I} - \frac{1}{m}\mathbf{ee}^T\right)\mathbf{x}.$$

Based on the data matrix, the estimate of the second moment of the random vector $E[\mathbf{X}\mathbf{X}^T]$ is given by

$$\frac{1}{m}\mathbf{x}^T\mathbf{x} = \frac{1}{m}\sum_{i=1}^{m}\mathbf{x}_i\mathbf{x}_i^T.$$

Then, the centered data matrix provides an estimate of the covariance matrix as follows:

$$\frac{1}{m}\mathbf{x_c}^T\mathbf{x_c} = \frac{1}{m}\sum_{i=1}^{m}(\mathbf{x}_i - \bar{\mathbf{x}})(\mathbf{x}_i - \bar{\mathbf{x}})^T$$

$$= \frac{1}{m}\mathbf{x}^T\left(\mathbf{I} - \frac{1}{m}\mathbf{e}\mathbf{e}^T\right)^2\mathbf{x}$$

$$= \frac{1}{m}\mathbf{x}^T\left(\mathbf{I} - \frac{1}{m}\mathbf{e}\mathbf{e}^T\right)\mathbf{x}$$

$$= \frac{1}{m}\mathbf{x}^T\mathbf{x} - \bar{\mathbf{x}}\bar{\mathbf{x}}^T.$$

J.4   A BRIEF LOOK AT M-MATRIX

Let $\mathbb{Z}_n$ denote the set of all $n \times n$ real matrices $\mathbf{A} = [a_{ij}]$ with $a_{ij} \leq 0$ for all $i \neq j, 1 \leq i, j \leq n$.

Definition J.15 *An $n \times n$ matrix $\mathbf{A}$ that can be expressed in the form $\mathbf{A} = s\mathbf{I} - \mathbf{B}$, with $b_{ij} \geq 0, 1 \leq i, j \leq n$, and $s \geq \rho(\mathbf{B})$, the spectral radius of $\mathbf{B}$, is called an M-matrix.*

It is easy to see that this class of M-matrices are non-singular from the Perron-Frobenius theorem by requiring $s > \rho(\mathbf{B})$. Note that if $\mathbf{A}$ is a non-singular M-matrix, then the diagonal elements a_{ii} of $\mathbf{A}$ must be positive. The applications of M-matrices include the establishment of convergence criteria for iterative methods for the solution of large sparse systems of linear equations (i.e., bounds on eigenvalues). The economists have studied M-matrices in connection with gross substitutability, stability of a general equilibrium and Leontief's input-output analysis in economic systems. This class of matrices have applications in stochastic modeling. For example, in Chapter 8, we utilize the oblique reflection mapping in modeling the general Jackson network in order to apply the diffusion process as the stochastic process limit. In this approach, the matrix $\mathbf{R}$ is an M-matrix, called a reflection matrix. No proofs are given in this review and the proofs of the various results can be found in the references (the original papers or textbooks). The following review is mainly taken from a reviw paper by Plemmons (1977). We first define the principle minors.

Definition J.16 *A principal sub-matrix of a square matrix* **A** *is the matrix obtained by deleting any k rows and the corresponding k columns. The determinant of a principal sub-matrix is called the principal minor of* **A**.

Furthermore, we can define the leading principal minors.

Definition J.17 *The leading principal sub-matrix of order k of an n × n matrix is obtained by deleting the last n − k rows and columns of the matrix. The determinant of a leading principal sub-matrix is called the leading principal minor of* **A**.

The main properties or conditions regarding the M-matrix can be grouped and their relations are identified. These groups are labeled by A to N. The properties or conditions in each group are equivalent. The relations among these groups are indicated by directed arrows (representing implications) as shown in the Figure for Theorem 1 in Plemmons (1977). Forty conditions are grouped into 14 categories and equivalent to the statement "**A** is a non-singular M-matrix". We state these conditions labeled by the group letter with the condition numner as the subscript under four general classes of properties. For example, A_3 represents the third condition in group A.

Positivity of Principal Minors:

A_1: All the principal minors of **A** are positive.

A_2: Every real eigenvalue of each principal sub-marix of **A** is positive.

A_3: $\mathbf{A} + \mathbf{D}$ is non-singular for each non-negative diagonal matrix **D**.

A_4: For each $\mathbf{x} \neq \mathbf{0}$ there exists a positive diagonal matrix **D** such that $\mathbf{x}^T \mathbf{A} \mathbf{D} \mathbf{x} > 0$.

A_5: For each $\mathbf{x} \neq \mathbf{0}$ there exists a non-negative diagonal matrix **D** such that $\mathbf{x}^T \mathbf{A} \mathbf{D} \mathbf{x} > 0$.

A_6: **A** does not reverse the sign of any vector. That is, if $\mathbf{x} \neq \mathbf{0}$, then for some subscript i $x_i(\mathbf{A}, \mathbf{x})_i > 0$.

A_7: For each signature matrix **S** there exists an $\mathbf{x} > \mathbf{0}$ such that $\mathbf{SASx} > \mathbf{0}$. Note that a signature matrix is a diagonal matrix whose diagonal elements are plus or minus 1.

B_8: The sum of all the $k \times k$ principal minors of **A** is positive for $k = 1, 2, ..., n$.

C_9: Every real eigenvalue of **A** is positive.

C_{10}: $\mathbf{A} + \alpha \mathbf{I}$ is non-singular for each scalar $\alpha \geq 0$.

D_{11}: All leading principal minors of **A** are positive.

D_{12}: There exist lower and upper triangular matrices **L** and **U** respectively, with positive diagonals, such that $\mathbf{A} = \mathbf{UL}$.

E_{13}: There exists a strictly increasing sequence of subsets $\emptyset \neq S_1 \subset \cdots \subset S_n = \{1, ...n\}$ such that the determinant of the principal sub-matrix of **A** formed by choosing row and columns indices from S_i is positive for $i = 1, ..., n$.

E_{14}: There exists a permutation matrix $\mathbf{P}$ and lower and upper triangular matrices $\mathbf{L}$ and $\mathbf{U}$ respectively, with positive diagonals, such that $\mathbf{PAP}' = \mathbf{LU}$.

Inverse-Positivity and Splittings

F_{15}: $\mathbf{A}$ is inverse-positive. That is, $\mathbf{A}^{-1}$ exists and $\mathbf{A}^{-1} > 0$.

F_{16}: $\mathbf{A}$ is monotone. That is, $\mathbf{Ax} \geq \mathbf{0} \Rightarrow \mathbf{x} \geq \mathbf{0}$ for all $\mathbf{x} \in \mathbb{R}^n$.

F_{17}: $\mathbf{A}$ has a convergent regular splitting. That is, $\mathbf{A}$ has a representation

$$\mathbf{A} = \mathbf{M} - \mathbf{N}, \quad \mathbf{M}^{-1} \geq 0, \ \mathbf{N} \geq 0$$

with $\mathbf{M}^{-1}\mathbf{N}$ convergent. That is, $\rho(\mathbf{M}^{-1}\mathbf{N}) < 1$.

F_{18}: $\mathbf{A}$ has a convergent weak regular splitting. That is, $\mathbf{A}$ has a representation

$$\mathbf{A} = \mathbf{M} - \mathbf{N}, \quad \mathbf{M}^{-1} \geq 0, \ \mathbf{M}^{-1}\mathbf{N} \geq 0$$

with $\mathbf{M}^{-1}\mathbf{N}$ convergent.

F_{19}: $\mathbf{A}$ has a weak regular splitting, and there exists $\mathbf{x} > 0$ with $\mathbf{Ax} > \mathbf{0}$.

F_{20}: There exist inverse-positive matrices $\mathbf{M}_l$ and $\mathbf{M}_2$, with $\mathbf{M}_1 < \mathbf{A} < \mathbf{M}_2$.

F_{21}: There exist an inverse-positive matrix $\mathbf{M}$, with $\mathbf{M} \geq \mathbf{A}$, and a nonsingular M-matrix $\mathbf{B}$ such that $\mathbf{A} = \mathbf{MB}$.

F_{22}: There exist an inverse-positive matrix $\mathbf{M}$ and a nonsingular M-matrix $\mathbf{B}$ such that $\mathbf{A} = \mathbf{MB}$.

G_{23}: Every weak regular splitting of $\mathbf{A}$ is convergent.

H_{24}: Every regular splitting of A is convergent.

Stability

I_{25}: There exists a positive diagonal matrix $\mathbf{D}$ such that $\mathbf{AD} + \mathbf{DA}^T$ is positive definite.

I_{26}: $\mathbf{A}$ is diagonally similar to a matrix whose symmetric part is positive definite. That is, there exists a positive diagonal matrix $\mathbf{E}$ such that for $\mathbf{B} = \mathbf{E}^{-1}\mathbf{AE}$, the matrix $(\mathbf{B} + \mathbf{B}^T)/2$ is positive definite.

I_{27}: For each nonzero positive semi-definite matrix $\mathbf{P}$, the matrix $\mathbf{PA}$ has a positive diagonal element.

I_{28}: Every principal sub-matrix of $\mathbf{A}$ satisfies condition I_{25}.

J_{29}: $\mathbf{A}$ is positive stable. That is, the real part of each eigenvalue of $\mathbf{A}$ is positive.

J_{30}: There exists a symmetric positive definite matrix $\mathbf{W}$ such that $\mathbf{AW} + \mathbf{WA}^T$ is positive definite.

J_{31}: $\mathbf{A} + \mathbf{I}$ is non-singular, and $\mathbf{G} = (\mathbf{A} + \mathbf{I})^{-1}(\mathbf{A} - \mathbf{I})$ is convergent.

J_{32}: $\mathbf{A} + \mathbf{I}$ is non-singular, and for $\mathbf{G} = (\mathbf{A} + \mathbf{I})^{-1}(\mathbf{A} - \mathbf{I})$ there exists a positive definite symmetric matrix $\mathbf{W}$ such that $\mathbf{W} - \mathbf{G}^T\mathbf{WG}$ is positive definite.

Semi-positivity and Diagonal Dominance

K_{33}: $\mathbf{A}$ is semi-positive. That is, there exists $\mathbf{x} > 0$ with $\mathbf{Ax} > \mathbf{0}$.

K_{34}: There exists $\mathbf{x} \geq \mathbf{0}$ with $\mathbf{Ax} > \mathbf{0}$.

K_{35}: There exists a positive diagonal matrix $\mathbf{D}$ such that $\mathbf{AD}$ has all positive row sums.

$L36$: There exists $\mathbf{x} > \mathbf{0}$ with $\mathbf{Ax} > \mathbf{0}$ such that if $(\mathbf{Ax})_{i_0} = 0$, then there exist indices $1 \leq i_1, ..., i_r \leq n$ with $a_{i_k, i_{k+1}}$ for $0 \leq k \leq r - l$ and $(\mathbf{Ax})_{i_r} > 0$.

M_{37}: There exists $\mathbf{x} > \mathbf{0}$ with $\mathbf{Ax} > \mathbf{0}$ and $\sum_{j=1}^{i} a_{ij}x_j > 0$ for each $i = 1, 2, ..., n$.

N_{38}: There exists $\mathbf{x} > \mathbf{0}$ such that for each signature matrix $\mathbf{S}$, $\mathbf{SASx} > \mathbf{0}$.

N_{39}: $\mathbf{A}$ has all positive diagonal elements, and there exists a positive diagonal matrix $\mathbf{D}$ such that $\mathbf{AD}$ is strictly diagonally dominant. That is,

$$a_{ii}d_i > \sum_{j \neq i} |a_{ij}| d_j$$

for $i = 1, 2, ..., n$.

N_{40}: $\mathbf{A}$ has all positive diagonal elements, and there exists a positive diagonal matrix $\mathbf{D}$ such that $\mathbf{D}^{-1}\mathbf{AD}$ is strictly diagonally dominant.

Here, we refer interested readers to the references for the details of proving the equivalence and implications. Below we put the reference number in square brackets (i.e., [#]'s) which are listed in the References in Plemmons (1997). conditions A_1 and A_2 can be found in Ostrowski [24]. Condition A_3 was shown to be equivalent to A_1 by Willson [40], while conditions A_4, A_5, and A_6 were listed by Fiedler and Ptak [10]. Condition A_6 is also given by Gale and Nikaido [14], and A_7, was shown to be equivalent to A_1, by Moylan [22].

Condition B_8 can be found in Johnson [16], and $C_9, C_{10}, D_{11}D_{12}, E_{13}$ and E_{14} are in Fiedler and Ptak [10]. Condition F_{15} is in the original paper by Ostrowski [24], condition F_{16} was shown to be equivalent to F_{15} by Collatz [7], condition F_{17} is implicit in the work of Varga [38] on regular splittings, condition F_{18} is in the work of Schneider, [32] and F_{19} and F_{20} are in the work of Price [29]. Conditions F_{21}, F_{22} and G_{23} are in Ortega and Rheinboldt [42] and H_{24} is given by Varga [38]. Condition I_{25} is in the work of Tartar [35] and of Araki [2]. The equivalence of I_{26}, I_{27}, and I_{28} to I_{25} is shown by Barker, Berman and Plemmons [3]. The stability condition J_{29} is in the work of Ostrowski [24]; its equivalence with J_{30} is the Lyapunov [19] theorem. The equivalence of J_{30} with J_{31} is given by Taussky [36], and the equivalence of J_{32} with J_{31} is the Stein [34] theorem. Conditions K_{33}, K_{34} and K_{35} are given by Schneider [30] and Fan [9]. Condition L_{36} is given in a slightly weaker form by Bramble and Hubbard [5]; in our form, by Varga [39]. Condition M_{37} is in the work of Beauwens [4], and N_{38} is in that of Moylan [22]. Conditions N_{39} and N_{40} are essentially given by Fiedler and Ptak [10].

The references for implications from one group of conditions to another are as follows. That $N \Rightarrow I$ was established in Barker, Berman and Plemmons [3]. That $I \Rightarrow J$ is shown by Lyapunov [19], and that $I \Rightarrow A$ is also shown by Barker, Berman and Plemmons [3]. The implications $A \Rightarrow B \Rightarrow C, A \Rightarrow D \Rightarrow E$ and $J \Rightarrow C$ are immediate. That $A \Rightarrow K$ can be found in Nikaido [23], and the implication $K \Rightarrow L$ is immediate. That $F \Rightarrow K$ can be found in Schneider [30].

Finally, the implication $F \Rightarrow G$ is given by Varga [38], and the implication $G \Rightarrow H$ immediate.

Below we give the necessary and sufficient conditions for an arbitrary $n \times n$ real matrix to be a non-singular M-matrix.

Theorem J.18 *Suppose that* $\mathbf{A}$ *is an* $n \times n$ *real matrix with* $n \geq 2$. *Then the following conditions are all equvalent to the statement* "$\mathbf{A}$ *is a non-singular M-matrix.*

> 1. $\mathbf{A} + \mathbf{D}$ *is inverse-positive for each non-negative diagonal matrix* $\mathbf{D}$.
> 2. $\mathbf{A} + \alpha\mathbf{I}$ *is inverse-positive for each scalar* $\alpha > 0$.
> 3. *Each principal sub-matrix of* $\mathbf{A}$ *is inverse-positive.*
> 4. *Each principal sub-matrix of* $\mathbf{A}$ *of orders 1, 2 and n is inverse-positive.*

For some characterization of non-singular M-matrices related to matrices over the complex field, interested readers are referred to Plemmons (1977).

J.5　DEFINITIONS OF DERIVATIVES IN MATRIX CALCULUS

Some definitions of derivatives in matrix calculus are summarized in this section. Let $\mathbf{y} = (y_1, y_2, ..., y_m)^T$, $\mathbf{x} = (x_1, x_2, ..., x_n)^T$ be two column vectors and $\mathbf{z} = (z_{ij})_{m \times n}$ be an $m \times n$ matrix. We simply call the derivative of function f with respect to x (i.e., df/dx) "f by x" below in each category and the boldface variables are vectors or matrices and non-boldface variables are scalars.

- **Vector by Scalar:**

$$\frac{\partial \mathbf{y}}{\partial x} = \left(\frac{\partial y_1}{\partial x}, \frac{\partial y_2}{\partial x}, ..., \frac{\partial y_m}{\partial x} \right)^T.$$

This type of derivative is for the case with m functions of x put as a column vector. The derivative matrix is also called Jacobian matrix.

- **Scalar by Vector:**

$$\frac{\partial y}{\partial \mathbf{x}} = \left(\frac{\partial y}{\partial x_1}, \frac{\partial y}{\partial x_1}, ..., \frac{\partial y}{\partial x_1} \right).$$

This type of derivative is for the case where y is an n variable function. Transposing this row vector into a column vector results in so-called the gradient of a multivariate function. That is

$$\nabla y = \left(\frac{\partial y}{\partial \mathbf{x}}\right)^T = \begin{pmatrix} \frac{\partial y}{x_1} \\ \frac{\partial y}{x_2} \\ \vdots \\ \frac{\partial y}{x_n} \end{pmatrix}.$$

Writing y as a function of $y = f(x_1, x_2, ..., x_n)$, if all second partial derivatives of f exist, we can also define the Hessian matrix, $\mathbf{H}$, of f as

$$\mathbf{H}_f = \begin{pmatrix} \frac{\partial^2 f}{\partial x_1^2} & \frac{\partial^2 f}{\partial x_1 \partial x_2} & \cdots & \frac{\partial^2 f}{\partial x_1 \partial x_n} \\ \frac{\partial^2 f}{\partial x_2 \partial x_1} & \frac{\partial^2 f}{\partial x_2^2} & \cdots & \frac{\partial^2 f}{\partial x_2 \partial x_n} \\ \vdots & \vdots & \ddots & \vdots \\ \frac{\partial^2 f}{\partial x_n \partial x_1} & \frac{\partial^2 f}{\partial x_n \partial x_2} & \cdots & \frac{\partial^2 f}{\partial x_n^2} \end{pmatrix}.$$

If furthermore the second partial derivatives are all continuous, the Hessian matrix is a symmetric matrix by Schwarz's theorem. These definitions are useful in the non-linear optimization theory.

- **Vector by Vector**:

$$\frac{\partial \mathbf{y}}{\partial \mathbf{x}} = \begin{pmatrix} \frac{\partial y_1}{\partial x_1} & \frac{\partial y_1}{\partial x_2} & \cdots & \frac{\partial y_1}{\partial x_n} \\ \frac{\partial y_2}{\partial x_1} & \frac{\partial y_2}{\partial x_2} & \cdots & \frac{\partial y_2}{\partial x_n} \\ \vdots & \cdots & \cdots & \vdots \\ \frac{\partial y_m}{\partial x_1} & \frac{\partial y_m}{\partial x_2} & \cdots & \frac{\partial y_m}{\partial x_n} \end{pmatrix}.$$

This type of derivative is for the case with m functions put in a column vector. Each function is a n variable function, i.e., $y_i = f_i(x_1, x_2, ..., x_n)$ with $i = 1, 2, ..., m$.

- **Matrix by Scalar**:

$$\frac{\partial \mathbf{z}}{\partial x} = \begin{pmatrix} \frac{\partial z_{11}}{\partial x} & \frac{\partial z_{12}}{\partial x} & \cdots & \frac{\partial z_{1n}}{\partial x} \\ \frac{\partial z_{21}}{\partial x} & \frac{\partial z_{22}}{\partial x} & \cdots & \frac{\partial z_{2n}}{\partial x} \\ \vdots & \cdots & \cdots & \vdots \\ \frac{\partial z_{m1}}{\partial x} & \frac{\partial z_{m2}}{\partial x} & \cdots & \frac{\partial z_{mn}}{\partial x} \end{pmatrix}.$$

This type of derivative is for the case where there are $m \times n$ functions put in a matrix. Each function is a function of a single variable x.

- **Scalar by Matrix**:

$$\frac{\partial y}{\partial \mathbf{z}} = \begin{pmatrix} \frac{\partial y}{\partial z_{11}} & \frac{\partial y}{\partial z_{21}} & \cdots & \frac{\partial y}{\partial z_{m1}} \\ \frac{\partial y}{\partial z_{12}} & \frac{\partial y}{\partial z_{22}} & \cdots & \frac{\partial y}{\partial z_{m2}} \\ \vdots & \cdots & \cdots & \vdots \\ \frac{\partial y}{\partial z_{1n}} & \frac{\partial y}{\partial z_{2n}} & \cdots & \frac{\partial y}{\partial z_{mn}} \end{pmatrix}.$$

This type of derivative is for the case where y is a scalar function of $m \times n$ variables put in a matrix.

REFERENCES

1. E. Kani, N. J. Pullman, and N. M. Rice, "Algebraic Methods", Queen's University, 2021. (Course Reader).
2. R. J. Plemmons, M-Matrix characterizations. I - nonsingular M-Matrices. *Linear Algebra and its applications*, 18, 175-188 (1977).

Index

$GI/M/1$ queue, 232
M-Matrix, 780
$M/G/1$ queue, 226
$M/M/\infty$ queue, 70
$M/M/s$ queue, 59
Q-learning, 536
ε-greedy policy, 517
σ-algebra, 588
c/μ rule, 434
$c\mu$ rule, 420

Absorbing states, 16
Action preference method, 527
Acyclic PH distribution, 95
Age replacement policy, 395
Almost sure convergence, 700
Arc since law, 275
Average cost Markov decision process, 480

Backward algorithm (β-pass), 563
Backward C-K equation, 10
Base-stock policy, 460
Basic Stochastic Ordering, 753
Batch Marovian arrival process (BMAP, 129
Bayesian updating, 567
Belief state in POMDP, 567
Binomial securities market model, 272
Birth-and-death process (BDP), 50
Blackwell's Theorem, 187
Borel-Cantelli Lemmas, 618
Branching processes, 41
Brownian motion, 288
Brownian motion network, 358
Busy period of $M/G/1$ queue, 405

Cauchy-Schwarz Inequality, 661
Change of probability measure, 691

Chapman-Kolmogorov equation, C-K equation, 10
Chebyshev's inequality, 656
Chernoff bounds, 663
Closure properties of PH distribution, 97
Compound Poisson process, 76, 206
Conditional Value at risk (CVaR), 440
Continuous mapping Theorem, 709
Continuous-time Markov chain, 47
Continuous-time QBD process, 147
Convergence in distribution, 697
Convergence in mean, 698
Convergence in probability, 698
Convergence of sequence of mapped processes, 326
Convergence-together Theorem, 712
Copula, 596
Customer assignment problem, 423

Decoding problem, 563
Decomposition approach to queueing network, 365
Dense property of PH distribution, 103
Diffusion process, 726
Discrete PH distribution, 95
Discrete QBD, 139
Discrete-time Markov chain, 9
Discrete-time Markov decison process (DTMDP), 450
Donsker's Theorem -FCLT, 713
Dynamic inventory control system, 458

Economic analysis of stable queueing systems, 406
Elementary Renewal Theorem, 181
Entropy, 682
Ergodic DTMC, 22

Erlang A, B, C models, 377
Existence of Brownian motion, 723
Expectation Maximization (EM) algorithm, 105
Expected sojourn time in a transient state, 36
Exponential distribution, 48
Exponential martingale, 307
Extinction probability of branching processes, 42

Failure rate ordering, 759
Fixed point theorem, 44
Flowing through a queue, 366
Fluid network, 355
Forward algorithm (α-pass), 562
Forward C-K equation, 10
Foster's criterion, 31
Functional central limit theorem (FCLT), 329
Functional law of the iterated logarithm (FLIL), 334
Functional strong law of large numbers (FSLLN), 327
Functions of square matrices, 767

Gaussian process, 291
Greedy policy, 517

Hölder's Inequality, 662
Halfin-Whitt regime, 376
Hidden Markov model (HMM), 558
Hitting time, 18
Hitting time probability, 248
Hoeffding's Inequality, 665

Increasing failure rate (IFR), 399
Independent process, 291
Individual optimization - self-interested equilibrium, 407
Inspection paradox, 195
Integration in probability measure, 631

Inventory process as the reflected Bronian motion, 341
Irreducible Markov chain, 16
Ito's Formula, 741

Jensen's Inequality, 660

Karush-Kuhn-Tucker (KKT) optimality conditions, 674
Key renewal theorem, 191
Kolmogorov's Backward Equation, 54
Kolmogorov's Forward Equation, 54
Kronecker product, 100
Kronecker product of matrices, 100
Kronecker sum of matrices, 100

Laplace–Stieltjes transform, 52
Laplace-Stieltjes transform (LST), 603
Large deviation estimate, 656
Law of the iterated logarithm, 666
Learning in stochastic games, 549
Level crossing method, 264
Likelihood ratio ordering, 759
Lindley recursion, 259

Marked Poisson process, 76
Markov inequality, 655
Markov Modulated Poisson Process (MMPP), 124
Markov property, 1, 9
Markov renewal equation, 218
Markov Renewal Process, 211
Markovian Arrival Process (MAP), 124
Maximum entropy principle, 684
Maximum Likelihood Estimation (MLE), 132
Maximum likelihood method, 105
Mean field method, 386
Mean value analysis, 366
Model-based learning, 531
Model-free learning, 533

Monte Carlo-based MDP model, 530
Multi-arm bandits problem, 515
Multi-class $M/G/1$ queue, 416
Multi-Dimensional Reflected Brownian Motion, 351
Multiple linear regression models, 675
Multivariate distribution, 596
Myopic policy, 464

Naive Bayes' Model, 688
Non-homogeneous Poisson process, 78
Null-recurrent state, 22

Oblique reflection mapping (ORM), 353
Off-policy algorithm, 534
On-policy algorithm, 534
One-step transition probability, 9
Open Jackson network, 60
Optimality of structured policy, 470
Optimization based on regenerative cycles, 395
Ornstein-Uhlenbeck process, 730

Partially observable Markov decision process (POMDP), 566
Performance potential, 426
Perturbation realization factor (PRF), 429
PH renewal process, 122
PH representation, 91
Phase type distribution, PH-distribution, 91
Phase-type (PH) distribution, 36
Poisson equation, 427
Poisson process, 48
Policy iteration method (PI), 475
Positive matrix, 761
Positive recurrent state, 22
Potential matrix, 20
Power of matrices, 763

Probability distribution of DTMC, 12
Probability Space, 588
Pure Birth Process, 79
Pure Death Processes, 86

Quality and efficiency driven (QED) regime, 377
Quasi-birth-and-death process (QBD), 137
Queue with server vacation, 402
Queueing systems with many servers, 373
Queues with time-varying parameters, 381

Radon-Nikodym Theorem, 632
Random Time-change Theorem, 711
Random variable and $\mathscr{F}$-measurable function, 592
Rate matrix, 142
Reducible DTMC, 39
Reflected Brownian motion, 319
Reflection mapping theorem, 322
Reflection principle, 247
Regenerative Process, 207
Renewal counting process, 172
Renewal equation, 174
Renewal function, 174
Renewal Process, 172
Renewal reward process, 205
Restricted Markov chain, 28
Risk-averse newsvendor problem, 442
Semi-Markov decision process, 495
Semi-Markov kernel, 212
Semi-Markov process, 222
Semi-regenerative process, 224
Sensitivity based optimization, 423
Service provider optimization, 411
Simple average method, 517
Simple random walk, 245
Skorohod Representation Theorem, 709

Social optimization, 408
Spitzer's identity, 256
Splitting a flow, 369
Stochastic differential equation (SDE), 730
Stochastic differentiation, 739
Stochastic game, 508
Stochastic Integration, 734
Stochastic matrix, 18
Strong Approximation Theorem, 719
Strong Law of Large Numbers (SLLN), 629
Strong Markov property, 1
Subgaussian random variable, 521
Submodular function, 686
Superposition of flows, 365
Supplementary variable method, 238

Taboo probability, 143
Tandem queue, 337
Temporal-difference learning, 540
The Yule Process, 81
Thompson sampling for Bernoulli bandit, 530

Threshold policy, 402
Time-homogeneous DTMC, 11
Time-nonhomogeneous DTMC, 10
Transforming distribution function, 601
Transient analysis of DTMC, 36
Transition probability, 9
Transition probability function, 47

Upper confidence bound method, 520

Value at risk (VaR), 438
Value function, 452
Value iteration method (VI), 471
Variability ordering, 759
Variation of Brownian motion, 732
Virtual waiting time, 265

Wald's equation, 183
Weak Law of Large Numbers (WLLN), 623

z-transform, 65

For Product Safety Concerns and Information please contact our EU
representative GPSR@taylorandfrancis.com
Taylor & Francis Verlag GmbH, Kaufingerstraße 24, 80331 München, Germany